Soil Biology

Volume 63

Series Editor

Ajit Varma, Amity Institute of Microbial Technology, Amity University, Noida, Uttar Pradesh, India

The Springer series *Soil Biology* publishes topical volumes in the fields of microbiology, environmental sciences, plant sciences, biotechnology, biochemistry, microbial ecology, mycology and agricultural sciences.

Special emphasis is placed on methodological chapters or volumes. This includes coverage of new molecular techniques relevant to soil biology research or to monitoring and assessing soil quality as well as advanced biotechnological applications. Leading international authorities with a background in academia, industry or government will contribute to the series as authors or editors.

Key Topics: microbial-plant interactions; microbial communities; root symbiosis, mycorrhiza; rhizosphere environment; soil fauna, e.g. termites etc.; biochemical processes, soil proteins, enzymes, nucleic acids; degradation of biomaterials, decomposition, nutrient cycles; soil genesis, mineralization; bioremediation of contaminated sites; biotechnological applications of soil microorganisms.

More information about this series at http://www.springer.com/series/5138

D. K. Choudhary • Arti Mishra • Ajit Varma
Editors

Climate Change and the Microbiome

Sustenance of the Ecosphere

Springer

Editors
D. K. Choudhary
Amity Institute of Microbial Technology
Amity University Uttar Pradesh
Noida, Uttar Pradesh, India

Arti Mishra
Amity Institute of Microbial Technology
Amity University Uttar Pradesh
Noida, Uttar Pradesh, India

Ajit Varma
Amity Institute of Microbial Technology
Amity University Uttar Pradesh
Noida, Uttar Pradesh, India

ISSN 1613-3382 ISSN 2196-4831 (electronic)
Soil Biology
ISBN 978-3-030-76862-1 ISBN 978-3-030-76863-8 (eBook)
https://doi.org/10.1007/978-3-030-76863-8

This Springer imprint is published by the registered company Springer Nature Switzerland AG.
The registered company address is: Gewerbestrasse 11, 6330 Cham, Switzerland

Contents

Part I
Impact of Climate Change on Soil Microbiome

Chapter 1
Impact of Climate Change on Functional Root-Derived Signals

Bartłomiej Świątek, Subodh Kumar Maiti, and Marcin Pietrzykowski

Abstract This chapter presents an analysis of current data on the impact of climate change on the biomass of fine roots. World forests store 44% of carbon in biomass, 11% in dead wood and litter, and 45% in soil. As a result of droughts and other extreme weather events, forest growth will be disrupted. It is important to understand the factors that control fine root production and death rates in order to comprehend the natural cycle of elements and progressive climate change. It is especially important to understand the role of fine roots in sequestration and emissions of carbon dioxide, the element primarily responsible for global warming. However, fine roots can only be a factor in climate change on a global scale. On a local scale, fine root biomass is influenced by many factors. Future research should link the physiological factors of fine roots with progressive climate change.

Keywords Climatic change · Nutrient cycle · Carbon sequestration · Fine roots · Forest adaptation · Cycle of elements

1.1 Introduction

Increases in the concentration of atmospheric carbon dioxide are the main anthropogenic cause of progressive climate change. Emissions from fossil fuel combustion amounted to 36.2 Gt CO_2 in 2017 and increased to 37.1 Gt CO_2 in 2018 (Le Quéré et al. 2018). Another anthropogenic cause of the increase in atmospheric CO_2 concentration is forest degradation. Deforestation is responsible for around 20% of total CO_2 emissions. The conversion of tropical forests into agricultural ecosystems

B. Świątek (✉) · M. Pietrzykowski
Department of Ecology and Silviculture, Faculty of Forestry, University of Agriculture in Kraków, Krakow, Poland

S. K. Maiti
Department of Environmental Science and Engineering, Indian Institute of Technology (Indian School of Mines), Dhanbād, India

D. K. Choudhary et al. (eds.), *Climate Change and the Microbiome*, Soil Biology 63,
https://doi.org/10.1007/978-3-030-76863-8_1

generates emissions into the atmosphere totalling about 1.7 PgC per year (Sullivan and O'Keeffe 2011; Vaughn 2010). Human activity is estimated to have increased the average annual air temperature on Earth by about 1 °C since preindustrial times (Allen et al. 2018). An increase in average annual air temperature may be accompanied by a change in total annual precipitation and a higher frequency of extreme weather events (Räisänen et al. 2004).

Forests play an important role in reducing greenhouse gas emissions and storing carbon. World forests store 44% of carbon in biomass, 11% in dead wood and litter, and 45% in soil (FAO 2010). The carbon content of forest ecosystems varies by latitude (Lal 2005). In boreal zone forests, carbon is stored mainly in the organic matter of soil, while it is stored in plants in tropical forests. The difference in the rate of accumulation of organic matter depends on the climate. In colder climates, organic matter is produced faster than it can decompose, whereas in warm climates, organic matter decomposes very quickly, leading to a rapid circulation of elements (Prentice 2001). The rise in temperature accelerates the decomposition rate of organic matter in arctic regions (i.e., those where the soil is rich in carbon and the rise in temperature is relatively faster). In areas where the initial amount of organic matter is low, the fast decomposition of organic matter can be offset by accelerated growth of aboveground plant biomass.

1.2 Impact of Climate Change on Plants

The increased accessibility of carbon dioxide and nitrogen, along with the extension of the vegetation period and higher air temperature, causes the acceleration of biomass growth along with a reduction in the natural mortality of trees. Moreover, very fast-growing and strongly densified stands are prone to drought (Rebetez and Dobbertin 2004). Periodic threats to forests directly affect the growth of trees by limiting photosynthesis and increasing susceptibility to pathogenic infections (Taeger et al. 2013). Progressive climate change may alter the range of some tree species, shifting the vertical forest boundary in the far north and the horizontal forest boundary in the mountains (Brooker et al. 2007).

One of the strongest indicators of plants' ecological responses to climate change is phenology. Many research results show that the phenological seasons of trees change with increasing temperatures. The phenology of roots and shoots is essential in determining the effect of global warming on carbon accumulation in plants (Radville et al. 2016).

1.3 Impact of Climate Change on Fine Roots

Understanding the life cycle and turnover of fine roots is important for determining the cycle of carbon and other nutrients in nature. While the impact of environmental changes on fine root turnover is difficult to quantify, studies show that roots, like leaves, have a set of functional characteristics related to their lifespan. The lifespan of roots is influenced by many factors related to their structure, including root diameter and density, nitrogen concentration, and colonization by mycorrhizal fungi. All these features are highly plastic and depend strongly on the concentration of nitrogen, carbon dioxide, and water in soils (Warren 2015).

The impact of global climate change on roots can be significant but difficult to describe. An increase in the concentration of carbon dioxide influences root length and leads to increased root colonization by mycorrhizal fungi and decreased nitrogen concentration in tissues (Pritchard and Rogers 2000; Tingey et al. 2000). Increasing the level of carbon dioxide, which lowers nitrogen concentrations in root tissues, leads to a slightly longer root life (Eissenstat et al. 2000). In turn, a decrease in soil moisture or an increase in soil temperature may contribute to an increase in root mortality, which may be counterbalanced by a reduced availability of nutrients (Wang et al. 2019).

Some studies indicate that the main element of the carbon cycle in nature is the growth and decomposition rate of fine roots (Alongi 2012). The rate of decomposition of fine roots is often close to—or even higher than—the rate of decomposition of litter. Fine roots have an impact on mitigating climate change through carbon sequestration. It has been proven that a higher concentration of carbon dioxide in the atmosphere affects the growth and density of fine roots. In addition, soil temperature, which is closely related to global warming, has a direct impact on the intensity of respiration and the early growth of fine roots observed in spring (Cao et al. 2020). Up to 33% of solar energy supplied to the Earth is used for the production of fine roots. Fine root turnover increases exponentially with higher mean annual air temperatures and is a key element in ecosystems' carbon and nutrient cycles (Eissenstat et al. 2000; Gill and Jackson 2000).

Despite the great variation in fine root morphology and physiology, little is known regarding which environmental factors alter root physiology or how they may be related to plant function. The chemical and physical stress caused by climate change is responsible for changes in root physiology (Hirano et al. 2006; Richter et al. 2006). Moisture deficiencies in soil resulting from drought reduce the availability of nutrients and decrease root cell division (North and Nobel 1997). Additionally, reductions in transpiration caused by drought contribute to an increase in the concentration of aluminum in the soil solution, which results in an increased concentration of aluminum in roots. Moreover, rainfall changes the concentration of nitrogen and phosphorus in fine roots, and the content of these elements decreases with increasing rainfall.

1.4 Influence of Temperature and Precipitation on the Biomass of Fine Roots

Temperature is an important factor regulating the growth rate of fine roots, and the biomass of fine roots differs in the forests of various climatic zones (Vogt et al. 1996; Jackson et al. 1997; Finér et al. 2007). The maximum monthly air temperature explains 65% of the variability of the root biomass of small coniferous stands. The increase in the biomass of fine roots in boreal zone forests, along with the increase in average annual air temperature, may at least partially result from the fact that the increase in the rate of mineralization of organic matter improves nitrogen availability and stimulates the growth of fine roots (Pregitzer et al. 2000). The turnover ratio of fine roots in areas with a southern slope is higher than on those with a northern slope, indicating that higher temperature may be the cause of the increased biomass of fine roots (Hendrick and Pregitzer 1993).

Coniferous forests are usually characterized by a higher fine root biomass than deciduous forests due to a low rate of fine root rotation and a longer period of photosynthesis (Jackson et al. 1996; Gordon and Jackson 2000; Yuan and Chen 2010). Results collected from 512 stands in 35 European countries indicate that the biomass of fine roots is higher in deciduous forests than coniferous forests (Finér et al. 2007, 2011). In China, fine root biomass of coniferous trees was higher than the fine root biomass of deciduous trees in areas where the average annual rainfall was about 500 mm. When the average annual rainfall increased to approximately 1000 mm, the biomass of fine roots was higher in deciduous forests (Wang et al. 2017). Moreover, in the forests of northeastern China, the biomass of fine roots with a diameter of less than 0.5 mm was positively correlated with soil temperature, water content in soil, and concentrations of nitrogen, potassium, and phosphorus. For roots with a diameter of 0.5 mm to 2.0 mm, biomass was correlated with soil temperature and phosphorus concentration. Slight differences in the biomass of fine roots may occur with changes in average annual rainfall from 500 to 1500 mm and average annual air temperature from 5 to 15 °C (Wang et al. 2017).

In temperate zone forests, no relationship was observed between the biomass of beech and spruce fine roots and average annual air temperature (Finér et al. 2007). In boreal zone forests, the root biomass of small trees increased with higher average annual air temperature and precipitation. It is estimated that only about 2% of the variation in the biomass of fine roots can be explained by the type of forest, the content of nitrogen and phosphorus carbon in the soil, and climatic factors (Finér et al. 2011, 2019).

1.5 Impact of Climate Change on Nutritional Status and Mycorrhiza of Fine Roots

A review of studies from 51 countries found that, on a global scale, average annual air temperature and average annual rainfall do not affect nitrogen concentration in fine roots (Yuan and Chen 2010; Freschet et al. 2017). On the other hand, the content of phosphorus in fine roots is negatively correlated with average annual rainfall (Yuan and Chen 2010; Yuan et al. 2011). Research conducted in China showed no relationship among average annual air temperature, average annual rainfall, and the content of nitrogen and phosphorus in fine roots. The discrepancies between the results obtained in China and those obtained globally may result from differences in the species composition of stands and heterogeneous nitrogen and phosphorus uptake by trees along the climatic gradient (Cao et al. 2020).

In addition to increasing the ability of plants to take up nutrients directly, fine roots are also the basis for colonization by mycorrhizal fungi (Tibbett et al. 1999; Brundrett 2002). In the northern Scots pine population, low-temperature roots developed larger cross-sectional areas of parenchyma bark, which were then available for fungal colonization, in order to maintain an efficient nutrient uptake. The relationship between the host plant and mycorrhizal fungi was visible on the basis of the increased surface area of the mantle as the temperature decreased.

1.6 Fine Root Biomass in the Extreme Site Conditions of Restored Post-mining Sites

Mining sites are example of large-scale environmental transformation and restoration of degraded land which have attracted a global interest. One of the main elements of a restored new ecosystem is soil process development as the foundation of terrestrial ecosystem functioning (Pietrzykowski 2014). The ecology of mine soils (technosoils) can be assayed using soil biota and soil microbial activity, which are closely connected to the study of the rhizosphere. Fine root biomass and accompanying nutrient cycling thus offer great possibilities to describe the rate of technosoils' development and the ecology of new phytocenoses (Świątek et al. 2019). The large-scale sandpits that remain after surface mining are characterized by low sorption and retention capacities. They are also characterized by a complete lack of soil organic matter and nutrient deficits—especially nitrogen, phosphorus, and potassium—as well as very low biological activity (Pietrzykowski and Krzaklewski 2007, 2009; Maiti 2013). The climate of regions in Poland with large-scale sandpits is characterized by an average annual air temperature of 8.7 °C and an average annual rainfall of 735 mm. The annual cycle of alder fine root biomass increase in sandy soils was over 1002 g m^{-2} (Świątek et al. 2019). This increase should be considered large compared to other data in the literature (Table 1.1). Under optimal conditions of natural habitats in central Poland, the

Table 1.1 Fine root biomass in temperate zone forest stands

Species	Fine root biomass g m^{-2}	References
Fagus sylvatica	389	Finér et al. (2007)
Picea abies	281	Finér et al. (2007)
Pinus sylvestris	377	Finér et al. (2007)
Cryptomeria japonica	117	Konôpka et al. (2006)
Chamaecyparis obtusa	666	Yamashita et al. (2004)
Abies amabilis	829	Leuschner and Hertel (2003)
Picea abies	302	Leuschner and Hertel (2003)
Pinus koraiensis	508	Leuschner and Hertel (2003)
Pinus radiata	232	Leuschner and Hertel (2003)
Pinus resinosa	478	Leuschner and Hertel (2003)
Pinus strobus	331	Leuschner and Hertel (2003)
Pinus sylvestris	416	Leuschner and Hertel (2003)
Pseudotsuga menziesii	477	Leuschner and Hertel (2003)
Betula sp.	318	Leuschner and Hertel (2003)
Fagus sylvatica	438	Leuschner and Hertel (2003)
Populus tremuloides	691	Leuschner and Hertel (2003)
Quercus alba	384	Leuschner and Hertel (2003)
Quercus petraea	309	Leuschner and Hertel (2003)
Quercus rubra	433	Leuschner and Hertel (2003)

biomass of alder fine roots reached 127 g m^{-2}, which is lower than the data obtained for European beech and English oak stands (Jagodziński et al. 2016). The dynamics of fine root growth clearly shows relationships with soil characteristics, in particular trophism and fertility. For example, in nutrient-poor soils, trees tend to develop increased fine root biomass, penetrating large volumes of soil to compensate for nutrient deficiencies (Hodge 2004). On the other hand, in the case of increased availability of supplied nutrients—for example, by mineral fertilization—the intensity of the fine root network and its biomass development are lower (Majdi and Kangas 1997).

1.7 Conclusions

To understand the natural circulation of elements and progressive climate change, it is important to identify the factors controlling fine root production and mortality. It is particularly important to investigate the role of fine roots in absorbing and emitting carbon dioxide, the element primarily responsible for global warming. Increases in fine root biomass with increasing air temperature may result in the use of a larger volume of soil for the uptake of water and minerals. Fine roots show wide plasticity in response to changing soil conditions, and controlling the dynamics of fine roots plays a key role in providing plants with water during a drought. However, fine roots can only be a factor in climate change on a global scale. On a local scale, the biomass of fine roots is influenced by many factors, including root diameter and density and colonization by mycorrhizal fungi. All of these features are highly plastic and depend strongly on the content of nitrogen, carbon dioxide, and water in the soil. Future research should address the physiological factors of fine roots together with progressive climate change.

Acknowledgments The study was financed by the National Science Centre, Poland, grant no. 2019/33/N/ST10/02509. The paper is a part PhD dissertation of Bartłomiej Świątek MSc. as a critical revival of current state of the art in the fine roots' ecology.

References

Allen MR, Dube OP, Solecki W, Aragón-Durand F, Cramer W, Humphreys S, Kainuma M, Kala J, Mahowald N, Mulugetta Y, Perez R, Wairiu M, Zickfeld K (2018) Framing and context. In: Masson-Delmotte P, Zhai H-O, Pörtner D, Roberts J, Skea PR, Shukla A, Pirani W, Moufouma-Okia C, Péan R, Pidcock S, Connors JBR, Matthews Y, Chen X, Zhou MI, Gomis E, Lonnoy T, Maycock M (eds) Global warming of 1.5°C. An IPCC special report on the impacts of global warming of 1.5°C above pre-industrial levels and related global greenhouse gas emission pathways, in the context of strengthening the global response to the threat of climate change, sustainable development, and efforts to eradicate poverty (in press)

Alongi DM (2012) Carbon sequestration in mangrove forests. Carbon Manage 3:313–322. https://doi.org/10.4155/cmt.12.20

Brooker R, Travis J, Clark EJ, Dytham C (2007) Modelling species range shifts in a changing climate: the impacts of biotic interactions, dispersal distance and the rate of climate change. J Theor Biol 245:59–65

Brundrett MC (2002) Coevolution of roots and mycorrhizas of land plants. New Phytol 154:275–304

Cao Y, Li YN, Zhang GQ, Zhang J, Chen M (2020) Fine root C:N:P stoichiometry and its driving factors across forest ecosystems in northwestern China. Sci Total Environ 737

Eissenstat DM, Wells CE, Yanai RD, Whitbeck JL (2000) Building roots in a changing environment: implications for root longevity. New Phytol 147:33–42

FAO (2010) Global forest resources assessment 2010. Main report. FAO Forestry Paper 163

Finér L, Helmisaari HS, Lohmus K, Majdi H, Brunner I, Borja I, Eldhuset T, Godbold D, Grebenc T, Konopka B, Kraigher H, Mottonen MR, Ohashi M, Oleksyn J, Ostonen I, Uri V, Vanguelova E (2007) Variation in fine root biomass of three European tree species: beech (*Fagus sylvatica* L.), Norway spruce (*Picea abies* L. Karst.), and Scots pine (*Pinus sylvestris* L.). Plant Biosyst 141:394–405

Finér L, Ohashi M, Noguchi K, Hirano Y (2011) Factors causing variation in fine root biomass in forest ecosystems. For Ecol Manag 261:265–277. https://doi.org/10.1016/j.foreco.2010.10.016

Finér L, Zverev V, Palviainen M, Romanis T, Kozlov MV (2019) Variation in fine root biomass along a 1000 km long latitudinal climatic gradient in mixed boreal forests of North-East Europe. For Ecol Manag 432:649–655. https://doi.org/10.1016/j.foreco.2018.09.060

Freschet GT, Valverde-Barrantes OJ, Tucker CM, Craine JM, McCormack ML, Violle C, Fort F, Blackwood CB, Urban-Mead KR, Iversen CM, Bonis A, Comas LH, Cornelissen JHC, Dong M, Guo DL, Hobbie SE, Holdaway RJ, Kembel SW, Makita N, Onipchenko VG, Picon-Cochard C, Reich PB, de la Riva EG, Smith SW, Soudzilovskaia NA, Tjoelker MG, Wardle DA, Roumet C (2017) Climate, soil and plant functional types as drivers of global fine-root trait variation. J Ecol 105:1182–1196. https://doi.org/10.1111/1365-2745.12769

Gill RA, Jackson RB (2000) Global patterns of root turnover for terrestrial ecosystems. New Phytol 147:13–31

Gordon WS, Jackson RB (2000) Nutrient concentrations in fine roots. Ecology 81:275–280. https://doi.org/10.2307/177151

Hendrick RL, Pregitzer KS (1993) The dynamics of fine-root length, biomass, and nitrogen-content in two northern hardwood ecosystems. Can J For Res 23:2507–2520

Hirano Y, Walthert L, Brunner I (2006) Callose in root apices of European chestnut seedlings: a physiological indicator of aluminium stress. Tree Physiol 26:431–440

Hodge A (2004) The plastic plant: root responses to heterogeneous supplies of nutrients. New Phytol 162:9–24

Jackson RB, Canadell J, Ehleringer JR (1996) A global analysis of root distributions for terrestrial biomes. Oecologia 108:389–411

Jackson RB, Mooney HA, Schulze ED (1997) A global budget for fine root biomass, surface area, and nutrient contents. Proc Nat Acad Sci Unit St Am 94:7362–7366

Jagodziński AM, Ziolkowski J, Warnkowska A, Prais H (2016) Tree age effects on fine root biomass and morphology over chronosequences of *Fagus sylvatica*, Quercus robur and Alnus glutinosa stands. PLoS One 11:1–25

Konôpka B, Noguchi K, Sakata T, Takahashi M, Konôpková Z (2006) Fine root dynamics in a Japanese cedar (Cryptomeria japonica) plantation throughout the growing season. For Ecol Manag 225:278–286

Lal R (2005) Forest soils and carbon sequestration. For Ecol Manage 220:242–258. https://doi.org/10.1016/j.foreco.2005.08.015

Le Quéré C, Andrew RM, Friedlingstein P, Sitch S, Pongratz J (2018) Global carbon budget 2017. Earth System Science Data 10:405–448. https://doi.org/10.5194/essd-10-405-2018

Leuschner C, Hertel D (2003) Fine root biomass of temperate forests in relation to soil acidity and fertility, climate, age and species. In: Esser K, Lüttge U, Beyschlag W, Hellwig F (eds) Progress in botany. Progress in botany, vol 64. Springer, Berlin. https://doi.org/10.1007/978-3-642-55819-1_16

Maiti SK (2013) Ecorestoration of the coalmine degraded lands. Springer Science and Business Media, pp 61–81. https://doi.org/10.1007/978-81-322-0851-8_1

Majdi H, Kangas P (1997) Demography of fine roots in response to nutrient applications in a Norway spruce stand in southwestern Sweden. Écoscience 4(2):199–205. https://doi.org/10.1080/11956860.1997.11682396

North GB, Nobel PS (1997) Root-soil contact for the desert succulent *Agave deserti* in wet and drying soil. New Phytol 135:21–29

Pietrzykowski M (2014) Soil quality index as a tool for pine (*Pinus sylvestris* L.) monoculture conversion planning on afforested, reclaimed mine land. J For Res 25:63–74

Pietrzykowski M, Krzaklewski W (2007) Soil organic matter, C and N accumulation during natural succession and reclamation in an opencast sand quarry (southern Poland). Arch Agron Soil Sci 53:473–483

Pietrzykowski M, Krzaklewski W (2009) Rekultywacja leśna terenów wyrobisk po eksploatacji piasków podsadzkowych na przykładzie Kopalni "Szczakowa". Katedra Ekologii Lasu UR, Kraków (in Polish)

Pregitzer KS, King JA, BurtonA J, Brown SE (2000) Responses of tree fine roots to temperature. New Phytol 147:105–115

Prentice IC (2001) The carbon cycle and atmospheric carbon dioxide. Climate change 2001: The scientific basis IPCC. Cambridge University Press, Cambridge, pp 183–237

Pritchard SG, Rogers HH (2000) Spatial and temporal deployment of crop roots in CO_2-enriched environments. New Phytol 147:55–71

Radville L, McCormack ML, Post E, Eissenstat DM (2016) Root phenology in a changing climate. J Exp Bot 67:3617–3628

Räisänen J, Hansson U, Ullerstig A, Döscher R, Graham LP, Jones C, Meier HEM, Samuelsson P, Willén U (2004) European climate in the late twenty–first century: regional simulations with two driving global models and two forcing scenarios. Clim Dyn 22:13–31

Rebetez M, Dobbertin M (2004) Climate change may already threaten Scots pine stands in the Swiss Alps. Theoret Appl Climatol 79(1–2):1–9. https://doi.org/10.1007/s00704-004-0058-3

Richter AK, Frossard E, Brunner I (2006) Polyphenols in woody roots of Norway spruce and European beech reduce TTC. Tree Physiol 27:155–160

Sullivan CA, O'Keeffe J (2011) Water, biodiversity and ecosystems: reducing our impact. Water resources planning and management. Cambridge University Press, Cambridge, pp 117–130. https://doi.org/10.1017/CBO9780511974304.009

Świątek B, Woś B, Chodak M, Maiti SK, Józefowska A, Pietrzykowski M (2019) Fine root biomass and the associated C and nutrient pool under the alder (Alnus spp.) plantings on reclaimed technosols. Geoderma 337:1021–1027. https://doi.org/10.1016/j.geoderma.2018.11.025

Taeger S, Zang C, Liesebach M, Schneck V, Menzel A (2013) Impact of climate and drought events on the growth of Scots pine (*Pinus sylvestris* L.) provenances. For Ecol Manage 307:30–42. https://doi.org/10.1016/j.foreco.2013.06.053

Tibbett M, Sanders FE, Cairney JWG, Leake JR (1999) Temperature regulation of extracellular proteases in ectomycorrhizal fungi (Hebeloma spp.) grown in axenic culture. Mycol Res 103:707–714

Tingey DT, Phillips DL, Johnson MG (2000) Elevated CO_2 and conifer roots: effects on growth, life span and turnover. New Phytol 147:87–103

Vaughn CC (2010) Biodiversity losses and ecosystem function in freshwaters: emerging conclusions and research directions. BioScience 60(1):25–35. https://doi.org/10.1525/bio.2010.60.1.7

Vogt KA, Vogt DJ, Palmiotto PA, Boon P, Ohara J, Asbjornsen H (1996) Review of root dynamics in forest ecosystems grouped by climate, climatic forest type and species. Plant Soil 187:159–219

Wang SZ, Wang ZQ, Gu JC (2017) Variation patterns of fine root biomass, production and turnover in Chinese forests. J For Res 28:1185–1194. https://doi.org/10.1007/s11676-017-0386-7

Wang N, Wang CK, Quan XK (2019) Variations in fine root dynamics and turnover rates in five forest types in northeastern China. J For Res. https://doi.org/10.1007/s11676-019-01065-x

Warren JM (2015) Root structural and functional dynamics in terrestrial biosphere models – evaluation and recommendations. New Phytol 205:59–78. https://doi.org/10.1111/nph.13034

Yamashita T, Kasuya N, Nishimura S, Takeda H (2004) Comparison of two coniferous plantations in central Japan with respect to forest productivity, growth, phenology and soil nitrogen dynamics. For Ecol Manag 200:215–226

Yuan ZY, Chen HYH (2010) Fine root biomass, production, turnover rates, and nutrient contents in boreal forest ecosystems in relation to species, climate, fertility, and stand age: literature review and meta-analyses. Crit Rev Plant Sci 29:204–221

Yuan ZY, Chen HYH, Reich PB (2011) Global-scale latitudinal patterns of plant fine root nitrogen and phosphorus. Nat Commun 2:344. https://doi.org/10.1038/ncomms1346

Chapter 2
Climate Change Alters Microbial Communities

Aliyu Dabai Ibrahim, Abdulbariu Ogirima Uhuami, Nafi'u Abdulkadir, and Ifeyinwa Monica Uzoh

Abstract Microbial communities are key players in regulating ecosystem processes. Climate change factors such as CO_2 and temperature alter the microbial composition which in turn influenced the activities of microbial communities in the ecosystem's settings. As a result of their activities in the ecosystems and their resultant effect, changes to climate do occur. The effects of global warming, extreme weather conditions and other biotic and abiotic factors on microbial community functioning and richness still remain unclear. The present study aimed to review the influential roles of climate change on structural composition and functionality of microbiomes in their ecological niche. We also discussed the impacts of climate change on microbial environments and how microbial communities are capable of responding to extreme climate changes. It is believed that knowledge of the interaction of climate change and microbiomes, including their adaptation, would play a major role in mitigation and combating of climate changes in different ways.

Keywords Climate change · Microbial communities · Ecosystems' sustainability · Global warming · Niche · Mitigation · Alterations

A. D. Ibrahim (✉) · A. O. Uhuami
Department of Microbiology, Usmanu Danfodiyo University, Sokoto, Nigeria
e-mail: ibrahim.aliyu@udusok.edu.ng

N. Abdulkadir
Department of Microbiology, Sokoto State University, Sokoto, Nigeria
e-mail: nafiu.abdulkadir@ssu.edu.ng

I. M. Uzoh
Department of Soil Science, University of Nigeria, Nsukka, Nigeria

D. K. Choudhary et al. (eds.), *Climate Change and the Microbiome*, Soil Biology 63,
https://doi.org/10.1007/978-3-030-76863-8_2

2.1 Introduction

Microbial communities are ubiquitous in nature and the major contributors and mediators of biogeochemical cycles and sustainability of the earth. As a result, they have great influence on the ecosystem and climate. That notwithstanding, climate changes affect microbial diversity either directly or indirectly (Nie et al. 2013). Microbial communities develop survival strategies in order to adapt to changes in climate, which increases their chances of survival in any ecosystems. Survival strategies by these microorganisms could be an alteration in microbial community (i.e. outcompeting of other species and primary succession of new species) or sudden changes in the physiology of individual species (Fierer et al. 2007; Philippot et al. 2010; Placella et al. 2012; Zimmerman et al. 2013).

Alteration in microbial community influenced changes of ecological features of microbial communities (Kvitek et al. 2008; Wang et al. 2011). For instance, microbes that have experienced harsh drought and rewetting during rainy season have higher chances of becoming more resistant in this type of unfavourable conditions than microbial species that are inexperienced to climate change challenges (Fierer et al. 2003; Bouskill et al. 2013; Evans and Wallenstein 2014). The composition of microbial communities in soil is very diverse, making them exposed to all sort of climate change factors such as high temperature, moisture fluctuations and nutrient availability (Bardgett and van der Putten 2014).

The variation in the functions and structures of the ecosphere leads to major climate changes like floods, drought, greenhouse gas emissions, ozone depletion and heat waves (Smith 2011; Reichstein et al. 2013). These have consequential effect on resistance and resilience of different microorganisms and their ability to recover from ecological changes that could have negative effects (Nimmo et al. 2015; Oliver et al. 2015; Ingrisch and Bahn 2018). Studies on climate changes in relation with alterations of microbial communities, especially the impact of climate changes in the transition of microbial diversity and population, are still limited. Therefore, there is need for holistic understanding of climate changes, and we need to understand their effect on microbial community variability, diversity and contributions as well as impacts of these microbial alterations to the ecosystems. The motivation behind this chapter is to shed light on the effects of climate change on microbial community composition and functions. We discussed adaptations of microbial communities to such effects.

2.1.1 Extreme Effects of Climate Change on Microbial Communities

Extreme conditions such as drought, high temperature and greenhouse gases can alter the structure of microbial distributions and growth in a specific ecological niche (Zhou et al. 2012). The microbial communities include virus, fungi, protista, archaea and bacteria that are inhabitants of different parts of ecosystems. These organisms play a vital role in carbon and nitrogen cycles which help in sustaining the ecosystem's processes (Bardgett and van der Putten 2014). Although studies on the transitional alteration of microbial communities are limited. proofs are beginning to unveil themselves that climate change could play a role in these transitions. Moreover, the reaction to transitional changes that occurred in microbial communities is unmeasurable. For instance, the reaction of soil microbial communities to greenhouse gas could either be positive or negative (Janus et al. 2005; Lipson et al. 2005, 2006; Lesaulnier et al. 2008; Austin et al. 2009). Furthermore, reduction in soil water retention, increase in carbon production and extreme drought could lead to alteration of microbial colonization of a particular habitat (Robinson et al. 2016).

Recent studies showed that drought serves as a key factor of climate change and has a stronger impact on bacterial community than fungi (Bapiri et al. 2010; Vries et al. 2012; Barnard et al. 2013; de Vries et al. 2018). The possible reason could be fungi hyphae penetrate deeper into soil profile for more access to water during drought. Therefore, they are able to resist drought condition than bacteria. Soil microbial community could also be altered in its functionality and compositions from their original state to a new transitional state when there is nutrient enrichment of a particular soil causing the elevated production of greenhouse gas especially CO_2 (Allison and Martiny 2008; Zhou et al. 2012; Leff et al. 2015). This effect could lead to increase in atmospheric CO_2 with concomitant occurrences of global warming. Then, the atmospheric CO_2 could also change the structure and composition of microbial communities and distributions in a given environment (Zhou et al. 2011).

Furthermore, a major trait for high resistivity to climate change had been traced back to fungi. This is characterized by high genomic potential. With this, they are able to withstand different forms of harsh conditions and weather (Egidi et al. 2019). Another important factor that also affects alteration of microbial community is nutrient availability. Nutrient availability also plays a very important role in microbial growth and population. Nutrient enrichment due to agricultural practices could have impacts on total mass of bacteria within a specific area and could bring about elevated population of bacteria (De Vries and Shade 2013; Bardgett and van der Putten 2014). Elevated population of bacteria affects the energy flow in the soil which have been connected with carbon and nitrogen cycling (Gordon et al. 2008; Vries et al. 2012).

2.1.2 Influence of Climate Change on Microbial Community's Functions and Compositions

2.1.2.1 Soil Microbial Communities

Soil microbial communities include all forms of microorganism found in the soil and terrestrial environments. The soil contains a large group of microbes making them the most complex diversified communities on earth (Flemming and Wuertz 2019).

They mostly include soil pathogens, symbionts, mutualists, producers, decomposers etc. Soil microbes play an essential role in shaping and regulating the amount of organic carbon stored in soil which is released back to the atmosphere in the ecosystem (Singh et al. 2010; Bardgett and van der Putten 2014). Climate changes play a vital role in the alteration of soil microbiome diversity and also their interaction with other organisms especially with plant. Plant interaction with soil fungi (mycorrhizae) is involved in plants' acquisition of phosphorus and nitrogen (Fellbaum et al. 2012). Various studies have shown how various climate extremes such as drought, flood and ice impact on the soil microbes.

The activities and the metabolic processes that are been carried out by soil microbes are of paramount importance because they help balance the elemental and chemical compounds within the earth's crust (Singh et al. 2010). Climate changes on soil microbiomes could have a positive impact when soil microbes enhance plant performance, for example, biomass production, organic matter decomposition and survival as in the case of legumes and nitrogen-fixing bacteria (NFB), and it could lead to negative impact within the microbial communities and even to plant when their effects are pathogenic, greenhouse gas production (Smith 2011).

Most soil bacteria and archaea are the major facilitators of biogeochemical cycling of essential elements such as nitrogen and carbon (de Vries et al. 2018). Organic matter decomposition is carried by soil fungi and bacteria, which plays a major role in carbon cycle and release of CO_2 into the atmosphere. Another important role played by soil microbes is the fixing of nitrogen by nitrogen-fixing bacteria (NFB) in the soil and plant during mutualistic relationship with leguminous plants (Fig. 2.1) (Hurd et al. 2018).

2.1.2.2 Marine Microbial Communities

The earth itself is made up of 70% waterbodies. Climatic factors such as temperatures affect the rate of biological and metabolic process, nutrient availability and marine microbiome dispersal (Jørgensen and Boetius 2007). Negative consequences of climate change such as shift in marine food webs and carbon export buried into the sea bed have been associated with the increase of greenhouse gas concentration on ocean acidification, nutrient supply, temperature and irradiation (Gao et al. 2012; Rintoul et al. 2018; Hurd et al. 2018). Marine phytoplanktons like cyanobacteria and algae are important in marine food chain and have been found out to perform half of the global photosynthesis CO_2 fixation and half of the oxygen production (Behrenfeld et al. 2016). Apart from marine phytoplanktons, chemolithoautotrophic (Fig. 2.1), marine archaea and bacteria could fix CO_2 under dark conditions in deep ocean waters (Pachiadaki et al. 2017). Cycling of elements is also contributed by marine archaea and bacteria (Bunse et al. 2016). A group of cyanobacteria known as *Prochlorococcus* and *Synechococcus* are very abundant photosynthetic microbes in the ocean that removed about 10 billion tons of carbon each year which is about

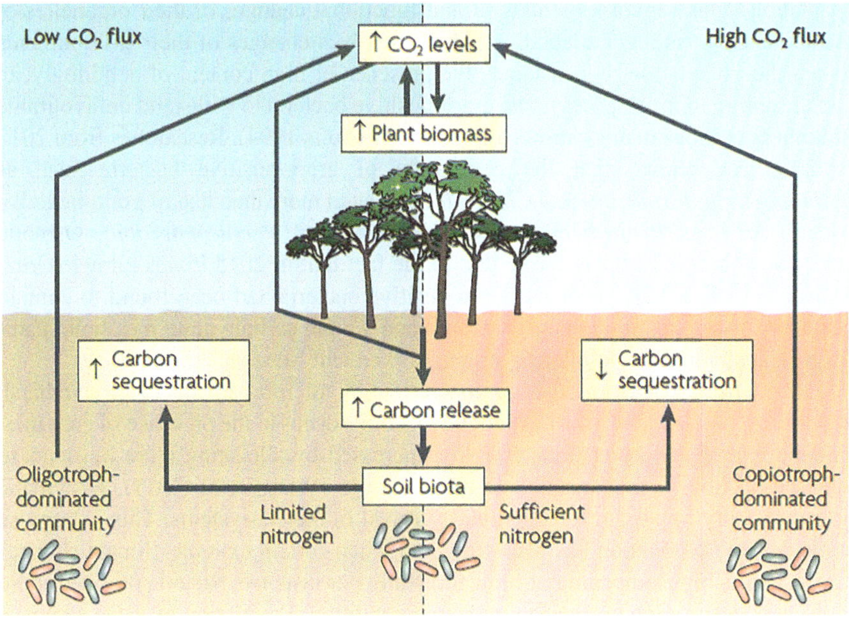

Fig. 2.1 A simple concept model illustrating the complex feedbacks climate change causes. Increased carbon dioxide (CO_2) levels resulting in a higher plant biomass and a higher carbon rhizodeposition, thereby increasing microbial biomass and activity in the short term. However, mineral nutrient limitation such as nitrogen may constrain this response in the long term. Such mineral limitation will affect oligotrophic and copiotrophic microorganism dominance in a given ecosystem, which in turn may influence the flux of CO_2 (with permission from Singh et al. (2010))

two-thirds of carbon fixation in ocean (Blount et al. 2008; Mariadassou et al. 2015; Youssef et al. 2015).

Thermal and latitudinal gradients and oceanic current are important factors for marine microbiome distributions (Wilkins et al. 2013; Cavicchioli 2015). These distributions could be affected by low pH which may lead marine archaea and bacteria to alter their gene expression to support cell maintenance (Bunse et al. 2016). Moreover, environmental and other factors influence the overall response and activities of marine microbes. For instance, reduction in cellular ribosomal concentration and increase in synthesis of protein in eukaryotic phytoplanktons occur in the presence of elevated temperature (Toseland et al. 2013).

2.1.3 Adaptation of Microbial Communities to Climate Change

Due to harmful environmental climate conditions, microorganisms have devised so many ways to adapt to unfavourable changes. These adaptations could be direct adaptation, which involves structural and functional changes of their organelles or metabolisms, or indirect adaptation, which involves changes of their environments to suit their habitation. For instance, the presence of high content of peptidoglycan and the ability to form spores make gram-positive bacteria to withstand unfavourable drought conditions than gram-negative bacteria (Potts 1994). Researches from 2013 to 2016 had shown that the population of gram-positive bacteria such as *Actinobacteria*, *Firmicutes* and *Chloroflexi* elevated more than that of gram-negative bacteria such as *Proteobacteria*, *Acidobacteria* and *Verrucomicrobia*. Osmotic stress was observed to play a role due to the fact that in 2015 it was rainier (Cruz-Martínez et al. 2012). The same gram-positive bacteria had been found to contain genes for producing amino sugar, alcohol and simple carbohydrate metabolic pathways which help them to tolerate stress (Borken and Matzner 2009).

Enzymatic activity is another approach used by microorganisms to survive harsh conditions. As we know, most metabolic reactions occur in the presence of enzymes. Therefore, enzyme productions could be increased by allocating more nutrients to their production in maintenance of the microbes (Wang et al. 2011). Enzymatic activity is very important for microbial survival in the ecosystems. This is because some enzyme production has been triggered when certain extreme climate changes occur such as high temperature or low moisture. For instance, bacteria produce spore to prevent desiccation during low moisture condition. Production of these spores is done with the activation of inert enzymes that help bacteria to survive unfavourable conditions. Despite all these, temperature and moisture fluctuation have impacts on enzyme productions and activities (Allison and Vitousek 2005).

Bacteria use two strategies for survival. The first is copiotrophic strategies involving the use of low resources or nutrients efficiently but with high growth which enable them to recover quickly from unfavourable conditions (resilience). The second is known as oligotrophic strategies that utilize high nutrients efficiently but have low growth rates making them to withstand unfavourable conditions (resistance) (Fierer et al. 2007; De Vries and Shade 2013). 'Ecological networking' has been shown as another approach that the microbial community could use for survival. This ecological networking involves interaction of a particular species with another which could affect their response to unconducive climate change (de Vries et al. 2018; Ramirez et al. 2018). Finally, some microbial community are known to possess 'traits'. These traits give them a special feature for survival during climate change. Some of these traits include dormancy genes (resuscitation promoting factors and sporulation) and operon count (Nemergut et al. 2016; Kearns and Shade 2018).

2.2 Contributors of Climate Change and Their Impacts on Microbial Community

2.2.1 *Temperature*

Temperature is one of the top contributors affecting the rate of metabolisms. Temperature plays a critical role for the success of metabolic processes. Instability of temperature in microbial metabolisms could bring about transitional alterations to the microbial community compositions. Moreover, high temperature contributes to the emission of atmospheric greenhouse gases that affect the survival of microbial community in many environmental settings (Fig. 2.2). Atmospheric greenhouse gas increments could be brought about by metabolic functionality changes in decomposers due to increase in temperature (Schindlbacher et al. 2011).

However, fungi play a crucial role in degradation of organic matters in the absence of nitrogen content in the soil, and temperature could cause warming which affects the amount of high nitrogen in the soil. Occurrence of this situation affects the activities of nitrogen bacteria such as nitrifying bacteria as they oxidize nitrogen and other nitrogen compounds due to high temperature that support their metabolism. Fungal decomposition is not only affected by sudden changes in the temperature but is also affected by other microbes that decompose organic matters and their diversity. Microbial community in water is also known to be affected by temperature change. High temperature change does not only affect their diversity or growth, but also it affects their metabolisms, population and resistivity. Algae species distribution all over the marine water bodies in the world especially the cyanobacteria is being affected by temperature (Beardall and Raven 2004). In the present twenty-first century, scientists have estimated that there may be continues temperature rise of surface of marine waters caused by global warming (Sarmento

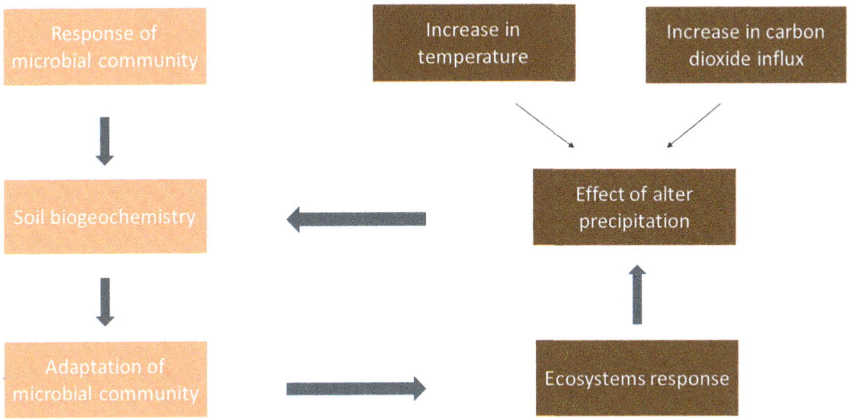

Fig. 2.2 An illustration of the interactions of climate change and some ecological factors that drives changes in microbial communities

et al. 2010). Therefore, increase in temperature could bring about negative impact on water chemistry which in turn influences microbial diversity, growth and populations (The USGS water science school 2015).

2.2.2 Water Content

Water is indispensable means of sustenance and functionality for all forms of life. Microbial community needs water to carry out their day-to-day metabolism and activities. The absence or availability of water affects the alteration or changes of microbes in the ecological systems. Microbial activities and composition are being affected by the presence of water. Furthermore, it stimulates these microbes to respond to soil respiration in regard to moisture and temperature (Aanderud et al. 2011). Changes in moisture content of terrestrial and soil niches determine the nature of microbial community in a particular ecological niche and also decomposition of organic materials (Fierer et al. 2003; Singh et al. 2010). The most intense consequence of different climate changes or any other forms of climate extremes on fungi, bacteria and any other microbial community is much higher when there is an alteration in water precipitation (Fig. 2.2). Therefore, increase or decrease in water precipitation regulates the microbial community of the ecosystem, their functions and structures and most importantly their metabolic processes (Schimel et al. 1999; Williams 2007; Castro et al. 2010).

Microbial activities could also be suppressed in environments such as soil and saltwater when there is low water availability and reduced enzymatic activity and hydration in the microbes. CO_2 emissions and productions to the atmosphere and the ecosphere could also be affected by soil moisture as it regulates soil respiration (Aanderud et al. 2011). Change in moisture and ecological factors is crucial for microbial lives, and processes depend on the regulation of these ecological factors (Smith et al. 2008).

2.2.3 Plant

Plant interaction with microbial community has been observed as the factor that alters microbial community diversity. One of the mechanisms is the distribution of plant–root absorbed carbon to soil microbial communities when plants are responding to climate changes. For instance, during dry weather conditions, there is reduction of photosynthetic processes due to the absence of water required for photosynthesis. This in turn reduces carbon allocation to soil microbes from plant which will result in low substrate for these microorganisms to carry out metabolisms. Fungi living in mutualistic association with plants are normally affected, for example, *mycorrhizae* (Hasibeder et al. 2015; Canarini and Dijkstra 2015; Fuchslueger et al. 2016; Bakhshandeh et al. 2019; Chomel et al. 2019).

Furthermore, bacteria community in the soil could also be affected by plant activities. Carbon emissions by plants to soil bacteria during rainy season increase population and growth of these microbes. Plant–soil relationship is able to be sustained by these bacteria due to the activities that they carry out in the soil such as soil organic matter decomposition or degradation, plant–microbial mutualistic relationships (nitrifying bacteria in legume) and oxidation of toxic compounds to nontoxic compounds which the plant could absorb and utilize (Karlowsky et al. 2018). Bacteria soil community could increase its size when there is increase in microbial activities and respiration which support decomposition of soil organic carbon. Drought-induced changes may trigger this process especially in the root exudates of plants (Chomel et al. 2019).

Plant association with soil microbes especially with fungi (*mycorrhizae*) and some mutualistic bacteria is a very important factor that needs to be studied as it affects alterations or transitions of microbial communities in the soil. Studies have shown that these mutualistic relationships could support some microbes living on a drought-tolerant plant where they could derive water, shelter and nutrients. Also, plant types and compositions could affect microbial community from recovery due to drought when there is low moisture content in the soil. These could cause nitrogen competition between plants and microbes (Orwin and Wardle 2005; Bloor and Bardgett 2012). Therefore, more researches should continue to be conducted in this area to understand plant interaction with microbes, their impacts on each other, their roles in climate changes and influence of plant alteration to microbial diversity.

2.3 Alteration of Microbial Community due to Climate Change in Other Aspects

2.3.1 Agriculture

Agriculture cannot be fully discussed without mentioning the roles of microorganisms. Microorganisms play an important role in the development and sustainability of crop growth and development as well as animal productions. Agriculture practices and methods and farmer activities also have an impact on microbial diversity (Table 2.1) and the ecosystem as well. Fertilizer applications have really contributed to the pollution of the environments, increase in nitrogen and distortion of biogeochemical cycles leading to threatening of the ecosystem (Steffen et al. 2015; Greaver et al. 2016). Microorganisms' oxidation and reduction of nitrogen compounds especially N_2O have made the agriculture sector as the highest emitter of greenhouse gas. Nitrogenous transformations such as ammonification, nitrification, nitrogen fixation and denitrification are different ways in which N_2O gas could be released into the atmosphere by these microbial communities (Greaver et al. 2016). Moreover, fertilizer applications could bring about microbial competitions and diversity.

Table 2.1 Environmental change transmission parameters and diseases (Eisenberg et al. 2007)

Environmental change	Description	Disease
Hospitalization	Increased people and time spent in hospitals	Tuberculosis (TB) Enteric and respiratory diseases
Urbanization	Increasing migration to and growth within towns	Diseases caused by faecal–oral pathogens Diseases caused by TB
Antibiotic usage	Emergence of antibiotic-resistant strains of bacterial pathogens	Multidrug-resistant TB and salmonellosis *Salmonella typhimurium*
Water projects	Water flow changes due to dam construction and irrigation networks	Schistosomiasis Malaria
Agricultural intensification	Changing crop and animal management practices; fertilizer and biocide use; use of genetically modified organisms	Cryptosporidiosis Diseases caused by *E. coli*
	Increased interplay between humans and domesticated animals	Influenza, severe acute respiratory syndrome (SARS), avian flu
Deforestation	Loss of forest cover, changing water flow patterns, reforestation and human encroachment along and into forested areas	Malaria Lyme disease Haemorrhagic fever AIDS
Transportation project	Construction of roads, increasing access to remote areas	Malaria STDS
Natural perturbations	Large-scale climate and other changes such as El Niño events	Cholera and leptospirosis
Cataclysmic events	Localized landscape changes caused by earthquakes, tsunamis, large fires and others	Water-related diseases like cholera
Climate change	Changing temperature and precipitation	Malaria, dengue fever and schistosomiasis

For instance, soil enrichment with elemental nutrients could result to unwanted algae bloom (Posch et al. 2012).

Rice cultivation and farm ruminant animals as other aspects of agriculture have also played a significant role in climate change and microbial biodiversity. Based on data from the World Bank, agricultural land, it has been estimated that 40% of terrestrial land has been devoted for crop production and animal rearing (Lanz et al. 2018). Natural CH_4 emissions that contribute to global warming are released by agricultural practices. CH_4 emissions from ruminant animals are the largest single source of this gas with the help of microbial community of intestinal tract of ruminant animals (Ripple et al. 2014). A total of 20% of agricultural CH_4 emissions by rice paddling contributes to CH_4 greenhouse gas. Scientific prediction has shown that by the end of this century, they may be doubling of CH_4 emissions only from rice paddling and cultivation (Groenigen et al. 2013). Thus, there is an urgent need for more researches and studies on agricultural practices in relationship with microbial activities.

2.3.2 *Infections*

Susceptibility of vectors and pathogens could be due to climate changes (McIntyre et al. 2017). The dispersion of microbial vector-borne disease and their virulence factors depend on climate change (Table 2.1). Changes in the ecosystems could affect the functionality of human health and food availability in ways where these microbial communities especially fungal, bacteria and virus cannot adapt to abiotic and biotic factors (Giraud et al. 2017; Cavicchioli et al. 2019). Fluctuation of rainfall and temperature due to climate variability is strongly attributed to many communicable diseases such as vector-borne and waterborne diseases and other forms of diseases such as Zika virus disease, plague, cholera and many more (Bouma and Dye 1997; Baylis et al. 1999; Rohani 2009; Kreppel et al. 2014; Caminade et al. 2017). For instance, the distribution of dengue fever and malaria which are known to be climate dependent often shifts in response to climate change (Bhatt et al. 2013; Pecl et al. 2017). Shift in host response and parasite adaptation to host are health risks that could be caused by climate change (Raffel et al. 2013). Antibiotic resistance of human bacterial pathogen has been predicted that climate change could also be another contributing factor (MacFadden et al. 2018).

Lower salinity and high temperature in estuaries' habitat caused by increase in precipitation could be associated with the spread of *Vibrio cholerae* infections which promote their growth. This has been observed in Bangladesh, Baltic Sea Region, North Atlanta and North Sea including human pathogen of *Vibrio* spp. (Pascual et al. 2000; Baker-Austin et al. 2013; Vezzulli et al. 2016). Transport and introduction of pathogens are influenced by effects of weather dispersal, and growth of the environmental conditions contributes to the spread and emergence of diseases (Bebber et al. 2013). Global environmental changes on pathogens, ecology of pathogens and host relationships with pathogens are basic knowledge that must be understood for strategic and effective control and spread of diseases (Johnson et al. 2017).

2.4 Microbial Mitigation to Climate Change

To combat climate change, we need to understand microbial efficacy and functionality towards mitigation of climate change. These involve harnessing of microbial biochemical molecules and processes and inducing of advantageous genetic sequences or genes into a potential microbe. For instance, the roles of microbes in agriculture could be supported when fertilizers are used with reduced nitrification inhibitors. This will help support soil bacteria especially nitrifying bacteria to produce more nitrates for plants and prevent subsequent leaching. Another approach could be the use of considerable amount of fertilizers which will reduce the availability of elemental nitrogen to soil microbes and less production of nitrous oxide. This will help reduce the impact of global warming (Smith et al. 2008). Carbon sequestration could be a very important approach in reducing atmospheric CO_2 (Prosser et al. 2007). Forest soils have been considered as effective for carbon stroage due to abundance of bacteria and fungi and favorable environmental conditions that support the growth of microbial communities (Bailey et al. 2002; De Deyn et al. 2008; Busse et al. 2009; Castro et al. 2010).

Methane flux emission is mostly caused by microbes. It is theoretically possible to control microbial activities in a considerable amount of CH_4 emissions from terrestrial ecosystem. Ninety percent of CH_4 emissions in the soil are oxidized by methanotrophs before escaping into the atmosphere (Tate et al. 2007; Smith et al. 2008). With these studies, rice cultivation has improved flood management and could reduce net emission of CH_4 by increasing oxygen availability in soils when methanotrophs absorb a proportion of the CH_4 produced. Also, quality feed and use of antibiotics, vaccines and other forms of electron acceptors are ways that could be employed to reduce methane emissions in ruminant animals (Smith et al. 2008).

The use of biochar could be a mitigation option in the treatment of climate change. Microbes play a role in breaking down organic matters that support the growth of plants. This organic matter decomposition by soil microbes could be mixed with biochar which will help in organic matter retention and preventing other microbes from carrying out ammonification and releasing of carbon (Weng et al. 2017). Finally, the use and development of non-greenhouse gas emission technologies and biotechnologies could be a lasting solution to global warming and climate change. These will surely solve the crisis of clean energy, clean water and industrial waste management treatment (Timmis et al. 2017).

2.5 Conclusion

The role of microbes in the ecosystem is of utmost importance in the regulation of the abiotic and biotic factors affecting the ecosystem. Microbial communities are the major regulators of all life processes and occurrence of the global climatic changes. Other factors also contribute to climate change such as human activities and industrial revolutions. These regulations could one way affect the alteration of microbial community and diversity. The alterations of these microbiomes could bring about positive or negative feedback to the environment and the ecosystem at large. The feedback could result in either direct impacts to the microbial community and other macro-organisms or indirect impacts to the environment. There is an urgent need for more researches to link climate change and microbial community and to understand the consequential impacts of transitions of microbial community after and before the processes do occur.

References

Aanderud ZT, Schoolmaster DR, Lennon JT (2011) Plants mediate the sensitivity of soil respiration to rainfall variability. Ecosystems 14:156–167. https://doi.org/10.1007/s10021-010-9401-y

Allison SD, Martiny JBH (2008) Colloquium paper: resistance, resilience, and redundancy in microbial communities. Proc Natl Acad Sci USA 105(Suppl 1):11512–11519. https://doi.org/10.1073/pnas.0801925105

Allison S, Vitousek P (2005) Responses of extracellular enzymes to simple and complex nutrient inputs. https://doi.org/10.1016/J.SOILBIO.2004.09.014

Austin EE, Castro HF, Sides KE et al (2009) Assessment of 10 years of CO_2 fumigation on soil microbial communities and function in a sweetgum plantation. Soil Biol Biochem 41:514–520. https://doi.org/10.1016/j.soilbio.2008.12.010

Bailey VL, Smith JL, Bolton H (2002) Fungal-to-bacterial ratios in soils investigated for enhanced C sequestration. Soil Biol Biochem 34:997–1007. https://doi.org/10.1016/S0038-0717(02)00033-0

Baker-Austin C, Trinanes JA, Taylor NGH et al (2013) Emerging Vibrio risk at high latitudes in response to ocean warming. Nat Clim Chang 3:73–77. https://doi.org/10.1038/nclimate1628

Bakhshandeh S, Corneo PE, Yin L, Dijkstra FA (2019) Drought and heat stress reduce yield and alter carbon rhizodeposition of different wheat genotypes. J Agron Crop Sci 205:157–167. https://doi.org/10.1111/jac.12314

Bapiri A, Bååth E, Rousk J (2010) Drying–rewetting cycles affect fungal and bacterial growth differently in an arable soil. Microb Ecol 60:419–428. https://doi.org/10.1007/s00248-010-9723-5

Bardgett RD, van der Putten WH (2014) Belowground biodiversity and ecosystem functioning. Nature 515:505–511. https://doi.org/10.1038/nature13855

Barnard RL, Osborne CA, Firestone MK (2013) Responses of soil bacterial and fungal communities to extreme desiccation and rewetting. ISME J 7:2229–2241. https://doi.org/10.1038/ismej.2013.104

Baylis M, Mellor PS, Meiswinkel R (1999) Horse sickness and ENSO in South Africa. Nature 397:574. https://doi.org/10.1038/17512

Beardall J, Raven JA (2004) The potential effects of global climate change on microalgal photosynthesis, growth and ecology. Phycologia 43:26–40. https://doi.org/10.2216/i0031-8884-43-1-26.1

Bebber DP, Ramotowski MAT, Gurr SJ (2013) Crop pests and pathogens move polewards in a warming world. Nat Clim Chang 3:985–988. https://doi.org/10.1038/nclimate1990

Behrenfeld MJ, O'Malley RT, Boss E et al (2016) Revaluating ocean warming impacts on global phytoplankton. https://doi.org/10.1038/NCLIMATE2838

Bhatt S, Gething PW, Brady OJ et al (2013) The global distribution and burden of dengue. Nature 496:504–507. https://doi.org/10.1038/nature12060

Bloor JMG, Bardgett RD (2012) Stability of above-ground and below-ground processes to extreme drought in model grassland ecosystems: interactions with plant species diversity and soil nitrogen availability. Perspect Plant Ecol Evol Syst 14:193–204. https://doi.org/10.1016/j.ppees.2011.12.001

Blount ZD, Borland CZ, Lenski RE (2008) Historical contingency and the evolution of a key innovation in an experimental population of *Escherichia coli*. Proc Natl Acad Sci USA 105:7899–7906. https://doi.org/10.1073/pnas.0803151105

Borken W, Matzner E (2009) Reappraisal of drying and wetting effects on C and N mineralization and fluxes in soils. Glob Chang Biol 15:808–824. https://doi.org/10.1111/j.1365-2486.2008.01681.x

Bouma MJ, Dye C (1997) Cycles of malaria associated with El Niño in Venezuela. JAMA 278:1772–1774

Bouskill NJ, Lim HC, Borglin S et al (2013) Pre-exposure to drought increases the resistance of tropical forest soil bacterial communities to extended drought. ISME J 7:384–394. https://doi.org/10.1038/ismej.2012.113

Bunse C, Lundin D, Karlsson CMG et al (2016) Response of marine bacterioplankton pH homeostasis gene expression to elevated CO_2. Nat Clim Chang 6:483–487. https://doi.org/10.1038/nclimate2914

Busse MD, Sanchez FG, Ratcliff AW et al (2009) Soil carbon sequestration and changes in fungal and bacterial biomass following incorporation of forest residues. Soil Biol Biochem 41:220–227

Caminade C, Turner J, Metelmann S et al (2017) Global risk model for vector-borne transmission of Zika virus reveals the role of El Niño 2015. PNAS 114:119–124. https://doi.org/10.1073/pnas.1614303114

Canarini A, Dijkstra FA (2015) Dry-rewetting cycles regulate wheat carbon rhizodeposition, stabilization and nitrogen cycling. Soil Biol Biochem 81:195–203. https://doi.org/10.1016/j.soilbio.2014.11.014

Castro HF, Classen AT, Austin EE et al (2010) Soil microbial community responses to multiple experimental climate change drivers. Appl Environ Microbiol 76:999–1007. https://doi.org/10.1128/AEM.02874-09

Cavicchioli R (2015) Microbial ecology of Antarctic aquatic systems. Nat Rev Microbiol 13:691–706. https://doi.org/10.1038/nrmicro3549

Cavicchioli R, Ripple WJ, Timmis KN et al (2019) Scientists' warning to humanity: microorganisms and climate change. Nat Rev Microbiol 17:569–586. https://doi.org/10.1038/s41579-019-0222-5

Chomel M, Lavallee JM, Alvarez-Segura N et al (2019) Drought decreases incorporation of recent plant photosynthate into soil food webs regardless of their trophic complexity. Glob Chang Biol 25:3549–3561. https://doi.org/10.1111/gcb.14754

Cruz-Martínez K, Rosling A, Zhang Y et al (2012) Effect of rainfall-induced soil geochemistry dynamics on grassland soil microbial communities. Appl Environ Microbiol 78:7587–7595. https://doi.org/10.1128/AEM.00203-12

De Deyn GB, Cornelissen JHC, Bardgett RD (2008) Plant functional traits and soil carbon sequestration in contrasting biomes. Ecol Lett 11:516–531. https://doi.org/10.1111/j.1461-0248.2008.01164.x

De Vries FT, Shade A (2013) Controls on soil microbial community stability under climate change. Front Microbiol 4. https://doi.org/10.3389/fmicb.2013.00265

de Vries FT, Griffiths RI, Bailey M et al (2018) Soil bacterial networks are less stable under drought than fungal networks. Nat Commun 9:3033. https://doi.org/10.1038/s41467-018-05516-7

Egidi E, Delgado-Baquerizo M, Plett JM et al (2019) A few Ascomycota taxa dominate soil fungal communities worldwide. Nat Commun 10. https://doi.org/10.1038/s41467-019-10373-z

Eisenberg JNS, Desai MA, Levy K et al (2007) Environmental determinants of infectious disease: a framework for tracking causal links and guiding public health research. Environ Health Perspect 115:1216–1223. https://doi.org/10.1289/ehp.9806

Evans SE, Wallenstein MD (2014) Climate change alters ecological strategies of soil bacteria. Ecol Lett 17:155–164. https://doi.org/10.1111/ele.12206

Fellbaum CR, Mensah JA, Pfeffer PE et al (2012) The role of carbon in fungal nutrient uptake and transport. Plant Signal Behav 7:1509–1512. https://doi.org/10.4161/psb.22015

Fierer N, Schimel JP, Holden PA (2003) Influence of drying-rewetting frequency on soil bacterial community structure. Microb Ecol 45:63–71. https://doi.org/10.1007/s00248-002-1007-2

Fierer N, Bradford MA, Jackson RB (2007) Toward an ecological classification of soil bacteria. Ecology 88:1354–1364. https://doi.org/10.1890/05-1839

Flemming HC, Wuertz S (2019) Bacteria and archaea on Earth and their abundance in biofilms. Nat Rev Microbiol 17:247–260. https://doi.org/10.1038/s41579-019-0158-9

Fuchslueger L, Bahn M, Hasibeder R et al (2016) Drought history affects grassland plant and microbial carbon turnover during and after a subsequent drought event. J Ecol 104:1453–1465. https://doi.org/10.1111/1365-2745.12593

Gao K, Xu J, Gao G et al (2012) Rising CO_2 and increased light exposure synergistically reduce marine primary productivity. Nat Clim Chang 2:519–523. https://doi.org/10.1038/nclimate1507
Giraud T, Koskella B, Laine A-L (2017) Introduction: microbial local adaptation: insights from natural populations, genomics and experimental evolution. Mol Ecol 26:1703–1710. https://doi.org/10.1111/mec.14091
Gordon H, Haygarth PM, Bardgett RD (2008) Drying and rewetting effects on soil microbial community composition and nutrient leaching. Soil Biol Biochem 40:302–311. https://doi.org/10.1016/j.soilbio.2007.08.008
Greaver TL, Clark CM, Compton JE et al (2016) Key ecological responses to nitrogen are altered by climate change. Nat Clim Chang 6:836–843. https://doi.org/10.1038/nclimate3088
Groenigen KJV, Kessel CV, Hungate BA (2013) Increased greenhouse-gas intensity of rice production under future atmospheric conditions. Nat Clim Chang 3:288–291. https://doi.org/10.1038/nclimate1712
Hasibeder R, Fuchslueger L, Richter A, Bahn M (2015) Summer drought alters carbon allocation to roots and root respiration in mountain grassland. New Phytol 205:1117–1127. https://doi.org/10.1111/nph.13146
Hurd CL, Lenton A, Tilbrook B, Boyd PW (2018) Current understanding and challenges for oceans in a higher-CO_2 world. Nat Clim Chang 8:686–694. https://doi.org/10.1038/s41558-018-0211-0
Ingrisch J, Bahn M (2018) Towards a comparable quantification of resilience. Trends Ecol Evol 33:251–259. https://doi.org/10.1016/j.tree.2018.01.013
Janus LR, Angeloni NL, McCormack J et al (2005) Elevated atmospheric CO_2 alters soil microbial communities associated with trembling aspen (*Populus tremuloides*) roots. Microb Ecol 50:102–109. https://doi.org/10.1007/s00248-004-0120-9
Johnson CN, Balmford A, Brook BW et al (2017) Biodiversity losses and conservation responses in the Anthropocene. Science 356:270–275. https://doi.org/10.1126/science.aam9317
Jørgensen BB, Boetius A (2007) Feast and famine — microbial life in the deep-sea bed. Nat Rev Microbiol 5:770–781. https://doi.org/10.1038/nrmicro1745
Karlowsky S, Augusti A, Ingrisch J et al (2018) Drought-induced accumulation of root exudates supports post-drought recovery of microbes in Mountain Grassland. Front Plant Sci 9. https://doi.org/10.3389/fpls.2018.01593
Kearns PJ, Shade A (2018) Trait-based patterns of microbial dynamics in dormancy potential and heterotrophic strategy: case studies of resource-based and post-press succession. ISME J 12:2575–2581. https://doi.org/10.1038/s41396-018-0194-x
Kreppel KS, Caminade C, Telfer S et al (2014) A non-stationary relationship between global climate phenomena and human plague incidence in Madagascar. PLoS Negl Trop Dis 8:e3155. https://doi.org/10.1371/journal.pntd.0003155
Kvitek DJ, Will JL, Gasch AP (2008) Variations in stress sensitivity and genomic expression in diverse *S. cerevisiae* isolates. PLoS Genet 4:e1000223. https://doi.org/10.1371/journal.pgen.1000223
Lanz B, Dietz S, Swanson T (2018) The expansion of modern agriculture and global biodiversity decline: an integrated assessment. Ecol Econ 144:260–277. https://doi.org/10.1016/j.ecolecon.2017.07.018
Leff JW, Jones SE, Prober SM et al (2015) Consistent responses of soil microbial communities to elevated nutrient inputs in grasslands across the globe. PNAS 112:10967–10972. https://doi.org/10.1073/pnas.1508382112
Lesaulnier C, Papamichail D, McCorkle S et al (2008) Elevated atmospheric CO_2 affects soil microbial diversity associated with trembling aspen. Environ Microbiol 10:926–941. https://doi.org/10.1111/j.1462-2920.2007.01512.x
Lipson DA, Wilson RF, Oechel WC (2005) Effects of elevated atmospheric CO_2 on soil microbial biomass, activity, and diversity in a chaparral ecosystem. Appl Environ Microbiol 71:8573–8580. https://doi.org/10.1128/AEM.71.12.8573-8580.2005

Lipson DA, Blair M, Barron-Gafford G et al (2006) Relationships between microbial community structure and soil processes under elevated atmospheric carbon dioxide. Microb Ecol 51:302–314. https://doi.org/10.1007/s00248-006-9032-1
MacFadden DR, McGough SF, Fisman D et al (2018) Antibiotic resistance increases with local temperature. Nat Clim Chang 8:510–514. https://doi.org/10.1038/s41558-018-0161-6
Mariadassou M, Pichon S, Ebert D (2015) Microbial ecosystems are dominated by specialist taxa. Ecol Lett 18:974–982. https://doi.org/10.1111/ele.12478
McIntyre KM, Setzkorn C, Hepworth PJ et al (2017) Systematic assessment of the climate sensitivity of important human and domestic animals pathogens in Europe. Sci Rep 7:7134. https://doi.org/10.1038/s41598-017-06948-9
Nemergut DR, Knelman JE, Ferrenberg S et al (2016) Decreases in average bacterial community rRNA operon copy number during succession. ISME J 10:1147–1156. https://doi.org/10.1038/ismej.2015.191
Nie M, Pendall E, Bell C et al (2013) Positive climate feedbacks of soil microbial communities in a semi-arid grassland. Ecol Lett 16:234–241. https://doi.org/10.1111/ele.12034
Nimmo DG, Mac Nally R, Cunningham SC et al (2015) Vive la résistance: reviving resistance for 21st century conservation. Trends Ecol Evol 30:516–523. https://doi.org/10.1016/j.tree.2015.07.008
Oliver TH, Heard MS, Isaac NJB et al (2015) Biodiversity and resilience of ecosystem functions. Trends Ecol Evol 30:673–684. https://doi.org/10.1016/j.tree.2015.08.009
Orwin KH, Wardle DA (2005) Plant species composition effects on belowground properties and the resistance and resilience of the soil microflora to a drying disturbance. Plant Soil 278:205–221. https://doi.org/10.1007/s11104-005-8424-1
Pachiadaki MG, Sintes E, Bergauer K et al (2017) Major role of nitrite-oxidizing bacteria in dark ocean carbon fixation. Science 358:1046–1051. https://doi.org/10.1126/science.aan8260
Pascual M, Rodó X, Ellner SP et al (2000) Cholera dynamics and El Niño-southern oscillation. Science 289:1766–1769. https://doi.org/10.1126/science.289.5485.1766
Pecl GT, Araújo MB, Bell JD et al (2017) Biodiversity redistribution under climate change: impacts on ecosystems and human well-being. Science 355. https://doi.org/10.1126/science.aai9214
Philippot L, Andersson SGE, Battin TJ et al (2010) The ecological coherence of high bacterial taxonomic ranks. Nat Rev Microbiol 8:523–529. https://doi.org/10.1038/nrmicro2367
Placella SA, Brodie EL, Firestone MK (2012) Rainfall-induced carbon dioxide pulses result from sequential resuscitation of phylogenetically clustered microbial groups. Proc Natl Acad Sci USA 109:10931–10936. https://doi.org/10.1073/pnas.1204306109
Posch T, Koster O, Salcher M, Pernthaler J (2012) Harmful filamentous cyanobacteria favoured by reduced water turnover with lake warming. Nat Clim Chang 2:5. https://doi.org/10.1038/nclimate1581
Potts M (1994) Desiccation tolerance of prokaryotes. Microbiol Rev 58:755–805
Prosser JI, Bohannan BJM, Curtis TP et al (2007) The role of ecological theory in microbial ecology. Nat Rev Microbiol 5:384–392. https://doi.org/10.1038/nrmicro1643
Raffel TR, Romansic JM, Halstead NT et al (2013) Disease and thermal acclimation in a more variable and unpredictable climate. Nat Clim Chang 3:146–151. https://doi.org/10.1038/nclimate1659
Ramirez KS, Geisen S, Morriën E et al (2018) Network analyses can advance above-belowground ecology. Trends Plant Sci 23:759–768. https://doi.org/10.1016/j.tplants.2018.06.009
Reichstein M, Bahn M, Ciais P et al (2013) Climate extremes and the carbon cycle. Nature 500:287–295. https://doi.org/10.1038/nature12350
Rintoul SR, Chown SL, DeConto RM et al (2018) Choosing the future of Antarctica. Nature 558:233–241. https://doi.org/10.1038/s41586-018-0173-4
Ripple WJ, Smith P, Haberl H et al (2014) Ruminants, climate change and climate policy. Nat Clim Chang 4:2–5. https://doi.org/10.1038/nclimate2081
Robinson DA, Jones SB, Lebron I et al (2016) Experimental evidence for drought induced alternative stable states of soil moisture. Sci Rep 6:20018. https://doi.org/10.1038/srep20018
Rohani P (2009) The link between dengue incidence and El Niño southern oscillation. PLoS Med 6: e1000185. https://doi.org/10.1371/journal.pmed.1000185

Sarmento H, Montoya JM, Vázquez-Domínguez E et al (2010) Warming effects on marine microbial food web processes: how far can we go when it comes to predictions? Philos Trans R Soc B: Biol Sci 365:2137–2149. https://doi.org/10.1098/rstb.2010.0045
Schimel JP, Gulledge JM, Clein-Curley JS et al (1999) Moisture effects on microbial activity and community structure in decomposing birch litter in the Alaskan taiga. Soil Biol Biochem 31:831–838. https://doi.org/10.1016/S0038-0717(98)00182-5
Schindlbacher A, Rodler A, Kuffner M et al (2011) Experimental warming effects on the microbial community of a temperate mountain forest soil. Soil Biol Biochem 43:1417–1425. https://doi.org/10.1016/j.soilbio.2011.03.005
Singh BK, Bardgett RD, Smith P, Reay DS (2010) Microorganisms and climate change: terrestrial feedbacks and mitigation options. Nat Rev Microbiol 8:779–790. https://doi.org/10.1038/nrmicro2439
Smith MD (2011) An ecological perspective on extreme climatic events: a synthetic definition and framework to guide future research. J Ecol 99:656–663. https://doi.org/10.1111/j.1365-2745.2011.01798.x
Smith P, Fang C, Dawson JJC, Moncrieff JB (2008) Impact of global warming on soil organic carbon. Adv Agron 97:1–43. https://doi.org/10.1016/S0065-2113(07)00001-6
Steffen W, Richardson K, Rockström J et al (2015) Planetary boundaries: guiding human development on a changing planet. Science 347. https://doi.org/10.1126/science.1259855
Tate KR, Ross DJ, Saggar S et al (2007) Methane uptake in soils from *Pinus radiata* plantations, a reverting shrubland and adjacent pastures: effects of land-use change, and soil texture, water and mineral nitrogen. Soil Biol Biochem 39:1437–1449
The USGS Water Science School (2015) Water and temperature. https://www.usgs.gov/special-topic/water-science-school/science/temperature-and-water?qt-science_center_objects=0#qt-science_center_objects Accessed 18th October, 2020
Timmis K, de Vos WM, Ramos JL, et al (2017) The contribution of microbial biotechnology to sustainable development goals. Microb Biotechnol 10:984–987. https://doi.org/10.1111/1751-7915.12818
Toseland A, Daines SJ, Clark JR et al (2013) The impact of temperature on marine phytoplankton resource allocation and metabolism. Nat Clim Chang 3:979–984. https://doi.org/10.1038/nclimate1989
Vezzulli L, Grande C, Reid PC et al (2016) Climate influence on Vibrio and associated human diseases during the past half-century in the coastal North Atlantic. PNAS 113:E5062–E5071. https://doi.org/10.1073/pnas.1609157113
Vries FTD, Liiri ME, Bjørnlund L et al (2012) Land use alters the resistance and resilience of soil food webs to drought. Nat Clim Chang 2:276–280. https://doi.org/10.1038/nclimate1368
Wang S, Spor A, Nidelet T et al (2011) Switch between life history strategies due to changes in glycolytic enzyme gene dosage in *Saccharomyces cerevisiae*. Appl Environ Microbiol 77:452–459. https://doi.org/10.1128/AEM.00808-10
Weng Z (Han), van Zwieten L, Singh BP, et al (2017) Biochar built soil carbon over a decade by stabilizing rhizodeposits. Nat Clim Chang 7:371–376. https://doi.org/10.1038/nclimate3276
Wilkins D, Lauro FM, Williams TJ et al (2013) Biogeographic partitioning of Southern Ocean microorganisms revealed by metagenomics. Environ Microbiol 15:1318–1333. https://doi.org/10.1111/1462-2920.12035
Williams MA (2007) Response of microbial communities to water stress in irrigated and drought-prone tallgrass prairie soils. Soil Biol Biochem 39(11):2750–2757
Youssef NH, Couger MB, McCully AL et al (2015) Assessing the global phylum level diversity within the bacterial domain: a review. J Adv Res 6:269–282. https://doi.org/10.1016/j.jare.2014.10.005
Zhou J, Deng Y, Luo F et al (2011) Phylogenetic molecular ecological network of soil microbial communities in response to elevated CO_2. mBio 2. https://doi.org/10.1128/mBio.00122-11
Zhou J, Xue K, Xie J et al (2012) Microbial mediation of carbon-cycle feedbacks to climate warming. Nat Clim Chang 2:106–110. https://doi.org/10.1038/nclimate1331
Zimmerman AE, Martiny AC, Allison SD (2013) Microdiversity of extracellular enzyme genes among sequenced prokaryotic genomes. ISME J 7:1187–1199. https://doi.org/10.1038/ismej.2012.176

Chapter 3
The Potential Impact of Climate Change on Soil Health, Soil Biota, and Soil Properties: A Review

Shikha Sharma, Arti Mishra, Kartikeya Shukla, Pratiksha Kumari, Tanu Jindal, and Smriti Shukla

Abstract The most complex community biologically is the soil environment, and the organisms present in soil are extremely diverse. They also contribute to a wide range of functions which are essential for sustainable functioning of both managed and natural ecosystems. The land productivity is affected by soil organisms in both direct and indirect ways. The direct effects involve crop yield alterations, whereas indirect effects involve carbon and nitrogen cycle, modification in soil structure, productivity followed by food web interactions. The soil is capable of producing agronomic, economic services along with environmental maintenance. Key indicators of soil health, similar to that of soil quality, are soil structure, soil organic carbon concentration and quality, water retention and intake rate, and soil biodiversity. However, these properties must be maintained and enhanced above threshold level to sustain soil health and productivity.

Enhancing the soil organic carbon pool also improves agroecosystem resilience, eco-efficiency, and adaptation to climate change. Technical potential of soil C sequestration through improvement in soil health is ~3 Pg/year for about 50 years with a drawdown capacity of reducing atmospheric CO_2 concentration by 50 ppm over the twenty-first century.

S. Sharma · P. Kumari · S. Shukla (✉)
Amity Institute of Environmental Toxicology, Safety and Management, Amity University, Noida, India
e-mail: sshukla6@amity.edu

A. Mishra
Amity Institute of Microbial Technology, Amity University, Noida, Uttar Pradesh, India

K. Shukla
Amity Institute of Environmental Sciences, Amity University, Noida, India

T. Jindal
Amity Institute of Environmental Toxicology, Safety and Management, Amity University, Noida, India

Amity Institute of Environmental Sciences, Amity University, Noida, India

D. K. Choudhary et al. (eds.), *Climate Change and the Microbiome*, Soil Biology 63,
https://doi.org/10.1007/978-3-030-76863-8_3

Keywords Soil · Soil health · Biota · Microorganisms · Properties · Climate change

3.1 Introduction

Organic matter along with minerals, liquids, gases, and microorganisms forms the soil. Soil has a vital role in environment such as it acts as plant growth medium, prevents water loss, and provides habitat for numerous microorganisms to live in. Apart from this, the fertility of soil directly impacts the vegetation. The word soil biota means the presence of microorganisms such as bacteria, fungi, and algae along with protozoans, nematodes, insects, earthworms, etc. (Fortuna 2012). A part of soil biota which is microorganisms plays a vital role in biogeochemical transformation's regulations. Major roles and functions performed by soil biota (Fig. 3.1) are soil organic matter's turnover and formation (mineralization and sequestration of C), nutrient cycles, transmission and prevention of disease, pollution decline, and soil structure improvement (Gupta et al. 1997).

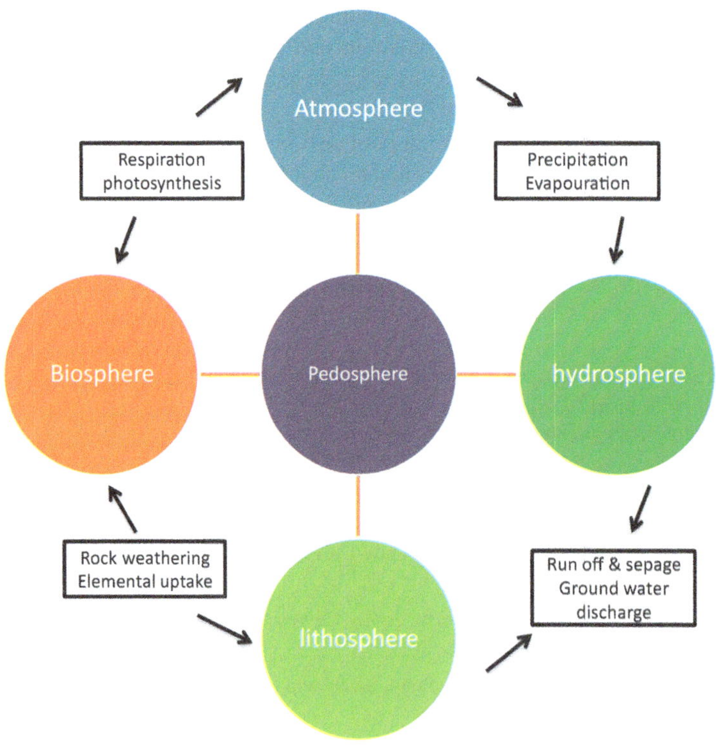

Fig. 3.1 Soil (pedosphere) interactions

The soil structure is influenced by soil organisms as they bind particles present in soil together, thus increasing the number and size of mass that favor microfauna's habitat. The oxidation and reduction of C and N compounds found in soil results in formation of byproducts including greenhouse gases consisting of carbon dioxide (CO_2), methane (CH_4), and nitrous oxide (N_2O).

Soil properties include physical, chemical, and biological characteristics. The physical properties of soil give data in relation to movement of air and water via soil (also with conditions affecting the germination, root growth, and processes of soil erosion). These physical properties are directly linked with soil's chemical and biological properties and might get affected by climatic conditions, position of landscape, and land use pattern (Allen et al. 2011). The climate can affect the physical properties of soil involving its structure, infiltration of water, density, rooting depth, and surface cover. Soil's chemical properties include pH, electrical conductivity, sorption capacity and cation exchange capacity, and available plant nutrients. The pH of soil is the main soil health indicator and is influenced by weathering time, vegetation, and climate and helps in identification of change in soil's biological and chemical functions (acidification, salinization, crop performance, nutrient availability along with biological activity). Soil health assessment is based on biological properties, or soil biota comprises soil organic matter, soil carbon, light fraction and macro-organic matter, potential mineralization of C and N, soil respiration, soil microbial biomass, and enzymatic activity. According to USDA, soil health or soil quality is soil's capacity to function properly to sustain life (plants, animals, humans). To nourish soil and its health, it must be provided with food and shelter along with water as it contains microorganisms and enhances its function for production of food and fiber. Physical, chemical, and biological characteristics of soil act as soil health indicators (Allen et al. 2011), and they may get affected with alteration in climate and surrounding environment.

Climate change can be defined as a change or shift in patterns of weather which directly affects production of food (agricultural impact), increasing sea levels (causing floods). Climate change is different than global warming (increase in temperature due to GHG). Climate change can show its potential impact on almost everything. According to UN, these impacts can include sea ice loss, rise in sea level (1–8 feet by 2100), intense heat waves, precipitate change patterns, stronger hurricanes, and more number of droughts (as summers are now longer than usual). In the next century, the temperature will rise by 2.5 to 10 degrees Fahrenheit majorly due to greenhouse gases (generated by anthropogenic activities) forecasted by IPCC.

The current chapter will focus on impacts of climate change on soil biota, soil health, and soil properties with its relative impact on food, agriculture, and ecosystems.

3.2 Soil Biota

A broad range of ecosystems gets contribution from soil organisms that play a major role in sustaining proper functioning of both natural and man-made ecosystems (Barrios 2007). The productivity of land gets affected by soil community either directly (poor crop yield, etc.) or indirectly (disturbances in C and other nutrient cycles, food web interactions). The plants may show negative and positive effects driven by soil microbe's pathogenic effects, mutualism of root and fungus (Mitchell and Power 2003; Packer and Clay 2000; Newsham et al. 1994). This generates feedback interactions of plant to microbial community at their root sides (Callaway et al. 2004). The feedback interactions are of two types: positive and negative feedback interactions. The accumulation of microbes by plant species is done near roots, it implements positive or beneficiary effects on plants (mycorrhizal fungi, nitrogen fixers, etc.), and this is called positive feedback but also causes local community diversity loss (Bever et al. 1997; Bever 2002), whereas negative feedback is observed when accumulation of pathogenic microbes is found in rhizospheres of plant species and helps in increasing community diversity. It is found that 1 g of soil contains 200 million fungal hyphae and 1 billion bacterial cells (Bender et al. 2016).

3.2.1 Impact of Climate Change on Biota

The increasing temperature or climate change can show changes in carbon cycling significantly as climate change directly affects the microbial breakdown in organic matter of soil, respiration of soil, and atmospheric greenhouse gas feedbacks (Bardgett et al. 2008; Craine et al. 2010; Schindlbacher et al. 2011). Climatic factors like increase in concentration of CO_2, rise in temperature, and change in precipitation patterns contribute to climate change, and these cause disturbances in soil microbial activity also known as soil microbiota (Mekala and Polepongu 2019). The activity of microbe might get enhanced with temperature change and may get deleted with precipitation change. These changes also show effects on plant growth. The related plant responses drive carbon entering type and amount into soil. The increase in temperature creates conditions like drought which causes stressful conditions in soil. Briones et al. in 2014 stated that drought enhances sensitivity of temperature of bacterial and fungal groups. The changes in soil moisture result in shifting of fungal communities, whereas it remains constant for bacteria (Mekala and Polepongu 2019). Arbuscular mycorrhizal fungi (AMF) colonization is reduced in drought conditions which particularly depend on strains in certain cases (Davies et al. 2002).

3.2.1.1 Impact of Alterations in CO_2 Levels on Soil Biota

Elevation in CO_2 will cause indirect effect to microbes via advanced root growth and rhizodeposition rates. The change of C availability in soil in relation to increased CO_2 concentrations is first seen through the microbial activity and nutrient availability in rhizosphere (Zak et al. 1993). Elevation in allocation of C to roots and elevated growth is often seen in plants grown with increased CO_2 concentrations (Curtis et al. 1990; Norby et al. 1986).

According to Smith and Paul in 1990, the magnitudes of C input to soil are regulated by plant production, and thus C is limited in soil for microbial populations which elevates allocation of microbial biomass. Organic acids play a vital role in the shape of soil bacterial communities, and hence impact on growth is observed with climate change (Shi et al. 2011). Colonization of plant growth-promoting fungi (PGPM) is observed to increase with increase in CO_2 levels. AMF (arbuscular mycorrhizal fungi) signifies plant's nutrient uptake and also provides other benefits for rhizosphere compounds' (carbohydrates) reciprocity. Hence, changes are observed in AMF community compositions because of elevated CO_2 levels (Klironomos et al. 2005).

The biomass showed no effect in response to increased CO_2 levels in Florida Scrub Ecosystem (Schortemeyer et al. 2000). Also increased CO_2 levels showed no effect on microbial biomass C, ergosterol elements, as well as fungal hyphae in conditions of lower nutrient supply in artificial tropical ecosystem (Zak et al. 2000). The change in climatic conditions results in induced changes in microbial activities by microbial communities. This might lead to formation of communities (that are already altered) and defining different responses by plants. The effects will affect diversity of plants and soil biota functioning. The higher levels of CO_2 can form link with symbiosis of plant and fungi, leading to higher frequency of endophyte infection but with limited production of toxin (Brosi et al. 2009).

3.2.1.2 Impact of Drought and Temperature on Soil Biota

Abiotic stresses on ecosystem are expected to increase due to climate change by way of extreme weather events causing intense drought and rainfall patterns (Meisner et al. 2013). Drought is a natural calamity which occurs when the availability of water below the ground is less, resulting in certain dryness and cracks above the ground. Temperature elevations or rise in temperature is the increase of atmospheric temperature. Temperature elevations may create a situation like drought. Drought and temperature elevations, together, eventually affect soil biota and its health and properties. Fungal and bacterial groups in soil have certain differential temperature sensitivity, and this sensitivity is amplified during drought conditions (Briones et al. 2014). The microbial composition also gets influenced by climate change (Mekala and Polepongu 2019). The microbial activities are seen to be reduced during drought, and when rainfall occurs, a small rise in activity occurs (Birch 1958; Butterly et al. 2009; Fierer and Schimel 2002).

The reduction of AMF colonization is observed during drought but this also depends upon the strains; Davies et al. in 2002 reported through a demonstration that *Glomus* sp. strain ZAC-19 showed enhanced arbuscular formation and hyphae development, whereas *Glomus fasciculatum* strain showed reduced colonization as an impact of drought. Soil respiration gets impacted indirectly due to change in climate which influences productivity and structure of plant community; hence, the C input's quality and quantity which is entering soil are determined (Whitaker et al. 2014).

Drought results in lower plant growths (in roots), and this increases plant's susceptibility to pathogenic attacks heading toward photosynthetic allocation changes in rhizosphere and ectomycorrhizal mycelium (ECM) formation alterations. This also affects the quality and quantity of litter below the ground with organic matter accumulation in soil (Mekala and Polepongu 2019).

Water stress also causes alterations in lignification of plant cell walls (Henry et al. 2005) and enhances productivity of grassland. Soil moisture along with high CO_2 levels causes drift in ammonium-oxidizing bacteria abundance and nitrogen cycle. Soil respiration might increase with temperature and elevated CO_2. The plant mediated indirect effect on activity of microbes and respiration of soil are non-understandable and this reflects knowledge gap in determining the response of terrestrial C cycling to future climate change (Whitaker et al. 2014).

Temperature change is the increase of the atmospheric temperature by $1 \pm \underline{2}$ °C. This also affects the availability of moisture in soil and may create a situation like drought and microbial communities may respond to temperature alterations. The temperature elevations can disturb various regulations of cycles. The temperature rise is regulated with increase in greenhouse gases such as carbon dioxide, methane, water vapor, nitrous oxide, and ozone. Drought as well as extreme rainfall can directly influence soil biota. Climate change promotes exotics (have specific traits like faster growth) and demotes native plant species via different mechanisms (Meisner et al. 2013). The temperature change globally is causing microbial soil respiration rate alterations due to the presence of soil microorganisms (temperature-mediated processes). Temperature change also leads to increase in decomposition of soil matter and its growth and respiration (Bradford et al. 2008). The microbial community composition is altered initially during experimental temperature, and also abundance of both gram-positive and gram-negative bacteria is observed to be shifted (Zogg et al. 1997). Temperature elevations are also seen to support the growth of plant and enhance tolerance to both biotic and abiotic stresses (Bent 2006). Some of the microbes entering host's root system enhance beneficial effects with endophytic lifestyle.

3.3 Soil Health and Properties

Soil health is an integral concept of sustainable agriculture and all the properties of soil acts as an indicator of soil health. Soil properties are divided into three main divisions which are physical properties, chemical properties, and biological properties relating all the functional processes of soil (Fig. 3.2). These properties or attributes are affected by climate change and management. Soil organic matter consists of carbon and nitrogen as major components (Brevik 2013). For many soil properties, soil structure formation, water holding capacity, exchange capacity of cations, and nutrient supply to ecosystem, organic matter is the most important (Brevik and Verheye 2009; Brevik and Burgess 2013). The climate change effects involve (potentially) changes in C and N cycles and influence soils. With change in CO_2 concentration, CO_2 fertilization effect is seen which enhances the growth of plants. However, with increased ozone concentration, the plant growth is restricted, and the enhanced effect is nullified (Long et al. 2005), and with limitation of nitrogen availability, plant growth is affected negatively (Brevik 2013). The change in rainfall affects soil moisture and crop stand and its germination. When soil moisture is adequate (like in Punjab), higher nitrogen is seen with yield benefits. The limitation in maximum crop yield is seen despite all the factors being normal.

3.3.1 Impact of Climate Change on Physical Properties of Soil

Physical properties of soil give water and air movement information and details along with the conditions that affect germination, root growth, and soil erosion processes. And thus, the physical properties form the foundation for other processes which are chemical and biological processes (Allen et al. 2011), and all these properties determines soil health and can also be termed as soil health indicators. The physical properties of soil are soil structure, water infiltration, bulk density, rooting depth, and soil surface cover, whereas soil physical processes are gains

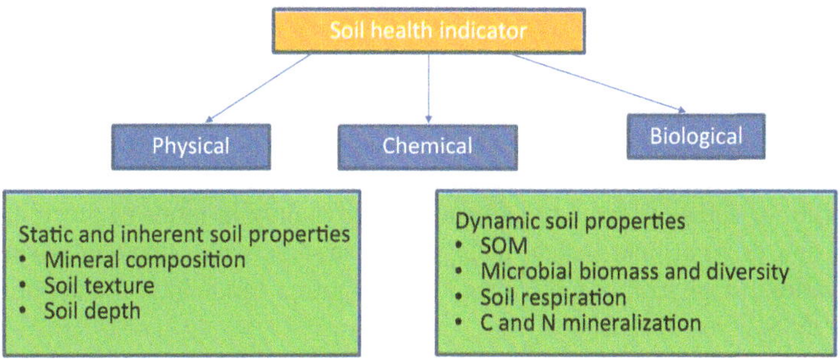

Fig. 3.2 Indicators of soil health

(addition of organic matter, oxygen, water via oxidation, and hydration cycles), losses (materials running off the surface), transfers, movements (air and water), organic matter, soluble salts, carbonates, silicate clay minerals, and silica (Patil and Lamnganbi 2018).

3.3.1.1 Impact of Climate Change on Soil Structure

Soil structure combines aggregate stability and porosity of soil together. Soil structure is defined as the primary and secondary particle arrangement and organization in soil mass (Patil and Lamnganbi 2018). It determines soil aggregate resistance against some external change such as climate change involving high rainfall, cultivation, etc. It is also a useful health indicator because its role involves maintaining major soil ecosystem function such as accumulation of organic carbon, infiltrating capability, water storage and movement, and activities of roots and microbial community (Allen et al. 2011). The amount and quality of organic matter, inorganic components in soil matrix, cultivating techniques and methods, and natural processes influence the soil's structure nature and quality. The declining levels of soil matter decrease soil aggregate stability and rates of infiltration while increasing compaction susceptibility (Bot and Benites 2005). The measurement of aggregate stability is done in different ways, but to measure the range of soil health monitoring framework (in climate changes), standardized procedures are required (Dalal and Moloney 2000; Sanchis et al. 2008). Porosity of soil is a basic void space measure for a material as the ratio of volume of voids and total volume. Distribution of pore size gives direct and quantitative estimate for soil's ability to store water and air. Development of root and activities of soil enzymes are in close proximity to soil porosity and its pore size distribution, and these are all affected by climate changes along with soil functions. These factors are also the governing factors for methane fluxes, emission and uptake, and nitrous oxide emissions from soil. The texture (sand silt and clay) of soil and soil processes shows direct impact of climate changes.

3.3.1.2 Impact of Climate Change on Soil Infiltration

The rate of water which enters the soil surface and goes to soil depth (Dalal and Moloney 2000) and availability of water to plants for their growth and processes are controlled by soil properties (porosity, field capacity, etc.). This is used as a section for soil health tests which helps in assessing management impacts. The water availability of soil is affected by climate change mainly in terms of precipitation or rainfall and drought, and hence management strategies like planting cover crops, conservation tillage, and organic matter incorporation can help in impact mitigation (Lal 1995; Sanchis et al. 2008).

3.3.1.3 Impact of Climate Change on Soil Bulk Density

Bulk density in agricultural systems is assessed on a routine basis for soil compactness characterizing in response to use of land and its management (Håkansson and Lipiec 2000). Bulk density is an indicator of soil functions like aeration, infiltration, etc. and is used to assess the soil health. Soil organic matter is obstructively correlated with soil bulk density (Allen et al. 2011), and with elevated temperature, increased loss of organic carbon is manifested (Davidson and Janssens 2006) which probably elevates bulk density, and soil gets prone towards compaction as a result of management of land and climatic alterations.

3.3.1.4 Impact of Climate Change on Rooting Depth

Any change in rooting depth is seen to affect plant and its water capacity, salinity, organic matter content, and subsoil directly and provides with indication to physiochemical constraints in a soil profile. When drought-like conditions appear, then more impact of constraints (salinity, chloride concentrations) is seen on plant water that is available along with its productivity (Patil and Lamnganbi 2018). The rooting depth is included as one of the parameters of soil health by Birkás et al. in 2008 to monitor soil conditions and growth of plant under unfavorable conditions like drought and alter rainfall which might indicate mitigation steps via rooting depth alterations (Allen et al. 2011).

3.3.1.5 Impact of Climate Change on Soil Surface Cover

Most of the important ecological functions are provided by soil surface cover like protecting soil surface from dissipating raindrop impact energy, stabilization of soil, erodible surface area reduction, nutrient and water retention, etc. including C and N fixations (Patil and Lamnganbi 2018). The conditions of soil (soil crust, soil seal formation) which are related with sodicity are some of the soil health indicators that can be used to understand the impact of climate change along with soil physical management, input of organic matter (linked with erosion), runoffs, etc. Soil crust's formation and seal formation affect processes performed by soil (involving infiltration, oxygen diffusion, surface water evaporation, and wind erosion).

3.3.1.6 Impact of Climate Change on Soil Temperature

Soil temperature exists in close proximity with air temperature, and with the change in climate, it is seen significantly affected. However, when air temperature increases, so does the soil temperature (Karmakar et al. 2016). Soil temperature is followed by various processes like gains and losses of radiations, evaporation, heat conduction, transfer via gas, and water movement. In 2011, Qian, B. et al. studied in Canada about trends related with soil and air temperature, and it was found that the relation between the two caused lower snow cover depth during winter and spring seasons when air temperature is increased. Soil temperature is directly linked with most of the soil processes which means if soil temperature increases, soil processes will increase (fast decomposition of organic matter, high microbial activity, enhanced nitrification, etc.) and there will be changes in vegetation types.

3.3.2 Impact of Climate Change on Chemical Properties of Soil

3.3.2.1 Impact of Climate Change on pH of Soil

pH is one of the most important and dominant characteristics of chemical properties or an indicator of weathering time, climate, and vegetation and identifies a range of functions such as acidification, salinization, performance of crop, and available nutrient. Hence, it is used as a parameter to assess soil health, land use change, and impact of agricultural practices (Gil et al. 2009; Idowu et al. 2009; Pattison et al. 2008; Schindelbeck et al. 2008). However, soil pH of most soil are not affected by temperature elevations, CO_2 fertilizations, precipitation changes, and deposition of atmospheric N, but these all can affect organic matter status, nutrient cycling (C and N), moisture and water content, as well as plant productivity, which eventually changes soil pH (Allen et al. 2011). The rainfall increase can cause increased leaching, nutrient loss, and high acidification, and the impact will depend upon intensity of change in climate. The pH (high salinization and alkalization) change is observed in areas with increased evaporation or decreased rainfall (Várallyay 1994). As capillary rise dominates, transient salinity is increased which allows salts to enter the root zone. Between different intervals of time, limited leaching is there, and when drying of subsoil increases, salt concentration in soil solution also increases. Reduced rainfalls also lead to reduction in groundwater levels and show significant impacts in semiarid zones. The salinity impact is reduced on plant growth with increase in CO_2 concentrations (Karmakar et al. 2016). The factors of management of land like drainage (man-made) will regulate how predicted climate change impacts will affect related landscapes.

3.3.2.2 Impact of Climate Change on Electrical Conductivity

EC or electrical conductivity is a salt concentration measure and a worthy indicator for soil health. Characteristics like salinity, performance of crop, nutrient cycling, and biological activity with pH are determined with soil electrical conductivity and inform declining soil structure (mostly in sodic soils) (Patil and Lamnganbi 2018). During crop management practices, EC is used to carry information to biological quality. As temperature is elevated and rainfall is declined, Smith et al. in 2002 found using elevated gradient in semiarid environment, pH is increased, and EC is decreased. Also, in 2001, Pariente examined salt (soluble) concentration dynamics from four different regions which were Mediterranean, semi-arid, mildly arid, and arid, and no linear relationship was found between rainfall and salt concentration. The climate change scenarios cause impact on N and C cycling (both are linked to one another) and other nutrient cycling like phosphorus, sulfur, etc.

3.3.2.3 Impact of Climate Change on Soil Sorption and Cation Exchange and Plants' Nutrient Availability

This is an important property for major nutrient cation retention like Ca^{2+}, Mg^{2+}, K^{+}, and potentially toxic cation's (Al^{3+}, Mn^{3+}) immobilization. This provides a clue about soil's capacity for plant's growth support and also helps in determining critical values of EHA (environmental hazard assessment). This property is also used as soil health parameter and gives information about nutrient absorption capacity and pesticide and chemical absorption capacity of soil (Ross et al. 2008). When atmospheric temperature is elevated, decomposition and loss of soil organic matter are increased and lead to cation exchange capacity loss with increased leaching as a response against high intense precipitation.

3.3.3 Impact of Climate Change on Biological Properties of Soil

The soil biota consists of all the biological properties, and under climate change, it is an important component for soil health. The major biological properties or indicators of soil are soil organic matter with its constituents, carbon of soil, soil respiration, and microbial mass. The impact of climate change on each is discussed below:

3.3.3.1 Impact of Climate Change on Soil Organic Matter

Soil organic matter (SOM) consists of a broad variety of both living and nonliving components, and hence it is the most complex and heterogeneous component differing in its properties and related functions. It is the source of C and N, regulates phosphorus and sulfur cycling, and provides habitat for micro- and macrofauna involving microbes. SOC or soil organic carbon is about 50% total SOM and organic N and C, and important plant productivity nutrients together form soil organic matter (SOM) (Allen et al. 2011). The SOM indicators are widely used in assessment of climate change experimentally, but the responses are controversial. Decline in SOM causes fertility and biodiversity decrease, soil structure loss, reduction in water

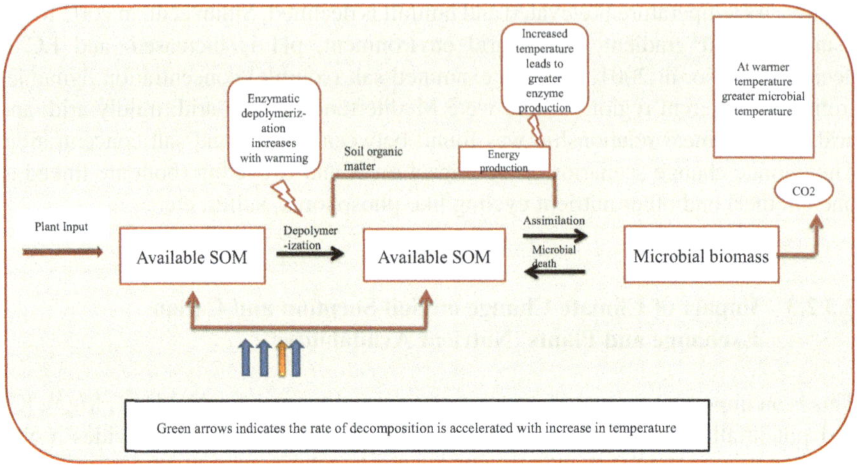

Fig. 3.3 SOM dynamics during climate change

holding capability, high erosion risk, high bulk density, and finally compaction (Patil and Lamnganbi 2018). Increased or buildup of SOM can help in absorption of atmospheric CO_2 by taking land use and management practices to mitigate the climatic impact on soil. However, increased temperature causes high SOM decomposition. High productivity of plant and organic matter input may be seen with temperature elevations, rainfall or precipitation, CO_2 fertilization, and C and N (atmospheric) which eventually increases soil organic matter. Flooding impacts of extreme rainfall events can be mitigated by increasing water storage during drought events; thus, soil resilience is enhanced (Fig. 3.3).

3.3.3.2 Impact of Climate Change on Soil Carbon

Soil carbon is a change indicator and is linked with performance of ecosystems as it is also a part of nutrient cycling. Potential mineralizable C and N indicate the quality of soil organic matter and acts as an interface among autotroph and heterotroph microbes in the process of nutrient cycling. Soil carbon is predictively decreased during elevated temperatures and generally gets modified with anthropogenic activities and together impacts the processes and functions of soil eventually affecting the soil health (Lal et al. 2007). Mineralization and decomposition are stimulated by high temperature and altered rainfalls which cause reduction in biomass accumulation, soil carbon depletion, and C/N ratio decline (Rosenzweig and Hillel 2000; Anderson 1992; Lal 2004). Efficiency of plant water used is increased with increase in CO_2 (atmospheric) concentration, and hence production of biomass per mm is increased (Kimball 2003). Reduction of organic carbon is seen in conditions where water deficit is increased. Increased water deficit causes more decomposition rate than that of NPP (net primary production) (Fig. 3.4).

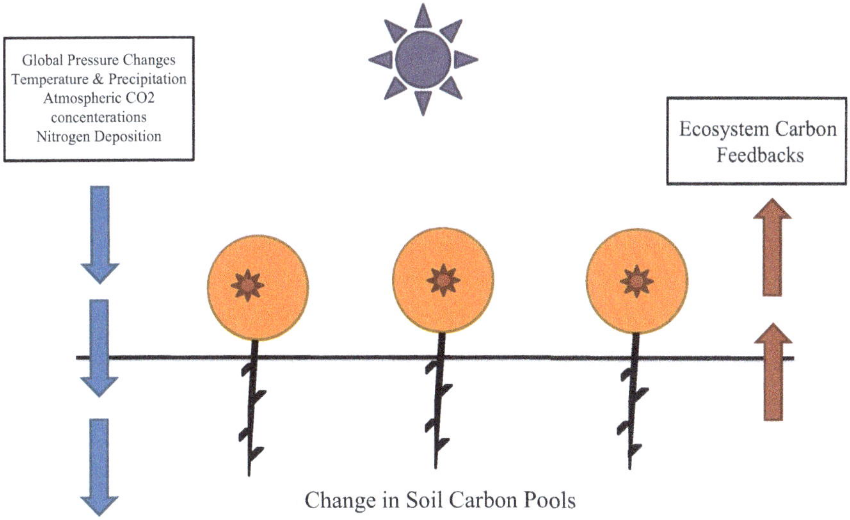

Fig. 3.4 Elevated temperature and SOC

3.3.3.3 Impact of Climate Change on Soil Flora and Fauna

Soil flora and fauna consist of hundreds and thousands of species that are found in 1 m^2 of soil area. These play a major role in retention, plant remain incorporations, breakdowns, and nutrient cycling and influence the structure and porosity of soil. Soil flora and fauna re not directly affected by climate change especially with temperature elevations as they have broad temperature optimum (Karmakar et al. 2016). When CO_2 level is increased, it causes increased growth of plant and carbon allocation down the ground which exhibits microbes to enhance rates of fixation, immobilization, and denitrification of nitrogen, elevated associations of mycorrhiza, soil aggregation elevations, and high mineral weathering. Soil microbes are affected in both ways: direct and indirect. Direct effects influence microbes, and indirect effects involve changes in plant productivity and diversity. This affects physiochemical conditions of soil, carbon supply, soil structure and activity, decomposition, and carbon release. The first mechanism that concerns the elevated CO_2 concentrations on soil is via photosynthesis along with photosynthate carbon transfer to roots and (mycorrhizal) fungi (Johnson et al. 2005; Högberg and Read 2006; Keel et al. 2006) and heterotrophic microorganisms. The elevated levels of CO_2 cause increased plant growth and photosynthesis (under nutrient-rich conditions) and carbon flux to roots and microbes by root exudation of organic acids, sugars, and amino acids.

3.3.3.4 Impact of Climate Change on Soil Respiration

It is a biological indicator of soil health and is related with soil organic matter, and determination is made as production and consumption of CO_2 and O_2, respectively. The nature of relationship between soil respiration, climate change, and global C cycle is currently under scientific debate. However, studies have shown soil respiration as responsive against climatic changes like rainfall and its timing, and it is predicted to alter on the basis of global and regional climate models (Chou et al. 2008).

3.3.3.5 Impact of Climate Change on Soil Microbial Biomass

Microbial biomass is soil organic matter's living component and is a pool of carbon in soil. It is a sensitive change indicator in processes of soil linking with soil energy and nutrient dynamics. The change in environment such as warming temperatures causes biomass decline, and it has been studied with the help of experiments by Rinnan R. et al. in 2007 and when it is combined with isotope 13C labeling technique, the shift in micro biomass 13C provides changes measure in carbon processes as response to climate change and land use change than total microbial biomass (Paterson et al. 2009). Due to differences in physiology, sensitivity to temperature, and growth rates of soil community, the effects of climate change is differentiable. In a temperate forest, warming by 58 °C can alter bacterial abundances and enhance the fungal ratio in the community, and these might lead to ecosystem's functional changes such as nitrogen fixation, nitrification, and denitrification, methanogenesis, etc.

3.3.3.6 Impact of Climate Change on Microbial and Metabolic Quotient

Physiological performance indicators such as respiration, growth and death, uptake of C, etc., against total organic carbon and total biomass C of microbes per unit of time are termed as microbial quotient and metabolic quotient (CO_2-C respired per hour per microbial C), respectively. It is used for management impact assessment of organic carbon dynamics (Moscatelli et al. 2005). Both microbial quotient and metabolic quotient are sensitive to climate change like elevated CO_2 and nitrogen fertilization, and this was reported on research by Moscatelli MC et al. in 2005 under controlled air CO_2 enrichment and nitrogen fertilization.

3.3.3.7 Impact of Climate Change on Soil's Enzymatic Activity

Activities of soil are linked with nutrient cycling and these are measured easily. These enzymes respond to changes such as in soil management. Enzymes integrate information of microbial status with physiochemical conditions of soil. Hence, it is used as an indicator of change in plant-soil system. An experiment done by Dorodnikov M et al. in 2009 showed that when the quality and quantity of carbon (belowground) input through plants are altered, elevation in CO_2 stimulates (stimulation extent is dependent on soil aggregate size) enzymatic activities, microbe abundance, and carbon turnover. This affects the normal functioning of microbial community in soil. Extracellular enzymes that help in decomposing SOC and are involved in nutrient cycling get affected with deposition of atmospheric N (Allen et al. 2011). Involvement processes of soil enzymatic activities in turning over organic carbon, nutrient cycling, emissions of GHG and their responses to climate change and land use change are still not well-known.

3.4 Conclusion

The change in climate such as elevated temperatures and changes in CO_2 concentrations and rainfall patterns may either affect the soil biota, health, and properties or enhance them. The degradation of land is also related with climate change's adverse effects. The impact of land degradation is mitigated by conservation farming. To overcome the impact on physical, chemical, and biological properties of soil, integrated nutrient management, crop improvement, site management practices of soil, and water conservation must be identified. The impact of soil is possibly understood using indicators or properties of soil to ecological functions, and continuous monitoring against climate change is done. The properties of soil which get affected are soil biota, microbial community, soil aggregate, SOM, SOC, nutrient cycling, microbial biomass, and diversity. It is however not always negatively impacted but some climate change may also benefit the soil. Climate change shows interlinked challenges which have not been anticipated in the last century. The main changes in soil like improvement in fertility, soil physical conditions, soil-forming processes, etc. are a result of climatic change. These changes are also mixed with mineral composition, organic matter content, and structural stability. The impact is monitored using various soil health indicators continuously.

References

Allen DE, Singh BP, Dalal RC (2011) Soil health indicators under climate change: a review of current knowledge. In: Soil health and climate change. Springer, Berlin, pp 25–45

Anderson JM (1992) Responses of soils to climate change. In: Advances in ecological research, vol 22. Academic, pp 163–210

Bardgett RD, Freeman C, Ostle NJ (2008) Microbial contributions to climate change through carbon cycle feedbacks. ISME J 2(8):805–814

Barrios E (2007) Soil biota, ecosystem services and land productivity. Ecol Econ 64(2):269–285

Bender SF, Wagg C, van der Heijden MG (2016) An underground revolution: biodiversity and soil ecological engineering for agricultural sustainability. Trends Ecol Evolut 31:440–452

Bent E (2006) Induced systemic resistance mediated by plant growth-promoting rhizobacteria (PGPR) and fungi (PGPF). In: Tuzun S, Bent E (eds) Multigenic and induced systemic resistance in plants. Springer, Berlin, pp 225–258

Bever JD (2002) Negative feedback within a mutualism: host–specific growth of mycorrhizal fungi reduces plant benefit. Proc R Soc Lond Ser B: Biol Sci 269(1509):2595–2601

Bever JD, Westover KM, Antonovics J (1997) Incorporating the soil community into plant population dynamics: the utility of the feedback approach. J Ecol 85:561–573

Birch HF (1958) The effect of soil drying on humus decomposition and nitrogen availability. Plant Soil 10(1):9–31

Birkás M, Stingli A, Szemők A, Kalmár T, Bottlik L (2008) Soil condition and plant interrelations in dry years. Cereal Res Commun 36(Suppl):15–18

Bot A, Benites J (2005) The importance of soil organic matter: key to drought-resistant soil and sustained food production (No. 80). Food & Agriculture Org

Bradford MA, Davies CA, Frey SD, Maddox TR, Melillo JM, Mohan JE, Reynolds JF, Treseder KK, Wallenstein MD (2008) Thermal adaptation of soil microbial respiration to elevated temperature. Ecol Lett 11:1316–1327

Brevik EC (2013) The potential impact of climate change on soil properties and processes and corresponding influence on food security. Agriculture 3(3):398–417

Brevik EC, Burgess LC (2013) The 2012 fungal meningitis outbreak in the United States: connections between soils and human health. Soil Horizons 54(2):1–4

Brevik EC, Verheye W (2009) Soil health and productivity. Soils, plant growth and crop production. In: Encyclopedia of life support systems (EOLSS), developed under the auspices of the UNESCO. EOLSS, Oxford

Briones MJI, McNamara NP, Poskitt J, Crow SE, Ostle NJ (2014) Interactive biotic and abiotic regulators of soil carbon cycling: evidence from controlled climate experiments on peatland and boreal soils. Glob Chang Biol 20:2971–2982

Brosi GB, Nelson JA, McCulley RL, Classen AT, Norby R (2009) PS 45-40: global change factors interact with fungal endophyte symbiosis to determine tall fescue litter chemistry. In: The 94th ESA annual meeting

Butterly CR, Bunemann EK, McNeill AM, Baldock JA, Marschner P (2009) Carbon pulses but not phosphorus pulses are related to decreases in microbial biomass during repeated drying and rewetting of soils. Soil Biol Biochem 41(7):1406–1416

Callaway RM, Thelen GC, Rodriguez A, Holben WE (2004) Soil biota and exotic plant invasion. Nature 427(6976):731–733

Chou WW, Silver WL, Jackson RD, Thompson AW, ALLEN-DIAZ BARBARA (2008) The sensitivity of annual grassland carbon cycling to the quantity and timing of rainfall. Glob Chang Biol 14(6):1382–1394

Craine JM, Fierer N, McLauchlan KK (2010) Widespread coupling between the rate and temperature sensitivity of organic matter decay. Nat Geosci 3(12):854–857

Curtis PS, Baldmann LM, Drake BG, Whigman DF (1990) Elevated atmospheric CO_2 effects on belowground processes in C_3 and C_4 estuarine communities. Ecology 71:2001–2006

Dalal RC, Moloney D (2000) Sustainability indicators of soil health and biodiversity. Management for sustainable ecosystems. Centre for Conservation Biology, Brisbane, pp 101–108

Davidson EA, Janssens IA (2006) Temperature sensitivity of soil carbon decomposition and feedbacks to climate change. Nature 440:165–173

Davies FT Jr, Olalde-Portugal V, Aguilera-Gómez L, Alvarado MJ, Ferrera-Cerrato RC, Boutton TW (2002) Alleviation of drought stress of Chile ancho pepper (*Capsicum annuum* L. cv. San Luis) with arbuscular mycorrhiza indigenous to Mexico. Sci Hortic 92:347–359

Dorodnikov M, Blagodatskaya E, Blagodatsky S, Marhan S, Fangmeier A, Kuzyakov Y (2009) Stimulation of microbial extracellular enzyme activities by elevated CO_2 depends on soil aggregate size. Glob Chang Biol 15(6):1603–1614

Fierer N, Schimel JP (2002) Effects of drying-rewetting frequency on soil carbon and nitrogen transformations. Soil Biol Biochem 34(6):777–787
Fortuna A (2012) The soil biota. Nat Educ Knowl 3(10):1
Gil SV, Meriles J, Conforto C, Figoni G, Basanta M, Lovera E, March GJ (2009) Field assessment of soil biological and chemical quality in response to crop management practices. World J Microbiol Biotechnol 25(3):439–448
Gupta VVSR, Neate SM, Leonard E, Cooperative Research Centre for Soil and Land Management (Australia) (1997) Life in the soil: the relationship between agriculture and soil organisms. Cooperative research Centre for Soil & Land Management, Glen Osmond
Håkansson I, Lipiec J (2000) A review of the usefulness of relative bulk density values in studies of soil structure and compaction. Soil Tillage Res 53(2):71–85
Henry HA, Clelad EE, Field CB, Vitousek PM (2005) Interactive effects of CO_2, N deposition, and climate change on plant litter quality in a California annual grassland. Oecologia 142:465–473
Högberg P, Read DJ (2006) Towards a more plant physiological perspective on soil ecology. Trends Ecol Evol 21(10):548–554
https://climate.nasa.gov/
https://climate.nasa.gov/effects/
https://www.un.org/en/sections/issues-depth/climate-change/
Idowu OJ, Van Es HM, Abawi GS, Wolfe DW, Schindelbeck RR, Moebius-Clune BN, Gugino BK (2009) Use of an integrative soil health test for evaluation of soil management impacts. Renew Agric Food Syst 24:214–224
Johnson D, Krsek M, Wellington EM, Stott AW, Cole L, Bardgett RD, Read DJ, Leake JR (2005) Soil invertebrates disrupt carbon flow through fungal networks. Science 309(5737):1047–1047
Karmakar R, Das I, Dutta D, Rakshit A (2016) Potential effects of climate change on soil properties: a review. Sci Int 4(2):51–73
Keel SG, Siegwolf RT, Körner C (2006) Canopy CO_2 enrichment permits tracing the fate of recently assimilated carbon in a mature deciduous forest. New Phytol 172(2):319–329
Kimball B (2003) Response of plants to elevated atmospheric CO_2. In: International congress of plant physiology
Klironomos JN, Allen MF, Rillig MC, Piotrowski J, MakvandiNejad S, Wolfe BE, Powell JR (2005) Abrupt rise in atmospheric CO_2 overestimates community response in a model plant–soil system. Nature 433:621–624
Lal R (1995) Tillage systems in the tropics: management options and sustainability implications (No. 71). Food & Agriculture Org
Lal R (2004) Soil carbon sequestration to mitigate climate change. Geoderma 123(1–2):1–22
Lal R, Follett RF, Stewart BA, Kimble JM (2007) Soil carbon sequestration to mitigate climate change and advance food security. Soil Sci 172(12):943–956
Long SP, Ainsworth EA, Leakey AD, Morgan PB (2005) Global food insecurity. Treatment of major food crops with elevated carbon dioxide or ozone under large-scale fully open-air conditions suggests recent models may have overestimated future yields. Philos Trans R Soc B: Biol Sci 360(1463):2011–2020
Meisner A, De Deyn GB, de Boer W, van der Putten WH (2013) Soil biotic legacy effects of extreme weather events influence plant invasiveness. Proc Natl Acad Sci 110(24):9835–9838
Mekala S, Polepongu S (2019) Impact of climate change on soil microbial community. In: Plant biotic interactions. Springer, Cham, pp 31–41
Mitchell CE, Power AG (2003) Release of invasive plants from fungal and viral pathogens. Nature 421(6923):625–627
Moscatelli MC, Lagomarsino A, Marinari S, De Angelis P, Grego S (2005) Soil microbial indices as bioindicators of environmental changes in a poplar plantation. Ecol Indic 5(3):171–179
Newsham KK, Fitter AH, Watkinson AR (1994) Root pathogenic and arbuscular mycorrhizal fungi determine fecundity of asymptomatic plants in the field. J Ecol 805–814
Norby RJ, O'Neill EG, Luxmoore RJ (1986) Effects of CO_2 enrichment on the growth and mineral nutrition of *Quercus alba* seedlings in nutrient-poor soil. Plant Physiol 82:83–89

Packer A, Clay K (2000) Soil pathogens and spatial patterns of seedling mortality in a temperate tree. Nature 404(6775):278–281

Paterson E, Midwood AJ, Millard P (2009) Through the eye of the needle: a review of isotope approaches to quantify microbial processes mediating soil carbon balance. New Phytol 184 (1):19–33

Patil A, Lamnganbi M (2018) Impact of climate change on soil health: a review. Int J Chem Stud 6 (3):2399–2404

Pattison AB, Moody PW, Badcock KA, Smith LJ, Armour JA, Rasiah V, Cobond JA, Gulino L-M, Mayer R (2008) Development of key soil health indicators for the Australian banana industry. Appl Soil Ecol 40(1):155–164

Qian B, Gregorich EG, Gameda S, Hopkins DW, Wang XL (2011) Observed soil temperature trends associated with climate change in Canada. J Geophys Res Atmos 116(D2)

Rinnan R, Michelsen A, Bååth E, Jonasson S (2007) Fifteen years of climate change manipulations alter soil microbial communities in a subarctic heath ecosystem. Glob Chang Biol 13(1):28–39

Rosenzweig C, Hillel D (2000) Soils and global climate change: challenges and opportunities. Soil Sci 165(1):47–56

Ross DS, Matschonat G, Skyllberg U (2008) Cation exchange in forest soils: the need for a new perspective. Eur J Soil Sci 59(6):1141–1159

Sanchis MS, Torri D, Borselli L, Poesen J (2008) Climate effects on soil erodibility. Earth Surf Process Landf 33(7):1082–1097

Schindelbeck RR, van Es HM, Abawi GS, Wolfe DW, Whitlow TL, Gugino BK, Idowu OJ, Moebius-Clune BN (2008) Comprehensive assessment of soil quality for landscape and urban management. Landsc Urban Plan 88(2–4):73–80

Schindlbacher A, Rodler A, Kuffner M, Kitzler B, Sessitsch A, Zechmeister-Boltenstern S (2011) Experimental warming effects on the microbial community of a temperate mountain forest soil. Soil Biol Biochem 43(7):1417–1425

Schortemeyer M, Dijkstra P, Johnson DW, Drake BG (2000) Effects of elevated atmospheric CO2 concentration on C and N pools and rhizosphere processes in a Florida scrub oak community. Glob Chang Biol 6:383–391

Shi S, Condron L, Larsen S, Richardson AE, Jones E, Jiao J, O'Callaghan M, Stewart A (2011) In situ sampling of low molecular weight organic anions from rhizosphere of *Pinus radiata* grown in a rhizotron system. Environ Exp Bot 70:131–142

Smith JL, Halvorson JJ, Bolton H Jr (2002) Soil properties and microbial activity across a 500 m elevation gradient in a semi-arid environment. Soil Biol Biochem 34(11):1749–1757

Várallyay G (1994) Climate change, soil salinity and alkalinity. In: Soil responses to climate change. Springer, Berlin, pp 39–54

Whitaker J, Ostle N, Nottingham AT, Ccahuana A, Salinas N, Bardgett RD, Meir P, McNamara NP (2014) Microbial community composition explains soil respiration responses to changing carbon inputs along an Andes-to-Amazon elevation gradient. J Ecol 102(4):1058–1071

Zak DR, Pregitzer KS, Curtis PS, Teeri JA, Fogel R, Randlett DL (1993) Elevated atmospheric CO_2 and feedback between carbon and nitrogen cycles. Plant Soil 151(1):105–117

Zak DR, Pregitzer KS, Curtis PS, Holmes WE (2000) Atmospheric CO_2 and the composition and function of soil on soil microbes 277 microbial communities. Ecol Appl 10:47–59

Zogg GP, Zak DR, Ringelberg DB, MacDonald NW, Pregitzer KS, White DC (1997) Compositional and functional shifts in microbial communities due to soil temperature. Soil Sci Soc Am J 61:475–481

Chapter 4
Impact of Climate Change on Soil Fertility

Abhilasha Shourie and Anamika Singh

Abstract Soil health and its fertility are an integral part of soil that provides a medium for plant growth and crop production. Climate is another most important aspect of nature and it is also a neutral thing similar to soil, but changes in climate occurs due to atmospheric environmental changes. Direct and indirect impacts of soil fertility are observed due to changes in climate. Although changes in soil are not an instant process, it takes years, and often sudden climatic changes destroy the vegetation of specific area and it may have many other effects. Human activity is directly or indirectly affecting climate change and further affecting soil health. Changes in soil health can affect food production such as crop, vegetables, etc. There are many factors in soil which can change due to effects of climate change; it can be physical, chemical, and biological. Climate change affects soil from its formation to its level of plant production. Soil affects climatics changes directly and indirectly as it is the main place of different biological and biochemical cycles.

Keywords Plant community · Ecosystem · Biodiversity · Geographical area · Species richness · Climate change

4.1 Introduction

Climate is generally defined as the average weather at a specific place, and it includes temperature, windiness, humidity and precipitation. Weather changes from day to night, while climatic changes occurs in years, significantly influencing the

A. Shourie
Department of Biotechnology, Faculty of Engineering and Technology, Manav Rachna International Institute of Research and Studies, Faridabad, India

A. Singh (✉)
Department of Botany, Maitreyi College, University of Delhi, New Delhi, India
e-mail: asingh@maitreyi.du.ac.in

D. K. Choudhary et al. (eds.), *Climate Change and the Microbiome*, Soil Biology 63,
https://doi.org/10.1007/978-3-030-76863-8_4

vegetation, productivity, chemical characteristics, etc. of soil (Brinkman and Brammer 1990; Scharpenseel et al. 1990). In general, soil structure can be influenced by changing climatic conditions in various directions, on different time scales and with different intensities. Change in soil organic carbon (SOC) content is a very slow process and detected by climatological time scales.

Nowadays, scientific studies are more focused on adverse effects of climate changes (Sinha and Swaminathan 1991; Saseendran et al. 2000; Aggarwal and Sinha 1993; Rao and Sinha 1994). Soil plays a major role in supplying micro- and macronutrient, which are used by different types of crops. Physical, chemical and biological properties of soil were changed with respect to climatic changes. Few important climatic factors such as atmospheric carbon dioxide (CO_2) levels, elevated temperature, altered precipitation (rainfall) and atmospheric nitrogen (N_2) deposition alter soil chemical, physical and biological functions (Arias et al. 2005; Moebius et al. 2007; Reynolds et al. 2009; Wixon and Balser 2009). Agriculture is mainly dependent upon rainfall, or uncertainty of monsoon in short climate change is linked to food security and food stability (Pimentel 2006; Lal 2010; Blum and Nortcliff 2013; Brevik 2013). Agriculture provides food for people to consume and also as a source of livelihood (Brevik 2013). The Intergovernmental Panel on Climate Change (IPCC) latest report shows that the rise in temperature between years 1980 and 1999 was 1.8–4.0 °C, and it is predicted that between years 2090 and 2099, it will be around 1.1–6.4 °C (Solomon et al. 2007). Definitely with these climatic changes, there will be change in environment, including the soil (Brevik 2012). Soil erosion, an important event, occurs in response to climatic changes. Soil quality has implications in ecosystem services and also influences food security. So, precipitation intensities or seasonal temperature variation highly influences soil hydro-physical properties. Changes affect the soil water regime, which may ultimately impact the environmental and economic development of a given area.

Solar radiation and precipitation are the two factors and the main sources of energy and moisture for Earth. Solar radiation is absorbed and re-emitted by the Earth's surface, which gradually heats the atmosphere. Precipitation is absorbed by the soil, used by plants and subsequently returned to the atmosphere via the processes of evaporation and transpiration. Thus, a continuous latent and sensible heat flux is present between the soil and the atmosphere, influencing the hydrothermal regime of the soil. Any changes in climatic factors can modify soil properties on the characteristic time scale of the driving processes. Soil energy and water balance components can respond rapidly to changes in atmospheric conditions. Changes in climatic temperature lead to global warming, which affects the agriculture, directly or indirectly. This may lead to decrease in food production, so ultimately any climatic change changes the soil quality and affects human life. So, impact of climate change and its effect on soil are major concerns and are discussed here in detail.

4.2 Causes of Climatic Changes

Both human and environment are interlinked together and have impact on each other. Their interaction may cause changes in environment as well as human life pattern. Climatic changes are mainly linked with sun and its refractivity of light on Earth. Sun is the major source of energy on Earth, and incoming and outgoing sun radiation affects Earth's climate (IPCC report 2007). Other than this, there are many other factors that cause changes in Earth's climate like natural, human and few local factors (Fig. 4.1). Natural factors like ice cores, volcanoes, ocean currents, continental drift, sediment microbes and trees are directly or indirectly linked to climatic changes. Earth's natural cycle is dynamic and changes frequently. Human is the most important factor for mis-utilisation of natural resources which leads to global warming and results in climatic changes. Over-industrialisation, deforestation and pollution are the main factors, and all these cause increase in carbon dioxide, methane and water vapour. Greenhouse gases actually helps to maintain the temperature of the Earth's surface by trapping heat from sun. Pollution adds harmful gases in the atmosphere, and it persists in the environment and causes destruction of the Earth's climate.

Concentrations of carbon dioxide, chlorofluorocarbons (CFCs), methane, nitrous oxide and most importantly water vapour are all known to be increasing in recent years. These increased concentrations cause a rise of the Earth's temperature from the last few years (Mitchell 1989). The radiation that comes from sun is not completely absorbed or reflected by the atmosphere (clouds) but it reaches the surface of the Earth. Only a small fraction (approximately 30%) of absorbed energy is reflected back. Seven percent of incoming radiation is absorbed by the atmosphere

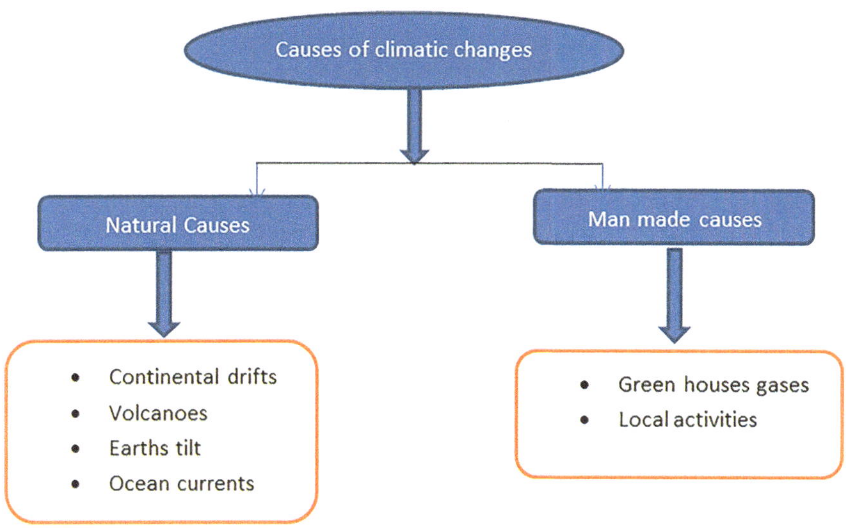

Fig. 4.1 Different causes of climatic change

and due to reflected energy, surface of Earth does not heat much. The radiated energy of Earth is having longer wavelengths than the Sun because it is colder. Most of the part of this longwave radiation is absorbed by these greenhouse gases that circulates and radiates energy into all directions, this is called Greenhouse effect.

Changes on land surface come under local activities. Local climatic changes often lead to global environment change, rainfall, drought and destruction of plants and vegetation in any area causing a rise in temperature. Lower plantation causes low rainfall and affects the Earth locally or regionally (Roy and Avissar 2002; Sage 1995).

4.3 Soil Fertility and Soil Nutrients

Soil quality is the measure of soil fertility and closest to plant nutrient management. Soil productivity mainly deals with its available nutrient status and productivity of soil. Soil productivity and fertility depend upon soil biological, physical and chemical components. All these factors directly or indirectly affect availability of nutrient and its dynamics. Soil fertility is actually a property of soil that helps in enhancing crop productivity level. Soil fertility is also the only most important aspect of soil productivity, e.g. if soil is very fertile, but little vegetation is observed, a situation may arise due to lack of some supporting factors responsible for fertility like water or unfavourable temperature. However, it has been observed that under most suitable environmental conditions, soil varies to generate perfect situations for plant root growth. Naturally healthy or improved soil health by different means is the main requirement of successful cropping. Good natural or improved soil fertility is essential for successful cropping. Soil is a natural medium and crop production is based on it.

There is no replacement of soil system for forest, grasslands and large level of crop production. Soils are the topmost layer of the Earth's crust, formed by weathering of rocks. Different types of soils are found throughout different regions of the Earth, and they vary by its origin, appearance, characteristics and production capacity. A well-developed soil is differentiated into different profiles or with different layers. Topsoil or a horizon is the topmost layer, rich in organic matter, soil microorganism and nutrients (Oliveira et al. 1992). Plants mainly use this layer or topsoil to obtain nutrient and water along with using B horizon or subsoil (Fig. 4.2). Physical degradation of rocks is called weathering, and ultimately it leads to soil formation, after addition of decomposed organic matter and minerals. Soil also contains many micro- and macronutrients (Table 4.1) and few beneficial metals which makes the soil fertile, and its absence causes loss of fertility and crop productivity. There are many other factors like organic matter, mineral, water, perfect pH and moisture, etc. that help in making the soil fertile. Soil nutrient is another aspect associated with soil fertility and it is mainly due to presence of micronutrients, macronutrients and benificial mineral elements.

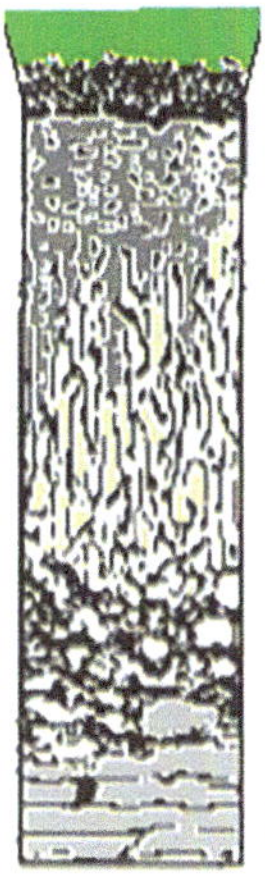

Fig. 4.2 Different soil layers

The first important factor that determines soil fertility is pH, which can be measured by chemical testing of soil. The second factor is soil organic matter which is mainly due to microorganism activity (decomposition), and it determines the quality of soil. The third factor is soil moisture content, which is the amount of moisture that resides in the soil that can also influence soil fertility. It is more desirable to use soil with higher moisture content for good fertility of soil. The fourth factor is clay content which is actually the cation exchange capacity (CEC). Low CEC indicates the possibility of easily losing nutrients by leaching, and the last and most important factor is bulk density of soil as it provides a passage for root penetration. It also blocks the path of roots so that they will not reach at the level of water and nutrients and also creates an obstacle for it and blocks path of roots and roots not able to reach at the level of water and nutrient. These few factors affects soil fertility drastically. Due to climatic changes there are changes occurs in these factors resulting variation in soil fertility.

4.4 Effect of Climate on Soil

Soil formation is a complex and slow process. The effect of atmospheric temperature on soil helps in decomposition of organic matter present in soil. Changes in soil moisture content depend upon rainfall, groundwater in the soil and the water consumption from the soil by evaporation, transpiration and run-off. So, soil moisture is mainly determined by climate, and few other factors like moisture are determined by climate, type of vegetation, human activities and the seasons. Climatic factors affect soil directly and indirectly. Organic matter in soil is the main source of mineral and carbon. Life of soil community members depends upon climatic changes because they have different physiologies, temperature sensitivities and growth rates (Whitaker et al. 2014). Temperature and moisture are the two

Table 4.1 List of micro- and macronutrient and beneficial elements required for soil fertility

S. no	Component (Symbol)	Function	Deficiency
Micronutrient			
1.	Iron (Fe)	Enzyme function, required for chlorophyll production	Vein yellowing in younger leaves. Occurs mainly at high pH
2.	Zinc (Zn)	Enzyme component	Vein yellowing in younger leaves. Rosette terminal leaves, occurs often at high pH
3.	Manganese (Mn)	Enzyme component	Yellowing of veins in younger leaves. Patches or freckled pattern found at high pH situation
4.	Copper (Cu)	Enzyme function	Curled leaves often bend downwards with stunted leaves often yellowing of leaves found at high pH
5.	Molybdenum (Mo)	Enzyme function	Usually an N deficiency due to the role in nitrate assimilation and in legumes in N-fixing bacteria. Yellowing of older leaves and light green in colour for the rest of the plant, at low pH
6.	Chlorine (Cl)	Osmotic balance, plant compounds	Generally deficient situation not found but yellowing and wilting of young leaves
7.	Boron (B)	Cell wall	Death of terminal buds, light general yellowing. Requirements are very plant specific
8.	Nickel (Ni)	Enzyme component	Almost never deficient
Macronutrients			
1.	Nitrogen (N)	Protein and enzyme component	Older leaves affected mainly by yellowing of leaves with stunted growth
2.	Phosphorus (P)	Membranes, energy, DNA	Difficult to visualise until severe. Dwarfed or stunted plants. Older leaves turn dark green or reddish purple
Beneficial mineral elements			
1.	Silicon (Si)	Provides resistance for pest-pathogen and drought	
2.	Sodium (Na)	Required for photosynthesis in C4 and CAM species of warm climatic-adapted plants	
3.	Cobalt (Co)	Heavy metal required for N-fixation by bacteria associated with legume tolerance, higher quality and yield of crop	

important factors linked to microbial growth, leading to soil quality improvement; global warming has effects on soil temperature and moisture influences microbial life. Climatic changes effects temperature and moisture that affect specific microbial functions in methanogen bacterial groups. It also effects microbial growth and changes in microbial ecosystem that ultimately leads to change in microbial soil decomposition and soil quality.

A direct effect of climate on soil is mainly due to rise of temperature, and in this condition, the concentration of carbon gets minimised. High peat formation and methane gas production are directly linked to high rainfall, while lower rainfall in any area experiencing loss of CO_2 and peat and increased moisture deficit for crops of shallow soils. High rainfall could increase atmospheric N deposition to soils; may promote soil disturbances, flooding and subsidence which changes in wetland and waterlogged habitats; and also enhance soil erosion, potentially leading to pollution of surface waters. Changes in soil properties like soil structural stability, soil biodiversity, plant–soil interactions and nutrient cycling were important changes observed due to climatic changes. Soil alteration due to climatic changes can be listed on the basis of many physical, chemical and biological changes, which actually affect soil formation and modify soil properties.

1. Physical Factors

a. *Soil surface*: It is an important factor that provides protection to soil surface water, nitrogen fixation, seed germination and soil nutrient (Box and Bruce 1996). Climatic changes directly affects soil structural conditions like soil crust formation and soil seal formation. It also affects oxygen diffusion with in soil, water infilteration in soil, surface water evaportaion and soil erosion.
b. *Soil water*: Soil water is the natural water present in soil and available for plants. Soil water is also called rhizic water. The three types of soil water are gravitational water, capillary water and hygroscopic water. This classification of soil water is based on functions of water in soil. Freely moving water through soil due to gravitational force is called gravitational water. It is found mainly in macrospores of soil and for plants its availability is very less. Gravitational water drains down due to the force of gravity thus unavailable. In micropores of soil water, capillary water is found, and it is highly available to roots of plants. Water found around soil particles is called hygroscopic water. Very less amount of this water is absorbed by the plants, but it helps mainly to bound soil particles tightly by adhesion. Soil water fluctuations can be observed through climatic changes, like precipitation affects soil water rapidly. Higher temperature causes high rate of evaporation causing loss of water from soil (Varallyay 1990a, b; Varallyay and Farkas 2008). Density, dynamics, species composition, biomass production, litter and root characteristics of crop plants are some features affecting soil water level. Human activity like higher human population, different cropping pattern, irrigation and drainage are few other factors that affect soil water directly. These few facts offer efficient measurement methods of climatic change prediction and its effects on soil (Lang 2006; Varallyay 2007; Birkas 2008; Harnos and Csete 2008). Structures of soil, organic matter, weathering and clay transformation are mainly affected by change in soil moisture content.
c. *Soil temperature*: Soil temperature is directly linked with sun radiation. There is a strong relationship between air temperature and soil temperature as increase in air temperature directly increases the soil temperature. In the process of evaporation, heat conduction occurs, through soil profile and movement of gas and water that take place (Karmakar et al. 2016). Warm soil temperature enhances soil process, like decomposition of organic matter, microbial activity, nitrification rate and chemical weathering of minerals due to accentuation. Soil temperatures affect the

type of vegetation of any area, which can be changed as per change in climate (Kalra and Aggarwal 1994).

d. *Rooting depth*: Length of plant roots depends upon available water, water capacity, subsoil salinity and carbon content. These properties indicate major constraints in the soil profile (Birkas et al. 2009; Dalal and Moloney 2000). Prolonged drought situation causes high chloride concentration and saline constraints (Dang et al. 2008; Rengasamy 2010) which affect productivity of soil.
e. *Soil texture and structure*: It is the proportion of sand clay and silt in soil. It is linked to climatic changes, crop productivity and vegetation type. Arid, semi-arid, subhumid and humid vegetation types have different textures in the soil profile (Brinkman and Brammer 1990; Scharpenseel et al. 1990). Organisation and combination of primary and secondary particles in a soil mass is known as soil structure. Soil structure controls the amount of water and air in soil. It is mainly concerned with movement of gases, water, pollutants, nutrients, etc. and directly affects plant growth, fauna and crop productivity, etc. The quality of soil depends upon organic matter present and inorganic content of soil matrix. Decrease of soil organic matter causes decrease of soil aggregate stability, while it increases soil infiltration and run-off causes soil erosion (Bot and Benites 2005; Karmakar et al. 2016). Soil aggregate stability provides an external energy for high intensity rainfall. Soil structure also determined chemical and biological properties of soil (Dalal and Moloney 2000; Moebius et al. 2007).
f. *Porosity*: It is the void spaces in a material pore size which mainly provides a space for water and air necessary for plant growth (Reynolds et al. 2002). Pore is linked to soil physical quality, bulk density, micro porosity and functions of pore volume. It provides aeration to soil and aeration capacity of soil depends upon pore size. Increased CO_2 and temperature causes variation in pore size distribution, and it changes root development and biological activity around the root. Development of root and soil enzymatic activities is closely related to each other, and both depend upon pore size and distribution.

2. Soil Biological Parameters

Many biological changes occur in soil with the influence of climatic changes. Microbial flora and fauna and soil organic matter are the most important biological factors. Soil microorganisms generally adapt to change of climate as it is not at all a sudden process, and microorganisms present in soil are indicators of soil health.

a. *Soil organic matter (SOM) and other components*: Normally, soil is full of organic and inorganic matter. SOM is the most complex and heterogeneous (Weil and Magdoff 2004). Soil plays an important role in cycling of carbon, nitrogen, phosphorus and sulphur. It is also full of multivalent ions and organic compounds. Soil provides shelter to many microorganisms, flora, algae, fungi and many more creatures. Growth of plants, microorganism and fauna affects soil properties like soil stability, water retention ability and hydraulic properties (Haynes 2008; Weil and Magdoff 2004). Decrease of SOM is observed to decrease soil fertility and biodiversity and water holding capacity and to increase the risk of erosion. SOM helps to absorb CO_2 from atmosphere and help to mitigate global warming. It also helps in water storage and minimises the risk of flood and high rainfall. Along with microorganic matter, soil is also full of macro organic matter and macro organic components mainly mineral-free particulate plant and animal residues, which serve as easily decomposable substrates for soil microorganisms (Post and Kwon 2000; Wagai et al. 2009). Rise of atmospheric temperature causes depletion of labile soil organic component. (Knorr et al. 2005).

Rise in atmospheric temperature and rainfall enhances mineralisation and decomposition (microbial activity) in soil. It causes biomass accumulation and decrease of soil carbon and nitrogen (C/N) ratio (Anderson 1992; Lal 2004). In cycling of nutrient, carbon and nitrogen, mineralizable organic matter acts as an interface between autotrophic and heterotrophic organisms (Gregorich et al. 1994). Under changing climatic conditions, mineralizable organic matter acts as indicator and affects nutrient dynamics.

b. *Soil respiration*: It is a measurement of released carbon dioxide (CO_2) from the soil. This CO_2 is released due to microbial decomposition of soil organic matter (SOM) and respiration from fauna and plant roots. It acts as a biological indicator of soil health, as it is linked to SOM content of soil. It is a link between global carbon cycle and climate change (Wixon and Balser 2009).
c. *Soil flora and soil fauna*: Soil microbial population is the main fauna of soil and responsible for SOM. They are sources of soil nutrient supply. Microbial biomass is the living component of SOM. Soil flora and fauna also participate in energy dynamics of soil (Saha and Mandal 2009). Flora of the soil includes plant varieties found in soil along with algal biomass and fungal filaments. Soil plants' root system interacts at a very close level and its leads to enzymatic activities. It is the indicator of physicochemical soil condition (Garcia-Ruiz et al. 2009).

3. Soil Chemical Parameters

a. *Soil pH*: It is an indicator of soil health and fertility. It has high effect on agriculture as rapid changes of pH result due to climatic changes such as temperature, CO_2, precipitation, etc. changes. It affects carbon and nutrient cycle and availability of water and also soil fertility ultimately productivity (Reth et al. 2005).
b. *Electrical conductivity*: It is the measure of salt concentration in soil, and it directly affects fertility of soil biological cycle and nutrient cycle. Increasing temperatures and decreasing precipitation increase the electrical conductivity under climate change scenarios (Smith et al. 2002). pH affects oil structure, especially in sodic soil (Arnold et al. 2005). All these biological, chemical and physical factors change due to climatic changes, and ultimately they affect soil fertility. Soil temperature of 80–90 °C and soil pH of 6 to 7 is best for good microbial growth, and it leads to high soil fertility. Climatic changes indirectly and directly affect soil fertility by changing the above-discussed factors.

Climate changes drive many changes in soil and they have been discussed in the chapter. Temperature and CO_2 are the main factors and are showing different effects on soil fertility. All these factors directly or indirectly affect the nutrient level of soil,

Table 4.2 Summary of expected effects of individual climate change variables on soil processes

S. no	Factors under increasing or decreasing concentration	Effects
1.	At increasing CO_2 concentration	Provides high concentration of carbon to soil, microorganism accelerates nutrient cycling in soil and water use efficiency also gets increased
2.	Increasing temperature	Soil organic matter (SOM) reduction and loss, reduction of moisture, increase of mineralisation rate, increase in soil respiration and soil structure, destruction was observed
3.	Reduction in rainfall	Reduction of SOM and nutrient availability and soil salinisation
4.	Increasing rainfall	It causes enhancement in rate of surface run-off and erosion. Increased soil organic matter and soil moisture, increased leaching of soil nutrient, reduction of nitrate and Fe

soil fertility, productivity of crops and soil moisture. Moreover, these factors are interlinked, and changes in any of two will also leads to change in soil fertility. Factors affecting climatic changes specifically temperature, CO_2 and rainfall were summarised in Table 4.1. Soil chemical properties are directly linked to the nutrient level of soil increasing temperature of soil and largely changing the precipitation level near the soil-root surface. Soil moisture and temperature are primary determinants of root growth and available nutrient. Climate changes have high impact on nutrient present in soil and its direct impacts on area near the root surface and influx rate (Brouder and Volenec (2008). And Table 4.2 summarised the list of factors that change due to climatic changes and its effects on soil.

4.5 Methods Utilised for Enhancing Soil Fertility in Changing Climatic Situation

There are many general methods to improve soil fertility level, which are mainly practised throughout word for the enhancement of crop productivity. These practices mainly contribute in improving soil structure and fertility. Few of these practices are as follows:

- *Mulching and residue management*: Mulch is a layer of material applied to the surface of soil. This method improves soil structure and also minimises soil erosion, reduces evaporation and increases soil aggregate formation. It also modifies soil temperature and moisture level and ultimately affects soil fauna.
- *Fertiliser*: Fertilisers may be chemical or biological in nature (cyanobacteria mainly). Application of fertilisers in soil enhances soil aggregation. It may have certain harmful effects under some circumstances, and then its application causes

decrease in soil organic component (SOC) and reduces soil aggregation and microbial communities. Correct application of fertiliser will improve the nutrient level in soil, increase plant productivity, increase SOC and promote microbial activity. All of these combinedly effects aggregation of soil particles and soil density.

- *Tillage*: It is a mechanical manipulation of the soil applied for good crop production. It affects soil water conservation, soil temperature, infiltration and evapotranspiration processes. It disrupts soil aggregates and also the factors which are helping soil (plants and animals) for aggregation. It is responsible for the decrease in soil organic matter (SOM), nutrient and microbial activities.
- *Manuring*: It is actually decomposed plants and animals which helps in improving soil fertility and crop yield and reducing soil erosion. It is used for the improvement of soil structure, soil density and aggregation of soil particles. Soil organic component (SOC) increase is a sign of good microbial and biological activity in the soil, which provides porosity of soil. Manured soil has higher earthworm population compared to unmanured soil with low SOC. Increased microbial activity due to increases in soil carbon from manure applications results in increases in aggregate stability.
- *Compost*: *Compost* is an organic material that helps plants in their growth. Food scraps and yard waste, fallen leaves, kitchen waste and left-over food can be easily converted into good compost. It is used to improve soil structure and to lower bulk density of soil. Composting materials can increase aggregation and aggregate stability. Environmental conditions such as drought can limit the effectiveness of compost. The effects of compost additions on soil structure may be short-lived although outcomes are generally positive.

4.6 Conclusion

Vegetation completely depends on the quality of soil and it directly effects the health of the human population. Changes in climate not only change the soil quality and soil fertility, but also affects the human and animal population as well as the whole ecosystem of any specific area. Changes in soil health can cause food loss directly affecting the human population. There are soil fertility enhancement methods that are widely accepted worldwide. Climate change can be natural or man made, but in both the cases it leads to damage the soil. In addition, soil-specific management practices will also help in retaining soil health, water conservation, crop productivity and crop improvements. These practices can overcome the impact of climatic changes in soil at physical, chemical and biological levels.

References

Aggarwal PK, Sinha SK (1993) Effect of probable increase in carbon dioxide and temperature on wheat yields in India. J Agric Meteorol 48:811–814
Anderson JM (1992) Responses of soils to climate change. Adv Ecol Res 22:163–210
Arias ME, Gonzalez-Perez JA, Gonzalez-Villa FJ, Ball AS (2005) Soil health-a new challenge for microbiologists and chemists. Int Microbiol 8:13–21
Arnold SL, Doran JW, Schepers J, Wienhold B (2005) Portable probes to measure electrical conductivity and soil quality in the field. Commun Soil Sci Plant Anal 36:2271–2287
Birkas M (2008) Environmentally-sound adaptable tillage. Akademiai Kiado, Budapest, p 351
Birkas M, Dexter A, Szemok A (2009) Tillage induced soil compaction as a climate threat increasing stressor. Cereal Res Commun 37:379–382
Blum WEH, Nortcliff S (2013) Soils and food security. In: Brevik EC, Burgess LC (eds) Soils and human health. CRC, Boca Raton, pp 299–321
Bot A, Benites J (2005) The importance of soil organic matter: key to drought-resistant soil and sustained food and production. FAO soils bull. No. 80. Food and Agriculture Organization of the United Nations, Rome, p 78
Box JE, Bruce RR (1996) The effect of surface cover on infiltration and soil erosion. In: Agassi M (ed) Soil erosion, conservation and rehabilitation. Marcel Dekker, New York, pp 107–118
Brevik EC (2012) Soils and climate change: gas fluxes and soil processes. Soil Horiz 53. https://doi.org/10.2136/sh12-04-0012
Brevik EC (2013) Climate change, soils, and human health. In: Brevik EC, Burgess LC (eds) Soils and human health. CRC, Boca Raton, pp 345–383
Brinkman R, Brammer H (1990) The influence of a changing climate in soil properties. In: Trans. 14th ISSS congress, Kyoto, pp 283–287
Brouder SM, Volenec JJ (2008) Impact of climate change on crop nutrient and water use efficiencies. Physiol Plant 133:705–724
Dalal RC, Moloney D (2000) Sustainability indicators of soil health and biodiversity. In: Hale P, Petrie A, Moloney D, Sattler P (eds) Management for sustainable ecosystems. Centre for Conservation Biology, Brisbane, p 101108
Dang YP, Dalal RC, Mayer D, McDonald M, Routley R, Schwenke GD et al (2008) High subsoil chloride concentrations reduce soil water extraction and crop yield on vertisol in northeastern Australia. Aust J Agric Res 59:321330
Garcia-Ruiz R, Ochoa V, Vinegla B, Hinojosa MB, PenaSantiago R, Liebanas G et al (2009) Soil enzymes, nematode community and selected physico-chemical properties as soil quality indicators in organic and conventional olive oil farming: influence of seasonality and site features. Appl Soil Ecol 41:305–314
Gregorich EG, Carter MR, Angers DA, Monreal CM, Ellert BH (1994) Towards a minimum data set to assess soil organic matter quality in agricultural soils. Can J Soil Sci 74:367–385
Harnos ZS, Csete L (2008) Climate change: environment-risk-society. In: Research results. Szaktudas Kiado Haz, Budapest, p 380
Haynes RJ (2008) Soil organic matter quality and the size and activity of the microbial biomass: their significance to the quality of agricultural soils. In: Huang Q, Huang PM, Violante A (eds) Soil mineral-microbe-organic interactions: theories and applications. Springer, Berlin, pp 201–230
IPCC (2007) Climate change mitigation. Contribution of working group III to the fourth assessment report of the intergovernmental panel on climate change. Cambridge University Press, Cambridge
Kalra N, Aggarwal PK (1994) Evaluating water production functions for yield assessment in wheat using crop simulation models. In: ten Berge HFM, Wopereis MCS, Shin JC (eds) Nitrogen economy of irrigated rice: field of simulation studies. SARP research proceedings. ABDLO, Wageningen, pp 254–266

Karmakar R, Das I, Dutta D, Rakshit A (2016) Potential effects of climate change on soil properties: a review. Sci Int 4:51–73
Knorr W, Prentice IC, House JI, Holland EA (2005) Long term sensitivity of soil organic carbon turnover to warming. Nature 433:298–301
Lal R (2004) Soil carbon sequestration to mitigate climate change. Geoderma 123:1–22
Lal R (2010) Managing soils and ecosystems for mitigating anthropogenic carbon emissions and advancing global food security. Bioscience 60:708–721
Lang I (2006) Global climate change: Hungarian impacts and reactions. Summarization of the VAHAVA project, Coordinated by the ministry of environmental protection and rural development and the hungarian academy of sciences, Budapest
Mitchell JFB (1989) The greenhouse effect and climate change. Rev Geophys 27:115–139
Moebius BN, Van EHM, Schindelbeck RR, Idowu OJ, Clune DJ, Thies JE (2007) Evaluation of laboratory measured soil properties as indicators of soil physical quality. Soil Sci 172:895–912
Oliveira JB, Jacomine, Camargo Amargo MN (1992) General soil classes in Brazil. Jaboticabal, FUNEP, 201 p
Pimentel D (2006) Soil erosion: a food and environmental threat. Environ Dev Sustain 8:119–137
Post WM, Kwon KC (2000) Soil carbon sequestration and land use change: processes and potential. Glob Chang Bio 6:317–327
Rao D, Sinha SK (1994) Impact of climate change on simulated wheat production in India. In: Rosenzweig C, Iglesias A (eds) Implications of climate change for international agriculture: crop modelling study. U.S. Environmental Protection Agency, Washington, DC, pp 1–10
Rengasamy P (2010) Soil processes affecting crop production in salt-affected soils. Func Plant Bio 37:613–620
Reth S, Reichstein M, Falge E (2005) The effect of soil water content, soil temperature, soil pH-value and root mass on soil CO_2 efflux-A modified model. Plant Soil 268:21–33
Reynolds WD, Bowman BT, Drury CF, Tan CS, Lu X (2002) Indicators of good soil physical quality: density and storage parameters. Geoderma 110:131–146
Reynolds WD, Drury CF, Tan CS, Fox CA, Yang XM (2009) Use of indicators and pore volume-function characteristics to quantify soil physical quality. Geoderma 152:252–263
Roy SB, Avissar R (2002) Impact of land use/land cover change on regional hydrometeorology in Amazonia. J Geophys Res Atmos 107:D20. https://doi.org/10.1029/2000JD000266
Sage RF (1995) Was low atmospheric CO_2 during the Pleistocene a limiting factor for the origin of agriculture? Glob Chang Biol 1:93–106
Saha N, Mandal B (2009) Soil health-a precondition for crop production. In: Khan MS, Zaidi A, Musarrat J (eds) Microbial strategies for crop improvement. Springer, Heidelberg, pp 161–168
Saseendran SA, Singh KK, Rathore LS, Singh SV, Sinha SK (2000) Effects of climate change on rice production in the tropical humid climate of Kerala, India. Climatic Tao, F., M. Yokozawa, Y. Hayashi and E. Lin, 2003. Future climate change, the agricultural water cycle and agricultural production in China. Agric Ecosyst Environ 95:203–215
Scharpenseel HW, Schomaker M, Ayoub A (1990) Soils on a warmer earth: effects of expected climate change on soil processes, with emphasis on the tropics and sub tropics. Elsevier, Amsterdam, p 273
Sinha SK, Swaminathan MS (1991) Deforestation, climate change and sustainable nutrition security: a case study of India. Clim Chang 19:201–209
Smith JL, Halvorson JJ, Bolton H (2002) Soil properties and microbial activity across a 500 m elevation gradient in a semi-arid environment. Soil Biol Biochem 34:1749–1757
Solomon S, Qin D, Manning M, Chen Z, Marquis M, Averyt KB, Tignor M, Miller HLIPCC (2007) Summary for policymakers. In: Climate change: the physical science basis; Contribution of working group I to the fourth assessment report of the intergovernmental panel on climate change. Cambridge University Press, Cambridge, pp 1–18
Varallyay G (1990a) Influence of climatic change on soil moisture regime, texture, structure and erosion. In: Scharpenseel HW, Schomaker M, Ayoub A (eds) Soils on a warmer earth: effects of

expected climate change on soil processes, with emphasis on the tropics and sub-tropics. Elsevier, Amsterdam, pp 39–49. ISBN: 9780080869889

Varallyay G (1990b) Potential impact of climate change on soil moisture regime and soil degradation processes. In: Proceedings of the international seminar on future research trends in MAB, August 20–29, Tokyo, pp 256–267

Varallyay G (2007) Potential impacts of climate change on agro-ecosystems. Agric Conspectus Scient 72:1–8

Varallyay G, Farkas CS (2008) Impact of climate change on soils. In: Harnos ZS, Csete L (eds) Climate change: environment-risk-society. Szaktudas Kiado Haz, Budapest, pp 91–129. (In Hungarian)

Wagai R, Mayer LM, Kitayama K (2009) Nature of the "occluded" low-density fraction in soil organic matter studies: a critical review. Soil Sci Plant Nutr 55:13–25

Weil RR, Magdoff F (2004) Significance of soil organic matter to soil quality and health. In: Magdoff F, Weil RR (eds) Soil organic matter in sustainable agriculture. CRC, Boca Raton, pp 1–43

Whitaker J, Ostle N, Nottingham AT, Ccahuana A, Salinas N, Bardgett RD, Meir P, McNamara NP (2014) Microbial community composition explains soil respiration responses to changing carbon inputs along an Andes-to-Amazon elevation gradient. J Ecol 102:1058–1071

Wixon DL, Balser TC (2009) Complexity, climate change and soil carbon: a systems approach to microbial temperature response. Syst Res Behav Sci 26:601–620

Chapter 5
Impact of Climate Change on Soil Microbes Involved in Biogeochemical Cycling

Anjali Singhal, Soumya Pandey, Neeta Kumari, D. K. Chauhan, and Pawan Kumar Jha

Abstract Anthropogenic activities have led to the emission of greenhouse gases which have accumulated in the earth's atmosphere over a period of time. The increased concentration of greenhouse gases has increased earth's temperature and has changed weather patterns. The enhanced CO_2 level, warming effect and changing soil moisture conditions have influenced soil microorganism. The microbial communities present in soil and the interactions taking place in terrestrial environment are extremely diverse and complex. The effect of climate change on soil microbial communities includes changes in microbial community composition, species abundance, diversity, survival and resilience, changes in enzyme production, and changes in interactions of microbes with roots of plants, production and sequestration of atmospheric gases (e.g. CO_2, CH_4, N_2O), utilization of soil nutrients and organic matter, etc. Further, the bidirectional nature of interactions where physical environment influences microorganisms and microorganisms in turn can impact environmental conditions, making it difficult to understand the effect of climate change. These microorganisms are involved in various biological processes associated with biogeochemical cycle. Thus, any change in microbial communities also affects the nutrient cycling through biogeochemical cycles. This chapter focuses on the effect of climate change on soil microorganisms and the impact on various microbial processes associated with carbon and nitrogen cycle.

Keywords Biological processes · Carbon cycle · Climate change · Extreme weather events · Nitrogen cycle · Soil microorganisms

---

A. Singhal · D. K. Chauhan
Department of Botany, University of Allahabad, Prayagraj, India

S. Pandey · N. Kumari
Department of Civil and Environmental Engineering, Birla Institute of Technology, Ranchi, India

P. K. Jha (✉)
Centre of Environmental Studies, University of Allahabad, Prayagraj, India

D. K. Choudhary et al. (eds.), *Climate Change and the Microbiome*, Soil Biology 63,
https://doi.org/10.1007/978-3-030-76863-8_5

5.1 Introduction

Due to the increased emission of greenhouse gases, there is a change in the climate, and it is predicted that the earth can become warmer by 5 °C by the end of this century. This changing climate has a profound impact on natural environment and human wellbeing. Climate change can strongly influence both abiotic and biotic interactions taking place in the environment. Biogeochemical cycling of nutrients is no exception (Frank et al. 2015). Climate change also affects terrestrial ecosystem. The microorganisms play an important role in maintaining the structure and functioning of terrestrial ecosystem. Climate change can affect soil microorganisms in many ways, including changes in species composition, abundance and distribution. It can also have direct and/or indirect effect on microorganisms causing changes in microbial community composition and/or enhanced or reduced physiological functions. Microbial communities in nature are complex and the interaction among different microorganisms is also varied. Thus, the response of soil microbes to changing environmental conditions is influenced by all these factors. Studies have shown that there is a change in biodiversity and function of ecosystem due to the impact of climate change. Soil microorganism interacts not only among themselves but also with flora and fauna of the region; thus, the interactions are extremely complex, making prediction of climate change-induced alteration very difficult. At the same time, the importance of soil microbes, especially those involved in biogeochemical cycles, stresses the need to do detailed analysis of the impact of climate change (Classen et al. 2015).

The interaction in between climate change and soil microbes is bidirectional. The soil acts as a sink for CO_2, while microbial processes taking place in soil lead to the emission of greenhouse gases like CO_2, CH_4, N_2O, etc. The microbial communities present in soil participate actively in various processes of biogeochemical cycles and regulate the movements of essential nutrients like carbon and nitrogen (French et al. 2009). Emission of CO_2 by various soil microbial processes can contribute up to 10% of total atmospheric CO_2; thus, soil microbial processes can influence the atmospheric factors at a global level (French et al. 2009; Mandal and Neenu 2012; Gougoulias et al. 2014). Microorganisms differ in their physiology, sensitivity, resilience and abundance; thus, changing physical conditions in terrestrial ecosystem has both direct and indirect impacts on them. When microorganism involved in key ecological processes like denitrification, nitrification, lignin degradation, etc. are affected, the functioning of ecosystem is also affected (Classen et al. 2015).

Various biological processes occurring in the terrestrial ecosystem have significant impact on the earth systems at a global scale. Carbon and nitrogen are the two most important nutrients required to sustain life. Their cycling among the various compartments of earth, atmosphere, lithosphere and hydrosphere is very important to sustain life and its processes. These biogeochemical cycles involve a number of biological processes mainly driven by soil microorganisms. The various processes for carbon cycle are photosynthesis, respiration, methanogenesis, fermentation, decomposition, etc. Nitrogen cycle involves biological processes like nitrification, denitrification, ammonification, etc. The climatic conditions have strong influence on the response and survival of soil microorganisms. These microbes, in turn, can have significant impact on environment by controlling the source and sink activities associated with carbon and nitrogen. Studies have shown that terrestrial ecosystem serves as a sink for CO_2. However, carbon sequestration also depends on the nitrogen content of the soil. Soils poor in nitrogen show reduced carbon fixation, and addition of nitrogen fertilizers promotes carbon sequestration as formation of soil organic matter requires a suitable C/N ratio along with other nutrients (French et al. 2009).

This chapter reviews the effect of climate change on microbial communities associated with carbon and nitrogen biogeochemical cycles. The chapter deals with the issue of the impact of changing abiotic conditions like increase in temperature, change in precipitation or increased CO_2 on structure and functioning of microbial communities. Studies related to the effect of climate change on microbial diversity, abundance, resilience and functioning are discussed. The changes in plant–microbe interaction, soil enzymes, rhizosphere, plant–microbe symbiotic relationships, pathogens and the associated changes in carbon and nitrogen cycles are also discussed.

5.2 Carbon Cycle and Microorganisms

Carbon is the essential nutrient for all life forms. In nature, it exists in both inorganic and organic forms. Microbes and plants interconvert the two forms of carbon and bring about its circulation among different compartments of environment, i.e. hydrosphere, atmosphere and lithosphere. The global carbon cycle is mainly driven by microbial communities, involved in the processes of fixing atmospheric C, plant growth and transformation and degradation of soil organic matter. Carbon is present in the atmosphere as CO_2 and CH_4. It is also present in the earth's crust in many inorganic forms like limestone and kerogens and in organic forms in soil. The process of converting C present in the atmosphere into organic form is called carbon fixation. In aerobic environment, photosynthesis is the dominant process for fixing atmospheric carbon. In this process, atmospheric CO_2 is converted into organic compounds and sunlight is used as a source of energy. Photosynthesis is performed mainly by plants and photosynthetic algae. Apart from photosynthesis, chemoautotrophic microorganisms (cyanobacteria, bacteria and some protozoa) also convert inorganic C compounds into organic compounds. The organic matter thus produced is consumed by animals and microbes for growth and maintaining their metabolic processes. As a result of these metabolic activities, CO_2 is generated and released in the environment. This process is called respiration. Terrestrial carbon cycle is a balance in between CO_2 fixed during photosynthesis and CO_2 released during respiration and organic matter decomposition. When living organisms die, their cells are transformed and decomposed by heterotrophs and carbon is released (mineralization). In anaerobic environment, microorganisms use organic compounds for obtaining energy and the process is called fermentation. Some of the commonly occurring fermenters are green and purple sulphur bacteria, *Thiobacillus ferrooxidans*, *Bacteroides succinogenes*, *Clostridium butyricum*, *Syntrophomonas* sp., etc. Fermentation is responsible for the release of CO_2 and CH_4 in environment. CO_2 is the major source of carbon followed by methane (CH_4). Methane exists in anaerobic environments. Methanogens are anaerobic archaebacteria that convert organic matter into methane by methanogenesis. Another group of bacteria, methanotrophs or methane-oxidizing bacteria, is a special group of aerobic bacteria capable of utilizing methane as an only source to satisfy carbon and energy requirements. Methanotrophs live at the boundary of aerobic and anaerobic environment so that they can have easy access to methane from anaerobic side and oxygen is available to them from aerobic side. The major microbial processes involved in carbon cycle are CO_2 fixation, methane production and utilization, respiration and decomposition of organic matter (Abatenh et al. 2018). Figure 5.1 shows the details of carbon cycle.

5.3 Effect of Climate Change on Soil Microorganisms of Carbon Cycle

Climate has a strong influence on the abiotic factors in the ecosystem. The growth, survival and activity of microbes are strongly regulated by abiotic conditions. Thus, climate change-induced variation in abiotic conditions can regulate and alter dynamics of microbial populations present. The two most important abiotic factors are temperature and moisture. The changes they can induce include abundance, composition and function of microorganisms. The growth and activity of any microorganism are its individual characteristics and can vary independently. Say, a change in abiotic condition induced higher activity; however, the growth of microorganism might reduce or might show lower biomass. Thus, growth and activity are two independent aspects of microbes and can respond differently to same changes in abiotic conditions (Mandal and Neenu 2012).

5.3.1 Effect of Enhanced CO_2 on Carbon Cycle Microbes

The amount of carbon locked in soil in organic form is almost three times the carbon available in the atmosphere. Annually, about 8% of carbon is circulated by carbon biogeochemical cycle in between the atmosphere and lithosphere. If the process of respiration and decomposition stops, then 100% of carbon present in the atmosphere will be fixed to organic matter in soil in about 12 years (Gougoulias et al. 2014). At present, the amount of carbon fixed by photosynthesis and autotrophic

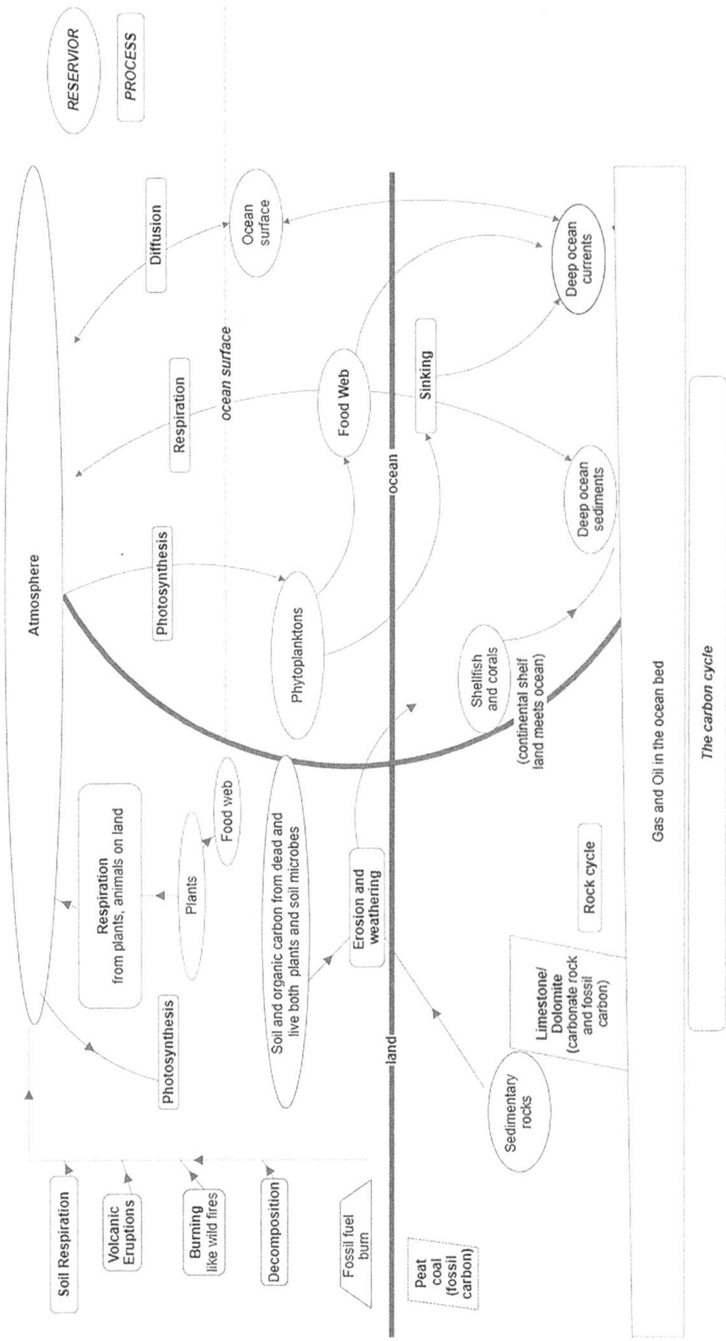

Fig. 5.1 The biogeochemical cycle of carbon

microorganism is almost 25% more than the carbon liberated during respiration. Thus, terrestrial carbon sequestration acts as a sink for CO_2.

Anthropogenic activities have disturbed the natural balance of the biogeochemical cycles. Burning of fossil fuels and cultivation of land have increased the emission of CO_2, CO, etc. Increase in atmospheric CO_2 content acts as fertilizer for photosynthesis, thereby stimulating growth. More root exudates are produced by plants and these organic substrates in turn enhance microbial activities in rhizosphere. Studies have reported changes in microbial community composition and activity. A study reported 121% increase in biomass of microorganism after treatment with CO_2 (690 ppm) in open top chambers for 22 weeks (French et al. 2009). Some studies have reported increased dominance of *Pseudomonas* spp. at elevated CO_2, some studies using molecular techniques have confirmed changes in community structure for bacteria as well as fungi, but some studies have reported no change (French et al. 2009). Arbuscular mycorrhizal fungi (AMF) form symbiotic relationship in between plants and fungi. Fungi colonize the roots of host plants and create vast connections between roots of plant and surrounding soil. This increases the surface area for nutrient uptake. An enhanced level of CO_2 promotes photosynthesis. More carbohydrates are available for roots. This promotes the growth of AMF. Thus, enhanced ambient CO_2 levels indirectly increase AMF development and promote symbiotic relationship (Choi et al. 2005; French et al. 2009).

The microbial decomposition and respiration also increased (French et al. 2009; Gougoulias et al. 2014). This in turn increases carbon mineralization, and more CO_2 is released in the environment. These changes will disturb the balance between carbon fixation and carbon mineralization. The processes are very diverse and complex. Thus, the net effect of the increase in CO_2 content in the atmosphere might favour carbon sink (photosynthesis) as well as carbon sources (decomposition and respiration). It is difficult to predict the outcome in the future. However, few models have predicted that increase in atmospheric CO_2 will lead to increased carbon emission (Gougoulias et al. 2014). Studies have suggested that increase in CO_2 will lead to enhanced photosynthesis by plants. The amount of litter generated by plants will also increase. This litter may alter the soil chemical and physical properties. Such changes can alter not only composition but also function of the microbial communities present there. Elevated CO_2 also promotes root growth, thus including changes in rhizosphere (Mandal and Neenu 2012). The published studies have also reported that the increasing CO_2 had no significant effect on microbial growth and activity (Kandeler et al. 2006; Pinay et al. 2007). Drissner et al. (2007) conducted a study to see the effect of elevated CO_2 on the soil enzymes commonly involved with biogeochemical cycles. In spring season, the activity of enzymes increases, urease (23.8%), xylanase (22.9%), protease (40.2%), invertase (36.2%) and alkaline phosphomonoesterase (54.1%) activities. However, in autumn season, enzyme activity decreased by 3–12%.

The effect on climate change on microbes can be evaluated at individual, community or global level. A study by Collins et al. (2008) evaluated the effect of CO_2 enrichment on phytoplankton community. Different microbial species respond differently to changes in environment. Also, by analysing evolutionary or physiological traits, it is difficult to predict the microbial response. Same microbial species may respond differently for single strain community and multi-strain community. For single strain communities, long-term CO_2 enrichment experiments enhanced biomass production. However, in multi-strain communities, the long-term CO_2 enrichment experiments lead to decline in CO_2 fixation. The difference in responses might be due to competitive interactions present in multispecies communities (Collins et al. 2008).

5.3.2 Effect of Drought and Increased Moisture on Carbon Cycle Microbes

The impact of climate change on biogeochemical cycle not only depends on the diversity and abundance of microorganisms but also on the prevalent environmental conditions in the ecosystem (Bardgett et al. 2008). In forest ecosystem, increase in severity and frequency of droughts will make the soil water deficient and dry. Various studies have shown that the moisture content may decrease to the level that it negatively affects microbial activity. The rate of decomposition and respiration reduces. The activity of phenol oxidase and amount of fungal and bacterial biomass also reduce (Nardo et al. 2004; Krivtsov et al. 2006). However, if similar changes occur in wetland or peatland, the effect will be opposite. Increased dry conditions will lower the water table in the region, thereby converting anaerobic soil into aerobic. This change will favour degradation and microbial respiration. Studies have shown that the activity of phenol oxidases increases (Freeman et al. 2004; Zibilske and Bradford 2007). Wetlands and peatlands have huge stock of organic matter. If the dry condition prevails, the level of oxygen in soil increases and CO_2 efflux increases. At the same time, the activity of methanogens is inhibited. These changes might have significant impact on global carbon cycle (Bardgett et al. 2008).

Global warming is leading to abrupt climate changes like increase in severity and frequency of drought, increased rainfall and increased episodes of extreme climate. Such changes in climate affect the global pattern for production and decomposition of organic matter. Drought can change the carbon allocation in between roots and foliage and thus can affect below-ground cycling of carbon and other nutrients. Rhizosphere is a zone where interactions in between roots and root-associated microorganisms take place. Drought can disrupt the various processes taking place in rhizosphere. Sanaullah et al. (2011) studied the changes in the microbial biomass and enzyme activity (xylanase, β-cellobiosidase, β-glucosidase, chitinase) in rhizosphere of grasses, grown as monoculture and mixed culture. It was observed that lesser carbon was allocated to shoot as compared to root. No trend was observed for changes in microbial biomass in monoculture conditions. However, in mixed culture, there was an increase in microbial biomass. Unplanted soil showed most adverse drought response with severe decline in enzyme activity of all the enzymes studied. The enzyme activity was lower in mixed plantations as compared with monoculture. The enzymes studied (xylanase, β-cellobiosidase, β-glucosidase, chitinase) are involved in decomposition of organic matter, and we can say that due to drought, the decomposition process will be slower in mixed plantations as compared with monoculture.

The response of microbes for change in moisture content also varies from region to region. The microbes inhabiting dry and arid place will have lower capability to respond to higher moisture content (Meisner et al. 2013). Due to climate change, there will be not only increasing episodes of drought and flood but also pulsed rain events where wet and dry spell will alternate. Studies have shown that microbial community and their functioning change with wet and dry spell with transition phases. The microbial activity is high during wet spell and lower during dry spell. However, some studies have indicated that dry spell exposes old C pools, and during wet spell, the microbial activity becomes so high that it is able to compensate the reduction in activity during dry spell (Collins et al. 2008; Evans et al. 2012; Meisner et al. 2013).

The duration of the study also has profound influence on the results. A study was conducted in Mediterranean-type grassland ecosystem. The effect of natural process of rainfall and dry period was studied. Short-term effects were studied for changes in the structure and function of microbial community. It was observed that microbial communities, especially bacteria, respond quickly for the rewetting for soil after a dry period. Rainfall was quickly followed by pulses of release of nutrients like carbon and nitrogen. Most of the abundant microbial communities (*Actinobacteria*, *Acidobacteria*, *Proteobacteria*, *Bacteroidetes*, *Firmicutes*, *Cyanobacteria*, *Verrucomicrobia*) showed strong correlation for moisture and available carbon. Many studies have reported that microbial communities are resilient to changes in moisture; however, most of them involve longer durations. Results can vary significantly with duration of the study (Cruz-Martínez et al. 2012).

5.3.3 Effect of Rise in Temperature on Carbon Cycle Microbes

One of the effects of climate change predicted is the overall temperature of earth will increase by 1–5 °C (IPCC 2007). It is generally believed that global warming will increase the degradation of organic matter, and thus the C flux from terrestrial ecosystems to atmosphere will increase. Increase in temperature can affect microbial community composition as well as physiological functioning. Studies have shown that bacteria play more important role in determining the rate of respiration as compared with fungi (Keiblinger et al. 2010). Most of the warming studies have shown that there is not much effect on microbial biomass due to rise in temperature. However, the changes in microbial community composition are varied. They can be change in fungal abundance, change in abundance of gram-positive bacteria, decrease in gram-negative bacteria or no change in microbial community structure

(Schindlbacher et al. 2011). A 5-year warming study was conducted in forest of Achenkirch, Austria. No change was observed in terms of microbial biomass or community structure over the entire period of the study (Schindlbacher et al. 2011). Zhang et al. (2005) studied warming effect on tall grass prairie for 2 years. The increase in fungal abundance was observed. Similar results were also reported by Castro et al. (2010). However, the increase in fungal abundance was an indirect effect caused due to changes in plant community. Warming may also lead to gradual decline in available decomposable matter. This change may also alter community composition. Thus, there are many direct and indirect factors affecting microbial processes (Schindlbacher et al. 2011). The soil respiration rate and thus CO_2 release increased as a result of warming of soil. This change can be linked to higher turnover rate and changes in substrate utilization by microorganisms. It was also observed that microbes were under stress and their respiration rate increased (Schindlbacher et al. 2011). Another study was conducted for 7 years (2002–2009) at deciduous forest in New England to evaluate the heating effect on CO_2 emission and sequestration. In warmed plots, temperature was maintained 5 °C more than the ambient temperature. Ambient temperature varied from 20 °C in summer to −6 °C in winter. Rainfall was evenly distributed. Increase in temperature promoted microbial activity, higher degradation of organic matter was observed and thus CO_2 emission increased. Warming also increased the storage of carbon in plants as compared with control plots. By the end of 7 years, the increased amount of carbon stored in plants was able to offset the increased CO_2 emission effect (Melillo et al. 2011).

Rise in atmospheric temperature or global warming has greatly affected the microbial functioning as the microbial processes are temperature sensitive. Many studies have explored the impact of rising temperature on soil microbes. However, there is no clear trend. Results also vary with experimental conditions like duration of study (short term or long term), or lab study or field study, single factor or multiple factors, etc. (Classen et al. 2015). Most of the studies have reported that the decomposition of organic matter and microbial respiration increase with rising temperature (Bradford et al. 2008; Sistla and Schimel 2013). Possible reasons for these changes can be changes in the structure of microbial community, substrate availability, quality and quantity of litter and relative abundance of labile carbon versus soil organic carbon (French et al. 2009). However, these changes can be for short duration. With time, as the labile C pool decreases in soil, the microbial activity also reduces. There might be change in microbial composition and functioning. Microbes respond to these changes by adaptation, evolution and interactions. The changes are diverse and complex (Bradford et al. 2008; French et al. 2009; Mandal and Neenu 2012; Sistla and Schimel 2013; Gougoulias et al. 2014).

An important indirect effect of rising temperature is greater loss of moisture from soil, creating drought-like conditions. Lack of moisture may negatively impact the availability of nutrients. The fast-growing bacterial community is more prone to adverse effect as compared with slow-growing fungal community. The changes in fungal community composition are usually more evident as compared with bacteria, showing better adaptability of fungi (Blankinship et al. 2011).

5.3.4 *Combined Effect of Different Climate Change Factors on Carbon Cycle Microbes*

Most of the studies have evaluated the effect of single factor while studying climate change and biological systems. However, in real scenario, there will be more than one factor affecting the environment. The net effect might be additive, antagonistic or no change. In nature, the changes in climatic conditions will not be individualistic. That is, changes in temperature will be overlapped with changes in CO_2 concentration, precipitation and so on. Thus, it is important to study the effect of microorganisms with variation in more than one abiotic condition. Effects of variation in temperature (ambient, 3 °C), precipitation (wet and dry) and CO_2 concentration (ambient, 300 ppm) were studied on bacterial and fungal diversity in oil field ecosystem (Mandal and Neenu 2012). Bacterial diversity increased in case of high temperature and high CO_2 concentration. The plots with high temperature and ambient CO_2 concentration showed decreased bacterial diversity. Fungal diversity increased in plots with high temperature (Mandal and Neenu 2012).

Studies combining the effect of stress due to climate change have often reported that microbial communities experiencing stress often trade growth for stress tolerance traits. Combined effect of warming (5 °C above ambient) and four freeze–thaw cycles on soil microorganisms was studies at Hubbard Brook Experimental Forest in the northeastern United States. The brown rot fungi and plant pathogens were favoured by rise in temperature, while growth of arbuscular mycorrhizal fungi (AMF) especially *Glomus* reduced. Warming and freeze–thaw cycle together promoted animal pathogens (genera *Trichosporon* and *Metarhizium*) and AMF recovered. The copiotrophic and cellulose degrading bacteria were also suppressed (Garcia et al. 2020). A study by Sheik et al. (2011) evaluated the combined effect of high temperature and drought on microbial communities of Oklahoma prairie soil, USA. It was observed that during the periods of normal rainfall and increase in temperature (2 °C above ambient), the microbial biomass increased by 40–150% but diversity decreased; thus, the composition changed. During the period of drought, the slight increase of 2 °C lead to severe drying of the soil and microbial population decreased by 50–80%. However, there was no long-term effect on community composition as species diversity, richness and evenness improved. May be under stressed conditions, fewer phylotypes were active (Sheik et al. 2011). All the physiological functions of the microorganisms are mediated through enzymes. Studies have shown that during stress, there might be a change in resource allocation, preference being given to enzyme production. This will help in optimizing the use of limited resources available (Steinweg et al. 2013). Microbial enzymes and microbial biomass were measured in a study conducted in an old abandoned field at Boston-Area Climate Experiment (BACE), USA. Combined effect of higher temperature (4 °C above ambient), lower rainfall (50% of normal) and higher rainfall (150% of normal) was evaluated. It was observed that microbial enzyme production increased with increasing temperature, while microbial biomass decreased. Possible reason can be allocation of resources for enzyme production changed. Thus, the popular view that changes in microbial biomass are followed by changes in microbial enzyme production was challenged (Steinweg et al. 2013).

5.3.5 Effect of Extreme Climatic Events on Carbon Cycle Microbes

Studies have shown that the frequency and severity of extreme weather events have increased due to climate change. This is in addition to the gradual effect climate change has on environmental conditions. The extreme weather events include heat waves, frosts, extreme drought, heavy precipitation, wind storms, etc. The past few years have witnessed many incidences of extreme weathers worldwide and also in India. Tables 5.1 and 5.2 summarize few events related to climate change and extreme events that occur worldwide and in India, respectively. Extreme weather events are considered as disturbances or pulse events that last for a short duration but have a strong impact on the surroundings. The effect can be categorized into four types: direct and concurrent like reduced productivity due to drought, indirect and concurrent like change in organic matter composition of soil due to forest fire caused by lightening, direct and lagged effect like reduced flowering and fruiting due to loss of fertile soil during flash floods and indirect and lagged effect like reduced productivity due to increased pest and pathogen population (Frank et al. 2015).

The ecosystems may experience huge fluctuations in their structure and function due to exposure to extreme climate events. These disturbances can sometimes be strong enough to cause abrupt change from one ecosystem state to another. Sometimes, ecosystems show good resistance and resilience (recovery) and are able to maintain their original state after extreme climate pulse disturbance. If these disturbances reoccur, then changes are inevitable. Different microorganisms adapt to different strategies to deal with disturbances. Members of phylum *Actinobacteria*, commonly found in soils of dry regions, have high tolerance for desiccation or are resistant to drying, while bacteria belonging to phylum *Acidobacteria* survive drought as they are more resilient and recover fast owing to fast growth strategy (Bardgett and Caruso 2020). *Actinobacteria* has oligotrophic characters and shows low growth rate and higher efficiency for resource utilization, but is resistant to change. *Acidobacteria* is copiotrophic, characterized by higher growth rate and lower efficiency for using resources, being resilient (Bardgett and Caruso 2020). Since there are many different types of extreme weather events, their effects also vary accordingly. For example, drought has direct concurrent effect on reduced enzyme activity of microbes. Recurrent droughts might alter the regional microbial community composition favouring drought-resistant species. This in turn will impact CO_2 sequestration and emission. If the extreme event is flash flood, then the top fertile soil is washed off, and the change in ecosystem can be so drastic that it might not be able to recover to its original state (Frank et al. 2015).

Table 5.1 Some of the natural disasters occurring worldwide due to climate change

Event	Year	Location	Details of event	Damage occurred	References
World					
Floods	2020	Indonesia	Coastal cities of Indonesia see frequent floods due to rising water level and excessive rainfall across local watersheds	Capital city Jakarta suffers from floods killing 67 people and displacing 300,000 people	Karyono TH, Vale R, Vale B (2019) Sustainable Building and Built Environments to Mitigate Climate Change in the Tropics: Conceptual and Practical Approaches. Springer
Bushfires	1995–2005, 2020	Australia	Most fire-prone continents suffered extreme fire in 2020. Changing climate and warming can result in dry weather, making the spread of small fires	Many species got burned, massive loss of life (animal and human) occurred, ecosystem was completely destroyed	Russell-Smith J, Yates C & Whitehead P et al. (2007) Bushfires 'down under': Patterns and implications of contemporary Australian landscape burning. Inter J Wildland Fire. 16. https://doi.org/10.1071/WF07018
Cyclone	2019	Southern Africa	Cyclones Idai and Kenneth hit Zimbabwe, Malawi, Mozambique of southeastern Africa simultaneously. Idai caused massive rainfall resulting in mudslides, extensive flooding and collapsing of dam	Destruction from wind and high-intensity rainfall, dam failure all costed lives and livelihood destruction and economic loss too	Cook N (2019) Cyclones Idai and Kenneth in Southeastern Africa: Humanitarian and Recovery Response in Brief. Congressional Research Service
Drought	2010–2011	East Africa	Drought caused by natural variability shifts in ocean temperature and air pressure in 2010 got more enhanced in 2011 with addition of Anthropocene activities. Climate change was held responsible for drought which has destroyed these areas of Africa	Majorly influenced by La Niña which impacted many countries brought drought to Africa. Everything from livestock to dairy and vegetation got destroyed resulting in famines	Lott FC, Christidis N, Stott PA (2013) Can the 2011 East African drought be attributed tohuman-induced climate change?. Geophys Res Lett. 40:1177–1181. doi:https://doi.org/10.1002/grl.50235
Ocean acidification	2016, 2017, 2020	Great Barrier Reef, Australia	As the ocean temperature rises, it results in coral bleaching. The coral in the Great Barrier Reef are very	Loss of various important species and food for other marine lives are all	The Guardian: https://www.theguardian.com/environment/2020/apr/07/great-barrier-reefs-

			sensitive to temperature and turn white under heat stress from water. This is frequently occurring and destroying the whole barrier reef	destroyed in repetitive bleaching of corals	third-mass-bleaching-in-five-years-the-most-widespread-ever
Algal bloom	2003	Antarctica	Within the 5 km of penguin colonies, algal bloom has been recorded and the nitrates of seabird and seal excrement are believed to be their food sources. This results in potential warming as green algae do not reflect sunlight completely, and this could cause temperature rise and melting of glaciers	Melting of Antarctica glaciers could lead to overall rise in water level drowning major coastline cities and destroying natural habitat of Antarctica	Kol E, Flint EA (1968) Algae in green ice from the Balleny Islands, Antarctica, New Zealand Journal of Botany, 6:3, 249–261, DOI: https://doi.org/10.1080/0028825X.1968.10428810
Eutrophication	2016	Australia	It resulted in unprecedented water level decline in lower lakes of Murray–Darling basin. Due to droughts, water acidification has increased along with salinity and eutrophication processes	This impacts the quality of water and aquatic biomes severely. And generally the water cannot be used for any other purposes	Li S, Bush RT, Maoo R et al. (2016) Extreme drought causes distinct water acidification and eutrophication in the Lower Lakes (Lakes Alexandrina and Albert), Australia. J Hydrol, doi: https://doi.org/10.1016/j.jhydrol.2016.11.015
Wildfires	2019	Amazon	In Brazil and Bolivia, large fires took place exceeding community and country capacity	Extensive loss of homes, loss of communities residing in the deep interior, crops, livestock and water supply system	Brando P, Macedo M, Silverio D et al. (2020) Amazon wildfires: Scenes from a foreseeable disaster
Hurricane	2020	Atlantic hurricane	A series of 20 tropical storms and 20 tropical hurricanes and 1 major hurricane occurred in the Atlantic region in the first half of 2020	More than 1 billion dollar of economic loss has occurred. 100 people have lost their lives	National hurricane center and Central Pacific Hurricane center (2020) 2020 Atlantic Hurricane season. Accessed on 17 September

Table 5.2 Some of the natural disasters occurring in India due to climate change

Event	Year	Location	Details of event	Damage occurred	References
India					
Drought	2004–2010	Bundelkhand	Historically droughts occurred every 16 years in the nineteenth century in Bundelkhand region, but with passing times, Anthropocene activities have brought out climatic changes and unorganized development has led to frequent droughts in the region, that is, from 2004 to 2010, this region faced drought frequently and severely. Although surrounded by Yamuna River and its tributaries, because of temperature rise, increased greenhouse gases, urbanization, etc., disasters have occurred	Droughts have caused crop failure and extreme debts, resulting in the death of more than 400 farmers because of suicide and starvation. Many residents had to mass relocate. From extremely rich, lush greenery, abundant water and glorious history, the region has converted into its poorest state with lack of food, water, fodder fuel, etc. with sparse vegetation. Along with climatic disaster, this is also a socioeconomic disaster	Gupta AK, Nair SS, Ghosh O et al. (2014) Bundelkhand Drought: Retrospective Analysis and Way Ahead. National Institute of Disaster Management. New Delhi, P 148
North Indian cold waves	2010–2011	Uttar Pradesh, Haryana, Punjab	A rapid decline in temperature within 24 h called the cold waves was felt severely in north Indian region. A temperature dropped to minus 26 degree in Kashmir and Leh region, while Mount Abu, Rajasthan, became the coldest desert with temperature drop to 4.2 degrees. Due to lack of any western disturbances, the cold waves persisted, resulting in the death of 650 people	Cold waves damage the agriculture and lives killing 650 people. Many viruses and influenza are believe to prevail in cold places more actively, so these waves pose potential threat for epidemic	Bhatla R, Pant M, Singh D et al. (2020) Evaluation of cold wave events over Indo-Gangetic Plain in India. J Agrometeorology 22 (2): 233–238

Cloud bursting, flash floods and landslides	2013	Uttarakhand	Cloud bursting (extreme rainfall >100 mm/h) due to loaded monsoon cloud coming from the Bay of Bengal and Arabian Sea over the Himalayan belt resulted in extreme flash floods on the hills of Uttarakhand causing massive destruction. The Chorabari Glacier melted abnormally fast flooding the Mandakini River causing this destruction. The debris from all the flooded area blocked the routes and cut-off the area from aid reach. The floods also triggered severe landslides adding further to the miseries	More than 5000 people went missing, and the death toll was around 5748. Huge loss to life and property making people homeless and creating potential endemic situation health-wise	Pradeepkumar AP, Behr FJ, Illiyas FT et al. (2014) Proc. 2nd Disaster, Risk and Vulnerability Conference 2014 (DRVC2014) 24–26 April 2014 Dept of Geology, Uni Kerala, Trivandrum, India pp. 57–64 ISBN 9788192344928
Floods	2014	Jammu and Kashmir	Due to severe rain for almost a week and overflowing channels, it caused floods on 6 September 2014 in J&K. The possible cause behind this massive disaster is believed to be climate change caused by Anthropocene activities over centuries	30% of the state was submerged in flood causing thousands of people stranded in their home. 400 villages were fully submerged and 2225 villages were partially submerged and 300 were completely cut-off from other parts of the area. Hundreds of lives were taken and rescue was even more difficult as flood water won't go down at all. Jhelum River was flowing 1 m above its embankment and swelled to a larger scale in hundreds of square km	Tabish SA, Nabil S (2015) Epic Tragedy: Jammu & Kashmir Floods: A Clarion Call. Emerg Med (Los Angel) 5: 233. Doi: https://doi.org/10.4172/2165-7548.1000233
Forest fires	2015–2016	Uttarakhand	Wildfires affected the 13 districts of the state of Uttarakhand. The causes behind these forest fires are believed to be dry weather, low moisture and	This has damaged nearly 4000 hectares of forest cover, destroying biomes and forest ecosystems hugely and killing 9 people and	Dalei N (2016) Forest Fires in Indian State of Uttarakhand. Eurasia Review

(continued)

Table 5.2 (continued)

Event	Year	Location	Details of event	Damage occurred	References
			humidity and deficit vapour pressure that allow fires to spread unconditionally. The heat waves in the summer in the state had promoted the conditions even more. Carelessness of locals as well unsupervised burning of pine needles, litter, etc. in forest with dry climatic conditions have costed the state a lot	injuring many. Glaciers of Himalayas have also been affected by carbon black deposits from smoke, and because ashes have high temperature-absorbing capacity, they may have enhanced the melting of glaciers. Flowing water may have toxic pollutants. 10,000 people had to relocate to avoid the fire	
Avalanche	2016	Siachen Glacier avalanche	In February, an avalanche hit the Siachen Glacier region which is an extremely important region between India and Pakistan border	10 soldiers were killed trapped inside the snow with temperature below −45 degrees. From 1984, 870 soldiers have died due to these major avalanches hitting the region	Wikipedia contributors. Siachen conflict. Wikipedia, the free encyclopedia. Available at: https://en.wikipedia.org/w/index.php?title=Siachen_conflict&oldid=978007208. Accessed September 16, 2020
Storm dust	2018	North India	High-velocity dust storms swept across parts of North India. Dust storm initiated in monsoon season and thunderstorm forecast with high winds were predicted, yet storm came in period of high temperature and increased weather intensity	Many people were killed in UP, Rajasthan, and other states of North India. Huge amount of debris brought along with the storm caused more destruction to lives, infrastructure, etc.	Gupta S (2018) More than 110 killed by high-intensity dust storm in India. CNN. Retrieved 3 May 2018.Accessed 16 September2020
Flash floods	2018	Kerala	Massive destruction caused due to heavy precipitation resulting in flash floods and spilling of dams in Kerala in 2018	Thousands of lives and livestock and animals were washed away or dead. The catastrophe is extreme in the past 100 years	Srivastava, Saikia P, Pandey AC et al. (2020) Evaluating the 2018 extreme flood hazard events in Kerala, India. Remote Sensing Letters 11(5):436–445

Invasion of desert locusts	2020	Rajasthan, Punjab and Gujarat	Invasion of locusts occurred in April 2020 in Rajasthan, Gujarat, Punjab and some part of Delhi. Locust swarm in group and are gluttonous insects consuming 2 g/ each of the crops. They usually bred in high humidity with wind and erratic rainfall. As the country had many rainfalls due to various low pressure disturbances from the west (mid-west Atlantic Ocean), it creates a perfect breeding ground for locusts. These insects travel at 100–200 km/day and consist of 50–100 b locusts per group	Massive destruction to crop fields and huge damage to food production industries. Severe losses to farmers financially. These outbreaks have been proving a great threat to food security and are frequently occurring because of severe climate change especially frequent rainfall, usually as a result of high humidity	Sankar PM, Shreedevasena S (2020) Desert Locusts (Schistocerca gregaria) – A Global Threatening Transboundary Pest for Food Security. Research Today Spl. 2(5): 389–391

5.3.6 *Impact of Climate Change on Plant and Soil Microbe Interactions*

Climate change has led to increase in temperature and thus, at some places, the plants are migrating to higher latitudes. The changes are drastic that the whole ecosystem has changed. For example, in the Arctic, woody shrubs have been replaced by grasses (Pearson et al. 2013). In soil, many microbial communities are closely associated with plants. This relationship is mutually beneficial and sometimes can have a great influence in terms of adaptation and survival. Changes in plant community in a region can affect microbial communities in many different ways. The microbial community might follow the migration of plant species, or it might migrate deep in soil, or there might also be redistribution of microorganism. In some cases, the microbial community composition changes and the new species emerge as dominant. Such changes in microbial communities affect not only soil carbon balance but also functioning and survival of plants. Changes in microbial community might also play some role in controlling the plant community structure and its resistance to disturbances and resilience (Classen et al. 2015).

5.4 Nitrogen Cycle and Microorganisms

Nitrogen reserves are available in abundance in air, in rock deposits and from living and dead organic matter. It is an important element required for synthesis of cellular components for all living beings. Nitrogen cannot be utilized directly by plants in its atmospheric form, hence requiring a more reactive form of nitrogen (Buresh et al. 1980).

The nitrogen cycle is one of the most important biogeochemical cycles on earth. It cycles the flow of nitrogen from atmosphere into ecosystems, both marine and terrestrial, through nitrogen fixation and finally returned to the atmosphere through denitrification (Wan et al. 2005). The fixed nitrogen is subsequently converted into a wide range of proteins and nucleic acids and oxidized compounds by microbes (Arnone 1999; Wan et al. 2016). Nitrogen cycle involves six distinct processes, mediated by microbes that proceed in an orderly fashion. Various processes like nitrogen fixation, nitrification, assimilation, ammonification and denitrification form the whole nitrogen cycle (Pajares and Bohannan 2016). Figure 5.2 shows details of nitrogen cycle.

Nitrogen fixation may be natural or industrial. In the natural process, nitrogen-fixing bacteria play a major role in nitrogen cycle as about 90% of nitrogen fixation happens due to them (Hu et al. 2016). These microbes are divided majorly into two groups. First, a symbiotic species that use root nodules of selective plants to live, mainly legumes, for example, *Rhizobium*, *Frankia* and certain species of *Azospirillum*. The second species live without host and freely and are found in soil systems and aquatic biomes, for example, *Cyanobacteria*: *Anabaena*, *Nostoc*,

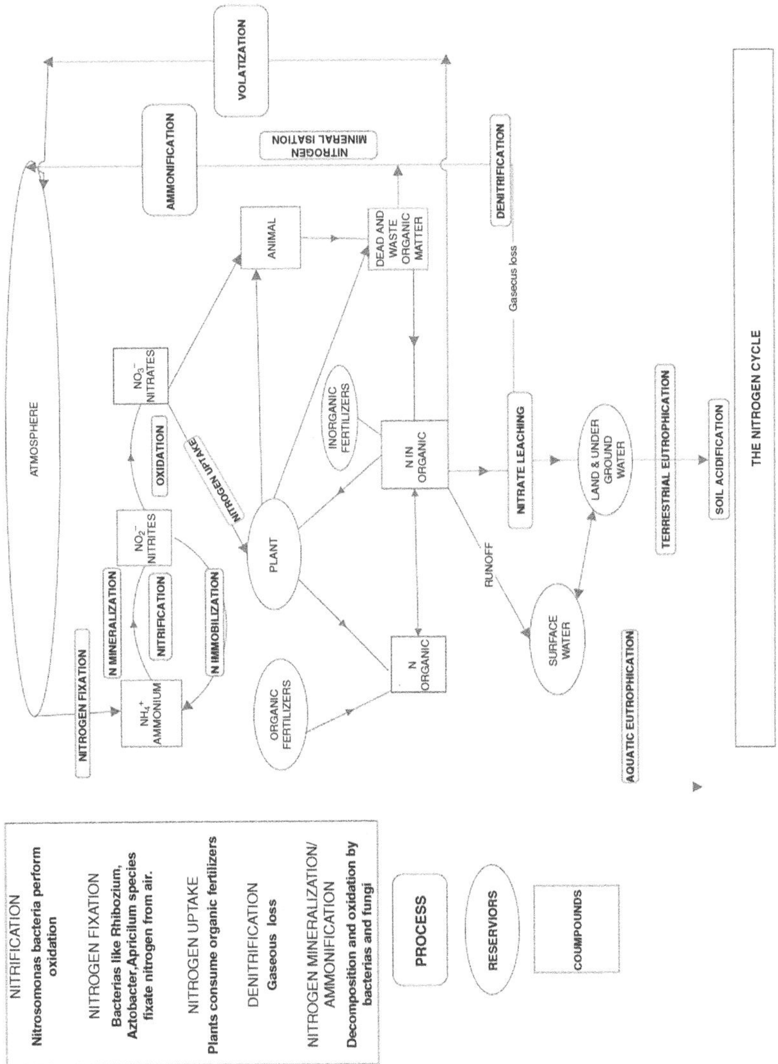

Fig. 5.2 The biogeochemical cycle of nitrogen

Azotobacter, etc. (Allison and Treseder 2008). Nitrogen fixation and nitrification are the processes leading to the formation of nitrate and ammonia. Nitrogen in the form of nitrate and ammonia is utilized by the plants and animals and the process is called assimilation. Plants absorb nitrogen through their roots and integrate them as proteins and nucleic acids. Animal use these by eating plants (Barnard et al. 2005a, b). Ammonification is the process where the organic nitrogen formed in the process of assimilation is converted into ammonia and hence becomes available for further nitrification and assimilation (Manning and Tiedemann 1995).

Nitrification is the biological conversion of ammonia to nitrate nitrogen. This usually takes place in two steps. In the first step, the microbes called *Nitrosomonas* convert ammonia and ammonium to nitrate, and then in the second step, the microbes called *Nitrobacter* convert nitrite to nitrate. The process is very rapid as these bacteria are aerobic and require dissolved oxygen of 1.0 mg/L or more for conversion. Denitrification is an anaerobic process in which nitrates are biologically reduced to nitrogen gas and released in air. The facultative and heterotrophic microbes are required for the process. This process occurs when oxygen is almost depleted (less than 0.5 mg/L). Nitrates act as oxygen source, are broken to gain oxygen and are converted to nitrous oxide released into air (Jiang et al. 2019).

5.4.1 Effect of Human Activities on Nitrogen Cycle

Anthropogenic activities have influenced the nitrogen cycle. Alteration in available nitrogen for plants has limited the growth of the plants as well as decreases its nutrient content in some region, mainly temperate and boreal. This problem is resolved by applying fertilizers into soil, which had initially increased the production of crops and plants. However, the bulk (80%) of N fertilizer applied in the field is washed off with run-off water or is lost as gas emissions and goes into the environment. At present, industrial fertilizers play a major role in providing nutrients to crop plants and about 50% food production depends on them. Use of industrial fertilizers and legume cultivation has increased the nitrogen addition to the environment to double. These increased quantities of plants sequestered the atmospheric carbon into the system and are believed to be the only positive effect of human activities on nitrogen cycle (Rakshit et al. 2012). The dumping of nitrogen especially in the form of fertilizers has led to the build-up of reactive N species in the environment and can have a toxic effect on humans as well as plants and animals. Thus, it is important to understand the various processes of nitrogen cycle so that agricultural practices can be improved, thereby minimizing detrimental effect of dumping of N in environment (Wallenstein and Hall 2012).

5.4.2 Effect of Enhanced CO_2 on N Cycle Microorganisms

In the process of plant growth and species diversity, CO_2 plays a crucial role. Studies reveal that rise in atmospheric CO_2 concentration has led to increased carbon from the atmosphere into the plants (Nie et al. 2014). It also enhanced the carbon content into the soil through rhizodeposition, leading to an increase in the organic matter content of soil. This would lead to further carbon sequestering causing simultaneous increase in nitrogen sequestration. Sequestered nitrogen will not be available for plant absorption and hence will gradually limit the plant productivity in terrestrial ecosystem (Hoosbeek et al. 2004). However, some studies also reveal that the soil carbon is not affected by change in CO_2 despite higher C inputs (Jensen et al. 2003). The impact of rising CO_2 is hard to predict without a good knowledge of interaction between carbon (C) and nitrogen (N) cycles (Phillips et al. 2012; Zang et al. 2015). There have been cases in which CO_2 has been responsible for enhanced N retention and decrease in leaching of nitrates and denitrification process (Phillips et al. 2006, 2009). In some other cases, increase in CO_2 has enhanced the process of leaching of nitrates and denitrification (Phillips et al. 2006). This increased CO_2 also increases the length and density of roots of the plants, hence also improving the N intake (nitrate and ammonium) of plants (Barnard et al. 2005a; Castro et al. 2010; Das and Mangwani 2015) and altering the N pool in soils (Björsne et al. 2014). N cycle processes like nitrification and denitrification are influenced by elevated CO_2 concentrations and in turn impact inorganic N concentrations in soil, leaching of nitrate and emission of N_2O (Cantarel et al. 2011; De Vries and Shade 2013). It is also important to understand the effect of elevated CO_2 on microbial N biomass because N immobilization in microbial biomass can have impact on plant productivity especially in N-limited ecosystems (De Vries and Shade 2013). Elevated CO_2 may have a good effect or no effect on soil microbial biomass of N. Change in microbial biomass is seen due to addition of fertilizers and hence the fertilizer-free soil has microbes insensitive to elevated CO_2 (Hartwig et al. 2002; Nowak et al. 2004; Fuchslueger et al. 2014). Studies also reveal that elevation in CO_2 is also responsible for increased root exudation which leads to more N immobilized in microbial biomass (Touceda-González et al. 2017). When the demand for N increases by heterotrophic bacteria, it tends to decrease the ammonium availability for nitrifiers and availability of soil nitrates (Cao et al. 2016). Root exudation results in mineralization of N as microbial cells and ammonium content in the soil increase (Paterson et al. 1997). Increased rate of mineralization can also promote nitrification and as a result soil nitrate concentration is modified (Zheng et al. 2008). Water availability is also affected by change in CO_2 concentration due to decrease in the rate of passage of CO_2 entering, or water vapour exiting from plants, also known as stomatal conductance. When this occurs for long period, it results in increase in denitrification process and loss in N reserves of soil (Zheng et al. 2008).

Due to human intervention and plants grown in fields, agricultural soils have a huge impact on global carbon and nitrogen cycles. For example, emission of nitrous oxide increases tenfold in cultivated soils as compared with conventional tillage practices (Robertson et al. 2000). From an agricultural viewpoint, elevated CO_2 concentration can alter the crop productivity and sustainability by improving N-use efficiency (Gamper et al. 2004; Nowak et al. 2004; Chen et al. 2017a, b). Due to increased CO_2, increase in fungal abundance and diversity has been observed in semi-arid regions. In dryland agricultural systems, the N cycle processes continue even under dry conditions, and the low N content of soil is the limitation for N availability (Schimel 2018; Li et al. 2020).

5.4.3 Effect of Enhanced Temperature on N Cycle

Microbial growth and enzyme activity are influenced by changes in temperature. Thus, change in temperature affects physiological processes like mineralization rate, growth of plants especially roots, diversity of plants and their distribution. Water occupies 70–90% of cell mass of microbes. Due to the warming and fluctuations in water level, it affects the soil microbial community in arid and semi-arid ecosystems. Warming induces water stress in soil microorganism and reduces their biomass. It is also responsible for decrease in the growth of plants which hugely impact the growth of soil microbes following which the nutrient content is reduced significantly (Pendall et al. 2004; Abbasi and Müller 2011; Xu et al. 2019). Impact of nitrogen on soil microbial communities also depends on water, as water and nitrogen have collective impact on microbes of nitrogen cycle. Higher water availability can improve the response of N cycle microbes. This is more prominently seen in temperate grasslands as water is limited in such ecosystems. In a case study, it is revealed that annual precipitation can drastically modify warming effects on microbe's community soil fungi in soil of meadows–steppe (Eckersten et al. 2001; Arcand et al. 2013). The microbes were stimulated by N addition or warming only in the presence of water and showed no response in the absence of it. This proves that water is primarily a limiting factor, and the warming effect of the functioning of microbes is dependent on the amount of water available (Kool et al. 2011; Rütting and Andresen 2015). Water stresses offer a very adverse growing condition and hinder activities in most organisms (Diao et al. 2020; Lafuente et al. 2020). Soil microbes constantly undergo a water stress environment and may get better adapted to drought environment. Hence, they become resistant to water stress. In semi-arid and steep desert sites soil microbes do not respond to warming much. Also, if microbes are not killed in the process of increased warming, it is seen that microorganisms lead to higher enzyme activities and increase in assimilation of nutrients (Chen et al. 2017a, b; Zhang et al. 2017). With the advancing world, high-latitude biomes, such as boreal and temperate ecosystems, experience the swiftest rates of impact of warming due to increased emission of harmful greenhouse gases. The impact of warming is also seen in snow-prone areas (Magill et al. 2000; Garrett et al. 2006; Caldwell et al. 2007; Butterly et al. 2015).

Microbial taxa show resistance to climate change conditions such as warming (Zak et al. 2011; Eldridge et al. 2020). Wood decay fungi which decompose the components of dead plants, such as cellulose, hemicellulose and lignin, have higher sensitivity to changes in temperature (Ainsworth and Long 2005; Choi et al. 2005; Maestre et al. 2013). Mycorrhizal fungi which live in living plant roots, exchanging nutrients from plant to soil, are comparatively less sensitive and can have both positive and negative responses to rise in temperature depending upon how this influences the soil and plant factors like nutrient and moisture present in soil and physiology of plants (Iversen 2010). Decomposition rate of microbes also increases with warming resulting in more soil carbon content (Garcia et al. 2020).

In snow-free months of a year, increases in carbon and nitrogen concentration take place as organic matters of soil decompose. But these effects tend to reduce during winter months under the increased freeze and thaw. It is also found that increase in freeze and thaw cycle disturbs the microbial plant interaction in N cycle processes and inorganic N availability is enhanced. Like warming, few microbial taxa can acclimate to freezing conditions. Different species have different levels of tolerance towards freezing. Since there are multiple functional groups of microbes involved in various processes of N cycle, like decomposition, nitrification and denitrification, it is hard to predict the behaviour of overall functional groups in the influence of change in climate (Yergeau and Kowalchuk 2008; Dooley and Treseder 2012). Most of the studies conducted are on biomes from artic, boreal or temperate regions as they usually have more impact from global climatic change like increase in temperature at higher elevation (French et al. 2009; Wan et al. 2016). A study on shrub land ecosystems has showed that there is an increase in soil respiration due to warming (French et al. 2009).

A high variability of mineralization of N is observed under the influence of warming. But as there is a lack of direct connection among temperature and N mineralization, the occurrence of processes on nitrogen cycle is terribly slow and has little impact on the N cycle. The influence of temperature on soil respiration controls carbon balance more in the short term rather than N mineralization controlled by water. In extreme temperature as N mineralization becomes unresponsive due to lack or excess of water, this promotes the N limitation process impacting both plant and microbial growth and also limits carbon sequestering as mentioned before. Higher increase in temperature may influence the soil moisture impacting the water-dependent process of N mineralization which would further cause increased N leaching and C sequestration.

Warming induces stress in microbial communities under various biogeochemical cycling; thus, there are physiological trade-offs and there is reallocation of resources in between growth and survival mechanisms. During growth seasons, the composition of soil microbial community tends to move towards fast-growing species that use less carbon (CO_2). These species are decomposers of cellulose and polysaccharides from plants as C sources and release CO_2 in huge amount in the atmosphere. Microbial activity may also decline with the drop in soil moisture as the growth of microbes is dependent on moisture availability. Higher temperature during growing season and freeze–thaw cycles of winter combined together negatively impact the biogeochemical cycles, by decreasing the amount of extractable organic C and N in soils. Moreover, a reduction in enzyme activities, respiration and biomass of microbes is also seen. A compounded suppression under stressful temperature conditions can occur, if there is biomass decrease in active microbes or if they exchange their traits which help them decompose with traits allowing them to adapt to temperature fluctuations. The traits which allow them to be more tolerant in these stress conditions include dehydration capacity, osmolyte production, thick cell walls and shock resistance proteins, C-storage vesicles. In the coming years, high grassland ecosystems are going to face more temperature rise, and by 2100 the temperature is going to rise by 3–8 °C.

5.4.4 *Effect of Drought and Increased Precipitation on N Cycle*

Droughts or lack of precipitation causes immense stresses in all living organisms especially soil microbes. Sometimes, droughts or lack of precipitation is also responsible for making them extinct (Fierer et al. 2005). The presence or lack of water plays an important role in physiology of plant communities and in regulating soil microbial activities. Drought and wet–dry cycles create immense challenge and bring out physiological stress in microbes as microbial population vitality and composition are altered by various factors like reduction in water and nutrient due to warming. Soil microbes have tendency to adapt to their immediate surrounding by undergoing osmosis when stressed due to drought, in the process of which they tend to retain water in their cells as the surrounding dries up. Fungi, although more drought-prone than bacteria, were found to be more repressed than bacteria in a study of grassland ecosystem. This was seen mainly due to increase in salinity and alkalinity of soil present as this change favours fungi growth. Also bacteria tend to be better at tolerating high salt concentrations and hence are more resistant to drought caused by warming (Niklaus et al. 2001; Bai et al. 2013). Other studies have revealed that microbes have positive reaction in response to increase in nitrogen when there is availability of water or lack of water stress; hence, this shows that although nitrogen have power of limiting the growth of microbes, its effects are highly dependent on change in precipitation. Increase in precipitation can incorporate nitrogen in soil and enhance the enzyme activities of microbes as water is necessary for nutrient distribution and renewal of soil (Rengel and Marschner 2005; Wan et al. 2016). Precipitation tends to release the microbes from there tensed state in drought conditions by replenishing the soil with resources. Hence, precipitation and drought take place alternatively.

The effect of drought can be seen on both nitrogen and carbon cycles although the impact is different. Mineralization increases in nitrogen cycle as the dry soil is wetted due to precipitation. The rewetted soils are rich in nitrogen and fuel the re-growing microbes with excess nitrogen, leading to nitrogen mineralization. Bacterial osmolytes and dead microbes are responsible for nitrogen-rich substrates that enhance the nitrogen content in soil. Fungi produce trehalose and polyols, which are nitrogen-free osmolytes that hinder the mobilization on rewetting (Robertson et al. 2000; Garrett et al. 2006; Butterly et al. 2015). Nitrification is sensitive to drought conditions, and the available ammonia is constrained in dry soils, but with the occurrence of precipitation, the rewetted soil generates a mass of nitrogen, showing a saturated state in the soil surrounding. In dry soil, ammonia is the dominant form of nitrogen, but post rewetting, a swift increase in nitrification is seen which allows a flux of nitrogen in gaseous form. Hence, drying–rewetting changing aspects appear to have disproportional effects on nitrogen losses (Ainsworth and Long 2005; Zak et al. 2011; Delgado-Baquerizo et al. 2013; Eldridge et al. 2020).

Drylands (arid, semi-arid and dry–subhumid ecosystems) provide ecosystem services like cattle raising and wool, meat and food production. Due to prevalent dry conditions, these ecosystems are more vulnerable to climate change. The change in precipitation and temperature rise has encouraged expansion of dryland and is expected to cover 10% of earth's surface by the end of this century (Smucker et al. 2007; Dong et al. 2010; Mueller et al. 2015). In these ecosystems, soils are generally deficient in nutrients; thus, nitrogen concentration plays an important role in determining net primary production and decomposition of organic matter. Dryland ecosystems are major contributor for gaseous N emissions and account for 30% of global emissions. Surface soil communities of drylands are comprised of mosses, lichens and cyanobacteria as they occupy open spaces between plant canopies (Gruza et al. 1999; McMichael et al. 2006; Rajkumar et al. 2013). Water availability is considered as an important parameter along with temperature for N cycle microbial-mediated processes. The various processes of N biogeochemical cycle like N fixation, production of dissolved organic N, nitrification and emission of gases are mediated and influenced by microbes growing in dryland soils.

5.4.5 Effect of Extreme Weather Events on N Cycle Microorganisms

Extreme weather events like waterlogging and extreme droughts put a severe impact on biomes by changing patterns of water availability to plants and microbial communities and also the physiochemical properties of soil. Changes in soil structure and pH brought by these weather events affect the availability of soil nutrients and cause changes in microbially mediated processes in biogeochemical cycles (Rosenzweig et al. 2001; Kumar et al. 2003; Coelho et al. 2013). According to recent researches, the community and functions of microbes show variable response to varying weather phenomena. Microbial community might be resistant to the various changes brought out by extreme weather events, and the ecosystem functioning is not believed to be affected by community changes in microbes. It is important to understand the microbial responses in terms of both community and functioning as these play a major role in the working of nutrient cycles and their sinking and pooling of the nutrient compounds (Zepp et al. 2007; Bowker et al.

2011; Castillo-Monroy et al. 2011). Along with weather events like drought and floods, one extreme weather event is wildfires or fire in general. Wildfire caused by extreme dry weather or any other anthropogenic activities also impacts the N cycle processes channelized by microbes. Severe fire tends to modify the properties of soil such as its biological, physical and chemical parameters, depending on temperature peak and its duration and soil's initial conditions, and negatively impacts the soil microbes too. Ash accumulated post fire changes the pH of soil and nitrogen gets volatized at temperature above 200 °C (Neary et al. 1999; Dooley and Treseder 2012). Nutrient availability is equally impacted and stays affected for many years post fire destruction. In some researches, it is also seen that N mineralization actually increases initially, increasing the inorganic N content in soil, but then tends to decrease approx. after 6 months into its original state. N mobilizes, causing leaching of nitrite oxide (NO^{3-}) through soil later post any fire event (Moreno-Jiménez et al. 2020).

5.5 Conclusions

Microbial processes associated with biogeochemical cycles play an important role in global fluxes of key greenhouse gases like CO_2, CH_4 and N_2O. These microbial processes are influenced greatly by climate change. These changes can be either positive (increased cell biomass and/or enhanced physiological functioning) or negative (decreased cell biomass/or reduced physiological functioning). Depending upon the response of the microorganisms, they either can help in maintaining the ecological balance and mitigating the effect of climate change or can aggravate the problem. Thus, it is necessary to study the changes caused due to climate change on microbial processes associated with biogeochemical cycles. This aspect must be incorporated in the models predicting the impact of climate change and mitigation measures, only then the results will be more realistic and meaningful. Most of the studies conducted have taken into account the effect of one factor. However, in nature, all the physical factors exert their influence at any given point of time. Thus, it is necessary to conduct more studies that mimic natural conditions as much as possible as the interactive effect of various climatic factors will be different from single factor effect. Moreover, due to climate change, incidences of extreme weather events have increased, but very few studies have been conducted in this direction. Thus, future studies should also take into the account the effect of extreme weather event.

References

Abatenh E, Gizaw B, Tsegaye Z et al (2018) Microbial function on climate change – a review. Environ Pollut Climate Change 2:1

Abbasi MK, Müller C (2011) Trace gas fluxes of CO_2, CH_4 and N_2O in a permanent grassland soil exposed to elevated CO_2 in the Giessen FACE study. Atmos Chem Phys Discuss 11:4199–4227
Ainsworth EA, Long SP (2005) What have we learned from 15 years of free-air CO_2 enrichment (FACE)? A meta-analytic review of the responses of photosynthesis, canopy properties and plant production to rising CO_2. New Phytol 165:351–372
Allison SD, Treseder KK (2008) Warming and drying suppress microbial activity and carbon cycling in boreal forest soils. Glob Chang Biol 14:2898–2909
Arcand MM, Knight JD, Farrell RE (2013) Estimating belowground nitrogen inputs of pea and canola and their contribution to soil inorganic N pools using 15N labeling. Plant Soil 371:67–80
Arnone JA (1999) Symbiotic N_2 fixation in a high alpine grassland: effects of four growing seasons of elevated CO_2. Funct Ecol 13:383–387
Bai E, Li S, Xu S et al (2013) A meta-analysis of experimental warming effects on terrestrial nitrogen pools and dynamics. New Phytol 199:441–451
Bardgett RD, Freeman C, Ostle NJ (2008) Microbial contributions to climate change through carbon cycle feedbacks. ISME J 2:805–814
Bardgett RD, Caruso T (2020) Soil microbial community responses to climate extremes: resistance, resilience and transitions to alternative states. Philos Trans R Soc B 375:20190112
Barnard R, Leadley PW, Hungate BA (2005a) Global change, nitrification, and denitrification: a review. Glob Biogeochem Cyc 19:1–13
Barnard R, Leadley PW, Lensi R et al (2005b) Plant, soil microbial and soil inorganic nitrogen responses to elevated CO_2: a study in microcosms of Holcus lanatus. Acta Oecol 27:171–178
Bhatla R, Pant M, Singh D et al (2020) Evaluation of cold wave events over Indo-Gangetic plain in India. J Agrometeorol 22:233–238
Björsne AK, Rütting T, Ambus P (2014) Combined climate factors alleviate changes in gross soil nitrogen dynamics in heathlands. Biogeochemistry 120:191–201
Blankinship JC, Niklaus PA, Hungate BA (2011) A meta-analysis of responses of soil biota to global change. Oecologia 165(3):553–565
Bowker MA, Mau RL, Maestre FT et al (2011) Functional profiles reveal unique ecological roles of various biological soil crust organisms. Funct Ecol 25:787–795
Bradford MA, Davies CA, Frey SA, Maddox TR, Melillo JM, Mohan JE, Reynolds JF, Treseder KK, Wallenstein MD (2008) Thermal adaptation of soil microbial respiration to elevated temperature. Ecol Lett 11:1316–1327
Brando P, Macedo M, Silverio D et al (2020) Amazon wildfires: scenes from a foreseeable disaster
Buresh RJ, Casselman ME, Patrick WH (1980) Nitrogen fixation in flooded soil systems: a review. Adv Agron 33:149–192
Butterly CR, Armstrong R, Chen D et al (2015) Carbon and nitrogen partitioning of wheat and field pea grown with two nitrogen levels under elevated CO_2. Plant Soil 391:367–382
Caldwell MM, Bornman JF, Ballare CL et al (2007) Terrestrial ecosystems increased solar ultraviolet radiation, and interactions with other climate change factors. Photochem Photobiol Sci 6:252–266
Cantarel AAM, Bloor JMG, Pommier T, Guillaumaud N, Moirot C, Soussana JF, Poly F (2011) Effects of climate change drivers on nitrous oxide fluxes in an upland temperate grassland. Ecosystems 14:223–233. https://doi.org/10.1007/s10021-010-9405-7
Cao Y, Lu C, Quan Z et al (2016) Elevated O_3 decreased N Rhizodeposition of spring wheat and its availability to subsequent buckwheat. Soil Tillage Res 162:18–25
Castillo-Monroy AP, Bowker MA, Maestre FT et al (2011) Relationships between biological soil crusts, bacterial diversity and abundance, and ecosystem functioning: insights from a semi-arid Mediterranean environment. J Veg Sci 22:165–174
Castro HF, Classen AT, Austin EE et al (2010) Soil microbial community responses to multiple experimental climate change drivers. App Environ Microbiol 76:999–1007
Chen J, Nie Y, Liu W et al (2017a) Ammonia-oxidizing archaea are more resistant than denitrifiers to seasonal precipitation changes in an acidic subtropical forest soil. Front Microbiol 8:1384

Chen J, Xiao G, Kuzyakev Y et al (2017b) Soil nitrogen transformation responses to seasonal precipitation changes are regulated by changes in functional microbial abundance in a subtropical forest. Biogeosciences 14:2513–2525
Choi DS, Quoreshi AM, Maruyama Y et al (2005) Effect of ectomycorrhizal infection on growth and photosynthetic characteristics of *Pinus densiflora* seedlings grown under elevated CO_2 concentrations. Photosynthetica 43:223–229
Classen AT, Sundqvist MK, Henning AJ et al (2015) Direct and indirect effects of climate change on soil microbial and soil microbial-plant interactions: what lies ahead? Ecosphere 6(8):130. https://doi.org/10.1890/ES15-00217.1
Coelho F, Santos A, Coimbra J et al (2013) Interactive effects of global climate change and pollution on marine microbes: the way ahead. Ecol Evolut 3:1808–1818
Collins SL, Sinsabaugh RL, Crenshaw C et al (2008) Pulse dynamics and microbial processes in aridland ecosystems. J Ecol 96:413–420
Cook N (2019) Cyclones Idai and Kenneth in southeastern Africa: humanitarian and recovery response in brief. Congressional Research Service
Cruz-Martínez K, Rosling A, Zhang Y, Song M, Andersen GL, Banfield JF (2012) Effect of rainfall-induced soil geochemistry dynamics on grassland. Appl Environ Microbiol 78:7587–7595
Dalei N (2016) Forest fires in Indian state of Uttarakhand. Eurasia Review
Das S, Mangwani M (2015) Ocean acidification and marine microorganisms: responses and consequences. Oceanologia 57:349–361. https://doi.org/10.1016/j.oceano.2015.07.003
De Vries FT, Shade A (2013) Controls on soil microbial community stability under climate change. Front Microbiol 4:1–16
Delgado-Baquerizo M, Gallardo A, Wallenstein M et al (2013) Vascular plants mediate the effects of aridity and soil properties on ammonia-oxidizing bacteria and archaea. FEMS Microbiol Ecol 85:273–282
Diao T, Peng Z, Niu X et al (2020) Changes of soil microbes related with carbon and nitrogen cycling after long-term CO_2 enrichment in a typical Chinese maize field. Sustainability (Switzerland) 12
Dong H, Jiang H, Yu B et al (2010) Impacts of environmental change and human activity on microbial ecosystems on the Tibetan plateau, NW China. GSA Today 20:4–10
Dooley SR, Treseder KK (2012) The effect of fire on microbial biomass: a meta-analysis of field studies. Biogeochemistry 109:49–61
Drissner D, Blum H, Tscherko D et al (2007) Nine years of enriched CO_2 changes the function and structural diversity of soil microorganisms in a grassland. Eur J Soil Sci 58:260–269
Eckersten H, Blombäck K, Kätterer T et al (2001) Modelling C, N, water and heat dynamics in winter wheat under climate change in Southern Sweden. Agric Eco Environ 86:221–235
Eldridge DJ, Travers SK, Delgado-Baquerizo M et al (2020) Grazing regulates the spatial heterogeneity of soil microbial communities within ecological networks. Ecosystems 23:932–942
Evans CD, Jones TG, Burden A, Ostle N, Zielinski P, Cooper MDA et al (2012) Acidity controls on dissolved organic carbon mobility in organic soils. Glob Chang Biol 18:3317–3331
Frank DC et al (2015) Effects of climate extremes on the terrestrial carbon cycle: concepts, processes and potential future impacts. Glob Chang Biol. https://doi.org/10.1111/gcb.12916
Fierer N, Craine JM, Mclauchlan K et al (2005) Litter quality and the temperature sensitivity of decomposition. Ecology 86:320–326
Freeman C, Ostle NJ, Fenner N et al (2004) A regulatory role for phenol oxidase during decomposition in peatlands. Soil Biol Biochem 36:1663–1667
French S, Levy-Booth D, Samarajeewa KE et al (2009) Elevated temperatures and carbon dioxide concentrations: effects on selected microbial activities in temperate agricultural soils. World J Microbiol Biotechnol 25:1887–1900
Fuchslueger L, Bahn M, Fritz K et al (2014) Experimental drought reduces the transfer of recently fixed plant carbon to soil microbes and alters the bacterial community composition in a mountain meadow. New Phytol 201:916–927

Gamper H, Peter M, Jansa J et al (2004) Arbuscular mycorrhizal fungi benefit from 7 years of free air CO_2 enrichment in well-fertilized grass and legume monocultures. Glob Chang Biol 10:189–199

Garcia MO, Templer PH, Sorensen PO et al (2020) Soil microbes trade-off biogeochemical cycling for stress tolerance traits in response to year-round climate change. Front Microbiol 11:616

Garrett KA, Dendy SP, Frank E et al (2006) Climate change effects on plant disease: genomes to ecosystems. Annu Rev Phytopathol 44:489–509

Gougoulias C, Clark JM, Shaw LJ (2014) The role of soil microbes in the global carbon cycle: tracking the below-ground microbial processing of plant-derived carbon for manipulating carbon dynamics in agricultural systems. J Sci Food Agric 94:2362–2371

Gruza G, Rankova E, Razuvaev V et al (1999) Indicators of climate change for the Russian Federation. Clim Chang 42:219–242

Gupta S (2018) More than 110 killed by high-intensity dust storm in India. CNN. Retrieved 3 May 2018. Accessed 16 Sept 2020

Gupta AK, Nair SS, Ghosh O et al (2014) Bundelkhand drought: retrospective analysis and way ahead. National Institute of Disaster Management. New Delhi, p 148

Hartwig UA, Wittmann P, Braun R et al (2002) Arbuscular mycorrhiza infection enhances the growth response of *Lolium perenne* to elevated atmospheric PCO_2. J Exp Bot 53:1207–1213

Hoosbeek MR, Lukac M, Dam DV et al (2004) More new carbon in the mineral soil of a poplar plantation under free air carbon enrichment (POPFACE): cause of increased priming effect? Glob Biogeochem Cycles 18:GB1040

Hu HW, Macdonald CA, Trivedi P et al (2016) Effects of climate warming and elevated CO_2 on autotrophic nitrification and nitrifiers in dryland ecosystems. Soil Biol Biochem. 92:1–15

IPCC (2007) Climate change 2007: the physical science basis. Contribution of working group I to the fourth assessment report of the intergovernmental panel on climate change. Cambridge University Press, Cambridge, p 996

Iversen CM (2010) Digging deeper: fine-root responses to rising atmospheric CO_2 concentration in forested ecosystems. New Phytol 186:346–357

Jensen KD, Beier C, Michelsen A et al (2003) Effects of experimental drought on microbial processes in two temperate heathlands at contrasting water conditions. Appl Soil Ecol 24:165–176

Jiang L, Shao J, Shi Z et al (2019) Responses of grasslands to experimental warming. In: Ecosystem consequences of soil warming: microbes, vegetation, fauna and soil biogeochemistry. Elsevier, pp 347–384

Kandeler E, Mosier AR, Morgan JA, Milchunas DG, King JY, Rudolph S, Tscherko D (2006) Response of soil microbial biomass and enzyme activities to the transient elevation of carbon dioxide in a semi-arid grassland. Soil Biol Biochem 38:2448–2460

Karyono TH, Vale R, Vale B (2019) Sustainable building and built environments to mitigate climate change in the tropics: conceptual and practical approaches. Springer

Keiblinger KM, Hall EK, Wanek W et al (2010) The effect of resource quantity and resource stoichiometry on microbial carbon-use-efficiency. FEMS Microbiol Ecol 73:430–440

Kol E, Flint EA (1968) Algae in green ice from the Balleny Islands, Antarctica. NZ J Bot 6 (3):249–261. https://doi.org/10.1080/0028825X.1968.10428810

Kool D, Dolfing J, Wrage N (2011) Nitrifier denitrification as a distinct and significant source of nitrous oxide from soil. Soil Bio Biochem 43:174–178

Krivtsov V, Bezginova T, Salmond R, Liddell K, Garside A, Thompson J et al (2006) Ecological interactions between fungi, other biota and forest litter composition in a unique Scottish woodland. Forestry 79:201–216

Kumar A, Tyagi M, Jha P et al (2003) Inactivation of cyanobacterial nitrogenase after exposure to ultraviolet-B radiation. Curr Microbiol 46(5):380–384

Lafuente A, Duran J, Delgado-Baquerizo M et al (2020) Biocrusts modulate responses of nitrous oxide and methane soil fluxes to simulated climate change in a Mediterranean dryland. Ecosystems

Li S, Bush RT, Maoo R et al (2016) Extreme drought causes distinct water acidification and eutrophication in the lower lakes (lakes Alexandrina and Albert), Australia. J Hydrol. https://doi.org/10.1016/j.jhydrol.2016.11.015

Li L, Zheng Z, Wang W et al (2020) Terrestrial N_2O emissions and related functional genes under climate change: a global meta-analysis. Glob Chang Bio. 26:931–943

Lott FC, Christidis N, Stott PA (2013) Can the 2011 East African drought be attributed to human-induced climate change? Geophys Res Lett 40:1177–1181. https://doi.org/10.1002/grl.50235

Maestre FT, Escolar C, Guevara M et al (2013) Changes in biocrust cover drive carbon cycle responses to climate change in drylands. Glob Chang Bio. 19:3835–3847

Magill A, Aber G, Berntson G et al (2000) Long-term nitrogen additions and nitrogen saturation in two temperate forests. Ecosystems 3:238–253

Mandal A, Neenu S (2012) Impact of climate change on soil biodiversity – a review. Agric Rev 33:283–292

Manning W, Tiedemann AV (1995) Climate change: potential effects of increased atmospheric carbon dioxide (CO_2), ozone (O_3), and ultraviolet-B (UV-B) radiation on plant diseases. Environ Pollut 88:219–245

McMichael A, Woodruff R, Hales S (2006) Climate change and human health: present and future risks. Lancet 367:859–869

Meisner A, Bååth E, Rousk J (2013) Microbial growth responses upon rewetting soil dried for four days or one year. Soil Biol Biochem 66:188–192

Melillo JM, Butler SM, Johnson JE, Mohan JE, Lux H, Burrows E, Bowles FP, Smith RM, Vario CL, Hill T, Burton AJ, Zhou Y, Tang J (2011) Soil warming, carbon-nitrogen interactions and forest carbon budgets. Proc Natl Acad Sci USA 108:9508–9512

Moreno-Jiménez E, Ochoa-Huesa C, Plaza C et al (2020) Biocrusts buffer against the accumulation of soil metallic nutrients induced by warming and rainfall reduction. Commun Biol 3(1)

Mueller RC, Belnap J, Kuske CR (2015) Soil bacterial and fungal community responses to nitrogen addition across soil depth and microhabitat in an arid Shrubland. Front Microbiol 6

Nardo CD, Cinquegrana A, Papa S, Fuggi A, Fioretto A (2004) Laccase and peroxidase isoenzymes during leaf litter decomposition of Quercus ilex in a Mediterranean ecosystem. Soil Biol Biochem 36:1539–1544

National Hurricane Center and Central Pacific Hurricane center (2020) 2020 Atlantic Hurricane season. Accessed 17 Sept

Neary DG, Klopatek CC, DeBano L et al (1999) Fire effects on belowground sustainability: a review and synthesis. For Ecol Manage 122:51–71

Nie M, Pendall E, Bell C et al (2014) Soil aggregate size distribution mediates microbial climate change feedbacks. Soil Bio Biochem. 68:357–365. https://doi.org/10.1016/j.soilbio.2013.10.012

Niklaus PA, Kandeler E, Leadley P et al (2001) A link between plant diversity, elevated CO_2 and soil nitrate. Oecologia 127:540–548

Nowak RS, Ellsworth DS, Smith SD (2004) Functional responses of plants to elevated atmospheric CO_2 – do photosynthetic and productivity data from FACE experiments support early predictions? New Phytol 162:253–280

Pajares S, Bohannan JM (2016) Ecology of nitrogen fixing, nitrifying, and denitrifying microorganisms in tropical forest. Soils 7:1–20

Paterson E, Hall J, Rattray E et al (1997) Effect of elevated CO_2 on rhizosphere carbon flow and soil microbial processes. Glob Chang Biol 3:363–377

Pearson R, Phillips S, Loranty M et al (2013) Shifts in Arctic vegetation and associated feedbacks under climate change. Nat Clim Chang 3:673–677

Pendall E, Bridgham S, Hanson P et al (2004) Below-ground process responses to elevated CO_2 and temperature: a discussion of observations, measurement methods, and models. New Phytol 162:311–322

Phillips DA, Fox TC, Six J et al (2006) Root exudation (net efflux of amino acids) may increase rhizodeposition under elevated CO_2. Glob Chang Biol 12:561–567

Phillips RP, Meier IC, Bernhardt E et al (2009) Elevated CO_2 increases root exudation from loblolly pine (*Pinus taeda*) seedlings as an N-mediated response. Tree Physiol 29:1513–1523
Phillips RP, Meier IC, Bernhardt E et al (2012) Roots and fungi accelerate carbon and nitrogen cycling in forests exposed to elevated CO_2. Ecol Lett 15:1042–1049
Pinay G, Gumiero B, Tabacchi E, Gimenez O, Hefting MM, Burt TP, Black VA, Nilsson C, Iordache V, Bureau F, Vought L, Petts GE, Decamps H (2007) Patterns of denitrification rates in European alluvial soils under various hydrological regimes. 52(2):252–266
Rajkumar M, Prasad M, Swaminathan S et al (2013) Climate change driven plant-metal-microbe interactions. Environ Int 53:74–86. https://doi.org/10.1016/j.envint.2012.12.009
Rakshit R, Patra AK, Pal D et al (2012) Effect of elevated CO_2 and temperature on nitrogen dynamics and microbial activity during wheat (*Triticum aestivum* L.) growth on a subtropical Inceptisol in India. J Agro Crop Sci 198:452–465
Rengel Z, Marschner P (2005) Nutrient availability and management in the rhizosphere: exploiting genotypic differences. New Phytol 168:305–312
Robertson GP, Paul EA, Harwood RR (2000) Greenhouse gases in intensive agriculture: contributions of individual gases to the radiative forcing of the atmosphere. Science 289:1922–1925
Rosenzweig C, Iglesias A, Yan X et al (2001) Climate change and extreme weather events; implications for food production, plant diseases, and pests. Glob Chang Hum Heal 2:90–104
Russell-Smith J, Yates C, Whitehead P et al (2007) Bushfires 'down under': patterns and implications of contemporary Australian landscape burning. Inter J Wildland Fire 16. https://doi.org/10.1071/WF07018
Rütting T, Andresen LC (2015) Nitrogen cycle responses to elevated CO_2 depend on ecosystem nutrient status. Nutr Cycl Agroecosyst 101:285–294
Sanaullah M, Blagodatskaya E, Chabbi A, Rumpel C, Kuzyakov Y (2011) Drought effects on microbial biomass and enzyme activities in the rhizosphere of grasses depend on plant community composition. Appl Soil Ecol 48:38–44
Sankar PM, Shreedevasena S (2020) Desert locusts (Schistocerca gregaria) – a global threatening transboundary pest for food security. Res Today Spl 2:389–391
Schimel JP (2018) Life in dry soils: effects of drought on soil microbial communities and processes. Annu Rev Ecol Evol Syst 49:409–432
Schindlbacher A, Rodler A, Kuffner M, Kitzler B, Sessitsch A, Zechmeister-Boltenstern S (2011) Experimental warming effects on the microbial community of a temperate mountain forest soil. Soil Biol Biochem. 43(7):1417–1425. https://doi.org/10.1016/j.soilbio.2011.03.005
Sheik CS, Beasley WH, Elshahed MS et al (2011) Effect of warming and drought on grassland microbial communities. ISME J 5:1692–1700
Sistla SA, Joshua P, Schimel JP (2013) Seasonal patterns of microbial extracellular enzyme activities in an arctic tundra soil: Identifying direct and indirect effects of long-term summer warming. Soil Biol Biochem 66:119–129
Smucker AJM, Park E, Dorner J (2007) Soil micropore development and contributions to soluble carbon transport within macroaggregates. Vadose Zo J 6:282–290
Srivastava SP, Pandey AC et al (2020) Evaluating the 2018 extreme flood hazard events in Kerala, India. Remote Sens Lett 11:436–445
Steinweg JM, Jagadamma S, Joshua Frerichs J, Mayes M (2013) Activation Energy of Extracellular Enzymes in Soils from Different Biomes. PLoS ONE 8(3):e59943
Tabish SA, Nabil S (2015) Epic tragedy: Jammu & Kashmir floods: a clarion call. Emerg Med (Los Angel) 5:233. https://doi.org/10.4172/2165-7548.1000233
The Guardian. https://www.theguardian.com/environment/2020/apr/07/great-barrier-reefs-third-mass-bleaching-in-five-years-the-most-widespread-ever
Touceda-González M, Prieto-Fernández Renella G, Giagnoni L et al (2017) Microbial community structure and activity in trace element-contaminated soils phytomanaged by gentle remediation options (GRO). Environ Pollut 231:237–251
Wallenstein M, Hall EK (2012) A trait-based framework for predicting when and where microbial adaptation to climate change will affect ecosystem functioning. Biogeochemistry 109:35–47

Wan S, Hui D, Wallace L et al (2005) Direct and indirect effects of experimental warming on ecosystem carbon processes in a tallgrass prairie. Glob Biogeochem Cyc 19:1–13

Wan R, Chen Y, Zheng X et al (2016) Effect of CO_2 on microbial denitrification via inhibiting electron transport and consumption. Environ Sci Technol 50:9915–9922

Wikipedia contributors. Siachen conflict. Wikipedia, The Free Encyclopedia. Available at: https://en.wikipedia.org/w/index.php?title=Siachen_conflict&oldid=978007208. Accessed 16 Sept 2020

Xu Q, Sullivan JO, Wang X et al (2019) Elevated CO_2 alters the rhizosphere effect on crop residue decomposition. Plant Soil 436:413–426

Yergeau E, Kowalchuk GA (2008) Responses of Antarctic soil microbial communities and associated functions to temperature and freeze-thaw cycle frequency. Wiley Online Library 10:2223–2235

Zak D, Pregitzner K, Burton A et al (2011) Microbial responses to a changing environment: implications for the future functioning of terrestrial ecosystems. Fungal Ecol 4:386–395. https://doi.org/10.1016/j.funeco.2011.04.001

Zang H, Yang X, Feng X et al (2015) Rhizodeposition of nitrogen and carbon by Mung bean (*Vigna radiata* L.) and its contribution to intercropped oats (*Avena nuda* L.). PLoS One 10(3)

Zepp RG, Erickson D, Paul N et al (2007) Interactive effects of solar UV radiation and climate change on biogeochemical cycling. Photochem Photobio Sci 6:286–300

Zhang W, Parker KM, Luo Y et al (2005) Soil microbial responses to experimental warming and clipping in a tallgrass prairie. Glob Chang Biol 11:266–277

Zhang C, Shen J, Sun Y et al (2017) Interactive effects of multiple climate change factors on ammonia oxidizers and denitrifiers in a temperate Steppe. FEMS Microbiol Ecol 93

Zheng J, Han S, Ren F et al (2008) Effects of long-term elevated CO_2 on N_2-fixing, denitrifying and nitrifying enzyme activities in forest soils under *Pinus sylvestriformis* in Changbai Mountain. J Fores Res 19:283–287

Zibilske LM, Bradford JM (2007) Oxygen effects on carbon, polyphenols, and nitrogen mineralization potential in soil. Soil Sci Soc Am J 71:133–139

Chapter 6
Climate Change with Its Impacts on Soil and Soil Microbiome Regulating Biogeochemical Nutrient Transformations

Kristina Medhi, Richa Bhardwaj, and Raj Laxmi

Abstract Global climate change has displayed prominent impact on the distinct nutrient transformations carried out in the soil ecosystem. Climate change intensifies seasonal variations and could even exaggerate to extreme actions. Carbon, nitrogen, phosphorus, sulfur, and potassium are the essential macronutrients essential for the survival and development of living beings. These nutrients are being cycled within the biosphere through the biogeochemical networking supported by diverse and versatile soil microbial populations. The soil environment as well as the biological reactions concerning nutrient availability occurring in soil regimes is considerably disturbed under the collective effects of climate change, biological invasions, and anthropogenic alterations of the environment. The soil microbiota are the originators and extenuators of the greenhouse gases (GHGs) represented as the chief contributors to the ongoing global warming. Environmental nutrient variability is an integral part of unavoidable disturbances. Sustainable agriculture management strategies such as no tillage, crop rotations, and soil organic amendments could not only assist in preventing nutrient loss but also encourage nutrient management making farmers and their agricultural practices smart.

Keywords Climate change · Biogeochemical cycles · Soil microbiome · Omic studies · Greenhouse gases · Nutrient management

---

K. Medhi (✉) · R. Laxmi
Central Pollution Control Board (CPCB), Regional Directorate, Lucknow, Uttar Pradesh, India

R. Bhardwaj
Central Pollution Control Board (CPCB), New Delhi, India

D. K. Choudhary et al. (eds.), *Climate Change and the Microbiome*, Soil Biology 63,
https://doi.org/10.1007/978-3-030-76863-8_6

6.1 Introduction

Global climate change is the most burning and challenging issue worldwide as its characteristic feature depicts the burning of the planet Earth. It is immensely and strongly being discussed by environmentalist, microbiologists, politicians, businessman, artists, the society, and mass media worldwide. Even children of smaller age are aware of this impending truth and participate wholeheartedly to save their Mother Nature. Presently, anthropogenic activities related to industrialization and globalization have led to the excessive use of fossil fuels, the chief contributors to the ongoing global warming. Climate change might be naturally induced or humanly induced. Climate is a complex but an interactive system that describes the long-term average weather conditions prevailing at a particular region basically comprising temperature, precipitation, pressure, and wind (Abatenh et al. 2018). It connects the atmosphere, lithosphere, hydrosphere, and biosphere and is determined by the Earth's movement. Our planet is encircled by a blanket of gases to keep the planet warm permitting plants, animals, and microbes to thrive in it. However, in recent years, the gases known as GHGs such as carbon dioxide (CO_2), methane (CH_4), nitrous oxide (N_2O), halocarbon gases, ozone, water vapor, and aerosols are significantly increasing day by day, accumulating in the atmosphere, heating up the Earth's surface beyond its capacity, and heightening global warming.

Soil being the most vital fragment in our ecosystem possesses enormous potential of supporting microbial growth, nutrient enhancement, and contaminant degradation and improving plant–microbe interaction by circulating and transforming essential nutrients through decomposition and mineralization of organic matter. Microbial diversity in different ecosystem has been contributing toward climate change from eons regulating and combating its negative impressions due to their versatile metabolic rate to survive in a comprehensive range of environmental conditions. Microbes and nutrient biogeochemical cycles are the two faces of one coin (Abatenh et al. 2018) and provide long- and short-term responses to global warming as well as climate change. Soil microbiota play a fundamental part in the production of as well as exhausting GHGs by recycling and transforming essential elements such as carbon (C) and nitrogen (N) essential for growth and development (Joshi and Shekhawat 2014). Biogeochemical networks are the pillars on which a sustainable environment depends upon. Effectiveness of biogeochemical cycles encompassing nutrient availability and recycling through soil environment is influenced by microbial population that mainly controls GHG emissions (Gupta et al. 2018). Nature itself is doing an excellent job by aiding to keep a balance of the soil C and N via biogeochemical nutrient cycles.

To better understand the massive unknown information regarding their mechanism during nutrient transformation, we need to detect the microbial genetics and functional diversity present in the soil ecosystem. Thus, omic studies have been assisting researchers to characterize and quantify a set of biomolecules existing in nature to detect the function, structure, physiology, and molecular mechanisms of the biodiversity to become aware of the ecological reactions in the environment

(Biswas and Sarkar 2018; Yun-Feng 2013). The application of the next-generation sequencing techniques along with traditional molecular methods and microbiome analysis provides valuable information and could support in the all-inclusive evaluation of variations occurring within a complex microbial biodiversity interacting with the soil including the reliability of their environmental impact assessments (Miao and Liu 2018). Thus, this chapter was tried to provide an insight how climate change has impacted on the soil environment and its biodiversity basically focusing on the nutrient (carbon, nitrogen, phosphorus, sulfur, and potassium) transformations happening in soil ecosystem. These biogeochemical networks are being stimulated by microbial activity and assist in the generation as well as abatement of the gases which leads to global warming. Modifications in agricultural practices and managements could result in enhanced C sequestration resulting in minimizing climate change as well as help in improving the soil fertility through reducing nutrient runoff from agricultural soils. The present study has further tried to focus on the approaches to protect and preserve distinct nutrient transformation progressions in the soil.

6.2 Climate Change: Causes and Effects

6.2.1 Natural Causes of Climate Change

6.2.1.1 Solar Variation

The Sun provides the Earth with most of the energy needed to make it a habitable planet, and any modification in the radiation output of the Sun also affects the energy balance of the Earth's surface and atmosphere. Therefore, solar fluctuations over a period of time could affect our climate. The luminosity of the Sun at the present is roughly 20–30% higher than when it was born 4.55 billion years ago. The concentration of GHGs such as CO_2 was abundant in the atmosphere of the young Earth, and thus changes in solar energy and GHG concentrations over billions of years have equal share in shaping the Earth's climate evenly. Due to this fact, Earth is about the same temperature today as it was back then. The warming from global warming has reduced the amount of solar radiation. However, the average solar radiation reaching Earth has changed little over the past hundreds of millions of years (Haigh 2000).

The other is the Sun's magnetic field, or rather in the outer part of the Sun's interior and in the solar atmosphere. Sunspots are magnetic storms that appear as relatively dark and cool areas on the surface of the Sun that denote short-term fluctuations in solar radiation. With sunspot fluctuation, solar radiation also varies, making alterations to the effects in space, in Earth's atmosphere, and on Earth's surface. For more than 40 years, satellite data collected have observed the solar activity with a variation in solar radiation of only up to 0.1%, probably too little to affect Earth's climate as compared to the warming 50 times higher brought about by anthropogenic GHG emissions over the same time length (Fahey et al. 2017).

6.2.1.2 Earth's Position

Collective effects of changes in Earth's position relative to the Sun strongly drive Earth's *long-term* climate, thereby once triggering the beginning and end of glaciation periods (Ice Ages) (Hostetler and Mix 1999). This was first hypothesized in the 1930s by a Serbian scientist Milutin Milankovitch. He proposed *Milankovitch theory* explaining how changes in the amount of solar activity output (known as insolation) signaling the Earth could be due to the Earth's orbital movements relative to the Sun that has a profound influence on global climate throughout (Campisano 2012). These cyclical positions (known as the Milankovitch cycles) may cause up to 25% variation in the amount of incoming insolation at Earth's midlatitude regions (Buis 2020).

6.2.1.3 Plate Tectonics and Volcanism

Tectonics involves both horizontal displacements and vertical movements of the Earth's crust. The direct correlation between tectonics and climate is through altering the oblique distribution of continental masses and the effect of opening and closing gates between major ocean basins and marginal seas. The most pronounced influences, in terms of regional climate, are precipitation and rain shadow effects due to cooling and moisture loss to rising air and heating of descending air after it passes over the summit of a mountain range. Finally, altitudes and mountain formation at high latitudes may be responsible for initiating ice sheet formation (Bailer-Jones 2009).

The indirect links are through subduction, removal of material from contacting with the Earth's liquid matter via volcanism, introducing gases into the atmosphere, and chemical weathering, overriding atmospheric gases. An example for the recent Earth is the timescale associated with the breakup of the most recent supercontinent, Pangea, which has been argued to have had significant climatic effects (Lee et al. 2013; McKenzie et al. 2016). It caused variations in plate spreading proportions, associated tectonic activity, and volcanism that altered aspects of climate–tectonic timescale.

6.2.1.4 Climate and Weather Oscillations

El Niño is described as a warm-water current that periodically disrupted the local fishery through the region it flows. It has since been identified as a warming factor of the tropical Pacific Ocean. This oceanic event is connected with a fluctuation of a global-scale tropical and subtropical surface pressure pattern called the Southern Oscillation. This coupled atmosphere–ocean phenomenon, with preferred timescales of 2–7 years, is known as the El Niño–Southern Oscillation (ENSO) (IPCC 2014), responsible for the magnitudes of weather changes around the world especially linked with droughts, heavy rainfall, storms, and floods.

6.2.2 Human-Induced Climate Change

The most emerging factor is the anthropogenic activities' effect on climate. Most of them are comparable to the natural forces that impact the climate. Changes in land use through deforestation, the building of cities, the storage and use of water, and the use of energy are all important factors locally. However, significant global impact on climate occurs widely due to fossil fuel usage that upsurges gases such as carbon dioxide (CO_2), methane (CH_4), and some other gases in the atmosphere.

6.2.2.1 Greenhouse Gases

Greenhouse gases (GHGs) are thought to be the main contributor to global climate change. Despite a growing number of climate change mitigation policies, annual GHG emissions grew on average by 1.0 $GtCO_2$-eq (2.2%) per year, from 2000 to 2010, compared to 0.4 $GtCO_2$-eq (1.3%) per year, from 1970 to 2000 (IPCC 2014). Similar to natural processes, human activities remove carbon from long-term sequestered reservoirs and let it out into the atmosphere. Combustion of fossil fuels releases CO_2 that has been stored in the Earth for a very longer period of time. Burning rain forests to create farmlands, through slash-and-burn agricultural practice prevalent in the tropics, liberates CO_2 stored in forests. Deforestation also reduces the amount of CO_2 that plants absorb from the atmosphere. CO_2 quantity in the atmosphere has amplified by far more than 30% since the beginning of the Industrial Revolution due to industrialization and deforestation, witnessed to be growing at the fastest decadal rate of change (2.0 ± 0.1 ppm/year) for 2002–2011 (IPCC 2014; Trenberth et al. 2000).

Like CO_2 levels, CH_4 levels have been rising swiftly for the past many decades due to population explosion. About 60% of CH_4 entering the atmosphere presently comes from anthropogenic activities. Rice production, feeding a large proportion of the planet's population, contributes to the largest share of CH_4 production. Another source is the huge increment in the meat production that releases this gas which has a highly significant impact in recent decades. CH_4 is also emitted via landfill decomposition, waste treatments, and incomplete forest material burning.

Other GHG levels are also aggregating because of human actions. Concentrations of ozone (O_3) in the troposphere, considered as a pollutant and GHG, have been consistently rising since 1976. Tropospheric O_3 is created by the action of sunlight on nitrogen oxide and hydrocarbon pollutants such as carbon- and hydrogen-based emissions from car exhaust. Nitrous oxides (NO and N_2O) are themselves GHG that come from fossil fuel combustion, forests, and crop wastes and also from the manufacture and usage of fertilizers. N_2O concentrations have steadily increased at a rate of 0.73 ± 0.03 ppb/year over the last three decades. Chlorofluorocarbons (CFCs) are extremely potent GHG in the troposphere (IPCC 2014), destroying the ozone layer in the stratosphere and primarily accountable for the springtime O_3 hole over Antarctica.

6.2.2.2 Land-Use Changes

When folks modify the method of how they are going to use the land, they unknowingly alter their climate. The foremost dramatic example of this is often the "urban heat island effect." The development whereby in urban areas is hotter than the encompassing rural area throughout the day and particularly at nighttime. The primary cause is the excess heat generated by the running of engines and given off by buildings. The second is the lower reflective power of synthetic surfaces, like concrete and asphalt, compared to natural surfaces. Man-made surfaces store the alternative energy that strikes and stores throughout the day and rerelease it into the atmosphere at nighttime. This is often why during nighttime temperatures over cities have move up vividly within the past few decades.

6.2.2.3 Global Dimming

There are also indirect effects such as altering the droplet size in clouds, reducing precipitation rates, and lengthening the clouds' lifetime (Geresdi et al. 2006). Clouds are made when water vapor condenses around natural particles to form tiny water droplets; these droplets coalesce and fall from the sky as rain when they become large enough. Significant escalations in fossil fuel combustion have greatly increased aerosol prevalence in the atmosphere in the form of sulfates, organic carbon, and black carbon (Dawson and Spannagle 2008). When water vapor condenses around these pollutant particles, they remain scattered throughout the cloud, a phenomenon known as "global dimming" that eventually increases the amount of sunlight reflected back to space which brings about a decrease in temperature, thereby having an overall cooling effect on the climate (Desonie 2008).

6.3 Climate-Sensitive Distinct Soil Ecosystems

6.3.1 The Arctic

In view of climate change and global warming, the Arctic, aka the permafrost of the poles being one of the most climate-sensitive regions on Earth, is the biggest terrestrial reservoir of carbon compounds (Weiman 2015). Even though the permafrost occupies 9% of the landmass of the Earth, it contains 25–50% of global soil organic carbon (European Commission 2015). The average temperatures growing at nearly two times the global rate have implicated in drastic changes in the landscape such as permafrost thawing, precipitation pattern variations, and vegetation variations that affect all living beings including us (Jansson and Hofmockel 2019). Arctic ecosystems lack nutrients as the thick layer of permafrost covers huge quantities of organic materials in the soil and just a thin layer of soil exposed with nutrients is available for the plants. However, biological processes in the Arctic are considerably hampered by temperature fluctuations and permafrost. Increasing temperature due to global warming results in permafrost thawing, activating soil microbes to decompose the huge pool of stored carbon (Schuur et al. 2015) and organic materials that are frozen-bound to soil. The thawing process also leads to soil thickening and releasing considerable quantities of nutrients mostly nitrogen and phosphorus. Recently it was suggested that ~5–15% of carbon currently contained in permafrost susceptible to microbial decomposition, consequences as a considerable foundation for CO_2 emissions (Schuur et al. 2015). These emissions along with the increasing CO_2 concentration augment plant biomass production; consequently, species adapted to the specific climate will be dominated by a wider but different species distribution resulting in overall biodiversity reduction and subsequently will reduce the biodiversity linked to other food chains. Even though the emitted CO_2 is fewer than that of current fossil fuel emissions, addition of microbial CH_4 and N_2O released due to melting and degradation of permafrost, to the persisting calculation, greatly speeds up the warming impact since both these GHGs pose higher global warming potential (GWP) or climate-forcing influence than CO_2 (Jansson and Hofmockel 2019). Warming of climate (1.5–2 °C) has been predicted to reduce permafrost by 28–53%, thereby making large C reservoirs accessible for microbial respiration and GHG emissions.

6.3.2 *Forests*

Forest soils together with boreal to temperate to tropical cover, which are ~30% of the total land surface, make up ~50% of terrestrial primary production and are crucially important as soil carbon sinks containing enormous quantities of stable organic matter (Lladó et al. 2017; Cavicchioli et al. 2019). Boreal forests are the key participants in the global C cycle as they accumulate 30–40% of terrestrial carbon in the form of organic soils crucial to sustain the productivity and biodiversity of these forests (Rumpel 2019). Climate change affects forests, and forests affect climate change as both of them are interlinked. Forests help to sequester and store up to 25% of anthropogenic CO_2 facilitating mitigation of climate change but become stressed through elevated temperatures, altered precipitation patterns, and extreme weather events owing to global climate change. Plant nutrient availability affects the net C balance in forests, with nutrient-poor forests releasing more C than nutrient-rich forests (Cavicchioli et al. 2019). A variety of routes underlie the exchange of CO_2 between the forests and the atmosphere such as soil respiration, microbial decomposition, and wildfire frequency, releasing substantial amount of CO_2 from the soil (Osman 2013) with the potential for converting from net sinks to net sources of CO_2 in the future. Both forest soils' fungal and bacterial communities have shown to react to climate changes, but the biodiversity with their specific biochemical reactions differs among forest ecosystems.

6.3.3 Grasslands

Grassland cover comprises ~29% of the terrestrial surface (Cavicchioli et al. 2019) and stores an estimated 20% of the total soil carbon stock (Jansson and Hofmockel 2019). The C pool of grassland soil is large due to the deep and abundant rhizosphere that deposits carbon in the soil. With the geographical shifting of climate conditions, the distributions of many plants and animals will also alter (Bagne et al. 2012). Climate change in grasslands is intimately correlated to plant–microorganism interactions taking place in the rhizosphere and to bulk soil processes that cycle carbon and other nutrients. Grassland microbiota will also likely be affected by climate change. Climate even being a vital driver of grassland ecosystems, increasing conflicts such as extensive period of drought, frequent fires, extreme precipitation events, and grazing also play a key role in influencing soil moisture for both the aboveground plant growth and the microbial community compositions and soil functioning in grasslands. With respect to differences in soil types and plant cover across variant grassland ecosystems, the elevated CO_2 effect on plant growth will vary according to the local climate patterns and species adaptations to water availability and nitrogen availability consequently impacting microbial community functions.

6.3.4 Drylands

Global arid regions (one-third of Earth's surface) collectively store ~27% of the total terrestrial organic C stocks. Water deficiency restricting plant and microbial mechanisms characterizes deserts and dryland soils. Arid soil ecosystems range from hot and cold deserts with limited annual precipitation as compared to grasslands, and thus it is difficult to generalize microbial responses to aggregating drought with climate change (Jansson and Hofmockel 2019). Studies indicate that nutrient depletion might take place faster in drier regions, with factors responsible such as plant community composition and grazing (Morgan et al. 2008). Changes in water availability could also have an intense and long-term impact on soil microbiomes. Climate change will likely accelerate land degradation (desertification) and aggravate poverty, food, and water insecurity in drylands (Laban et al. 2018). Soil desertification (20–35% drylands) is anticipated to intensify by 11–23% by this century's end.

6.4 Impact of Climate Change on Soil Environment and Soil Microbiome

Climate change is a slow process that relatively changes temperature and precipitation rather gradually over a length of time; however, these slow ongoing and unseen climatic changes influence various soil processes, predominantly soil fertility. Natural regimes are drastically getting disturbed under the cumulative effects of both climate interruptions and biological invasions. Perturbations in the soil environment initiate microbiome ecological modifications as well as alter microbial metabolic action involved in soil nutrient cycling and preserving soil productivity.

6.4.1 Soil Formation and Development

The process of soil formation is lengthy and requires ages to form just 1 cm of soil layer. Soil formation is determined through a complex cycle between the soil minerals, its environment, and the biota contained by the soil ecosystem (Mandal and Neenu 2012). Soil-forming factors govern and regulate the formation of terrestrial soil cover. Soil is formed particularly through weathering with due course of time and also acts as an essential element source for microorganisms present within the soil. Weathering of materials on Earth's surface such as mechanical breakage of rocks or minerals enduring to chemical processes could generally work together in soil development. Weathering process also aids in soil biota growth as it could release the essential elements, and in turn the biota helps in rapidly increasing the weathering process (Mandal and Neenu 2012). Numerous parameters such as temperature and precipitation along with their variation patterns are involved in the soil formation but are vigorously impacted by climate change as it involves the amount and frequency of the soil determining factors and alterations in both biogeochemical and hydrological cycles (Gelybó et al. 2018). Temperature and precipitation ratio could also influence soil formation by providing fluctuating weathering conditions. External factors' fluctuations involved in soil formation could lead to transformation of internal factors' inclusive energy stored and requirement and water or moisture availability and most particularly could affect the biological soil content (Pareek 2017). Climate change disrupting soil minerals would make them more tolerant to weathering and result in losing of soil function to maintain fertility and becoming more dependent on synthetic fertilizers influencing nutrient cycles within the soil system. Climate is one of the main factors in determining soil's development accompanied by parent materials (rocks and minerals) and type of vegetation involved. Water or better known as soil moisture determines the extent of chemical weathering taking place as rainfall is essential for accelerating weathering. Thereby, changes in soil moisture fluxes could induce alterations in vegetation types and patterns, plant growth extents, water withdrawal by plants from soil, and the effect of enhanced CO_2 levels on plant transpiration (Pareek 2017). Moreover, fluctuating climatic conditions not only deteriorate soil formation and development but hamper weathering giving rise to dissimilar soil profiles within a same area.

6.4.2 Soil Fertility and Nutrient Availability

Soil-forming factors would be potentially affected from the ever-expanding global climatic change due to loss of biomass organic matter supply that assists in maintaining the soil fertility and improves its productivity to sustain various forms of life on the Earth. Soil physical properties greatly influence the chemical and biological processes inclusive of water absorption and adsorption, energy transportation, and biological interactions involved in nutrient regime that ultimately works in preserving soil quality and upholding the soil system fertility. Soil temperature regime regulates organic matter decomposition rate and nutrient transformation within the soil. Higher temperatures or elevated CO_2 concentrations would be able to increase nutrient availability and accelerate nutrient cycling but eventually would impact in the running out of the available organic matter content diminishing soil fertility. Temperature and precipitation fluctuations significantly impact temperature and moisture regimes related to the plant root zone as they are well-known contributors of governing nutrient accessibility, root growth and development, as well as nutrient acquisition, and the changed climate will reflect unnecessary outcomes. Nutrient cycling represents an important aspect in determining soil fertility and soil quality especially soil N cycle since it can be coupled to both water and C cycles (Mandal and Neenu 2012). The deterioration of potential and effective indicators of soil fertility is somewhat an irreversible process such as a study reporting the decrease in the amount of humus, an important indicator of fertility in Azerbaijan soils due to the global warming on ongoing soil degradation creating a negative warning as an ecological crisis (Jafarov and Mustafayev 2020). Amouzou et al. (2019) observed that climate change scenarios decreased N and P uptake by crops irrespective of CO_2 fertilization effects, and improvement in soil–crop management usage for enhancing yields and water- and N-use efficiencies was not able to compensate the total yield loss due to climate change. Global climate change has strongly impacted soil functioning focused on biological transformation and influences nitrogen and sulfur solution concentrations in soils.

6.4.3 Impact Mechanisms of Climate Change on Soil Microbes

Microbes are expected to prone toward climate change and bring about environmental, economic, and societal impacts (French et al. 2009). Climate change is described mostly by fluctuations in temperature and precipitation (Smith et al.

2008). Thus, its impact on soil microbes is demonstrated primarily via temperature and moisture change. Factors such as atmospheric CO_2 increment, altered temperature patterns, and warming pinpoint to the direct and indirect concerns on soil microbiomes (Castro et al. 2010).

6.4.3.1 Changes in Soil Microbial Diversity

Climate warming wields abiotic stress causing alterations in soil microbial diversity and their mechanisms (Shade et al. 2012). Various microbial groups have their own optimal temperature ranges for growth and activity; thus, temperature rise influences the microbial community structure. With temperature elevation, the rate of fungal and bacterial organic matter degradation, nutrient cycling, and soil aggregate formation is accelerated resulting in a shift of microbes that are better acclimatized to higher temperatures and show higher growth rate (Pareek 2017; Castro et al. 2010). Thus, climatic change precisely alters the relative soil microorganisms' richness as well as purpose due to their difference in respect to physiology, temperature sensitivity, and growth rates. Subsequently, the regulation of the specific routes of nutrient cycling supported by these microbes is directly affected (Classen et al. 2015).

6.4.3.2 Mechanisms Acting Through Physiological Changes

Climate warming accelerates soil microbial activity, thereby increasing soil respiration (Wu et al. 2011) which in turn is also influenced by variations in microbiota composition (Balser et al. 2006), substrate availability (Davidson and Janssens 2006), plant litter quality and quantity (Rustad et al. 2001), and relative abundance of available carbon (Fierer et al. 2005). Thus, global climatic changes such as rise in temperature can directly alter the rates of soil microbe's respiration in response to the temperature sensitiveness of microbial metabolism and the activities they carry out (Classen et al. 2015). Also, soil respiration is positively correlated with temperature but could be suppressed with alternating moisture rates or precipitation rates (Yiqi and Zhou 2006; Aanderud et al. 2013). Another important aspect of soil microbial activity affecting soil respiration is enzymatic activity, but climatic changes could also result to short-term enzymatic activity disbalance.

6.4.3.3 Mechanisms Acting Through Plants

Plants being essential biotic factors could alter the soil microbial respiration exchanges by liberating carbon substrates via roots (Scott-Denton et al. 2006), by adjusting soil moisture and temperature thru transpiration, and by providing shade (Lauenroth and Bradford 2006) as well as by shifting the rainfall volume reaching the soil. A fundamental application that influences in vegetation structure shifts, driving microbes and their metabolism, is by altering the quality and quantity of organic matter incoming into the soil as plant remains. Warming and varied precipitation regimes are capable of altering the distribution of vegetation composition at both local and global scales (Woodward et al. 2004) that can modify the soil physical surroundings, through changes in root design and rooting depth (Jackson et al. 1996).

6.4.3.4 Mechanisms Acting Through Moisture Fluctuations

Microbial activity, soil respiration patterns, and decomposition of SOM are mostly disturbed by factors associated with varying water and moisture content such as water transportation, gaseous and solute diffusion, and endurance of microorganisms (Aanderud et al. 2011). Water content is capable of suppressing microbial action in many environments inclusive of soils. Low moisture declines the intracellular water potential and causes reduction in the hydration and enzyme mechanism (Stark and Firestone 1995) as well as intensifies the aspects and production of CO_2 (Aanderud et al. 2011).

6.5 Factors Regulated by Climate Change

6.5.1 Temperature

Nowadays, issues related with the atmosphere have become the biggest global issues facing mankind as an outcome of scientific facts about the growing concentration of greenhouse gases in the surrounding and the changing climate of the Earth (Cubasch et al. 2001). Climate change, an environmental term, means any consistent, measurable, and long-term changes in the average global temperature that can be caused by numerous natural as well as anthropogenic activities (Eneji et al. 2017; VijayaVenkataRaman et al. 2012). Currently, Earth is undergoing rapid heating due to rising intensity of GHGs and accumulation and overloading of tremendous amount of CO_2 in the Earth's atmosphere (Allison 2009). Globally, concentration of GHGs in the atmosphere is increasing day by day through industrialization, transportation through vehicles, etc. The level of CO_2 rises due to anaerobic decay of buried lifeless remnants and natural gas, power stations, industrial units, factories, and deforestation. Since globalization and industrialization, the temperature of the Earth's surface has steeped by 0.7 degrees, and we must take effective measures soon because by 2100 temperatures might rise by as much as 5 degrees. This temperature rise will have a destructive effect globally, leading to further extreme weather conditions and additionally widespread extinction of many fauna and flora (UN 2007). When temperature increases, many different changes take place on the Earth, resulting in more floods, longer droughts, intense rains and storms, and more frequent and severe heat waves as well as alteration in the life cycle of flora and fauna which will be challenging to our humanity in the future decades.

6.5.2 *Moisture*

Decline in relative humidity over land is a dominant feature of future climate projections, with implication for impacts of worldwide climate change (Byrne and O'Gorman 2018). Humidity denotes the existence of vapors of water in the ambient air and affects how warm the air feels to us. Generally, warm temperatures feel even warmer as the moisture rises since as it rises, the rapidity upon which water evaporates at a certain temperature reduces. As the environment keeps on warming, the vision of progressively more moist heat waves will escalate. That will make heat indexes of inferior quality. Moreover, higher soil moisture, frequently stimulated by agricultural practices, can supplement more moisture to the air (Cappucci 2019). Variations in moisture also impact the pathogens of contagious infections and diseases. The infectious microorganisms of airborne transmittable disease, for example, influenza, have a tendency to be receptive to humidity conditions (Shaman and Kohn 2009; Xu et al. 2014).

6.5.3 *Precipitation*

Changes in precipitation are also one of the most serious aspects to determine the overall effect of climate change. A warmer environment can grasp more wetness, and globally vapors rise by 7% for each degree Celsius (°C) of warming. Globally, Earth's temperature is rapidly increasing and the quantity and distribution of precipitation are being distorted (Cubasch et al. 2001). Precipitation is any type of water that forms in the Earth's atmosphere and then drops onto the surface of the Earth, encompassing a wide series of forms, from rain to hail. Recent research work confirms that even if crops would respond positively to prominent CO_2 in the nonappearance of climate change, the connected effects of high temperatures, changed patterns of rainfall and precipitation, and probably increased occurrence of intense events like drought and floods will probably unite to depress soil characteristics that impact on yields and also amplify production risks in numerous areas of the world expanding the gap between wealthy and deprived countries. Climate change might cause alterations in precipitation that affects the spread of waterborne infectious microorganisms. The variations in rainfall and precipitation altered the relative abundance of *Acidobacteria* and *Proteobacteria*, where *Acidobacteria* declined with an affiliated enlargement in the *Proteobacteria* in moist relative to parched treatments (Mandal and Neenu 2012). Rainfall plays an imperative role in the expansion of waterborne pathogens. Thus, unusual rainfall after extensive drought can result in disease outbreak due to increase of disease-causing pathogens (Wilby et al. 2005). Considerably enlarged amount of precipitation has been detected on the east of Northern Europe, South and North America, and Central and Northern Asia. Precipitation is greatly changeable geographically and temporally. Westerly winds in midlatitude have become stronger in both hemispheres ever since the 1960s (IPCC 2007).

6.5.4 C/N Ratio

Uptake of carbon concentration in land ecosystems depends upon the availability of nutrients like nitrogen to sustain new development, and fertilization studies show that the accessibility of nitrogen confines prime production in numerous innate and managed ecosystems (Elser et al. 2007; Vitousek and Howarth 1991; LeBauer and Treseder 2008). The large bulk of nitrogen (N) in new plant growth obtained from the disintegration of plant wastes and organic matter from the soil results in a sturdy coupling where heterotrophic respiration depends on the organic (natural) substances produced by plants and growth of plants depends on the mineral (e.g., nitrogen) released from organic substances for the duration of decomposition (Vitousek and Howarth 1991; Melillo et al. 2002). In agriculture, N usage significantly alters universal nitrogen cycle with short- as well as long-term effects on climate change and global warming. It increases emanation of NOx (nitrous oxide), which donates 6.2%, while CO_2 and CH_4 contribute 76% and 16%, with respect to global warming. Universally, N is the majorly used nutrient in crop growing. Nitrogen fertilizer contributes to N_2O emission because it acts as a resource of global warming. However, it also plays a foremost role to global cooling with emissions of NOx and NH_3. According to Fagodiya et al. (2017), the net global temperature change potential is 6.9% lower and 2.4% lower in 20-year timescale and in 100-year timescale when warming and cooling effects of N use in cultivation were considered compared to considering warming due to emission of N_2O alone.

6.6 The Interlinked Influences of Soil Microbes on Biogeochemical Nutrient Cycles

Microbes are an important component of carbon biogeochemical cycling as they are involved in the emission as well as the removal of greenhouse gases like CO_2 and CH_4 which in turn are responsible for climate change (Singh et al. 2010; Dutta and Dutta 2016). Microbial role with respect to climate change requires attention (Walsh 2015) as they have resulted in several changes in world climate which has also, in turn, influenced them (Zimmer 2010). Microbes are critical in decomposition process and converting matter into forms that can be reused by other organisms making the microbial enzyme as a key "instrument" that drives the Earth's biogeochemical cycles (Abatenh et al. 2018).

Soil is one of the highly diverse ecosystems on our planet with an interacting community of bacteria, archaea, viruses, fungi, and protozoa, collectively referred to as the "soil microbiome" (Kuzyakov and Blagodatskaya 2015). The amount of C present in soil is more than two-thirds of the total terrestrial carbon (Hibbard et al. 2005; Davidson and Janssens 2006). Terrestrial soil efflux of CO_2 is an important contributor to global carbon cycling (Raich and Schlesinger 1992). An understanding of soil microbial ecology is crucial to our ability to assess terrestrial *carbon cycle–climate change* impact relationship as microbes are responsible for the main processes causing GHG fluxes between soils and the atmosphere (Falkowski et al. 2008).

6.6.1 Role of Soil Microbes with Respect to Carbon Cycling

The terrestrial soil carbon cycle is dominated by the balance between photosynthesis and respiration (includes autotrophic root respiration and heterotrophic soil microbial respiration). Microbes contribute greatly to net carbon exchange through the processes of heterotrophic soil microbial respiration which is mediated by soil microbes (Singh et al. 2010). Carbon is transferred from the atmosphere to soil via "carbon-fixing" autotrophic organisms such as photosynthesizing plants and photo- and chemoautotrophic microorganisms (including algae, cyanobacteria, and some other forms of bacteria) that synthesize atmospheric CO_2 into organic matter. In the carbon cycle, "primary producers" are grazed upon by "primary consumers" such as herbivores moving up fixed C thru trophic levels, and these in turn can be predated by "secondary consumers" and so on. The fixed carbon is then returned to the soil organic matter (SOM) when the organisms die, or through excreta which is used by microbes as their metabolism substrate that generates atmospheric CO_2 or CH_4. Therefore, one of the most significant contributions of soil microbes to climate change is their role in SOM decomposition, a fundamental process regulating the carbon cycle. In terrestrial soil ecosystems, microbial decomposition converts SOM into CO_2 and releases nutrients for plant growth (Glassman et al. 2018). In fact, the amount of atmospheric carbon produced due to soil microbial decomposition is 7.5–9 times higher as compared to the anthropogenic emissions worldwide annually (Crowther et al. 2015). However, there are several mechanisms which can increase the residence time of carbon in soils. Short-chain carbohydrates are highly fragile and do not generally persist in soils for long. However, more complex molecules, especially lignins and tannins, are much more recalcitrant and can remain in soils for many years. Therefore, soil microorganisms carry out the dichotomous roles of mineralization of soil organic carbon (SOC) and stabilization of carbon inputs into organic forms (Heath et al. 2005). The balance between these two processes governs the net efflux of CO_2 to the atmosphere. The proportion of carbon substrate that is retained in the microbial biomass compared with that respired as CO_2 is referred to as the "microbial carbon use efficiency" (Bond-Lamberty et al. 2018).

6.6.2 *Role of Soil Microbes with Respect to Nitrogen Cycling*

Nitrogen (N) is the most abundant atmospheric gas constituting 78% of the global atmosphere. Being a colorless, odorless element, nitrogen could be found in the air we breathe, in the soil beneath our feet, as well as in the water we drink. Nitrogen biogeochemical cycle includes two large N pools on Earth, atmospheric, unreactive, and stable elemental dinitrogen (N_2) and biologically reactive ammonia, nitrate, nitrite, and organic nitrogen. N cycle is a repeating cycle of processes inclusive of internal interactions within the reactive nitrogen pool and in- and out-flow between reactive nitrogen and atmospheric N_2 pools as well as its movement through both living and nonliving things: the atmosphere, soil, water, plants, animals, and bacteria (Gruber and Galloway 2008). N is considered to be a macronutrient used by living organisms including us necessary for growth and survival (Medhi et al. 2017). Nitrogen is one of the six major essential elements of CHNOPS for life putting together the building blocks and integrating molecules for metabolism (amino acids and proteins), heredity (nucleotides and nucleic acids), and other important biological functions (Takai 2019). Nitrogen transformation in soil is a very crucial process (Fig. 6.1), and unlike other soil nutrients, nitrogen does not originate from the soil but is gained from the air. Some of the nitrogen accumulates in the soil when rainfall absorbs nitrates from the atmosphere, while some nitrogen is fixed by the soil

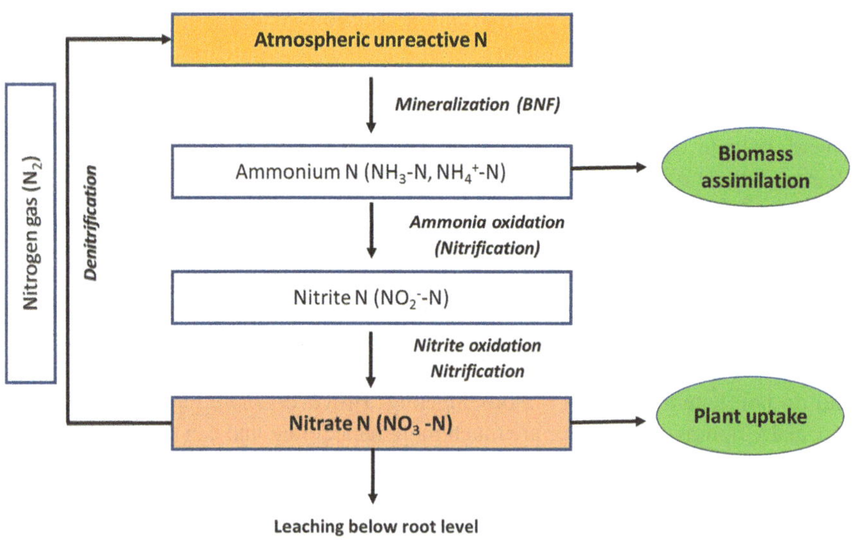

Fig. 6.1 Microbial nitrogen transformation pathways in soil

microbes, and without its fixation, it is unusable by plants or animals. N_2 enters into the biosphere specifically via biological fixation and is carried out in a bidirectional manner. Human activities additionally have a profound effect on the amount of bioavailable nitrogen, mainly owing to the high input of industrial nitrogen-based fertilizers (Kuypers et al. 2018). The bioavailability of N is rare in many environments, making it a growth-limiting nutrient primarily controlled by microbial equations that alter the oxidation as well as reduction state of N. The N-transforming processes have much diverse fluxes due to the presence of distinct microbial enzymes that perform 14 redox reactions involving eight key inorganic nitrogen species of different oxidation states (Kuypers et al. 2018). Conversion of nitrogen from one state to another state involving a network of interactions with N-transforming microbes to harvest and accumulate N-energy for their survival and evolvement is briefly described in the following N-transforming metabolic pathways:

Nitrogen Fixation Biological nitrogen fixing (BNF) is a unique way of converting the largest inventory of freely accessible nitrogen (N_2) into different reactive forms to be utilized by other organisms for growth and survival. It is the first and the foremost step which enables the movement of nitrogen from the atmosphere into the soil. The bioavailability of the inert N_2 is made possible by a highly diverse but rare nitrogen-fixing bacteria and archaea known as diazotrophs (key in the global N-cycle) that carry the nitrogenase metalloenzyme that combines gaseous nitrogen with hydrogen and assists in the fixing of dinitrogen into ammonia. This process is also regarded as the primary route of transforming nitrogen usable to be taken up by plants. Nitrogenase enzyme includes three different types such as iron–iron (Fe–Fe), vanadium–iron (V–Fe), and molybdenum–iron (Mo–Fe) nitrogenases that offer bacteria and archaea with a competitive advantage in environments that are N-depleted (Kuypers et al. 2018). Two kinds of nitrogen-fixing bacteria are recognized: the first one are the free-living (non-symbiotic) bacteria, including cyanobacteria, *Anabaena*, *Nostoc*, *Azotobacter*, *Beijerinckia*, and *Clostridium* and the second one are the mutualistic (symbiotic) bacteria mainly *Rhizobium* associated with leguminous plants such as soybeans and alfalfa. During symbiotic N_2 fixation, the plant converts the "fixed" ammonium ion transformed by *Rhizobium* (energy through photosynthesis) to nitrogen oxides and amino acids to form proteins and other molecules like alkaloids, thus providing an important N input in many plant communities and having major impacts on N dynamics and ecosystem productivity in a good manner (Abatenh et al. 2018; Thomas et al. 2015). A minor amount of N can also be fixed through lightning providing the energy needed for N_2 to react with oxygen for producing NO (nitrogen oxide) and NO_2 (nitrogen dioxide). However, this abiotic nitrogen fixation input has been estimated to account for only $<1/10$ of BNF (Takai 2019).

Nitrification Nitrification is the transformation of ammonia (NH_3), the foremost product of mineralization, to nitrites (NO_2^-) and nitrates (NO_3^-); the nitrification products are also carried out in soils biologically and aerobically. Ammonia gas and nitrite are toxic to plants and animals; thus, their immediate conversion to ammonium ion and nitrate is essential. Moreover, ammonium ion provides energy source to microorganisms involved in the system. The nitrification process is comprised of two metabolic pathways—ammonia oxidation to nitrite via hydroxylamine oxidation and nitrite oxidation to nitrate, carried out by ammonia-oxidizing bacteria (AOB), ammonia-oxidizing archaea (AOA), and nitrite-oxidizing bacteria (NOB). Most of the natural and anthropogenic habitats comprising nitrifying microbial communities are dependent on the close cooperation between AOB and NOB metabolism as well as between archaeal and bacterial populations (Stahl and de la Torre 2012). AOA include phylogenetically and physiologically diverse members and possess a significant role in natural environments. In the primary stage of nitrification, NH_3 is oxidized into NO_2^- involving enzymes ammonia monooxygenase (AMO) and hydroxylamine oxidoreductase (HAO) performed by autotrophic AOB, *Nitrosomonas* and *Nitrosococcus*, and secondly, oxidation of NO_2^- to NO_3^- in the presence of enzyme nitrite oxidase performed by NOB, *Nitrobacter* and *Nitrospira* (Thakur and Medhi 2019; Fagodiya et al. 2020). Recently, genus *Nitrospira* previously presumed to be capable of only nitrite oxidation has also been found to hold the ability for ammonia oxidation all the way to nitrate containing both AMO and HAO enzymes (Daims et al. 2015). Nitrite oxidation is the main biochemical pathway generating nitrate catalyzed by the enzyme nitrite oxidoreductase (NXR), encoded by versatile aerobic NOB such as *Alphaproteobacteria*, *Betaproteobacteria*, *Gammaproteobacteria*, *Chloroflexi*, *Nitrospinae*, and *Nitrospirae* phyla, anoxygenic phototrophs including *Thiocapsa* sp. KS1 and *Rhodopseudomonas* sp. LQ17, and anaerobic AOB (Kuypers et al. 2018).

Assimilation The phase of nitrogen uptake by plants or in other words the assimilation of mineralized NH_4^+-N and NO_3^--N by plants through their roots transporting to the shoot via xylem where nitrogen assimilation is mostly carried out and transforming them into amino acids, nucleic acids, and chlorophyll for their needs. NO_3^--N is predominantly present in aerobic soils as a plant-available nitrogen form, whereas in anaerobic and flooded soils (paddy soils), NH_4^+-N is the predominant form of plant-available nitrogen (Fagodiya et al. 2020). In legume plants, some nitrogen is assimilated in the form of ammonium ions directly from the nodules with the help of *Rhizobium* bacteria. Just like plants, soil microbes requiring nitrogen as an energy source take up nitrogen from the soil through the decomposing plant residues and incorporate in their biomass, but sometimes when microorganisms take in NH_4^+ and NO_3^-, these N are no longer available to the plants triggering nitrogen deficiency in soils.

Denitrification It is the reverse process of nitrification likely occurring anaerobically in the presence of nitrates and organic matter. It is a microbially mediated sequential process involving the dissimilatory reduction of one or both ionic nitrogen oxides, NO_3^- and NO_2^-, to gaseous nitrogen oxides, nitric oxide (NO), and nitrous oxide (N_2O) and finally reduces to the ultimate end-product, dinitrogen (N_2), thus ending the N cycle and removing biologically available nitrogen in the soil and returning it to the atmosphere (Thakur and Medhi 2019). During low conditions or absence of oxygen, most of the soil microorganisms extract oxygen from nitrates and use organic matter as the carbon source for growth and development. Denitrification is half of a nitrogen buffering action in the soil, the other half being nitrogen fixation by free-living organisms, and hence if NO_3^--N are low, N will be fixed, but if high, it will be denitrified. Nitrate reduction to nitrite is used for respiration as well as for nitrogen assimilation into biomass reaction catalyzed by either a membrane-bound nitrate reductase (NAR) at cytoplasm or periplasmic nitrate reductase (NAP) at periplasm performed by special and unique group of denitrifiers like *Pseudomonas* and *Clostridium*, including the model organism *Paracoccus denitrificans*, containing both NAP and NAR enzymes (Medhi et al. 2018). Many microorganisms have the ability to reduce nitrite to nitric oxide, such as *Proteobacteria*, anaerobic AOB, and *Bacteroidetes* catalyzed by two unrelated enzymes: a heme-containing cd1 nitrite reductase (nirS) and a Cu-containing nitrite reductase (nirK). Microorganisms such as *P. denitrificans* and *Pseudomonas stutzeri* capable of N_2O reduction possess nitrous oxide-producing nitric oxide reductases (NOR) for detoxification or respiration of nitric oxide that belongs to a diverse group of enzymes ranging from flavoproteins to heme copper oxidases. AOB could also produce N_2O in a process termed as nitrifier–denitrification. Diverse bacteria, including members of the *Proteobacteria, Bacteroidetes* and *Chlorobi* phyla, and archaea from Crenarchaeota and Halobacteria utilize nitrous oxide reductase (NOS) to reduce N_2O to N_2 (Kuypers et al. 2018).

6.6.3 Role of Soil Microbes with Respect to Phosphorus Cycling

Both nitrogen and phosphorus (P) help in strengthening photosynthetic processes, cell growth, metabolism, and protein synthesis in living organisms, but their natural sources and rates of supply are very different from each other. Phosphorus is one of the most important plant nutrients present in soils and supports in the transfer of energy from sunlight to plants, stimulates early root and plant growth, and hastens maturity. P originates from phosphate rock, renewed with the uplift of continental rock, but today the primary source of P emanates from the considerable amount of anthropogenic inputs of fertilizers onto agricultural soils (Guignard et al. 2017). Phosphorus biogeochemical cycle depicts the movement of P through weathering of rocks, water, soil, and organisms. While soil N is "leaky," soil P tends to be "clingy." Even though there is plenty of phosphorus present in the soil in both organic and inorganic forms, it is mostly in unavailable inorganic forms; thus, most of the plants obtain it only as orthophosphate ions (soluble inorganic forms) and is considered as a nutrient-limiting factor. Phosphorus can be found dissolved in the soil solution in very low amounts or associated with soil minerals or organic materials. Precipitation and weathering additionally cause rocks to release phosphate ions, and other minerals consequently get distributed in soils and water.

Mineralization is the process via which the tightly held organic P not available for plant uptake becomes available when the P-bound organic materials are decomposed. In layman terms, it is the conversion of organic phosphorus into insoluble inorganic phosphates and is carried out by phosphatase enzyme mainly of soil microbial origin that attacks many of the organic phosphorus compounds in the soil and release inorganic phosphate. Increased activity of phosphatases occurs in response to P deficiency as part of P starvation responses that occur in soils (Richardson and Simpson 2011). When soil suspensions or soil extracts are treated with an excess of phosphatase activity (phytase activity), organic P substrates get hydrolyzed and appreciable amounts of orthophosphate are released. Increased mineralization of soil organic matter is directly proportional to microbial activity associated with plant rhizosphere. Certain plant–symbiotic soil microbes, especially mycorrhizal fungi, are efficient at helping plants to access P-bound forms of soil. The rate of P release is affected by various factors such as soil moisture, organic material composition, oxygen concentration, and pH. A microbial-assisted reverse process known as immobilization also occurs in the soil where microbes use phosphorus for their own nutritional needs by competing with plants. Both mineralization and immobilization occur simultaneously in soil, and if the organic P content is high enough to fulfill the requirements of the microbial population, then mineralization takes place. Conversion of insoluble inorganic phosphates into soluble inorganic phosphates leads to solubilization process. A wide range of microorganisms able to solubilize inorganic P such as fungi (e.g., *Aspergillus*, *Penicillium*, *Fusarium*) and bacteria, namely, *Bacillus*, *Pseudomonas*, *Micrococcus*, *Flavobacterium*, *Actinomycetes*, etc. are commonly isolated from plant rhizosphere where sugars from root exudates are converted into organic acids biologically. The P availability largely depends on the solubilization degree by the substantial amounts of organic and inorganic acids such as sulfuric acid and nitric acid produced by these microorganisms in soil. This solubilized form of inorganic phosphates is subsequently taken up by plants.

6.6.4 Role of Soil Microbes with Respect to Sulfur Cycling

Individual's activities are the prime reason of significant upsurge of heat-trapping greenhouse gases (e.g., CO_2, CH_4 and N_2O) and aerosols in the surroundings driven by secretion from combustion of fossil fuels, manufacturing units, agriculture activities, and deforestation. Dimethyl sulfide (DMS) is a significant semi-volatile organic compound (Kloster 2006) produced mainly by marine organisms which is the most vital source of oceanic sulfate aerosol precursor, that is, dimethylsulfoniopropionate (Andreae 1990; Kiene et al. 2000). In the lower atmosphere, DMS is largely oxidized, resulting in the formation of sulfur dioxide, which reacts with HOx oxidants, and exerting a direct radiative effect that condenses to make sulfate aerosol particles (Schwinger et al. 2017). These sulfur-containing droplets have a direct cooling effect on climate by scattering solar radiation. They

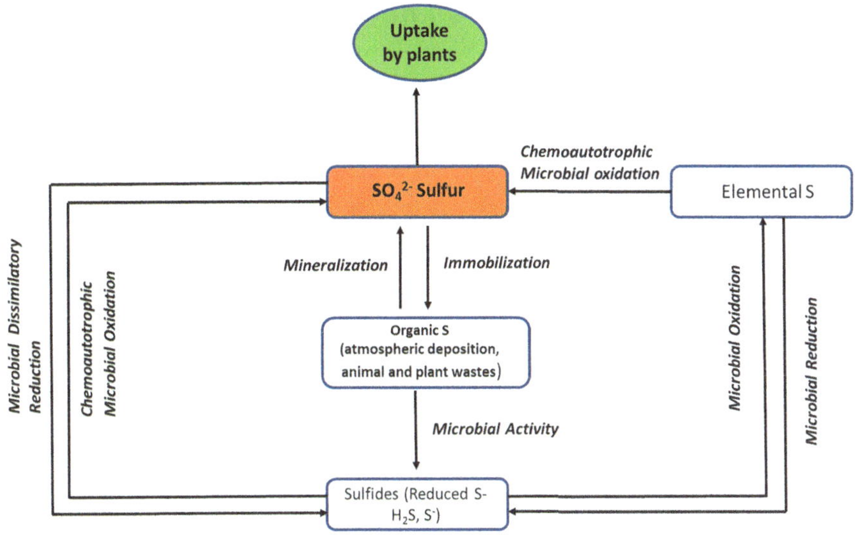

Fig. 6.2 Microbial sulfur transformation pathways in soil

also play an important role in cloud condensation nuclei (CCN), modifying properties of clouds and thereby affecting cloud formation and development and influencing clouds' radiative effects on climate (Andreae et al. 1995; Wang et al. 2018; Kloster 2006).

Sulfur is an imperative component of essential vitamins and metabolites. It is released by the breaking of rocks and minerals from the Earth's surface. Cysteine and methionine are the two amino acids where sulfur occurs. Despite of its inimitability in cells, it is an utterly vital element for existing living systems. In the similar way with nitrogen and carbon, the microorganisms can convert sulfur from the most oxidized state of sulfate (SO_4) to the most reduced form of sulfide (H_2S) as shown in Fig. 6.2. Various distinctive groups of prokaryotes and their processes are required in the sulfur cycle. Two diverse assemblages of prokaryotes oxidize H_2S (hydrogen sulfide) to S (sulfur) and from S (sulfur) to SO_4 (sulfate). This process does not occur readily in poorly aerated soils. Sulfate is the form of sulfur which is assimilated by plants. Since SO_4 and sulfur can be utilized as electron receivers for respiration process, sulfate-lessening bacteria generate H_2S during a process of respiration under anaerobic conditions corresponding to denitrification. The method results in the distinct odor of hydrogen sulfide that occurs in anaerobic bogs, sediments, and soils. Sulfur is acquired by bacteria and plant species as sulfate (SO_4) for utilization and diminution to sulfide (Mandal and Neenu 2012).

6.6.5 Role of Soil Microbes with Respect to Potassium Cycling

Potassium (K) is a necessary nutrient and a foremost constituent present in all living cells. In contrast to N, P, and S, K reserves are available in soil solution in the form of positively charged cation (K^+), and its behavior is influenced mainly by soil cation exchange and mineral weathering, least by biological activity. Gradually weathered K is more prone to be bound within the soil solution or taken up by plant roots rather than being leached out as shown in Fig. 6.3. Naturally, soils surround potassium in superior amounts than other nutrients, though the majority of the potassium concentration is out of stock for flora uptake. However, potassium-solubilizing bacteria (KSB) can solubilize potassium-bearing mineral deposits and convert the unsolvable potassium to solvable state of potassium accessible for flora uptake. Several bacteria, for example, *Paenibacillus* spp., *Acidithiobacillus ferrooxidans*, *Bacillus mucilaginosus*, *Bacillus circulans*, and *Bacillus edaphicus*, have the ability to solubilize potassium minerals (e.g., feldspar, biotite, illite, orthoclase, muscovite, and mica). Potassium-solubilizing bacteria are generally available in all types of soils, even though their numeral, multiplicity, and capability for potassium solubilization fluctuate lying depending upon the climatic and soil conditions. KSB are capable of dissolving the silicate minerals and liberate potassium all the way throughout the production of acidolysis, inorganic and organic acids, polysaccharides, chelation, and exchange reactions (Etesami et al. 2017). Plant's nutrient

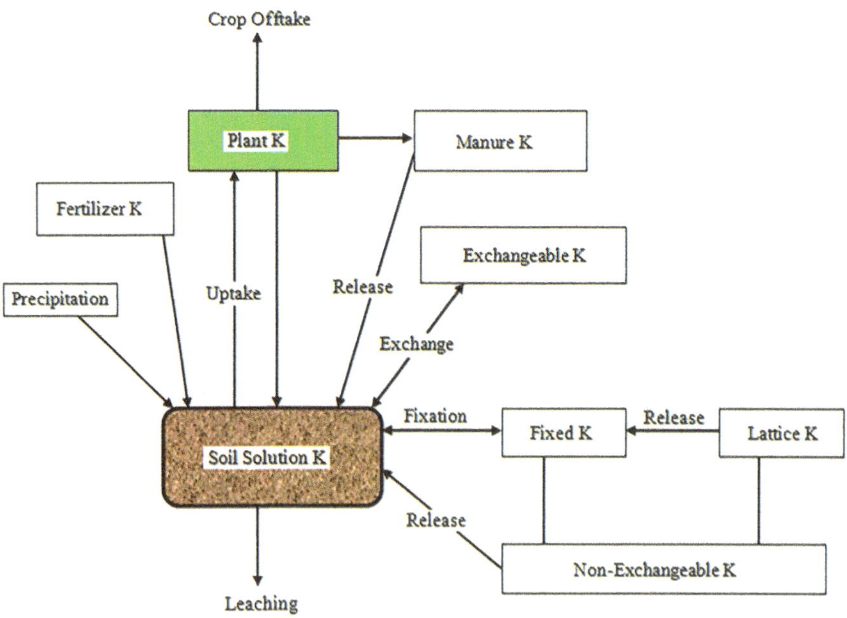

Fig. 6.3 Microbial potassium transformation pathways in soil

availability by complexolysis depends on the length of the pathway the nutrients should pass through the soil to get to the root surface.

6.7 Molecular Strategies on Soil Microbes' Functioning During Climate Change

Most of the soil ecosystem investigation is based on the measurement of physico-chemical parameters such as temperature, pH, and concentrations of carbon, nitrogen, and phosphorus, but the characterization of the biological entity of soil would also support in the comprehensive understanding of the impact of scaling climate change on soil health. The timeline of microbial studies predates back from Leeuwenhoek (1676), to Koch (1888) describing microbiota and their growth and isolation on nutrient media to Sanger (1977), Kary Mullis (1980), and Ventor (1995) developing DNA sequencing, polymerase chain reaction (PCR), and whole genome analysis and advancing toward the next-generation sequencing (NGS) (Nair and Raja 2017). Understanding of soil ecosystem that hosts an immensely diversified and evolving microbiota demands clarifications such as who are they, what are they doing, and how are they doing? The soil presents the most heterogeneous and dynamic environment for the growth of different organisms including prokaryotes as well as archaea, viruses, algae, and fungi (Biswas and Sarkar 2018). These microbial groups are independent as well as dependent on each other to perform a crucial role of influencing each and every activity carried out in the soil environment. However, from many decades, limitations in the field of microbiology in terms of laboratory culture-based methods has permitted $<1\%$ of microbes to be isolated from the environment. It has been suggested that 80% of soil bacteria could be dormant as microbiota dormancy is found to be a common phase in soil environments (Edge et al. 2020; Tecon and Or 2017). The approaches to study and understand the microbial life thriving in various distinct environments have undergone a remarkable transformation since the period of culturing bacteria using synthetic media with significant implications contributing adequately to environmental assessment, monitoring, and remediation programs (Nair and Raja 2017).

The modern molecular tools intend to answer the central questions revolving around soil biodiversity with the help of its genome-enabled omics studies such as genomics, metagenomics, metaproteomics, metatranscriptomics, and metabolomics, whose individual as well as combined efforts have tried to uncover and bring out new and novel information concerning the microbial world. Omics studies have revolutionized our understanding about how organisms have evolved and are gradually adapting to the progressing climate change. Microbial nucleic acids record both how the environment affects these microorganisms and how they respond to the changing environmental conditions (Mock et al. 2016). In recent years, researchers have tried to focus their projects related to environmental monitoring using both culture-dependent and culture-independent molecular approaches (Kumar et al. 2016; Rathour et al. 2020; Mishra et al. 2018).

Genomics is the study of an entire genome of an organism allowing defined sets of genes involved in various metabolic network systems to regulate entire but distinct cellular functions. Transcriptomic and proteomic approaches establish the structure and function of different proteins and RNA molecules and the coupling of these metabolic networks together. Moreover, they are also responsible for the change/adaptation in accordance with the availability of resources or environmental stress. Metagenomics is the study of community genomics that involves the genetic content of the entire microbiota by directly assessing environment-recovered samples and is advantageous for studying distinct ecological role of a particular microbial community present in different ecosystems. Presently, most of the research laboratories have been applying NGS for metagenomic studies (labeling DNA fragments with fluorescent adapters) using a NGS machine (Illumina Hi-seq, PacBio) since its usage removes the biases and artifacts that were present in clone-based approach (cloning the DNA fragments into a vector). Metabolomics study enables the mapping of the entire metabolic profile of an organism and can be regarded as "microbial metabolomics" in microbiological context reflecting novel biosynthetic and degradative pathways utilizing the organic content present in soil ecosystem (Nair and Raja 2017).

However, the integration of multi-omics tactics would also permit a researcher to disclose the contribution of active, dormant, or dead microbes to distinct ecosystem functions. A study revealed the importance of pathways' linking processes by conducting a genome-centered metagenomics across a permafrost thaw regime that illuminated novel fungal pathways for plant polysaccharide degradation and syntrophic interactions resulting in CH_4 production (Edwards et al. 2020). Whole genome analysis along with metagenomics could be used to assess the functional gene capabilities of the bacterial communities for nutrient transformations that are being widely influenced by other relevant environmental factors that assist in shaping up the soil microbial populations. Metatranscriptomic approach used by Masuda et al. (2017) predicted that nitrogen fixation in paddy fields was majorly driven by deltaproteobacterial populations, such as *Anaeromyxobacter* and *Geobacter*, and not by any other *Proteobacteria* or *Cyanobacteria*, the major diazotrophs considered to be present in most paddy soils. Shotgun metagenomics is utilized to decipher unbiased interpretation on the phylogenetic and functional changes in the microbial community in biochar-amended soil microbiomes (Yu et al. 2019). The significant shifts exhibited the key metabolic pathways such as the utilization of plant-derived carbohydrates and denitrification which was likely stimulated by the increased available nutrients associated with biochar amendment, findings conferring enhanced nutrient cycling and a minor decline in CO_2 emissions but potentially abating N_2O emissions. Omics tools not only help in studying the analysis of metabolic footprint of different soil biota in response to nutrient transformations but also provide guidance to innovate smart and sustainable agriculture environment.

6.8 Impact of Soil Microbes on Climate Change with Respect to GHGs

6.8.1 CO_2 Emissions

6.8.1.1 Effects of Temperature

Soil's ability to sequester carbon is the outcome of the soil microbial activity that in due course controls the rate-limiting steps in the decomposition process, thereby ultimately influencing abiotic factors on decomposition (Classen et al. 2015). As explained previously, climate warming modifies the structural makeup of soil microbes involved in decomposition process, thereby inducing CO_2 efflux from soil (Schindlbacher et al. 2011). Increasing temperatures speed up the efflux of CO_2 from soil to atmosphere due to enhanced heterotrophic microbial breakdown of SOM. In comparison to primary production, soil respiration rate of SOM is more sensitive to temperature (Schimel et al. 1994). SOM can vary greatly in terms of carbon compounds with divergent chemical composition, and therefore temperature-dependent microbial decomposition of the same is a complex phenomenon (Glassman et al. 2018). There are extensive environmental constraints, such as physical and chemical protection of SOM, to decrease substrate availability for microbial attack, thereby relegating microbial responses to global warming (Davidson and Janssens 2006). Climate warming likely increases the net transfer of carbon from soil to atmosphere, thereby creating a positive feedback on climate change as depicted in Fig. 6.4. The effects of increasing temperature initiated via decomposition accompanying CO_2 release are unequal among different soils (Dutta and Dutta 2016).

6.8.1.2 The Permafrost Issues

Temperature increment, on expenses of climate change, results in the melting of inert permafrost layers. Subsequently, the dormant microbiome within these layers gets activated, and biotic activities such as respiration, fermentation, and methanogenesis are prompted (Weiman 2015). As thawing proceeds, the soil begins transitioning more into decomposition (Svoboda 2015) leading to the transformation of soil carbon into GHGs (Weiman 2015), thus creating a positive feedback on climate change (Davidson and Janssens 2006). Arctic and alpine carbon is viewed as potential indicator as well as regulator of global warming (European Commission 2015) because when it becomes warmer and drier, the permafrost microbes, initially predicted to yield CO_2, start releasing CH_4 instead (Atkin 2015), thereby further heating up because of climate change.

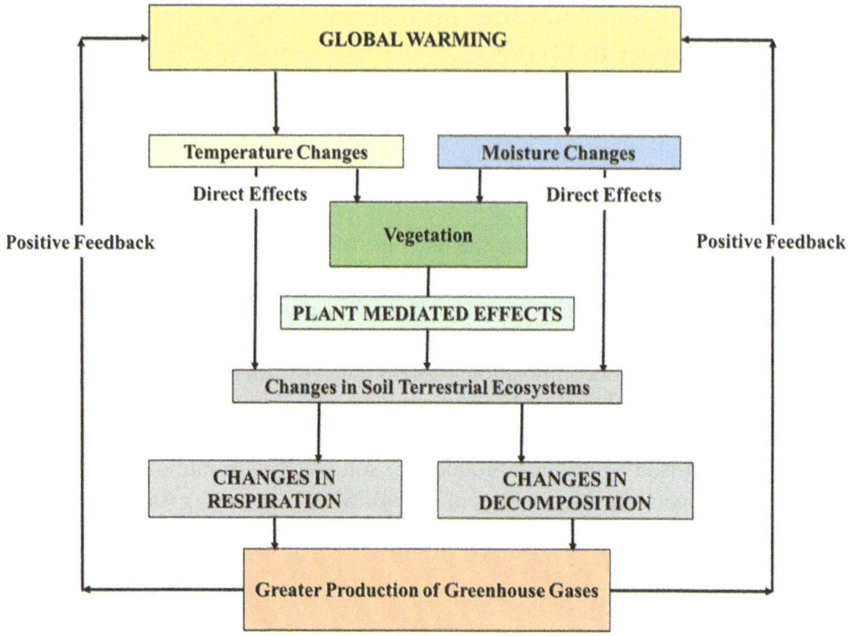

Fig. 6.4 Positive feedback between global warming and soil microbial effects

6.8.1.3 Effects of Changes in Precipitation

Extreme occasions of precipitation such as drought and flooding have increased due to climate change, thereby affecting the availability of water. Precipitation-induced fluxes are significant factors that determine whether ecosystems can serve as sources or sinks of CO_2 (Shim et al. 2009). In fact, precipitation is critical in determining the changeability of soil moisture and respiratory undertakings in soils (Aanderud et al. 2011). Shifts in precipitation establishments are quite substantial since the moisture level determines soil microbiota and the magnitude of soil decomposition. On the contrary, drought in wet ecosystems, such as peatlands and wetlands, stimulates carbon cycling as drying up of soils maximizes oxygen accessibility, accordingly raising CO_2 flux (Fierer and Schimel 2003; Singh et al. 2010).

6.8.1.4 Effects of Elevated Carbon Dioxide Levels

Prominent intensities of atmospheric CO_2 can result in substantial increase in microbial soil respiration and cause soil microbes to emanate more potent GHGs such as CH_4 and N_2O (Pathak and Pathak 2012). In fact, raised CO_2 levels not only proliferates CH_4 efflux but also declines the uptake of CH_4 by soil microbes (up to 30%) (Phillips et al. 2001; Ineson et al. 1998). Moreover, higher CO_2 altitudes also lead to discrete and significant amendments in the soil microbial communities of leaves of trees in the form of litter. This could have widespread consequences on the food chain as such microorganisms are sources of nutrients for the small phytophagous animals (Dutta and Dutta 2016). Elevated levels of CO_2 quantitatively and qualitatively modify the discharge of labile sugars, organic acids, and amino acids from plant roots that can encourage microbial growth and activity, thereby changing the CO_2 flux (Singh et al. 2010). Use of labile carbon over complex carbon by soil microbes slows down the rates of litter decomposition, which in turn may minimize CO_2 emissions by respiration and favor carbon sequestration in the soil.

6.8.1.5 Effects Mediated Through Plants

Warming and reformed precipitation establishments potentially alter the distribution of plant species and functional groups at both local and global scales (Woodward et al. 2004). A variety of indirect impacts of climate change are intermediate through plants that are allied to soil microbial communities (Kardol et al. 2010). Plants minimize the respiration sensitivity under high rainfall changes but offer reverse action under conflicting conditions in mesic habitats (Aanderud et al. 2011). When thermal stress occurs, there are changes in the carbon-rich secretions released by plants, leading to shifts in the soil microbial secretions (Ngumbi 2015). The phonologies of roots and shoots are altered due to climate change, subsequently altering rhizosphere interactions that distinctly stimulate seasonal assemblages of soil microbiomes (Classen et al. 2015). Raised temperature along with nutrient availability can result in vegetation changes (Hobbie 1996). This in turn influences composition of soil microbes and their metabolism through changes in the quality and quantity of organic matter entering the soil as plant matter that is known to differ consistently across plant functional groups (in terms of C_3 and C_4 plants) (Dorrepaal et al. 2005) and correlates strongly with rates of decomposition and hence heterotrophic soil microbial respiration (Engelkes et al. 2008; Bardgett et al. 2009).

6.8.2 *N_2O Emissions*

Along with CO_2 flux, soil microbiota can also govern other GHG fluxes, i.e., methane (CH_4), carbon cycle byproduct, and nitrous oxide (N_2O) and nitrogen cycling byproduct. Even though these GHGs represent smaller fluxes compared to those of CO_2, they are much more potent than GHGs with GWP approximately 21 times CO_2 eq and 300 times CO_2 eq, respectively, implicating they could linger on for much longer in the atmosphere than CO_2 and actively participate in maximizing the global warming issue. Each year, the global N_2O emissions are increasing at an alarming rate that contributes approximately to 6%–8% of the overall greenhouse effect, and agricultural soils represent as the major contributors due to the excessive but inefficient use of N fertilizers (Thakur and Medhi 2019; Fagodiya et al. 2020). Nitrogen is one of the key nutrient controlling elements for sustaining life at an optimum concentration, and its transformation in soil is mainly controlled by mineralization, immobilization, and assimilation carried out by the highly indigenous and diverse soil microbiota. During microbial elimination of ammonium, both nitrification and denitrification initiate N_2O production and emissions into the environment contributing to climate change. N_2O occurs during fertilizer usage and fossil fuel combustion but naturally occurs and released from soil and ocean. Of the cumulative GHGs emitted, 80% of N_2O is being emanated via soil processes in distinct managed ecosystems (Mandal and Neenu 2012).

6.8.2.1 Effect of CO_2 Increase and Water Response

The culprits, both fungi and certain classes of bacteria living in soils, take up inorganic chemicals like NO_3^- and CO_2 as part of their respiratory processes for N_2O production and energy generation. Anaerobic habit helps these microorganisms to flourish in the presence of increased atmospheric CO_2 concentrations. Extra CO_2 concentrations reduce water usability by plants, creating abundant moisture in soils as well as favoring anaerobic/anoxic conditions for these microorganisms affecting how soils release the N_2O gas. Robust microbial responses to freeze–thaw process due to increased GHG emission and changing patterns of microbial substrate usage have enhanced soil N transformation, and expression of denitrifying genes has been detected in studies prominently leading to additional N_2O emission to the global warming budget (Zhe et al. 2018; Bardgett et al. 2008). Drylands or extreme drought conditions also have shown noticeable effects on the substantial rate of N_2O emissions. Wetting of dry soils not only improves soil hydraulic conditions but also stimulates bursts of respiration and mineralization via microbial activity through high flux of nutrients into soils that rapidly depletes soil O_2 levels, allowing generous N_2O emissions to occur (Hu et al. 2017). Aeration and higher irrigation along with temperature rise affected both mean soil nitrifier and denitrifier abundance proving them as the primary factors influencing soil N_2O fluxes (Chen et al. 2019).

6.8.2.2 Effect of Plant Influences

Composition and types of plant species govern the litter quality influencing N availability, and the rhizosphere exudation of the mineralized N organic compounds speeds up microbial activity for N_2O production (Wrage-Mönnig et al. 2018). Biological nitrogen-fixing forests can fuel CO_2 sequestration but can also encourage soil N_2O. Symbiotic N-fixing trees' ability to either abate or aggravate climate change compared to non-fixing trees depends on their N-fixing strategy and enrichment of nitrogen deposition substantially upsetting CO_2 sequestration and potentially elevating soil N_2O emissions (Kou-Giesbrecht and Menge 2019). Increase in CO_2 level makes plants growth faster, eventually supplying energy to soil microbes boosting their metabolism that could slow down climate change, but with more CO_2, there is a chance of leaking the extra carbon fueling microbial byproduct, N_2O, that ends up in the atmosphere counteracting the ecological climate change abatement strategy (van Groenigen and Trinity College Dublin 2011) as more uptake of CO_2 releases other potent GHGs.

6.8.3 CH_4 Emissions

Soil biodiversity referring to the variety of numerous organisms existing below and above the soil surface are mostly responsible for the 60–80% of the biological activities via nutrient cycle regulations and decomposition of organic residues, preserving the soil quality, fertility, and productivity. Methane production is the result of anthropogenic routine activities, for example, production of fossil fuels, circulation and burning, landfills and squander, cattle's farming, decomposition of plant matter, and burning of biomass and rice paddies. Of the cumulative GHGs emitted, 50% of CH_4 is being emanated via soil processes in distinct managed ecosystems (Mandal and Neenu 2012). CH_4-eating bacteria called methanotrophs are significant to sustaining a healthy climate on Earth. Bacteria utilize methane for assimilation as power supply (Semrau et al. 2010; Nikiema et al. 2005; Bousquet et al. 2006) and convert it to CO_2 through their digestive process. These bacteria can devour huge quantity of methane which is supportive in dropping methane release from methane-producing manufacturing units and landfills (Gupta et al. 2014; Shindell et al. 2012). Production of methane (CH_4) also takes place as a fraction of the carbon cycle underneath anaerobic state by the action of microbiota. In peatlands, decay-resistant trash (e.g., *Sphagnum* mosses polysaccharides and anti-microbial phenolics) restrains decomposition of microbes, and water infiltration confines exchange of oxygen and promotes the expansion of anaerobes and emission of CO_2 and CH_4. Thawing of permafrost directs to rises in water-permeated soils, which endorse anaerobic methane production by methanogenic bacteria (Cavicchioli et al. 2019). Increasing atmospheric temperature coupled with liquefying permafrost enhances the activity rates of soil microbes (Yvon-Durocher et al. 2014), which straightly lead to larger production of CH_4 soils (Cooper et al. 2017). Prominent CO_2 effects on methane fluxes were extremely uneven in highlands soils and were principally net sinks for methane (via oxidation by methanotrophs). An increase in soil humidity beneath upraised CO_2 possibly has either decreased diffusion of methane into the soil (thereby dropping the concentration of CH_4 oxidation by methanotrophs) or elevated methane production by methanogenic bacteria (Phillips et al. 2001). Raised CO_2 frequently increased methane release in peatlands, wetlands, and rice agricultural fields. The anoxic situations in rice paddies, wetlands, and peatlands endorse the production of CH_4 by methanogenic bacteria. Increased CH_4 production, when subjected to eminent CO_2 conditions, has been accredited to enlarged C input into the soil (Tokida et al. 2010; Dijkstra et al. 2012).

6.9 Methods to Protect Nutrient Transformations in Soil

6.9.1 *Reduce Soil Tillage*

Soil microbial communities are the primary organisms responsible for catalyzing a range of events in nutrient cycling, essential to the productivity and sustainability of soil ecosystems (Bissett et al. 2013). Soil microbial biomass is the living part of SOM that generally comprises <10% of SOM and is performing an important ecological functions in soil ecosystem (Salinas-Garcia et al. 2002). Soil microbial biomass and soil enzyme activities are regarded as sensitive indicators of dissimilarities between sustainable cropping systems (Kennedy and Papendick 1995; Eivazi et al. 2003) implying soil quality. Soil microbial biomass often reacts quickly to soil management fluctuations.

Tillage is regarded as the soil mechanical manipulation for crop production purpose affecting considerably the soil characteristics such as soil water conservation, soil temperature, infiltration, and evapotranspiration (Busari et al. 2015). However, excessive or inappropriate tillage practices contribute to the degradation of soil properties and in doing so impact the nutrient-related processes. Tillage disrupts SOM, vital for refining soil fertility and soil water-holding capacity (Brust 2019). It also disturbs soil structure heterogeneity, thereby affecting the diversity of dominant soil microbial community that in due course alters the relationships among the members of the soil microbiome within the soil ecosystem (Altieri 1999), ultimately impacting decomposition of SOM. It has been reported that about one-third of the global GHGs is attributed to agricultural practices (Gattinger et al. 2014) such as tillage. Tillage modifies edaphic factors, thereby impacting the rate of carbon mineralization (Curtin et al. 2012). Conventional tillage disrupts soil aggregates contributing to erosion and increased atmospheric CO_2 and ultimately to global warming (Roldan et al. 2003). Therefore, use of more sustainable agricultural practices such as "conservation tillage" to improve residue cover with "minimum tillage" or "no tillage" has been established in many environments recently and can advance the substrate as well as nutrient availability to soil microbes. In fact, high carbon sequestration has been given as one of the credits of no tillage (Lal et al. 2007). Gambolati et al. (2005) observed that conservation tillage practices decreased the exposure of unmineralized organic substances to the soil microbial processes, thus reducing SOM decomposition and CO_2 emission.

Conservation tillage can also increase water retention in drought conditions and help to restore carbon in soils (Busari et al. 2015). Conservation tillage soil has reduced C concentrations from oxidation of labile SOM due to tillage. With no-till practices, an increase of the total soil nitrogen by 38–68% with depth has been observed in arid wheat-based systems (Dou and Hons 2006). Total nitrogen in zero tillage was 51–60% higher than in minimum and conventional tillage practices, respectively (Lopez-Fando and Pardo 2009). Therefore, potential benefits of conservation tillage practices in reducing carbon and nitrous oxide emissions to the atmosphere have become more important now than ever.

6.9.2 Landscape Management

Land is both a source and a sink of GHGs and chiefly participates in the nutrient cycling. Land ecosystems are vulnerable to ongoing climate change and climate extents. However, land degradation is itself a major reason of climate change (Scherr and Sthapit 2009). Global nutrient cycles have been greatly altered by land-use/land-cover changes over the last century that have a major implication in climate change (Downing et al. 1999). In general, land-use pattern change affects soil physical and chemical properties such as pH, moisture, organic matter content, and nutrient availability especially with respect to soil carbon and other nutrient cycling (Potthast et al. 2010; Paustian et al. 2000).

With population explosion, regularly lands are being converted to agricultural or degraded landscapes often acting as inefficient sinks for carbon and other elements which have dramatically altered regional hydrology and are subject to large losses of nutrient elements such as carbon, nitrogen, or phosphorus (Downing et al. 1999). Land-use changes can also disrupt carbon, nitrogen, or phosphorus dynamics as well as SOM storage across a wide range of habitats (Monkiedje et al. 2006), which are generally regarded as major factors causing shifts in microbial community composition (Schimel and Bennett 2004; Cookson et al. 2007). Land-cover changes leave the soil unprotected making way to CO_2, CH_4, and N_2O release. CH_4 flux from soil to atmosphere is the net result of bacterial processes that are strongly influenced by land use, land management, and the type of soil (IPCC 2014). Other unsustainable land management practices degrading soils include tillage-based crop production systems; simplified crop rotations that lead to soil nutrient mining; improper fertilizer application; unsuitable irrigation practices; overstocking, overgrazing, and burning of rangelands; inefficient grazing methods; and overexploitation or clearance of wooded lands and forestlands. Therefore, sustainable land management can contribute to lower the negative impacts associated with land use, thereby having implications in nutrient cycling as well as climate change.

Soil nutrients are closely related to land-use types and their associated sustainable management practices (Reijneveld et al. 2009; Wang et al. 2010). The conversion of any vulnerable land for unsustainable uses must be crucially prevented for sequestering nutrients and preventing nutrient loss. SOC content can be conserved, restored, and increased through appropriate land uses and agricultural management practices that can be applied at the landscape level (Corsi et al. 2012). Evaluations of land resources could categorize hotspots where sustainable land management practices might help in generating multiple ecosystem benefits.

6.9.3 Crop Management

Crop production is highly sensitive to climate change wherein both affect each other. Climate change affects crop production in terms of CO_2 concentrations, temperature, and precipitation, whereas crop production produces GHGs that have an impact on climate change (Smith and Almaraz 2004) in particular CO_2, CH_4, and N_2O. The SOC stock may either increase or decrease in response to changes in climate and cropping practices (Smith et al. 2015). High SOC content in the soil improves nutrient storage. C sequestration in soils therefore may be an efficient way to store C for climate regulation (Pan et al. 2013) by improving SOM content and thereby soil quality (Lal 2004) whose degradation otherwise is a major driver of climate change.

Specific cropping practices encourage soil capacity to conserve and accumulate SOC. These practices can reduce GHGs at the source and maintain or improve yields and enable crop systems to adapt to the projected impacts of climate change. Shifting crop variety production is a sustainable option for reducing fluxes. Perennial crops also reduce the need for annual tillage by increasing nutrient supply and also enhancing water-holding capacity (Lal 2004) which in turn can increase soil microbial community. Use of perennial crops or agroforestry could also result in similar remunerations (Mbow et al. 2014). Also, an intensive and diversified crop rotation practice, eliminating fallow periods and returning crop residues to the soil with an average C/N ratio in the 25–30 range, could be used. In this process, carbon accumulates in the soil and facilitates nitrogen in the decaying surface residues to be released slowly to the next crop. Cropping practices should also include greater diversity of crops that also contributes to integrated pest management by preventing weed growth (FAO 2010). Weed species competes for nutrients, water, and light, and removing them by tilling, pulling, mowing, and/or using herbicide ultimately leads to low primary productivity environments where the exposed soil leads to losses in SOM and soil microbial diversity (Louis et al. 2016).

6.9.4 Amendments of Organic Residues

Soil quality index is generally governed by the soil organic carbon since it has always been a vital factor in sustaining crop productivity and rejuvenating environmental quality through soil water availability, nutrient cycling, gas flux, plant root growth, and soil conservation (Patra et al. 2010). Plants acquire nutrients from natural sources inclusive of organic matter (soils carry 2–10%) and minerals. Soil organic matter (SOM) not only aggregates soil property and improves water availability but also acts as a nutrient buffering bank to uphold soil quality as represented in Fig. 6.5. Organic soil amendments extend a huge prospective for improvement of soil microbial biomass influencing soil enzymatic activities since the amended residues generally hold intra- and extracellular enzymes as well as assist in carbon, nitrogen, and other nutrient transformation for sustainable productivity. Application of animal wastes, sludge wastes, poultry manure, other carbon-rich wastes, or compost improves the soil organic matter facilitating better soil fertility and periodical decomposition of organic residues before field application inputs C-rich compounds, immobilizes available soil N, as well as substitutes for N required in the soil

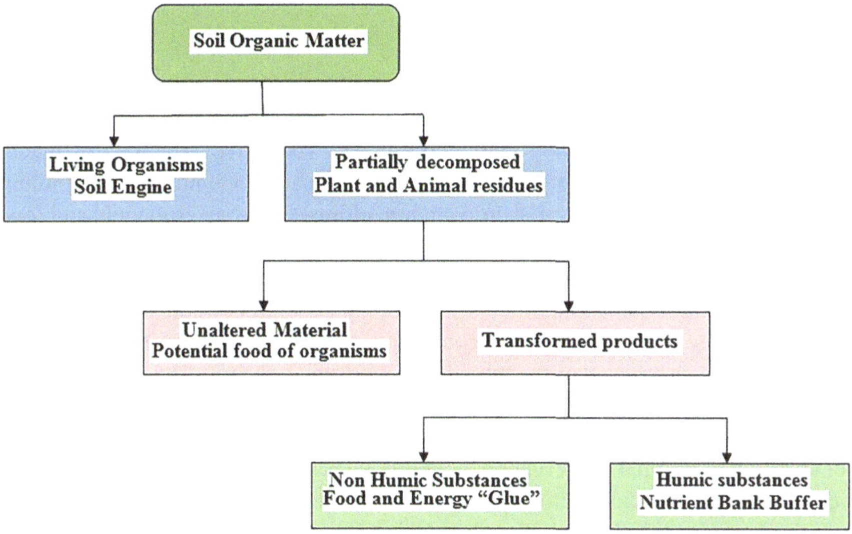

Fig. 6.5 Soil organic matter importance with its distinct components partaking biological functions (adapted from Mohammadi et al. 2011)

(Khursheed 2016). Composting practice conserves nutrients and emits less CO_2 and N_2O (Shrestha et al. 2018).

During carbon sequestration, it is necessary to transfer C to a stable or inert carbon pool to avoid C emission to the atmosphere and biochar application could easily provide the needed direction. The biochar synthesized from biomass pyrolysis could persist in the soils for longer period possessing resistance to chemical and biological degradation, thus owning the capability of amending C in terrestrial ecosystems. Biochar cycle being a slower process would appreciably act as a carbon sink greatly impacting the atmospheric CO_2 concentration uptake by plants as photosynthesis is eightfold higher than GHGs emissions mitigating almost 10% of current anthropogenic CO_2 release (Qambrani et al. 2017). Biochar application in soils involved increasing soil aeration and reducing the frequency of methanogenesis occurrence resulting in CH_4 emissions close to zero (Verheijen et al. 2010). Biochar containing low N concentrations and high C/N ratios aids in the biological immobilization of inorganic N by efficiently adsorbing NH_3 from soil and consequently declining ammonia volatilization from agricultural soils and reducing N_2O emission by 3% to 84% (Qambrani et al. 2017; Shrestha et al. 2018). Biochar participation has been widely documented in improvising agricultural soil properties precisely for increasing soil porosity, soil moisture retention, and aggregate stability and additionally influencing organic carbon stock, cation exchange capacity, and availability of P and K reserves. It structures the microbial community in forest soils, by enhancing soil microbial biomass, and significantly reflects in minimizing soil N_2O emissions, increasing soil CH_4 uptake, and complex changes of soil CO_2 emissions (Li et al. 2018).

6.9.5 Nutrient Management

Agriculture is wholly dependent on the climatic conditions and that's how it easily falls prey to climate change. Increased temperature on soils is accountable for the stresses, including interference with germination, vegetative growth, dry matter partitioning, reproductive processes, loss of organic matter, and nutrient and biological functions drastically affecting crop quality, ultimately losing crop yield and food quality (Agrimonti et al. 2020). Moreover, currently, the modern intensive agricultural practices, including land clearance, extreme and ineffective fertilizer usage, irrigation variations, and agricultural machines run by nonrenewable resources, have made agriculture sector a prominent contributor to GHG emissions (Heidecke et al. 2018). Thus, it is necessary to uphold the nutritional security, to come up with innovative initiatives for fostering soil eminence for future use, and one such approach is nutrient management. Nutrient management science is the study of the combination of optimal conservation strategies interlinking soil utilization, crop and weather factors, irrigation factors, and essential nutrient (CNPSK) input with diverse cultures that are directly or indirectly helping in optimizing and exploiting ideal nutrient use efficiency to crops to refine plant value as well as soil health environment (Farooqi et al. 2018; Bhattacharyya et al. 2016). Dependence and pressurization for more and quality food with population explosion have raised the bars for traditional farming practices, to move toward sustainable and smart approaches such as usage of beneficial microbes for higher agricultural byproducts. Soil microbes especially bacteria and fungi in the form of biofertilizers could play a central and vibrant role in the nutrient management by decomposing and recycling soil organic residues as well as supplying important nutrients (nitrogen, phosphorus, and potassium) to the associated plants (Bhattacharyya et al. 2016). Mulching (organic, inorganic, or mixed) is one of the soil moisture conservation techniques practiced on agricultural fields nowadays that consists of covering the soil surface defending against soil erosion to compensate the nutrient loss and enhance soil fertility as well as positively impact soil temperature and microbiology (Kader et al. 2017). Synthetic N fertilizers become prone to loss due to leaching, mineralization, volatilization, gaseous emissions, and runoff, less or not available for the plants to assimilate. Thus, use of enhanced efficiency fertilizers (EEFs) could be applied to improve the crop's N-use efficiency (Uchida and von Rein 2018). By selecting the right type of EEFs, crops demand of N and the N release by the fertilizers could be synchronized. Stabilized fertilizers like nitrification and urease inhibitors extend the N timing in original form by directly slowing down microbial activity providing higher chance of assimilating nitrates by plants for an extended period of time. Endophytic fungi could be helpful for improving N-use efficiency by supplying NO_3^--N to plant roots (Bhowmik et al. 2017). Thus, these agriculture services could not only help in conserving nutrients but could also contribute to combat climate change.

6.10 Conclusion and Future Perspectives

For supporting various forms of life on Earth, the most pivotal role is usually being played by the soil ecosystem. However, climate change possesses a dynamic and strong impact on soil functions directly via elevated temperature, changes in precipitation patterns, and moisture regime changes and indirectly associated with anthropogenic activities encouraged with agricultural practice adaptations such as management of nutrients, irrigation systems, crop rotation changes, and tillage practices. It is a well-known fact that indigenous soil microbiota or the bioaugmentation of soil microbes in soils could prove to be a promising technique in stabilizing and maintaining various nutrient reserves in soil and eventually making it available to the plants, promoting plant growth and curtailing the application of synthetic fertilizers. Many studies have tried to develop our understanding the relationships between soil properties and climate change, but still there has been a lack of comprehensive explanation investigating this issue. With rapid industrialization and population explosion, anthropogenic pollution along with climate change is pressurizing on the limited water resources. Any ecosystem could not afford to become debt with water scarcity as climate change results in greater soil moisture evaporation. Thus, sustainable farming practices such as mulching should be greatly practiced on agricultural soils to help in conserving both soil moisture and soil nutrients. Organic mulches under suitable temperature and water content not only add nutrients to the soil after microbial biodegradation but also input in nitrogen mobilization, support carbon sequestration, and add plant nutrients to soils. Therefore, future research need to investigate the effects of biodegradable mulching materials on soil biota, soil fertility, adaptability to climate change, and crop yields. Legacy phosphorus defined as the cumulative P accumulated in soils over the years from past fertilizer contributions and manures is the second largest global source of P that could substitute synthetic fertilizers. Thus, integrating the legacy P in the holistic nutrient management approach could provide a sustainable and smart agriculture for future food, bioenergy, and water security. Biochar applicability in forest soils has not only resulted in the improvement of soil physicochemical and biological properties focusing more on soil fertility but has also retained its impression for abatement of soil GHG emissions. The study of phosphate-solubilizing bacteria under field conditions is very rare as it has huge potential in mineralizing soil K reserves for plant growth, but its part in sustaining soil nutrient has not yet being exploited. Management of terrestrial microbial processes incurs the capability of abating climate change by decreasing GHG emissions projecting a tempting prospect for the future research. When microbial communities and biogeochemical cycles are interlinked, they could become a good mechanism to resolve maximum proportion of this environmental mechanism for solving climate change. Therefore, more research on the impacts of climate change threats on soil microbiome services need to be thoroughly understood, and fundamental applicability of multi-omics studies needs to be incorporated for gaining awareness and maximum perception into the unexplored microbial community to fight the negativity of climate change and restore nature's sustainability.

Acknowledgment The authors are much thankful to the Central Pollution Control Board (CPCB), Lucknow, and Central Pollution Control Board (CPCB), New Delhi, for providing necessary support for this publication.

References

Aanderud ZT, Schoolmaster DR, Lennon JT (2011) Plants mediate the sensitivity of soil respiration to rainfall variability. Ecosystems 14:156–167

Aanderud ZT, Jones SE, Schoolmaster DR, Fierer N, Lennon JT (2013) Sensitivity of soil respiration and microbial communities to altered snowfall. Soil Biol Biochem 57:217–227

Abatenh E, Gizaw B, Tsegaye Z, Tefera G (2018) Microbial function on climate change – a review. Environ Pollut Clim Change 2:147. https://doi.org/10.4172/2573458X.1000147

Agrimonti C, Lauro M, Visioli G (2020) Smart agriculture for food quality: facing climate change in the 21st century. Crit Rev Food Sci Nutr 1–11

Allison I (2009) The Copenhagen diagnosis, 2009: updating the world on the latest climate science. The University of New South Wales Climate Change Research Centre (CCRC), Sydney. Retrieved 8 June 2014

Altieri MA (1999) The ecological role of biodiversity in agroecosystems. Agric Ecosyst Environ 74 (1–3):19–31. https://doi.org/10.1016/S0167-8809(99)00028-6

Amouzou KA, Lamers JP, Naab JB, Borgemeister C, Vlek PL, Becker M (2019) Climate change impact on water-and nitrogen-use efficiencies and yields of maize and sorghum in the northern Benin dry savanna, West Africa. Field Crop Res 235:104–117

Andreae MO (1990) Ocean-atmosphere interactions in the global biogeochemical sulfur cycle. Mar Chem 30:1–29

Andreae MO, Elbert W, de Mora SJ (1995) Biogenic sulfur emissions and aerosols over the tropical South Atlantic: 3. Atmospheric dimethylsulfide, aerosols and cloud condensation nuclei. J Geophys Res Atmos 100:11335–11356

Atkin E (2015) Why this new study on arctic permafrost is so scary? http://thinkprogress.org/climate/2015/04/08/3643953/greenlandpermafrostthawmicrobes. Accessed 07 Aug 2020

Bagne K, Ford P, Reeves M (2012) Grasslands and climate change. US Department of Agriculture, Forest Service, Climate Change Resource Center. http://www.fs.fed.us/ccrc/topics/grasslands/index.shtml

Bailer-Jones CAL (2009) The evidence for and against astronomical impacts on climate change and mass extinctions: a review. Int J Astrobiol 8:213–239

Balser TC, McMahon KD, Bart D et al (2006) Bridging the gap between micro- and macro-scale perspectives on the role of microbial communities in global change ecology. Plant Soil 289:59–70

Bardgett RD, Freeman C, Ostle NJ (2008) Microbial contributions to climate change through carbon cycle feedbacks. ISME J 2:805–814

Bardgett RD, DeDeyn GB, Ostle NJ (2009) Plant-soil interactions and the carbon cycle. J Ecol 97:838–839

Bhattacharyya PN, Goswami MP, Bhattacharyya LH (2016) Perspective of beneficial microbes in agriculture under changing climatic scenario: a review. J Phytol 26–41

Bhowmik A, Cloutier M, Ball E, Bruns MA (2017) Underexplored microbial metabolisms for enhanced nutrient recycling in agricultural soils. AIMS Microbiol 3:826

Bissett A, Richardson AE, Baker G, Kirkegaard J, Thrall PH (2013) Bacterial community response to tillage and nutrient additions in a long-term wheat cropping experiment. Soil Biol Biochem 58:281–292

Biswas R, Sarkar A (2018) 'Omics' tools in soil microbiology: the state of the art. In: Advances in soil microbiology: recent trends and future prospects. Springer, Singapore, pp 35–64

Bond-Lamberty B, Bailey V, Chen M, Gough C, Vargas R (2018) Globally rising soil heterotrophic respiration over recent decades. Nature 560:80–83

Bousquet P, Ciais P, Miller JB, Dlugokencky EJ, Hauglustaine DA, Prigent C, Van der Werf GR, Peylin P, Brunke EG, Carouge C, Langenfelds RL (2006) Contribution of anthropogenic and natural sources to atmospheric methane variability. Nature 443:439–443

Brust GE (2019) Management strategies for organic vegetable fertility. In: Safety and practice for organic food, pp 193–212. https://doi.org/10.1016/B978-0-12-812060-6.00009-X

Buis A (2020) Milankovitch (orbital) cycles and their role in Earth's climate. Features. https://climate.nasa.gov/

Busari MA, Kukal SS, Kaur A, Bhatt R, Dulazi AA (2015) Conservation tillage impacts on soil, crop, and the environment. Int Soil Water Conserv Res 3:119–129

Byrne MP, O'Gorman PA (2018) PNAS 115(19):4863–4868; first published April 23, 2018. https://doi.org/10.1073/pnas.1722312115

Campisano CJ (2012) Milankovitch cycles, paleoclimatic change, and hominin evolution. Nat Educ Knowl 4(3):5

Cappucci M (2019) Increasing humidity, driven in part by climate change, is making even modest heat waves unbearable. The Washington Post

Castro HF, Classen AT, Austin EE, Norby RJ, Schadt CW (2010) Soil microbial community responses to multiple experimental climate change drivers. Appl Environ Microbiol 76 (40):999–1007

Cavicchioli R, Ripple WJ, Timmis KN, Azam F, Bakken LR, Baylis M, Behrenfeld MJ, Boetius A, Boyd PW, Classen AT, Crowther TW (2019) Scientists' warning to humanity: microorganisms and climate change. Nat Rev Microbiol 17(9):569–586

Chen H, Shang Z, Cai H, Zhu Y (2019) Response of soil N_2O emissions to soil microbe and enzyme activities with aeration at two irrigation levels in greenhouse tomato (*Lycopersicon esculentum* Mill.) fields. Atmosphere 10(2):72

Classen AT, Sundqvist MK, Henning JA et al (2015) Direct and indirect effects of climate change on soil microbial and soil microbial-plant interactions: what lies ahead? Ecosphere 6(8):130. https://doi.org/10.1890/ES15-00217.1

Cookson WR, Osman M, Marschner P et al (2007) Controls on soil nitrogen cycling and microbial community composition across land use and incubation temperature. Soil Biol Biochem 39:744–756

Cooper MD, Estop-Aragonés C, Fisher JP, Thierry A, Garnett MH, Charman DJ, Murton JB, Phoenix GK, Treharne R, Kokelj SV, Wolfe SA (2017) Limited contribution of permafrost carbon to methane release from thawing peatlands. Nat Clim Chang 7(7):507–511

Corsi S, Friedrich T, Kassam A, Pisante M, de Moraes Sa JC (2012) Soil organic carbon accumulation and greenhouse gas emission reductions from conservation agriculture: a literature review. Integr Crop Manage 16:101

Crowther TW, Thomas SM, Maynard DS et al (2015) Biotic interactions mediate soil microbial feedbacks to climate change. Proc Natl Acad Sci 112(22):7033–7038

Cubasch U, Meehl GA, Boer GJ, Stouffer RJ, Dix M, Noda A, Senior CA, Raper S, Yap KS (2001) Projections of future climate change. In: Ding Y et al (eds) Climate change 2001: the scientific basis. Contribution of working group I to the third assessment report of the intergovernmental panel on climate change, pp 526–582

Curtin D, Beare MH, Hernandez-Ramirez G (2012) Temperature and moisture effects on microbial biomass and soil organic matter mineralization. Soil Sci Soc Am J 76:2005–2067

Daims H, Lebedeva EV, Pjevac P, Han P, Herbold C, Albertsen M, Jehmlich N, Palatinszky M, Vierheilig J, Bulaev A, Kirkegaard RH (2015) Complete nitrification by Nitrospira bacteria. Nature 528(7583):504–509
Davidson EA, Janssens IA (2006) Temperature sensitivity of soil carbon decomposition and feedbacks to climate change. Nature 440:165–173
Dawson B, Spannagle M (2008) The complete guide to climate change is an indispensable resource. Routledge, Oxon
Desonie D (2008) Climate: causes and effects of climate change. Chelsea House, New York
Dijkstra FA, Prior SA, Runion GB, Torbert HA, Tian H, Lu C, Venterea RT (2012) Effects of elevated carbon dioxide and increased temperature on methane and nitrous oxide fluxes: evidence from field experiments. Front Ecol Environ 10(10):520–527
Dorrepaal E, Cornelissen JHC, Aerts R, Wallen B, van Logtestijn RSP (2005) Are growth forms consistent predictors of leaf litter quality and decomposability across peatlands along a latitudinal gradient? J Ecol 93:817–828
Dou F, Hons FM (2006) Tillage and nitrogen effects on soil organic matter fractions in wheat-based systems. Soil Sci Soc Am J 70:1896–1905
Downing J, McClain M, Twilley R et al (1999) The impact of accelerating land-use change on the N-cycle of tropical aquatic ecosystems: current conditions and projected changes. Biogeochemistry 46(1):109–148
Dutta H, Dutta A (2016) The microbial aspect of climate change. Energy Ecol Environ 1 (4):209–232
Edge TA, Baird DJ, Bilodeau G, Gagné N, Greer C, Konkin D, Newton G, Séguin A, Beaudette L, Bilkhu S, Bush A (2020) The Ecobiomics project: advancing metagenomics assessment of soil health and freshwater quality in Canada. Sci Total Environ 710:135906
Edwards A, Cameron KA, Cook JM, Debbonaire AR, Furness E, Hay MC, Rassner SM (2020) Microbial genomics amidst the Arctic crisis. Microb Genom mgen000375
Eivazi F, Bayan MR, Schmidt K (2003) Select soil enzyme activities in the historic Sanborn field as affected by long-term cropping systems. Commun Soil Sci Plant Anal 34:2259–2275
Elser JJ, Bracken MES, Cleland EE, Gruner DS, Harpole WS, Hillebrand H, Ngai JT, Seabloom EW, Shurin JB, Smith JE (2007) Global analysis of nitrogen and phosphorus limitation of primary producers in freshwater, marine, and terrestrial ecosystems. Ecol Lett 10:1135–1142. https://doi.org/10.1111/j.1461-0248.2007.01113.x
Eneji VCO, Inyang-Abia ME, Ekpo CG, Isa AM (2017) A review of global warming/climate change, causes, effects and mitigations. Environ Stud 1(1):44–71
Engelkes T, Morrien E, Verhoeven KJF et al (2008) Successful range-expanding plants experience less aboveground and below-ground enemy impact. Nature 456:946–948
Etesami H, Emami S, Alikhani HA (2017) Potassium solubilizing bacteria (KSB): mechanisms, promotion of plant growth, and future prospects: a review. J Soil Sci Plant Nutr 17(4):897–911
European Commission (2015) The role of Arctic microbes in climate change. Cordis: News and Events. http://cordis.europa.eu/news/rcn/127717_en.html. Accessed 02 Aug 2020
Fagodiya RK, Pathak H, Kumar A, Bhatia A, Jain N (2017) Global temperature change potential of nitrogen use in agriculture: a 50-year assessment. Sci Rep 7:44928
Fagodiya RK, Kumar A, Kumari S, Medhi K, Shabnam AA (2020) Role of nitrogen and its agricultural management in changing environment. In: Contaminants in agriculture. Springer, Cham, pp 247–270
Fahey D, Doherty S, Hibbard KA, Romanou A, Taylor P (2017) Physical drivers of climate change
Falkowski PG, Fenchel T, Delong EF (2008) The microbial engines that drive earth's biogeochemical cycles. Science 320:1034–1039
Farooqi ZUR, Sabir M, Zeeshan N, Naveed K, Hussain MM (2018) Enhancing carbon sequestration using organic amendments and agricultural practices. Carbon Capture, Utilization and Sequestration
Fierer N, Schimel JPA (2003) A proposed mechanism for the pulse in carbon dioxide production commonly observed following the rapid rewetting of a dry soil. Soil Sci Soc Am J 67:798–805

Fierer N, Craine JM, McLauchlan K et al (2005) Litter quality and the temperature sensitivity of decomposition. Ecology 86:320–326

Food and Agriculture Organization (2010) Climate-smart agriculture: policies, practices and financing for food security, adaptation and mitigation. Report prepared for The Hague Conference on Agriculture, Food Security, and Climate Change, Rome

French S, Levy-Booth D, Samarajeewa A et al (2009) Elevated temperatures and carbon dioxide concentrations: effects on selected microbial activities in temperate agricultural soils. World J Microbiol Biotechnol 25:1887–1900

Gambolati G, Putti M, Teatini P, Camporese M, Ferraris S, Gasparetto Stori G (2005) Peatland oxidation enhances subsidence in the Venice watershed. Eos Trans AGU 86:217–220

Gattinger A, Jawtusch J, Muller A, Mader P (2014) Climate change and agriculture: no-till agriculture – a climate smart solution? Report no. 2. Misereor, Aachen, p 24

Gelybó G, Tóth E, Farkas C, Horel Á, Kása I, Bakacsi Z (2018) Potential impacts of climate change on soil properties. AgrokémiaésTalajtan 67(1):121–141

Geresdi I, Meszaros E, Molnar A (2006) The effect of chemical composition and size distribution of aerosol particles on droplet formation and albedo of stratocumulus clouds. Atmos Environ 40:1845–1855

Glassman SI, Weihe C, Li J et al (2018) Decomposition responses to climate depend on microbial community composition. Proc Natl Acad Sci 115(47):11994–11999

Gruber N, Galloway JN (2008) An earth-system perspective of the global nitrogen cycle. Nature 451:293–296

Guignard MS, Leitch AR, Acquisti C, Eizaguirre C, Elser JJ, Hessen DO, Jeyasingh PD, Neiman M, Richardson AE, Soltis PS, Soltis DE (2017) Impacts of nitrogen and phosphorus: from genomes to natural ecosystems and agriculture. Front Ecol Evol 5:70

Gupta C, Prakash DG, Gupta S (2014) Role of microbes in combating global warming. Int J Pharm Sci Lett 4:359–363

Gupta N, Vats S, Bhargava P (2018) Sustainable agriculture: role of metagenomics and metabolomics in exploring the soil microbiota. In: In silico approach for sustainable agriculture. Springer, Singapore, pp 183–199

Haigh JD (2000) Solar variability and climate. Weather 55:399–405

Heath J, Ayres E, Possell M et al (2005) Rising atmospheric CO_2 reduces soil carbon sequestration. Science 309:1711–1713

Heidecke C, Montgomery H, Stalb H, Wollenberg L (eds) (2018) International conference on agricultural GHG emissions and food security – connecting research to policy and practice – volume of abstracts, Braunschweig. Johann Heinrich von Thunen Institut Berlin

Hibbard K, Law B, Reichstein M, Sulzman J (2005) An analysis of soil respiration across northern hemisphere temperate ecosystems. Biogeochemistry 73:29–70

Hobbie SE (1996) Temperature and plant species control over litter decomposition in Alaskan tundra. Ecol Monogr 66:503–522

Hostetler SW, Mix AC (1999) Re-assessment of ice-age cooling of the tropical ocean and atmosphere. Nature 399:673–676

Hu HW, Trivedi P, He JZ, Singh BK (2017) Microbial nitrous oxide emissions in dryland ecosystems: mechanisms, microbiome and mitigation. Environ Microbiol 19(12):4808–4828

Ineson P, Coward PA, Hartwig UA (1998) Soil gas fluxes of N_2O, CH_4 and CO_2 beneath *Lolium perenne* under elevated CO_2: the Swiss free air carbon dioxide enrichment experiment. Plant Soil 198:89–95

Intergovernmental Panel on Climate Change (2014) Contribution of working groups I, II and III to the fifth assessment report of the Intergovernmental Panel on Climate Change. IPCC, Geneva, p 151

IPCC (2007) Climate change 2007: the physical science basis. Contribution of working group I to the fourth assessment report of the intergovernmental panel on climate change. Cambridge University Press, New York

Jackson RB, Canadell J, Ehleringer JR et al (1996) A global analysis of root distributions for terrestrial biomes. Oecologia 108:389–411

Jafarov FT, Mustafayev MG (2020) The impact of climate change on soil fertility in the central region of Azerbaijan. Екологічнабезпекатаприродокористування 34(2):81–87

Jansson JK, Hofmockel KS (2019) Soil microbiomes and climate change. Nat Rev Microbiol 1–12

Joshi PA, Shekhawat DB (2014) Microbial contributions to global climate changes in soil environments: impact on carbon cycle (short review). Ann Appl Bio-Sci 1(1):R7–R9

Kader MA, Senge M, Mojid MA, Ito K (2017) Recent advances in mulching materials and methods for modifying soil environment. Soil Tillage Res 168:155–166

Kardol P, Cregger MA, Campany CE, Classen AT (2010) Soil ecosystem functioning under climate change: plant species and community effects. Ecology 91:767–781

Kennedy AC, Papendick RI (1995) Microbial characteristics of soil quality. J Soil Water Conserv 50:243–248

Khursheed S (2016) Soil biodiversity and climate change. Adv Plants Agric Res 3(5):00113

Kiene RP, Linn LJ, Bruton JA (2000) New and important roles for DMSP in marine microbial communities. J Sea Res 43(3–4):209–224

Kloster S (2006) DMS cycle in the ocean-atmosphere system and its response to anthropogenic perturbations. Doctoral dissertation, University of Hamburg, Hamburg

Kou-Giesbrecht S, Menge D (2019) Nitrogen-fixing trees could exacerbate climate change under elevated nitrogen deposition. Nat Commun 10(1):1–8

Kumar M, Gazara RK, Verma S, Kumar M, Verma PK, Thakur IS (2016) Genome sequence of *Pandoraea* sp. ISTKB, a lignin-degrading betaproteobacterium, isolated from Rhizospheric soil. Genome Announcements 4(6)

Kuypers MM, Marchant HK, Kartal B (2018) The microbial nitrogen-cycling network. Nat Rev Microbiol 16(5):263

Kuzyakov Y, Blagodatskaya E (2015) Microbial hotspots and hot moments in soil: concept and review. Soil Biol Biochem 83:184–199

Laban P, Metternicht G, Davies J (2018) Soil biodiversity and soil organic carbon: keeping drylands alive. Gland: International Union for Conservation of Nature and Natural Resources. https://portals.iucn.org/library/sites/library/files/documents/2018-004-En

Lal R (2004) Soil carbon sequestration impacts on global climate change and food security. Science 304:1623–1627. https://doi.org/10.1126/science.1097396

Lal R, Reicosky DC, Hanson JD (2007) Evolution of the plow over 10,000 years and the rationale for no-till farming. Soil Tillage Res 93:1–12

Lauenroth WK, Bradford JB (2006) Ecohydrology and the partitioning AET between transpiration and evaporation in a semiarid steppe. Ecosystems 9:756–767

LeBauer DS, Treseder KK (2008) Nitrogen limitation of net primary productivity in terrestrial ecosystems is globally distributed. Ecology 89:371–379

Lee CA, Shen B, Slotnick BS et al (2013) Continents-island arc fluctuations, growth of crustal carbonates, and long-term climate change. Geosphere 9(1):21–36. https://doi.org/10.1130/GES00822.1

Li Y, Hu S, Chen J, Müller K, Li Y, Fu W, Lin Z, Wang H (2018) Effects of biochar application in forest ecosystems on soil properties and greenhouse gas emissions: a review. J Soils Sediments 18(2):546–563

Lladó S, López-Mondéjar R, Baldrian P (2017) Forest soil bacteria: diversity, involvement in ecosystem processes, and response to global change. Microbiol Mol Biol Rev 81(2)

Lopez-Fando C, Pardo MT (2009) Changes in soil chemical characteristics with different tillage practices in a semi-arid environment. Soil Tillage Res 104:278–284

Louis BP, Maron P, Vlaud V, Leterme P, Menasseri-Aubry S (2016) Soil C and N models that integrate microbial diversity. Environ Chem Lett 14(3):331–344

Mandal A, Neenu S (2012) Impact of climate change on soil biodiversity – a review. Agric Rev 33 (4):283–292

Masuda Y, Itoh H, Shiratori Y et al (2017) Predominant but previously-overlooked prokaryotic drivers of reductive nitrogen transformation in paddy soils, revealed by metatranscriptomics. Microbes Environ 32:180–183

Mbow C, Smith P, Skole D, Duguma L, Bustamante M (2014) Achieving mitigation and adaptation to climate change through sustainable agroforestry practices in Africa. Curr Opin Environ Sustain 6:8–14

McKenzie NR, Horton BK, Loomis SE et al (2016) Continental arc volcanism as the principal driver of icehouse-greenhouse variability. Science 352(6284):444–447. https://doi.org/10.1126/science.aad5787

Medhi K, Singhal A, Chauhan DK, Thakur IS (2017) Investigating the nitrification and denitrification kinetics under aerobic and anaerobic conditions by *Paracoccus denitrificans* ISTOD1. Bioresour Technol 242:334–343

Medhi K, Mishra A, Thakur IS (2018) Genome sequence of a heterotrophic nitrifier and aerobic denitrifier, *paracoccus denitrificans* strain ISTOD1, isolated from wastewater. Genome Announc 6(15)

Melillo JM, Steudler PA, Aber JD, Newkirk K, Lux H, Bowles FP, Catricala C, Magill A, Ahrens T, Morrisseau S (2002) Soil warming and carbon-cycle feedbacks to the climate system. Science 298:2173–2176

Miao L, Liu Z (2018) Microbiome analysis and-omics studies of microbial denitrification processes in wastewater treatment: recent advances. Sci China Life Sci 61(7):753–761

Mishra A, Jha G, Thakur IS (2018) Draft genome sequence of *Zhihengliuella* sp. strain ISTPL4, a psychrotolerant and halotolerant bacterium isolated from Pangong Lake, India. Genome Announcements 6(5)

Mock T, Daines SJ, Geider R, Collins S, Metodiev M, Millar AJ, Moulton V, Lenton TM (2016) Bridging the gap between omics and earth system science to better understand how environmental change impacts marine microbes. Glob Chang Biol 22(1):61–75

Mohammadi K, Heidari G, Khalesro S, Sohrabi Y (2011) Soil management, microorganisms and organic matter interactions: a review. Afr J Biotechnol 10(86):19840–19849

Monkiedje A, Spiteller M, Fotio D, Sukul P (2006) The effect of land use on soil health indicator in peri-urban agriculture in the humid forest zone of southern Cameroon. J Environ Qual 35:2402–2409

Morgan JA, Derner JD, Milchunas DG, Pendall E (2008) Management implications of global change for Great Plains rangelands. Rangelands 30(3):18–22

Nair GR, Raja SSS (2017) Decoding complex soil microbial communities through new age "omics". J Microb Biochem Technol 9(6):301–309

Ngumbi E (2015) Turning to bacteria to fight the effects of climate change. http://blogs.scientificamerican.com/guestblog/turningtobacteriatofighttheeffectsofclimatechange/. Accessed 25 July 2020

Nikiema J, Bibeau L, Lavoie J, Brzezinski R, Vigneux J, Heitz M (2005) Biofiltration of methane: an experimental study. Chem Eng J 113(2–3):111–117

Osman KT (2013) Forest soil and climate change. In: Forest soils. Springer, Cham, pp 173–182

Pan G, Huang ZQ, Wang JK et al (2013) Soil organic matter dynamics beyond carbon; a report of the 4th international symposium on soil organic matter dynamics. Carbon Manage 4:485–489

Pareek N (2017) Climate change impact on soils: adaptation and mitigation. MOJ Eco Environ Sci 2 (3):00026

Pathak A, Pathak R (2012) Microorganisms and global warming. Int J Appl Microbiol Sci 1:21–23

Patra PK, Saha N, Mukherjee R, Chakraborty A, Sarkar S, Mukherjee D (2010) Influence of tillage techniques and organic matter on carbon and nitrogen transformation in the rice rhizosphere in an alluvial soil of West Bengal. Appl Ecol Environ Res 8(4):313–327

Paustian K, Six J, Elliott ET, Hunt HW (2000) Management options for reducing CO_2 emissions from agricultural soils. Biogeochemistry 48:147–163

Phillips RL, Whalen SC, Schlesinger WH (2001) Influence of atmospheric CO2 enrichment on methane consumption in a temperate forest soil. Glob Chang Biol 7(5):557–563

Potthast K, Hamer U, Makeschin F (2010) Impact of litter quality on mineralization processes in managed and abandoned pasture soils in southern Ecuador. Soil Biol Biochem 42:56–64
Qambrani NA, Rahman MM, Won S, Shim S, Ra C (2017) Biochar properties and eco-friendly applications for climate change mitigation, waste management, and wastewater treatment: a review. Renew Sust Energ Rev 79:255–273
Raich JW, Schlesinger W (1992) The global carbon-dioxide flux in soil respiration and its relationship to vegetation and climate. Tellus B 44:81–99
Rathour R, Gupta J, Mishra A, Rajeev AC, Dupont CL, Thakur IS (2020) A comparative metagenomic study reveals microbial diversity and their role in the biogeochemical cycling of Pangong lake. Sci Total Environ 139074
Reijneveld A, van Wensem J, Oenema O (2009) Soil organic carbon contents of agricultural land in the Netherlands between 1984 and 2004. Geoderma 152(3–4):231–238. https://doi.org/10.1016/j.geoderma.2009.06.007
Richardson AE, Simpson RJ (2011) Soil microorganisms mediating phosphorus availability update on microbial phosphorus. Plant Physiol 156(3):989–996
Roldan A, Caravaca F, Hernandez MT et al (2003) No-tillage, crop residue additions, and legume cover cropping effects on soil quality characteristics under maize in Patzcuaro watershed (Mexico). Soil Tillage Res 72:65–73
Rumpel C (2019) Soils linked to climate change. Nature
Rustad L, Campbell J, Marion G et al (2001) A meta-analysis of the response of soil respiration, net nitrogen mineralization, and aboveground plant growth to experimental ecosystem warming. Oecology 126:543–562
Salinas-Garcia JR, Velazquez-Garcia J, Gallardo-Valdez M et al (2002) Tillage effects on microbial biomass and nutrient distribution in soils under rainfed corn production in Central-Western Mexico. Soil Tillage Res 66:143–152
Scherr SJ, Sthapit S (2009) Farming and land use to cool the planet in: state of the world- into a warming world. Worldwatch Institute, Cambridge
Schimel JP, Bennett J (2004) Nitrogen mineralization: challenges of a changing paradigm. Ecology 85:591–602
Schimel DS, Braswell BH, Holland EA et al (1994) Climatic, edaphic, and biotic controls over storage and turnover of carbon in soils. Glob Biogeochem Cycles 8:279–293
Schindlbacher A, Rodler A, Kuffner M et al (2011) Experimental warming effects on the microbial community of a temperate mountain forest soil. Soil Biol Biochem 43(7):1417–1425
Schuur EA, McGuire AD, Schädel C, Grosse G, Harden JW, Hayes DJ, Hugelius G, Koven CD, Kuhry P, Lawrence DM, Natali SM (2015) Climate change and the permafrost carbon feedback. Nature 520(7546):171–179
Schwinger J, Tjiputra J, Goris N, Six KD, Kirkevåg A, Seland Ø, Heinze C, Ilyina T (2017) Amplification of global warming through pH dependence of DMS production simulated with a fully coupled earth system model. Biogeosciences 14(15):3633
Scott-Denton LE, Rosenstiel TN, Monson RK (2006) Differential controls by climate and substrate over the heterotrophic and rhizospheric components of soil respiration. Glob Chang Biol 12:205–216
Semrau JD, DiSpirito AA, Yoon S (2010) Methanotrophs and copper. FEMS Microbiol Rev 34 (4):496–531
Shade A, Peter H, Allison SD et al (2012) Fundamentals of microbial community resistance and resilience. Front Microbiol-Aquat Microbiol 3:417
Shaman J, Kohn M (2009) Absolute humidity modulates influenza survival, transmission, and seasonality. Proc Natl Acad Sci 106(9):3243–3248
Shim JH, Pendall E, Morgan JA, Ojima DS (2009) Wetting and drying cycles drive variations in the stable carbon isotope ratio of respired carbon dioxide in semi-arid grassland. Oecology 160:321–333
Shindell D, Kuylenstierna JC, Vignati E, van Dingenen R, Amann M, Klimont Z, Anenberg SC, Muller N, Janssens-Maenhout G, Raes F, Schwartz J (2012) Simultaneously mitigating near-

term climate change and improving human health and food security. Science 335 (6065):183–189
Shrestha BM, Chang SX, Bork EW, Carlyle CN (2018) Enrichment planting and soil amendments enhance carbon sequestration and reduce greenhouse gas emissions in agroforestry systems: a review. Forests 9(6):369
Singh BK, Bardgett RD, Smith P, Reay DS (2010) Microorganisms and climate change: terrestrial feedbacks and mitigation options. Nat Rev Microbiol 8:779–790
Smith DL, Almaraz JJ (2004) Climate change and crop production: contributions, impacts, and adaptations. Can J Plant Pathol 26(3):253–266
Smith P, Fang C, Dawson JJC, Moncrieff JB (2008) Impact of global warming on soil organic carbon. Adv Agron 97:1–43
Smith P, Cotrufo MF, Rumpel C et al (2015) Biogeochemical cycles and biodiversity as key drivers of ecosystem services provided by soils. Soil 1:665–685
Stahl DA, de la Torre JR (2012) Physiology and diversity of ammonia-oxidizing archaea. Annu Rev Microbiol 66:83–101
Stark JM, Firestone MK (1995) Mechanisms for soil moisture effects on activity of nitrifying bacteria. Appl Environ Microbiol 61(1):218–221
Svoboda E (2015) How wetland microbes impact global climate? http://discovermagazine.com/2015/june/22smallwonders. Accessed 25 July 2020
Takai K (2019) The nitrogen cycle: a large, fast, and mystifying cycle. Microbes Environ 34 (3):223–225
Tecon R, Or D (2017) Biophysical processes supporting the diversity of microbial life in soil. FEMS Microbiol Rev 41(5):599–623
Thakur IS, Medhi K (2019) Nitrification and denitrification processes for mitigation of nitrous oxide from waste water treatment plants for biovalorization: challenges and opportunities. Bioresour Technol 282:502–513
Thomas RQ, Brookshire EJ, Gerber S (2015) Nitrogen limitation on land: how can it occur in earth system models? Glob Chang Biol 21(5):1777–1793
Tokida T, Fumoto T, Cheng W, Matsunami T, Adachi M, Katayanagi N, Matsushima M, Okawara Y, Nakamura H, Okada M, Sameshima R (2010) Effects of free-air CO_2 enrichment (FACE) and soil warming on CH 4 emission from a rice paddy field: impact assessment and stoichiometric evaluation. Biogeosci Discuss 7(2)
Trenberth K, Miller K, Mearns L, Rhodes S (2000) Effects of changing climate on weather and human activities. University Science Books, California
Uchida Y, von Rein I (2018) Mitigation of nitrous oxide emissions during nitrification and denitrification processes in agricultural soils using enhanced efficiency fertilizers. In: Soil contamination and alternatives for sustainable development. IntechOpen
UN (2007) United Nations expert group meeting on population distribution, urbanization, internal migration and development, United Nations population division, UN/POP/BGM-URB/2007/01
van Groenigen J, Trinity College Dublin (2011) Soil microbes accelerate global warming. Science Daily. Retrieved August 27, 2020 from www.sciencedaily.com/releases/2011/07/110713131423.htm
Verheijen F, Jeffery S, Bastos AC, Van der Velde M, Diafas I (2010) Biochar application to soils. A critical scientific review of effects on soil properties, processes, and functions. EUR, 24099, p 162
VijayaVenkataRaman S, Iniyan S, Goic R (2012) A review of climate change, mitigation and adaptation. Renew Sust Energ Rev 16(1):878–897
Vitousek PM, Howarth RW (1991) Nitrogen limitation on land and in the sea: how can it occur? Biogeochemistry 13(87–115):1991
Walsh DA (2015) Consequences of climate change on microbial life in the ocean. Microbiology Today. Microbiology Society, London

Wang L, Tang LL, Wang X et al (2010) Effects of alley crop planting on soil and nutrient losses in the citrus orchards of the three gorges region. Soil Tillage Res 110(2):243–250. https://doi.org/10.1016/j.still.2010.08.012

Wang S, Maltrud M, Elliott S, Cameron-Smith P, Jonko A (2018) Influence of dimethyl sulfide on the carbon cycle and biological production. Biogeochemistry 138(1):49–68

Weiman S (2015) Microbes help to drive global carbon cycling and climate change. Microbe Mag 10(6):233–238

Wilby RL, Hedger M, Orr H (2005) Climate change impacts and adaptation: a science agenda for the Environment Agency of England and Wales. Weather 60(7):206–211

Woodward FI, Lomas MR, Kelly CK (2004) Global climate and the distribution of plant biomes. Proc R Soc Lond Biol Sci 359:1465–1476

Wrage-Mönnig N, Horn MA, Well R, Müller C, Velthof G, Oenema O (2018) The role of nitrifier denitrification in the production of nitrous oxide revisited. Soil Biol Biochem 123:A3–A16

Wu Z, Dijkstra P, Koch GW, Penuelas J, Hungate BA (2011) Responses of terrestrial ecosystems to temperature and precipitation change: a meta-analysis of experimental manipulation. Glob Chang Biol 17:927–942

Xu B, Jin Z, Jiang Z, Guo J, Timberlake M, Ma X (2014) Climatological and geographical impacts on the global pandemic of influenza A (H1N1) 2009. In: Global urban monitoring and assessment through earth observation. CRC, pp 254–269

Yiqi L, Zhou X (2006) Soil respiration and the environment. Academic, London

Yu J, Deem LM, Crow SE, Deenik J, Penton CR (2019) Comparative metagenomics reveals enhanced nutrient cycling potential after 2 years of biochar amendment in a tropical oxisol. Appl Environ Microbiol 85(11)

Yun-Feng Y (2013) Omics breakthroughs for environmental microbiology. Omics Environ Microbiol l40:18–33

Yvon-Durocher G, Allen AP, Bastviken D, Conrad R, Gudasz C, St-Pierre A, Thanh-Duc N, Del Giorgio PA (2014) Methane fluxes show consistent temperature dependence across microbial to ecosystem scales. Nature 507(7493):488–491

Zhe CHEN, Yang SQ, Zhang AP, Xin JING, Song WM, Mi ZR, Zhang QW, Wang WY, Yang ZL (2018) Nitrous oxide emissions following seasonal freeze-thaw events from arable soils in Northeast China. J Integr Agric 17(1):231–246

Zimmer C (2010) The microbe factor and its role in our climate future. http://e360.yale.edu/feature/the_microbe_factor_and_its_role_in_our_climate_future/2279/. Accessed 15 July 2020

Chapter 7
Climate Change and Its Impact on Soil Properties

Ashutosh Tripathi, Vanya Pandey, and Manju Rawat Ranjan

Abstract Increased industrialization and urbanization are resulting in changes in the global environment, often with severe consequences. Changes in the gas composition of the atmosphere—majorly due to CO_2 and other greenhouse gases—are leading to a rise in global temperature with high spatial and temporal climate variability, changing precipitation patterns, and altered global circulation processes. These modifications are reflected sensitively by ecosystems. Soils are linked to the atmospheric/climatic system through the carbon (C), nitrogen (N), and hydrological cycles. Hence, several biogeochemical and hydrological cycles are altered due to increased temperature and changing precipitation (frequency as well as amount), in turn altering pedogenesis (modifying the rate of organic matter formation, soil water regime, mineral composition, etc.). These soil properties play a central part in the quality and productivity of agricultural/forestland. The chapter attempts to review the potential changes in the properties (physical and chemical) of soil due to changing climatic variable, soil degradation as a result of altered temperature and precipitation patterns, and studying soil as a part of some of the major biogeochemical cycles (C and N).

Keywords Climate change · Soil properties · Climate–soil interaction · Biogeochemical cycles

7.1 Introduction

7.1.1 Background: Climate Change

The global environment is changing at unprecedented rates with disastrous potential consequences to the future generations owing to the mindless human activities.

A. Tripathi (✉) · V. Pandey · M. R. Ranjan
Amity Institute of Environmental Sciences, Amity University, Noida, Uttar Pradesh, India
e-mail: atripathi1@amity.edu

D. K. Choudhary et al. (eds.), *Climate Change and the Microbiome*, Soil Biology 63,
https://doi.org/10.1007/978-3-030-76863-8_7

According to the Massachusetts Institute of Technology, Cambridge, we are currently on track for an increase of temperature between 6.3 and 13.3 degrees Fahrenheit, with a high probability of 9.4 degrees Fahrenheit by 2100. There have been/will be changes in all spheres of the world as a result of climate change, including the atmosphere, hydrosphere, biosphere, and lithosphere, which includes soil.

This needs a clear understanding of the whole Earth's system, the relationship between climate and soils, and the knowledge of how climate change can affect soils.

With rising temperatures, an increased number of mountain glaciers, polar ice caps, as well as other permafrost regions may melt, as expected. Changes in patterns of water flow, such as flood waves and surface rush, would contribute to an increased level of eustatic sea and endanger low-lying human protected lands, cities, farming regions, and low-lying extended seas. The increased expansion of the areas affected by salt would be further impacted by direct seawater flooding or by a rise in the saline or brackish water level associated with the sea.

A changing climate can lead to significant changes to the effects of the climate feedback effect on natural vegetation and the practice of land use, such as changed albedo, surface roughness, and near-surface atmosphere's heat and energy balance, and the pattern of temperature and precipitation significantly affects the processes of field water cycle and soil formation/degradation.

7.2 Climate: A Soil-Forming Factor

The various factors influencing the process of formation of soil are called as soil-forming factors. The parent material, topography, biological factors, and time all play an essential role in the formation of soil, climate being one of them (United States Department of Agriculture 2020). Soil-forming factors influence as well as determine the composition of soil. The relationship between soil and the major soil-forming factors was first identified by Dokuchaev in 1899 and by Jenny in 1941. Various physical and chemical processes occur during soil formation, each requiring a different time scale. When soil enters a "maturity" phase, soil processes either stabilize or enter a state of quasi-equilibrium with the surrounding environment; and the changing environment, thus, can dramatically alter the soil development pathways.

Climate, as one of the critical influencing factors in soil formation, has both direct and indirect consequences. The direct effect of climatic variables (precipitation and increased radiation leading to more heat) on the environment of soil can be interpreted.

Some of the direct effects influencing soil properties are surface runoff and filtrating water, especially during heavy rains and thunderstorms (intensity and frequency of which are characteristic features of climate change) and destructing the role of raindrops.

Climate change has an indirect impact on the biosphere as a whole, which affects the properties and development of soils. As a consequence, combinations of temperature and humidity will decide the processes of conversion of mineral compounds in the soil, such as weathering and its rate and deposition of soil-forming products (vegetation, the rate of development and degradation of organic matter, soil microflora, its functions and intensity of its functions, and water erosion processes).

In hot, dry regions of the desert, for example, there is little vegetation and therefore little organic soil content. Failure to precipitate prevents chemical weathering in arid regions, leading to rough texture. The cold temperatures that allow organic matter to grow in the tundra limit the work of bacteria. In the warm and wet tropics, bacterial development in the leaf litters persists at a high rate. The trees under the lush tropical canopy of forest can easily retrieve useful nutrients. Any organic matter is also eliminated from the Earth by the high annual rainfall. These factors combine to set soils that are deficient in organic matter in their upper horizons.

The atmosphere, which interacts with plants, also influences the chemistry of the soil. Pine forests dominate cool, humid climates. The decomposition of the pine needles in the presence of water produces a weak acid that removes soluble soil bases, leaving it acidic. Pine trees often have a low nutrient demand, so that the trees retrieve small soil nutrients and then recycle them through a rotting needle litter. The soil beneath pine trees has a very less water retention capacity, and hence the water does not get absorbed and glides over the surface, increasing the risk of mudslides and landslides. The soil beneath oak trees has a high-water retention capacity, and rainwater easily gets absorbed into the soil, thus keeping the water table alive.

The chemical reaction within the soil is another significant influence, so does the chemical reaction as the temperature increases. In the evolution of soils, chemical processes play a dominant role. The weathering of the parent rock in tropical areas is estimated to be three times faster than in the temperate zone.

7.3 Climate and Soil: The Interaction

Changes in climate variables such as precipitation intensity or seasonal temperatures may have a direct impact on the hydrophysical properties of the soil. These adjustments have an impact on the soil water regime, which in turn has an impact on the environmental and economic development of a country. Figure 7.1 illustrates the relationship of climate with soil and, as a result, agriculture.

The primary energy and humidity sources for biological and soil processes are solar radiation and precipitation. The Earth's surface heats the atmosphere and absorbs and re-emits solar radiation. Precipitation is absorbed by the soil, used by crops, and accompanied by evaporation and transpiration to the atmosphere. As a result, the soil and the atmosphere exchange of heat is constant.

However, the study and evaluation of possible collective future changes is far easier than done due to uncertainties in estimation of the longer-term temperature and precipitation trends (temporal and spatial change), the evolving water cycle, and the dynamic impacts of natural vegetation and Earth use patterns (change in socio-economic conditions).

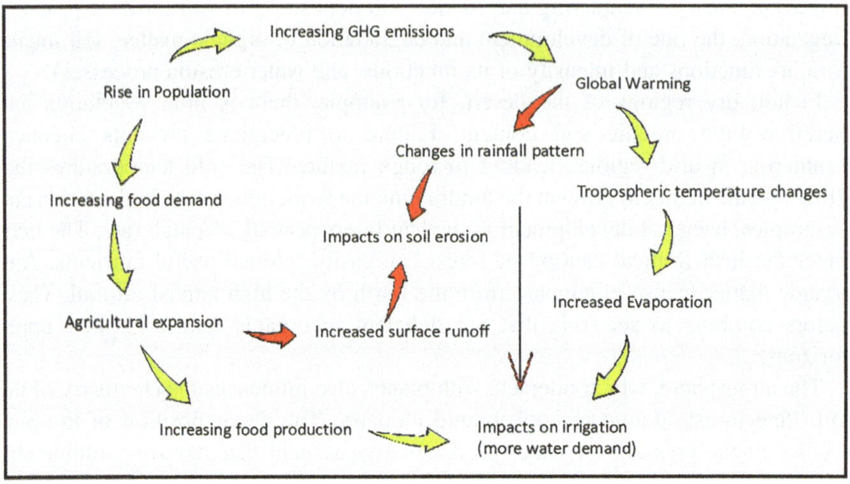

Fig. 7.1 Interaction between changing climate, soil, and agriculture

7.3.1 *Soil Physical Properties*

Physical properties of soil are the properties associated with size and arrangement of soil particles. The effect of solar particle size and arrangement on the movement of liquids and gases through it is also taken into account. This includes texture, soil structure, and dynamic characteristics such as capacity to hold water, infiltration, bulk density, porosity, etc. The physical characteristics of soil have a great effect on its biological (biological activity, nutrient supply, adsorption, water, heat transport) and chemical processes, which in turn affects the soil fertility.

Theoretically, four cases can arise due to changing temperature and precipitation:

(a) Decreased temperature with decreased precipitation
(b) Increased temperature with decreased precipitation
(c) Increased temperature with increased precipitation
(d) Decreased temperature with increased precipitation

7.3.1.1 Soil Texture

The most important and basic determining factor for soil characteristics is its texture. It is defined as the mineral particle's size or the correlated proportions of several groups of mineral size present in a given soil sample. The soil texture is of three types, that is, clay, silt, and sand (Climate and Soil Considerations 2020).

The change in soil texture in accordance with geologic time scale is very slow; therefore, it is not susceptible to be influenced with time, hence is not much of an importance to climate change studies. However, it still remains as a crucial determinant of the sensitivity degree of soil to the climatic variations.

However, the clay soil phenomenon of shrinking and cracking with respect to increase in drying and wetting cycles made this soil type vulnerable to climatic changes. This result in significant soil crack development. The direct and rapid water flow movement from the surface of soil through the bypass flow (or the drainage installations) is a result of deep cracks, which can reduce the filtering mechanism of soil, and the risk of nutrient loss and water contamination increases. These processes are common in clay soils, but the loss of material and water increases when there are frequent droughts which are followed by intense precipitation events.

Silt soils are known to have high-water retention capacity and clay soils have lowest sensitivity toward climatic changes. The soil texture remains the dominating influencing factor despite soil sensitivity being a function of various properties such as vegetation density, transpiration, and groundwater depth.

7.3.1.2 Structure: Shape and Stability

The fundamental determining factors for the aeration and moisture status of soil are pore size and porosity together with the soil stability (structure) and size. The change in the soil porosity can directly impact the water storing ability and can also change the carbon dioxide (aerobic condition) and methane (anaerobic condition) emissions of the soil. The soil aggregate to an extent is directly affected by the raindrops. There is decrease in soil aggregate stability and size with less soil organic and biomass content due to increase in temperature and lower water availability.

7.3.1.3 Porosity and Bulk Density

The change on soil form, structure, and spatial distribution with the soil aggregate stability with respect to climatic change has been a very complex procedure. Soil structures are directly impacted with the increase in temperature and changes in the temporal distribution and volume of precipitation via processes of dispersion, slaking, compaction, and mechanical disruption.

The bulk density which is closely linked to the soil texture properties and organic matter quality is known to be dependent on climate. There is increase in bulk density with the soil erosion or decomposition rate which results in the loss of organic matter present in the soil and leads to the soil compaction with all of the implications like root growth inhibition due to formation of compact layer and decrease in soil porosity.

7.3.1.4 Soil Hydrothermal Regime

The integrated influence on the functioning of soils of changes in environment and related hydrological, vegan, and land use is shown by the water balance in the field and soil humidity regime (Gelybó et al. 2018).

The dynamisms of soil moisture combine and represent seasonal changes in soil moisture supply, plant water availability, deeper-layered liquidation, and weathering, making the environment a valuable descriptor. The supply of soil water and the conservation of water, which have a strong influence on the functioning of soil ecosystems on the possible effects of climate change, are dependent on the rate of inflation. According to some, infiltration can be one of the most critical soil tools, helping to improve soil water store and minimize soil erosion and the risk of flash flooding and dryness. The hydrothermal regime of the soil (water mode, heat mode) therefore constitutes a significant determinant of the moisture system of the soil and an important soil climate descriptor. Figure 7.2 shows the various ecological conditions of soil and its indirect linkage to soil fertility and biomass production.

7.3.1.5 Soil Organic Carbon

The ratio of the soil to the organic carbon concentration of the soil determines the impact on the water's ability of the organic carbon content of the soil. The effect of

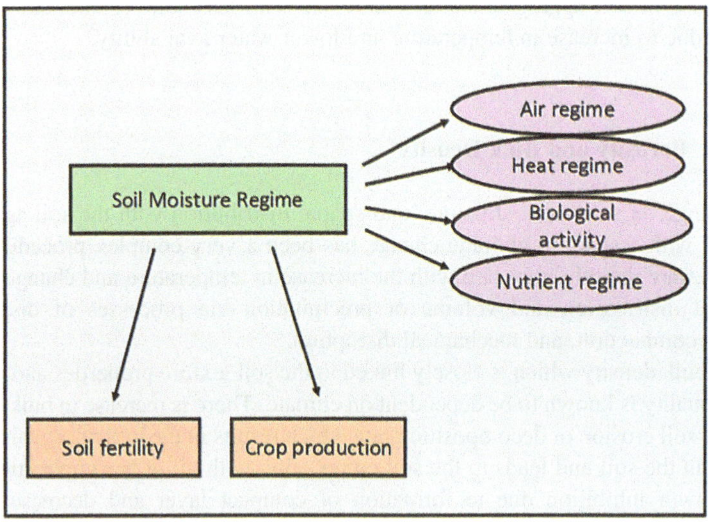

Fig. 7.2 Ecological conditions of soil (air regime, heat regime, biological activity, nutrient regime) directly affecting the soil moisture regime, which in turn affects the soil fertility and biomass (crop) production

climate change on the rate of decomposition in temperate grasslands was studied in an experiment conducted by Walter et al. 2014.

Thus, when litterbags were exposed to drought conditions for about 6 weeks, the rate of decay decreased. This thus showed that decomposition was more prone to drought or climate change. Extreme droughts can dramatically alter the composition of soil bacterial communities and decomposition processes. It was found that even a very short drought accompanied by heavy rain pulses decreases the rate of litter mass loss by 5% relative to the exposure duration.

Climate change is caused by changes to soil humidity or temperature, including the composition and operation of microbial and soil communities, which affect soil biological processes directly, which change decomposition processes. Thus, drier climate conditions under global warming could reduce nutrient cycling and alter the carbon balance of the soil in more intensively maintained grasslands. Changes in climatic factors have a significant effect on decomposition through changes in the biotic behavior of the soil and not through changes in litter content.

Stress of soil moisture will significantly decrease healthy soil function, thereby affecting the productivity of plants. A rise in organic carbon turnover can be caused by drier and warmer soil conditions, whereas changes in precipitation can affect the availability of water. In the long term, the detrimental effects of variations in precipitation on hydrophysical soil properties are attributed to the rapid decomposition of organic matter under wet conditions.

7.3.2 Soil Chemical Properties

The pH, carbonate concentration and other nutrient contents and distribution of these nutrients in the soil profile, soluble salt content and base saturation (BS) value, and cation exchange capacity (CEC) are some of the greatest essential soil chemical properties.

7.3.2.1 Soil pH

Since the pH of soil is determined by the parent material, climate, the vegetation, and rate of weathering, it is unlikely to change rapidly as a result of direct climate change effects. Increased precipitation, on the other hand, can exacerbate leaching and contribute to soil acidification.

Low pH values can mobilize toxic elements, and this can lead to heavy metal leaching in well-drained, structurally stable soil that receives a lot of rain. The changes in soil chemistry are the result of acidification of soil.

Crocker and Major (1955) performed a study that looked at glaciated soil surfaces and how they evolved over time due to vegetation and surface age. The predicted 50-year changes revealed that if vegetation begins to expand, soil pH will drop dramatically, while pH shift on a bare soil surface will be negligible for decades (Gelybó et al. 2018).

7.3.2.2 Salt Content

High concentrations of H^+ and Al^{3+} may be found in very acidic soils, whereas Ca^{2+} and Mg^{2+} predominate in neutral and moderately alkaline soils. The way which adsorbed cations affect the development of soil aggregate structure, hence soil water management, is important. The high amount of sodium in the adsorbed cations causes weak, diffluent structure, while calcium can preserve and protect the aggregate structure of salt-affected soils. David and Dimitrios (2002) studied saline structured soils with various Na^+ and Ca^{2+} solutions as well as SAR values.

7.3.2.3 Electrical Conductivity

The effects of electrical conductivity (EC), organic soil (SAR), and organic soil matter-related variables on tensile strength have been examined in a study by Rahimi et al. (2000). The authors found that the treatments with higher electron donating abilities had higher tensile strength and the treatments with higher electron accepting abilities had lower tensile strength. The study of the soil aggregation responses in soil was used for the identification of four climatic conditions (along a climate transect). This study found a nonlinear relationship between changes in soil structure and climate in several ways (Gelybó et al. 2018).

By transitioning into weather conditions, the surface chemical properties of drier soils will significantly change, as weathering depletes primary minerals, which can further be lost by liquefaction and leads to a major change in surface chemistry.

Lavee et al. (1998) reported that organic matter content, and aggregate size and stability decrease with aridity, while the sodium adsorption ratio and the runoff coefficient increases with aridity; with respect to the experiment conducted along a climatic transect, from the Mediterranean climate to the arid zone in Israel. They also stated that the rate of change of these variables along the climatic transect was nonlinear. According to their study, a threshold existed at the semi-arid area which separated arid and the Mediterranean ecogeomorphic systems. This existence of threshold meant that only a relatively small climatic change is needed to shift the border between the Mediterranean and the arid ecogeomorphic systems. And since several Mediterranean climate and semiarid regions lied adjacent to the arid ones, even a small fluctuation in climate can make them susceptible to desertification.

7.3.2.4 Cycle of Nutrients

It should be remembered that nitrogen is one of the nutrients in the soil which is closely related to the water cycle. The biological and physical behavior of the soil, as well as the interactions with the physical and biological soil processes, depends on the chemical properties of the soil. They combine soil pH monitoring with biological activity, to determine the soil's nutrient regime and fertility. Therefore, factors that influence the water delivery cycle as well as the carbon cycle will have an impact on nutrient availability. Loss of carbonate in the soil profile is accelerated by a rise in annual rainfall (cases 2 and 3 in Sect. 7.3.1), and percolation is one of the most important factors involved in carbonate leaching (Gelybó et al. 2018).

Rising annual precipitation will increase the rate of down-flooding and leaching, thereby causing greater acidification. Acidification will increase the mobilization of toxic elements (such as heavy metals), which leads to unhealthy and uninhabitable environments for plants and other living organisms. Kopittke et al. (2012) analyzed factors that cause soil acidification due to anthropogenic CO_2 release. Drought event scenarios were simulated in the summer during the germination process, and control scenarios were also simulated. In both treatments, the soil solution had a lower pH value than that of the control plot. However, the soil solution pH in the control plots was more acidic than those in the treatment plots. The drought evidently has had an inhibitory effect on the ozone hole. Growing sea levels from climate change would cause more extreme flooding and salt accumulation on coastal regions. Rising temperatures caused less severe leaching of soil, allowing some plants to regenerate faster. Salinization and carbonate deposition could have adverse effects on plants and animals if there is acidic water.

An increase in irrigation water use and inadequate irrigation practice could result in extensive secondary salinization. In recent regions where the source of salt is shallow groundwater and decreases are required, upward capillary transportation is significant. Climate change's possible impacts on soil properties would be decreased, resulting in fewer soluble salts being transported to the surface.

7.4 Climate Change Leading to Soil Degradation

Climate-related land degradation refers primarily to the soil quality changes caused by wind erosion, water, and vegetation degradation. In areas in which the soil-water budget has been balanced previously, increasing evaporation will lead to evaporation and an increasing evaporation risk. Groundwater that is salty and close to the surface is usually toxic. Soil erosion is a major factor that can have a significant effect on soil properties. Impacts of climatic change on soil leading to soil degradation are as follows:

7.4.1 Soil Erosion

For a variety of reasons, the rate of soil erosion is likely to change as a result of climate change. Soil erosion is a relatively short-term process in terms of time. Climate change has a significant influence on erosion rates because of the erosive force of precipitation, but changes in plant biomass may also play a role. Since it affects elements of runoff and soil erosion, the relationship between biomass and climate change can be a very complicated process. The rate, shape and degree, form of soil erosion, the amount and intensity of precipitation, relief, vegetation (nature, quality, density), and soil erosion characteristics all play a role.

The main influences on soil erosion due to potential climatic changes are as follows:

7.4.1.1 High Precipitation

Higher rainfall, particularly heavy rainfall and thunderstorms, will increase the rate of erosion (runoff) if it is not offset by the increased impact of denser and more permanent vegetation on soil conservation as a result of better water supply.

After 3 years of tracking of soil structural variables under arid and Mediterranean climates, Sarah (2005) reported a high correlation between the mean annual precipitation and annular precipitation variability and several aggregation variables. With declining precipitation, the percentage of microparticles decreased. The overall stability increased. The aggregate destructive role of the raindrops, surface rushes, and water filters are one of the most critical direct impacts of soil erosion caused by precipitation change. The rate of structural damage depends on the intensity and resistance of the destructive factor to such soil aggregates (Jakab et al. 2016).

7.4.1.2 Decreased Vegetation and Inappropriate Land Use

Lower precipitation normally decreases the erosion rate, but due to moisture constraints, it can be counterbalanced by the weaker vegetation. Climate change can cause changes in plant biomass and land use by changing the natural vegetation cover or agricultural practices. Vegetation changes can have positive or negative consequences on soil structure (desertification, watershed, and salinity-alkalinity changes) (Farkas et al. 2014). Overgrazing, irrational land use, misguided agricultural use (cropping pattern, crop rotation), and inappropriate technologies have unfavorable, sometimes permanent, and almost impossible to correct consequences (heavy machinery, over-tillage, and over-irrigation). It may, on the other hand, assist in the preservation or restoration of good soil structure by wise land use, appropriate technologies, and improved practices.

7.4.1.3 Lower Precipitation

Wind erosion may also be intensified by lower precipitation. Water erosion can be minimized by less precipitation itself, while the cohesion between soil particles can decrease due to loss of moisture, resulting in accelerated wind erosion.

Moreover, the average annual rainfall, surface runoff, and denudation/erosion rates are not linearly connected. Based on six GCM models, Shiono et al. (2013), who determined that the predicted climate change will enhance soil erosion on farmland in Japan by more than 20% by 2100, have investigated rainfall erosion factor. Mullan et al. (2012) modeled (three scenarios) soil erosion and potential effects on soil properties using UK, Canada, and Australia climate change forecasts and concluded that both increases and decreases in rates of erosion may occur depending on the types of scenarios used.

7.4.2 Acidification

Decreasing rainfall may decrease downward filtration and leaching. In this indirect way, climate determines the dominant types of vegetation, their productivity, and the decomposition rate of their litter deposits and affects soil reactions.

7.4.3 Salinization/Sodification

The increase in the eustatic sea level is the consequence of a projected global warming that will lead to more flooded areas (in particular in heavy-duty delta and river valleys) and areas threatened by seawater intrusion. Growing precipitation (increasing rates of filtering downward) decreases, and decreased precipitation and elevated temperatures exacerbate salinization and solidification processes: higher evaporation improves the transfer of solutes from groundwater into the root region through a capillary transport of water + no to no leaching. In plains and poorly drained lowlands (evaporative basins, i.e., the Carpathian lowlands), where shallow salt and brackish soil water are the main source of salt, the groundwater deposition can be offset by groundwater table (but by the negative water balance: $G > Gi + ET$). The formation of compact (petrocalcic) accumulation horizons should be presumed to create the same patterns by carbonate leaching or deposits.

7.4.4 Structure Destruction: Compaction

The aggregate-destructing effect of raindrops, surface runoff, and filtering water (see earlier) is the most significant direct impact. By means of the vegetation pattern and land use practices, the indirect factors function.

7.4.5 Biological Degradation

Changes in temperature, precipitation, and vegetation have a substantial effect on biological soil processes, but there is little evidence available on these effects.

7.4.6 Unfavorable Changes in the Biogeochemical Cycles of Plant Nutrients and Pollutants

All of these processes are connected to the regime of soil moisture as well as the abiotic and biotic phenomena of transformation (fixation, release of immobilization, engagement, modifications in solubility, redox status, etc.). High precipitation accelerates the leaching, loss of filtration (potential pollution and reduction processes in groundwater). Low precipitation can affect soluble, mobile, and accessible compounds and elements (dry conditions).

7.5 Soil as a Part of Carbon and Nitrogen Cycles

Soil is an essential part of several biogeochemical cycles. The carbon and nitrogen cycles are the two biogeochemical cycles that are important from the perspective of soils and climate change interactions. This is because C and N are both essential components of organic matter present in the soil, and CH_4, CO_2, and N_2O are the utmost important long-lasting greenhouse gases.

Due to low population and technological levels prior to the Industrial Revolution, anthropogenic greenhouse gas emissions were negligible, and C and N cycles were in equilibrium. However, after industrialization, the natural balance was broken by increased fossil fuel combustion, soil tilling, and other human activities.

Tillage is also correlated to the amount of organic carbon in the soil. More carbon and nitrogen are released into the atmosphere each year than global sinks can absorb. Human land management has a significant effect on the nitrogen and carbon cycle, as it impacts global climate change by way of C and N gas emissions.

Management choices can limit a soil's ability to sequester C. The extensive use of heavy machinery, for example, has made soil compaction an important problem in modern agricultural production that has demonstrated that the sequestration of C was restricted.

Organic soils, which form in damp environments and must be drained for agricultural purposes, can pose a special C management issue. This runoff transforms anaerobic soil to aerobic soil, hastening the decomposition of organic matter and releasing greenhouse gases into the atmosphere. A net source of CO_2, with mean annual emissions of CO_2 ranging from 1500 kg to 7500 kg, were all cultivated systems in a 2013 study in Finland surveys.

The introduction of more intensive types of tillage and drainage wetlands to the agricultural production system is an example of a change in soil management that increases CO_2 emissions. It is also true that if management decisions leading to their sequestration are reversed, sequestered C may in the future sometime be released back to the environment.

In short, depending on how they are handled, regulated soils may be either net CO_2 sinks or sources. Another element of the carbon cycle that is connected to soil is methane. Rice production is the most common source of soil-derived methane emissions, accounting for 47% of global anthropogenic CH_4 emissions per year. Since different plants grow on similar soil, there may be variances in CH_4 emissions or consumption. According to Hu, R. et al., the soil under forest vegetation acted as a net sink of CH_4, while the same soil in a neighboring field planted with maize (*Zea mays*) was essentially CH_4 neutral, and a third field planted with grass cover was a net source of CH_4 to the atmosphere (2001). It was caused by differences in methanotrophic survival rates and gas exchange due to different physical properties in soils, which were dominated by vegetation (Brevik 2013).

It makes a difference to monitor CH_4 and CO_2 flow in soil. CH_4 production is reduced by dry tillage of land, dry seed, and other methods to reduce soil saturation during rice production. Wassmann et al. (1993) found that potassium fertilizers based on minerals (K) had no effect on the rice paddy soils' generation of CH_4, but Lu et al. (1999) found CH_4 emission of the phosphorus (P) fertilizers had declined. The increase of CH_4 emissions in P-deficient soil was caused by increased root exudates, according to Lu et al., while the plant was trying to harness the soil environment to increase the uptake of P. The addition of oxidizing mineral fertilizers, according to Stepniewski et al. (2011), will reduce CH_4 emissions by 20% to 70%. Ducks have been introduced into the rice system and CH_4 emissions have been reduced, mixing oxygen into paddy water and increasing CH_4 oxidation and reducing CH_4 emissions as a result of duck swimming, according to Zhang et al. (2011).

Soils are important components of C and N cycles, which are also important components of the global climate system. Agriculture emits a disproportionately large amount of anthropogenic CH_4 into the environment per year. Decisions on agricultural management have a significant impact on whether soil is a net CO_2, CH_4, or N_2O source of greenhouse gases, which indicates that management systems could affect climate change.

Climate change, in turn, is projected to have an effect on soil erosion and food security, as discussed in the following pages.

If climate change and its possible implications for food security are to be considered through the erosion cycle, we need to be aware of climate changes. Our understanding of the effect of climate change on the C and N cycles is incomplete, and more research on these topics is required.

In aerated soils, CO_2 emissions are prevalent, while in anaerobic conditions, production of CH_4 is associated. The balance of C in the soil relative to soil C is used to assess if the soil raises or decreases its total C levels. The soil C levels increase, with C removed from the atmosphere and the levels of the atmosphere decreased; the atmosphere levels increase when C is introduced to the atmosphere when soil C levels decrease.

The impact of human activity on C balance in controlled soils is important. While recent studies have shown that non-freeze systems will lead simply to higher accumulation of C of the top 15–20 cm of soil without C growth taking into consideration the whole soil profile, soil management strategies like non-freeze systems will result in lower carbon emissions and a more extensive C sequence in soil compared to intensive tillage management systems.

Additional adjustments to management, such as the use of cover crops, the rotation instead of monocropping, and the reduction or elimination of fallow times, can also help to increase land C sequestration by returning land to local woods or grasslands from agricultural use. C sequestration tends to be quick at first, but it slows down over time. For 50–150 years following changes in management, mainly agricultural soils can only sequester carbon until C is saturated.

7.6 Conclusion

The Earth's climate system is changing because of the changing levels of greenhouse gases in the Earth's atmosphere, the C and N dependency being the major ones. Since the soils are part of the cycles C and N, and both soils C and N are essential components of the soil organic matter, the content of soils can be affected by climate change. The state of any environment is also largely determined by the conditions of the soil, depending in particular on the climate in general and precipitation. This will in turn have an impact on essential soil features such as aggregate formation and stability, water holding capacities, and soil nutrient cation capacity and quality.

Because of many global climate change uncertainties (directional, rate, seasonal), and in the prediction of their environmental, ecological, economical, and even social consequences, major assumptions cannot be ruled out in the prediction of how a single climatic variable (either T or P) affects the soil properties. However, we know that the risk is that soils will add increasing amounts of greenhouse gas to the atmosphere as global temperatures rise and lose their ability to serve as a sink for C.

As a result, understanding how nutrient limitations, such as N and P, affect soil C sequestration is important. We also need to learn about the effect on the nitrogen cycle of climate change, which has received significantly fewer scientific attentions than the carbon cycle. At this time, little is known about the impact of climate change on soil species, which are crucial in driving parts of the soil's C and N cycles. Changes in CO_2 levels in the atmosphere can affect plant metal absorption, resulting in food products with insecure levels of those metals in their tissues, but there has been no conclusive research in this area.

Better knowledge of these areas is therefore important to give us an idea of how soil erosion and food protection can be affected by changes in processes and properties.

Hence, we require more detailed, integrated multidisciplinary studies to study the complex interactions taking place in the environment to give an exact scientific basis for the consequences of climate change.

References

Brevik EC (2013) The potential impact of climate change on soil properties and processes and corresponding influence on food security. Agriculture

Climate and Soil Considerations (2020). https://smallfarms.cornell.edu/guide/2-climate-and-soilconsiderations/. Accessed 25 Oct 2020

Crocker L, Major J (1955) Soil development in relation to vegetation and surface age at Glacier bay, Alaska. J Ecol 43(2):427–448

David R, Dimitrios P (2002) Diffusion and cation exchange during the reclamation of saline-structured soils. Geoderma 107(3-4):271–279

Farkas C, Gelybó G, Bakacsi Z, Horel A, Hagyó A, Dobor L, Kása I, Tóth E (2014) Impact of expected climate change on soil water regime under different vegetation conditions. Biologia 69:1510–1519

Gelybó G, Tóth E, Farkas C, Horel A, Kása I, Bakacsi Z (2018) Potential impacts of climate change on soil properties. Agrochemistry and Soil science

Hu R, Kusa K, Hatano R (2001) Soil respiration and methane flux in adjacent forest, grassland, and cornfield soils in Hokkaido, Japan. Soil Sci Plant Nutr 47(3):621–627

Jakab G, Szabó J, Szalai Z, Mészáros E, Madarász B, Centeri CS, Szabó B, Németh T, Sipos P (2016) Changes in organic carbon concentration and organic matter compound of erosion delivered soil aggregates. Environ Earth Sci 75:144–155

Kopittke GR, Tietema A, Verstraten JM (2012) Soil acidification occurs under ambient conditions but is retarded by repeated drought: results of a field-scale climate manipulation experiment. Sci Total Environ 42:332–439

Lavee H, Imeson AC, Sarah P (1998) The impact of climate change on geomorphology and desertification along a mediterranean-arid transect. Wiley Online Library

Lu Y, Wassmann R, Neue H-U, Huang C (1999) Impact of phosphorus supply on root exudation, aerenchyma formation and methane emission of rice plants. Biogeochemistry 47:203–218

Mullan D, Favis-Mortlock D, Fealy R (2012) Addressing key limitations associated with modelling soil erosion under the impacts of future climate change. Agric For Meteorol 156:18–30

Rahimi H, Pazira E, Tajik F (2000) Effect of soil organic matter, electrical conductivity and sodium adsorption ratio on tensile strength of aggregates. J Water Soil Sci 6(3):151–161

Sarah P (2005) Soil aggregation response to long- and short-term differences in rainfall amount under arid and Mediterranean climate conditions. Geomorphology 70(1–2):1–11

Shiono T, Ogawa S, Miyamoto T, Kameyama KN (2013) Expected impacts of climate change on rainfall erosivity of farmlands in Japan. Ecol Eng 61(Part C):678–689

Stepniewski W, Stepniewski Z, Rożej A (2011) Gas exchange in soils. In: Hatfield JL, Sauer TJ (eds) Soil management: building a stable base for agriculture. Soil Science Society of America, Madison, pp 117–144

United States Department of Agriculture (2020) National Resources Conservation Service Soil. https://www.nrcs.usda.gov/wps/portal/nrcs/detail/soils/edu/?cid=nrcs142p2_054278. Accessed 25 Oct 2020

Walter J, Hein R, Beierkuhnlein C, Hammerl V, Jentsch A, Schädler M, Schuerings J, Kreyling J (2014) Combined effects of multifactor climate change and land use on decomposition in temperate grassland. Soil Biol Biochem 60:10–18

Wassmann R, Schütz H, Papen H, Rennenberg H, Seiler W, Aiguo D, Renxing S, Xingjian S, Mingxing W (1993) Quantification of methane emissions from Chinese rice fields (Zhejiang Province) as influenced by fertilizer treatment. Biogeochemistry 20:83–101

Zhang J-E, Ouyang Y, Huang Z-X, Quan G-M (2011) Dynamic emission of CH4 from a rice-duck farming ecosystem. J Environ Prot 2:537–544

Rowell David, Pateras Dimitrios, (2002) Diffusion and cation exchange during the reclamation of saline-structured soils. Geoderma 107 (3-4):271-279

Chapter 8
Climate Change Impacts on Plant–Microbe Interactions

Anamika Singh and Abhilasha Shourie

Abstract Climate change has impacted all forms of life and its consequences are evident from altered ecological functions. Plant–microbe interactions are crucial in maintaining the ecosystem structure and driving important functions such as geochemical cycles, soil formation, carbon sequestration, and greenhouse gas emission. Microbial communities are very sensitive toward environmental stressors like temperature, moisture, pH, CO_2, and any changes in the abundance, diversity, and activity of microbes which are likely to have profound impact on the associated plant communities. The plant–microbe interactions range from positive such as symbiosis to negative such as parasitism; however, benefits of microbial association certainly outnumber the negative relationships. Numerous vital physiological functions of plants supported by microbes such as nitrogen fixation, phosphate solubilization, sequestration of minerals, and conferment of stress tolerance and plant immunity can be compromised due to climate change-induced alterations in microbial community structure and dynamics. Climate change effects are also observed as shifts in phonological patterns of both plants and microbes, which can lead to serious disturbances in interspecific phenomena such as pollination, herbivory, predation, etc., threatening community stability and leading to changes in demographic processes. Plant–soil feedbacks (PSFs) play a significant role in shaping the community structure and regulating ecosystem processes. Shifts in microbial community dynamics also feedback to affect plant performance, coexistence, and community composition. PSFs also potentially modify the process of succession as the positive PSFs favoring colonization of symbiotic nitrogen fixers and mycorrhizal fungi help in establishment of primary successors on nutrient devoid substratum and the PSFs favoring decomposer community facilitate carbon cycling, soil mineralization, and nutrient mobilization, accelerating the secondary succession. The chapter discusses

A. Singh
Department of Botany, Maitreyi College, Delhi University, Delhi, India

A. Shourie (✉)
Department of Biotechnology, FET, Manav Rachna International Institute of Research and Studies, Faridabad, India

D. K. Choudhary et al. (eds.), *Climate Change and the Microbiome*, Soil Biology 63,
https://doi.org/10.1007/978-3-030-76863-8_8

different aspects of plant–microbe interactions which are directly or indirectly affected by the abiotic and biotic factors influenced by climate change.

Keywords Climate change · Microbial diversity · Plant–microbe interactions · Abiotic factors · Plant phenology · Plant–soil feedbacks

8.1 Introduction

The effects of climate change on animal and plant worldwide are largely visible and thus well studied and well documented; on the contrary, the effects on microbial communities being less obvious are not fairly characterized. However, climate change has profound impact on almost all aspects of microbial dynamics such as community composition, physiological responses, ecological functions, and evolutionary aspects. Microbes also play a crucial role in regulating climate change due to their huge diversity, adaptive capabilities, and varied responses under changing environment. This makes microbes the key drivers of ecosystem functions enabling them to impact the resilience of other organisms towards climate change. Thus, microbial response to climate change itself can potentially determine the structure and function of an ecosystem.

Soil environment is richly populated with huge diversity of microorganisms and is a hot spot of microbial activity. Microbes residing in the soil interact not only with each other but also with the plants. These interactions greatly influence the ecosystem structure and function including plant and animal abundance, diversity, and composition. The effects of such interactions can be positive or negative depending upon the beneficial or hostile nature of interaction. In general, the interactions such as symbiosis and mutualism that enhance the growth and productivity of plants are said to be positive, while the interactions such as parasitism or host–pathogen relationship that tend to decrease the performance of plants are called negative. There are a number of beneficial plant–microbe interactions which not only enhance plant growth but also directly or indirectly improve the host plant resistance to abiotic and biotic stress factors such as drought, salinity, heavy metals, toxins, extreme temperature, nutrient scarcity, and diseases. Such interactions strongly help the plants to withstand the adverse effects of climate change.

Further, many other essential functions of ecosystem are also largely driven by microbes such as biogeochemical cycles including water cycle and nutrient cycles, determination of soil composition, carbon sequestration, and greenhouse gas emission. Microbes modify the soil properties and water regimes adding to the nutrient acquisition efficiency of the plants and contributing in their growth and biomass. This has far-reaching consequences as the plants are at the base of trophic levels serving as producers and providing nutrition to all other organisms.

8.2 Plant Microbiota and Key Interactions

The association of plants and microbes is inseparable, intimate, vital, and evolutionary. Their relationships are ecologically diverse and complex ranging from antagonistic to mutualistic, and both the plant host and microbes seem to have coevolved in response to climate change. All the microorganisms associated with a plant are together known as plant microbiota. Plant microbiota comprises diverse organisms including archaea, bacteria, fungi, microbial eukaryotes, etc. that interact not only with the host plant but also with the surrounding environment. While pathogens are harmful to their plant hosts, most other plant–microbe relationships often have functional advantages to the plants such as growth promotion, nutrient uptake, stress tolerance, induction of secondary metabolism, and biocontrol of pathogens. There are a number of factors that play a role in shaping the structure of plant microbiota such as geographical location, climate, soil characteristics, plant community structure, and environmental stresses.

8.2.1 Functional Zones of Plant Ecosystem

Microorganisms can inhabit in any of the four functional zones of the plant ecosystem, viz., rhizosphere, rhizoplane, phyllosphere, and endosphere.

1. Rhizosphere—Rhizosphere refers to the brief zone of soil immediately surrounding the plant roots. The rhizosphere microbiota is present at the root–soil interface, is directly affected by the root exudates and in turn also affects other organisms such as soil invertebrates in the microenvironment. They also contribute to the soil formation by facilitating mineralization.
2. Rhizoplane—Rhizoplane is the region where the root surface is in contact with soil and corresponds to the inner limit of the rhizosphere.
3. Phyllosphere—Phyllosphere comprises the aerial parts of plant which are colonized by a large number of epiphytes including bacteria, fungi, algae, protozoans, and nematodes. Phyllosphere microbes face the most challenging environment conditions due to rapid atmospheric changes in the factors like temperature, humidity, intensity of solar radiation, etc. However, most of the phyllosphere microbes have developed certain adaptations to survive through these hostile conditions, for example, epiphytic bacteria have pigments to protect from UV rays and *Pseudomonas* species can find way through the cuticle by producing biosurfactants to reach the nutrient source on the leaf surface (Hirsch and Fujishige 2008; Warriner 2005).
4. Endosphere—The endosphere comprises the internal tissues of various plant parts such as leaf, stem, and roots. It harbors the endophytic microbes in intercellular as well as intracellular spaces. Root endosphere is the most vulnerable and susceptible towards microbial infection and endospheric colonization.

8.2.2 *Rhizospheric Microbial Diversity*

The plant–microbe interactions happening below the ground are usually more complex as compared to those taking place on the soil surface (Bais et al. 2004) or those happening in the aerial parts of plant. The most active zone of plant–microbe interaction is the rhizosphere where the plant roots are in close proximity to a large number and variety of microbes. Many studies have shown that the properties of rhizospheric soil differ from the bulk soil to a great extent, especially in terms of microbial diversity and activity. The rhizosphere microbiota is affected by soil factors such as soil pH, salinity, soil type, soil structure, soil moisture, and soil organic matter and root exudates. Besides, the recruitment of soil microbiota in rhizosphere is influenced largely by plant roots through physical interaction, secretion, and plant immunity. There seem to exist some evolutionarily conserved mechanisms of selection in plants and adaptation in microbes that are responsible for microbiota recruitment and cohabitation.

Roots biochemically attract certain microbial species to populate in the rhizosphere. Plant root exudates such as organic acids, amino acids, fatty acids, phenolics, plant growth regulators, nucleotides, sugars, putrescine, sterols, and vitamins largely affect the microbial community structure in the rhizosphere (Mendes et al. 2013).

8.2.3 *Plant Growth-Promoting Rhizobacteria*

Rhizosphere microbiota is also referred to as plant growth-promoting rhizobacteria (PGPR) due to their contribution in plant growth enhancement through various means like phytohormone production, protection against pathogens, and alleviation from environmental stresses. The plant growth-promoting rhizobacteria (PGPR) can be found in free-living or endophytic modes.

8.2.3.1 Free-Living Bacteria

The PGPR community in soil microbiota often consists of *Pseudomonas*, *Azospirillum*, *Azotobacter*, *Klebsiella*, *Bacillus*, *Serratia*, *Burkholderia*, *Enterobacter*, *Alcaligenes*, *Rhizobium*, *Erwinia*, *Mycobacterium*, *Mesorhizobium*, and *Flavobacterium* (Ahmad et al. 2008). Most of the decomposers living freely in the soil belong to the genera *Azotobacter*, *Bacillus*, *Beijerinckia*, *Burkholderia*, *Clostridium*, *Desulfovibrio*, *Enterobacter*, *Klebsiella*, and *Serratia*. The free-living diazotrophic bacteria include *Firmicutes*, *Rhizobia*, *Bradyrhizobium*, *Rhodobacteria* (*Alphaproteobacteria*), *Burkholderia*, *Nitrosospira* (*Betaproteobacteria*), *Pseudomonas*, *Xanthomonas* (*Gammaproteobacteria*), and *Nostoc* (*Cyanobacteria*) (Morris and Schniter 2018).

Tropical rainforest soil was found to be dominated by *Heliobacterium*, *Gluconacetobacter*, *Methylobacterium*, *Azospirillum*, and *Zymomonas* in Costa Rica (Reed et al. 2010) and *Azospirillum*, *Azorhizobium*, *Bradyrhizobium*, *Methylobacterium*, *Burkholderia*, *Paenibacillus*, *Heliobacterium*, *Nostoc*, *Anabaena*, and methanogenic archaea, *Methanoregula*, *Methanosphaerula*, and *Methanocella* in Western Amazon Basin of Brazil (Mirza et al. 2014).

8.2.3.2 Endophytic Bacteria

Endophytic PGPR specialize in invading the plant tissues as they reside in the plant endosphere and develop a deep mutualistic relationship with the host. They can be found in all the plant parts, aboveground and belowground; however, the infection mostly occurs through roots. Roots are colonized by endophytic bacteria following the invasion through root tissues and systemic spread through xylem to other plant parts. Root exudates play an important role in recognition and communication, facilitating infection and colonization of rhizobacteria. Endophyte diversities are different in endosphere and phyllosphere due to anatomical and ecological differences in the aboveground and belowground plant parts.

The predominant phyla of root endophytes found in grapevine were *Proteobacteria*, *Acidobacteria*, *Actinobacteria*, *Bacteroidetes*, *Verrucomicrobia*, *Planctomycetes*, *Chloroflexi*, *Firmicutes*, and *Gemmatimonadetes* (Burns et al. 2015; Zarraonaindia et al. 2015; Samad et al. 2017; Faist et al. 2016), while the most abundant phyllosphere genera were *Pseudomonas*, *Sphingomonas*, *Frigoribacterium*, *Curtobacterium*, *Bacillus*, *Enterobacter*, *Acinetobacter*, *Erwinia*, *Citrobacter*, *Pantoea*, and *Methylobacterium* (Zarraonaindia et al. 2015; Kecskemeti et al. 2016). Compartment-specific localization of microbes suggested a strong functional relationship between the host and microbes.

Endophytic bacteria *Burkholderia and Enterobacter* lived intracellularly in root tissue of cereals (Wakelin and Ryder 2004), while diazotrophic bacterial genera associated with maize were *Rahnella*, *Pantoea*, *Rhizobium*, *Pseudomonas*, *Herbaspirillum*, *Enterobacter*, *Brevundimonas*, and *Burkholderia* (Montanez et al. 2012). *Klebsiella and Burkholderia* were also found to be dominant in roots and rhizosphere of maize (Arruda et al. 2013). *Pseudomonas stutzeri*, an endophytic diazotroph, can switch over to different nitrogen-fixing processes like denitrification under anaerobic conditions, nitrification under aerobic conditions, and nitrogen fixation under microaerobic conditions. The root endophyte *Herbaspirillum seropedicae* is a *Betaproteobacterium* that fixes nitrogen and efficiently colonizes the roots of several plants like wheat, rice, corn, sorghum, and sugarcane (Tkacz and Poole 2015).

Azoarcus is a plant growth-promoting bacterium as it fixes nitrogen that its host appears to be able to access but lacks the usual genetic components involved in plant pathogenicity (e.g., type III and IV secretion) (Krause et al. 2006). *Azoarcus* along with the nitrogen-fixing *Azospirillum* has been found to be a common root colonizer of rice. Plants were found to exert some control of the endophytic N-fixing community as wild rice species were preferably colonized by *Azoarcus*, while modern cultivars selected *Azospirillum* (Engelhard et al. 2000).

The endosymbiotic association is well exemplified by the relationship between actinorhizal plants and *Frankia* (Gram-positive bacteria) and between legumes and *Rhizobia* (Gram-negative bacteria). Both these diazotrophs are the nodule-forming bacteria. Besides legumes and actinorhizal plants, there is another example of colonization of *Rhizobium* bacteria in a nonlegume plant which is *Parasponia andersonii* that belongs to the family Cannabaceae (Sytsma et al. 2002).

Members of the genera *Bacillus*, *Georgenia*, *Mycobacterium*, *Bosea*, *Microbacterium*, *Psychrobacillus*, *Roseomonas*, *Chitinophaga*, and *Leifsonia* are found in the root endosphere of wheat (Rilling et al. 2018). Soil actinobacterium *Frankia*, a Gram-positive, heterotrophic bacterium, is a common endosymbiont of *Casuarina* trees, which lives inside the root nodules. There are many species of *Casuarina* that are actinorhizal such as *C. glauca* and *C. equisetifolia*. Clusters of different strains of *Frankia* infect different plants of different families, such as Betulaceae and Myricaceae (Cluster 1a); Casuarinaceae (Cluster 1c); Coriariaceae, Datiscaceae, Rosaceae, and *Ceanothus* belonging to Rhamnaceae (Cluster 2); and Myricaceae, Rhamnaceae, Elaeagnaceae, and *Gymnostoma* belonging to Casuarinaceae (Cluster 3) (Ngom et al. 2016).

Nostoc, a cyanobacterium, occurs widely in free-living form and is able to enter into symbiotic association with several lower and higher plant groups, such as bryophytes (liverworts and hornworts), *Azolla* (water fern), cycads (gymnosperm), and the herbaceous angiosperm *Gunnera*. The relationship of *Gunnera and Nostoc* is facultative endosymbiotic. *Nostoc* resides in the stem glands of *Gunnera* and fixes nitrogen (Bergman et al. 1992). Endophytic diazotrophs found in sugarcane stems are *Gluconacetobacter diazotrophicus*, *Herbaspirillum seropedicae*, *Herbaspirillum rubrisubalbicans*, and *Burkholderia silvatlantica* (Lery et al. 2011; Dong et al. 1994).

8.2.4 Plant-Associated Fungi

Plant community structure is strongly influenced by the fungal diversity and their mutual relationships. The three major roles played by soil fungi are decomposition, biological control of pests and pathogens, and ecosystem regulation. Fungi act as decomposers and stabilize the soil organic matter contributing to nutrient cycling and availability. Fungi also act as biological controllers by regulating diseases, pests, and the growth of other organisms. They actively participate in physiological processes like nitrogen fixation, hormone production, biological control against root pathogens, and protection against drought, conferring numerous advantages to the coinhabitants. They also function as ecosystem regulators since they regulate the dynamics of soil structure and functions by modifying the physiological processes in the soil environment which change the microhabitats for other organisms.

On the basis of beneficial relationship of fungi with plants, they can be broadly grouped into two categories.

8.2.4.1 Decomposers

These are saprophytic fungi which serve to decompose the dead organic matter and release minerals and nutrients into the soil. Since the plant litter decomposition is necessary for recycling of carbon and minerals, the saprophytic fungi are indispensable because of their capability of decomposing lignocellulosic material in the litter. Surface litter horizon is largely controlled by saprotrophic fungi of the group Basidiomycetes and Ascomycetes (Chanda et al. 2020).

Fungal diversity in senescent leaves and submerged litter was explored, and the fungal taxa found in senescent leaves were *Bullera*, *Dioszegia*, *Cryptococcus*, *Aspergillus*, *Aureobasidium*, *Glomeromycota* sp., *Tremellomycetes* sp., and *Wallemia sebi.* The fungi found exclusively in stream-immersed litter were *Alatospora* sp., *Amanita wadjukiorum*, *Basidiomycota* sp., *Ciboria carunculoides*, *Circinaria esculenta*, *Cryptococcus* sp., *Cyphellostereum* sp., *Ditopella aseptatospora*, *Kockovaella prillingeri*, and *Paramicrothyrium* sp. (Koivusaari et al. 2019).

8.2.4.2 Mutualistic or Symbiotic

Mycorrhizal fungi form symbiotic association with plant roots extending a number of benefits to their host such as mineral solubilization, support biological nitrogen fixation, and improve water availability. In return, the fungi derive carbohydrates from the host plant. Ectomycorrhiza colonize the root surface and are mostly found on trees, while endomycorrhiza grow intercellularly and intracellularly within the root tissues.

Arbuscular mycorrhizal fungi (AMF) are obligate symbiont endomycorrhizal fungi. They facilitate the plant growth by improving nutrient uptake and resistance to several abiotic and biotic stress factors. Four orders of AMF, namely, *Glomerales*, *Archaeosporales*, *Paraglomerales*, and *Diversisporales*, have been identified in this subphylum that consists of 25 genera (Redecker et al. 2013). Many other endophytic fungi which exist as microsymbionts in different plant parts belong to genera *Atkinsonella*, *Balansia*, *Balansiopsis*, *Echinodothis*, *Epichloë*, *Myriogenospora*, *Parepichloe*, and *Neotyphodium* (Chanda et al. 2020).

8.3 Physiological Aspects of Plant–Microbe Interactions

Microbes extend numerous benefits to the plant community as they are involved in vital physiological functions that support the growth and development of plants. As much as the plants depend upon microbes for life functions, microbial community structure and functions are also influenced by plant functions. Diverse genera of plant growth-promoting bacteria are listed in Table 1.1. Some of the prominent physiological activities of plants which depend largely on microbial association are discussed further.

8.3.1 Nitrogen Fixation

Nitrogen is not available to plants and animals for utilization in its free form, and the transformations and conversions of its various forms are carried out typically by microbes. Although the biological N cycle is quite complex, its main processes can be enumerated as:

1. Release of nitrogen in the form of ammonia (NH_4^+) from dead and decomposed organic matter
2. Assimilation of NH_4^+ and oxides of nitrogen (NO_2^- and NO_3^-) by microorganisms and plants
3. Dissimilative processes such as nitrification, denitrification, and nitrate reduction to ammonium including nitrogen fixation

Biological nitrogen fixation contributes the most in reducing atmospheric nitrogen (N_2) to bioavailable forms with the help of free-living and symbiotic diazotrophic microorganisms. There are many free-living diazotrophic bacteria found in the rhizosphere, some of which enter into symbiotic relationship with certain plants. The nitrogen-fixing symbiotic bacteria include the genera *Rhizobium*, *Bradyrhizobium*, *Mesorhizobium*, and *Sinorhizobium* (Graham and Vance 2003). They colonize the roots of leguminous plants and form root nodules where they fix nitrogen with the help of nitrogenase enzyme. The *nifH* gene that encodes the reductase subunit of nitrogenase enzyme is recognized as the genetic marker to study the diversity and abundance of diazotrophs (Gaby and Buckley 2011).

8.3.2 Phosphate Solubilization

After nitrogen, phosphorus is the most essential macronutrient for plants that is required for their growth and development and can be a limiting nutrient. It is present in inorganic and organic forms in soil in abundance but is not readily available for plants to be absorbed as it is bound to aluminum (in the form of strengite), iron

Table 1.1 Diversity of plant growth-promoting bacteria

Physiological function	Genus/group	Species	References
Nitrogen fixation	*Azospirillum*	*A. brasilense, A. zeae, A. lipoferum, Azospirillum amazonense, and Azospirillum halopraeferens*	Venieraki et al. (2011), Magalhaes et al. (1983), Reinhold et al. (1987)
	Burkholderia	*B. tropica, B. xenovorans, B. silvatlantica, B. caballeronis, B. unamae, B. vietnamiensis*	Reis et al. (2004), Vandamme (2004), Perin et al. (2006), Martinez-Aguilar (2013), Estrada-De Los Santos et al. (2001)
	Azotobacter	*A. vinelandii*	Jacobson et al. (1989)
	Klebsiella	*K. pneumoniae*	Arnold et al. (1988)
	Pseudomonas	*P. graminis, P. odorifer, P. stutzeri, P. koreensis, and P. entomophila*	Venieraki et al. (2011), Li et al. (2017)
	Bacillus	*Bacillus brevis, B. cereus, B. circulans, B. firmus, B. licheniformis, B. megaterium, B. pumilus, and B. subtilis*	Xie et al. (1998)
Phosphate solubilization	*Pseudomonas*	*Pseudomonas aeruginosa, P. mosselii, P. monteilii, P. plecoglossicida, P. putida, P. fulva, and P. fluorescens*	Naik et al. (2008)
	Rhizobium	*Rhizobium leguminosarum bv. trifolii*	Abril et al. (2007)
	Vibrio	*Vibrio proteolyticus*	Vazquez et al. (2000)
	Xanthobacter	*Xanthobacter agilis*	Vazquez et al. (2000)
	Actinomycetes	*Actinomyces, Micromonospora, and Streptomyces*	Sharma et al. (2013)
Iron sequestration and siderophore production	*Sphingobacterium*	*Sphingobacterium*	Tian et al. (2009)
	Pseudomonas	*Pseudomonas poae*	Tian et al. (2009)
	Enterobacter	*Enterobacter*	Tian et al. (2009)
	Delftia	*Delftia acidovorans*	Tian et al. (2009)
	Achromobacter	*Achromobacter xylosoxidans*	Tian et al. (2009)
Plant growth production			
IAA production	*Actinobacteria*	*Actinobacteria*	Kumar et al. (2020)
	Nocardia	*Nocardia*	Kumar et al. (2020)
	Frankia	*Frankia*	Kumar et al. (2020)
	Kitasatospora	*Kitasatospora*	Kumar et al. (2020)
	Streptomyces	*Streptomyces*	Kumar et al. (2020)

(continued)

Table 1.1 (continued)

Physiological function	Genus/group	Species	References
	Actinomycetes	*Streptomyces, Nocardia, Nocardiopsis, Spirillospora, Microbispora, and Micromonospora*	Shutsrirung et al. (2013)
Ethylene regulating ACC deaminase enzyme production	*Pseudomonas*	*Pseudomonas syringae, Pseudomonas fluorescens, Pseudomonas putida*	Ahmad et al. (2011), Saravanakumar and Samiyappan (2007), Cheng et al. (2007)
	Rhizobium phaseoli	*Rhizobium phaseoli*	Ahmad et al. (2011)
Gibberellin production	*Serratia*	*Serratia nematodiphila*	Kang et al. (2015)
	Bacillus	*Bacillus pumilus, Bacillus licheniformis, Bacillus* sp.	Gutierrez-Manero et al. (2001), Bottini et al. (2004)
	Acetobacter	*Acetobacter* sp.	Gutierrez-Manero et al. (2001), Bottini et al. (2004)
	Azospirillum	*Azospirillum* sp.	Gutierrez-Manero et al. (2001), Bottini et al. (2004)
Cytokinin production	*Bacillus*	*Bacillus* sp.	Naz et al. (2009)
	Pseudomonas	*Pseudomonas* sp.	Naz et al. (2009)
	Acetobacter	*Acetobacter* sp.	Naz et al. (2009)
	Azospirillum	*Azospirillum* sp.	Naz et al. (2009)
ABA production	*Proteus*	*Proteus mirabilis*	Karadeniz et al. (2006)
	Klebsiella	*Klebsiella pneumoniae*	Karadeniz et al. (2006)
	Bacillus	*B. megaterium, and B. cereus*	Karadeniz et al. (2006)
Other growth promotion activities	*Bacillus*	*B. megaterium, B. safensis, B. simplex, B. pumilus, B. circulans, B. cereus, B. laevolacticus, B. amyloliquefaciens*	Akinrinlola et al. (2018), Tilak and Reddy (2006), Egamberdiyeva (2005)
	Paenibacillus	*Paenibacillus graminis*	Akinrinlola et al. (2018)
	Arthrobacter	*Arthrobacter simplex*	Egamberdiyeva (2005)
	Pseudomonas	*Pseudomonas denitrificans, P. rathonis*	Egamberdiyeva (2005)

(in the form of variscite), or calcium (in the form of apatite) (Tkacz and Poole 2015). Plants acquire phosphate in soluble forms like $H_2PO_4^-$ or HPO_4^-, and the bioavailability of these forms depends upon the soil pH. Phosphorus is made bioavailable through solubilization and mineralization by phosphorus-solubilizing microorganisms (PSM) which reside in the soil and rhizosphere. These microbes chelate the cations bound to phosphate via production of organic acids and siderophores, thus releasing it into the soil. PSM is a diverse group of microbes comprising of bacteria, fungi, actinomycetes, and algae which solubilize insoluble inorganic phosphorus and mineralize insoluble organic phosphorus (Sharma et al. 2013).

8.3.3 Iron Sequestration

Iron is an essential microelement for plant growth but cannot be readily absorbed due to low solubility and low bioavailability of its ionic forms. Its absorption is facilitated by the soil bacteria and fungi which produce specialized iron-chelating molecules called siderophores. Siderophores are low molecular weight peptides that can bind to ferric ions (Fe_3^+) with very high affinity to form iron–siderophore complex. The plant roots and other soil microbes take up this complex through specific outer membrane receptor proteins and utilize the iron after cleavage of the siderophore molecule (Saha et al. 2013).

There exist more than 500 types of siderophores which can be broadly categorized into three groups that differ in their complexation mode and stability constant with Fe_3^+ ions. The siderophore groups classified on the basis of the functional groups that bind the ferric ions (Fe_3^+) are:

1. Catecholates (including phenolates)
2. Hydroxamates
3. Carboxylates (or hydroxycarboxylates)

Phyllobacterium strain produced siderophores that promoted the growth and quality of strawberries (Flores-Felix et al. 2015). Catechol and hydroxamate siderophore-producing bacteria *Bacillus* spp. and *Pseudomonas* spp. were isolated from plant roots (Grobelak and Hiller 2017). Iron-chelating activity of bacteria *Azotobacter vinelandii*, *Bacillus megaterium*, *Bacillus subtilis*, *Pantoea allii*, and *Rhizobium radiobacter* was ascertained. *B. megaterium* showed the highest iron-chelating capacity followed by *B. subtilis* and *A. vinelandii*. The bacteria were also evaluated for their siderophore production kinetics, level of siderophore production, and types of siderophores produced. *B. subtilis* and *R. radiobacter* produced catecholates, B. *megaterium* and *P. allii* produced hydroxamates, while *A. vinelandii* produced both types of siderophores (Ferreira et al. 2019).

Plants also secrete iron-chelating molecules called phytosiderophores which are efficient in binding to the ferric ions (Fe_3^+). They act as hexadentate ligands that can coordinate Fe(III) with the help of their amino and carboxyl groups (Singh et al. 2011). The iron–phytosiderophore complex is taken up through the root surface with the help of membrane receptors. Different phytosiderophores have been identified in plants of family Gramineae, for example, avenic acid A in oats (*Avena sativa*) and distichonic acid in barley (*Hordeum vulgare*) (Nomoto et al. 1981).

Detailed study of tobacco rhizosphere revealed the presence of siderophore-producing bacteria which mainly belonged to the divisions *α-Proteobacteria*, *β-Proteobacteria*, *γ-Proteobacteria*, *Sphingobacteria*, *Bacilli*, and *Actinobacteria*.

The dominant genera of γ-*Proteobacteria* group were *Pseudomonas, Enterobacter, Serratia, Pantoea, Erwinia*, and *Stenotrophomonas* (Tian et al. 2009).

8.3.4 Plant Growth Hormone (PGH) Production

Many root-associated microbes synthesize plant growth regulators that not only facilitate growth of plants but also alleviate them from several environmental stresses. Auxins play a vital role in growth and development of plants by influencing cellular-level events like cell division, elongation, and differentiation. Auxins are also reported to improve abiotic stress tolerance in plants. The most common auxin is indoleacetic acid (IAA) which can be produced through different biosynthetic pathways. IAA production is a very common growth-promoting phenomenon among PGPR, and these bacteria often have two or sometimes three functional IAA biosynthesis pathways, suggesting the efficiency of their role. Aragon et al. (2014) reported that the auxin-producing genes could be located on the plasmid as found in *P. savastanoi* pv. *savastanoi*, or on the chromosomal DNA as seen in *P. syringae* pv. *syringae*. PGPR isolated from cucumber plants *Pseudomonas stutzeri*, *Bacillus subtilis*, *Stenotrophomonas maltophilia*, and *Bacillus amyloliquefaciens* were found to produce IAA (Islam et al. 2016).

Another PGR cytokinin (CK) controls cellular proliferation and differentiation in plants. Cytokinins are essentially produced by plants and much information is not available regarding production of cytokinins by PGPR. Cytokinin zeatin is produced by various PGPR like *Bradyrhizobium japonicum* (Cassan et al. 2009), *Pseudomonas fluorescens* (Garcia de Salamone et al. 2001), *Bacillus licheniformis* (Hussain and Hasnain 2009), *Bacillus subtilis* (Arkhipova et al. 2005), *Paenibacillus polymyxa* (Timmusk et al. 1999), *Arthrobacter giacomelloi*, and *Azospirillum brasilense* (Cacciari et al. 1989).

Ethylene is another endogenously produced plant growth regulator which is known as a ripening hormone as it ripens the fruits and promotes senescence. It is also produced in response to several abiotic and biotic stresses. It is reported to be produced by PGPR *Azospirillum brasilense* (Perrig et al. 2007). Gibberellins are reported to be produced by several PGPR such as *Acinetobacter calcoaceticus*, *Azospirillum* spp., *Azotobacter* spp., *Herbaspirillum seropedicae*, *Achromobacter xylosoxidans*, *Gluconacetobacter diazotrophicus*, *Bacillus* spp., and *Rhizobia* (Egamberdieva et al. 2015; Dodd et al. 2010). Abscisic acid (ABA), the stress hormone in plants, was reported to be produced in vitro by *Bacillus pumilus* and *Pseudomonas* sp. (Kudoyarova et al. 2019). Microbial ABA is also produced by *Azospirillum lipoferum* which further increases the ABA concentrations in *Arabidopsis thaliana* and alleviates it from drought stress (Cohen et al. 2015).

8.4 Effect of Climatic Change on Diversity and Functions of Plant-Associated Microbes

There is a huge diversity of microbes in soil which exhibit a variety of morphology, physiology, and metabolic capabilities and growth dynamics. The distribution, relative abundance, and function of soil microbial communities are affected by climate change to a large extent. Ecological factors attributed to climate change can potentially act as selection forces causing spatial and temporal changes in plant–microbe interactions leading to adaptive changes. Since plant–microbe relationships may involve reciprocal manipulation of phenotypes, including morphology, physiology, and life cycle of either of the two or both, the adaptive changes might get manifested in the form of different kinds of interaction patterns. For example, the ergot fungus (*Claviceps purpurea*), which is primarily a plant pathogen, serves as a conditional defensive mutualist for its host grass when exposed to herbivory, by producing toxic alkaloids. Asexual *Epichloë* endophyte acts as a parasite by reducing host fitness in the absence of the enemy, but shows mutualism on exposure to enemy by improving host fitness (Wali et al. 2013; Saikkonen et al. 2016). Figure 8.1 depicts the impact of climate change on relationships of plants and microbes and their feedbacks.

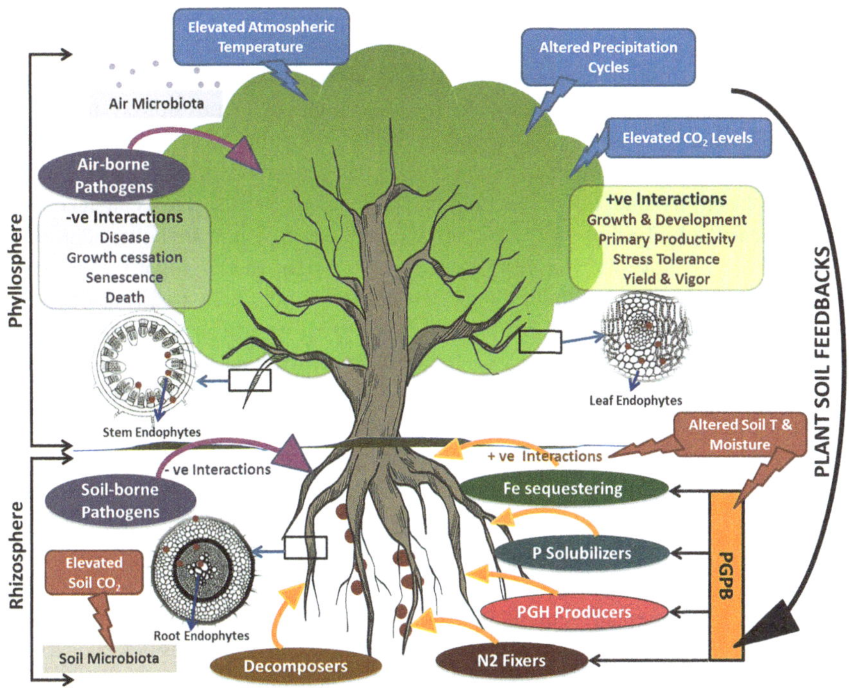

Fig. 8.1 Dynamics of plant–microbe interactions under climate change

Rhizosphere plays a significant role in shaping the plant community structure. Rhizospheric microbial communities are sensitive toward the changes in soil temperature, moisture, and pH, the factors which are the most likely to be responsive toward climate change. The effect of each of these parameters on microbial diversity is discussed further.

8.4.1 Effect of Soil Temperature

According to the assessment of climate change at 1.5 °C of global warming, the increase in global mean surface temperature (GMST) has reached 0.87 °C (+0.10 °C likely range) above pre-industrial values in the 2006–2015 decade. One-quarter of the land has experienced an intensification of hot extremes (maximum temperature on the hottest day of the year) by more than 1 °C and a reduction in the intensity of cold extremes by at least 2.5 °C (minimum temperature on the coldest night of the year) (Hoegh-Guldberg et al. 2018). A number of soil processes such as soil respiration, microbial decomposition and formation of soil organic matter, mineralization, and pedogenesis are sensitive to diurnal temperature changes. Increase in temperature affects the microbial respiration which modulates the decomposition of organic matter in soil. Soil temperature change has a significant impact on the terrestrial ecosystem functions and is thus considered as one of the major indicators of climate change. Alterations in soil temperature directly affect the distribution and abundance of soil microbial community which further changes the soil structure and composition, subsequently influencing the individual plant physiology and plant community structure, having a profound impact on the entire vegetation of a region.

Increase in temperature usually enhances the microbial turnover and rate of respiratory activity and other biochemical processing events. This can lead to a shift in the microbial community structure where the taxa which are better adapted to increased temperature are favored (Castro et al. 2010). In this regard, it must be added that abundances of both fungi and bacteria are likely to be affected by warming (Schindlbacher et al. 2011). This holds great significance because ecosystem functions such as nitrogen fixation, nitrification, denitrification, and methanogenesis are regulated by specific microorganisms. Consequently, alterations in their relative abundances directly impact the rate of these processes.

Overall microbial diversity and abundance both increase with the increase in temperature. Certain bacterial populations like *Alphaproteobacteria* and *Actinobacteria* show a surge on soil warming (Hayden et al. 2012). Many Gram-negative bacteria, *Acidobacteria*, *Planctomycetes*, *Verrucomicrobia* (Sheik et al. 2011), and *Firmicutes* (Gray et al. 2011) are found to increase with rise in soil temperature. Some species of archaea also increase in soil with warming like *Crenarchaeota* (Sheik et al. 2011) and methanotrophs (Zheng et al. 2012). Ammonia-oxidizing bacteria (Long et al. 2012) and ammonia-oxidizing archaea (AOA) (Jung et al. 2011) are observed to decrease with rise in soil temperature.

The fungal diversity and relative abundance of many fungal species in soil are affected with increase in temperature. A study reported 2.75-fold increase in *Mucoromycotina* with a temperature increase of 5 °C (Andrade-Linares et al. 2016). The effect of climate change on plant-associated fungi was studied in a soil solarization experiment, and it was found that a number of fungi exhibited changes which were *Alternaria alternata*, *Cochliobolus spicifer*, some varieties of *Emericella nidulans*, *E. rugulosa*, *Rhizopus stolonifer*, *Fusarium chlamydosporum*, *F. dimerum*, *Gibberella fujikuroi*, *Cochliobolus sativus*, *Gliocladium roseum*, *Melanospora zamiae*, *Nectria haematococca* (anamorph), *Aspergillus fumigatus*, *A. flavus*, and *A. niger* (El-Shanawany et al. 2004).

There are several studies reporting the effect of temperature on colonization of ectomycorrhizal fungi on their host which can alter the community structure. Many ectomycorrhizal fungi like *C. geophilum*, *Suillus intermedius*, and *Lactarius* cf. *pubescens* show reduced respiration under increased temperature. Colonization of ectomycorrhizal fungi *C. geophilum* in *Quercus myrsinifolia* was found to decrease due to increase in temperature (Compant et al. 2010).

8.4.2 Effect of Soil Moisture

Soil moisture varies with temperature, hydrogeological factors, and precipitation. Warming of soil causes loss of moisture subsequently making the soil dry. Microbial activities have been often correlated with frequencies of seasonal rains, while microbial community composition and function are reported to be influenced by drought events. Large variations in soil moisture can impact microbial distribution, abundance, and activity. Patterns of soil respiration are also affected by soil moisture in many terrestrial ecosystems. The moisture content influences the rate of decomposition of organic matter and in turn alters the soil properties such as percolation of water, diffusion of gases and solutes, bioavailability of minerals, etc. which are further responsible for shaping the ecological community structure. Moisture is reported to suppress the microbial activity in environments like saltwater and soils. Increase in humidity due to precipitation can increase the abundance of *Proteobacteria* while decreasing *Acidobacteria* (Mandal and Neenu 2012).

Soil hydration after rain favored the increase in *Firmicutes* followed by increase in *Alphaproteobacteria*, *Betaproteobacteria*, and *Gammaproteobacteria* and a rapid decline in *Actinobacteria* population in a successive manner (Placella et al. 2012) probably due to hydration-mediated competition, predation, or phage bloom. In a study, the effect of hydration–desiccation cycles on dynamics of arid soil bacterial communities was examined by manipulating rain intensity and incubation conditions. Under heavy rain at the lower temperature, a significant decrease in the ribosomal abundance of total soil bacteria, *Actinobacteria* and *Firmicutes*, was seen. Bacterial diversity was also affected and there was a drop in both population richness and evenness. It was observed that there was a community shift upon application of heavy rain at the lower temperature entailing a decrease in *Actinobacteria* and increase in *Bacteroidetes* and *Firmicutes* (Stovicek et al. 2017).

The impact of moisture can be much greater in extreme environments such as hot deserts and Arctic desert and salt lakes. Interestingly, such environments harbor unique and rare diversity of microbes which specialize to survive and thrive through the extremities of these climates. However, the effects of climate change are far more pronounced on these extreme environments than the moderate environments. A study explored the effects of addition of water and organic matter microbial community structure and function in a polar desert along a naturally occurring salinity gradient. Water addition showed a steep decline in relative abundance of *Actinobacteria* while increase in *Proteobacteria* and *Bacteroidetes* at low-salinity soil sites. In the moderate-salinity soil, addition of water favored the dominance of *Acidobacteria* and *Actinobacteria*, which on addition of organic matter were replaced by *Firmicutes* and *Bacteroidetes*. In the high salinity, all bacterial communities declined on addition of water and organic matter except *Firmicutes*. It was suggested that persistent increase in moisture due to permafrost thaw and other reasons, along with organic matter inputs, could stimulate microbial activity, shifting the community composition to copiotrophic organisms (Van Horn et al. 2014).

Another study investigated the effects of environmental variables on the bacterial and fungal communities of permafrost soils in the Beilu River of Tibetan Plateau and found the soil moisture to be the most important factor in determining microbial diversity. The soil with higher moisture content was dominated by *Proteobacteria* followed in relative abundance by *Acidobacteria*, *Actinobacteria*, and *Bacteroidetes*. In soil with lesser moisture content, abundance of *Proteobacteria* and *Acidobacteria* decreased, whereas that of *Bacteroidetes* increased greatly. The most abundant fungal population in low moisture content soil was *Ascomycota* followed by *Basidiomycota*, *Blastocladiomycota*, *Mucoromycotina*, and *Chytridiomycota* (Zhang et al. 2013).

8.4.3 Effect of Soil pH

Soil pH is a critical factor that influences the chemical, physical, and biological properties of soil and therefore plays a critical role in shaping soil community structure. It directly affects the microbial activity and can impose a selective pressure on microbial communities. Soil pH is controlled by several factors such as leaching of ions (both cations and anions), formation of carbonic acid due to dissolution of CO_2 in soil water, nitrification and denitrification processes, humification of soil organic matter, etc. The pH of soil affects the solubility of minerals and availability of nutrients such as carbon, nitrogen, and phosphate, further affecting their uptake by plants and soil microbes. The soil pH also affects the release of plant root exudates potentially impairing the host–microbe communication. This might in turn influence the process of microbial recruitment by plant community causing marked ecological impacts.

Influence of pH on soil bacterial diversity and community composition has been demonstrated in several studies. Pronounced shifts in bacterial community composition and relative abundance were observed across the pH gradient using denaturing gradient gel electrophoresis (DGGE). The DGGE fingerprints showed the most pronounced changes for *Acidobacteria*, *Verrucomicrobia*, and *Gammaproteobacteria* pH 4.5 and 5.0 clusters, while the pattern changes across the gradient were less pronounced for *Firmicutes*, *Actinobacteria*, *Alphaproteobacteria*, and *Betaproteobacteria*. On the basis of NMF (nonnegative matrix factorization) using 16S rRNA gene data analysis, *Dyella* and *Rhodanobacter* (*Gammaproteobacteria*) were identified as low-pH-associated groups and *Anaerolineae* (a class under the phylum *Chloroflexi*) and *Verrucomicrobia* were among the medium-pH cluster. *Lysobacter* (*Gammaproteobacteria*) was also identified in the medium- and high-pH NMF cluster. No considerable changes in *Firmicutes* patterns were observed across the pH gradient (Bartram et al. 2014).

A comparative study of soil bacterial diversity in acidic (pH = 5.2) and neutral soil (pH = 7.7) was performed, and higher degree of bacterial diversity, evenness, and richness was found in acidic than neutral soil, although it represented a nonoptimal pH. Phyla occurring commonly in both types of soils were *Proteobacteria*, *Actinobacteria*, *Cyanobacteria*, and *Acidobacteria*. *Proteobacteria* and *Actinobacteria* were the most abundant in both soils, while the frequently occurring *Acidobacteria* were very less in abundance. The bacteria found commonly in both soils were from the order *Burkholderiales (Betaproteobacteria)* and *Rhizobiales (Alphaproteobacteria)*. *Planctomycetes* were found in acidic soil, while *Bacteroidetes* were present only in neutral soil. Among different classes of *Proteobacteria*, which was the most abundant phylum, the alpha, beta, and gamma *Proteobacteria* dominated with proportions of 28.7, 5.3, and 13.6% in acidic soil and 18.3, 5.2, and 2.0% in neutral soil, respectively. The orders found abundant in acidic soil were *Actinomycetales*, *Caulobacterales*, *Legionellales*, *Planctomycetales*, *Rhodospirillales*, *Solirubrobacterales*, and *Xanthomonadales*, while those dominated at neutral pH were *Acidobacteriales*, *Clostridiales*, and *Sphingomonadales* (Cho et al. 2016).

8.4.4 Effect of Elevated Atmospheric CO_2

Increase in the levels of atmospheric CO_2 is a prominent consequence of climate change and global warming. Soil microbiota plays a major role in regulation of all the biogeochemical cycles including carbon cycle in the ecosystems. Microbial communities transform the organic matter, thereby contributing to the soil fertility and controlling carbon fluxes to a great extent. Under increased atmospheric CO_2, soil microbes release the greenhouse gases methane (CH_4) and nitrous oxide (NO) in more amounts, while methane uptake by soil microbes decreases.

Altered CO_2 levels in atmosphere are also expected to have a profound impact on plant carbon fluxes. Elevated CO_2 concentrations are also expected to increase the leaf photosynthetic rate by making the CO_2 substrate available to the enzyme RuBisCO for fixation. However, this relationship is not linear due to other effects posed by enhanced CO_2 such as suppression of photorespiration, reduction in stomatal conductance, and impairment of nitrogen uptake in certain cases. Consequently, the net primary productivity and carbon fluxes in the food chain are expected to get affected (Dusenge et al. 2019). However, under elevated CO_2 conditions, an increase in plant productivity and detritus production would certainly increase the organic matter in the soil, accelerating the rate of microbial respiration (De Graff et al. 2006).

Responses of plants and soil microbiota were studied in *Populus tremuloides* rhizosphere, and it was observed that there was no change in abundance of total bacterial and eukaryotic populations while an increase was recorded in heterotrophic decomposers and ectomycorrhizal fungi. Significant changes in *Proteobacteria* included decrease in OTUs of *Bradyrhizobium*, *Rhodobium*, and *Duganella (Alphaproteobacteria)*, *Chitinophaga (Bacteroidetes)*, *Acidobacteria*, *and Verrucomicrobia* while an increase in *Rhodoplanes*, *Thiomonas*, *Polaromonas (Alphaproteobacteria)*, *Pseudomonas (Gammaproteobacteria)*, *Arthrobacter*, *Lechevalieria (Actinobacteria)*, and *Niastella (Bacteroidetes)*. Among fungi, the abundance of *Basidiomycota* increased but their diversity decreased. Ectomycorrhiza *Inocybe*, a dominant genus belonging to *Homobasidiomycetes*, increased in abundance, while *Heterobasidiomycetes* decreased in abundance at elevated CO_2. *Urediniomycetes*, which are predominantly plant pathogenic fungi and OTUs associated with *Zygomycota*, decreased in abundance (Lesaulnier et al. 2008).

Increased root biomass due to elevated CO_2 could stimulate the fungal species that thrive on both dead and live root biomass. Changes in composition of arbuscular mycorrhizal (AM) fungi associated with the woody shrub *Adenostoma fasciculatum* were recorded under increased CO_2 conditions where the AM genera *Scutellospora* and *Acaulospora* dominated indicating a shift in community composition and thereby in carbon allocation (Treseder et al. 2003).

8.5 Effect of Climate Change on Phenology

Phenology encompasses the study of the recurrent timing or seasonality of biological phenomena in natural communities and the influences of biotic and abiotic drivers on such events. Global warming has played a consequential role in disrupting the phenological patterns of a large number of naturally occurring communities including plants, animals, and microbes. Thus, changing phonological profiles serve as effective indicators of climate change (Workie and Debella 2018). Phenological shifts have been observed as a response to climate change across several taxa of plant kingdom; however, the magnitude and direction of phenological responses may vary among different genera (Rafferty et al. 2013).

Temperature and rainfall have the most profound effects on plant phenophases such as vegetative growth period, maturity onset, flowering, seed set, dormancy onset, and senescence. Any shifts in these life events of vegetation have considerable impact on the nature and strength of interspecific interactions in the ecosystem such as plant–pollinator mutualism, butterfly emergence, predator–prey relationship, plant–herbivore relationship, migratory bird arrival to breeding sites, etc. This further imposes the risk of phenological mismatches causing temporal overlap with mutualists and antagonists, threatening community stability and leading to changes in demographic processes (Rafferty et al. 2013).

In a study conducted across eco-regions in Ethiopia, the temporal changes in the phenology of vegetation were examined in relation to the shifting patterns of temperature and rainfall. It was observed that over a 14-year period, there was an early onset of growing period along with its prolongation in most of the eco-regions. Temperature variability weakly affected the photosynthetic activities as compared to the water availability due to rainfall. Temperature related inversely to vegetation greenness as it enhanced evapotranspiration causing the vegetation to lose the leaf moisture content and eventually become either pale green or wilt. The prolonged growth period was correlated with an increase in gross primary productivity (GPP) and net primary productivity (NPP) due to modification of nutrient cycling and energy fluxes (Workie and Debella 2018).

8.5.1 Effect of Microbial Community Shift on Plant Phenology

Soil microbiomes were studied to affect the flowering phenology and reproductive fitness of *Boechera stricta*, a wild relative of *Arabidopsis*, in which they were found to potentially contribute to its phenotypic plasticity of flowering time and differential selection observed between habitats. The flowering time of the plant was not only sensitive to abundance of microbes but also to the altered soil chemistries (Wagner et al. 2014).

Rhizosphere microbiota can contribute to phenotypic plasticity by producing phytohormones like auxins which may regulate plant growth and development. A novel metabolic network was reported in which soil microbiota modulated flowering time of *Arabidopsis thaliana*. Rhizosphere microorganisms induced nitrification delayed flowering through a cascade of events that involved production of auxin indoleacetic acid (IAA) from tryptophan, increased and prolonged nitrogen bioavailability, downregulation of flowering genes, and continuation of vegetative growth (Lu et al. 2018).

The influence of soil microbial community on phenology and growth of *Ipomea purpurea* was studied under controlled greenhouse experiment to determine the role of soil community as a selection agent for important plant traits. Comparing the complex and simplified soil communities, the flowering time was found to experience positive and stabilizing selection under both soil environments, and early flowering plants were at a fitness advantage irrespective of the soil microbial composition. However, the complex soil community strongly favored early flowering, and the change in soil environment altered the linear selection differential on flowering time by 55% between environments (Chaney and Baucom 2020).

8.5.2 Impact of Plant Phenological Shifts on Microbial Communities

Phenological patterns in plants and animals are easily observable and predictable, but such patterns are not well documented for microorganisms. However, there are some studies reporting effects of environmental factors and seasonal variations on microbial phenology. For example, bacterioplanktons in a high-elevation lake were studied over an extended period of 3 years and were found to exhibit consistent phenological patterns which were largely predictable from inter- and intra-annual variation. The bacterioplankton community composition changed with seasonal variation exemplified by emergence of distinct, inter-annually recurring community types. The communities appearing each summer were found to be associated with thermal stratification. Increased heterogeneity of bacterioplankton community was also observed throughout the water column, which might be due to the extended physical isolation and decreased community mixing at different depths (Nelson 2009).

Marked effects of drought stress and phenological stages of *Enteropogon macrostachyus*, an indigenous African grass, were studied on rhizosphere microbial biomass. Drought stress controlled the soil organic matter decomposition and nutrient mobilization, and the nutrient uptake competition between plant and microbes led to increase in seedling root biomass as well as the microbial biomass and activities of enzymes β-glucosidase, cellobiohydrolase, and chitinase (Mganga et al. 2019).

Forest ecosystem processes like productivity and nutrient cycling were hypothesized to be affected by two factors—tree leaf phenology and plant–microbe interactions. The increase in temperature affects the tree leaf phenology by extending the length of the growing season. In a simulation model, interactive effects of the length of the growing season and microbe-mediated plant–soil feedback were studied under increased temperature, and the two factors together were found to intensify the negative effects of increasing temperature on the size of soil carbon stock as compared with the length of the growing season alone (Miki and Doi 2016).

The spatiotemporal dynamics of soil bacterial community composition and functional diversity were studied in a tropical lowland evergreen rain forest in relation to the precipitation seasonality and spatial variability in soil characteristics that arose from litterfall-derived nutrient input. Precipitation seasonality-related changes in soil nutrients and moisture regimes gave rise to high spatial heterogeneity and large species turnover. During high precipitation, the abundance of decomposers of complex polymers increased of which the anaerobic saprophytes and N_2 fixers dominated. Increase in abundance of phyllosphere-derived bacteria *Beijerinckiaceae* at the end of the dry season was observed with the increase in litter inputs suggesting the changes in soil microbial composition with respect to canopy phenology patterns (Buscardo et al. 2018).

8.6 Plant–Soil Feedbacks (PSFs)

Plant roots form an intricate network within the soil, harboring a huge diversity of microbes, with which they establish deep relationships. Plants not only affect the microbial community structure and functions, but by virtue of these interactions with microbiota, they also affect the soil properties. These effects of plants on the biotic and abiotic properties of soil last for longer durations, even after they have depleted from the soil, influencing the performance of successively colonizing plants. These effects are known as plant–soil feedbacks and play a significant role in shaping the plant community structure and regulating ecosystem processes. Shifts in microbial community dynamics also feedback to affect plant performance, coexistence, and community composition (Lau and Lennon 2011). Plant–soil feedback is not a unidirectional process; instead, it is the net effect that results from a number of interactions ranging from mutualistic to antagonistic. The strength and direction of the PSFs are determined by several abiotic (such as temperature, soil pH, soil moisture, nutrient availability, and organic matter) and biotic (such as roots and shoots, microbiota, and herbivores) drivers of the ecosystem processes. Since most of the factors have already been discussed in the chapter, some key factors, processes, and effects in terms of plant–soil feedbacks are discussed here.

8.6.1 Direct Influence of Abiotic Factors on PSFs

Warming or increased temperature exerts direct effect on microbial processes as well as the physiological aspects of plants. Also the effects of climate change on ecosystem carbon and nutrient dynamics are mediated via plants and their interactions with soil microbes (De Vries et al. 2012). Warming can enhance the microbial activity in rhizosphere, increasing the rate of decomposition of organic matter and nutrient availability. This may lead to increase in plant productivity, litter production, and rhizodeposition, altering the plant-derived soil inputs. The climate extremes such as drought and heavy rainfall also affect the structure and function of soil microbial communities impacting the carbon dynamics. Increased aridity and persistent drought conditions may lead to reduced aboveground biomass, less litter production, low litterfall quality, and plant species turnover. The feedback thus generated can be observed in the form of slow rate of mineralization, disturbed nutrient cycling, effect on plant diversity, and effect on carbon dynamics (Pugnaire et al. 2019).

In a study, the plant–soil feedback responses of paired native and non-native warm- and cool-season perennial grasses to temperatures and soil moisture were examined. The soil microbial dynamics was found to be stable in either ambient or elevated temperatures, but was disrupted in variable environments. The PSFs were strongly negative at the elevated temperatures in combination with the soil legacy of elevated temperatures and at the constant training and testing temperatures in the cool-season grasses tested under drought conditions. It could be concluded from the study that the year-to-year inconsistency in environmental conditions like temperature and moisture may undermine the stabilizing forces of negative PSF and favor non-native grasses (Duell et al. 2019).

8.6.2 Soil Community Dynamics and PSFs

The abiotic and biotic drivers of climate change affect the microbial functions which in turn influence the direction and magnitude of PSFs, increasing the complexity of the process. There are many examples which show the microbial effect on PSFs turning them either into positive or negative direction depending on the net effect. Beneficial plant–microbe relationships such as symbiosis and mutualism lead to positive PSFs, while pathogenic and parasitic interactions give rise to negative PSFs. Symbiotic nitrogen fixers, PGPR, free-living decomposers, and mycorrhizal fungi are all known to enhance the plant growth and soil health, therefore leading to positive PSFs. Although climatic changes like increase in temperature and altered precipitation cycles impact the PSFs directly, the adaptation of soil microbial communities to climatic extremes can be of immense help in improving plant fitness. For example, symbiotic soil bacteria and mycorrhizal fungi adapted to drought can enhance the drought tolerance of plants and affect the PSFs accordingly (Pugnaire et al. 2019).

Besides, there are some studies exploring the role of belowground trophic interactions on PSFs. A meta-analysis revealed that the top-down control on soil biota through belowground predators results in positive feedbacks for the performance of both fast- and slow-growing plants. Although the consumption of mutualists, symbionts and litter consumers by higher trophic-level organisms is likely to dampen positive PSFs, and the consumption of soil pathogens is expected to reduce negative PSFs, the overall positive feedbacks have been observed. This was attributed to the enhanced nutrient cycling and thereby speeding up of the plant–litter feedback pathway (De Long et al. 2019).

8.6.3 Climate Change and Range Expansion

Plant species show range expansion in response to climate change due to their rapid and farther dispersal efficiency. This enables them to escape from their native enemies like pathogens, parasites, herbivores, etc. Changes are also expected in microbial community and diversity across the range expansion gradient. However, the range expanding plant species may not be accompanied by the same associative microbial species in the new range, which were present earlier. The negatively interacting species are usually not present in their new range and the range shifting plant species are at the competitive advantage with respect to the native plants of the new range.

As already discussed, the soil microbial community structure and function are determined by both abiotic and biotic factors, and the range shifts of these communities also depend upon the interplay of a number of these factors. The rhizosphere microbiota composition is strongly influenced by abiotic factors like temperature, soil pH, moisture, nutrients, and carbon dynamics and biotic factors like root exudates, architecture, and species-specific affinity. These influencing features are instrumental in building species-specific microbiomes through selection across spatial and temporal scales. The dispersal capabilities of rhizosphere microbes are different from plants; therefore, the rhizosphere species might not expand their range with the same rate as the associated plant community, which disrupts the soil community structure in the event of range expansion.

The range expander plant species not only experience the benefit of encountering less pathogenic microbes but also lose their certain specialized soil pathogens (Ramirez et al. 2019), which makes them more successful in the new range. The PSFs due to range shifts thus normally range from lesser negative to positive in the new range. These PSFs can be dynamic in longer duration and turn negative due to coevolution of species-specific pathogens or adaptive development of pathogenesis in native microbial species.

8.6.4 PSFs and Succession

PSFs also play an important role in modifying the process of primary and secondary succession via several mechanisms. Positive plant–soil feedbacks are crucial for early successional stages such as colonization of symbiotic nitrogen fixers and mycorrhizal fungi which are extremely helpful in establishment of primary successors on nutrient devoid substratum. Further, the PSFs involving soil microbes can also have positive or negative effects on the secondary succession. If PSFs favor the decomposer community that facilitates the carbon cycling, soil mineralization, and nutrient mobilization, the succession can be accelerated. Else the negative feedback may be generated if the plants produce recalcitrant litter that can only be metabolized by their own mycorrhizal symbionts and not by other species, retarding the succession (Pugnaire et al. 2019). In a study, it has been shown that in some temperate regions, warming-induced feedbacks from arbuscular mycorrhizal (AM) fungi lead to dominant inhabitation of trees, instead of the feedbacks from ectomycorrhizal (EM) and saprophytic fungi (Tedersoo et al. 2014).

A negative soil feedback is exemplified by symbiotic relationship of early successors with nitrogen fixers, which in the early stages facilitate the plants with nitrogen fixation, but eventually reduce their competitive ability against later successional plant species due to increased pathogenicity and decreased symbiosis. The successive plant species are tolerant to the pathogens of their predecessors and easily replace the early successors. This also exemplifies the indirect positive PSF for later successional plants (Van der Putten et al. 2013).

8.6.5 Climate Change and Legacy Effects

Legacy effects refer to the long-lasting effects of existing plant species on abiotic and biotic conditions of the soil which impact the future plant communities even after their disappearance. The soil legacy effects vary greatly in their spatial and temporal extent, depending upon several factors such as type and abundance of the effector plant community, soil characteristics, and influence on soil microbiota. The legacy effects on soil microbial community dynamics can further affect plant–plant interactions, altering the PSFs. However, the legacy effects on soil microbes are expected to persist for very long durations, sometimes for decades, mainly because of greater dependency of microbial communities on soil characteristics which itself change very slowly over the time (Ladau et al. 2018). Soil pathogens harboring the rhizosphere of monospecific plant communities produce conspecific legacy effects, while those which are less species-specific tend to create heterospecific soil legacy effects (Kaisermann et al. 2017).

In a microcosm study, the microbial soil legacies of early-successional grasses were observed to be more pronounced on mid-successional grass community than on any other plants (Kardol et al. 2007). Extremes of climatic change like drought and altered rainfall patterns are likely to change the plant community composition. There are evidences of strong legacy effects of drought on soil microbiota which have far-reaching consequences on plant–plant interactions potentially altering the PSFs. In arid soil, the microbes can neutralize the legacy effects of drought and promote seed germination and plant growth of a particular species. In another mesocosm study, exotic species dominated over natives due to better performance under drought conditions because they experience weaker negative and positive feedbacks from interactions with soil biota (Engelkes et al. 2008). The soil legacy effects favoring exotics were found to be mediated through changes in the soil microbiota and processes like N mineralization and nutrient availability (Meisner et al. 2013).

8.7 Conclusion

Disruptive impacts of climate change are evident on all the ecological communities. The huge diversity of plant-associated microorganisms ranging from archaea, bacteria, and fungi to insects and nematodes share a common ecological niche and function together to contribute toward the stability of the ecosystem. The biotic and abiotic stresses consequent to climate change potentially affect all these organisms in some or the other way. Plant–microbe interactions are multidimensional and exert effect on all the trophic levels and tend to change the food web dynamics. The overall impact of climate change on an ecosystem depends upon the net effect of the individual responses of ecological communities. The effects of climate change on microbial alpha diversity (diversity within a sample, from the same site experiencing the same environment) may also be different from those on the beta diversity (diversity variation between samples from different sites experiencing different environments). The spatial and time scales of microbial and plant functions also vary greatly, and therefore manifold mechanisms could be responsible for shaping the community structures across the environmental gradients. Prominently visible aspects of plant responses toward climate change include alterations in phonological patterns and range expansion, which have profound impact on microbial diversity and activity. Such ecological shifts generate strong feedbacks, which further influence the biotic and abiotic drivers of climate change, making it a seemingly cyclic process. The complexity of plant–microbe interactions thus cannot be overlooked, and all the aspects are to be necessarily considered for predicting the climate change effects in their entirety, in order to draw the sustainable ecological strategies.

References

Abril A, Zurdo-Pineiro JL, Peix A, Rivas R, Velázquez E (2007) Solubilization of phosphate by a strain of Rhizobium leguminosarum bv. trifolii isolated from *Phaseolus vulgaris* in El Chaco Arido soil (Argentina). In: First international meeting on microbial phosphate solubilization. Springer, Dordrecht, pp 135–138

Ahmad F, Ahmad I, Khan MS (2008) Screening of free-living rhizospheric bacteria for their multiple plant growth promoting activities. Microbiol Res 163(2):173–181

Ahmad M, Zahir ZA, Asghar HN, Asghar M (2011) Inducing salt tolerance in mung bean through coinoculation with rhizobia and plant-growth-promoting rhizobacteria containing 1-aminocyclopropane-1-carboxylate deaminase. Can J Microbiol 57:578–589

Akinrinlola RJ, Yuen GY, Drijber RA, Adesemoye AO (2018) Evaluation of Bacillus strains for plant growth promotion and predictability of efficacy by *in vitro* physiological traits. Int J Microbiol

Andrade-Linares DR, Veresoglou SD, Rillig MC (2016) Temperature priming and memory in soil filamentous fungi. Fungal Ecol 21:10–15

Aragon IM, Pérez-Martínez I, Moreno-Pérez A, Cerezo M, Ramos C (2014) New insights into the role of indole-3-acetic acid in the virulence of *Pseudomonas savastanoi* pv. savastanoi. FEMS Microbiol Lett 356:184–192

Arkhipova TN, Veselov SU, Melentiev AI, Martynenko EV, Kudoyarova GR (2005) Ability of bacterium Bacillus subtilis to produce cytokinins and to influence the growth and endogenous hormone content of lettuce plants. Plant Soil 272:201–209

Arnold W, Rump A, Klipp W, Priefer UB, Pühler A (1988) Nucleotide-sequence of a 24,206-base-pair DNA fragment carrying the entire nitrogen-fixation gene-cluster of Klebsiella-Pneumoniae. J Mol Biol 203

Arruda L, Beneduzi A, Martins A, Lisboa B, Lopes C, Bertolo F, Passaglia LMP, Vargas LK (2013) Screening of rhizobacteria isolated from maize (*Zea mays* L.) in Rio Grande do Sul State (South Brazil) and analysis of their potential to improve plant growth. Appl Soil Ecol 63:15–22

Bais HP, Park SW, Weir TL, Callaway RM, Vivanco JM (2004) How plants communicate using the underground information superhighway. Trends Plant Sci 9:26–32

Bartram AK, Jiang X, Lynch MD, Masella AP, Nicol GW, Dushoff J, Neufeld JD (2014) Exploring links between pH and bacterial community composition in soils from the Craibstone experimental farm. FEMS Microbiol Ecol 87:403–415

Bergman B, Johansson C, Soderback E (1992) Tansley review no. 42. The Nostoc-Gunnera symbiosis. New Phytol 379–400

Bottini R, Cassán F, Piccoli P (2004) Gibberellin production by bacteria and its involvement in plant growth promotion and yield increase. Appl Microbiol Biotechnol 65:497–503

Burns KN, Kluepfel DA, Strauss SL, Bokulich NA, Cantu D, Steenwerth KL (2015) Vineyard soil bacterial diversity and composition revealed by 16S rRNA genes: differentiation by geographic features. Soil Biol Biochem 91:232–247

Buscardo E, Geml J, Schmidt SK, Freitas H, da Cunha HB, Nagy L (2018) Spatio-temporal dynamics of soil bacterial communities as a function of Amazon forest phenology. Sci Rep 8:1–13

Cacciari I, Lippi D, Pietrosanti T, Pietrosanti W (1989) Phytohormone-like substances produced by single and mixed diazotrophic cultures of Azospirillum and Arthrobacter. Plant Soil 115:151–153

Cassan F, Perrig D, Sgroy V, Masciarelli O, Penna C, Luna V (2009) *Azospirillum brasilense* Az39 and *Bradyrhizobium japonicum* E109, inoculated singly or in combination, promote seed germination and early seedling growth in corn (*Zea mays* L.) and soybean (*Glycine max* L.). Eur J Soil Biol 45:28–35

Castro HF, Classen AT, Austin EE, Norby RJ, Schadt CW (2010) Soil microbial community responses to multiple experimental climate change drivers. Appl Environ Microbiol 76:999–1007

Chanda A, Maghrawy H, Sayour H, Gummadidala PM, Gomaa OM (2020) Impact of climate change on plant-associated Fungi. In: Climate change impacts on agriculture and food security in Egypt. Springer, Cham, pp 83–96

Chaney L, Baucom RS (2020) The soil microbial community alters patterns of selection on flowering time and fitness-related traits in *Ipomoea purpurea*. Am J Bot 107:186–194

Cheng Z, Park E, Glick BR (2007) 1-Aminocyclopropane-1-carboxylate deaminase from Pseudomonas putida UW4 facilitates the growth of canola in the presence of salt. Can J Microbiol 53:912–918

Cho SJ, Kim MH, Lee YO (2016) Effect of pH on soil bacterial diversity. J Ecol Environ 40:10

Cohen AC, Bottini R, Pontin M, Berli FJ, Moreno D, Boccanlandro H, Travaglia CN, Piccoli PN (2015) *Azospirillum brasilense* ameliorates the response of *Arabidopsis thaliana* to drought mainly via enhancement of ABA levels. Physiol Plant 153:79–90

Compant S, Van Der Heijden MG, Sessitsch A (2010) Climate change effects on beneficial plant–microorganism interactions. FEMS Microbiol Ecol 73(2):197–214

De Graff MA, Van Groenigen KJ, Six J, Hungate B, van Kessel C (2006) Interactions between plant growth and soil nutrient cycling under elevated CO_2: a meta-analysis. Glob Chang Biol 12 (11):2077–2091

De Long JR, Fry EL, Veen GF, Kardol P (2019) Why are plant–soil feedbacks so unpredictable, and what to do about it? Funct Ecol 33(1):118–128

De Vries FT, Liiri ME, Bjørnlund L, Bowker MA, Christensen S, Setälä HM, Bardgett RD (2012) Land use alters the resistance and resilience of soil food webs to drought. Nat Clim Chang 2 (4):276–280

Dodd IC, Zinovkina NY, Safronova VI, Belimov AA (2010) Rhizobacterial mediation of plant hormone status. Ann Appl Biol 157(3):361–379

Dong Z, Canny MJ, McCully ME, Roboredo MR, Cabadilla CF, Ortega E, Rodes R (1994) A nitrogen-fixing endophyte of sugarcane stems (a new role for the apoplast). Plant Physiol 105 (4):1139–1147

Duell EB, Zaiger K, Bever JD, Wilson GWT (2019) Climate affects plant-soil feedback of native and invasive grasses: negative feedbacks in stable but not in variable environments. Front Ecol Evol 7:419. https://doi.org/10.3389/fevo

Dusenge ME, Duarte AG, Way DA (2019) Plant carbon metabolism and climate change: elevated CO_2 and temperature impacts on photosynthesis, photorespiration and respiration. New Phytol 221(1):32–49

Egamberdieva D, Shrivastava S, Varma A (eds) (2015) Plant-growth-promoting rhizobacteria (PGPR) and medicinal plants. Springer International, Cham

Egamberdiyeva D (2005) Plant-growth-promoting rhizobacteria isolated from a Calcisol in a semi-arid region of Uzbekistan: biochemical characterization and effectiveness. J Plant Nutr Soil Sci 168(1):94–99

El-Shanawany AA, El-Ghamery AA, El-Sheikh HH, Bashandy AA (2004) Soil solarization and the composition of soil fungal community in upper Egypt. Ass Univ Bull Environ Res 7:137–152

Engelhard M, Hurek T, Reinhold-Hurek B (2000) Preferential occurrence of diazotrophic endophytes, Azoarcus spp., in wild rice species and land races of *Oryza sativa* in comparison with modern races. Environ Microbiol 2(2):131–141

Engelkes T, Morriën E, Verhoeven KJ, Bezemer TM, Biere A, Harvey JA, McIntyre LM, Tamis WL, van der Putten WH (2008) Successful range-expanding plants experience less above-ground and below-ground enemy impact. Nature 456(7224):946–948

Estrada-De Los Santos P, Bustillos-Cristales R, Caballero-Mellado J (2001) Burkholderia, a genus rich in plant-associated nitrogen fixers with wide environmental and geographic distribution. Appl Environ Microbiol 67(6):2790–2798

Faist H, Keller A, Hentschel U, Deeken R (2016) Grapevine (*Vitis vinifera*) crown galls host distinct microbiota. Appl Environ Microbiol 82(18):5542–5552

Ferreira CM, Vilas-Boas Â, Sousa CA, Soares HM, Soares EV (2019) Comparison of five bacterial strains producing siderophores with ability to chelate iron under alkaline conditions. AMB Express 9(1):78

Flores-Felix JD, Silva LR, Rivera LP, Marcos-García M, García-Fraile P, Martínez-Molina E, Mateos PF, Velázquez E, Andrade P, Rivas R (2015) Plants probiotics as a tool to produce highly functional fruits: the case of Phyllobacterium and vitamin C in strawberries. PLoS One 10(4):e0122281

Gaby JC, Buckley DH (2011) A global census of nitrogenase diversity. Environ Microbiol 13 (7):1790–1799

Garcia de Salamone IE, Hynes RK, Nelson LM (2001) Cytokinin production by plant growth promoting rhizobacteria and selected mutants. Can J Microbiol 47(5):404–411

Graham PH, Vance CP (2003) Legumes: importance and constraints to greater use. Plant Physiol 131(3):872–877

Gray SB, Classen AT, Kardol P, Yermakov Z, Mille RM (2011) Multiple climate change factors interact to alter soil microbial community structure in an old-field ecosystem. Soil Sci Soc Am J 75(6):2217–2226

Grobelak A, Hiller J (2017) Bacterial siderophores promote plant growth: screening of catechol and hydroxamate siderophores. Int J Phytoremediation 19(9):825–833

Gutierrez-Manero FJ, Ramos-Solano B, Probanza AN, Mehouachi J, Tadeo R, F. and Talon, M. (2001) The plant-growth-promoting rhizobacteria *Bacillus pumilus* and *Bacillus licheniformis* produce high amounts of physiologically active gibberellins. Physiol Plant 111 (2):206–211

Hayden HL, Mele PM, Bougoure DS, Allan CY, Norng S, Piceno YM, Brodie EL, DeSantis TZ, Andersen GL, Williams AL, Hovenden MJ (2012) Changes in the microbial community structure of bacteria, archaea and fungi in response to elevated CO_2 and warming in an Australian native grassland soil. Environ Microbiol 14(12):3081–3096

Hirsch AM, Fujishige NA (2008) Commensalisms. In: Encyclopedia of ecology. Elsevier, Oxford, pp 679–683

Hoegh-Guldberg O, Jacob D, Bindi M, Brown S, Camilloni I, Diedhiou A, Djalante R, Ebi K, Engelbrecht F, Guiot J, Hijioka Y (2018) Impacts of 1.5 C global warming on natural and human systems. Global warming of 1.5° C. An IPCC Special Report

Hussain A, Hasnain S (2009) Cytokinin production by some bacteria: its impact on cell division in cucumber cotyledons. Afr J Microbiol Res 3(11):704–712

Islam S, Akanda AM, Prova A, Islam MT, Hossain MM (2016) Isolation and identification of plant growth promoting rhizobacteria from cucumber rhizosphere and their effect on plant growth promotion and disease suppression. Front Microbiol 6:1360

Jacobson MR, Brigle KE, Bennett LT, Setterquist RA, Wilson MS, Cash VL, Beynon J, Newton WE, Dean DR (1989) Physical and genetic map of the major nif gene cluster from *Azotobacter vinelandii*. J Bacteriol 171(2):1017–1027

Jung J, Yeom J, Kim J, Han J, Lim HS, Park H, Hyun S, Park W (2011) Change in gene abundance in the nitrogen biogeochemical cycle with temperature and nitrogen addition in Antarctic soils. Res Microbiol 162(10):1018–1026

Kaisermann A, de Vries FT, Griffiths RI, Bardgett RD (2017) Legacy effects of drought on plant–soil feedbacks and plant–plant interactions. New Phytol 215(4):1413–1424

Kang SM, Khan AL, Waqas M, You YH, Hamayun M, Joo GJ, Shahzad R, Choi KS, Lee IJ (2015) Gibberellin-producing Serratia nematodiphila PEJ1011 ameliorates low temperature stress in *Capsicum annuum* L. Eur J Soil Biol 68:85–93

Karadeniz A, Topcuoğlu ŞF, Inan S (2006) Auxin, gibberellin, cytokinin and abscisic acid production in some bacteria. World J Microbiol Biotechnol 22(10):1061–1064

Kardol P, Cornips NJ, van Kempen MM, Bakx-Schotman JT, van der Putten WH (2007) Microbe-mediated plant–soil feedback causes historical contingency effects in plant community assembly. Ecol Monogr 77(2):147–162

Kecskemeti E, Berkelmann-Löhnertz B, Reineke A (2016) Are epiphytic microbial communities in the carposphere of ripening grape clusters (*Vitis vinifera* L.) different between conventional, organic, and biodynamic grapes? PLoS One 11(8):e0160852

Koivusaari P, Tejesvi MV, Tolkkinen M, Markkola A, Mykrä H, Pirttilä AM (2019) Fungi originating from tree leaves contribute to fungal diversity of litter in streams. Front Microbiol 10:651

Krause A, Ramakumar A, Bartels D, Battistoni F, Bekel T, Boch J, Böhm M, Friedrich F, Hurek T, Krause L, Linke B (2006) Complete genome of the mutualistic, N2-fixing grass endophyte Azoarcus sp. strain BH72. Nat Biotechnol 24(11):1384–1390

Kudoyarova G, Arkhipova TN, Korshunova T, Bakaeva M, Loginov O, Dodd IC (2019) Phytohormone mediation of interactions between plants and non-symbiotic growth promoting bacteria under edaphic stresses. Front Plant Sci 10:1368

Kumar A, Singh S, Gaurav AK, Srivastava S, Verma JP (2020) Plant growth-promoting Bacteria: biological tools for the mitigation of salinity stress in plants. Front Microbiol 11

Ladau J, Shi Y, Jing X, He JS, Chen L, Lin X, Fierer N, Gilbert JA, Pollard KS, Chu H (2018) Existing climate change will lead to pronounced shifts in the diversity of soil prokaryotes. Msystems 3(5)

Lau JA, Lennon JT (2011) Evolutionary ecology of plant–microbe interactions: soil microbial structure alters selection on plant traits. New Phytol 192(1):215–224

Lery LM, Hemerly AS, Nogueira EM, von Krüger WM, Bisch PM (2011) Quantitative proteomic analysis of the interaction between the endophytic plant-growth-promoting bacterium *Gluconacetobacter diazotrophicus* and sugarcane. Mol Plant-Microbe Interact 24(5):562–576

Lesaulnier C, Papamichail D, McCorkle S, Ollivier B, Skiena S, Taghavi S, Zak D, Van Der Lelie D (2008) Elevated atmospheric CO_2 affects soil microbial diversity associated with trembling aspen. Environ Microbiol 10(4):926–941

Li HB, Singh RK, Singh P, Song QQ, Xing YX, Yang LT, Li YR (2017) Genetic diversity of nitrogen-fixing and plant growth promoting Pseudomonas species isolated from sugarcane rhizosphere. Front Microbiol 8:1268

Long X, Chen C, Xu Z, Linder S, He J (2012) Abundance and community structure of ammonia oxidizing bacteria and archaea in a Sweden boreal forest soil under 19-year fertilization and 12-year warming. J Soils Sediments 12(7):1124–1133

Lu T, Ke M, Lavoie M, Jin Y, Fan X, Zhang Z, Fu Z, Sun L, Gillings M, Peñuelas J, Qian H (2018) Rhizosphere microorganisms can influence the timing of plant flowering. Microbiome 6 (1):1–12

Magalhaes FM, Baldani JI, Souto SM, Kuykendall JR, Dobereiner J (1983) New acid-tolerant Azospirillum species. Anais-Academia Brasileira de Ciencias

Mandal A, Neenu S (2012) Impact of climate change on soil biodiversity-a review. Agric Rev 33 (4):283–292

Martinez-Aguilar L, Salazar-Salazar C, Méndez RD, Caballero-Mellado J, Hirsch AM, Vásquez-Murrieta MS, Estrada-De Los Santos P (2013) *Burkholderia caballeronis* sp. nov., a nitrogen fixing species isolated from tomato (*Lycopersicon esculentum*) with the ability to effectively nodulate *Phaseolus vulgaris*. Antonie Van Leeuwenhoek 104(6):1063–1071

Meisner A, De Deyn GB, de Boer W, van der Putten WH (2013) Soil biotic legacy effects of extreme weather events influence plant invasiveness. Proc Natl Acad Sci 110(24):9835–9838

Mendes R, Garbeva P, Raaijmakers JM (2013) The rhizosphere microbiome: significance of plant beneficial, plant pathogenic, and human pathogenic microorganisms. FEMS Microbiol Rev 37:634–663

Mganga KZ, Razavi BS, Sanaullah M, Kuzyakov Y (2019) Phenological stage, plant biomass, and drought stress affect microbial biomass and enzyme activities in the rhizosphere of *Enteropogon macrostachyus*. Pedosphere 29(2):259–265

Miki T, Doi H (2016) Leaf phenological shifts and plant–microbe–soil interactions can determine forest productivity and nutrient cycling under climate change in an ecosystem model. Ecol Res 31(2):263–274

Mirza BS, Potisap C, Nüsslein K, Bohannan BJ, Rodrigues JL (2014) Response of free-living nitrogen-fixing microorganisms to land use change in the Amazon rainforest. Appl Environ Microbiol 80(1):281–288

Montanez A, Blanco AR, Barlocco C, Beracochea M, Sicardi M (2012) Characterization of cultivable putative endophytic plant growth promoting bacteria associated with maize cultivars (*Zea mays* L.) and their inoculation effects in vitro. Appl Soil Ecol 58:21–28

Morris JJ, Schniter E (2018) Black queen markets: commensalism, dependency, and the evolution of cooperative specialization in human society. J Bioecon 20(1):69–105

Naik PR, Raman G, Narayanan KB, Sakthivel N (2008) Assessment of genetic and functional diversity of phosphate solubilizing fluorescent pseudomonads isolated from rhizospheric soil. BMC Microbiol 8(1):230

Naz I, Bano A, Ul-Hassan T (2009) Isolation of phytohormones producing plant growth promoting rhizobacteria from weeds growing in Khewra salt range, Pakistan and their implication in providing salt tolerance to *Glycine max* L. Afr J Biotechnol 8(21)

Nelson CE (2009) Phenology of high-elevation pelagic bacteria: the roles of meteorologic variability, catchment inputs and thermal stratification in structuring communities. ISME J 3(1):13–30

Ngom M, Gray K, Diagne N, Oshone R, Fardoux J, Gherbi H, Hocher V, Svistoonoff S, Laplaze L, Tisa LS, Sy MO (2016) Symbiotic performance of diverse Frankia strains on salt-stressed *Casuarina glauca* and *Casuarina equisetifolia* plants. Front Plant Sci 7:1331

Nomoto K, Mino Y, Ishida T, Yoshioka H, Ota N, Inoue M, Takagi SI, Takemoto T (1981) X-ray crystal structure of the copper (II) complex of mugineic acid, a naturally occurring metal chelator of graminaceous plants. J Chem Soc Chem Commun 7:338–339

Perin L, Martínez-Aguilar L, Paredes-Valdez G, Baldani JI, Estrada-De Los Santos P, Reis VM, Caballero-Mellado J (2006) *Burkholderia silvatlantica* sp. nov., a diazotrophic bacterium associated with sugar cane and maize. Int J Syst Evol Microbiol 56(8):1931–1937

Perrig D, Boiero ML, Masciarelli OA, Penna C, Ruiz OA, Cassán FD, Luna MV (2007) Plant-growth-promoting compounds produced by two agronomically important strains of *Azospirillum brasilense*, and implications for inoculant formulation. Appl Microbiol Biotechnol 75(5):1143–1150

Placella SA, Brodie EL, Firestone MK (2012) Rainfall-induced carbon dioxide pulses result from sequential resuscitation of phylogenetically clustered microbial groups. Proc Natl Acad Sci 109 (27):10931–10936

Pugnaire FI, Morillo JA, Peñuelas J, Reich PB, Bardgett RD, Gaxiola A, Wardle DA, Van Der Putten WH (2019) Climate change effects on plant-soil feedbacks and consequences for biodiversity and functioning of terrestrial ecosystems. Sci Adv 5(11):eaaz1834

Rafferty NE, Cara Donna PJ, Burkle LA, Iler AM, Bronstein JL (2013) Phenological overlap of interacting species in a changing climate: an assessment of available approaches. Ecol Evol 3 (9):3183–3193

Ramirez KS, Snoek LB, Koorem K, Geisen S, Bloem LJ, Ten Hooven F, Kostenko O, Krigas N, Manrubia M, Caković D, van Raaij D (2019) Range-expansion effects on the belowground plant microbiome. Nat Ecol Evol 3(4):604–611

Redecker D, Schüßler A, Stockinger H, Stürmer SL, Morton JB, Walker C (2013) An evidence-based consensus for the classification of arbuscular mycorrhizal fungi (Glomeromycota). Mycorrhiza 23(7):515–531

Reed SC, Townsend AR, Cleveland CC, Nemergut DR (2010) Microbial community shifts influence patterns in tropical forest nitrogen fixation. Oecologia 164(2):521–531

Reinhold B, Hurek T, Fendrik I, Pot B, Gillis M, Kersters K, Thielemans S, De Ley J (1987) *Azospirillum halopraeferens* sp. nov., a nitrogen-fixing organism associated with roots of Kallar grass (*Leptochloa fusca* (L.) Kunth). Int J Syst Evol Microbiol 37(1):43–51

Reis VM, Estrada-De Los Santos P, Tenorio-Salgado S, Vogel J, Stoffels M, Guyon S, Mavingui P, Baldani VLD, Schmid M, Baldani JI, Balandreau J (2004) Burkholderia tropica sp. nov., a novel nitrogen-fixing, plant-associated bacterium. Int J Syst Evol Microbiol 54(6):2155–2162

Rilling JI, Acuña JJ, Sadowsky MJ, Jorquera MA (2018) Putative nitrogen-fixing bacteria associated with the rhizosphere and root endosphere of wheat plants grown in an andisol from southern Chile. Front Microbiol 9:2710

Saha R, Saha N, Donofrio RS, Bestervelt LL (2013) Microbial siderophores: a mini review. J Basic Microbiol 53(4):303–317

Saikkonen K, Young CA, Helander M, Schardl CL (2016) Endophytic Epichloë species and their grass hosts: from evolution to applications. Plant Mol Biol 90(6):665–675

Samad A, Trognitz F, Compant S, Antonielli L, Sessitsch A (2017) Shared and host-specific microbiome diversity and functioning of grapevine and accompanying weed plants. Environ Microbiol 19(4):1407–1424

Saravanakumar D, Samiyappan R (2007) ACC deaminase from *Pseudomonas fluorescens* mediated saline resistance in groundnut (Arachis hypogea) plants. J Appl Microbiol 102(5):1283–1292

Schindlbacher A, Rodler A, Kuffner M, Kitzler B, Sessitsch A, Zechmeister-Boltenstern S (2011) Experimental warming effects on the microbial community of a temperate mountain forest soil. Soil Biol Biochem 43(7):1417–1425

Sharma SB, Sayyed RZ, Trivedi MH, Gobi TA (2013) Phosphate solubilizing microbes: sustainable approach for managing phosphorus deficiency in agricultural soils. SpringerPlus 2(1):587

Sheik CS, Beasley WH, Elshahed MS, Zhou X, Luo Y, Krumholz LR (2011) Effect of warming and drought on grassland microbial communities. ISME J 5(10):1692–1700

Shutsrirung A, Chromkaew Y, Pathom-Aree W, Choonluchanon S, Boonkerd N (2013) Diversity of endophytic actinomycetes in mandarin grown in northern Thailand, their phytohormone production potential and plant growth promoting activity. Soil Sci Plant Nutr 59(3):322–330

Singh JS, Pandey VC, Singh DP (2011) Efficient soil microorganisms: a new dimension for sustainable agriculture and environmental development. Agric Ecosyst Environ 140 (3–4):339–353

Stovicek A, Azatyan A, Soares MIM, Gillor O (2017) The impact of hydration and temperature on bacterial diversity in arid soil mesocosms. Front Microbiol 8:1078

Sytsma KJ, Morawetz J, Pires JC, Nepokroeff M, Conti E, Zjhra M, Hall JC, Chase MW (2002) Urticalean rosids: circumscription, rosid ancestry, and phylogenetics based on rbcL, trnL-F, and ndhF sequences. Am J Bot 89(9):1531–1546

Tedersoo L, Bahram M, Põlme S, Kõljalg U, Yorou NS, Wijesundera R, Ruiz LV, Vasco-Palacios AM, Thu PQ, Suija A, Smith ME (2014) Global diversity and geography of soil fungi. Science 346(6213):1078–1088

Tian F, Ding Y, Zhu H, Yao L, Du B (2009) Genetic diversity of siderophore-producing bacteria of tobacco rhizosphere. Braz J Microbiol 40(2):276–284

Tilak KVBR, Reddy BS (2006) Bacillus cereus and *B. circulans*–novel inoculants for crops. Curr Sci 90(5):642–644

Timmusk S, Nicander B, Granhall U, Tillberg E (1999) Cytokinin production by *Paenibacillus polymyxa*. Soil Biol Biochem 31(13):1847–1852

Tkacz A, Poole P (2015) Role of root microbiota in plant productivity. J Exp Bot 66(8):2167–2175

Treseder KK, Egerton-Warburton LM, Allen MF, Cheng Y, Oechel WC (2003) Alteration of soil carbon pools and communities of mycorrhizal fungi in chaparral exposed to elevated carbon dioxide. Ecosystems 6(8):786–796

Van der Putten WH, Bardgett RD, Bever JD, Bezemer TM, Casper BB, Fukami T, Kardol P, Klironomos JN, Kulmatiski A, Schweitzer JA, Suding KN (2013) Plant–soil feedbacks: the past, the present and future challenges. J Ecol 101(2):265–276

Van Horn DJ, Okie JG, Buelow HN, Gooseff MN, Barrett JE, Takacs-Vesbach CD (2014) Soil microbial responses to increased moisture and organic resources along a salinity gradient in a polar desert. Appl Environ Microbiol 80(10):3034–3043

Vandamme P (2004) Classification of the PCB-and biphenyl-degrading strain LB400 and relatives as *Burkholderia xenovorans* sp. nov. Int J Syst Evol Microbiol 54:16771681Green

Vazquez P, Holguin G, Puente ME, Lopez-Cortes A, Bashan Y (2000) Phosphate-solubilizing microorganisms associated with the rhizosphere of mangroves in a semiarid coastal lagoon. Biol Fertil Soils 30(5–6):460–468

Venieraki A, Dimou M, Pergalis P, Kefalogianni I, Chatzipavlidis I, Katinakis P (2011) The genetic diversity of culturable nitrogen-fixing bacteria in the rhizosphere of wheat. Microb Ecol 61 (2):277–285

Wagner MR, Lundberg DS, Coleman-Derr D, Tringe SG, Dangl JL, Mitchell-Olds T (2014) Natural soil microbes alter flowering phenology and the intensity of selection on flowering time in a wild Arabidopsis relative. Ecol Lett 17(6):717–726

Wakelin SA, Ryder MH (2004) Plant growth-promoting inoculants in Australian agriculture. Crop Manage 3(1):1–5

Wali PP, Wali PR, Saikkonen K, Tuomi J (2013) Is the pathogenic ergot fungus a conditional defensive mutualist for its host grass? PLoS One 8(7):e69249

Warriner K (2005) Pathogens in vegetables. In: Improving the safety of fresh fruit and vegetables. Woodhead, pp 3–43

Workie TG, Debella HJ (2018) Climate change and its effects on vegetation phenology across ecoregions of Ethiopia. Glob Ecol Conserv 13:e00366

Xie G, Su B, Cui Z (1998) Isolation and identification of N_2-fixing strains of Bacillus in rice rhizosphere of the Yangtze River valley. Wei Sheng wu xue bao= Acta Microbiologica Sinica 38(6):480–483

Zarraonaindia I, Owens SM, Weisenhorn P, West K, Hampton-Marcell J, Lax S, Bokulich NA, Mills DA, Martin G, Taghavi S, van der Lelie D (2015) The soil microbiome influences grapevine-associated microbiota. MBio 6(2)

Zhang XF, Zhao L, Xu SJ Jr, Liu YZ, Liu HY, Cheng GD (2013) Soil moisture effect on bacterial and fungal community in Beilu River (Tibetan Plateau) permafrost soils with different vegetation types. J Appl Microbiol 114(4):1054–1065

Zheng Y, Yang W, Sun X, Wang SP, Rui YC, Luo CY, Guo LD (2012) Methanotrophic community structure and activity under warming and grazing of alpine meadow on the Tibetan plateau. Appl Microbiol Biotechnol 93(5):2193–2203

Chapter 9
Climate Changes in Soil Microorganism–Plant Interactions

Ashutosh Dubey

Abstract As earth's climate is constituted by land, water bodies, atmosphere within and around the earth, and living organisms including humans, it is a very intricate yet collaborative system (Abatenh et al., J Bacteriol Infec Dis 2(1):8–27, 2018), which is primarily changing due to human activities. Soil is a medium of growth and development for microbes and plants, and collaborations between plants and soil microorganism constitute part of land-based terrestrial ecosystems (Hawkins and Crawford, AoB Plants 10:ply039, 2018). With changing global climate, proper understanding of possible feedbacks from climate to the soil system and its microbial population and vice versa is necessary. Microbes have vast diversity in type and function and considered crucial for survival of life on earth, as these organisms participate in nutrient recycling for soil fertility, detoxify xenobiotics, and regulate carbon storage due to their exceedingly adaptable metabolism. The microbial populations determine key soil functions; hence, they have straightforward influence on soil properties. Terrestrial ecosystems have interacting biotic and abiotic components where soilborne microbes influence plant growth by symbiotic or nonsymbiotic interfaces. Advantageous plant-affiliated microbes, i.e., plant growth-promoting rhizobacteria (PGPR), not only stimulate plant growth but also strengthen plant resistance to biotic and abiotic stresses (Tyagi et al., Eur J Mol Biotechnol 5:149–156, 2014).

Although different factors are related to climate change such as rise in temperature and atmospheric CO_2, scarcity of water causing drought conditions is nowadays a crucial point of research to understand altering interaction dynamics between plants and microorganisms (Compant et al. 2010). As these changing climate scenarios not only have direct impact on microbial populations in soil but also have many potential consequences for soil parameters as loss of soil carbon, variations in soilborne greenhouse gas levels are crucial for plant–microbe interactions.

A. Dubey (✉)
Department of Biochemistry, College of Basic Sciences and Humanities, G. B. Pant University of Agriculture and Technology, Pantnagar, India
e-mail: ashutoshdubey.cbsh@gbpuat-tech.ac.in

D. K. Choudhary et al. (eds.), *Climate Change and the Microbiome*, Soil Biology 63,
https://doi.org/10.1007/978-3-030-76863-8_9

Keywords Plant–microbe interaction · Symbiosis · Secondary metabolite · PGPR · Stress · Climate change

9.1 Soil, Microbes, and Plants

Several fungal and bacterial species found in the soil environment have a vital role to alter and evolve functional ecosystem (Binyamin et al. 2019). Rhizospheric microorganisms have biochemical machinery to produce, consume, and decompose various metabolites. These soil microorganisms are distributed in diverse phylogenetic groups carrying exceptional genetic diversity (Torsvik et al. 1990). Environmental factors play a key role during influence and interaction of biotic soil communities and plants and provide significant motivation for plant–soil–microbe interactions (Reynolds et al. 2003).

For the synergetic ecological conduct, complementary resources transference among host plant and beneficial microbes and subsequent recovery from disturbance are necessary for ecosystem maintenance (Reynolds et al. 2003). The tripartite interaction between abiotic conditions of soil; biotic constituents of soil, e.g., microorganisms; and plants is collectively known as plant–soil feedbacks (PSFs), and these complex feedbacks are the prime motivational stimulus for ecosystem processes which influences diversity, performance, and community structure of plants and microbes (Bever 1994, 2002; Bever et al. 1997; Ehrenfeld et al. 2005; Kulmatiski et al. 2008; van der Putten et al. 2013). The plant–microbe interactions also constitute important ingredient for these feedbacks that influences ecosystem (Pugnaire et al. 2019). PSFs studies have specifically examined the exclusive positive or negative ramifications of microbial population on plant performance (Wagg et al. 2015). The plant growth depends on the equilibrium among beneficial or antagonistic interactions with the host plant and associated terrestrial microbial population. This plant–microbe association depends upon abiotic (viz., soil pH, and EC, physical structure, and nutrient availability) as well as biotic (plant and bacterial specific) components (Bennett and Klironomos 2019). Although soil microorganisms have complicated influence on PSFs, free-living PGPR or PGPB and PGPF capable of providing growth-limiting nutrients to plants (Kaisermann et al. 2017) or suitable for organic matter decay (Veen et al. 2015) accomplish more positive or less negative PSFs. Host-specific pathogenic fungi found in rhizosphere add burden on plant species for cohabitation and steer to negative PSFs (Maron et al. 2011; Bever et al. 2015; Mommer et al. 2018), while rhizospheric aggregation of beneficial microorganisms, e.g., mycorrhizal fungi (Bever et al. 2012; Cortois et al. 2016) and nitrogen-fixing symbionts (Carter et al. 2019; Siefert et al. 2019), is responsible for positive PSFs.

9.2 Beneficial Plant–Microbe Interactions

Symbiotic relation between almost all terrestrial plants and range of microorganisms is well recognized (Brundrett 2009). Soil zone around the plant roots, i.e., rhizosphere, has maximum microbial activity and several plant root-allied soil microorganisms depend on root exudates for their existence (Whipps 1990; Bent 2006; Lugtenberg and Kamilova 2009). Rhizospheric microbiota not only interacts with each other but also collaborates with plant roots (van der Heijden et al. 1998; Binyamin et al. 2019). Although plant–microbe interactions take place not only within soil domain but also in the troposphere, they are more intricate in the rhizosphere than above the soil surface (Bais et al. 2004). Microbial responses toward plants are environment dependent, e.g., functions and interactions with plants of nitrogen-fixing bacteria and mycorrhizal fungi are dependent on mineral availability in the surrounding soils (Bethlenfalvay et al. 1982; van der Heijden et al. 2008). Soil microfloras have a wide range of collaboration with higher plants that span from synergistic to antagonistic. The microorganisms provide the nutritional supply to host plants as nitrogen-fixing bacteria directly provide mineral resources to the plants or mycorrhizal fungi, nonmycorrhizal fungi, various rhizosphere bacteria, protozoa, and nematodes provide protection to plants against other pests like fungi, bacteria, actinobacteria, protozoa, nematodes, and viruses in the rhizosphere (Schlesinger 1991; Schimel 1995; Lavelle and Spain 2001; Reynolds et al. 2003; Reynolds and Haubensak 2008; Eom et al. 2000) while photosynthesizing plants with use of solar energy fix carbon that is used by various cohabitating organisms. Competition for nutrients between microbes and plants persists in soil (Hodge et al. 2000; Burns et al. 2013), making it a codependent yet a combative alliance (Harte and Kinzig 1993).

Besides these competition for nutrients, plant–microbe interactions may be favorable or detrimental for the host plant (Bever 1999; Saharan and Nehra 2011; Nadeem et al. 2013; Schikora 2018; Ding et al. 2019; Wille et al. 2019). Some of plant root-associated soil microorganisms can be neutral or pathogenic for their hosts according to the de Bary definition (de Bary 1879). Whereas plant growth and tolerance against biotic and abiotic stresses are favorably supported by plant growth promoting rhizobacteria or bacteria (PGPR or PGPB) and plant growth-promoting fungi (PGPF) (Kloepper and Schroth 1978; Bashan and Holguin 1998; Bent 2006). Some microorganisms can also enter the root system of their hosts and enhance their beneficial effects with an endophytic lifestyle (Stone et al. 2000) and adapt plants for inhospitable environment (Yang et al. 2008; de Zelicourt et al. 2013; Pieterse et al. 2014; Pii et al. 2015; Verbon and Liberman 2016).

The affirmative plant–microbe interactions stimulate plant growth and for that plant growth-promoting rhizobacteria (PGPR) are prominent group of organisms that includes the following genera: *Pseudomonas*, *Bacillus*, *Serratia*, *Enterobacter*, *Erwinia*, *Beijerinckia*, *Klebsiella*, *Flavobacterium*, *Burkholderia*, and *Gluconacetobacter* (Podile and Kishore 2006). The microbes of these genera comprise multitudinous growths promoting attributes that aid and facilitate plant expansion (Mehmood et al. 2018). The microbes assist plants in the form of various inputs such as nutrient solubilization, siderophore production, formation of exopolysaccharides, hormone production, supply of metabolically advantageous enzymes such as chitinase and ACC deaminase, and last but not the least nitrogen fixation and sulfur and carbon cycling (Nadeem et al. 2010a, b; Glick 2014). Accessibility of soil phosphorus and potassium to the plant is viable by activity of phosphate- and potassium-solubilizing bacteria (Archana et al. 2012; Panhwar et al. 2014). Equivalently, calcium and other nutrients are also readily made available to plants by PGPR activity (Lee et al. 2010). Under water-deficient conditions, the PGPR produce exopolysaccharides to safeguard plants from dehydration (Sandhya et al. 2009). The plant growth is shunned as a result of the abiotic stresses, viz., salinity and drought, because of root growth inhibition by heightened ethylene concentrations (Nadeem et al. 2010b). These effects of stresses, i.e., increased ethylene, are mitigated by ACC deaminase activity of plant growth-promoting rhizobacteria (Glick et al. 2007).

The rhizospheric symbiotic microbes not only influence plant nutrition and growth but can also shield the host plant from detrimental responses of various plant pathogens by inducing resistant mechanism in the entire plant against pathogens (Binyamin et al. 2019). The plant defenses are upgraded either by disintegrating the pathogen cell wall by activity of proteases, cellulases, glucanases, lipases, chitinases, and other enzymes. (Beneduzi et al. 2012) or by causing scarcity of pathogen-specific nutrients, e.g., PGPR-produced siderophores make iron unavailable to the deleterious organisms (Bhattacharyya and Jha 2012).

Beneficial microbes of rhizosphere also instigate induced systemic resistance (ISR) in host plants to enhance plant resistance (Saravana and Samiyappan 2007) against biotic stresses caused by biological agents as different pathogens and insects (Romera et al. 2019). In bacteria, lipopolysaccharides, salicylic acid, and siderophores are mobilized for induced systemic resistance (ISR) in host (Lugtenberg and Kamilova 2009), and in response, several hormones and signaling molecules, like ethylene, auxin, and nitric oxide (NO), and many transcription factors are induced in host plants (Romera et al. 2019). These ISR-related signals perceived in plant activate and strengthen plant defense systems by different mechanisms to counter the pathogens (Van Loon 2007).

9.3 Microbes and Climate Change

Various indirect impacts of climate changes on biotic and abiotic constituents of terrestrial ecosystems include risk of physical injury as well as emergence / spread of diseases and bereavement from natural disasters such as heat waves, wildfires, storms, floods, poor air quality, and drought (Wallenstein and Hall 2012; Compant et al. 2010; de Vries and Griffiths 2018; Endeshaw et al. 2018). Under such environmental conditions, not only abundance and composition of microbial communities in the form of microbial ecology and ecosystem structure as well as microbial processes like respiration, fermentation, and methanogenesis are also accelerated/altered. Furthermore, substantial changes may also be observed in the functional genes and traits of microbes under the stimulus of biogeochemical changes (Yergeau et al. 2011; Sayer et al. 2017).

Simultaneously, rhizospheric microorganisms such as bacteria, fungus, algae, and archaea also speed up global warming like climate changes through organic matter decomposition and finally increase the flux of CO_2 in the atmosphere (Fierer and Schimel 2003; Bardgett et al. 2008; Castro et al. 2010; Tyagi et al. 2014; Weiman 2015).

Soil microbial groups are hypersensitive to environmental changes and not able to promptly recuperate after disturbance (Allison and Martiny 2008). In the changed environmental conditions, microbes get acquainted in new conditions by varying resource use efficacy and subsequent growth kinetics as microorganisms have distinctive resistance and acclimatization attribute (Bradford et al. 2008; Schimel et al. 2007). In strained terrestrial surroundings, fungi have superior carbon use competence and C/N biomass stoichiometry yet slower growth rate to facilitate surpassing of adverse conditions, while rapid growth and higher turnover make bacteria better equipped for survival (Waring et al. 2013). These common provisions for resource use and turnover further verify that in soil fungi dominate food webs and hence more resistant to climate changes, whereas bacterial systems are more robust (De Vries et al. 2012a, b). Nonetheless, there are also substantial differences in the physiologies, adaptive capacities, and resource use of organisms within a given taxonomic group, which will shape community-level responses to climate change.

9.4 Effect of Climate Change on Plant–Microbe Interaction

Climate change leads to alteration in abundance, diversity, proportion, and activity of rhizospheric microorganisms (Drigo et al. 2008). Evidently, soil microbial community composition is immensely affected due to changing environment, but every microbial group responds differently to climate change (Hawkins and Crawford 2018; de Vries et al. 2018). Due to climatic fluctuations, soil microbial population has direct impact on their respective growth and other activities, while indirect changes in the flora lead to alteration in resource availability for microbes. Consequences of change in vegetation composition lead to long-term impact on microbial population, while microbial growth and activities are affected immediately (Sayer et al. 2017). In the event of climatic changes like temperature rise, drought, flooding, and increased CO_2 in the atmosphere, K-strategists or oligotrophic organisms like fungi and gram-positive bacteria fair better due to slow growth rates, thick cell walls, and the ability to form spores. These physiological assets provide stress resistance to these microbial groups when subjected to changed climate conditions (Schimel et al. 2007; De Vries and Shade 2013). Under the influence of changes in favorable environment, these K-strategist microbes trade their resources for defense rather than infusing in growth and acquiring resistance (De Vries and Shade 2013). On the basis of response to changed climate, bacteria can be segregated into two groups: (1) copiotrophic, e.g., *Betaproteobacteria* and *Bacteroidetes*, and (2) oligotrophic, e.g., *Acidobacteria* (Fierer et al. 2007). Copiotrophic bacterial groups cannot face the adverse environmental changes, while oligotrophic organisms thrive even in the antagonistic environmental circumstances (DeAngelis et al. 2015). Although oligotrophic organisms have better survival than copiotrophic organisms under harsh or changed environmental conditions, if environmental conditions are optimized, again copiotrophic organisms can regain their previous opulence (de Vries et al. 2018). Thus, resource accessibility undoubtedly supports the microbial community to overcome environmental disruptions (De Vries et al. 2012; De Vries and Shade 2013).

Despite the fact that impact of climate change on plant physiology and phenology is not yet fully inferred, it is well studied that any change in environment also cause alterations in plant in terms of plant growth, biomass allocation, chlorophyll content, stomatal activity, photosynthetic rate, litter quality, phytohormone concentration, their distribution, and water use efficiency, and all of these may likely influence/disturb soil microbial communities and their synergy with plants (Gutknecht et al. 2012). Plants have detailed metabolic network to dissipate the abiotic stresses (Dubey and Verma 2019). Increment in CO_2 concentrations due to climate changes causes enhanced carbon availability to roots that changes root exudate constitution (Compant et al. 2010). The modifications in plant discharge may alter nutrition availability, change in C/N ratio, availability of chemoattractants or signaling molecules (Kandeler et al. 2006; Haase et al. 2007) and mutually or exclusively these changes can also impact the microbial communities to revise microbe-mediated PSFs (de Vries et al. 2018; Chomel et al. 2019). Transformations transpired in microbial population and plants due to climate change, beneficial microbes might be impaired to colonize the host, or colonizing efficacy of host may also be hampered or totally altered in favor of non-beneficial microbes (Compant et al. 2010).

Since climate has major influence on growth, development, and distribution of species, any change in climatic condition has direct and potent impact on interactions amid plant and microbial communities. Pinpointing the changes in microbial behavior leading to alterations in plant–microbe interactions such as availability of nutrient supplying saprophytic taxa to plants or changed ratio of mutualistic to pathogenic taxa is difficult, but such climate change-induced adaptations in microbial activities cause major shifts and deviations in PSFs (Allison et al. 2010; van der Putten et al. 2016). Evidently, biodiversity of soil microbes and their function are intensely linked to climate and climate change, for example, grasslands' rhizospheric microflora is prominently associated with timing and intensity of rainfall (Sheik et al. 2011) and drought influences the distribution of soil fungal and bacterial communities (Meisner et al. 2018). Studies have indicated that microbial metabolic functions including microbial gene expressions are too connected with climate (Bahram et al. 2018).

Microorganism-mediated plant–soil feedbacks (PSFs) can have a significant impact due to the excessive change in climate, e.g., prolonged water scarcity causes adjustments in prevailing microbial communities which affect invasiveness and interactions among plant–plant and plants–microbes (Meisner et al. 2013; Kaisermann et al. 2017; Fry et al. 2018). Drought-adapted soil microflora not only enhances plant acclimatization toward water deficiency (Lau and Lennon 2012) but also instigates colonization and symbiotic interaction with drought tolerance promoting fungi and bacteria (Mariotte et al. 2017) and sways the PSFs. Abiotic stress like high temperature induces alterations with mutualistic symbionts and also causes loss of connection with host-specific soil communities, together with pathogenic microbes and plants (van der Putten et al. 2010).

However, PSFs' shifting microbial activities are unique for each ecosystem that reflect all involved components, viz., plants, soils, and climate. Climate change has robust unwavering effects on plants and soil organisms, as well as secondary effects through changes in plant physiology and the quality and quantity of resources entering the soil (i.e., as litter and rhizodeposits). It is not easy to predict fate of the plant–microbe interaction with change in climate, but consequential PSF change will occur due to various processes of ecosystem and altered dynamics of vegetation along with feedback effects to climate at local and even global levels. Any irreversible alteration in plant–microbe interactions caused by climate change is disastrous for the global ecosystem.

References

Abatenh E, Gizaw B, Tsegay Z, Tefera G, Aynalem E (2018) Health benefits of probiotics. J Bacteriol Infec Dis 2(1):8–27

Allison SD, Martiny JBH (2008) Resistance, resilience, and redundancy in microbial communities. Proc Natl Acad Sci USA 105:11512–11519

Allison SD, Wallenstein MD, Bradford MA (2010) Soil-carbon response to warming dependent on microbial physiology. Nat Geo Sci 3:336–340

Archana DS, Nandishm MS, Savalagi VP, Alagawadi AR (2012) Screening of potassium solubilizing bacteria (KSB) for plant growth promotional activity. Bioinfolet-A Q J Life Sci 9:627–630

Bahram M, Hildebrand F, Forslund SK, Anderson JL, Soudzilovskaia NA, Bodegom PM, Bengtsson-Palme J, Anslan S, Coelho LP, Harend H, Huerta-Cepas J, Medema MH, Maltz MR, Mundra S, Olsson PA, Pent M, Põlme S, Sunagawa S, Ryberg M, Tedersoo L, Bork P (2018) Structure and function of the global topsoil microbiome. Nature 560:233–237

Bais HP, Park SW, Weir TL, Callaway RM, Vivanco JM (2004) How plants communicate using the underground information superhighway. Trends Plant Sci 9:26–32

Bardgett RB, Freeman C, Ostle NJ (2008) Microbial contributions to climate change through carbon cycle feedbacks. ISME J 2:805–814

Bashan Y, Holguin G (1998) Proposal for the division of plant growth-promoting rhizobacteria into two classifications: biocontrol-PGPB (plant growth-promoting bacteria) and PGPB. Soil Biol Biochem 30:1225–1228

Beneduzi A, Ambrosini A, Passaglia LMP (2012) Plant growth-promoting rhizobacteria (PGPR): their potential as antagonists and biocontrol agents. Genet Mol Biol 35:1044–1051

Bennett JA, Klironomos J (2019) Mechanisms of plant-soil feedback: interactions among biotic and abiotic drivers. New Phytol 222:91–96

Bent E (2006) Induced systemic resistance mediated by plant growth-promoting rhizobacteria (PGPR) and fungi (PGPF). In: Tuzun S, Bent E (eds) Multigenic and induced systemic resistance in plants. Springer, Berlin, pp 225–258

Bethlenfalvay GJ, Brown MS, Pacovsky RS (1982) Relationships between host and endophyte development in mycorrhizal soybean. New Phytol 90:537–543

Bever JD (1994) Feedback between plants and their soil communities in an old field community. Ecology 75:1965–1977

Bever JD (1999) Dynamics within mutualism and the maintenance of diversity: inference from a model of interguild frequency dependence. Ecol Lett 2:52–61

Bever JD (2002) Negative feedback within a mutualism: host-specific growth of mycorrhizal fungi reduces plant benefit. Proc R Soc Lond 269:2595–2601

Bever JD, Westover KM, Antonovics J (1997) Incorporating the soil community into plant population dynamics: the utility of the feedback approach. J Ecol 85:561–573

Bever JD, Platt TG, Morton ER (2012) Microbial population and community dynamics on plant roots and their feedbacks on plant communities. Annu Rev Microbiol 66:265–283

Bever JD, Mangan SA, Alexander HM (2015) Maintenance of plant species diversity by pathogens. Annu Rev Ecol Evol Syst 46:305–325

Bhattacharyya PN, Jha DK (2012) Plant growth-promoting rhizobacteria (PGPR): emergence in agriculture. World J Microbiol Biotechnol 28:1327–1350

Binyamin R, Nadeem SM, Akhtar S, Khan MY, Anjum R (2019) Beneficial and pathogenic plant-microbe interactions: a review. Soil Environ 38:11–33

Bradford MA, Davies CA, Frey SD, Maddox TR, Melillo JM, Mohan JE, Reynolds JF, Treseder KK, Wallenstein MD (2008) Thermal adaptation of soil microbial respiration to elevated temperature. Ecol Lett 11:1316–1327

Brundrett MC (2009) Mycorrhizal associations and other means of nutrition of vascular plants: understanding the global diversity of host plants by resolving conflicting information and developing reliable means of diagnosis. Plant Soil 320:37–77

Burns RG, DeForest JL, Marxsen J, Sinsabaugh RL, Stromberger ME, Wallenstein MD, Weintraub MN, Zoppini A (2013) Soil enzymes in a changing: current knowledge and future directions. Soil Biol Biochem 58:216–234

Carter DR, Slesak RA, Harrington TB, Peter DH, D'Amato AW (2019) Scotch broom (*Cytisus scoparius*) modifies microenvironment to promote nonnative plant communities. Biol Invasions 2:1055–1073

Castro HF, Classen AT, Austin EE, Norby RJ, Schadt CW (2010) Soil microbial community responses to multiple experimental climate change drivers. Appl Env Microbiol 76:999–1007

Chomel M, Lavallee JM, Alvarez-Segura N, de Castro F, Rhymes JM, Caruso T, de Vries FT, Baggs EM, Emmerson MC, Bardgett RD, Johnson D (2019) Drought decreases incorporation of recent plant photosynthate into soil food webs regardless of their trophic complexity. Glob Change Biol 25:3549–3561

Compant S, van der Heijden MGA, Sessitsch A (2010) Climate change effects on beneficial plant – microorganism interactions. FEMS Microbiol Ecol 73:197–214

Cortois R, Schröder-Georgi T, Weigelt A, van der Putten WH, De Deyn GB (2016) Plant–soil feedbacks: role of plant functional group and plant traits. J Ecol 104:1608–1617

de Bary AH (1879) De la symbiose. Rev Int Sci 3:301–309

de Vries FT, Griffiths RI (2018) Impacts of climate change on soil microbial communities and their functioning. In: Horwath WR, Kuzyakov Y (eds) Climate change impacts on soil processes and ecosystem properties. Elsevier, The Hague, pp 111–129

de Vries FT, Shade A (2013) Controls on soil microbial community stability under climate change. Front Microbiol 4:265

de Vries FT, Liiri M, Bjørnlund L, Setälä H, Christensen S, Bardgett RD (2012) Legacy effects of drought on plant growth and the soil food web. Oecologia 170:821–833

de Vries FT, Liiri ME, Bjørnlund L, Bowker MA, Christensen S, Setälä HM, Bardgett RD (2012a) Land use alters the resistance and resilience of soil food webs to drought. Nat Clim Chang 2:276–280

de Vries FT, Manning P, Tallowin JRB, Mortimer SR, Pilgrim ES, Harrison KA, Hobbs PJ, Quirk H, Shipley B, Cornelissen JHC, Kattge J, Bardgett RD (2012b) Abiotic drivers and plant traits explain landscape-scale patterns in soil microbial communities. Ecol Lett 15:1230–1239

de Vries FT, Griffiths RI, Bailey M, Craig H, Girlanda M, Gweon HS, Hallin S, Kaisermann A, Keith AM, Kretzschmar M, Lemanceau P, Lumini E, Mason KE, Oliver A, Ostle N, Prosser JI,

Thion C, Thomson B, Bardgett RD (2018) Soil bacterial networks are less stable under drought than fungal networks. Nat Commun 9:3033
de Zelicourt A, Al-Yousif M, Hirt H (2013) Rhizosphere microbes as essential partners for plant stress tolerance. Mol Plant 6:242–245
DeAngelis KM, Pold G, Topcuoglu BD, van Diepen LTA, Varney RM, Blanchard JL, Melillo J, Frey SD (2015) Long-term forest soil warming alters microbial communities in temperate forest soils. Front Microbiol 6
Ding LJ, Cui HL, Nie SA, Long XE, Duan GL, Zhu YG (2019) Microbiomes inhabiting rice roots and rhizosphere. FEMS Microbiol Ecol 95. https://doi.org/10.1093/femsec/fiz040
Drigo B, Kowalchuk GA, van Veen JA (2008) Climate change goes underground: effects of elevated atmospheric CO_2 on microbial community structure and activities in the rhizosphere. Biol Fert Soils 44:667–679
Dubey A, Verma AK (2019) Salinity stress-dependent coordination of metabolic networks in relation to salt tolerance in plants. In: Giri B, Varma A (eds) Microorganisms in saline environments: strategies and functions. Springer Nature, Cham, p 466
Ehrenfeld JG, Ravit B, Elgersma K (2005) Feedback in the plant-soil system. Annu Rev Environ Resour 30:75–115
Endeshaw A, Birhanu G, Zerihun T, Genene T (2018) Microbial function on climate change – a review. Open J Environ Biol 3:001–007
Eom AH, Hartnett DC, Wilson GWT (2000) Host plant species effects on arbuscular mycorrhizal fungal communities in tallgrass prairie. Oecologia 122:435–444
Fierer N, Schimel JP (2003) A proposed mechanism for the pulse in carbon dioxide production commonly observed following the rapid rewetting of a dry soil. Soil Sci Soc Am J 67:798–805
Fierer N, Bradford MA, Jackson RB (2007) Toward an ecological classification of soil bacteria. Ecology 88:1354–1364
Fry EL, Johnson GN, Hall AL, Pritchard WJ, Bullock JM, Bardgett RD (2018) Drought neutralises plant–soil feedback of two mesic grassland forbs. Oecologia 186:1113–1125
Glick BR (2014) Bacteria with ACC deaminase can promote plant growth and help to feed the world. Microbiol Res 169:30–39
Glick BR, Todorovic B, Czarny J, Cheng Z, Duan J, McConkey B (2007) Promotion of plant growth by bacterial ACC deaminase. Crit Rev Plant Sci 26:227–242
Gutknecht JL, Field CB, Balser TC (2012) Microbial communities and their responses to simulated global change fluctuate greatly over multiple years. Glob Chang Biol 18:2256–2269
Haase S, Neumann G, Kania A, Kuzyakov Y, Römheld V, Kandeler E (2007) Elevation of atmospheric CO_2 and nutritional status modify nodulation, nodule-carbon supply, and root exudation of *Phaseolus vulgaris* L. Soil Biol Biochem 39:2208–2222
Harte J, Kinzig AP (1993) Mutualism and competition between plants and decomposers: implications for nutrient allocation in ecosystems. Am Nat 141:829–846
Hawkins AP, Crawford KM (2018) Interactions between plants and soil microbes may alter the relative importance of intraspecific and interspecific plant competition in a changing climate. AoB Plants 10(4):ply039. https://doi.org/10.1093/aobpla/ply039
Hodge A, Robinson D, Fitter A (2000) Are microorganisms more effective than plants at competing for nitrogen? Trends Plant Sci 5:304–308
Kaisermann A, de Vries FT, Griffiths RI, Bardgett RD (2017) Legacy effects of drought on plant–soil feedbacks and plant–plant interactions. New Phytol 215:1413–1424
Kandeler E, Mosier AR, Morgan JA, Milchunas DG, King JY, Rudolph S, Tscherko D (2006) Response of soil microbial biomass and enzyme activities to the transient elevation of carbon dioxide in a semi-arid grassland. Soil Biol Biochem 38:2448–2460
Kloepper JW, Schroth MN (1978) Plant growth-promoting rhizobacteria on radishes. In: Station de Pathologie Végétale et Phyto-Bactériologie (ed) Proceedings of the 4th international conference on plant pathogenic bacteria, vol II. Gilbert-Clarey, Tours, pp 879–882
Kulmatiski A, Beard KH, Stevens JR, Cobbold SM (2008) Plant-soil feedbacks: a metaanalytical review. Ecol Lett 11:980–992
Lau JA, Lennon JT (2012) Rapid responses of soil microorganisms improve plant fitness in novel environments. Proc Natl Acad Sci USA 109:14058–14062

Lavelle P, Spain AV (2001) Soil ecology. Kluwer Academic, Boston

Lee MH, Chiu-Min C, Tatiana R, Chien-Ming C, Richard MB (2010) Over expression of a redox-regulated cutinase gene, MfCUT1, increases virulence of the brown rot pathogen *Monilinia fructicola* on Prunus spp. Mol Plant Microbe Interact 23:176–186

Lugtenberg B, Kamilova F (2009) Plant-growth-promoting rhizobacteria. Annu Rev Microbiol 63:541–556

Mariotte P, Canarini A, Dijkstra FA (2017) Stoichiometric N:P flexibility and mycorrhizal symbiosis favour plant resistance against drought. J Ecol 105:958–967

Maron JL, Marler M, Klironomos JN, Cleveland CC (2011) Soil fungal pathogens and the relationship between plant diversity and productivity. Ecol Lett 14:36–41

Mehmood U, Inam-ul-Haq M, Saeed M, Altaf A, Azam F, Hayat S (2018) A brief review on plant growth promoting Rhizobacteria (PGPR): a key role in plant growth promotion. Plant Prot 2:77–82

Meisner A, De Deyn GB, de Boer W, van der Putten WH (2013) Soil biotic legacy effects of extreme weather events influence plant invasiveness. Proc Natl Acad Sci USA 110:9835–9838

Meisner A, Jacquiod S, Snoek BL, ten Hooven FC, van der Putten WH (2018) Drought legacy effects on the composition of soil fungal and prokaryote communities. Front Microbiol 9:294

Mommer L, Cotton TEA, Raaijmakers JM, Termorshuizen AJ, van Ruijven J, Hendriks M, van Rijssel SQ, van de Mortel JE, van der Paauw JW, Schijlen EGWM, Smit-Tiekstra AE, Berendse F, de Kroon H, Dumbrell AJ (2018) Lost in diversity: the interactions between soil-borne fungi, biodiversity and plant productivity. New Phytol 218:542–553

Nadeem SM, Zahir ZA, Naveed M, Asghar HN, Arshad M (2010a) Rhizobacteria capable of producing ACC-deaminase may mitigate salt stress in wheat. Soil Sci Soc Am J 74:533–542

Nadeem SM, Zahir ZA, Naveed M, Ashraf M (2010b) Microbial ACC deaminase: prospects and applications for inducing salt tolerance in plants. Crit Rev Plant Sci 29:360–393

Nadeem SM, Naveed M, Zahir ZA, Asghar HN (2013) Plant-microbe interactions for sustainable agriculture: fundamentals and recent advances. In: Arora N (ed) Plant microbe symbiosis: fundamentals and advances. Springer, Delhi, pp 53–103

Panhwar QA, Naher UA, Shamshuddin J, Othman R, Latif MA, Ismail MR (2014) Biochemical and molecular characterization of potential phosphate-solubilizing bacteria in acid sulfate soils and their beneficial effects on rice growth. PLoS One 9:e97241

Pieterse CMJ, Zamioudis C, Berendsen RL, Weller DM, Van Wees SCM, Bakker PAHM (2014) Induced systemic resistance by beneficial microbes. Annu Rev Phytopathol 52:347–375

Pii Y, Mimmo T, Tomasi N, Terzano R, Cesco S, Crecchio C (2015) Microbial interactions in the rhizosphere: beneficial influences of plant growth promoting rhizobacteria on nutrient acquisition process. A review. Biol Fertil Soils 51:403–415

Podile AR, Kishore GK (2006) Plant growth promoting rhizobacteria. In: Gnanamanickam SS (ed) Plant associated bacteria. Springer, Dordrecht, pp 195–230

Pugnaire FI, Morillo JA, Peñuelas J, Reich PB, Bardgett RD, Gaxiola A, Wardle DA, van der Putten WH (2019) Climate change effects on plant-soil feedbacks and consequences for biodiversity and functioning of terrestrial ecosystems. Sci Adv 5:eaaz1834

Reynolds HL, Haubensak KA (2008) Soil fertility, heterogeneity, and microbes: towards an integrated understanding of grassland structure and dynamics. Appl Veg Sci 12:33–44

Reynolds HL, Packer A, Bever JD, Clay K (2003) Grassroots ecology: plant-microbe-soil interactions as drivers of plant community structure and dynamics. Ecology 84:2281–2291

Romera FJ, García MJ, Carlos L, Ainhoa M-M, Aparicio MA, José R, Esteban A, Macarena A, Rafael P-V (2019) Induced systemic resistance (ISR) and Fe deficiency responses in dicot plants. Front Plant Sci 10:287

Saharan B, Nehra V (2011) Plant growth promoting rhizobacteria: a critical review. Int J Stat Med Res 21:1–30

Sandhya V, Ali SKZ, Grover M, Reddy G, Venkateswarlu B (2009) Alleviation of drought stress effects in sunflower seedlings by the exopolysaccharides producing Pseudomonas putida strain GAP-P45. Biol Fertil Soils 46:17–26

Saravana KD, Samiyappan R (2007) ACC deaminase from *Pseudomonas fluorescens* mediated saline resistance in groundnut (Arachis hypogea) plants. J Appl Microbiol 102:1283–1292
Sayer EJ, Oliver AE, Fridley JD, Askew AP, Mills RTE, Grime JP (2017) Links between soil microbial communities and plant traits in a species-rich grassland under long-term climate change. Ecol Evol 7:855–862
Schikora A (2018) Plant-microbe interactions in the rhizosphere. Caister Academic, Braunschweig
Schimel J (1995) Ecosystem consequences of microbial diversity and community structure. In: Chapin F, Korner C (eds) Arctic and alpine biodiversity: patterns, causes, and ecosystem consequences. Springer, Berlin, pp 239–254
Schimel J, Balser TC, Wallenstein M (2007) Microbial stress response physiology and its implications for ecosystem function. Ecology 88:1386–1394
Schlesinger WH (1991) Biogeochemistry: an analysis of global change. Academic, San Diego, 443 pp
Sheik CS, Beasley WH, Elshahed MS, Zhou X, Luo Y, Krumholz LR (2011) Effect of warming and drought on grassland microbial communities. ISME J 5:1692–1700
Siefert A, Zillig KW, Friesen ML, Strauss SY (2019) Mutualists stabilize the coexistence of congeneric legumes. Am Nat 193:200–212
Stone JK, Bacon CW, White JF (2000) An overview of endophytic microbes: endophytism defined. In: Bacon CW, White JF (eds) Microbial endophytes. Marcel Dekker, New York, pp 3–29
Torsvik V, Goksøyr J, Daae FL (1990) High diversity in DNA of soil bacteria. Appl Environ Microbiol 56:782–787
Tyagi S, Singh R, Javeria S (2014) Effect of climate change on plant-microbe interaction: an overview. Eur J Mol Biotechnol 5:149–156
van der Heijden MGA, Klironomos JN, Ursic M, Moutoglis P, Streitwolf-Engel R, Boller T, Wiemken A, Sanders IR (1998) Mycorrhizal fungal diversity determines plant biodiversity, ecosystem variability and productivity. Nature 396:69–72
van der Heijden MGA, Bardgett RD, van Straalen NM (2008) The unseen majority: soil microbes as drivers of plant diversity and productivity in terrestrial ecosystems. Ecol Lett 11:296–310
van der Putten WH, Macel M, Visser ME (2010) Predicting species distribution and abundance responses to climate change: why it is essential to include biotic interactions across trophic levels. Philos Trans R Soc B Biol Sci 365:2025–2034
van der Putten WH, Bardgett RD, Bever JD, Bezemer TM, Casper BB, Fukami T, Kardol P, Klironomos JN, Kulmatiski A, Schweitzer JA, Suding KN, van de Voorde TFJ, Wardle DA (2013) Plant-soil feedbacks: the past, the present and future challenges. J Ecol 101:265–276
van der Putten WH, Bradford MA, Brinkman EP, van de Voorde TFJ, Veen GF (2016) Where, when and how plant–soil feedback matters in a changing world. Funct Ecol 30:1109–1121
Van Loon LC (2007) Plant responses to plant growth-promoting rhizobacteria. Eur J Plant Pathol 119:243–254
Veen GF, Freschet GT, Ordonez A, Wardle DA (2015) Litter quality and environmental controls of home-field advantage effects on litter decomposition. Oikos 124:187–195
Verbon EH, Liberman LM (2016) Beneficial microbes affect endogenous mechanisms controlling root development. Trends Plant Sci 21:218–229
Wagg C, Boller B, Schneider S, Widmer F, van der Heijden MGA (2015) Intraspecific and intergenerational differences in plant-soil feedbacks. Oikos 124:994–1004
Wallenstein MD, Hall EK (2012) A trait-based framework for predicting when and where microbial adaptation to climate change will affect ecosystem functioning. Biogeochemistry 109:35–47
Waring BG, Averill C, Hawkes CV (2013) Differences in fungal and bacterial physiology alter soil carbon and nitrogen cycling: insights from meta-analysis and theoretical models. Ecol Lett 16:887–894
Weiman S (2015) Microbes help to drive global carbon cycling and climate change. Microbe Mag 10:233–238

Whipps JM (1990) Carbon economy. In: Lynch JM (ed) The rhizosphere. Wiley, Chichester, pp 59–99

Wille L, Messmer MM, Studer B, Hohmann P (2019) Insights to plant–microbe interactions provide opportunities to improve resistance breeding against root diseases in grain legumes. Plant Cell Environ 42:20–40

Yang J, Kloepper JW, Ryu CM (2008) Rhizosphere bacteria help plants tolerate abiotic stress. Trends Plant Sci 14:1–4

Yergeau E, Bokhorst S, Kang S, Zhou J, Greer CW, Aerts R, Kowalchuk GA (2011) Shifts in soil microorganisms in response to warming are consistent across a range of Antarctic environments. ISME J 6:692–702

Chapter 10
Microbial Sequestration of Atmospheric Carbon Dioxide

Manju Rawat Ranjan, Pallavi Bhardwaj, and Ashutosh Tripathi

Abstract Increased industrialization and urbanization have made humans rely on the usage of fossil fuels which leads to the generation of carbon dioxide (CO_2) and causes global warming. The concentration of greenhouse gases majorly carbon dioxide is rapidly increasing in the atmosphere because of various anthropogenic as well as natural activities. The loss of carbon from the terrestrial ecosystem into the atmosphere has both short- and long-term negative effects on our planet earth. Carbon cycling in a terrestrial as well as an aquatic ecosystem is a major concern in climate change; hence, it is required to comprehend the role of biota in the management of carbon fixation and its cycling. This chapter mainly focuses on the potential of different organisms that have the capability of CO_2 assimilation. With this, the process by which organism proceeds sequestration of CO_2 is explained in brief. Different types of microorganisms and their adaptations for effective CO_2 sequestration are also highlighted. It is important to implement biological methods to manage CO_2 because they synthesize useful by-products such as bioplastics, medicinal compounds, and biofuels.

The current chapter will help us to understand how biological diverse organisms can work as a sustainable technology to mitigate the increasing CO_2 levels in the atmosphere.

Keywords CO_2 sequestration · Assimilation · Climate change · Microorganisms · Greenhouse gases · Carbon cycling

M. R. Ranjan (✉) · P. Bhardwaj · A. Tripathi
Amity Institute of Environmental Sciences (AIES), Sector-125, Amity University, Noida, Uttar Pradesh, India
e-mail: mrranjan@amity.edu

D. K. Choudhary et al. (eds.), *Climate Change and the Microbiome*, Soil Biology 63,
https://doi.org/10.1007/978-3-030-76863-8_10

10.1 Introduction

The present human-induced change in climate globally increases the greenhouse gases which has become one of the biggest scientific challenges of the twenty-first century. An increase in the concentration of greenhouse gas like carbon dioxide (CO_2) in the atmosphere is a major concern. According to the Intergovernmental Panel on Climate Change (IPCC), it has been estimated that carbon dioxide is a potent greenhouse gas, accounting for 76.7% (v/v) of the total GHG emissions, and its concentration has increased exponentially since the onset of industrialization (Ramanathan 1988). Concentrations of CO_2 which started from 280 parts per million (ppm) during pre-industrial times have now exceeded 400 ppm (http://www.esrl.noaa.gov/gmd/ccgg/trends) and are expected to reach up to 600–800 ppm by the end of the century (Knohl and Veldkamp 2011). The world's maximum greenhouse gas emissions come from small countries, whereas China, the United States, and the European Union are the three largest emitters. Global carbon dioxide emissions since 1850 are shown in Fig. 10.1. Per capita greenhouse gas emissions are highest in the United States and Russia, and the share of each country is described in Fig. 10.2 provided by the International Energy Agency, after estimating the sources of CO_2 emissions which include coal combustion; natural gas, oil, and other fuels; industrial waste, and nonrenewable municipal waste.

The major sources of CO_2 emissions are fossil fuel burning and higher deforestation rate which directly increases the concentration of atmospheric CO_2. As per Global Carbon Budget 2018, global fossil CO_2 emissions have increased from an average of 3.1 ± 0.2 GtC year^{-1} (gigatons carbon per year) in the 1960s to an average of 9.4 ± 0.5 GtC year^{-1} during 2008–2017, whereas CO_2 emissions from land use, land-use change, and forestry have remained relatively constant, at around 1.3 ± 0.7 GtC year^{-1} over the past half-century (Quéré et al. 2018). Along with this, it is important to notice that the rate of fossil fuel burning has not decreased since the 1940s (Bastos et al. 2016).

The increased concentration of carbon dioxide in the atmosphere leads to harmful environmental effects such as global warming and climate change. Some of these observed negative effects are (1) increased frequency and duration of heat waves, (2) warming of ocean and atmosphere, (3) rise in sea level, and (4) decrease in mass of Greenland and Antarctic ice sheets. Such effects can deteriorate human health as we know that biological systems are the most climate sensitive. For example, increased heat wave frequency may result in heat-related illnesses and varying temperatures, and precipitation patterns may cause certain diseases. Thus, the unpredictability of infectious diseases may become the biological expressions of unstable climate (Dev et al. 2019).

That is why it becomes an emergent challenge in front of our society, and an area of interest to reduce atmospheric CO_2 which accounts for 76% of the total greenhouse gas (GHG) emissions (Rossi et al. 2015). Many anthropogenic actions, like fossil fuel combustion, deforestation, forest fire, automobile combustion, etc. increase the CO_2 concentrations. Soil, plant, and the ocean are the major natural

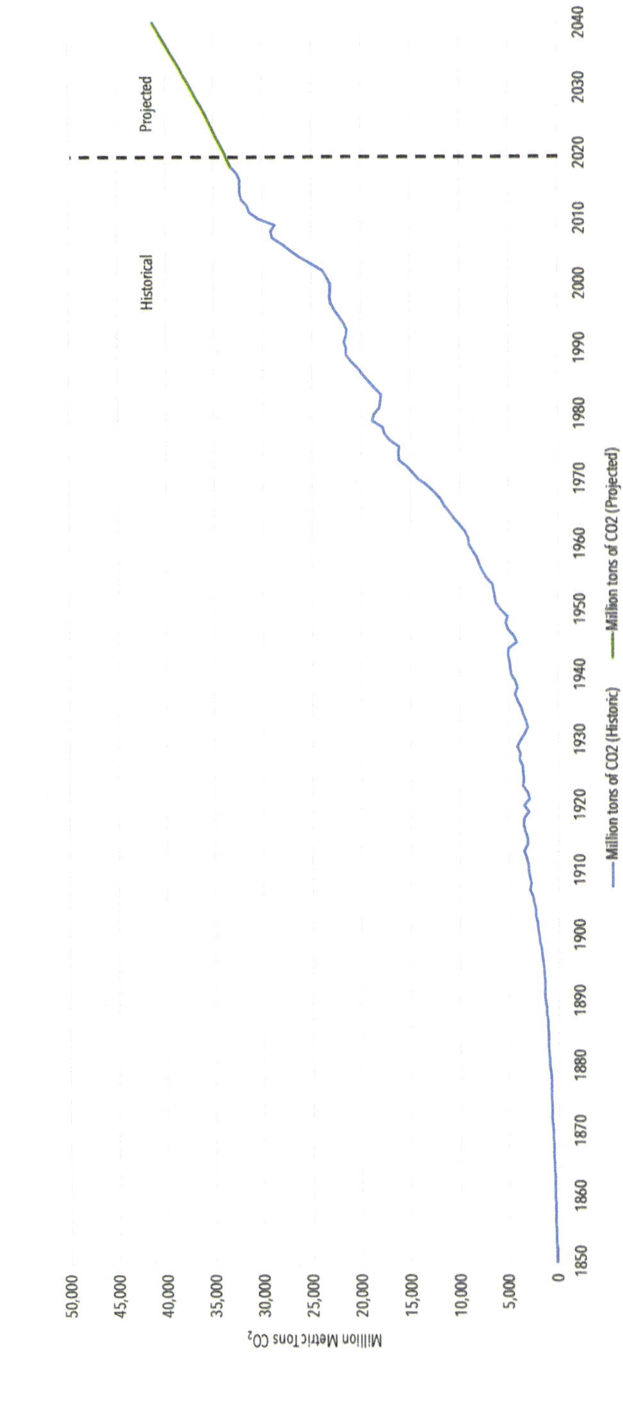

Fig. 10.1 Global carbon dioxide emissions through different years. Source: Carbon dioxide Information analysis Centre (Oak Ridge National Laboratory, 2017) and World Energy outlook (International Energy Agency, 2019)

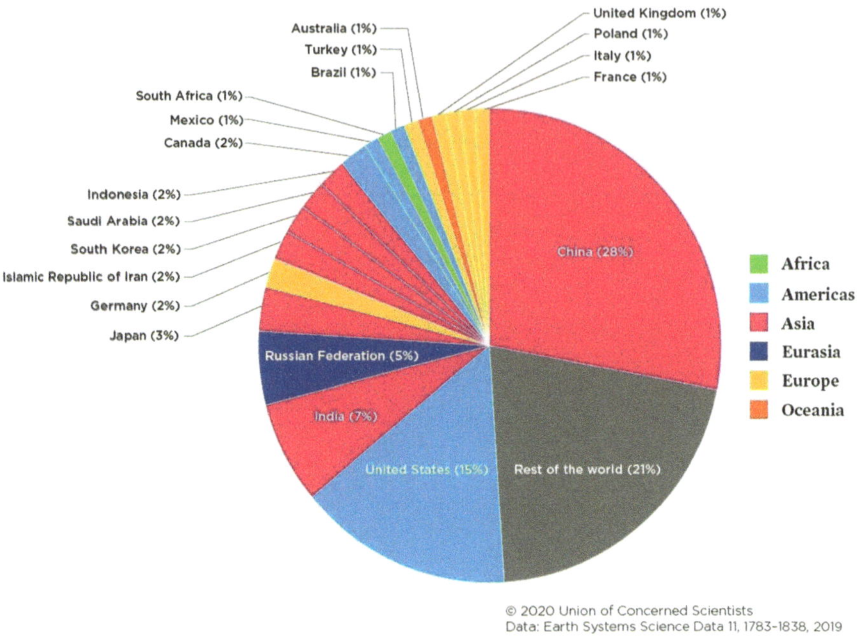

Fig. 10.2 Contribution of individual countries in CO_2 emissions

sinks for atmospheric CO_2. These changing atmospheric levels of CO_2 can be managed using different ways like by increased efficiency of energy conversion technologies and by using energy sources of low carbon content or free of carbon. The scientific community is continuously working on mitigating climate change and global warming by reducing atmospheric CO_2 levels through the process of carbon sequestration (Bhattacharyya et al. 2018). In these ways, we need sustainable and effective technologies to capture, store, and sequester CO_2, so that adverse effects of CO_2 emissions on the environment can be prevented (Ho et al. 2012). One such sustainable and effective approach to balance the levels of CO_2 in the atmosphere is carbon sequestration (Litynski et al. 2006). Even though we have technologies to utilize CO_2 as a chemical feedstock, at the industrial level, it is still limited. Rare industries are capable of utilizing CO_2 for producing various chemicals such as urea (~70 Mt. CO_2 year^{-1}, megatons carbon dioxide per year), inorganic carbonates and pigments (~30 Mt. CO_2 year^{-1}), methanol (~6 Mt. CO_2 year^{-1}), salicylic acid (~20 kt CO_2 year^{-1} (kilotons carbon dioxide per year)) and propylene carbonate (a few kt CO_2 year^{-1}) (Wong 2014). Compared to this, photosynthetic organisms fix around 100 Gt of carbon into biomass annually (Field et al. 1998).

10.2 Overview on Carbon Sequestration

Carbon sequestration is a process that involves capturing and securing storage of atmospheric CO_2 to reduce global warming (Herzog 2001). There are majorly three types of routes to carbon sequestration, i.e. geological, biological, and physiochemical.

10.2.1 Geologic Carbon Sequestration

Geologic carbon sequestration involves capturing of carbon dioxide (CO_2) in geologic formations. In this process, CO_2 is pressurized to convert it into a liquid form, which is further injected into porous rocks in geologic basins. This method of carbon storage helps in enhancing oil recovery. The concept of geologic sequestration can be described in a better way using Fig. 10.3.

10.2.2 Biologic Carbon Sequestration

Geologic carbon sequestration means the storage of atmospheric carbon in vegetation, soils, woody products, and aquatic environments (oceans) by encouraging afforestation or by using certain CO_2 assimilating microbes, etc. Biological (or terrestrial) sequestration involves maximum removal *of atmospheric* CO_2. It is the long-term storage of carbon in soils and vegetation using microbes and is explained in Fig. 10.4.

Removal of CO_2 from atmosphere by increasing its concentration in soils and vegetation by implementing afforestation or in the ocean by proceeding iron fertilization are the forms of carbon sequestration. Geologic storage means deep saline formations (subterranean and sub-seabed), exhausted oil and gas reservoirs, formations for oil recovery operations, and un-minable coal seams. In deep ocean storage, direct injection of liquid carbon dioxide into the water column is performed at intermediate depths (1000–3000 m), or at depths greater than 3000 m, where it becomes heavier than seawater and drops to the ocean bottom forming a "CO_2 lake" (Herzog 2001). With this, the natural process of CO_2 sequestration is mainly performed by photosynthetic organisms present in the terrestrial and aquatic ecosystems and maintains the global carbon cycle (Mistry et al. 2019). Some microorganisms like autotrophic bacteria can fix CO_2 by non-photosynthetic pathway. Few studies have witnessed efficient CO_2 sequestering organisms are present in soil and belowground which can manage atmospheric carbon levels (Cole et al. 2007). Iron fertilization experiments in the Southern Ocean are evident to show that increased net phytoplankton growth has emerged as one method for mitigating rising atmospheric CO_2 levels. Because of the high reproduction and faster growth rates, diverse

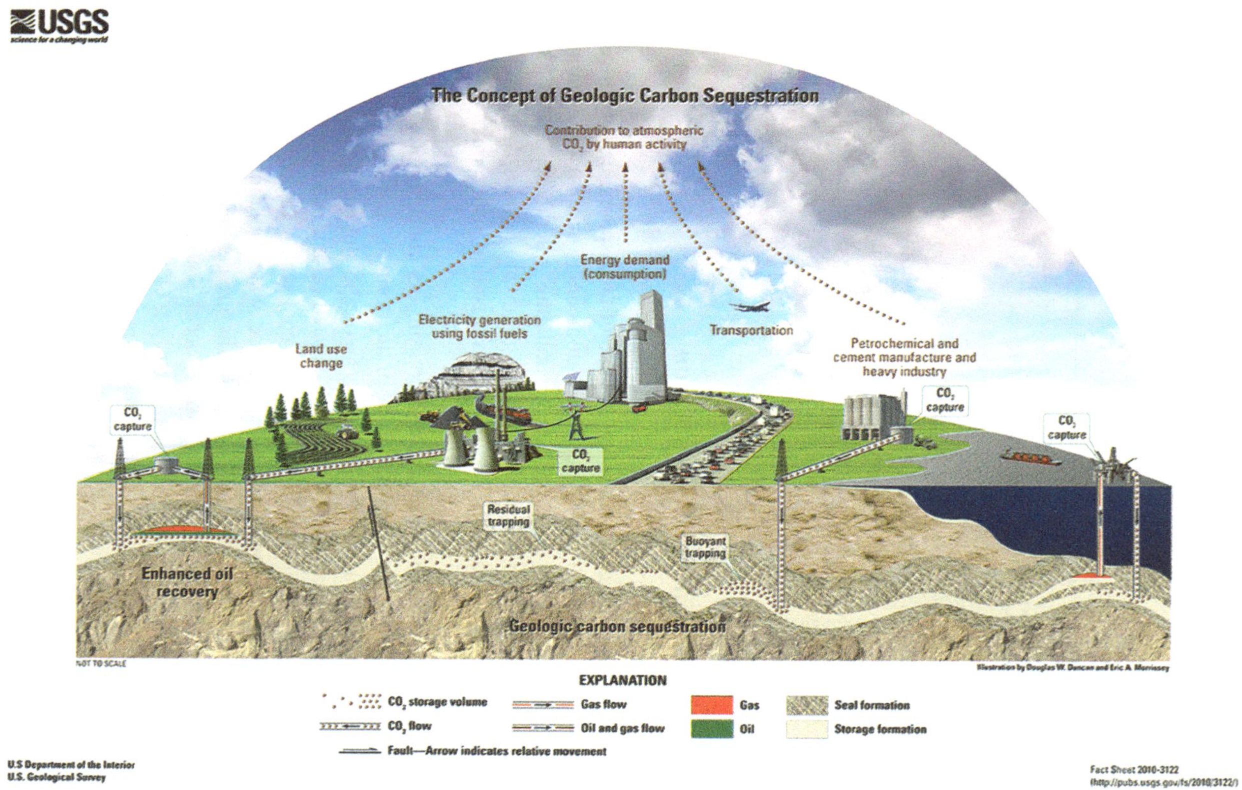

Fig. 10.3 Concept of geologic carbon sequestration in biosphere

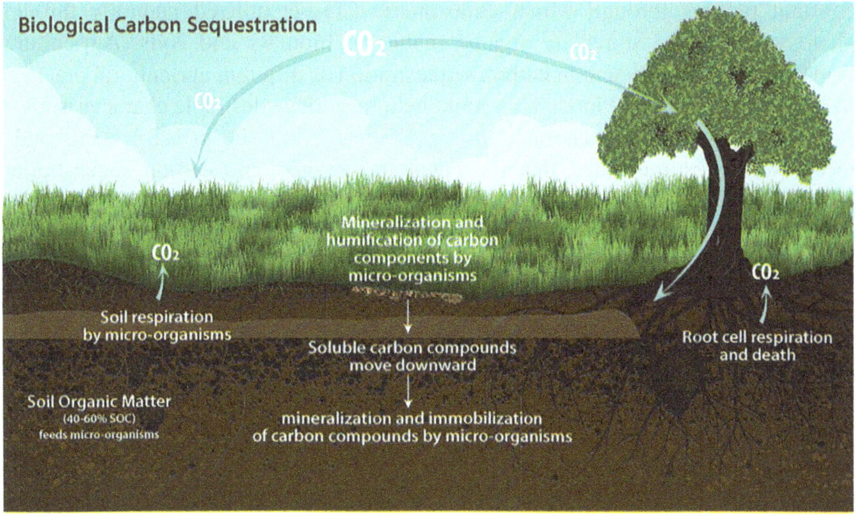

Fig. 10.4 Concept of biological carbon sequestration in environment. Source: Organics. https://www.calrecycle.ca.gov/organics/compostmulch/toolbox/carbonsequest/. Accessed 28 Oct 2020

distribution, and ubiquitous nature, soil microorganisms have the potential to work as bio-monitor to detect the effects of global change on ecosystems (Foissner 1999).

In biological sequestration of CO_2, the organism produces biomolecules such as proteins, lipids, and carbohydrates after consuming CO_2. Terrestrial ecosystems like orchards, forests, and agricultural land help in carbon dioxide sequestration (Guo et al. 2010).

10.3 Terrestrial CO_2 Sequestration

Carbon (C) is a crucial and life-supporting element on the planet. Naturally, it exists in both organic and inorganic forms. About 18% (about 50% on a dry basis), the concentration of carbon in living matter is almost 100 times greater than the average concentration of C in the earth's crust (0.19%) (Kumar et al. 2018). Thus, a continuous carbon cycle should be managed properly so that life can sustain on earth. Different terrestrial ecosystems like forests, orchard, and agricultural fields play different roles in sequestering CO_2 (Mistry et al. 2019). Forests are the most important for balancing levels of carbon. The type of vegetation and present soil varies from forest to forest; thus, the sequestration varies too. For example, as per World Bank, tropical forests have carbon stock of approximately 109 tonnes/acre and wetlands have around 306 tonnes/acre.

Plants sequester atmospheric CO_2 via the process of photosynthesis in the presence of chlorophyll where the conversion of CO_2 to glucose takes place for their growth in the presence of solar energy. This carbon returns to the atmosphere from fixed organic matter as CO_2 by respiration, combustion, and decay. Thus, it is a constant process through which atmospheric CO_2 is utilized naturally through photosynthesis and stored as organic matter in biomass and soils. Agricultural lands can be more efficient in carbon sequestration as different agricultural practices like the use of microbial inoculants could help to achieve desirable characteristics in soil. A variety of microbial communities exist in terrestrial subsurface systems, and a recent study recommended that the energetics in geologic carbon sequestration reservoirs can be advantageous to microbial metabolism under specific conditions (West et al. 2011). Though the "Microbial Carbon Pump" is a well-established theory for marine open ocean environments (Jiao et al. 2010). Bacteria acquire considerable benefits over photosynthetic organisms, e.g., higher plants and microalgae, which have gained the attention to capture CO_2. Keepig this view in mind the role of microbes in the process of atmospheric carbon sequestration is described to use them as a potential approach to mitigate the CO_2 levels in the atmosphere.

10.4 Microbial CO_2 Sequestration

This is also known as biological sequestration where the CO_2 capturing process is enhanced using algae, bacteria, cyanobacteria, and archaea. Microbes help in increasing the carbon inputs and decrease the levels of carbon outputs in the soils. Thus, soil microbial communities have important roles in carbon sequestration and soil carbon emission (Fang et al. 2014). Soil microorganisms choose different mechanisms to contribute to the process of sequestration like possessing metabolic activities resulting in the capture of atmospheric CO_2, the ability to sediment carbonates, forming recalcitrant vegetative tissues and products, and the ability to form stable forms such as soil aggregates that protect carbon soil organic forms (Ahmed et al. 2019). Few studies are evident enough to show that some bacterial strains can reduce CO_2, and they could contribute to carbon sequestration and help in the mitigation of atmospheric CO_2 levels (Nie et al. 2015). Figure 10.5 describes the advantages of using microbes for carbon sequestration. Dissolved inorganic carbon (DIC) which include CO_2 and bicarbonate are vital elements for microorganisms. The CO_2 requirement of a bacterial cell is fulfilled by transporting it into the cell by hydration reaction on the basis of bicarbonate and CO_2 concentration inside and outside the cell. In this process, carbon dioxide gets converted into bicarbonate at a very low rate. That is why being the fastest enzyme, carbonic anhydrase catalyzes the reaction with typical catalytic rates (Heck et al. 1994). The reaction can be read as:

$$CO_2 + H_2O \overset{CA}{\rightarrow} HCO_3^- + H^+$$

Carbonic anhydrase enzyme is stated to be present in animals, plants, and microorganisms, and it existed in them even before the division of Archaea and

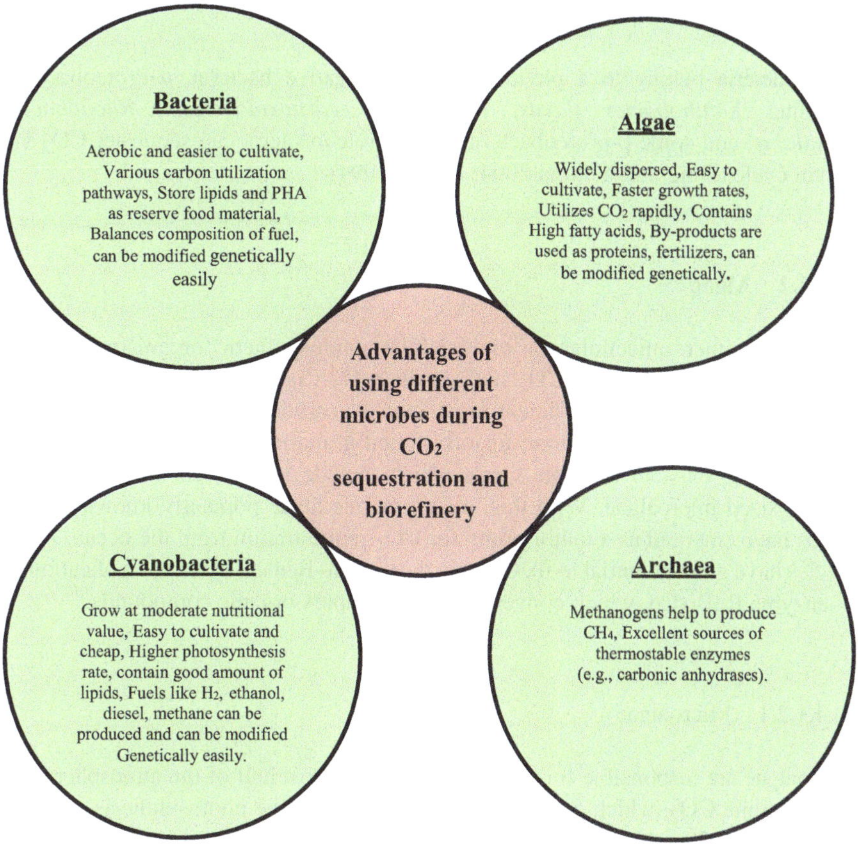

Fig. 10.5 Advantages of using microbes during the sequestration of carbon dioxide and biorefinery

Bacteria (Karlsson et al. 1998). According to Thauer, autotrophic organisms can fix CO_2 in four ways which are:

1. Calvin cycle or Calvin–Benson–Bassham (CBB) pathway or reductive pentose pathway
2. Reductive TCA cycle or reverse citric acid cycle
3. Reductive acetyl CoA pathway or Wood–Ljungdahl pathway
4. 3-Hydroxypropionate cycle

10.4.1 Calvin Cycle/Calvin–Benson–Bassham (CBB) Pathway/Reductive Pentose Pathway

A number of prokaryotic microorganisms perform the Calvin cycle for CO_2 fixation (Saini et al. 2011) using RuBisCO (Bharti et al. 2014). Organisms like plants, algae, cyanobacteria, proteobacteria, and mycobacteria follow this pathway.

10.4.1.1 Proteobacteria

Proteobacteria belong to a phylum of Gram-negative bacteria. α-Proteobacteria including *Xanthobacter flavus*, *Oligotropha carboxidovorans*, *Rhodobacter capsulatus*, and some β-proteobacteria and γ-proteobacteria can sequester CO_2 via Calvin cycle (Ding and Yokota 2004; Meijer 1994).

10.4.1.2 Algae

Algae are either unicellular or multicellular photosynthetic organisms and are primary producers that fix CO_2 and produce O_2 (Sahoo and Elangbam 2012). These are one of the most efficient organisms for carbon sequestration. They exist throughout the biosphere but are majorly found in marine and freshwater environments. Algae exist in multiple forms which include large-sized macroalgae and smaller-sized microalgae. With this, macro marine algae popularly known as seaweeds have emerged as a major group for CO_2 sequestration from the ocean. Algal species have great potential to fix CO_2 via the Calvin–Benson cycle by application of an enzyme RuBisCo, which converts CO_2 to complex organic compounds.

10.4.1.2.1 Microalgae

Microalgae are responsible for the production of almost half of the atmospheric O_2 and consume CO_2, which accounts for almost 50% of the photosynthesis on earth (Paul et al. 2020). About 1.0 kg of cultivated microalgae can fix 1.83 kg of CO_2 (Cheah et al. 2015). Microalgae species such as *Anabaena* sp. and *Chlorella vulgaris* can fix CO_2 at rates of 1.45 and 6.24 g/L/d, respectively (Ghorbani et al. 2014). According to Beardall and Raven (2004), marine photosynthesis contributes 54–59 Pg C $year^{-1}$ (picogram carbon per year) of the total primary productivity of the planet, and out of this, ~1 Pg C $year^{-1}$ is contributed by seaweeds and sea grasses (Beardall and Raven 2004). CO_2 fixation by algae is more advantageous over terrestrial plants. CO_2 captured as bicarbonates by sparging in water profusely gets utilized by aquatic microalgae to produce biomass. Algae have a better CO_2-concentrating mechanism, a higher CO_2 fixation efficiency, and a higher growth rate (Zhang 2011).

10.4.1.2.2 Marine Macroalgae

Marine macroalgae such as *Macrocystis*, *Laminaria*, *Sargassum*, *Ascophyllum*, *Fucus*, *Porphyra*, *Palmaria*, *Ulva*, and *Enteromorpha* also achieve high rates of CO_2 assimilation per gram fresh weight (Gao and McKinley 1994). Macroalgae can incorporate an average of 0.26 × 106 tonnes C into the harvested algae annually (Chung et al. 2011); thus, seaweeds also have a good potential in capturing carbon.

10.4.2 Reductive TCA Cycle or Reverse Citric Acid Cycle

This pathway is used as an alternative pathway to the Calvin cycle and works in the reductive direction of the TCA cycle. Microorganisms in extreme conditions, such as high temperature, anaerobic conditions, and acidic conditions, usually acquire this way to fix CO_2. Microorganisms like proteobacteria and green sulfur bacteria follow this pathway.

10.4.3 Wood–Ljungdahl Pathway

This is also known as the reductive acetyl-CoA pathway used by microbes to fix CO_2. Ragsdale and Wood in 1985 stated that in this metabolic activity, autotrophic acetogens and methanogens synthesize acetic acid and methane from CO_2. Anaerobic organisms including *Proteobacteria*, *Planctomycetes*, *Spirochaetes*, and Euryarchaeota use this pathway. In anaerobic conditions, Archaea like *Methanobrevibacter aboriphilus*, *Methanothermobacter thermautotrophicus*, and *Methanosarcina barkeri* act as methanogens and produce methane by consummating CO_2 as a carbon source and H_2 as an energy source (Rittmann et al. 2015). Clostridia, predominantly *Clostridium* spp., being an obligate anaerobe and gram-positive bacteria balances the carbon cycle and can fix CO_2 as well as CO. For example, *Clostridium autoethanogenum* captures CO_2 (in the presence of H_2) and CO into central metabolite acetyl-CoA via the Wood–Ljungdahl pathway (Liew et al. 2016). This pathway is the most efficient thermodynamically in synthesizing acetate and is the only linear pathway for CO_2 fixation (Fast and Papoutsakis 2012).

10.4.4 Hydroxypropionate Cycle

Some green non-sulfur bacteria of the family *Chloroflexaceae* have found to use this pathway. Archaea, the single-celled prokaryotic microorganisms, are termed as extremophiles because they are found in extreme conditions like acidic environments, high salinity, anaerobic environment, and high and low temperatures. *Metallosphaera*, *Sulfolobus*, and *Cenarchaeum* spp. are such members of autotrophic archaea that can sequester CO_2 by the 3-hydroxypropionate-4-hydroxybutyrate cycle (Berg et al. 2007).

Few more examples are *Pseudomonas fluorescens* which being a plant growth-promoting bacterium increases the productivity of plants and has a high potential to increase terrestrial carbon sequestration (Pemencilan et al. 2014). *Bacillus mucilaginosus* produces carbonic anhydrase which captures the atmospheric CO_2 and fixes it through bacterial metabolism and forms carbonate to form carbonated minerals which lead to decreased atmospheric CO_2 levels (Han et al. 2013). Bacterial strains like *Bacillus pumilus* (Stocks-Fischer et al. 1999) *Bacillus cereus* (Li et al. 2015), and *Bacillus pasteurii* (Ekblad et al. 2013) are known to reduce CO_2 as they can sediment carbonates and can contribute to the carbon cycle and sequestration.

10.5 Fungi and Carbon Sequestration

It is observed from previous studies that fungi contribute more than bacteria in carbon sequestration (Espinoza Pérez et al. 2017). The photosynthetic rates and respiratory losses of the autotrophs and symbionts like mycorrhizal fungi have direct impact on the carbon sequestration. These fungi utilize 5–20% of the net primary productivity of the symbiotic system (Liao et al. 2010). A fungus is a non-chlorophyllous organism and is heterotrophic (requires an organic source of carbon) in nature, i.e., it obtains its food from either dead organic matter or the autotrophic/heterotrophic associates (Chung et al. 2011). *Agaricus*, *Aspergillus*, *Morchella*, *Mucor*, *Penicillium*, *Rhizopus*, *Saprolegnia*, etc. are a few examples of saprophytic nutrition. The role of fungi in the process of carbon sequestration is very substantial because fungi symbiosize with plant roots and help the plants in utilizing the nutrients from the soil. It helps in the sequestration of soil carbon by maximizing the quantity of carbon allocation to the soil by synthesizing biomolecules that enhance aggregate stability (Govindarajulu et al. 2005). There is a different pathway of converting the atmospheric CO_2 to plant biomass (Fellbaum et al. 2012). Carbon accumulation varies with different fungal species (Abraham et al. 2016). Fungal mycelium grows into soils to provide water and nutrients to the fungi. Carbon being the main constituent of fungal mycelia captures maximum carbon to grow. Therefore, fungal mycelia are considered a vital carbon sink to the soils.

10.5.1 Mycorrhizal Fungi

Mycorrhizal fungi stimulate plant growth which results in the faster removal of atmospheric CO_2 through its conversion into plant biomass. Mycorrhizal fungi dominate in the microbial community of soils. Changes in their abundance and their contribution to carbon sequestration could have global effects. Two groups of mycorrhizal fungi, ericoid and ectomycorrhizal fungi, are especially common in high latitude systems (Read et al. 2004). Ectomycorrhizal fungi proliferate when subjected to variable concentrations of atmospheric CO_2 and decline after nitrogen enrichment (Treseder and Allen 2000). With this, Clemmensen et al. found that mycorrhizal fungi make smaller contributions to soil carbon in more frequently burned boreal forests, potentially because dead fungal tissues decay faster in these soils (Clemmensen et al. 2013). Fungi are a potential CO_2 sequester because they secrete glomalin which forms soil aggregates, and it is the recalcitrant nature of glomalin, which makes it hard to decay. Stable aggregates in the soil contribute to carbon sequestration preventing degradation of carbonaceous compounds and keep carbon stored and sequestered in soils for long periods.

10.6 Interlinkage Between Carbon Sequestering Biological Systems

The most followed metabolic pathway by trees, plants, algae, and some proteobacteria for sequestration of carbon is Calvin–Benson cycle. Few microorganisms like archaea, clostridia, and proteobacteria (delta- and epsilon-proteobacteria) follow biochemical pathways such as 3-hydroxypropionate-4-hydroxybutyrate cycle, Wood–Ljungdahl pathway, and reductive TCA cycle. Organisms ranging from eukaryotic plants to prokaryotic microorganisms are found to be active in terms of carbon sequestration and achieve the same using different metabolic routes. Sequestration using algae is found to be more beneficial as it removes CO_2 from different sources in an eco-friendly way. And it can be further used to produce several biofuels such as biodiesel, bioethanol, etc. which is a positive economic aspect of it.

10.7 Importance of CO_2 Sequestration by Microbes in Biorefinery

Biorefinery means the conversion of biomass into various commercially valuable products and fuels. A biorefinery is similar to an oil refinery, where energy, fuels, chemicals, and materials are produced by using organic matter as raw material (Mata et al. 2010). The overuse of fossil fuels by society generates the issue of managing natural resources to achieve sustainable development goals. The biorefineries simultaneously produce biofuels and biomaterials and minimize environmental degradation by waste utilization (Ghimire et al. 2017). For example, the microalgal biomass constitutes lipids (7–23%), proteins (6–71%), and carbohydrates (5–64%) (Bhati et al. 2010) and has the potential to be used as feedstock for the production of biofuels like biodiesel, biogas, biohydrogen, bioethanol, and biobutanol. The lipids in microalgae are converted into FAMEs (fatty acid methyl esters) via the process of transesterification to produce biodiesel and glycerol as by-products. Microalgal biomass is efficient enough to produce biogas which includes methane, hydrogen, and biohythane (Koller et al. 2014). The carbohydrate component of microalgae mainly constitutes glucose, starch, cellulose, and hemicellulose which can be used for transforming into bioethanol via fermentation. The residual part of microalgae after oil extraction can be processed further for the production of biobutanol which is more suitable than biomethanol or bioethanol as a biofuel because of its high energy density. These also produce extracellular polymeric substances (EPSs), and cyanobacterial genera are capable of accumulating polyhydroxyalkanoates (PHAs) (Olaizola 2003). EPS has several industrial applications, whereas PHAs are used for manufacturing biodegradable bioplastics (Forjan et al. 2007). Microalgal biomass is been used as a nutrition source too. India predominantly cultivates cyanobacteria *Arthrospira* sp. and *Spirulina* sp. for their high protein content. The pigments of algae contain high contents of pro-vitamin A and vitamin E (Forjan et al. 2007). The advantages of using microbes using biorefineries are summarized in Fig. 10.5.

10.8 Way Forward

All three types of carbon sequestration are employed for CO_2 mitigation. Using the geologic sequestration process, CO_2 emissions from anthropogenic sources can be injected into the oceans or deep inside the earth's surface. Biologically, CO_2 gets sequestered either by terrestrial plants or microorganisms. Physical and geological methods have certain limitations to perform the process, that is why biological methods are more effective in the active application. Microbes utilize the carbon and produce biomass which gets further used for the production of value-added products, such as biofuels. These biofuels serve as a substitute for dependence on fossil fuel, synthetic plastics, surfactants, and chemically synthesized flocculent. India being an agriculture-based nation can manage CO_2 levels by opting for a variety of soil management strategies. Soil microbes play a major role in soil inorganic carbon (SIC) dynamics via carbonate precipitation. Several growth-promoting bacteria have the potential for carbon sequestration, and they could be studied further to use them as microbial inoculants so that carbon levels in soil can be increased. And the composition of microbial populations and their metabolic pathways for microbial mineral formation are correlated positively with the rate of microbial CO_2 incorporation into SIC. Thus, soil present in agricultural lands acts as a great carbon sink due to the diverse symbiotic relationships between crops and soil microorganisms.

References

Abraham A, Mathew AK, Sindhu R, Pandey A, Binod P (2016) Potential of rice straw for bio-refining: an overview. Bioresour Technol 215:29–36. https://doi.org/10.1016/j.biortech.2016.04.011

Ahmed AAQ, Odelade KA, Babalola OO (2019) Microbial inoculants for improving carbon sequestration in agroecosystems to mitigate climate change. In: Handbook of climate change resilience. Springer International, pp 1–21. https://doi.org/10.1007/978-3-319-71025-9_119-1

Bastos A, Ciais P, Barichivich J, Bopp L, Brovkin V, Gasser T, Trudinger CM (2016) Re-evaluating the 1940s CO_2 plateau. Biogeosciences 13(17):4877–4897. https://doi.org/10.5194/bg-13-4877-2016

Beardall J, Raven JA (2004) The potential effects of global climate change on microalgal photosynthesis, growth and ecology. Phycologia 43:26–40

Berg IA, Kockelkorn D, Buckel W, Fuchs G (2007) A 3-hydroxypropionate/4-hydroxybutyrate autotrophic carbon dioxide assimilation pathway in archaea. Science 318(5857):1782–1786. https://doi.org/10.1126/science.1149976

Bharti RK, Srivastava S, Thakur IS (2014) Production and characterization of biodiesel from carbon dioxide concentrating chemolithotrophic bacteria, Serratia sp. ISTD04. Bioresour Technol 153:189–197. https://doi.org/10.1016/j.biortech.2013.11.075

Bhati R, Samantaray S, Sharma L, Mallick N (2010) Poly-β-hydroxybutyrate accumulation in cyanobacteria under photoautotrophy. Biotechnol J 5:1181–1185. https://doi.org/10.1002/biot.201000252

Bhattacharyya R, Bhatia A, Das TK, Lata S, Kumar A, Tomer R, Biswas AK (2018) Aggregate-associated N and global warming potential of conservation agriculture-based cropping of maize-wheat system in the north-western Indo-Gangetic Plains. Soil Tillage Res 182:66–77. https://doi.org/10.1016/j.still.2018.05.002

Cheah WY, Show PL, Chang JS, Ling TC, Juan JC (2015) Biosequestration of atmospheric CO_2 and flue gas-containing CO_2 by microalgae. Bioresour Technol 184:190–201. https://doi.org/10.1016/j.biortech.2014.11.026

Chung IK, Beardall J, Mehta S, Sahoo D, Stojkovic S (2011) Using marine macroalgae for carbon sequestration: a critical appraisal. J Appl Phycol 23(5):877–886. https://doi.org/10.1007/s10811-010-9604-9

Clemmensen KE, Bahr A, Ovaskainen O, Dahlberg A, Ekblad A, Wallander H, Lindahl BD (2013) Roots and associated fungi drive long-term carbon sequestration in boreal forest. Science 340(6127):1615–1618. https://doi.org/10.1126/science.1231923

Cole JJ, Prairie YT, Caraco NF, McDowell WH, Tranvik LJ, Striegl RG, Melack J (2007) Plumbing the global carbon cycle: integrating inland waters into the terrestrial carbon budget. Ecosystems 10(1):171–184. https://doi.org/10.1007/s10021-006-9013-8

Dev S, Lalli D, Price S, Peters L (2019) Impacts of climate change: spread of communicable diseases in humans. Retrieved from www.sciencedirect.com/science/article/pii/S0160412015300489.Diseases

Ding L, Yokota A (2004) Proposals of *Curvibacter gracilis* gen. nov., sp. nov. and *Herbaspirillum putei* sp. nov. for bacterial strains isolated from well water and reclassification of [Pseudomonas] huttiensis, [Pseudomonas] lanceolata, [Aquaspirillum] delicatum and [Aquaspirillum] autotrophicum as *Herbaspirillum huttiense* comb. nov., *Curvibacter lansceolatus* comb. nov., *Curvibacter delicatus* comb. nov. Int J Syst Evolut Microbiol 54:2223–2230. https://doi.org/10.1099/ijs.0.02975-0

Ekblad A, Wallander H, Godbold DL, Cruz C, Johnson D, Baldrian P, Plassard C (2013) The production and turnover of extramatrical mycelium of ectomycorrhizal fungi in forest soils: role in carbon cycling. Plant Soil 366:1–27. https://doi.org/10.1007/s11104-013-1630-3

Espinoza Pérez AT, Camargo M, Narváez Rincón PC, Alfaro Marchant M (2017) Key challenges and requirements for sustainable and industrialized biorefinery supply chain design and management: a bibliographic analysis. Renew Sustain Energy Rev 69:350–359. https://doi.org/10.1016/j.rser.2016.11.084

Fang H, Cheng S, Wang Y, Yu G, Xu M, Dang X, Wang L (2014) Changes in soil heterotrophic respiration, carbon availability, and microbial function in seven forests along a climate gradient. Ecol Res 29(6):1077–1086. https://doi.org/10.1007/s11284-014-1194-6

Fast AG, Papoutsakis ET (2012) Stoichiometric and energetic analyses of non-photosynthetic CO_2-fixation pathways to support synthetic biology strategies for production of fuels and chemicals. Curr Opin Chem Eng 1:380–395. https://doi.org/10.1016/j.coche.2012.07.005

Fellbaum CR, Mensah JA, Pfeffer PE, Toby Kiers E, Bücking H (2012) The role of carbon in fungal nutrient uptake and transport: implications for resource exchange in the arbuscular mycorrhizal symbiosis. Plant Signal Behav 7. https://doi.org/10.4161/psb.22015

Field CB, Behrenfeld MJ, Randerson JT, Falkowski P (1998) Primary production of the biosphere: integrating terrestrial and oceanic components. Science 281:237–240. https://doi.org/10.1126/science.281.5374.237

Foissner W (1999) Soil protozoa as bioindicators: pros and cons, methods, diversity, representative examples. Ecosyst Environ 74

Forjan E, Garbayo I, Casal C, Vilchez C (2007) Enhancement of carotenoid production in nanochloropsis by phosphate and sulphur limitation. Commun Curr Res Educ Top Trends Appl Microbiol 356–364

Gao K, McKinley KR (1994) Use of macroalgae for marine biomass production and CO_2 remediation: a review. J Appl Phycol 6:45–60. https://doi.org/10.1007/BF02185904

Ghimire A, Kumar G, Sivagurunathan P, Shobana S, Saratale GD, Kim HW, Munoz R (2017) Bio-hythane production from microalgae biomass: key challenges and potential opportunities for algal bio-refineries. Bioresour Technol 241:525–536. https://doi.org/10.1016/j.biortech.2017.05.156

Ghorbani A, Rahimpour HR, Ghasemi Y, Zoughi S, Rahimpour MR (2014) A review of carbon capture and sequestration in Iran: microalgal biofixation potential in Iran. Renew Sustain Energy Rev 35:73–100. https://doi.org/10.1016/j.rser.2014.03.013

Govindarajulu M, Pfeffer PE, Jin H, Abubaker J, Douds DD, Allen JW, Shachar-Hill Y (2005) Nitrogen transfer in the arbuscular mycorrhizal symbiosis. Nature 435(7043):819–823. https://doi.org/10.1038/nature03610

Guo JH, Liu XJ, Zhang Y, Shen JL, Han WX, Zhang WF, Zhang FS (2010) Significant acidification in major Chinese croplands. Science 327(5968):1008–1010. https://doi.org/10.1126/science.1182570

Han J, Lian B, Ling H (2013) Induction of calcium carbonate by *Bacillus cereus*. Geomicrobiol J 30 (8):682–689. https://doi.org/10.1080/01490451.2012.758194

Heck RW, Tanhausers SM, Mandas R, Tu C, Laipiss PJ, Silverman DN (1994) Catalytic properties of mouse carbonic anhydrase V. J Biol Chem 269:24742–24746

Herzog HJ (2001) Peer reviewed: what future for carbon capture and sequestration? Environ Sci Technol 35:148A–153A. https://doi.org/10.1021/es012307j

Ho SH, Chen CY, Chang JS (2012) Effect of light intensity and nitrogen starvation on CO_2 fixation and lipid/carbohydrate production of an indigenous microalga *Scenedesmus obliquus* CNW-N. Bioresour Technol:244–252. https://doi.org/10.1016/j.biortech.2011.11.133

Jiao N, Herndl GJ, Hansell DA, Benner R, Kattner G, Wilhelm SW, Azam F (2010) Microbial production of recalcitrant dissolved organic matter: long-term carbon storage in the global ocean. Nat Rev Microbiol 8:593–599. https://doi.org/10.1038/nrmicro2386

Karlsson J, Clarke AK, Chen Z-Y, Hugghins SY, Park Y-I, David Husic H, Ran Samuelsson G (1998) A novel α-type carbonic anhydrase associated with the thylakoid membrane in *Chlamydomonas reinhardtii* is required for growth at ambient CO_2. EMBO J 17

Knohl A, Veldkamp E (2011) Global change: indirect feedbacks to rising CO_2. Nature 475:177–178

Koller M, Muhr A, Braunegg G (2014) Microalgae as versatile cellular factories for valued products. Algal Res 6:52–63. https://doi.org/10.1016/j.algal.2014.09.002

Kumar M, Sundaram S, Gnansounou E, Larroche C, Thakur IS (2018) Carbon dioxide capture, storage and production of biofuel and biomaterials by bacteria: a review. Bioresour Technol 247:1059–1068. https://doi.org/10.1016/j.biortech.2017.09.050

Li N, Xu YZ, Han XZ, He HB, Zhang, X. dong, & Zhang, B. (2015) Fungi contribute more than bacteria to soil organic matter through necromass accumulation under different agricultural practices during the early pedogenesis of a Mollisol. Eur J Soil Biol 67:51–58. https://doi.org/10.1016/j.ejsobi.2015.02.002

Liao C, Luo Y, Fang C, Li B (2010) Ecosystem carbon stock influenced by plantation practice: implications for planting forests as a measure of climate change mitigation. PLoS One 5. https://doi.org/10.1371/journal.pone.0010867

Liew F, Henstra AM, Winzer K, Köpke M, Simpson SD, Minton NP (2016) Insights into CO_2 fixation pathway of *Clostridium autoethanogenum* by targeted mutagenesis. MBio 7. https://doi.org/10.1128/mBio.00427-16

Litynski JT, Klara SM, McIlvried HG, Srivastava RD (2006) The United States Department of Energy's regional carbon sequestration partnerships program: a collaborative approach to carbon management. Environ Int 32:128–144. https://doi.org/10.1016/j.envint.2005.05.047

Mata TM, Martins AA, Caetano NS (2010) Microalgae for biodiesel production and other applications: a review. Renew Sust Energ Rev 14:217–232. https://doi.org/10.1016/j.rser.2009.07.020

Meijer WG (1994) The Calvin cycle enzyme phosphoglycerate kinase of *Xanthobacter flavus* required for autotrophic CO_2 fixation is not encoded by the CBB Operon Downloaded from (vol 176). Retrieved from http://jb.asm.org/

Mistry AN, Ganta U, Chakrabarty J, Dutta S (2019) A review on biological systems for CO_2 sequestration: organisms and their pathways. Environ Progress Sustain Energy 38:127–136. https://doi.org/10.1002/ep.12946

Nie M, Bell C, Wallenstein MD, Pendall E (2015) Increased plant productivity and decreased microbial respiratory C loss by plant growth-promoting rhizobacteria under elevated CO_2. Sci Rep 5. https://doi.org/10.1038/srep09212

Olaizola M (2003) Commercial development of microalgal biotechnology: from the test tube to the marketplace. Biomol Eng 20:459–466. https://doi.org/10.1016/S1389-0344(03)00076-5

Paul V, Chandra Shekharaiah PS, Kushwaha, Sapre A, Dasgupta S, Sanyal D (2020) Role of algae in CO_2 sequestration addressing climate change: a review. https://doi.org/10.1007/978-981-32-9578-0_23

Pemencilan P, Karbon Dioksida Oleh Bacillus B, Komala T, Khun TC (2014) Biological carbon dioxide sequestration potential of *Bacillus pumilus*. Sains Malaysiana 43:1149–1156

Quéré C, Andrew R, Friedlingstein P, Sitch S, Hauck J, Pongratz J, Pickers PA, Korsbakken JI, Peters GP, Canadell JG, Arneth A, Arora VK, Barbero L, Bastos A, Bopp L, Chevallier F, Chini LP, Ciais P, Doney SC, Gkritzalis T, Goll DS, Harris I, Haverd V, Hoffman FM, Hoppema M, Houghton RA, Hurtt G, Ilyina T, Jain AK, Johannessen T, Jones CD, Kato E, Keeling RF, Klein Goldewijk K, Landschützer P, Lefèvre N, Lienert S, Liu Z, Lombardozzi D, Metzl N, Munro DR, Nabel JEMS, Nakaoka S, Neill C, Olsen A, Ono T, Patra P, Peregon A, Peters W, Peylin P, Pfeil B, Pierrot D, Poulter B, Rehder G, Resplandy L, Robertson E, Rocher M, Rödenbeck C, Schuster U, Schwinger J, Séférian R, Skjelvan I, Steinhoff T, Sutton A, Tans PP, Tian H, Tilbrook B, Tubiello FN, van der Laan-Luijkx IT, van der Werf GR, Viovy N, Walker AP, Wiltshire AJ, Wright R, Zaehle S, Zheng B (2018) Global carbon budget 2018. Earth Syst Sci Data 10:2141–2194. https://doi.org/10.5194/essd-10-2141-2018

Ramanathan V (1988) The greenhouse theory of climate change: a test by an inadvertent global experiment. Science 240:293–299. https://doi.org/10.1126/science.240:4850.293

Read DJ, Leake JR, Perez-Moreno J (2004) Mycorrhizal fungi as drivers of ecosystem processes in heathland and boreal forest biomes. Can J Bot 82:1243–1263. https://doi.org/10.1139/B04-123

Rittmann S, Seifert A, Herwig C (2015) Essential prerequisites for successful bioprocess development of biological CH_4 production from CO_2 and H_2. Crit Rev Biotechnol 35:141–151. https://doi.org/10.3109/07388551.2013.820685

Rossi F, Olguín EJ, Diels L, de Philippis R (2015) Microbial fixation of CO_2 in water bodies and in drylands to combat climate change, soil loss and desertification. New Biotechnol 32:109–120. https://doi.org/10.1016/j.nbt.2013.12.002

Sahoo D, Elangbam G (2012) Using algae for carbon dioxide capture and bio-fuel production to combat climate change Understanding the impact of microbial transformation during food fermentation on health View project Seabuckthorn View project. Retrieved from https://www.researchgate.net/publication/267408503

Saini R, Kapoor R, Kumar R, Siddiqi TO, Kumar A (2011) CO_2 utilizing microbes – a comprehensive review. Biotechnol Adv 29:949–960. https://doi.org/10.1016/j.biotechadv.2011.08.009
Stocks-Fischer S, Galinat JK, Bang SS (1999) Microbiological precipitation of $CaCO_3$. Soil Biol Biochem 31:1563–1571. https://doi.org/10.1016/S0038-0717(99)00082-6
Treseder KK, Allen MF (2000) Mycorrhizal fungi have a potential role in soil carbon storage under elevated CO # and nitrogen deposition. Rev New Phytol 147
West JM, McKinley IG, Palumbo-Roe B, Rochelle CA (2011) Potential impact of CO_2 storage on subsurface microbial ecosystems and implications for groundwater quality. Energy Procedia 4:3163–3170. https://doi.org/10.1016/j.egypro.2011.02.231
Wong TS (2014) Carbon dioxide capture and utilization using biological systems: opportunities and challenges. J Bioprocess Biotech 04. https://doi.org/10.4172/2155-9821.1000155
Zhang (2011) Bacillus mucilaginosus can capture atmospheric CO_2 by carbonic anhydrase. Afr J Microbiol Res 5:106–112. https://doi.org/10.5897/AJMR10.690

Chapter 11
Direct and Indirect Impacts of Climatic Change on Soil Communities and Plants

Rajiv Kumar, Moni Kumari, Mayank Pathak, and Jagpreet Singh Kamboj

Abstract Climate change is one of the most important environmental issues that may lead to changes in the biotic and abiotic components of the planet. The soil communities and plants are affected directly and indirectly by the changing climate, such as increasing average temperatures, increasing CO_2 content in the atmosphere, and disturbed rainfall patterns. Climate plays a significant role, right from the development of the soil to its maintenance. The changes in the terrestrial ecosystems that dwell in the soil are a result of the global climate change. The species distribution as well as their interaction with other species in the ecosystem is being altered due to the global climate change. The natural communities are composed of organisms with varying traits and abilities. This chapter aims at comprehensive description of the direct and indirect impacts of climate change on the soil communities and plants and how climate change disturbed or changed soil structure and the microbial communities present in the soil. Further, the chapter also deals about how climate change affects the interaction of soil microbial communities among themselves and with the plant. The discussion also throws light on the extent of effects climate has on the soil degradation, biodiversity in soil community, and other soil dynamics.

Keywords Climate change · CO_2 fertilization · Plant–microbe interactions · soil respiration · FACE · Food security

R. Kumar (✉) · M. Pathak · J. S. Kamboj
Department of Civil Engineering, Dr. B R Ambedkar National Institute of Technology, Jalandhar, Punjab, India
e-mail: kumarr@nitj.ac.in

M. Kumari
Department of Botany, Gaya College, Gaya, Bihar, India

D. K. Choudhary et al. (eds.), *Climate Change and the Microbiome*, Soil Biology 63,
https://doi.org/10.1007/978-3-030-76863-8_11

11.1 Introduction

Biological, chemical, and physical processes occurring in the land, ocean, and atmosphere determine the climate of the planet (Denman et al. 2007). The wide variety of organisms at each tropic level play an indispensable role. Any external disturbance either natural or anthropogenic may disturb the well-established equilibrium within the ecosystem. If the disruption continues with time, eventually it will result in complete ecosystem destruction and biodiversity loss.

Climate change is a major environmental challenge of the twenty-first century facing mankind. Anthropogenic activities play a vital role in climate change around the world (Mishra 2016). Industrial revolution, burning of fossil fuel, increasing population, deforestation, and land use changes are responsible for the changing climate (Salam and Noguchi 2005). Worldwide, it may impose dramatic effects on the society and the ecosystem. Shifting precipitation, rising of temperatures, occurrence of more frequent wildfires, melting of the glaciers and snow, and the rising sea level all indicate that climate change is already happening. Climate change alters or disturbs the natural ecosystem, posing dangerous effects on human health across the globe. It also threatens the biodiversity and severely affects global food production. Decline soil fertility, disturbed rainfall patterns, floods, and increasing population are now raising questions about food security. On the other hand, temperature increase due to climate change favors microbial growth, thus reducing food safety (Hammond et al. 2015). The constant increase in population in this stressful environment is raising an important question on how we will provide adequate food to the entire population in the future (Kang et al. 2009).

The interaction between plant and soil is called plant–soil feedbacks (PSFs) (Kulmatiski and Kardol 2008). Basically, the two major steps of PSFs are (1) different soils are created by different plant species and (2) growth of different plant species is affected by the presence of different soil types. The plant ecosystem processes and the dynamics of the plant community are significantly governed by the plant–soil feedbacks. Since growth of organism and distribution of organism are driven by climate, the changing climate may prominently affect the plant composition, soil microbial communities, and the interactions between them. However, very little is understood how climate change affects plant–soil feedbacks (Pugnaire et al. 2019).

The inter-community interactions, soil microbial biodiversity, and ecosystem function could be constructive or destructive and might even have no significant impacts due to the changing climate. How PSFs cope themselves from the adverse environment effects arises due climate change. Despite a number of studies present, climate change triggered changes in these interactions that add up to alter the biodiversity, with a few of them specializing in soil communities. Soil organisms act with each other like plants in a number of ways that shape and maintain the properties of the ecosystem. Soil–microbe interactions can shape landscape patterns of plant and animal abundance, diversity, and composition (Classen et al. 2015).

Several theoretical, mathematical, and conceptual models are there providing insightful information about how climate change alters PSFs (Bellard et al. 2012). However, applied testing in the laboratory and field studies is very necessary to explain the underlining mechanisms behind the changes in the PSFs in response to climate change (Van Der Putten et al. 2013). The effect of climate change on the PSFs may be direct or indirect. A study at the ClimGrass site indicates that the indirect effects due to climate change in the future predominate over the direct effects (Deltedesco et al. 2020). Botkin et al. (2007) predict the effect of global warming on plant migration and genetic diversity. This study shows that due to global warming plant migration as well as loss of genetic diversity will take place. A survey conducted in the Amarkantak region shows that due to climate change and its effects on PSFs, there is loss of microbial biodiversity and diversity of medicinal plants. Decrease in fishery catches and productivity of maize, wheat, and soybean has also been reported. This negative aspect of climate change on PSFs is risky for the local tribal people of that region who are dependent on forest and agriculture for their livelihood (Ahirvar et al. 2020). Thus, it is very necessary to study the impact of changing climate on the agrological regions in a regular basis (Karmakar et al. 2016).

11.2 General Overview of Climate Change

The natural ecosystems are significantly affected by the increasing global temperatures with global warming (Chao and Feng 2018). The emissions of greenhouse gases from the natural or anthropogenic sources contribute significantly in the global climate change. It leads to changes in parameters like disturbed precipitation, snow cover, humidity, and sea level rise, increasing the mean temperature. Besides that, climate change severely affects the ecosystem services, biodiversity, species composition, plant growth, and productivity (Amedie 2013). The cropland area and cropping frequency are more sensitive to climate change than crop yield (Chao and Feng 2018; Cohn et al. 2016). Due to climate change, it is predicted that the global bio-production may reduce by around 17% by 2050. By the end of this century, the annual food production would decline by 0% to 2%, and the demand for food would increase by 14% by 2050 (Nelson et al. 2014). Based on a mathematical modeling, it is specified that grain production in Southeast Asia and Southern Africa will be more severely affected by climate change (Fischer and Edmeades 2010; Rai 2020).

The impact of climate change can be mainly divided into two categories:

1. Biophysical impact: This category takes into account the effects that directly affect the physical environment due to climate change like drought and flooding, causing an effect on physical environment such as (1.1) effects on quality and amount of crops, pasture, forest, and livestock; (1.2) change in natural resources quality and quantity of soil, land, and water resources, (1.3) increased weed and bug blighter challenges, and (1.4) sea level rise and change in ocean salinity.

2. Socioeconomic impacts: It includes decline in yield and production, reduced marginal gross domestic product from agriculture sector, fluctuation of world market value, change in geographical distribution of trade regimes, hunger, and migration (Amedie 2013; Fao 2017).

11.3 Impact of Climate Change on Plants

Plants are autotrophs, capable of synthesizing organic compounds using simple inorganic molecules. The energy generated by the plant through photosynthesis is utilized as food. Climate change creates abiotic stress environment; thus, the plant growth and yield are greatly influenced. Impact of the changing climate on plants is shown in Fig. 11.1. Several stressful environments like high temperature or heat, drought, waterlogging, cold, and salinity under natural climate conditions are often experienced by the plants. Thus, the physiology of plants is greatly affected by several means. Climate change also affects plants at the level of molecular function, developmental processes, and morphological traits.

The climatic factors influencing plant growth and yields are discussed below.

Fig. 11.1 Impact of climate change on plants

11.3.1 Light

Light is vital for the plants right after their germination. The intensity of light, brightness, photoperiod, and wavelength have their own role in the growth of plants. Light is also inevitably required by plants for chlorophyll production and to carry out photosynthesis. The movements of stomata are significantly governed by light. These movements are responsible for inhibition or producing various processes like abscission and translocation. Various properties that play a crucial role are light intensity, quality, and photoperiod. Variance in this climatic factor causes variations in the seed production, maturity, and flowering of the plants (Pandey 2020).

11.3.2 Temperatures

By the end of this century, it is predicted that the global average surface temperature will increase by around 1.0–3.7 °C (IPCC 2014). Variations in the diurnal temperatures have a significant effect on the maturity and expansion of the plants. Balasubramanian et al. (2006) reported that due to increasing atmospheric temperature both vegetative and reproductive phase get shifted earlier in *Arabidopsis thaliana*. In response to increase in temperature, the allocation of food to the roots may increase, and the root function also gets affected including nutrient uptake by the root and rate of respiration (Awal et al. 2003; Atkin et al. 2000). Different plants need different favorable day-and-night temperatures for their growth, and yields of different crops vary with different atmospheric temperatures (Table 11.1) (Pandey 2020; Warren 2011). By means of global warming, different latitudes of the earth's surface experienced different degrees of temperature stress. Thus, plants growing in different regions face different levels of temperature stress. Thus, in the Southern USA, elevated temperature leads to 2.4% yield loss of soybean opposing 1.7% increase in yield of the same in the Midwestern USA (Hatfield et al. 2011). The negative relationship between crop yield and temperature has been reen reported in the Midwestern USA. For both soybean and maize for every 1 °C rise in temperature, growing season yield loss decreases by 17% (Lobell and Asner 2003). The reproductive structure of the plant is greatly affected by the elevated temperature. Hedhly (2011) reported that plant male reproductive part is more sensitive to the temperature stress when compared with pistil or female gametophyte. In response to temperature stress, there is decrease in the number of ovule, and ovule abortion rate is increased in *Arabidopsis* (Whittle et al. 2009). Elevated temperature strongly altered the leaf developmental processes like rate of leaf initiation and leaf expansion. Leaf morphology and the emergence of new leaf are also affected by the increase in temperature (Granier et al. 2002). Plant dry matter content determines the physical structure or toughness of the plant. Elevated temperature increases the carbon content of the plant in the plant tissue, thus enhancing toughness of the plant

Table 11.1 Quantitative impact of climate change at 2 °C and 4 °C on agriculture and food, ecosystems and feedback processes in the earth system (Warren 2011)

Impact sector	Projections at 2 °C above preindustrial levels	Projections at 4 °C above preindustrial levels
Agriculture and food	32% of currently cultivable land becomes unfit for crop growth	50% of currently cultivable land becomes unfit for crop growth; threats to food security
Ecosystems	20% species face risk of extinction, damage to arctic ecosystem functioning, some increases in fire frequency	40% species face risk of extinction, disruption to functioning of major global ecosystems, large increase in boreal and Mediterranean fire frequency
Feedback processes in the earth system	Much reduced risk of loss of terrestrial carbon sink	Loss of terrestrial carbon sink causing large-scale loss of forests via desiccation and fire

(Gross and Lombardo 2018). Warmer temperature accelerates plant growth by extending growing season and increasing net photosynthesis (Berggren et al. 2009). Temperature not only influences the quantity but also the quality of net production of the plant.

11.3.3 Rainfall

Rainfall is the basic requirement of plants and it affects crop production worldwide. Regular and adequate rainfall is very crucial for good crop productivity. One side variation in rainfall pattern due to climate change either causes extreme precipitation or floods. On another side, climate change may lead to drought due to complete absence or very little rainfall for a longer time. By the next 50 years across the world, transformed rainfall pattern or drought will hinder crop productivity in more than 50% of the total arable land (Table 11.2) (Dhankher 2018). The function of photosystem II (PSII) gets affected more strongly and decreased when exposed to combined stress of heat and drought in *Leymus chinensis* (Xu and Zhou 2006). Plant faces multiple stresses under drought condition and elevated atmospheric CO_2 because it initiates the generation of reactive oxygen species (ROS) in the plants (Raza et al. 2019). The enhanced production of ROS will further lead to break down of the cell membrane; thus, eventually the plant development frequency, plant respiration, and photosynthesis get negatively affected (Ahmad et al. 2018). The net benefits from the plant mycorrhizal mutualistic relationship to the plant are also decreased due to decreased rate of photosynthesis in response to drought stress (Johnson et al. 2015). Based on data collected on mathematical modeling experiments conducted by Easterling et al. (2007), it is proposed that variation in rainfall pattern, increase in atmospheric CO_2, and increase in temperature by 1–3 °C have positive impacts on crop yields in mid- to high-latitude regions. But even moderate increase in temperature (1–2 °C) has deleterious impact on the yield of major cereals

Table 11.2 Consequences of drought on different plant species

S. no.	Name of plant	Description
1	*Triticium aestivum* L.	Reduction in total growth and yield, delay in pre-anthesis
2	*Chenopodium quinoa* Wild	Delay in pre-anthesis
3	*Oryza sativa* L.	Delay in flowering
4	*Glycine max* L.	Drought at the time of grain filling accelerates the maturity and decreases the yield
5	*Pennisetum glaucum* L.	Increase in the rate of abortion

in low-latitude regions. In soybean, the number and density of root nodule have been decreased which houses nitrogen-fixing bacteria (*Bradyrhizobium*) in response to elevated atmospheric CO_2 and decreased precipitation (Gray et al. 2013, 2016). The variation in precipitation pattern poses impact on the moisture content of the soil or soil humidity. Walck et al. (2011) reported that this will affect the time and rate of success of germination and regeneration of the seedling.

Due to global warming, the temperature of the earth's atmosphere is increased which leads to melting of the snow and ultimately sea level rise. This will possibly increase the ratio of rainfall to snowfall. The rate of rainfall has been increased especially in the winter (Sheffield and Wood 2008). Due to increase in soil moisture content, the number of pre-germination seed mortality has been increased in response to enhanced fungal pathogen infection (Walck et al. 2011).

11.3.4 Raising CO_2

Increasing atmospheric concentrations of CO_2 due to climate change are possible which have profound direct effects on the plant chemistry, plant growth, and plant physiology (Ziska 2008). Plant responds to increase in atmospheric concentration of CO_2 by closing their stomata and reducing the water loss through transpiration. The chemical composition of the plant tissue is also changed in response to elevated atmospheric CO_2. Free-air CO_2 enrichment (FACE) experiments conducted showed that the plant photosynthetic activity is increased due to elevated CO_2 concentrations. Carbohydrate content of the nonstructural leaf is also increased by 30–40% (Ainsworth 2008; Ainsworth and Long 2005). The nitrogen content of the leaf is decreased by 13% per unit leaf mass in FACE experiments (Ainsworth and Long 2005). In India, the yield of wheat, rice, oilseeds, and legumes has been increased by 10–20% in response to increase in atmospheric CO_2 to 500 ppm (Mahato 2014).

Generally, it is well studied that climate change reduces the crop yields. The CO_2 fertilization effects tend to increase crop yields. The increasing concentration of CO_2 in the atmosphere leads to CO_2 fertilization where the rate of photosynthesis is increased. It gives a positive feedback in terms of increasing crop yield (Miller 2001; Pandey 2020). The CO_2 fertilization effect is predominantly seen in C3 species because in these plants with rise in atmospheric CO_2 the rate of carbon assimilation is increased. The RuBisCo enzyme of C3 species is not completely saturated with the atmospheric concentration of CO_2. On the other hand, increased atmospheric CO_2 inhibit the oxygenation of the enzyme; thus, the loss of CO_2 via photorespiration is decreased. In both C3 and C4 species, elevated atmospheric CO_2 decline stomatal conductance; thus, the use of canopy water has been decreased. This will upsurge the moisture content of the soil and thus plant yields are maintained even in low water availability (Leakey 2009; Leakey et al. 2006).

Fertilization effect of CO_2 is also dependent on other factors like availability of water and nutrients for the plant and outbreak of pests and diseases that may also be affected by climate change (Erda et al. 2005). In the next 20–80 years without considering CO_2 fertilization effect, it is also studied by the same group using modeling approach that in China only climate change could reduce the yields of wheat, rice, and maize by up to 18–37%. The quality of wheat is also affected very seriously if the atmospheric concentration of CO_2 is more than 460 ppm. 100 ppm increase in atmospheric CO_2 resulted in 5–17% increase in average crop production over regions including Russia, Central Asia, Africa, Brazil, etc. It is also well noted that the positive response of plant with increasing concentration of CO_2 is more in dry regions when compared with wet regions (McGrath and Lobell 2013). This study report also mentions that the response of CO_2 fertilization is especially large in Russia, due to high production of potatoes, a tuberous species. It is projected that by 2050 or midcentury with increase in CO_2 concentration, free-air CO_2 enrichment (FACE) experiments conducted in the US Midwest expected a 100% increase in the dry mass of cassava tubers (Rosenthal et al. 2012; Rosenthal and Ort 2012).

11.4 Impact of Climate Change on the Soil Communities

The relative abundance and function of soil communities will possibly change with the climatic condition. This is because different soil microbial community members differ in their temperature sensitivity, growth rate, and physiology (Whitaker et al. 2014; Briones et al. 2014).

Global climate change alters the distribution of species and interaction among them (Van Der Putten et al. 2013). Understanding climate change impacts on the soil dynamics is very important. The functioning of soil and ecosystem is significantly governed by the composition of the soil microbial communities (Kardol et al. 2010; Schimel and Schaeffer 2012). An alteration in the soil physical and chemical properties in response to climate change directly affects the soil microbial community. The interactions of the soil microbial communities among each other as well as with plants shape the plant structures and biodiversity (Van Der Putten et al. 2013). Alteration in ecosystem function and biodiversity happened due to climate change.

There are very few study reports focusing on the impact of climate change on the soil microbial communities (de Vries and Griffiths 2018).

11.4.1 Temperature

Soil temperature change is always linked with moisture content of the soil that may further explain how microbial community responds to climate change. Due to climate change, the atmospheric temperature has been increased. Global warming promotes growth of microbial biomass, decomposition of soil organic matter, and soil respiration. The effects of increased atmospheric temperature on microbial physiology are very complex, and microbial system of the soil probably adapted or evolved with time. The rate at which the temperature of the atmosphere is increased is with the same rate the moisture content of the soil is decreased. It is reported that as compared to fungal communities, soil bacterial communities react rapidly to the moisture content of soils (Hagerty et al. 2014). However, Classen et al. (2015) stated that when the water holding capacity of the soil is decreased by 30% in response to climate change, the dominant fungal communities of the soil shift from one to another, while the bacterial soil communities remain unaffected. At the present time, how microbial metabolism reacted with elevated temperature is gaining attention (Hagerty et al. 2014). An increase in the average temperature increased the abundance of sagebrush at cold sites, and decreased the same at hot sites (Kleinhesselink 2017). DeAngelis et al. (2015) reported that in temperate forest, a temperature increase of 5 °C altered the relative abundance of soil bacterial community and also increased the ratio of bacterial community to fungal community. Increase in temperature and soil moisture content poses special effects on the microbial community involved in ecosystem functions like nitrogen metabolism such as nitrification, nitrogen fixation, denitrification, and methanogenesis (Isobe et al. 2011; Bakken et al. 2012; Bodelier et al. 2000).

Soil microbial communities adapted with the warmer soil by altering their physiology mainly carbon use efficiency (CUE). The amount of carbon utilized by the microbes for their growth is called CUE. Since microbial respiration is increased with the soil temperature, the CUE of the microbes is reduced, thus inhibiting carbon allocation to the microbial growth. Soil microbial production of extracellular enzymes is significantly changed due to global warming or climate change because long-term warming shifts microbial community of the soil organic matter content of the soil that determines its health and productivity. It is also important for the enormous number and variety of microflora and fauna present in the soil. In the upper layer of the soil in the forest, fungi composed a large proportion of microbial biomass. It plays a major role in nutrient recycling and degradation of soil carbon. Complex organic substrates of the soil including lignin and chitin are degraded by the enzymes extracellularly produced by the fungi (Kellner and Vandenbol 2010).

11.4.2 Elevated CO_2

The rising levels of global CO_2, for example, have remarkable impacts on soil ecosystems (Janus et al. 2005). Various types of soil processes govern these soil dynamics, at different rates and ways to make the soil a complex system to be studied. The major changes in the soils over the past 50 years are reflected on the dynamicity of the soil processes. The soil processes are influenced directly by climate factors such as precipitation, global temperatures, and atmospheric CO_2 changes and indirectly by the climate-induced changes in the land use and management.

In this section, we aim to study the impacts of climate change on the soil communities with different parameters such as physical parameters (water, temperature, and structure and texture differentiation), biological parameters (organic matter), chemical parameters, and soil degradative parameters. These areas holistically help us to visualize the impacts of the climate changes on the soil communities with various parameters. Under elevated atmospheric CO_2 conditions the plant yield is increased due to CO_2 fertilization effect and enhanced belowground carbon allocation. This will further support the soil microbial communities to increase the rate of nitrogen fixation, immobilization of nitrogen, denitrification, increased aggregation of the soil, and mutualistic relationship between roots of higher plants and fungi (Jones et al. 2009). Increased CO_2 also accelerate the colonization of plant growth-promoting bacteria.

11.5 Climate Change Impact on Plant–Microbe Interactions

Various changes in the land use and land cover (LULC) due to the climatic changes are responsible for the effects on the plant–microbe interaction. There are a wide variety of microbes present in the rhizosphere or associated with the plant root. Some of them are neutral or may be pathogenic for the plants on which they reside. However, some are very beneficial for the plants, accelerating their growth and enhancing their tolerance to abiotic or biotic stress. Plant roots communicate via chemicals with the microbial community present in the rhizosphere. Root exudates play a vital role in the plant–microbe interactions. Several external factors including photosynthetic activity of the plants, size of plants, and soil condition determine the chemical nature of the root exudates (Mhlongo et al. 2018).

Climate change alters the environmental condition; thus, the physiology of plants and nature of its root exudation changed (Compant et al. 2010). Due to elevated atmospheric CO_2, the allocation of carbon to the plant root has been increased that will potentially change the chemical composition of the root exudates. Alteration resulted in changed C/N ratio, nutrient availability, and the availability of signaling molecules present in the root exudates (Haase et al. 2008). In the same manner, increased atmospheric temperature and drought due to climate change also resulted in changes in chemical composition of the root exudates in the same manner. Thus, climate change leads to alteration in microbial community associated with the plant root. As an effect of the changing climate, the soil communities may change their distributions in the soil profile (Jansson and Hofmockel 2019; Classen et al. 2015).

11.5.1 Effect of Global Changing Conditions on Arbuscular Mycorrhizal Fungi (AMF)

Roots of around 90% of the higher plants present on the earth are associated with the mycorrhiza. In the exchange for carbohydrate compounds present in the rhizosphere, AMF boost nutrient uptake by the plants especially phosphorus (P). It is reported by Leake et al. (2004) that in the terrestrial ecosystem, AMF contributes to 20–30% of the total microbial biomass. The hyphae of AMF may possibly occupy 100 m cm^{-3} of soil (Miller et al 1995).

Due to climate change effects such as soil warming or drought and increase in atmospheric CO_2 concentrations, several studies have reported the effects of climate change factors on these plant-associated fungi (Augé 2001). The effect of drought on the AMF colonization is a major section of concern. Augé (2001) reported that AMF protect plants from drought resistance and also support plant–water relation. The negative effect of drought on the AMF depends on the species of the AMF as well as the plant, and it is also determined by the hyphal growth within or outside the root (Wilson et al. 2016).

The hyphae of the AMF strains under the elevated atmospheric CO_2 (600μL L^{-1}) showed increase growth in the rhizosphere of *Prunella vulgaris* when compared with the CO_2 level of 350μL L^{-1} (Sanders et al. 1998). This is further related with the increased root biomass due to increased atmospheric CO_2 level. Tang et al. (2009) reported that due to increased CO_2 level, colonization of AMF is increased in C4 plants, whereas the rate of canonization is unaffected in C3 plants. Climate variables like elevated CO_2 level and atmospheric temperature have positive effects on the AMF colonization due to enhanced plant productivity, and thus the demand for plant nutrients and the production of root exudates have been increased (Zavalloni et al. 2012). The effect of elevated CO_2 on the AMF colonization is also influenced by other factors like fungal species, precipitation, and the availability of water (Veresoglou et al. 2016). The ratio of *Glomeraceae* to *Gigasporaceae* fungal species in the AMF community composition has been increased with increase in the concentration of atmospheric CO_2 (Cotton et al. 2015).

The effect on AMF due to increase in atmospheric temperature is studied. Mohan et al. (2014) mentioned in his study report about the effect of increased atmospheric temperature on the abundance of the mycorrhizal communities. He reported that in response to increased temperature, 17% of mycorrhizal communities shows decreased, 20% neutral, and 63% positive effects in their abundance. Majority of the study reports stated that AMF colonization and hyphal growth increase with increase in atmospheric temperature. There are a few reports on the neutral or negative effects of temperature on AMF colonization and hyphal growth. Augé et al. (2015) reported that when the atmospheric temperature is rise upto 27 °C, the growth of AMF is increased by 10%. In the Mediterranean region, an increase in atmospheric temperature by 3 °C decreases the rate of colonization of AMF (Wilson et al. 2016). At higher temperature, the activity of AMF decreases (Mohan et al. 2014). Hawkes et al. (2008) mentioned in his study report that the structure of AMF hyphae is altered in response to increased atmospheric temperature, and also a number of vesicles are increased. The rate by which carbon is allocated to the rhizosphere has been increased (Heinemeyer et al. 2006). Similarly, Zhang et al. (2016) reported that at high temperature, the diversity of AMF species and the growth of hyphae are increased.

Way Forward

The effect of the changing climate on the plant and soil communities may be unpredictable initially. With time as the temperature of the atmosphere is increased, atmospheric level of CO_2, frequent drought, and altered precipitation is increased to a greater extent. The consequences of changing climate are visible. The present chapter discussed about the different factors of the changing climate like how they affect either alone or all together the growth and development of plants. Due to the temperature sensitivity, the composition of soil microbial community also shifts from their natural form.

References

Ahirvar BP, Chaudhry S, Kumar M, Das P (2020) Climate change impact on forest and agrobiodiversity: a special reference to Amarkantak area, Madhya Pradesh. In: Contemporary environmental issues and challenges in era of climate change. Springer, Singapore, pp 65–76. https://doi.org/10.1007/978-981-32-9595-7_3

Ahmad Z, Anjum S, Waraich EA, Ayub MA, Ahmad T, Tariq RMS, Ahmad R, Iqbal MA (2018) Growth, physiology, and biochemical activities of plant responses with foliar potassium application under drought stress—a review. J Plant Nutr 1734–1743

Ainsworth EA (2008) Rice production in a changing climate: a meta-analysis of responses to elevated carbon dioxide and elevated ozone concentration. Glob Chang Biol 14:1642–1650

Ainsworth EA, Long SP (2005) What have we learned from 15 years of free-air CO_2 enrichment (FACE)? A meta-analytic review of the responses of photosynthesis, canopy properties and plant production to rising CO_2. New Phytol 165:351–372

Amedie FA (2013) Impacts of climate change on plant growth, ecosystem services, biodiversity, and potential adaptation measure. Master Thesis in Atmospheric Science, Department of Biological and Environmental Sciences, University of Gothenburg, Sweden

Atkin OK, Edwards EJ, Loveys BR (2000) Response of root respiration to changes in temperature and its relevance to global. New Phytol 147:141–154. https://doi.org/10.1046/j.1469-8137.2000.00683.x

Augé RM (2001) Water relations, drought and vesicular–arbuscular mycorrhizal symbiosis. Mycorrhiza 11:3–42

Augé RM, Toler HD, Saxton AM (2015) Arbuscular mycorrhizal symbiosis alters stomatal conductance of host plants more under drought than under amply watered conditions: a meta-analysis. Mycorrhiza 25:13–24

Awal MA, Ikeda T, Itoh R (2003) The effect of soil temperature on source-sink economy in peanut (*Arachis hypogaea*) environ. Exp Bot 50:41–50

Bakken LR, Bergaust L, Liu B, Frostegard A (2012) Regulation of denitrification at the cellular level: a clue to the understanding of N_2O emissions from soils. Philos Trans R Soc B 367:1226–1234

Balasubramanian S, Sureshkumar S, Lempe J, Weigel D (2006) Potent induction of *Arabidopsis thaliana* flowering by elevated growth temperature. PLoS Genet 2:0980–0989. https://doi.org/10.1371/journal.pgen.0020106

Bellard C, Bertelsmeier C, Leadley P, Thuiller W, Courchamp F (2012) Impacts of climate change on the future of biodiversity. Ecol Lett 154:365–377. https://doi.org/10.1111/j.1461-0248.2011.01736.x

Berggren A, Björkman C, Bylund H, Ayres MP (2009) The distribution and abundance of animal populations in a climate of uncertainty. Oikos 118:1121–1126

Bodelier PLE, Roslev P, Henckel T, Frenzel P (2000) Stimulation by ammonium-based fertilizers of methane oxidation in soil around rice roots. Nature 403:421–424

Botkin DB, Saxe H, Araujo MB, Betts R, Bradshaw RHW, Cedhagen T, Chesson P, Dawson TP, Etterson JR, Faith DP, Ferrier S, Guisan A, Hansen AS, Hilbert DW, Loehle C, Margules C, New M, Sobel MJ, Stockwell DRB (2007) Forecasting the effects of global warming on biodiversity. Bioscience 57:227–236

Briones MJI, McNamara NP, Poskitt J, Crow SE, Ostle NJ (2014) Interactive biotic and abiotic regulators of soil carbon cycling: evidence from controlled climate experiments on peatland and boreal soils. Glob Chang Biol 20:2971–2982

Chao Q, Feng A (2018) Scientific basis of climate change and its response. Glob Energy Interconnect 1:420–427. https://doi.org/10.14171/j.2096-5117.gei.2018.04.002

Classen AE, Sundqvist MK, Henning JA, Newman GS, Moore JA, Cregger MA, Moorhead LC, Patterson CM (2015) Direct and indirect effects of climate change on soil microbial and soil microbial-plant interactions: what lies ahead? Ecosphere 6(8). https://doi.org/10.1890/ES15-00217.1

Cohn AS, Vanwey LK, Spera SA, Mustard JF (2016) Cropping frequency and area response to climate variability can exceed yield response. Nat Clim Chang 6:601–604. https://doi.org/10.1038/NCLIMATE2934

Compant S, Van Der Heijden MGA, Sessitsch A (2010) Climate change effects on beneficial plant-microorganism interactions. FEMS Microbiol Ecol. https://doi.org/10.1111/j.1574-6941.2010.00900.x

Cotton TEA, Fitter AH, Miller RM, Dumbrell AJ, Helgason T (2015) Fungi in the future: interannual variation and effects of atmospheric change on arbuscular mycorrhizal fungal communities. New Phytol 205:1598–1607. https://doi.org/10.1111/nph.13224

De Vries FT, Griffiths RI (2018) Impacts of climate change on soil microbial communities and their functioning. Dev Soil Sci 35. https://doi.org/10.1016/B978-0-444-63865-6.00005-3

DeAngelis KMG, Pold BD, Topcuoglu L, van Diepen TA, Varney RM, Blanchard JL, Melillo J, Frey SD (2015) Long-term forest soil warming alters microbial communities in temperate forest soils. Front Microbiol 6

Deltedesco E, Keiblinger KM, Piepho H-P, Antonielli L, P€, E. M., Zechmeister-Boltenstern, S., Gorfer, M. (2020) Soil microbial community structure and function mainly respond to indirect effects in a multifactorial climate manipulation experiment. Soil Biol Biochem 142:107704. https://doi.org/10.1016/j.soilbio.2020.107704

Denman KL, Brasseur G, Chidthaisong A, Ciais P, Cox PM, Dickinson RE, Hauglustaine D, Heinze C, Holland E, Jacob D, Lohmann U, Ramachandran S, Leite da Silva Dias P, Wofsy SC, Zhang X (2007) Couplings between changes in the climate system and biogeochemistry (PDF). In: Solomon S, Qin D, Manning M, Chen Z, Marquis M, Averyt KB, Tignor M, Miller HL (eds)

IPCC. 2007. Climate change 2007: the physical science basis. Contribution of working group I to the fourth assessment report of the intergovernmental panel on climate change. Cambridge University Press, Cambridge. Accessed 31 Aug 2009

Dhankher A (2018) Climate resilient crops for improving global food security and safety, vol 2018. Wiley

Easterling WE, Aggarwal PK, Batima P, Brander LM, Erda L, Howden SM et al (2007) Food, fibre and forest products. In: Parry ML, Canziani OF, Palutikof JP, van der Linden PJ, Hanson CE (eds) Climate change 2007: impacts, adaptation and vulnerability. Contribution of working group II to the fourth assessment report of the intergovernmental panel on climate change. Cambridge University Press, Cambridge, pp 273–313

Erda L, Wei X, Hui J, Yinlong X, Yue L, Liping B, Liyong X (2005) Climate change impacts on crop yield and quality with CO_2 fertilization in China. Philos Trans R Soc B: Biol Sci 3601463:2149–2154. https://doi.org/10.1098/rstb.2005.1743

Fao (2017) Adaptation to climate change in agriculture, forestry and fisheries: perspective, framework and priorities. In: Food and agriculture organisation of the united nations rome. https://www.google.com/url?sa=t&source=web&rct=j&url=http://www.fao.org/3/a-au030e.pdf&ved=2ahUKEwiVjJiG6L3rAhVJyTgGHeDOC54QFjABegQIDBAG&usg=AOvVaw0-rB_qVizHBxya7hSXCZC3

Fischer RA, Edmeades GO (2010) Breeding and cereal yield progress. Crop Sci 50(S1):S-85-S-98. https://doi.org/10.2135/cropsci2009.10.0564

Granier C, Massonnet C, Turc O, Muller B, Chenu K, Tardieu F (2002) Individual leaf development in *Arabidopsis thaliana*: a stable thermal-time-based programme. Ann Bot 89:595–604

Gray SB, Strellner RS, Puthuval KK, Shulman R, Siebers M, Rogers A, Leakey ADB (2013) Nodulation of field-grown soybean is enhanced by free-air CO_2 enrichment only when combined with drought stress. Funct Plant Biol 40(2):137–147

Gray SB, Siebers M, Locke AM, Rosenthal DR, Strellner RS, Paul RE, Klein SP, Ruiz UM, McGrath VJ, Dermody O, Ainsworth EA, Bernacchi CJ, Long SP, Ort DR, Leakey ADB (2016) Intensifying drought eliminates the expected benefits of elevated [CO_2] for soybean. Nat Plan Theory. https://doi.org/10.1038/NPLANTS.2016.132

Gross EM, Lombardo P (2018) Limited effect of gizzard sand on consumption of the macrophyte *Myriophyllum spicatum* by the great pond snail *Lymnaea stagnalis*. Hydrobiologia 812:131–145. https://doi.org/10.1007/s10750-016-2890-8

Haase S, Philippot L, Neumann G, Marhan S, Kandeler E (2008) Local response of bacterial densities and enzyme activities to elevated atmospheric CO_2 and different N supply in the rhizosphere of *Phaseolus vulgaris* L. Soil Biol Biochem 40:1225–1234

Hagerty SB, van Groenigen KJ, Allison SD, Hungate BA, Schwartz E, Koch GW et al (2014) Accelerated microbial turnover but constant growth efficiency with warming in soil. Nat Clim Chang 4:903–906. https://doi.org/10.1038/nclimate2361

Hammond ST, Brown JH, Burger JR, Flanagan TP, Fristoe TS, Mercado-Silva N, Nekola JC, Okie JG (2015) Food spoilage, storage, and transport: implications for a sustainable. Future Biosci 65 (8):758–776

Hatfield JL, Boote KJ, Kimball BA, Ziska LH, Izaurralde RC, Ort D, Thomson AM, Wolfe D (2011) Climate impacts on agriculture: implications for crop production. Agron J 103:351–370. https://doi.org/10.2134/agronj2010.0303

Hawkes CV, Hartley IP, Ineson P, Fitter AH (2008) Soil temperature affects carbon allocation within arbuscular mycorrhizal networks and carbon transport from plant to fungus. Glob Chang Biol 14:1–10

Hedhly A (2011) Sensitivity of flowering plant gametophytes to temperature fluctuations. Environ Exp Bot 74:9–16

Heinemeyer A, Ineson P, Ostle N, Fitter AH (2006) Respiration of the external mycelium in the arbuscular mycorrhizal symbiosis shows strong dependence on recent photosynthates and acclimation to temperature. New Phytol 171:159–170

IPCC (2014) Climate change 2014: synthesis report. In: Core Writing Team, Pachauri RK, Meyer LA (eds) Contribution of working groups I, II and III to the fifth assessment report of the intergovernmental panel on climate change. IPCC, Geneva, 151 pp

Isobe KK, Koba S, Otsuka, Senoo. K. (2011) Nitrification and nitrifying microbial communities in forest soils. J For Res 16:351–362

Jansson JK, Hofmockel KS (2019) Soil microbiomes and climate change. Nat Rev Microbiol 18:35–46. https://doi.org/10.1038/s41579-019-0265-7

Janus LR, Angeloni NL, McCormack J, Rier ST, Tuchman NC, Kelly JJ (2005) Elevated atmospheric CO_2 alters soil microbial communities associated with trembling aspen (*Populus tremuloides*) roots. Microb Ecol 50:102–109. https://doi.org/10.1007/s00248-004-0120-9

Johnson NC, Wilson GWT, Wilson JA, Miller RM, Bowker MA (2015) Mycorrhizal phenotypes and the law of the minimum. New Phytol 205:1473–1484

Jones DL, Nguyen C, Finlay RD (2009) Carbon flow in the rhizosphere: carbon trading at the soil-root interface. Plant Soil 321:5–33

Kang Y, Khan S, Ma X (2009) Climate change impacts on crop yield, crop water productivity and food security – a review. Prog Nat Sci 19:1665–1674

Kardol P, Cregger MA, Campany CE, Classen AT (2010) Soil ecosystem functioning under climate change: plant species and community effects. Ecology 91(3):767–781. https://doi.org/10.1890/09-0135.1

Karmakar R, Das I, Dutta D, Rakshit A (2016) Potential effects of climate change on soil properties: a review. Sci Int 42:51–73. https://doi.org/10.17311/sciintl.2016.51.73

Kellner H, Vandenbol M (2010) Fungi unearthed: transcripts encoding lignocellulolytic and chitinolytic enzymes in forest soil. PLoS One 5:E10971. https://doi.org/10.1371/journal.pone.0010971

Kleinhesselink AR (2017) Direct and indirect effects of climate change on plant populations and communities in Sagebrush Steppe. In All Graduate Theses and Dissertations. https://digitalcommons.usu.edu/etd/5417

Kulmatiski A, Kardol P (2008) Getting plant–soil feedbacks out of the greenhouse: experimental and conceptual approaches. In: Luttge UE, Esser K, Beyschlag W, Murata J (eds) Progress in botany-69 (Review genetic physiology ecology). Springer, Berlin, pp 449–472

Leake J, Johnson D, Donnelly D, Muckle G, Boddy L, Read D (2004) Networks of power and influence: the role of mycorrhizal mycelium in controlling plant communities and agroecosystem functioning. Can J Bot 82(8):1016–1045. https://doi.org/10.1139/b04-060

Leakey ADB (2009) Rising atmospheric carbon dioxide concentration and the future of C4 crops for food and fuel. Proc R Soc B: Biol Sci 276:2333–2343

Leakey ADB, Bernacchi CJ, Ort DR, Long SP (2006) Long-term growth of soybean at elevated [CO_2] does not cause acclimation of stomatal conductance under fully open-air conditions. Plant Cell Environ 29:1794–1800

Lobell DB, Asner GP (2003) Climate and management contributions to recent trends in U.S. agricultural yields. Science 299:1032

Mahato A (2014) Climate change and its impact on agriculture. Int J Sci Res 4:1–6

McGrath JM, Lobell DB (2013) Regional disparities in the CO_2 fertilization effect and implications for crop yields. Environ Res Lett 8(1):014054. https://doi.org/10.1088/1748-9326/8/1/014054

Mhlongo MI, Piater LA, Madala NE, Labuschagne N, Dubery IA (2018) The chemistry of plant–microbe interactions in the rhizosphere and the potential for metabolomics to reveal signaling related to defense priming and induced systemic resistance. Front Plant Sci 9. https://doi.org/10.3389/fpls.2018.00112

Miller D (2001) Distributing responsibilities. J Polit Philos 9(4). https://doi.org/10.1111/1467-9760.00136

Miller RM, Jastrow JD, Reinhardt DR (1995) External hyphal production of vesicular-arbuscular mycorrhizal fungi in pasture and tallgrass prairie communities. Oecologia 103:17–23

Mishra T (2016) Climate change and production of secondary metabolites in medicinal plants: a review. Int J Herb Med 4:27–30

Mohan JE, Cowden CC, Baas P, Dawadi A, Frankson PT, Helmick K et al (2014) Mycorrhizal fungi mediation of terrestrial ecosystem responses to global change: mini-review. Fungal Ecol 10:3–19. https://doi.org/10.1016/j.funeco.2014.01.005

Nelson GC, Valin H, Sands RD, Havlík P, Ahammad H, Deryng D, Elliott J, Fujimori S, Hasegawa T, Heyhoe E, Kyle P, Von Lampe M, Lotze-Campen H, Mason D'croz D, Van Meijl H, Van Der Mensbrugghe D, Müller C, Popp A, Robertson R, Performed DW (2014) Climate change effects on agriculture: economic responses to biophysical shocks. Soc Sci 111:3274–3279. https://doi.org/10.1073/pnas.1222465110

Pandey D (2020) Agricultural sustainability and climate change nexus. In: Contemporary environmental issues and challenges in era of climate change. Springer Nature, Singapore. https://doi.org/10.1007/978-981-32-9595-7_4

Pugnaire FI, Morillo JA, Peñuelas J, Reich PB, Bardgett RD, Gaxiola A, Wardle DA, Van Der Putten WH (2019) Climate change effects on plant-soil feedbacks and consequences for biodiversity and functioning of terrestrial ecosystems. Sci Adv 5–11. https://doi.org/10.1126/sciadv.aaz1834

Rai R (2020) Heat stress in crops: driver of climate change impacting global food supply. In: Contemporary environmental issues and challenges in era of climate change. Springer Nature, Singapore, pp 99–117. https://doi.org/10.1007/978-981-32-9595-7_5

Raza A, Razzaq A, Mehmood SS, Zou X, Zhang X, Lv Y, Xu J (2019) Impact of climate change on crops adaptation and strategies to tackle its outcome: a review. Plan Theory 8:34. https://doi.org/10.3390/plants8020034

Rosenthal DM, Ort DR (2012) Examining cassava's potential to enhance food security under climate change trop. Plant Biol 5:30–38

Rosenthal DM, Slattery RA, Miller RE, Grennan AK, Cavagnaro TR, Fauquet CM, Gleadow RM, Ort DR (2012) Cassava about-FACE: greater than expected yield stimulation of cassava (*Manihot esculenta*) by future CO_2 levels. Glob Change Biol 18:2661–2675

Salam MA, Noguchi T (2005) Impact of human activities on carbon dioxide (CO_2) emissions: a statistical analysis. Environmentalist 25:19–30

Sanders IR, Streitwolf-Engel R, Van Der Heijden MGA, Boller T, Wiemken A (1998) Increased allocation to external hyphae of arbuscular mycorrhizal fungi under CO_2 enrichment. Oecologia 117:496–503

Schimel JP, Schaeffer SM (2012) Microbial control over carbon cycling in soil. Front Microbiol 3. https://doi.org/10.3389/fmicb.2012.00348

Sheffield J, Wood EF (2008) Projected changes in drought occurrence under future global warming from multi-model, multi-scenario, IPCC AR4 simulations. Clim Dyn 31:79–105

Tang J, Xu L, Chen X, Hu S (2009) Interaction between C_4 barnyard grass and C_3 upland rice under elevated CO_2: impact of mycorrhizae. Acta Oecol 35:227–235

Van Der Putten WH, Bardgett RD, Bever JD, Martijn Bezemer T, Casper BB, Fukami T, Kardol P, Klironomos JN, Kulmatiski A, Schweitzer JA, Suding KN, Van De Voorde TFJ, Wardle DA (2013) Plant-soil feedbacks: the past, the present and future challenges. J Ecol 101(2):265–276. https://doi.org/10.1111/1365-2745.12054

Veresoglou SD, Anderson IC, de Sousa NMF, Hempel S, Rillig MC (2016) Resilience of fungal communities to elevated CO_2. Microb Ecol 72:493–495

Walck JL, Hidayati SN, Dixon KW, Thompson K, Poschlod P (2011) Climate change and plant regeneration from seed. Glob Change Biol 17:2145–2161

Warren R (2011) The role of interactions in a world implementing adaptation and mitigation solutions to climate change. Trans R Soc A 369:217–241. https://doi.org/10.1098/rsta.2010.0271

Whitaker JN, Ostle AT, Nottingham A, Ccahuana N, Salinas RD, Bardgett P, Meirand NP, McNamara (2014) Microbial community composition explains soil respiration responses to changing carbon inputs along an Andes-to-Amazon elevation gradient. J Ecol 102:1058–1071

Whittle CA, Otto SP, Johnston MO, Krochko JE (2009) Adaptive epigenetic memory of ancestral temperature regime in *Arabidopsis thaliana*. Botany (Botanique) 87:650–657

Wilson H, Johnson BR, Bohannan B, Pfeifer-Meister L, Mueller R, Bridgham SD (2016) Experimental warming decreases arbuscular mycorrhizal fungal colonization in prairie plants along a Mediterranean climate gradient. Peer J 4:e2083. https://doi.org/10.7717/peerj.2083
Xu ZZ, Zhou GS (2006) Combined effects of water stress and high temperature on photosynthesis, nitrogen metabolism and lipid peroxidation of a perennial grass Leymus chinensis. Planta 224:1080–1090
Zavalloni C, Vicca S, Büscher M et al (2012) Exposure to warming and CO_2 enrichment promotes greater above-ground biomass, nitrogen, phosphorus and arbuscular mycorrhizal colonization in newly established grasslands. Plant Soil 359:121–136
Zhang T, Yang X, Gu R, Guo J (2016) Response of AM fungi spore population to elevated temperature and nitrogen addition and their influence on the plant community composition and productivity. Sci Rep 6:24749. https://doi.org/10.1038/srep24749
Ziska LH (2008) Rising atmospheric carbon dioxide and plant biology: the overlooked paradigm. In: Kleinman DL, Cloud-Hansen KA et al (eds) Controversies in science and technology, from climate to chromosomes. Liebert, New Rochele, 379–400

Chapter 12
How Climate Change Alters Soil Productivity

Moni Kumari and Rajiv Kumar

Abstract Climate change or global warming occurs constantly. Climate change drivers including temperature, moisture, altered rainfall pattern, and greenhouse gases (GHGs) are likely to have harmful effects on different soil properties and processes. The results of the changed climatic conditions are environmental problems like soil erosion, soil degradation, loss of soil fertility, and desertification. The development of several soil processes and properties depends on their soil organic matter (SOM) content. The presence of organic matter in the soil enhanced soil structure and its quality. A section in the present chapter especially emphasized on how climate change alters the organic content of the soil, resulting in reduced soil productivity. The present study deals about the physical, chemical, and biological parameters necessary for a healthy soil and how the changing climatic condition affects soil productivity. A little idea about CO_2 fertilization effect is also discussed here and how the other factors of climate change combat the effect of CO_2 fertilization, so that the overall outcome of climate change on soil productivity is adverse.

Keywords Climate change · Greenhouse gases · Soil organic carbon · Soil organic matter · Soil leaching · Soil acidification

12.1 Introduction

Climate of an area is determined by the temperature and moisture content of that area. Presently, climate change is acknowledged as the most serious issue facing the planet. Climate change or global warming occurs constantly. The rate of climate change is enhanced very rapidly from the last 150 years. By 2090–2099, the average

M. Kumari (✉)
Department of Botany, Gaya College, Gaya, Bihar, India

R. Kumar
Department of Civil Engineering, Dr. B R Ambedkar National Institute of Technology, Jalandhar, Punjab, India

D. K. Choudhary et al. (eds.), *Climate Change and the Microbiome*, Soil Biology 63,
https://doi.org/10.1007/978-3-030-76863-8_12

Table 12.1 Summary of expected effects of individual climate change variables on soil processes

1.	Increasing temperature	Loss of soil organic matter
		Reduction in labile pool of SOM
		Reduction in moisture content
		Increase in mineralization rate
		Damage of soil structure
		Enhanced soil respiration rate
2.	Increasing CO_2 concentration	Soil organic matter (SOM) is increased
		Water use efficiency is enhanced
		More carbon is easily available to soil microorganisms
		Nutrient cycling is accelerated
3.	Increasing rainfall	Soil moisture is increased
		Enhanced surface runoff and erosion
		Improved SOM
		Leaching of nutrients is noticed
		Increased reduction of Fe and nitrates
		More volatilization loss of nitrogen
		Arid regions' productivity is increased
4.	Reduction in rainfall	Decline SOM
		Soil alkalinity
		Reduction in nutrient availability

worldwide temperature will most likely rise between 1.1 and 6.4 °C (Brevik 2013). Based on the data presented by IPCC globally between 1970 and 2004, the emission of greenhouse gases rose by 70% (IPCC 2007). The adverse effects of climate change are linked to droughts, heavy rainfall, storms, frosts, and the increasing levels of the sea in coastal areas and temperature increase. It has been studied that the predicted change in climate is expected to have profound effects on the natural ecosystem as well as on soil water availability, carbon storage, and yields (Chao and Feng 2018; Cox et al. 2018). The results of the changed climatic conditions are environmental problems like soil erosion, soil degradation, loss of soil fertility, and desertification. Table 12.1 shows the expected effects of individual climate change variables on soil processes.

Climate change drivers including temperature, moisture, altered rainfall pattern, and greenhouse gases (GHGs) are likely to have profound effects on different soil properties and processes, thus having significant effects on soil fertility and productivity. However, the consequences of climate change factors cannot be assessed independently. They are interrelated and influence one over the other. The outcome of changing climatic condition on the soil properties and processes depends on the region, its magnitude, and the type of soil present in that region. In India, 9 types of soil are present out of the 12 types that exist in the world. India is subdivided into 15 agroclimatic zones, with different seasons, crop yields, and agriculture systems.

The impact of different climate change factors on the soil fertility is discussed in this chapter and also explained with the various study reports.

Soil is the sole vital natural resource available to all living systems present on the planet. It is the base of all factors, namely, food, feed, clothes, shelter, and medicine. Its multifunction role enormously supports the living system by biomass production, fiber, and fuel. It maintains biodiversity and acts as a reservoir of the gene pool. Besides that, other important ecological functions of the soil are detoxification of organic materials, nutrient recycling, carbon turnover, carbon sequestration, and water as well as air purifier of the terrestrial system. It also plays a major role in conserving the quality and quality of water. Nitrogen and carbon cycle together with hydrological cycle linked soils to the climate system (Brevik 2013). Either directly or indirectly, soil functions are affected by climate change. The direct consequences of climate change on soil include transformation of soil organic carbon (SOC), nutrient cycling, soil moisture content, soil temperature, and altered rainfall pattern. Climate change and soil management can change the ability of soils to perform soil functions. Various study reports are there dealing with effects of climate change on soil functions (Coyle et al. 2016; Ostle et al. 2009; Xiong et al. 2014). Due to global warming, it has been reported by Díaz et al. (1997) that there is a decrease in dry matter production and the rate of soil mineralization gets increased. These changes in the climatic conditions adversely affect soil functions like reduced soil fertility, soil compaction, and soil erosion, thus lowering agricultural productivity (Lal et al. 2011). In the future, threatening situations will be present for food due to climate change (Brevik 2013; Fanzo et al. 2018).

Global climate change is greatly influenced by the soil since it acts as a natural sink for the atmospheric CO_2. An increase in SOC positively influences soil fertility, soil structure, the water-holding capacity of the soil, nutrient retention, and decreased soil erosion (Hoyle 2013). Climate change has positively influenced the agriculture system via CO_2 fertilization effects. It is mentioned in Kulshreshtha (2019) that there is a possibility of CO_2 fertilization under the influence of increased atmospheric CO_2 that ultimately enhances crop yields especially in C3 plants like canola and wheat. Under climate change, water using the efficiency of crops also get increased that may further support crop yield. An increase in atmospheric CO_2 has ultimately supported the soil organic carbon (SOC) that may positively influence crop productivity. It has been demonstrated by Cheng and Johnson (1998) that under the elevated CO_2 environment, plant produced 15–26% more biomass.

The basic needs (physical, chemical, or biological) for plant growth are provided by the soil only (Abbott and Murphy 2007). For proper growth and development of plants, 14 soil-derived nutrients are required. If the soil has the potential to supply all the 14 nutrients, then the soil is fertile. The most important features of soil are its fertility and how it is managed for crop production in current agriculture system Due to less fertile soil, developing nations in continents like Asia and Africa face food security problem (St. Clair and Lynch 2010; Sanchez and Swaminathan 2005). It has been estimated by Lal (2008) that nearly 40% of agricultural land in Europe is prone to soil degradation.

This chapter focuses on the how changing climatic conditions alter or harm soil health with a special focus on soil fertility.

12.2 Soil Parameter Determines Soil Health

Soil health is also referred to as soil quality. Soil health of an area is influenced by several factors including climate of that area, agriculture system, urbanization, forestry, waste disposal, natural events, etc. Maintenance of soil health is very important. Further, soil health has impacts on atmospheric balance, plant health, soil microbial community, soil ecosystem health, and the most important human health. Doran (2002) initially discusses the concept of soil health. He stated that in order to sustain plant life, maintenance of air and water quality, promotion of animal and plant health, and the role of living soil are very important.

The basis of agriculture and natural plant communities is soil. Soil health depends upon the different physical, chemical, and biological parameters of the soil. Physical indicators of soil health include moisture, soil texture, aggregation, infiltration, porosity, and bulk density. Chemical indicators comprise cation exchange capacity; total C and N content; mineral nutrients; availability of N, P and K; and organic matter present in the soil, while the biological indicators of soil health are the microbial biomass C and N, soil respiration, soil enzyme, and the floral and faunal biodiversity (Girija Veni et al. 2020). The soil biological, physical, and chemical health depend on a number of factors; thus, determination of soil health status is a very complex process (Fig. 12.1).

Today, soil health is destroyed very rapidly in the form of soil erosion, soil alkalinity, soil acidity, loss of soil organic carbon (SOC), loss of soil organic matter (SOM), soil contamination, and soil compaction. Due to climate change, the pattern of rainfall (altered precipitation) has been changed, and the atmospheric temperature and carbon dioxide (CO_2) concentrations are elevated and the atmospheric nitrogen gets deposited. It is expected that climate change vigorously affects different physiochemical parameters of the soil including organic matter, moisture, pH, salinity, and bulk density. These altered climatic factors adversely affect soil health profile. Incidence of extreme climate variabilities including floods, drought, and occasional heavy storms has also become very frequent, destroying soil health and threatening soil fertility, agricultural productivity, and food security (Nikolskii et al. 2010). It has been observed that food insecurity is more likely to occur in regions of the world where soil degradation is more common, for example, in sub-Saharan Africa and South Asia (Abrahams 2002).

Conservation of soil health is very essential to fulfill food security for the future generations and to mitigate the harmful effects of the changing climate. In order to enhance economic profit and agronomic profit, maintenance of soil health is also important.

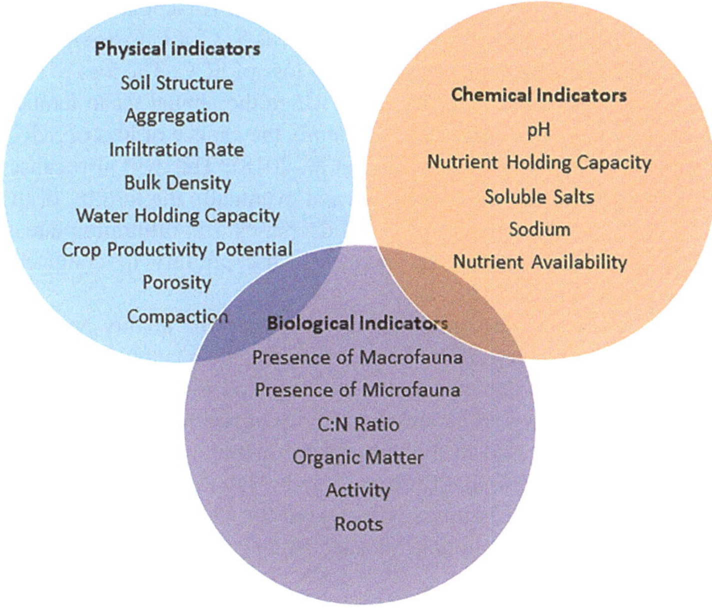

Fig. 12.1 Several physical, chemical and biological indicators that determining soil health

12.3 How Climate Change Influenced Soil Productivity

12.3.1 Impact of Climate Change on Soil Organic Carbon (SOC) and Soil Productivity

The development of several soil processes and properties depends on their SOM content. The presence of organic matter in the soil enhanced soil structure and its quality. Besides that, it also plays a major role in soil fertility, nutrient storage capacity, and water holding capacity (Obalum et al. 2017).

The nitrogen (N) and carbon (C) dynamics of the atmosphere are changed with the changing climatic conditions. Soil organic carbon and soil inorganic carbon together constitute the total carbon content of the soil. The changing climatic conditions mainly poor rainfall and increasing atmospheric temperature intensify the loss of organic carbon from the soil. Lehman et al. (2015) reported that soil health primarily depends on the soil organic matter (SOM) content. Fifty percent of the total SOM mass is composed of only soil organic carbon (SOC). Jenny (1941) reported SOC content of the soil is beneficial to vegetation pattern, inherent fertility of soil, soil moisture content, soil structure, soil erosion reduction, water holding potential of the soil, parent material, and soil biological activity (Sanderman et al. 2010; Hoyle 2013).

The deposition of organic matter to the soil is highly sensitive to the temperature. SOC content of the soil is inversely related with the atmospheric temperature. As atmospheric temperature increases, the SOC decreases. Kirschbaum (1995) reported that in regions where the annual mean temperature is 5 °C, a 1 °C rise in atmospheric temperature due to climate change causes 10% loss of SOC. The loss of SOC is 3% only with the same increase in temperature where the annual mean temperature is 30 °C. With increasing atmospheric temperature, the chance of loss of SOC is more in older soil than the younger one (Zhou et al. 2018). This will also cause loss of SOC. Thus, the soil health will deteriorate and eventually the fertility of the soil is lost (Lal. 2004). As SOC content of the soil decreases, soil infiltration rate, stability of soil aggregate, soil runoff, and increased susceptibility to compaction also decrease (Guo et al. 2019).

Under the elevated atmospheric CO_2 environment, the quality and quantity of root exudate have been changed. In the root exudate, the quantity N-rich metabolite is decreased, whereas C-rich metabolite is increased under the climatic change conditions. This will further stimulate the activity of soil microbial community. As a result of this, CO_2 released to the atmosphere or the amount of organic carbon stored in the soil is decreased (Tarnawski and Aragno 2006). Because of CO_2 fertilization effect, the plant biomass is increased by 15–26% and at the same time a 56–74% increase in C respiration released by the root rhizosphere (Cheng and Johnson 1998).

Changing climate causes alteration in rainfall pattern, and this may lead to either droughts or floods. Intense rainfall vigorously losses the organic carbon content of the soil. This will further result in increased bulk density of the soil, causing soil compaction and decreased soil porosity, and thus compact layer formation eventually inhibits root growth (Singh et al. 2011). Also, intense rainfall pattern may exaggerate soil leaching, reduce soil pH, and then cause soil acidification.

At present, important management practices have been used, leading to reduction of soil surface erosion, maintenance of soil total nutrient content, improvement of soil fertility, and overall preservation of soil health. By using modern technology if crop productivity is increased, than the amount of crop residues is also increased. Thus, the formation of soil organic matter or the level of SOC has been increased (Snyder et al. 2009). Any management practices that are able to preserve SOC ensure food security for the future generation.

12.3.2 Impact of Climate Change on Soil Temperature and Soil Productivity

Soil moisture and soil temperature are the main drivers of soil processes. Air temperature and soil temperature are complementary to each other. When the air temperature increases, the soil temperature also increases. The temperature system of the soil is governed by increases and misfortunes of radiation at the surface, the procedure of dissipation, heat conduction through the soil profile, and convective transfer by the movement of gas and water (Karmakar et al. 2016). By the end of the twenty-first century, due to the tremendous emission of greenhouse gases (GHGs), the global temperature is projected to increase by 2 °C–4 °C.

The increase in soil tempertaure will increase soil processes, make decomposition of organic matter rapid, increase microbiological activity, quicken nutrient release, increase nitrification rate, and generally accentuate chemical weathering of minerals. However, soil temperatures will also be affected by the type of vegetation occurring at its surface, which may change itself because of climate change or adaptation management. Besides, in many regions, climatic change conditions, mainly enhanced air temperature, would intensify evaporative demand (IPCC 2019). Eventually, this leads to deficit in soil moisture content, and if the drainage system is not well established, soil salinity will increase and the carbon storage capacity of the soil is reduced (Ostle et al. 2009; Xiong et al. 2014). Climate change consequences including elevated temperatures and atmospheric CO_2 concentrations resulted in decline in water content of the soil; thus, the risk of wind soil erosion has been increased. In climatic conditions of the Asian grassland, it is observed via simulation experiments that when atmospheric temperature is increased by 2 °C, soil erosion via wind will increase by approximately 25% (Gao et al. 2003). This is because at high temperatures, the stability of soil aggregates has been decreased and surface soil erosion is increased (Edwards 1991).

The enzymatic activity of microbial community present in the soil is an important factor to assess soil quality, which can reveal soil fertility (Guoju et al. 2012). Due to increasing atmospheric temperature, soil temperature has been increased. It directly or indirectly affects the structure of microbial community, its enzymatic activities, total microbial biomass, microbial respiration, microbial diversity, decomposition of soil organic matter, and mineralization rate of SOM (Xue et al. 2003).

As climate warms, microbial populations shrink or die. Numerous studies have demonstrated an expansion in microbial biomass in transient trials; however, over the long term under raised temperature, biomass is bound to diminish. This is because the effectiveness of microbial growth changes at higher temperatures. Some physiological changes are observed in microorganisms at higher temperatures including configuration of membrane-bound lipid and the expression of warmth shock proteins. In the in vivo environment, rise in atmospheric temperature caused shift of microbial community and functional gene field (De Vries and Griffiths 2018; Jansson and Hofmockel 2019). Catalase and urease enzymes of the soil microorganisms are linked with the conversion of SOC and total N content of the soil, respectively. Several study reports observed that if the atmospheric temperature is increased, it facilitates the decomposition of soil organic matter; thus, the fertility of the soil has been decreased (Hinojosa et al. 2004).

Thakur et al. (2016) reported that due to increase in soil temperature, interaction between plant and soil biota has been potentially modified. Temperature also influences the plant physiological processes including photosynthesis and root respiration because it affects the activities of both aboveground and belowground enzymes (Liang et al. 2013; Atkin et al. 2000). Heinze et al. (2016) observed that due to warmer temperature, the soil root structure is altered, whereas the biomass production is increased and the ratio of plant root to shoot is decreased.

Nutrient-use efficiencies and nutrient uptake by the plants are also positively influenced by increasing soil temperature (Gavito et al. 2001).

Possible vulnerabilities due to increase in atmospheric temperature are decrease in water table, increase in evapotranspiration, soil erosion, landslides, floods, inundation of standing crops, and reduction of soil fertility.

12.3.3 Effects of Higher CO_2 due to Climate Change on Soil Productivity

The global climate change has produced changes in the soil composition, since different types of soils are closely linked with the geological carbon and nitrogen cycles. Soils with a sufficient amount of organic matter are more productive than soils that are deficient in organic matter. One of the main impacts of climate change on soil processes and resources is how soils will be influenced by the future changes in the C and N cycles (Brevik 2013). The disturbance in the nitrogen and phosphorous cycles has produced significant changes in the soil organic matter. The flow of carbon and nitrogen from land to the atmosphere is increased due to higher concentration of atmospheric CO_2. Thus decreasing the soil carbon and nitrogen content.

This problem is further increased by soil tilling, which changes the soil structure and resulted in the decrease of carbon content (Brevik 2013). Other parameters such as the moisture content, global CO_2 levels, and global temperatures also affect the soil fertility. However, the changes in the soil caused by these parameters are highly region-specific along with being dependent on the magnitude of changes in climate (Pareek 2017). Early expectations were that increased atmospheric CO_2 would lead to increased productivity of plants coupled with increased soil C sequestration, which would help offset rising atmospheric CO_2 levels by raising plant growth and the soil–plant system. The CO_2 fertilization effect may not be as high as originally believed. However, recent studies suggest that rising ozone levels will potentially offset the effect of CO_2 fertilization that contributes to decreased plant growth under high CO_2, and the negative effects of elevated temperatures on plant growth can also cancel out any impact of CO_2 (Hattenschwiler et al. 2002; Poorter and Navas 2003; Zaehle et al. 2010). Plant growth may be adversely affected by nitrogen limitations, and modeling of C dynamics as influenced by N suggests less C sequestration by soil than originally expected providing CO_2 fertilization. Some researchers conclude that as opposed to a CO_2 fertilization effect, the increases in plant productivity they saw were largely attributable to the soil moisture status (Niklaus and Körner 2004). Experiments looking at the decomposition of plant tissues grown under elevated atmospheric SOC decrease due to increased microbial activity under elevated atmospheric CO_2 levels (Carney et al. 2007; Brevik 2013). It is observed with the help of FACE (free-air CO_2 enrichment) experiment that as compared to ambient environment in the high CO_2 environment, pH drop more rapidly in submerged soil (Kumeleh et al. 2009). In the same experiment, it is found

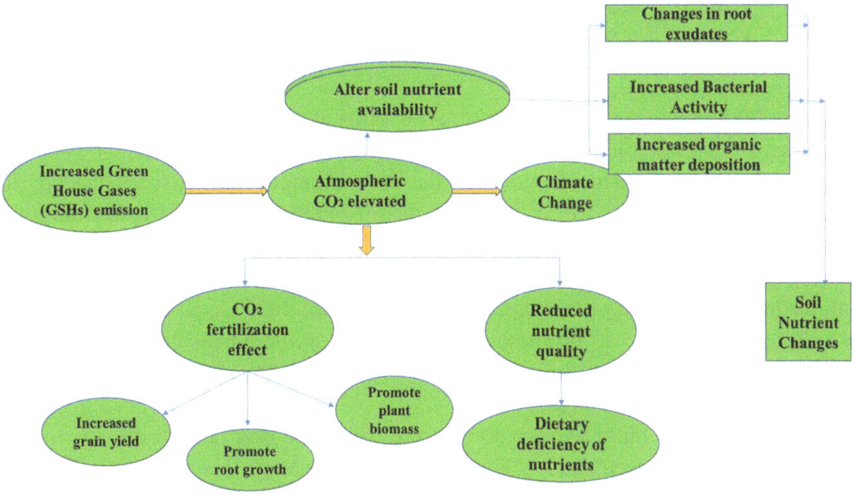

Fig. 12.2 Schematic diagram showing the impact of enhanced CO_2 concentrations on soil fertility

that elevated CO_2 environment causes deficiency of minerals like nitrogen, calcium, and magnesium in the soil. The minerals present in the soil form complexs that are not readily utilized by the plants at change soil pH. On the other hand, soil organic matter, potassium (K), phosphorus (P), iron (Fe), zinc (Zn), and manganese (Mn) are found to increase in soil concentration under the elevated atmospheric CO_2 level. The elevated atmospheric CO_2 alter the soil mineral equilibrium; thus, the nutrient-use efficiency and mineral uptake by the plants get badly affected (Nakandalage and Seneweera 2018). Thus, due to increasing atmospheric CO_2 concentration, the nutrient imbalance of minerals may occur in the future, and it will adversely affect soil fertility. Across a wide variety of plant species due to CO_2 fertilization, the effect of the elevated CO_2 atmosphere in plant growth may increase, but the quality of the grain is reduced (Dong et al. 2018).

It is expected that the drivers of climate change, such as humidity, temperature, and CO_2, would have considerable and variable effects on different soil processes and properties related to soil fertility and productivity. These consequences of the causes of climate change, however, cannot be considered independently, being one factor influencing the other, and the resulting impact will be complex. Furthermore, depending on the extent of climate change, soil properties and climatic conditions, all of these impacts may be extremely region-specific. Since climate change is a reality, it will have a direct and indirect effect on soil development processes and crop production-related resources that affect the livelihoods of millions of people worldwide (Fig. 12.2).

12.3.4 Rainfall Pattern/Altered Precipitation due to Climate Change Effect Soil Productivity

In the rainfed areas, it is soil moisture availability that determines the seed germination crop stand establishment. Alteration in rainfall patterns due to climate change largely affects the moisture content of the soil. Changed rainfall pattern leads to either flooding or drought. In both cases, it disturbs the water table and physiology of trees along with soil properties. In addition to deleterious effects on the carbon and water equilibrium of the plants, mineral nutrition will also be influenced badly in the future due to extreme water availability (Ciais et al. 2005). Soil N is lost due to waterlogging because it promotes nitrate leaching, denitrification, and runoff (Kopyra and Gwód 2004). Soil N mineralization rate is also greatly reduced due to waterlogging (Haddad et al. 2013).

Depending upon the buffering pool of the soil, increased rainfall due to climate change leads to enhanced soil leaching and soil acidification. There is a possibility of greater nutrient loss from the soil. Increased deposition of N to the soil, soil disturbances, enhanced soil erosion, flooding, and habitat changes also occur due to increased rainfall (Karmakar et al. 2016). Water holding capacity of the atmosphere has been increased by 7% due to 1 °C rise in atmospheric temperature. Thus, the water vapor content of the atmosphere is increased, and eventually it leads to more intense rainfall. In the United Kingdom, only due to 7% increase in rainfall intensity soil erosion is enhanced by 26% (Favis-Mortlock and Boardman 1995). It has been predicted that this altered rainfall pattern and rising atmospheric temperature have profound effect on soil productivity in the future (Gang et al. 2015). As the rate of soil erosion is increased due to climate change, it adversely affects food production.

Because of high intensity of rainfall due to climate change, there is a possibility that the rate of soil erosion will be affected. Soil erodibility, land use pattern, and vegetation cover will also get disturbed. Peat formation and methane release are also enhanced by increased rainfall (Chander 2012). A study report of an experiment conducted in Australia indicated that due to increased rainfall intensity, there is significant increase in the chances of surface runoff. The probabilities of surface runoff are much more likely in arid areas than in temperate and wet areas (Chiew et al. 1995). The chance of more surface runoff leads to enhanced soil erosion. UK water erosion model depicted that during wet years the rate of soil erosion is increased by 150% if rainfall in winter is increased by 10% (Favis-Mortlock and Boardman 1995). Teng et al. (2017) estimated that erosion rates of the Tibetan Plateau will increase by 14%–41% by the mid-twenty-first century. If the drainage system of an area is not working properly, then waterlogging may occur due to intense rainfall. It leads to oxygen deprivation in the soil. And due to lack of proper supply of oxygen from the soil to the plant roots and microbial community present in the soil, their growth is retarded.

Decreased rainfall could undergo peat and CO_2 loss and increased moisture deficit for arable crops (especially on shallow soils) and for forest soils, thereby affecting foraging patterns, reproduction, and survivability of the soil invertebrates (Chander 2012). Low-rainfall due to climate change resulted in enhanced soil salinity or soil alkalinity. Many study reports of soil erosion and climate change depicted that it depends on the change in the mean annual erosion rates, which in temperate zones are thought to exhibit a linear relation to annual precipitation amount (Obalum et al. 2017; Rounsevell et al. 1999). A decrease in the availability of water, caused by the climatic changes, in the eco-geomorphic system of the Mediterranean region may cause open bare soil patches in the future (Lavee et al. 1998). Salinization, acidification, and soil erosion caused by climate change are majorly responsible for the soil degradation (Rounsevell et al. 1999). In case of decreased rainfall due to climate change, soil erosion is also accelerated because of increased risk of wind erosion (Parry et al. 1998). If erosion rates either via water or wind are unchecked, degradation of soil will continue. Due to low rainfall in the Southern Europe, countries like Serbia, Hungary, Romania, and Bulgaria will face the most negative impact, i.e., very low crop productivity (Olesen et al. 2011).

12.3.5 Impact of Climate Change on Soil Microbial Communities and Soil Productivity

Soil microbial community is an indicator of soil biological health, and it plays a major role in retaining soil fertility (Zhu et al. 2020). Microbial community present in the soil transforms carbon (C) biomass present there into humus. Due to climate change, an increase in atmospheric temperature, altered precipitation, and modified vegetation have significant impact on the biological soil processes. It also leads to alternation in distribution of species across the globe. Climate of an area also determines the interactions among species. Factors of climate change both positively and negatively affect the microbial communities present in the soil (Varma et al. 2019). The physiology, temperature sensitivity, and growth rate of soil microbial communities are dependent on the factors of climate change (Whitaker et al. 2014). Due to global warming in temperate forest, an increase in temperature by 5 °C leads to alternation in soil bacterial abundance and the percentage of bacterial and fungal community in the soil (DeAngelis et al. 2015). The microbial activity, growth rate, and processing are increased with increase in temperature (Bradford et al. 2008). Climate change lowers moisture content of the soil; thus, it leads to cytoplasmic dehydration and slows down the diffusion of substrate within the soil-nitrifying bacteria; hence, its activity gets reduced (Stark and Firestone 1995). Due to global warming, the concentration of CO_2 in the atmosphere has been increased, indicating high degree of variability in microbial biomass and microbial immobilization. Soil wetness or soil moisture is a noble way to illustrate land use changes and climate–hydrology. Karmakar et al. (2016) mentioned in his study report that it is the moisture content of the soil which determines framing processes, soil weathering, and complex soil formation processes (podzolization), movement of clay, and gleying. The rate of water evaporation from the soil surface is enhanced with increasing atmospheric temperature. The moisture content of the soil system has decreased, and this would affect the soil microbial communities (De Vries and Griffiths 2018).

How soil microbial community is influenced and changed with its moisture level is highly variable. There is no such record that linked soil respiration and moisture and temperature together and its influence on each other.

12.4 Conclusion

Understanding the impact of climate change on soil productivity is very necessary. The mechanism behind alternation of soil physical, chemical and biological properties due to changing climatic conditions is very important. That eventually leads to depecit soil fertility. The impact of climate change on soil is changed with changing soil type and factors of climate change. The chapter concludes how the factors of climate change deteriorate soil quality. It is also very essential to study and opt for conservation practices in order to maintain and improve soil structure. This would help agriculture system to adapt to changing climatic conditions. Threats of food security due to climate change are only overcome by maintaining soil health.

References

Abbott LK, Murphy DV (2007) What is soil biological fertility? In: Abbott LK, Murphy DV (eds) Soil biological fertility – a key to sustainable land use in agriculture. Springer, New York, pp 1–15

Abrahams P (2002) Soils: their implications to human health. Sci Total Environ 291:1–32

Atkin OK, Edwards EJ, Loveys BR (2000) Response of root respiration to changes in temperature and its relevance to global change. New Phytol 147:141–54

Bradford MA, Davies CA, Frey SD, Maddox TR, Melillo JM, Mohan JE, Reynolds JF, Treseder KK, Wallenstein MD (2008) Thermal adaptation of soil microbial respiration to elevated temperature. Ecol Lett 11:1316–1327

Brevik EC (2013) The potential impact of climate change on soil properties and processes and corresponding influence on food security. Agriculture 3(3):398–417

Carney KM, Hungate BA, Drake BG, Megonigal JP (2007) Altered soil microbial community at elevated CO_2 leads to loss of soil carbon. PNAS USA 104(12):4990–4995

Chander S (2012) Impact of climate change on insects. In: Pathak H, Aggarwal PK (eds) Climate change impact, adaptation and mitigation in agriculture: methodology for assessment and application. Indian Agricultural Research Institute, New Delhi, p 131

Chao Q, Feng A (2018) Scientific basis of climate change and its response. Global Energy Interconnect 1(4):420–427

Cheng WX, Johnson DW (1998) Elevated CO_2, rhizosphere processes, and soil organic matter decomposition. Plant Soil 202:167–174

Chiew FHS, Whetton PH, McMahon TA, Pittock AB (1995) Simulation of the impacts of climate change on runoff and soil moisture in Australian catchments. J Hydrol 167:121–147

Ciais P, Reichstein M, Viovy N, Granier A, Ogée J, Allard V, Valentini R (2005) Europe-wide reduction in primary productivity caused by the heat and drought in 2003. Nature 437 (7058):529–533

Cox PM, Huntingford C, Williamson MS (2018) Emergent constraint on equilibrium climate sensitivity from global temperature variability. Nature 553(7688):319–322

Coyle C, Creamer RE, Schulte RPO, O'Sullivan L, Jordan P (2016) A functional land management conceptual framework under soil drainage and land use scenarios. Environ Sci Policy 56:39–48

De Vries FT, Griffiths RI (2018) Impacts of climate change on soil microbial communities and their functioning. Dev Soil Sci 35. https://doi.org/10.1016/B978-0-444-63865-6.00005-3

DeAngelis KM, Pold G, Topcuoglu BD, van Diepen LTA, Varney RM, Blanchard JL, Melillo J, Frey SD (2015) Long-term forest soil warming alters microbial communities in temperate forest soils. Front Microbiol 6:104

Díaz RA, Magrin GO, Travasso MI, Rodríguez RO (1997) Climate change and its impact on the properties of agricultural soils in the Argentinean rolling pampas. Clim Res 9(1–2):25–30

Dong J, Gruda N, Lam SK, Li X, Duan Z (2018) Effects of elevated CO_2 on nutritional quality of vegetables: a review. Front Plant Sci 9

Doran J (2002) Soil health and global sustainability: translating science into practice. Agric Ecosyst Environ 88:119–127

Edwards WM (1991) Soil structure: process and management. In: Lal R, Pierce FJ (eds) Soil management for sustainability. Soil and Water Conservation Society, Ankeny, pp 7–14

Fanzo J, Davis C, McLaren R, Choufani J (2018) The effect of climate change across food systems: implications for nutrition outcomes. Glob Food Secur 18:12–19

Favis-Mortlock D, Boardman J (1995) Nonlinear responses of soil erosion to climate change: a modeling study on the UK South Downs. Catena 25:365–387

Gang CC, Wang ZQ, Zhou W, Chen YZ, Li JL, Cheng JM, Guo L, Odeh I, Chen C (2015) Projecting the dynamics of terrestrial net primary productivity in response to future climate change under the RCP2.6 scenario. Environ Earth Sci 74:5949–5959

Gao X, Luo Y, Lin W, Zhao Z, Filippo G (2003) Simulation of effects of land use change on climate in China by a regional climate model. Adv Atmos Sci 6:224–230

Gavito ME, Curtis PS, Mikkelsen TN, Jakobsen I (2001) Interactive effects of soil temperature, atmospheric carbon dioxide and soil N on root development, biomass and nutrient uptake of winter wheat during vegetation growth. J Exp Bot 362(19):13–23

Girija Veni V, Srinivasarao C, Sammi Reddy K, Sharma KL, Rai A (2020) Soil health and climate change. Clim Change Soil Interact 751–767. https://doi.org/10.1016/b978-0-12-818032-7.00026-6

Guo Z, Han J, Li J, Xu Y, Wang X et al (2019) Effects of long-term fertilization on soil organic carbon mineralization and microbial community structure. PLoS One 14(4):e0216006

Guoju X, Qiang Z, Jiangtao B, Fengju Z, Chengke L (2012) The relationship between winter temperature rise and soil fertility properties. Air Soil Water Res 5:ASWR-S8599

Haddad SA, Tabatabai MA, Loynachan TE (2013) Biochemical processes controlling soil nitrogen mineralization under waterlogged conditions. Soil Sci Soc Am J 77:809–816

Hattenschwiler S, Handa IT, Egli L, Asshoff R, Ammann W, Korner C (2002) Atmospheric CO_2 enrichment of alpine treeline conifers. New Phytol 156(3):363–375

Heinze J, Gensch S, Weber E, Joshi J (2016) Soil temperature modifies effects of soil biota on plant growth. Plant Ecol 10:808–821

Hinojosa MB, Carreira JA, Ruíz RG, Dick RP (2004) Soil moisture pre-treatment effects on enzyme activities as indicators of heavy metal-contaminated and reclaimed soils. Soil Biol Biochem 36 (10):1559–1568

Hoyle F (2013) Managing soil organic matter: a practical guide. Central Queensland Soil Health, Kingston, ACT

IPCC (2007) Climate change 2007: mitigation. Contribution of Working Group III to the Fourth Assessment Report of the Intergovernmental Panel on Climate Change. Summary for policymakers. Cambridge University Press, Cambridge, New York

IPCC (2019) Climate change and land: an IPCC special report on climate change, desertification, land degradation, sustainable land management, food security, and greenhouse gas fluxes in terrestrial ecosystems—summary for policymakers

Jansson JK, Hofmockel KS (2019) Soil microbiomes and climate change. Nat Rev Microbiol 18:35–46. https://doi.org/10.1038/s41579-019-0265-7

Jenny H (1941) Factors of soil formation. McGraw-Hill, New York

Karmakar R, Das I, Dutta D, Rakshit A (2016) Potential effects of climate change on soil properties: a review. Sci Int 4(51):51–73

Kirschbaum MUF (1995) The temperature dependence of soil organic matter decomposition and the effect of global warming on soil organic carbon storage. Soil Biol Biochem 27:753–760

Kopyra M, Gwód EA (2004) The role of nitric oxide in plant growth regulation and responses to abiotic stresses. Acta Physiol Plant 26:459–473

Kulshreshtha S (2019) Resiliency of prairie agriculture to climate change. Clim Change Agric. https://doi.org/10.5772/intechopen.87098

Kumeleh AS, Sharmila P, Uprety DC, Saradhi PP (2009) Effects of elevated CO_2 on soil physico-chemical characteristics under free air CO_2 enrichment (FACE) technology. IOP Conf Ser: Environ Earth Sci 6(29):292045. https://doi.org/10.1088/1755-1307/6/29/292045

Lal R (2004) Soil carbon sequestration impacts on global climate change and food security. Science 304(5677):1623–1627

Lal R (2008) Soil carbon stocks under present and future climate with specific reference to European ecoregions. Nutr Cycl Agroecosyst 81(2):113–127

Lal R, Delgado JA, Groffman PM, Millar N, Dell C, Rotz A (2011) Management to mitigate and adapt to climate change. J Soil Water Conserv 66:276–285

Lavee H, Imeson AC, Sarah P (1998) The impact of climate change on geomorphology and desertification along a Mediterranean-arid transect. Land Degrad Dev 9:407–422

Lehman RM, Cambardella CA, Stott DE, Acosta-Martinez V, Manter DK, Buyer JS, Maul JE, Smith JL, Collins HP, Halvorson JJ, Kremer RJ, Lundgren JG, Ducey TF, Jin VL, Karlen DL (2015) Understanding and enhancing soil biological health: the solution for reversing soil degradation. Sustainability 7:988–1027

Liang J, Xia J, Liu L et al (2013) Global patterns of the responses of leaf-level photosynthesis and respiration in terrestrial plants to experimental warming. J Plant Ecol 6:437–447

Nakandalage N, Seneweera S (2018) Micronutrients use efficiency of crop-plants under changing climate. In: Hossain MA, Kamiya T, Burritt DJ, Phan Tran L-S, Fujiwara T (eds) Plant micronutrient use efficiency. Academic, New York

Niklaus PA, Körner C (2004) Synthesis of a six-year study of calcareous grassland responses to in situ CO_2 enrichment. Ecol Monogr 74(3):491–511

Nikolskii YN, Castillo-Alvarez M, Bakhlaeva OS, Gama-Castro J, Landeros-Sanchez C (2010) Assessing the effect of possible global climate changes on the fertility of Mexican soils and the prediction of crop yields. Eurasian Soil Sci 43:985–992

Obalum SE, Chibuike GU, Peth S, Ouyang Y (2017) Soil organic matter as sole indicator of soil degradation. Environ Monit Assess 176:988–1027

Olesen JE, Trnka M, Kersebaum KC, Skjelvåg A, Seguin B, Peltonen-Sainio P, Rossi F, Kozyra J, Micale F (2011) Impacts and adaptation of European crop production systems to climate change. Eur J Agron 34:96–112

Ostle NJ, Levy PE, Evans CD, Smith P (2009) UK land use and soil carbon sequestration. Land Use Policy 26:274–283

Pareek N (2017) Climate change impact on soils: adaptation and mitigation. MOJ Ecol Environ Sci 2(3):136–139

Parry M, Arnell N, Hulme M, Nicholls R, Livermore M (1998) Buenos Aires and Kyoto targets do little to reduce climate change impacts. Glob Environ Change 8(4):1–5

Poorter H, Navas M (2003) Plant growth and competition at elevated CO_2: on winners, losers and functional groups. New Phytol 157(3):175–198

Rounsevell MD, Evans S, Bullock P (1999) Climate change and agriculture soils impact and adaptation. Clim Chang 43(4):683–709. https://doi.org/10.1023/A:1005597216804
Sanchez PA, Swaminathan MS (2005) Hunger in Africa: the link between unhealthy people and unhealthy soils. Lancet 365:442–444
Sanderman J, Farquharson R, Baldock J (2010) Soil carbon sequestration potential: a review for Australian agriculture. CSIRO Report, Canberra
Singh BP, Cowie AL, Chan KY (2011) Soil health and climate change. Soil biology. Springer, Berlin, pp 1–414
Snyder CS, Bruulsema TW, Jensen JL, Fixen PE (2009) Review of greenhouse gas emissions from crop production systems and fertilizer management effects. Agric Ecosyst Environ 133:247–266
St. Clair SB, Lynch JP (2010) The opening of Pandora's Box: climate change impacts on soil fertility and crop nutrition in developing countries. Plant Soil 335:101–115
Stark JM, Firestone MK (1995) Mechanisms for soil moisture effects on activity of nitrifying bacteria. Appl Environ Microbiol 61:218–221
Tarnawski S, Aragno M (2006) The influence of elevated CO_2 on diversity, activity and biogeochemical function of rhizosphere and soil bacterial communities. In: Nösberger J, Long SP et al (eds) Managed ecosystems and CO_2 case studies, processes and perspectives. Ecological studies series, vol 187. Springer, Berlin, pp 393–409
Teng HT, Ma ZQ, Chappell A, Shi Z, Liang ZZ, Yu W (2017) Improving rainfall erosivity estimates using merged TRMM and gauge data. Remote Sens 9:1134
Thakur MP, Reich PB, Wagg C et al (2016) Effects of soil warming history on the performance of congeneric temperate and boreal herbaceous plant species and their associations with soil biota. J Plant Ecol. https://doi.org/10.1093/jpe/rtw066
Varma A, Tripathi S, Prasad R (2019) Plant biotic interactions. https://doi.org/10.1007/978-3-030-26657-8
Whitaker J, Ostle N, Nottingham AT, Ccahuana A, Salinas N, Bardgett RD, Meir P, McNamara NP (2014) Microbial community composition explains soil respiration responses to changing carbon inputs along an Andes-to-Amazon elevation gradient. J Ecol 102:1058–1071
Xiong X, Grunwald S, Myers DB, Ross CW, Harris WG, Comerford NB (2014) Interaction effects of climate and land use/land cover change on soil organic carbon sequestration. Sci Total Environ 493:974–982
Xue L, Kuang LG, Chen HY (2003) Soil nutrients, microorganisms and enzyme activities of different stands, Acta Pedologica Sinica. 40:280–285
Zaehle S, Friedlingstein P, Friend AD (2010) Terrestrial nitrogen feedbacks may accelerate future climate change. Geophys Res Lett 37(1)
Zhou X, Xu X, Zhou G, Luo Y (2018) Temperature sensitivity of soil organic carbon decomposition increased with mean carbon residence time: field incubation and data assimilation. Glob Change Biol 24:810–822
Zhu Z, Bai Y, Lv M, Tian G, Zhang X, Jia Z, Jiang Y, Ge S (2020) Soil fertility, microbial biomass, and microbial functional diversity responses to four years fertilization in an apple orchard in North China. Hortic Plant J. https://doi.org/10.1016/j.hpj.2020.06.003

Part II
Impact of Climate Change on Plant Health

Chapter 13
Crop Microbiome Engineering and Relevance in Abiotic Stress Tolerance

Mayur Naitam, Rajeev Kaushik, and Anjney Sharma

Abstract Agriculture is the panacea of human's very existence. Maintaining or increasing the agricultural productivity is needed to feed the ever-increasing global population, but we have limited resources like land and water, so effective management of available resources is need of the hour in ever-changing global environmental threats. Agriculture is exposed to vagaries of nature, and various abiotic stresses like drought, salinity, extremes of temperature and heavy metal toxicity are the major limiting factor for plant growth, threatening agricultural productivity and ecological sustainability. Microorganisms are the inhabitants of the most diverse environment and possess metabolic and physiological capabilities to tolerate various abiotic stresses. Such microorganisms naturally inhabiting plant rhizosphere can be exploited to ameliorate the effect of abiotic stresses in agriculture and enhancing growth and productivity. Having known the identity and characteristics of microbes involved in imparting tolerance to abiotic stress, provides us with the plethora of opportunities to modulate and restructure the microbial community composition at the roots of host plants through varied microbiome engineering techniques. Effect of abiotic stress on agriculture, microbes involved in stress mitigation and how to engineer plant microbiome to recruit ideal microbiome at host plants to alleviate effects of abiotic stress and to enhance overall agricultural productivity is reviewed here.

Keywords Abiotic stress · Plant rhizosphere · Microorganisms · Stress mitigation · Microbiome engineering

---

M. Naitam
Division of Microbiology, ICAR-Indian Agriculture Research Institute, New Delhi, India

R. Kaushik
Principal Scientist, Division of Microbiology, ICAR-Indian Agriculture Research Institute, New Delhi, India

A. Sharma (✉)
Key Laboratory of Sugarcane Biotechnology and Genetic Improvement (Guangxi), Ministry of Agriculture, Sugarcane Research Center, Chinese Academy of Agricultural Sciences, Guangxi Key Laboratory of Sugarcane Genetic Improvement, Sugarcane Research Institute, Guangxi Academy of Agricultural Sciences, Nanning, Guangxi, China

D. K. Choudhary et al. (eds.), *Climate Change and the Microbiome*, Soil Biology 63,
https://doi.org/10.1007/978-3-030-76863-8_13

13.1 Introduction

Agriculture is a game of uncertainties and is exposed to vagaries of nature, where various abiotic stresses affect the plant growth one or other way. Abiotic stresses specifically drought, salinity, elevated temperature and heavy metal toxicity influence the crop growth and thus reduce agricultural productivity drastically (Rolli et al. 2015). Enhancing the agricultural productivity under such abiotic stresses has come up as a challenge because of an incomplete understanding of how stress affects the plants and what stress tolerance mechanisms are employed to produce better growth by plants. Efforts are being taken to understand and develop the strategies for combating abiotic stress, but unfortunately, it has met with little success (Hussain et al. 2018). Development of improved crop varieties having tolerance to various abiotic stresses using current breeding technologies seems to have far-reaching implications, but still, we need to wait a bit longer for development and worldwide use of such crops (Ashraf and Foolad 2007). Development of transgenic crops seems a feasible and attractive option, which involves the transfer of genes imparting abiotic stress tolerance to the domesticated high-yielding varieties of economically important crops. Transgenics has got immense potential and can be used to solve the problem of abiotic stresses in agriculture, but the progress is limited due to problems faced in the identification of genes involved in abiotic stress tolerance and their use without harming the present yield potential of crop plants (Marasco et al. 2016; Thao and Tran 2016). Microorganisms are well adapted to varying ecological conditions and are found to grow at extremes of conditions like high and low temperature, salt concentration, pH, water activity and nutrient availability. This distinguishing property of microorganisms can be exploited to improve crop growth and productivity in agriculture while combating abiotic stresses. The potential of bacterial inoculants in alleviating drought, salinity and other abiotic stress in plants has been demonstrated earlier. It was reported that high-affinity potassium transporters (HKTs), small heat shock proteins (sHSPs) and drought-responsive element binding (DREB) genes play crucial roles in salinity and drought tolerance. "High-affinity potassium transporters (HKTs) are a large superfamily of transporters in plants, bacteria, and fungi which resist salinity via removal of Na^+ from the xylem during salinity stress" (James et al. 2006). "In general, little is known about Na^+ excluding genes in plants, particularly in wild genotypes, or their degree of genetic diversity. In wheat, the group I HKT genes confer salt tolerance through sodium exclusion mechanisms in leaves" (Huang et al. 2006). "Small heat shock proteins (sHSPs) of the molecular mass of 15–42 kDa, are a family of HSPs, or chaperons, which are synthesized upon heat shock and are responsible for proper folding of unfolded or partially-folded polypeptides" (AL-Quraan et al. 2002; Waters 2013). Drought tolerance in plants can be enhanced by overexpression of the gene coding for HSP17 protein (AL-Quraan et al. 2002; Zou et al. 2012). "Different members of HSPs play different roles under drought stress. Drought-tolerant cultivars exhibit accumulation of HSP17 proteins and their transcripts" (Grigorova et al. 2011; Svoboda et al. 2016). The drought-responsive element binding (DREB) genes activate multiple stress-responsive genes by interacting with the dehydration-responsive element (DRE) (Huang et al. 2012; Yamaguchi-Shinozaki and Shinozaki 1994). Microorganisms are ubiquitous in nature, occupying every single possible niche in the environment. They exist either singly or in communities comprised of the same species or different ones. Over the due course of evolution, these interesting creatures have coevolved with their host species resulting in the development of ecological communities on earth. Microbial communities interact effectively among themselves and with their host to form co-operative and mutualistic relationships, which determines the composition of the microbial community (microbiome) as well as affects the physiology of the host (Foo et al. 2017). Microbiomes, through their dynamic interactions, not only can influence the properties of the ecosystem in their vicinity but also affect the condition of local host as whole, i.e. holobiont, the organism together with its microbiome. The microbiome of the particular crop plants and the soil can have a considerable impact on the productivity of the concerned crop and the environment (Chaparro et al. 2012). Soil microbial communities are highly complex and are evolving continuously responding to environmental fluctuations. Manipulation and management of these soil and plant microbiomes have great potential towards improving plant growth and health, enhancing agricultural productivity, suppressing pathogenic microorganisms, recruiting beneficial microbiota to impart tolerance to various biotic and abiotic stresses and promoting environmentally friendly agriculture (Mueller and Sachs 2015). Thus the agricultural productivity and environmental sustainability would depend on the wise application and utilization of agriculturally important microorganisms, which may prove beneficial for improving agricultural productivity in the near future (Hussain et al. 2018). Microorganisms are at the heart of nutrient cycling, helping replenish the nutrients for plant growth; at the same time they play an important role in integrated management of various abiotic and biotic stresses; thus they can help in reducing the reliance on chemical fertilizers and pesticide for better plant growth and higher productivity. Looking at the considerable impact of microbiome on agriculture and environment, there is immediate need to understand microbiome and the factors contributing towards variability in microbiome to develop strategies for engineering the microbiome for altering microbiota to model ecosystem as per our interest. Microorganisms having abiotic stress tolerance abilities can be well utilized to alter the microbiome composition of target plants in prevailing conditions to impart stress tolerance abilities to the plants.

13.2 Abiotic Stress in Agriculture: Harnessing Rhizobacterial-Plant Interaction for Increased Resilience

13.2.1 Drought Stress

"Microorganisms could play an important role in adaptation strategies and increase of tolerance to abiotic stresses in agriculture, moreover plant growth-promoting rhizobacteria (PGPR) are associated with plant roots and mitigate most effectively the impact of abiotic stresses (drought, low temperature, salinity, metal toxicity, and high temperatures) on plants through the production of exopolysaccharides and biofilm formation" (Dimkpa et al. 2009). Plants increase the synthesis of osmolytes in response to a lack of water, increasing the osmotic capacity within cells (Farooq et al. 2009). "Incidentally, compounds exudated by root zone bacteria also include such osmolytes, like glycine betaine which is produced by osmotolerant bacteria and can act synergistically with plant produced glycine betaine in response to the stress, and this way, increase drought tolerance" (Yuwono et al. 2005). Another essential factor in these osmotolerant bacteria's ability to promote growth is their ability to produce IAA. This hormone is likely to cause an increase in root proliferation in inoculated drought-stressed rice plants, possibly for improved water uptake (Yuwono et al. 2005). *Achromobacter piechaudii* ARV8 isolated from arid and salty environments helped tomato and pepper plants to grow better in water-stressed conditions and also helped to recover faster and better once watering was resumed after drought imposition (Mayak et al. 2004). "Bacteria occurring on root surfaces containing ACC deaminase have been shown to modify the sensitivity of root and leaf growth to soil drying, apparently by influencing ethylene signalling" (Mayak et al. 2004). The drought stress tolerance imparted by the bacterium can be attributed to the production of ACC deaminase in rhizosphere, which can break down the ACC, the immediate precursor of ethylene into α-ketobutyrate and ammonia, thus eliminating the stress response. Inoculation of the plant growth-promoting rhizobacteria, *Pseudomonas* sp., was shown to ameliorate the effects of drought stress on pea plant. Certainly, the vegetative growth was affected due to drought, but the decrease in growth was much less in inoculated plants as compared to uninoculated. Moreover 40% to 62% increase in grain yield and delayed ripening of pod were observed in inoculated vs. uninoculated (Arshad et al. 2008). Additionally, inoculation of *Pseudomonas* sp. reduced the intensity of triple response which was induced through external application of ACC to the pea leaves. Timmusk et al. (2014) observed a noteworthy effect of bacterial priming (isolated from the rhizosphere of drought-stressed plants) on drought-stressed wheat which enhanced its tolerance towards drought. Plant biomass was increased up to 78% and supported fivefold more survival under such severe drought conditions. Moreover, the emission of stress-related volatile organic compounds was also observed. Inoculation of bacterium *Burkholderia phytofirmans* PsJN on drought-stressed wheat growing in field conditions increased transpiration rate, the relative water content of the plant cells, photosynthesis rate and total grain yield (18–21% higher than uninoculated ones), and it also improved ionic balance, level of antioxidants and N, P, K and total protein content of the wheat grains as compared to uninoculated wheat grains (Naveed et al. 2014a). Similar kind of results was obtained when drought-stressed maize was inoculated with *Burkholderia phytofirmans* PsJN and *Enterobacter* sp. FD17. Additionally, both the bacteria were found to establish an endophytic relation with the maize plant by colonizing endorhizosphere of maize (Naveed et al. 2014b). Sandhya et al. (2009) inoculated the bacterium *Pseudomonas* sp. GAP-45 on sunflower seedlings, resulting in the enhancement of biomass production and survival rate under severe drought conditions, suggesting that plant growth-promoting bacteria can be sustainably utilized for drought stress amelioration. Similarly, three bacterial strains, two *Pseudomonads* and a *Bacillus megaterium* were shown to sensitize plant growth under stress conditions. Inoculation with bacterial strains increased shoot and root biomass and also the water content under prevailing water-stressed condition. Bacterial production of Indol-3-acetic acid (IAA) under stress conditions can be accounted for increased tolerance towards drought stress (Marulanda et al. 2009). Kohler et al. (2008) evaluated the effect of inoculation of *Pseudomonas mendocina* and *Glomus intraradices* on *Lactuca sativa*. Highest level of stress alleviation was observed with fertilization and microbial PGPR inoculation together. PGPR inoculation resulted in increased nitrate reductase, phosphatase activity and antioxidant enzymes activity (the activity of peroxidase and catalase increased with increasing drought but not of superoxide dismutase), along with the accumulation of compatible solutes like proline in leaves (Kohler et al. 2008). This establishes the role of PGPR in alleviating oxidative damage under drought stress conditions. Marulanda et al. (2007) reported similar findings with lavender plants when inoculated with different arbuscular mycorrhizal (AM) fungus, *Glomus* species. Through various studies as mentioned above, now it has been well established that plant growth-promoting microbes can be utilized economically to alleviate drought stress in agriculture.

13.2.2 Salinity Stress

"Salinity is one of the most brutal environmental factor, since most crop plants are vulnerable to salinity, which is caused by high salt concentrations in the soil, and the area of land affected is growing by the day" (James et al. 2006; Huang et al. 2006). "For all-important crops, average yields are only a fraction between 20% and 50% of record yields, mostly due to drought and high soil salinity. Efficient resource management and crop improvement for the development of better breeds can help to overcome salinity stress" (Huang et al. 2006). "However, since such strategies are long-drawn and cost-intensive, there is a need to develop simple, low-cost, biological methods for managing salinity stress that can be used in the short term". Microorganisms could play a significant role in this respect if we take advantage of their unique properties, such as saline tolerance, plant growth-promoting hormone production and their interaction with crop plant. “Saline soil is generally defined as one in which the electrical conductivity (EC) of the saturation extract (ECe) in the root zone exceeds 4 dS m^{-1} (approximately 40 mM NaCl) at 25 °C and has exchangeable sodium of 15% and the yield of most crop plants is reduced at this ECe, though many crops exhibit yield reduction at lower ECes” (AL-Quraan et al. 2002; Waters 2013). “It has been estimated that worldwide 20% of total cultivated and 33% of irrigated agricultural lands are afflicted by high salinity. Furthermore, the salinized areas are increasing at a rate of 10% annually for various reasons. It has been estimated that more than 50% of the arable land would be salinized by the year 2050” (Jamil et al. 2011).

Besides drought, there is ample evidence to suggest that the low productivity in rain-fed agriculture is also due to decreased nutrient availability and soil salinity (Arshad et al. 2008; Timmusk et al. 2014). Yadav et al. (2015) suggested that rhizospheric archaea assist the plants in sustaining the effects of drought and salinity by altering the level of osmoprotectants (carbohydrates, amino acids, proline), which helps in maintaining their membrane integrity under water-deficient conditions as well as salinity stress. These rhizospheric archaea were also found to possess plant growth-promoting attributes that assist in nutrient management such as solubilization of P, K and Zn. Microorganisms use different mechanisms to alleviate salinity stress in crops. Accumulation of exopolysaccharide and biofilm formation by certain bacteria affects the growth and development of crop plants grown in saline environments by restricting sodium uptake (Afrasayab et al. 2010). Zhang et al. (2008) studied induced systemic tolerance to salinity stress in *Arabidopsis* using *Bacillus subtilis* GB03. Interestingly, some of the volatile organic compounds (VOCs) that are emitted from *B. subtilis* GB03 are bacterial determinants involved in salinity stress tolerance. These VOCs caused tissue-specific regulation of high-affinity K^+ transporter (HKT1) which controls Na^+ homeostasis under salt stress (Zhang et al. 2008). Halotolerant bacteria can help plants grow better in salt-affected areas through various mechanisms. Halotolerant bacteria isolated from the rhizosphere of halophytes growing near Yellow Sea, Korea, when inoculated with *Canola* plants, resulted in increase in root length and dry weight up to 40% as compared to uninoculated ones (Siddikee et al. 2010). This growth-enhancing property can be attributed to various PGP activities as observed in the halotolerant isolates like N_2 fixation, P and Zn solubilization and IAA production, along with the production of ammonia and other extracellular hydrolytic enzymes, but the most promising was ACC deaminase production. Similarly, Upadhyay et al. (2009) reported plant growth-promoting properties of bacteria isolates, isolated from the rhizosphere of wheat, showing tolerance to NaCl concentration up to 8%. Presence of these isolates in the wheat rhizosphere confirms their contribution to the enhanced growth and productivity of the host crop. A new group of diazotrophic halotolerant bacteria present at the roots of a halophyte *Salicornia brachiata* was shown to have plant growth-promoting attributes (IAA and ACC deaminase production and phosphorus solubilization) by Jha et al. (2012) at varying salt concentrations. Nitrogen is one of the most important nutrients for plant growth, and these nitrogen-fixing halotolerant bacteria can provide plants growing in salt stress with a readily available source of biologically fixed nitrogen helping them grow more efficiently. PGPR priming of bean plants can alleviate the negative effects of higher salt concentrations (Yildirim and Taylor 2005). Also, inoculation of *Azospirillum* on lettuce mitigated the negative effect of salinity (Barassi et al. 2006). Similar observations were also reported for sugarcane (Leite et al. 2014). In addition to bacteria and archaea, AM fungi were also observed to impart salinity tolerance in *Sesbania aegyptiaca* and *Sesbania grandiflora* (Giri and Mukerji 2004), moong bean (Jindal et al. 1993), tomato (Al-Karaki et al. 2001) and maize (Feng et al. 2002) through enhanced Mg^{2+} and suppressed levels of Na^{+} in plant tissue. This signifies the paramount importance of recruiting plant growth-promoting microorganisms as they help the plant to alleviate the salinity stress at least to some extent.

13.2.3 High- or Low-Temperature Stress

Extensive industrial growth in the last century has led to an unprecedented increase in global environmental temperature, owing to global warming. It is predicted that by the end of this century, the global temperature will rise by 3.5 °C (IPCC 2007; Masson-Delmotte et al. 2019), resulting in hampering the plant growth severely, threatening the ecosystem functioning and food security to the ever-increasing world population. The major effect of temperature rise is cellular damage, protein aggregation and improper folding which directly affects the protein function in addition to altered physiology and biochemistry of plants (Hassan et al. 2020). Not only plants but microorganisms growth also gets affected by elevated temperature levels, to alleviate which they use "complex multilevel regulatory processes", which involves higher-level expression of heat shock proteins like stress sigma factors (RpoS) (Srivastava et al. 2008). Thus priming of thermotolerant microorganisms on crop plants can help mitigate the heat stress. Khan et al. (2020a, b) reported the thermotolerance effect of plant growth-promoting *Bacillus cereus* SA1 on soybean growing under heat stress. Bacterium inoculation resulted in increased biomass, chlorophyll content and fluorescence under heat stress and normal conditions, which can be attributed to reduced abscisic acid (ABA) and increased salicylic acid (SA) production along with enhanced antioxidant enzyme activity and amino acid content of the plant. Additionally, soybean was found to overexpress heat shock proteins, *GmLAX3* and *GmAKT2*, which can be associated with decreased reactive oxygen species production and altered auxin and ABA stimulus of the plant critical in heat stress. Similar results were reported for wheat (Abd El-Daim et al. 2014; Zulfikar Ali et al. 2011), Dichanthelium lanuginosum, and tomato (De Zelicourt et al. 2013; Khan et al. 2020b). Ali et al. (2009) also reported the growth-enhancing and sustaining effects of inoculation of thermotolerant strain *Pseudomonas* sp. AKM-P6, isolated from pigeon pea growing in arid regions on *Sorghum* seedlings. The bacterium supported seedling growth for 15 days in contrast to 5 days in uninoculated at 47–50 °C, as a result of reduced membrane injury, biosynthesis of high molecular weight proteins and cellular metabolites like proline, sugars, chlorophyll and amino acids along with increased shoot and root biomass. As studies mentioned above indicates the potential role of thermotolerant microorganisms in sustaining and enhancing host plant growth at elevated temperature regimes, they can be well utilized as bioinoculants for increasing agricultural productivity in semi-arid and arid regions. Development and use of microbiome module using thermotolerant microbial strain as one of the components of the modular microbiome might have imperative implications for increasing agricultural productivity in the regions of higher-temperature stress.

Low temperature significantly limits the crop growth and productivity specifically in areas where annual temperature variation is high. Temperature governs the rate of photosynthesis and respiration and thus the yield potential of the plant. Drastic reduction in temperature during plant growth may also lead to chilling or freezing injury, forming ice crystals in the plant tissue and damaging or rupturing the membranes (de Langre 2008). *Burkholderia phytofirmans* PsJN a PGPR, inducing resistance to grey mould, has been shown to impart tolerance to non-freezing low-temperature (4 °C) stress unlike normal temperature (26 °C) in grapevine by Theocharis et al. (2012). As a result of PGPR priming, the stress-responsive metabolites and transcripts increased in plant tissue, reaching higher levels earlier than non-primed plantlets. *Burkholderia phytofirmans* PsJN also had growth-enhancing effects on switchgrass (Kim et al. 2012). Earlier Ait Barka et al. (2006) also reported a rise in amino acids, phenolics, starch and proline in grapevine inoculated with *Burkholderia phytofirmans* PsJN, strengthening the low-temperature tolerance. AM fungus *Glomus versiforme* and *Rhizophagus irregularis* forming symbiotic relation with barley enhanced plant performance and post freezing survival at −5 °C. Inoculated plants were observed to have improved phosphorus uptake, photosynthesis rate and plant growth. Improved post freezing tolerance of the barley plant can be accounted for lower cellular damage as a result of increased antioxidant defence mechanism and phenolics-related metabolism (Hajiboland et al. 2019). Similar results were also reported by Ma et al. (2019) on inoculation of *Rhizophagus irregularis* on cucumber. Inoculation of rhizobacterial consortium composing of *Bacillus amyloliquefaciens* Bk7 and *Brevibacillus laterosporus* B4 in rice resulted in imparting tolerance to dual stress of drought and cold (Kakar et al. 2016). A fungal isolate from Iran, *Trichoderma harzianum* (*Th*) AK20G, found to impart chilling resistance in tomato against the cold stress (Ghorbanpour et al. 2018). Molecular communications, improved growth, relative water content, electrolyte retention and increased proline biosynthesis relative to uninoculated controls can be accounted for increased chilling tolerance. PGPR imparting cold stress tolerance to the crop plant can significantly be utilized to improve crop growth and productivity at low-temperature regimes.

13.2.4 Heavy Metal Stress

Industrialization along with the geological and anthropogenic activities has led to extensive soil pollution due to heavy metal release. These heavy metals cannot be degraded; thus they persist through times and affect ecosystem functioning and agricultural productivity (Chibuike and Obiora 2014). "Essential and non-essential heavy metals generally produce common toxic effects on plants, such as low biomass accumulation, chlorosis, inhibition of growth and photosynthesis, altered water balance and nutrient assimilation, and senescence, which ultimately cause plant death" (Singh et al. 2016). Bioremediation is the most favoured and economical approach for treating the heavy metal polluted soils. Phytoremediation is being practised as routine at polluted sites as it accumulates heavy metals in plant parts and thus can be removed from the soils (Ashfaque et al. 2016). In addition to the use of plants, microorganisms associated with plants can successfully be exploited for bioremediation of heavy metals, as they improve uptake and accumulation by plants through various activities like increasing the bioavailability and mobility (Yang et al. 2012). Various studies have proved the effectiveness of utilizing microorganisms for bioremediation of heavy metals while simultaneously improving the plant growth and productivity. *Staphylococcus arlettae* strain NBRIEAG-6 an arsenic hypertolerant bacterium having PGP activity was observed to improve growth as well as arsenic uptake by Indian mustard. Bacterium possesses arsenic reductase activity in addition to IAA, siderophores and ACC deaminase production, as revealed by biochemical profiling (Srivastava et al. 2013). *Phyllobacterium myrsinacearum* RC6b having high heavy metal resistance successfully mobilized the cadmium (Cd), zinc (Zn) and lead (Pb), resulting in the higher accumulation of these metals by *Sedum plumbizincicola*. The bacterium also possesses PGP attributes, i.e. ACC deaminase, siderophore and IAA production and phosphorus solubilization, which improved plant growth at various metal concentrations (Ma et al. 2013). *Pseudomonas koreensis* AGB-1 inoculation enhanced solubility and bioavailability of Cd, copper (Cu), Pb and Zn in the rhizosphere of *Miscanthus sinensis* along with improved plant growth. The *Pseudomonas koreensis* inoculated plants promoted remediation and phytostabilization of heavy metal-polluted mine soils (Babu et al. 2015). A bio-surfactant-producing bacterium *Bacillus* sp. J119, exhibiting resistance to Pb, Cd, Cu, nickel (Ni) and Zn and antibiotics, showed plant growth promotion in heavy metal-contaminated pot soils. Additionally, it helped the tomato plant to accumulate higher concentration of Cd (39% to 70% in live bacterium inoculated plants against dead bacterium inoculated controls) (Sheng et al. 2008). Differential response to bacterial inoculation was observed in rapeseed, tomato, maize and sudangrass. This may reflect the host specificity of *Bacillus* sp. J119 for helping the plants to grow better and bio-remediate the contaminated soils. Similar results for plant growth promotion, bioremediation and phytostabilization were reported for *Rhizobium leguminosarum* strain TAL-102 and *Azotobacter chroococcum* in maize (Hadi and Bano 2010), *Rhizobium* sp. RP5 in pea (Wani et al. 2008), *Bradyrhizobium* sp. (*Vigna*) in moong bean (Wani et al. 2007) and *Enterobacter aerogenes* and *Rahnella aquatilis* in Indian mustard (Kumar et al. 2009). Therefore use of heavy metal-tolerant microorganisms in addition to other abiotic stress-tolerant ones in increasing the plant growth, productivity and overall ecological sustainability is imperative.

13.3 Engineering the Crop Microbiome: Recruiting Beneficial Microflora at the Host Roots

The microbiome is

the community of microorganisms (such as bacteria, archaea, fungi and viruses) that inhabit a particular environment and especially the collection of microorganisms living in or on the plant, human or any living organism's body. Plant microbiome comprises the microbial community which inhabits different plant organs like root (rhizobiome), leaf, stem, flower, etc. Plant along with its complete microbiome is regarded as phytobiome. However these microbial communities are labile to environmental fluctuations, and the community structure is always in a dynamic state, owing to the multitude of factors like microbe-microbe interaction, biotic and abiotic environmental stresses like drought and pathogen invasion and most important among all, the host genotype (Foo et al. 2017). Physiology of host can be influenced by modulating its microbiome composition, and thus crop or plant microbiome engineering has come up as a potential tool for improving plant fitness attributes, such as disease suppression, enhanced nutrient uptake, increase in photosynthetic rate and biomass accumulation, higher productivity and tolerance to abiotic stresses like drought. Engineering of plant microbiome occurs naturally at plant level through a variety of mechanisms, which we should know before developing strategies for engineering microbiome. Plants use strategies for recruiting their microbiome like production of specific root exudates, e.g. malic acid and succinic acid (Sasse et al. 2018); secretion of secondary metabolites, e.g. DIMBOA by maize plant (Neal et al. 2012); exudation of signalling molecules, e.g. AHL by *Medicago truncatula* (Veliz-Vallejos et al. 2014); plant genotype, e.g. glucosinolates production by transgenic *Arabidopsis* (Bressan et al. 2009); and defence activation and recruitment in response to infection, e.g. firmicutes against *Ralstonia solanacearum*. Plant microbiome is being continuously engineered by humans without proper scientific understanding since the advent of agriculture when humans started selecting and domesticating the crops of their interest. This unobserved microbiome engineering might have taken place due to selection during the domestication of crop plants, plant breeding to change quality and quantity of root exudates, development of transgenic varieties, biofertilizers introduction on seed, planting material or in soil, foliar spray of nutrient, organic compound and chemicals, through crop management practices and management of soil type and properties.

13.3.1 Microbiome Transfer and Transplantation

Plant microbiome can be engineered for desired characters through various ways, the most common of which is the transfer of microbiome itself; others are the transfer of synthetic microbial communities like inoculation of biofertilizer on planting materials and artificial selection of plant beneficial microbial groups based on host response through phenotypes. The most common and familiar example of microbiome transfer is the transfer of disease-suppressive soil to disease-conducive soil to tackle soil sickness as a measure of plant disease management. Many of the diseases like potato scab, black root rot of tobacco and infection in sugar beet have been impeded successfully using microbiome transfer through suppressive soils (Gopal et al. 2013). Along with soil sickness, microbiome transfer have been found effective in alleviating drought stress and enhancing biomass accumulation in water stresses areas. Moreover, the success of this strategy depends on the availability of native microbiota capable of suppressing a known disease of economic importance. Microbiome transfer was successfully used against *Pseudomonas syringae* a pathogen of bacterial canker of kiwifruit, where the microbiome from a medicinal plant *Leptospermum scoparium* was transferred to kiwifruit through wound infection. It was observed that kiwifruit produced the same antimicrobial effect as *Leptospermum scoparium* and was able to control *Pseudomonas syringae* infection (Wicaksono et al. 2018). Root microbiome of *Ralstonia solanacearum*-resistant and susceptible tomato varieties was analysed for its effect on disease invasion by pathogen *Ralstonia solanacearum* using metagenome analysis. Results showed an abundance of *Flavobacterium* in root microbiome of resistant variety. Transfer of this *Flavobacterium*-abundant microbiome from the rhizosphere of resistant variety to susceptible one was able to suppress the disease incidence (Kwak et al. 2018). Therefore, native microbiome has promising roles in protecting plants from the microbial pathogen and can be successfully used for engineering the microbiome of susceptible crop cultivars.

13.3.2 Synthetic Microbial Community

Engineering of crop microbiome requires in-depth knowledge of the composition of microbial community forming a mutualistic relationship with the particular crop plant. The microbiome structure of the crop can be analysed using standard laboratory procedures which will provide you with the culturable part of the community. However, the use of sequencing and omics techniques such as metagenomics will give you the complete picture of plant microbial community composition. This prior knowledge of microbiome composition comes handy in predicting and formulating synthetic microbial community which may represent the natural microbiome of a particular plant. The plant microbiome can be divided into core microbiome and the rest. The core microbiome is necessary for the plant to grow and function well, and this core microbiome can be found to be associated with the host plant in all cultivation sites and all climatic conditions. A similar approach was used to study and characterize the potential interspecies interaction in gnotobiotic maize rhizosphere and to study its beneficial effect on the host. Microbiome composition of maize rhizosphere and its abundance profile at maize roots was studied using both culture-based and culture-independent techniques which involve isolation of total DNA from roots and rhizosphere followed by 16S rRNA sequencing and identification of isolate and, on the other hand, culturing the microbes from dilutions prepared from rhizospheric soil. From the relative abundance of culture from day 0 to 15, a simplified synthetic microbial community was designed consisting of seven isolates, *Enterobacter cloacae*, *Stenotrophomonas maltophilia*, *Curtobacterium pusillum*, *Ochrobactrum pituitosum*, *Pseudomonas putida*, *Herbaspirillum frisingens* and *Chryseobacterium indologenes*. The association and effect of this synthetic community on maize roots were studies for 15 days, and it was found that the community was able to form a stable association with maize roots. Community composition and root association were stable up to 15 days in all inoculations, but when the *Enterobacter cloacae* was excluded from inoculation group, it was observed that the whole community got collapsed after initial growth. The collapsed community showed an overgrowth of *Curtobactrium pussilosum*, these suggest that *Enterobacter cloacae* plays a crucial role in maintaining the synthetic community on maize roots, and it is important for the establishment and continued association of the synthetic community on maize roots (Niu et al. 2017). Besides *Enterobacter cloacae* plays a key role in maintaining other species of the community onto the roots, and this seven species synthetic community protects maize seedling from *Fusarium moniliforme* and *Fusarium verticillioides* a seedling blight-causing fungus (Niu et al. 2017; Niu and Kolter 2018). Hence, microbial strain with desired functions can be assembled into synthetic microbial communities as per research interest and used for engineering the host plant's microbiome. It is the consortium of plant-associated microorganisms which helps plants tackle the stressed conditions as evident from *Nicotiana attenuata*. A consortium of five root-associated bacteria of native plants but not the individual ones was successful at rescuing from the sudden attack of wilt similar to symptoms produced by *Fusarium-Altarneria* disease complex in continuous cropping of *Nicotiana attenuata* (Santhanam et al. 2015). Thus, a native microbiome seems to impart resistance against the pathogen buildup and disease incidence during continuous cropping. The community composition of the synthetic microbiome can be controlled and modified in fulfilment of desired research goals, but all these advantages are limited by our understanding of core microbiome. Moreover, only culturable part of the plant-associated microbes can be used in the formulation of the synthetic microbiome, which may miss upon the more active and important but unculturable member of the community, thus hindering the exploitation of the full potential of these synthetic microbiomes.

13.3.3 Host-Mediated Artificial Selection

Another approach in the engineering of crop microbiome is host-mediated artificial selection (HMAS), which can be used to change ecology and plant microbiomes could be obliged to evolve. This host-mediated artificial selection is being practised for centuries to shape the properties of particular organisms, and it holds the promise for shaping properties of the whole ecosystem (Swenson et al. 2000). This approach involves propagating genetically constant cultivars in microbiome from the soil which showed promise in producing desired phenotypes in previous studies and can be successfully used to change plant phenotypes such as biomass accumulation and flowering specifically with *Arabidopsis thaliana* (Swenson et al. 2000). Studies with rice show that the microbiome varies with soil source and genotype in greenhouse conditions. On the other hand, geographical location and cultivation practice, i.e. organic vs. conventional, were the major contributing factors to microbiome variation. "Dynamic changes observed during microbiome acquisition, as well as steady-state compositions of spatial compartments, support a multistep model for root microbiome assembly from soil wherein the rhizoplane plays a selective gating role" (Edwards et al. 2015). In the case of *Brassica rapa* host-mediated artificial selection was used for engineering plant microbiome through repeated multigenerational cultivation of best-performing plants to impart resistance against drought stress and for better fitness. "The research findings suggest that when faced with environmental change, plants may not be limited to "adapt or migrate" strategies; instead, they also may benefit from association with interacting species, especially diverse soil microbial communities, that respond rapidly to environmental change" (Lau and Lennon 2012).

13.3.4 MAP-Assisted Microbiome Engineering and Modular Microbiome

However, unobserved microbiomes contribute to host growth and fitness, its physiology and its response to the prevailing ecological conditions. Oyserman and colleagues introduced a new concept in plant microbiome engineering referred to as microbiome-associated phenotype (MAP)-assisted microbiome engineering (Oyserman et al. 2018). Host-mediated microbiome selection based on phenotypic characterization has always been a qualitative and taxonomically driven approach, but to provide with a more reliable solution, Oyserman and group have introduced a MAPs first approach which is theoretical as well as incorporates experimental observations, and it involves quantitative profiling of microbiome-associated phenotypes across genetically variable host cultivars while subsequently identifying the mechanism lying underneath. Authors have also introduced the concept of modular microbiome, i.e. synthetic microbial consortia that can be engineered co-operatively with the host genotypes to give out different but at the same time mutually consistent microbiome-associated phenotypes to single or population of hosts. The main guiding principle in MAPs first approach is going back to the roots which involve the search for the missing plant microbes to restore the plant-microbe interactions lost during domestication or breeding for disease resistance. This MAPs first approach involves systematic quantification of most significant MAPs across wild and domesticated host, herein natural ecosystem; traditional and modern agriculture serves as a reservoir of genetic and ecological potential, for identification of microbiome-associated phenotypes. These MAPs are systematically screened for identification of plant-microbe and environmental combinations in which MAPs provide the largest fitness advantage. This framework will guide into the mechanism that drives MAPs, and this generated information can be used for targeted plant breeding and microbiome engineering in concert with the plant genotype referred to as next-generation agriculture. MAPs are defined quantitatively, for examples, salt tolerance, phosphorus solubilization, disease suppressiveness, etc., but the contribution of MAPs to fitness and the conditions in which maximum benefits from cumulative microbiome effect comes should be assessed quantitatively. These studies will inform us about where to target the mechanistic investigation and where to apply direct microbiome engineering efforts, and it will also guide the field applications for results generated and applications developed.

The distinct functional communities can be considered as modules, which are individually separate and discrete. "Module can be combined to provide new functional combinations designed to improve host fitness over the various dimensions of the niche" (Oyserman et al. 2018). Oyserman et al. (2017) provided the functional basis of modularity in the microbiome when he tried to develop a novel waste water treatment by combining photosynthetic-nitrifying group with polyphosphate-accumulating organisms group. The concept aimed to exploit the full potential of the unique metabolic potential of each functional guild. The need for mechanical aeration has been overcome, and polyphosphate cycling was not interrupted, by the design of such biogeochemically complementary communities or modules. From this experiment, it becomes clear that the impact of different communities or modules on the function is nonlinear. Some communities can be dominant with no significant change in cumulative microbiome effect, and some are recessive with a significant change in cumulative microbiome effect after mixing due to trade-off between the two functional MAPs, which constitutes a suboptimal space. Successful implementation of microbial communities as a modular component requires, firstly, minimizing the overlap in resource requirement between the functional guilds, example between endosphere and rhizosphere and secondly minimizing the trade-off between the functional guilds, which will be the key for developing customizable and modular microbiome. As no silver bullet organisms can provide optimum MAP under any ecological and environmental conditions, the modularity of microbiome would provide agriculture with the toolbox for rapid adaptation and maximization of crop output under different environmental and ecological conditions. This modularity can be achieved by dividing resources of root exudates, such as plant microbiome cross-feeding strategies in opinal plants which select microbial categorization preferentially opinal and minimize a crosstalk with non-target microbes. These modular microbiome modularities can be achieved through the resource division of the root exudates. Breeding for tailored root exudates composition and designing functional microbial modules depending on these distinct exudate profiles, exploiting natural compartmentalization produced by rhizosphere, endosphere and phyllosphere communities. Engineering MAP heterogeneity at the population level can lead to inter-compatibility between MAPs without necessitating individual host level modularity, for example, an approach analogous to intercropping can be adapted wherein neighbours or interacting microbes provide combinatorial effect for the population.

13.3.5 Manipulation of Root Exudates for Engineering Microbiome

Soil is home for the vast diversity of microorganisms, but the density of microflora increases as you move near the rhizosphere zone as compared to remaining bulk soil. Rhizosphere harbours only a limited diversity, i.e. subset of the microbial population in bulk soil on account of secretion of root exudates. This macro and microflora in the rhizosphere zone comprise the rhizobiome. As you look deeper into the rhizoplane and endosphere, they shelter a subset of the rhizospheric microbial population. The rhizobiome composition of every plant differs, and this selectivity is imparted by the root exudates. This very fact that rhizobiome composition depends on the type of compounds exuded from roots opens up the plethora of options and strategies for designing and engineering root microbiome composition. In *Arabidopsis thaliana*, it was reported that synthesis and foliar application of salicylic acid is the most determining factor on the assembly of the normal microbiome at the roots (Lebeis et al. 2015). Salicylic acid is also involved in imparting tolerance towards pathogen attack; thus being an important part of the central immune system of the plant, it harmonizes the assembly of specific bacterial taxa on plant roots and thus the whole colonization process. Therefore central immunity regulators, as well as specific metabolite compounds, can be applied for moulding plant microbial community structure for enhanced productivity, growth and fitness, tolerance to biotic and abiotic stresses and resilience towards changing environmental conditions. Unlike establishing symbiosis, recruitment of plant-associated microbial community is still unclear. Induction of plant-microbe symbiosis involves secretion of exudates into the rhizosphere attracting microbes. The well-known examples of plant microbes symbiosis are mycorrhizal and rhizobial association, which are attracted by strigolactones and flavonoids secreted by plants. Strigolactones were initially reported to stimulate germination of root parasitic weed seeds; later it was characterized as signalling molecule having a role in the initiation of symbiotic relationship with mycorrhiza fungi excreted through ABC transporters of root cells (Kretzschmar et al. 2012). Alfalfa roots secret flavonoids from infectable zones to the concentration of one micromolar which stimulates assembly of *Rhizobium meliloti* on roots, but nodulation minus mutant lines of Alfalfa exudates compounds, i.e. morin, naringenin and chrysin, which have an inhibitory effect on *Rhizobium meliloti* (Peters and Long 1988). Plant signalling molecules are not yet well characterized but are proved to have an important role in the assembly of the microbiome. Pathogen elicited *Arabidopsis thaliana* showed specific root colonization by the biocontrol agent *Bacillus subtilis* as a result of increased ALMT1 expression and exudation of malic acid (Rudrappa et al. 2008). Exudation of malic acid (signaling molecules in this case) in pathogen elicited plant helps better colonization by *Bacillus subtilis* as compared in absence of malic acid but, it also signifies the presence of other signaling molecules which needs to be characterized. The compounds of plant-microbe crosstalk can also be responsible for root colonization by microbiota. Moreover, plant-secreted cell wall polysaccharides of maize and other plant signalling molecules were reported to commence root colonization by microbes and biofilm formation (Benizri et al. 2007; Dombrowski et al. 2017). The genetic network that commands the phosphate stress response in *Arabidopsis thaliana* has been shown to regulate the root microbial community structure, even in the absence of phosphate stress (Castrillo et al. 2017). However, how the plant metabolites, signalling molecules or immune system moulds formation of microbiome remains a mystery.

13.3.6 Host Genotype, Crop Breeding and Transgenic Varieties in Microbiome Engineering

It is now well established that host plant genotype plays an indispensable role in assembly, function and determination of the structure of microbiome. If we look at the microbiome level, different cultivar genotypes recruit different microbiota extending to beneficials like plant growth-promoting rhizobacteria (PGPR), disease suppressing and immunity busting through variations in their root exudates (Arif et al. 2020). Genome-wide analysis studies in cultivars of *Arabidopsis thaliana* establishes the fact that there occurs a differential response of genotypes to get benefited from PGPR being tested. Only some of the plant genotypes showed good response to inoculation of *Pseudomonas simiae* WCS417r in the form of an increase in fresh shoot biomass and the number of lateral roots formed as compared to others which barely showed any response (Wintermans et al. 2016). Comparison of eight different wild and common modern bean also showed a strong genotype-dependent response in the assembly of rhizosphere microbiome along with a profound impact of the expansion of habitat from native soil to agricultural soils. There was reduced complexity of interaction in bean rhizosphere in agricultural soils as compared to native soils, and only a small fraction of all sequenced OUTs formed the core microbiome which was shared among all the eight bean cultivars in both native and agricultural soils (Pérez-Jaramillo et al. 2019). Plant phenotypic and functional genomics studies in concert with the plant microbiome interaction have provided with the means of manipulating the microbial community at plant roots (Stringlis et al. 2018). Tailoring the root exudates and hormones being released into the soil through crop breeding or developing the designer plants through transgenics provides us with the toolbox to manipulate and maintain beneficial community at plant roots. Wild relatives of the domesticated plants can provide newer point view about the role of genes in wild plants which are associated with microbiome assembly (Jaramillo et al. 2016), as emphasized by Oyserman et al. (2018) that going back to the roots with the search for missing plant microbes to restore the plant-microbe interactions lost during domestication or breeding for disease resistance can be a guiding principle in microbiome engineering (Oyserman et al. 2018). Breeding for resistance against pathogen also affects the microbiome composition and assembly. Comparison of different common bean cultivars, i.e. resistant to *Fusarium oxysporum* (fox) and susceptible ones, showed striking differences in their root microbiome composition. Fox-resistant cultivars had more microbial diversity and abundance of *Pseudomonadaceae*, *Bacillaceae*, *Solibacteraceae* and *Cytophagaceae* in their rhizosphere as compared to susceptible. Besides metagenome analysis suggested the abundance of genes responsible for protein secretion systems and biosynthesis of antifungal compounds like rhamnolipid and phenazines in rhizobiome of fox-resistant cultivars of bean (Mendes et al. 2018). Like root microbiome engineering, engineering the seed microbiome economically important crops might have importance in enhancing agricultural productivity of food, fodder and fibre crops through crop breeding and improvement. Mitter et al. (2017) introduced an approach to alter the microbial composition of seed through the introduction of an endophyte, *Paraburkholderia phytofirmans*, to the flowers of parent plants which could be included in the microbiome of seed. This approach may lead to the transfer of newly introduced microbe to future generations through vertical inheritance and have a potential role in determining the plant traits.

13.3.7 Effect of Biofertilizer Inoculation on Root Microbial Community Structure

The application of biofertilizer when inoculated on seeds, planting material or directly through soil, enhances crop growth and fitness, mineral nutrition, disease resistance and drought tolerance and increases environmental sustainability (Bhardwaj et al. 2014). Microbial inoculants grow in the rhizosphere and interact with the host plant roots. Similarly, they interact among themselves and other microbes in the rhizosphere to bring about changes in the composition of microbial community which may in turn benefit crop growth, productivity and immunity towards disease incidence (Dal Cortivo et al. 2020). Inoculation of arbuscular mycorrhizal fungi in various agricultural fields influenced the microbial community structure in a field-dependent manner. External inoculation changed the abundance of indigenous AM fungi along with other fungi. Additionally, enrichment of several bacterial OUTs either by changing abundance or introducing new bacterial members to the community was observed (Akyol et al. 2019). Xiong et al. (2017) reported that alteration in the microbial community structure and bacterial abundance on inoculation of *Bacillus* and *Trichoderma* spp. as biofertilizers played a key role in containing *Fusarium oxysporum* infection. This biocontrol activity is the result of suppressiveness induced in the soil through alteration of soil microbiome composition rather than direct antagonism. Endophyte-enriched root microbiome of rice showed restructuring in response to growth-promoting treatment of urea and *Rhizobium leguminosarum* compared to untreated ones. Treated root microbial community showed a reduction in the OUT richness and read abundances. Only 12.04% OTU represented the core microbiome associated with roots with the dominance of α- and γ-proteobacteria followed by *Firmicutes* and *Bacteriodetes* (Jha et al. 2020). No difference in the community composition and biodiversity in the rhizosphere was observed when inoculated with seed-applied biofertilizers. However, biofertilizer treatment resulted in increased microbial biomass in the rhizosphere along with the enhanced activity of enzymes involved in the decomposition of organic matter (Dal Cortivo et al. 2020). This suggests dynamic interactions of root-associated microbiome with externally inoculated biofertilizers may impact microbiome composition altering abundance and richness of bacterial communities of common occurrence, which may help microbiome evolve and benefit the host plant in various ways.

13.3.8 Soil Amendments: Organic and Inorganic

Indiscriminate use of inorganic fertilizers has disturbed the resident microflora of soil, thus affecting soil health in turn. This damage to soil microbial community can be resurrected through soil amendments with organic compounds which enhances the positive microbial activity and functional diversity. Incorporation of organic soil amendments like organic waste and manure, peat, lignite, biochar and various crop residues would have long-term positive effects on soil health and sustainability (Liu et al. 2020). In addition to organic amendments, inorganic amendments like, fine sand, vermiculite, lime, perlite, etc. can be used for alleviating the soil salinity and acidity (Qadir et al. 2019), which in turn increases colonization by the greater number of microbial groups, resulting in increased plant biomass and root exudates, which further strengthens the microbiome (Wang et al. 2015). Incorporation of organic manure in the soil is well-known for altering the physicochemical properties of soils, which causes a shift in microbiome towards beneficial ones and increases agricultural yields (Qian et al. 2016). Use of sheep manure in Alfalfa planting resulted in higher crop yields as a result of boost in microbial activity and removal of toxic metal ions like lead, zinc and cadmium from rhizosphere region (Elouear et al. 2016). Signalling molecules from plants can be applied purposefully to modulate soil microbiome. Plant root exudates like coumarins and benzoxazinoids are reported to induce disease suppression in soil, increase the bioavailability of iron compounds and provide protection against herbivore insects (Voges et al. 2019; Cotton et al. 2019). Incorporation of 1-aminocyclopropane-1-carboxylic acid (ACC) 2 weeks prior planting alters the microbiome composition to add ACC cleaving bacterial groups, which cleaves off ACC into ammonia and α-ketobutyrate through enzyme ACC deaminase, reducing ethylene levels in plants, thus enabling plants to grow better in salinity-stressed conditions (Liu et al. 2019). Therefore, optimal utilization of organic and inorganic soil amendments as mentioned above might help enhance positive microbial interaction and microbial ecosystem services to establish soil health and agricultural sustainability.

13.4 Conclusion

Abiotic stresses severely affect the plant growth and reduce the yield by 20% to 50%. Land affected by the abiotic stress is predicted to increase further in the future diminishing the total area under cultivation, threatening the global food security. Better cultivars of crop plants can be developed using modern breeding approaches and transgenics, but it is slow and time taking. Alternatively plant growth-promoting microorganisms exhibiting tolerance to various abiotic stresses can be utilized to alleviate effect of abiotic stress in agriculture. Through extensive studies numerous microorganisms have been identified and characterized for their ability to ameliorate abiotic stresses, but mechanisms behind how these microbes help to mitigate stress and their interaction with host plants are yet to be understood completely. Microbiome engineering can successfully be utilized for recruiting beneficial microbes either directly or through tailored root exudate, which can respond to specific abiotic or biotic stresses. Microbiome engineering can further facilitate development of modular microbiome through use of microbial consortium having varied metabolic and physiological activities, complementing each other to achieve a cumulative goal of helping plant to withstand the abiotic stress and enhance the productivity.

References

Abd El-Daim IA, Bejai S, Meijer J (2014) Improved heat stress tolerance of wheat seedlings by bacterial seed treatment. Plant Soil 379:337–350. https://doi.org/10.1007/s11104-014-2063-3

Afrasayab S, Faisal M, Hasnain S (2010) Comparative study of wild and transformed salt tolerant bacterial strains on *Triticum aestivum* growth under salt stress. Braz J Microbiol 41:946–955. https://doi.org/10.1590/S1517-83822010000400013

Ait Barka E, Nowak J, Clément C (2006) Enhancement of chilling resistance of inoculated grapevine plantlets with a plant growth-promoting rhizobacterium, *Burkholderia phytofirmans* strain PsJN. Appl Environ Microbiol 72:7246–7252. https://doi.org/10.1128/AEM.01047-06

Akyol TY, Niwa R, Hirakawa H et al (2019) Impact of introduction of arbuscular mycorrhizal fungi on the root microbial community in agricultural fields. Microbes Environ 34:23–32. https://doi.org/10.1264/jsme2.ME18109

Ali SZ, Sandhya V, Grover M et al (2009) *Pseudomonas sp.* strain AKM-P6 enhances tolerance of sorghum seedlings to elevated temperatures. Biol Fertil Soils 46:45–55. https://doi.org/10.1007/s00374-009-0404-9

Al-Karaki GN, Hammad R, Rusan M (2001) Response of two tomato cultivars differing in salt tolerance to inoculation with mycorrhizal fungi under salt stress. Mycorrhiza 11:43–47. https://doi.org/10.1007/s005720100098

AL-Quraan NA, Locy RD, Singh NK et al (2002) Small heat shock proteins and stress toler. Plant Physiol Biochem 1577:1–9

Arif I, Batool M, Schenk PM (2020) Plant microbiome engineering: expected benefits for improved crop growth and resilience. Trends Biotechnol 1–12. https://doi.org/10.1016/j.tibtech.2020.04.015

Arshad M, Shaharoona B, Mahmood T (2008) Inoculation with *Pseudomonas* spp. containing ACC-deaminase partially eliminates the effects of drought stress on growth, yield, and ripening of pea (*Pisum sativum* L.) project supported by the higher education commission, Islamabad, Pakistan (No. PIN 041 2). Pedosphere 18:611–620. https://doi.org/10.1016/S1002-0160(08)60055-7

Ashfaque F, Inam A, Sahay S, Iqbal S (2016) Influence of heavy metal toxicity on plant growth, metabolism and its alleviation by phytoremediation – a promising technology. J Agric Ecol Res Int 6:1–19. https://doi.org/10.9734/jaeri/2016/23543

Ashraf M, Foolad MR (2007) Roles of glycine betaine and proline in improving plant abiotic stress resistance. Environ Exp Bot 59:206–216. https://doi.org/10.1016/j.envexpbot.2005.12.006

Babu AG, Shea PJ, Sudhakar D et al (2015) Potential use of *Pseudomonas koreensis* AGB-1 in association with *Miscanthus sinensis* to remediate heavy metal (loid)-contaminated mining site soil. J Environ Manag 151:160–166

Barassi CA, Ayrault G, Creus CM et al (2006) Seed inoculation with *Azospirillum* mitigates NaCl effects on lettuce. Sci Hortic (Amsterdam) 109:8–14. https://doi.org/10.1016/j.scienta.2006.02.025

Benizri E, Nguyen C, Piutti S et al (2007) Additions of maize root mucilage to soil changed the structure of the bacterial community. Soil Biol Biochem 39:1230–1233

Bhardwaj D, Ansari MW, Sahoo RK, Tuteja N (2014) Biofertilizers function as key player in sustainable agriculture by improving soil fertility, plant tolerance and crop productivity. Microb Cell Factories 13:1–10. https://doi.org/10.1186/1475-2859-13-66

Bressan M, Roncato MA, Bellvert F et al (2009) Exogenous glucosinolate produced by *Arabidopsis thaliana* has an impact on microbes in the rhizosphere and plant roots. ISME J 3:1243–1257. https://doi.org/10.1038/ismej.2009.68

Castrillo G, Teixeira PJPL, Paredes SH et al (2017) Root microbiota drive direct integration of phosphate stress and immunity. Nature 543:513–518. https://doi.org/10.1038/nature21417

Chaparro JM, Sheflin AM, Manter DK, Vivanco JM (2012) Manipulating the soil microbiome to increase soil health and plant fertility. Biol Fertil Soils 48:489–499

Chibuike GU, Obiora SC (2014) Heavy metal polluted soils: effect on plants and bioremediation methods. Appl Environ Soil Sci 2014. https://doi.org/10.1155/2014/752708

Cotton TEA, Pétriacq P, Cameron DD et al (2019) Metabolic regulation of the maize rhizobiome by benzoxazinoids. ISME J 13:1647–1658. https://doi.org/10.1038/s41396-019-0375-2

Dal Cortivo C, Ferrari M, Visioli G et al (2020) Effects of seed-applied biofertilizers on rhizosphere biodiversity and growth of common wheat (*Triticum aestivum* L.) in the field. Front Plant Sci 11:1–14. https://doi.org/10.3389/fpls.2020.00072

de Langre E (2008) Effects of wind on plants. Annu Rev Fluid Mech 40:141–168

De Zelicourt A, Al-Yousif M, Hirt H (2013) Rhizosphere microbes as essential partners for plant stress tolerance. Mol Plant 6:242–245. https://doi.org/10.1093/mp/sst028

Dimkpa C, Weinand T, Asch F (2009) Plant-rhizobacteria interactions alleviate abiotic stress conditions. Plant Cell Environ 32:1682–1694. https://doi.org/10.1111/j.1365-3040.2009.02028.x

Dombrowski N, Schlaeppi K, Agler MT et al (2017) Root microbiota dynamics of perennial *Arabis alpina* are dependent on soil residence time but independent of flowering time. ISME J 11:43–55. https://doi.org/10.1038/ismej.2016.109

Edwards J, Johnson C, Santos-Medellín C et al (2015) Structure, variation, and assembly of the root-associated microbiomes of rice. Proc Natl Acad Sci USA 112:E911–E920. https://doi.org/10.1073/pnas.1414592112

Elouear Z, Bouhamed F, Boujelben N, Bouzid J (2016) Application of sheep manure and potassium fertilizer to contaminated soil and its effect on zinc, cadmium and lead accumulation by alfalfa plants. Sustain Environ Res 26:131–135. https://doi.org/10.1016/j.serj.2016.04.004

Farooq M, Wahid A, Kobayashi N, Fujita D, SMA B (2009) Plant drought stress: effects, mechanisms and management. Agron Sustain Dev 29:185–212

Feng G, Zhang FS, Li XL et al (2002) Improved tolerance of maize plants to salt stress by arbuscular mycorrhiza is related to higher accumulation of soluble sugars in roots. Mycorrhiza 12:185–190. https://doi.org/10.1007/s00572-002-0170-0

Foo JL, Ling H, Lee YS, Chang MW (2017) Microbiome engineering: current applications and its future. Biotechnol J 12:1–11

Ghorbanpour A, Salimi A, Ghanbary MAT et al (2018) The effect of *Trichoderma harzianum* in mitigating low temperature stress in tomato (*Solanum lycopersicum* L.) plants. Sci Hortic (Amsterdam) 230:134–141

Giri B, Mukerji KG (2004) Mycorrhizal inoculant alleviates salt stress in *Sesbania aegyptiaca* and *Sesbania grandiflora* under field conditions: evidence for reduced sodium and improved magnesium uptake. Mycorrhiza 14:307–312. https://doi.org/10.1007/s00572-003-0274-1

Gopal M, Gupta A, Thomas GV (2013) Bespoke microbiome therapy to manage plant diseases. Front Microbiol 4:10–13. https://doi.org/10.3389/fmicb.2013.00355

Grigorova B, Vaseva I, Demirevska K, Feller U (2011) Combined drought and heat stress in wheat: changes in some heat shock proteins. Biol Plant 55:105–111. https://doi.org/10.1007/s10535-011-0014-x

Hadi F, Bano A (2010) Effect of diazotrophs (*Rhizobium* and *Azotobacter*) on growth of maize (*Zea mays* L.) and accumulation of lead (Pb) in different plant parts. Pakistan J Bot 42:4363–4370

Hajiboland R, Joudmand A, Aliasgharzad N et al (2019) Arbuscular mycorrhizal fungi alleviate low-temperature stress and increase freezing resistance as a substitute for acclimation treatment in barley. Crop Pasture Sci 70:218–233

Hassan MU, Chattha MU, Khan I et al (2020) Heat stress in cultivated plants: nature, impact, mechanisms, and mitigation strategies—a review. Plant Biosyst 0:1–24

Huang S, Spielmeyer W, Lagudah ES et al (2006) A sodium transporter (HKT7) is a candidate for Nax1, a gene for salt tolerance in durum wheat. Plant Physiol 142:1718–1727. https://doi.org/10.1104/pp.106.088864

Huang GT, Ma SL, Bai LP et al (2012) Signal transduction during cold, salt, and drought stresses in plants. Mol Biol Rep 39:969–987. https://doi.org/10.1007/s11033-011-0823-1

Hussain SS, Mehnaz S, Siddique KHM (2018) Harnessing the plant microbiome for improved abiotic stress tolerance. In: Plant microbiome: stress response, pp 1–41

IPCC (2007) Summary for policymakers. In: Climate change 2007: the physical science basis

James RA, Davenport RJ, Munns R (2006) Physiological characterization of two genes for Na^+ exclusion in durum wheat, Nax1 and Nax2. Plant Physiol 142:1537–1547. https://doi.org/10.1104/pp.106.086538

Jamil A, Riaz S, Ashraf M, Foolad MR (2011) Gene expression profiling of plants under salt stress. CRC Crit Rev Plant Sci 30:435–458

Jaramillo JEP, Mendes R, Raaijmakers JM (2016) Impact of plant domestication on rhizosphere microbiome assembly and functions. Plant Mol Biol 90:635–644. https://doi.org/10.1007/s11103-015-0337-7

Jha B, Gontia I, Hartmann A (2012) The roots of the halophyte *Salicornia brachiata* are a source of new halotolerant diazotrophic bacteria with plant growth-promoting potential. Plant Soil 356:265–277. https://doi.org/10.1007/s11104-011-0877-9

Jha PN, Gomaa AB, Yanni YG et al (2020) Alterations in the endophyte-enriched root-associated microbiome of rice receiving growth-promoting treatments of urea fertilizer and *Rhizobium* biofertilizer. Microb Ecol 79:367–382. https://doi.org/10.1007/s00248-019-01406-7

Jindal V, Atwal A, Bs S, Singh R (1993) Effect of vesicular-arbuscular mycorrhizae on metabolism of moong plants under NaCl salinity. Plant Physiol Biochem 31:475–481

Kakar KU, Ren XL, Nawaz Z et al (2016) A consortium of rhizobacterial strains and biochemical growth elicitors improve cold and drought stress tolerance in rice (*Oryza sativa* L.). Plant Biol 18:471–483. https://doi.org/10.1111/plb.12427

Khan MA, Asaf S, Khan AL et al (2020a) Thermotolerance effect of plant growth-promoting *Bacillus cereus* SA1 on soybean during heat stress. BMC Microbiol 20:175. https://doi.org/10.1186/s12866-020-01822-7

Khan MA, Asaf S, Khan AL et al (2020b) Extending thermotolerance to tomato seedlings by inoculation with SA1 isolate of *Bacillus cereus* and comparison with exogenous humic acid application. PLoS One 15:1–25. https://doi.org/10.1371/journal.pone.0232228

Kim S, Lowman S, Hou G et al (2012) Growth promotion and colonization of switchgrass (*Panicum virgatum*) cv. alamo by bacterial endophyte *Burkholderia phytofirmans* strain PsJN. Biotechnol Biofuels 5:1–10. https://doi.org/10.1186/1754-6834-5-37

Kohler J, Hernández JA, Caravaca F, Roldán A (2008) Plant-growth-promoting rhizobacteria and arbuscular mycorrhizal fungi modify alleviation biochemical mechanisms in water-stressed plants. Funct Plant Biol 35:141–151

Kretzschmar T, Kohlen W, Sasse J et al (2012) A petunia ABC protein controls strigolactone-dependent symbiotic signalling and branching. Nature 483:341–344. https://doi.org/10.1038/nature10873

Kumar KV, Srivastava S, Singh N, Behl HM (2009) Role of metal resistant plant growth promoting bacteria in ameliorating fly ash to the growth of *Brassica juncea*. J Hazard Mater 170:51–57

Kwak MJ, Kong HG, Choi K et al (2018) Rhizosphere microbiome structure alters to enable wilt resistance in tomato. Nat Biotechnol 36:1100–1116. https://doi.org/10.1038/nbt.4232

Lau JA, Lennon JT (2012) Rapid responses of soil microorganisms improve plant fitness in novel environments. Proc Natl Acad Sci USA 109:14058–14062. https://doi.org/10.1073/pnas.1202319109

Lebeis SL, Paredes SH, Lundberg DS et al (2015) Salicylic acid modulates colonization of the root microbiome by specific bacterial taxa. Science (80) 349:860–864. https://doi.org/10.1126/science.aaa8764

Leite MCBS, Farias ARB, Freire FJ et al (2014) Isolation, bioprospecting and diversity of salt-tolerant bacteria associated with sugarcane in soils of Pernambuco, Brazil. Rev Bras Eng Agríc Ambient 3715–3725

Liu H, Khan MY, Carvalhais LC et al (2019) Soil amendments with ethylene precursor alleviate negative impacts of salinity on soil microbial properties and productivity. Sci Rep 9:1–13. https://doi.org/10.1038/s41598-019-43305-4

Liu H, Brettell LE, Qiu Z, Singh BK (2020) Microbiome-mediated stress resistance in plants. Trends Plant Sci 25:733–743. https://doi.org/10.1016/j.tplants.2020.03.014

Ma Y, Rajkumar M, Luo Y, Freitas H (2013) Phytoextraction of heavy metal polluted soils using *Sedum plumbizincicola* inoculated with metal mobilizing *Phyllobacterium myrsinacearum* RC6b. Chemosphere 93:1386–1392. https://doi.org/10.1016/j.chemosphere.2013.06.077

Ma J, Janoušková M, Ye L et al (2019) Role of arbuscular mycorrhiza in alleviating the effect of cold on the photosynthesis of cucumber seedlings. Photosynthetica 57:86–95. https://doi.org/10.32615/ps.2019.001

Marasco R, Mapelli F, Rolli E et al (2016) *Salicornia strobilacea* (synonym of *Halocnemum strobilaceum*) grown under different tidal regimes selects rhizosphere bacteria capable of promoting plant growth. Front Microbiol 7:1–11. https://doi.org/10.3389/fmicb.2016.01286

Marulanda A, Porcel R, Barea JM, Azcón R (2007) Drought tolerance and antioxidant activities in lavender plants colonized by native drought-tolerant or drought-sensitive *Glomus* species. Microb Ecol 54:543–552. https://doi.org/10.1007/s00248-007-9237-y

Marulanda A, Barea JM, Azcón R (2009) Stimulation of plant growth and drought tolerance by native microorganisms (AM Fungi and bacteria) from dry environments: mechanisms related to bacterial effectiveness. J Plant Growth Regul 28:115–124. https://doi.org/10.1007/s00344-009-9079-6

Masson-Delmotte V, Zhai P, Pörtner H-O, Roberts D, Skea J, Shukla PR, Pirani A, Moufouma-Okia W, Péan C, Pidcock R, Connors S, Matthews JBR, Chen Y, Zhou X, Gomis MI, Lonnoy E, Maycock T, Tignor M (2019) Global warming of 1.5°C. An IPCC special report on the impacts of global warming of 1.5°C above pre-industrial levels and related global greenhouse gas emission pathways, in the context of strengthening the global response to. IPCC, 2018 Summary policymakers 1–24. https://doi.org/10.1016/j.oneear.2019.10.025

Mayak S, Tirosh T, Glick BR (2004) Plant growth-promoting bacteria that confer resistance to water stress in tomatoes and peppers. Plant Sci 166:525–530. https://doi.org/10.1016/j.plantsci.2003.10.025

Mendes LW, Raaijmakers JM, De Hollander M et al (2018) Influence of resistance breeding in common bean on rhizosphere microbiome composition and function. ISME J 12:212–224. https://doi.org/10.1038/ismej.2017.158

Mitter B, Pfaffenbichler N, Flavell R et al (2017) A new approach to modify plant microbiomes and traits by introducing beneficial bacteria at flowering into progeny seeds. Front Microbiol 8:1–10. https://doi.org/10.3389/fmicb.2017.00011

Mueller UG, Sachs JL (2015) Engineering microbiomes to improve plant and animal health. Trends Microbiol 23:606–617. https://doi.org/10.1016/j.tim.2015.07.009

Naveed M, Hussain MB, Zahir ZA et al (2014a) Drought stress amelioration in wheat through inoculation with *Burkholderia phytofirmans* strain PsJN. Plant Growth Regul 73:121–131. https://doi.org/10.1007/s10725-013-9874-8

Naveed M, Mitter B, Reichenauer TG et al (2014b) Increased drought stress resilience of maize through endophytic colonization by *Burkholderia phytofirmans* PsJN and *Enterobacter* sp. FD17. Environ Exp Bot 97:30–39. https://doi.org/10.1016/j.envexpbot.2013.09.014

Neal AL, Ahmad S, Gordon-Weeks R, Ton J (2012) Benzoxazinoids in root exudates of maize attract *Pseudomonas putida* to the rhizosphere. PLoS One 7. https://doi.org/10.1371/journal.pone.0035498

Niu B, Kolter R (2018) Quantification of the composition dynamics of a maize root-associated simplified bacterial community and evaluation of its biological control effect. Bio-Protocol 8:1–21. https://doi.org/10.21769/bioprotoc.2885

Niu B, Paulson JN, Zheng X, Kolter R (2017) Simplified and representative bacterial community of maize roots. Proc Natl Acad Sci USA 114:E2450–E2459. https://doi.org/10.1073/pnas.1616148114

Oyserman BO, Martirano JM, Wipperfurth S et al (2017) Community assembly and ecology of activated sludge under photosynthetic feast-famine conditions. Environ Sci Technol 51:3165–3175. https://doi.org/10.1021/acs.est.6b03976

Oyserman BO, Medema MH, Raaijmakers JM (2018) Road MAPs to engineer host microbiomes. Curr Opin Microbiol 43:46–54

Pérez-Jaramillo JE, De Hollander M, Ramírez CA et al (2019) Deciphering rhizosphere microbiome assembly of wild and modern common bean (*Phaseolus vulgaris*) in native and agricultural soils from Colombia. Microbiome 7:1–16. https://doi.org/10.1186/s40168-019-0727-1

Peters NK, Long SR (1988) Alfalfa root exudates and compounds which promote or inhibit induction of *Rhizobium meliloti* nodulation genes. Plant Physiol 88:396–400. https://doi.org/10.1104/pp.88.2.396

Qadir M, Quillérou E, Nangia V et al (2019) Economics of salt-induced land degradation and restoration to cite this version: HAL id: hal-01954413. HAL Arch Ouvert 38:282–295

Qian M, Wu H, Wang J et al (2016) Occurrence of trace elements and antibiotics in manure-based fertilizers from the Zhejiang Province of China. Sci Total Environ 559:174–181. https://doi.org/10.1016/j.scitotenv.2016.03.123

Rolli E, Marasco R, Vigani G et al (2015) Improved plant resistance to drought is promoted by the root-associated microbiome as a water stress-dependent trait. Environ Microbiol 17:316–331. https://doi.org/10.1111/1462-2920.12439

Rudrappa T, Czymmek KJ, Paré PW, Bais HP (2008) Root-secreted malic acid recruits beneficial soil bacteria. Plant Physiol 148:1547–1556. https://doi.org/10.1104/pp.108.127613

Sandhya V, Ali S, Grover M et al (2009) *Pseudomonas* sp. strain P45 protects sunflowers seedlings from drought stress through improved soil structure. J Oilseed Res 26:600–601

Santhanam R, Luu VT, Weinhold A et al (2015) Native root-associated bacteria rescue a plant from a sudden-wilt disease that emerged during continuous cropping. Proc Natl Acad Sci USA 112: E5013–E5120. https://doi.org/10.1073/pnas.1505765112

Sasse J, Martinoia E, Northen T (2018) Feed your friends: do plant exudates shape the root microbiome? Trends Plant Sci 23:25–41. https://doi.org/10.1016/j.tplants.2017.09.003

Sheng X, He L, Wang Q et al (2008) Effects of inoculation of biosurfactant-producing *Bacillus sp.* J119 on plant growth and cadmium uptake in a cadmium-amended soil. J Hazard Mater 155:17–22

Siddikee MA, Chauhan PS, Anandham R et al (2010) Isolation, characterization, and use for plant growth promotion under salt stress, of ACC deaminase-producing halotolerant bacteria derived from coastal soil. J Microbial Biotechnol 20:1577–1584. https://doi.org/10.4014/jmb.1007.07011

Singh S, Parihar P, Singh R et al (2016) Heavy metal tolerance in plants: role of transcriptomics, proteomics, metabolomics, and ionomics. Front Plant Sci 6:1–36

Srivastava S, Yadav A, Seem K et al (2008) Effect of high temperature on *Pseudomonas putida* NBRI0987 biofilm formation and expression of stress sigma factor RpoS. Curr Microbiol 56:453–457. https://doi.org/10.1007/s00284-008-9105-0

Srivastava S, Verma PC, Chaudhry V et al (2013) Influence of inoculation of arsenic-resistant *Staphylococcus arlettae* on growth and arsenic uptake in *Brassica juncea* (L.) Czern. Var. R-46. J Hazard Met 262:1039–1047

Stringlis IA, Zhang H, Pieterse CMJ et al (2018) Microbial small molecules-weapons of plant subversion. Nat Prod Rep 35:410–433

Svoboda P, Janská A, Spiwok V et al (2016) Global scale transcriptional profiling of two contrasting barley genotypes exposed to moderate drought conditions: contribution of leaves and crowns to water shortage coping strategies. Front Plant Sci 7:1–17. https://doi.org/10.3389/fpls.2016.01958

Swenson W, Wilson DS, Elias R (2000) Artificial ecosystem selection. Proc Natl Acad Sci USA 97:9110–9114. https://doi.org/10.1073/pnas.150237597

Thao NP, Tran L-SP (2016) Editorial (thematic issue: enhancement of plant productivity in the post-genomics era). Curr Genomics 17:295–296. https://doi.org/10.2174/138920291704160607182507

Theocharis A, Bordiec S, Fernandez O et al (2012) *Burkholderia phytofirmans* PsJN primes *Vitis vinifera* L. and confers a better tolerance to low nonfreezing temperatures. Mol Plant-Microbe Interact 25:241–249. https://doi.org/10.1094/MPMI-05-11-0124

Timmusk S, Abd El-Daim IA, Copolovici L, et al (2014) Drought-tolerance of wheat improved by rhizosphere bacteria from harsh environments: enhanced biomass production and reduced emissions of stress volatiles. PLoS One 9. https://doi.org/10.1371/journal.pone.0096086

Upadhyay SK, Singh DP, Saikia R (2009) Genetic diversity of plant growth promoting rhizobacteria isolated from rhizospheric soil of wheat under saline condition. Curr Microbiol 59:489–496. https://doi.org/10.1007/s00284-009-9464-1

Veliz-Vallejos DF, Van Noorden GE, Yuan M, Mathesius U (2014) A *Sinorhizobium meliloti*-specific N-acyl homoserine lactone quorum-sensing signal increases nodule numbers in

Medicago truncatula independent of autoregulation. Front Plant Sci 5:1–13. https://doi.org/10.3389/fpls.2014.00551

Voges MJEEE, Bai Y, Schulze-Lefert P, Sattely ES (2019) Plant-derived coumarins shape the composition of an Arabidopsis synthetic root microbiome. Proc Natl Acad Sci USA 116:12558–12565. https://doi.org/10.1073/pnas.1820691116

Wang S, Tan Y, Fan H et al (2015) Responses of soil microarthropods to inorganic and organic fertilizers in a poplar plantation in a coastal area of eastern China. Appl Soil Ecol 89:69–75. https://doi.org/10.1016/j.apsoil.2015.01.004

Wani PA, Khan MS, Zaidi A (2007) Effect of metal tolerant plant growth promoting *Bradyrhizobium* sp. (vigna) on growth, symbiosis, seed yield and metal uptake by greengram plants. Chemosphere 70:36–45

Wani PA, Khan MS, Zaidi A (2008) Effect of metal-tolerant plant growth-promoting *Rhizobium* on the performance of pea grown in metal-amended soil. Arch Environ Contam Toxicol 55:33–42

Waters ER (2013) The evolution, function, structure, and expression of the plant sHSPs. J Exp Bot 64:391–403. https://doi.org/10.1093/jxb/ers355

Wicaksono WA, Jones EE, Casonato S et al (2018) Biological control of *Pseudomonas syringae* pv. actinidiae (Psa), the causal agent of bacterial canker of kiwifruit, using endophytic bacteria recovered from a medicinal plant. Biol Control 116:103–112

Wintermans PCA, Bakker PAHM, Pieterse CMJ (2016) Natural genetic variation in *Arabidopsis* for responsiveness to plant growth-promoting rhizobacteria. Plant Mol Biol 90:623–634. https://doi.org/10.1007/s11103-016-0442-2

Xiong W, Guo S, Jousset A et al (2017) Bio-fertilizer application induces soil suppressiveness against *Fusarium* wilt disease by reshaping the soil microbiome. Soil Biol Biochem 114:238–247. https://doi.org/10.1016/j.soilbio.2017.07.016

Yadav AN, Sharma D, Gulati S et al (2015) Haloarchaea endowed with phosphorus Solubilization attribute implicated in phosphorus cycle. Sci Rep 5:1–10. https://doi.org/10.1038/srep12293

Yamaguchi-Shinozaki K, Shinozaki K (1994) A novel cis-acting element in an *Arabidopsis* gene is involved in responsiveness to drought, low-temperature, or high-salt stress. Plant Cell 6:251–264. https://doi.org/10.1105/tpc.6.2.251

Yang Q, Tu S, Wang G et al (2012) Effectiveness of applying arsenate reducing bacteria to enhance arsenic removal from polluted soils by *Pteris vittata* L. Int J Phytoremediation 14:89–99

Yildirim E, Taylor AG (2005) Effect of biological treatments on growth of bean plants under salt stress. Science 123

Yuwono T, Handayani D, Soedarsono J (2005) The role of osmotolerant rhizobacteria in rice growth under different drought conditions. Aust J Agric Res 56:715–721. https://doi.org/10.1071/AR04082

Zhang Y, Zhao L, Wang Y et al (2008) Enhancement of heavy metal accumulation by tissue specific co-expression of iaaM and ACC deaminase genes in plants. Chemosphere 72:564–571. https://doi.org/10.1016/j.chemosphere.2008.03.043

Zou J, Liu C, Liu A et al (2012) Overexpression of OsHsp17.0 and OsHsp23.7 enhances drought and salt tolerance in rice. J Plant Physiol 169:628–635. https://doi.org/10.1016/j.jplph.2011.12.014

Zulfikar Ali S, Sandhya V, Grover M et al (2011) Effect of inoculation with a thermotolerant plant growth promoting *Pseudomonas putida* strain AKMP7 on growth of wheat (*Triticum* spp.) under heat stress. J Plant Interact 6:239–246. https://doi.org/10.1080/17429145.2010.545147

Chapter 14
Impact of Abiotic Stress on Plant Brassinosteroids

Zahra Dehghanian, Ali Bandehagh, Khashayar Habibi, Khosro Balilashaki, and Behnam Asgari Lajayer

Abstract Brassinosteroids (BRs) are a category of steroid hormones which have different functions on development and plant growth process, in addition to stress response. BR deficiencies disrupt physiological processes and induce phenotypic anomalies in plants. A wide number of studies indicate that BRs could have a beneficial impact on the plant tolerance to abiotic tensions, consisting of low temperature, sun, salinity, drought, pesticides, and heavy metals. The fundamental mechanisms of tension resistance caused by BR however remain still unknown. BR reception occurs in the cellular level through BR receiver, resulting in flow of phosphorylation incidents triggering main transcription factors Brassinazole-Resistant1 (BZR1) which regulate BR-responsive gene transcription in cell nucleus. BRs enhance photosynthetic performance affected by tension situation that lead in large part to enhanced growth and accumulation of biomass. Genetic experiments indicate established association between endogenous rates of BR and resistance to abiotic stress, but this conclusion contradicts the results of some BR mutants under stress. Importantly, the reaction of plants to BRs varies greatly depending on the species, stages of development, and environmental factors. Furthermore, some chemical messengers that engage in the effective regulating of BR efficacies often do a significant role in plant adjustment to tension. In this chapter, the role of BRs in response to abiotic stresses in plants was studied. These findings show the elaboration of BR function in interceding tension tolerance in plants.

Z. Dehghanian
Department of Biotechnology, Faculty of Agriculture, Azarbaijan Shahid Madani University, Tabriz, Iran

A. Bandehagh · K. Habibi
Department of Plant Breeding and Biotechnology, Faculty of Agriculture, University of Tabriz, Tabriz, Iran

K. Balilashaki
Department of Horticultural Science, Faculty of Agriculture, University of Guilan, Guilan, Iran

B. Asgari Lajayer (✉)
Health and Environment Research Center, Tabriz University of Medical Sciences, Tabriz, Iran
e-mail: h-asgari@tabrizu.ac.ir

D. K. Choudhary et al. (eds.), *Climate Change and the Microbiome*, Soil Biology 63,
https://doi.org/10.1007/978-3-030-76863-8_14

Keywords Environmental stress · Reactive oxygen species · Stress tolerance · Brassinolide · Hormones · Plants

14.1 Introduction

Associated with both daily and seasonal changes, plants have to tolerate the usual trends in environmental parameters. Moreover, pollution in the environment and abnormal weather events mostly lead to plant stress (Ahammed et al. 2015; Fang et al. 2019). Plant durability depends mainly on understanding the stress incentives and quick response to the impacts caused by stress, as they are unable to move (Nolan et al. 2020; Planas-Riverola et al. 2019); and also by the help of several signaling molecules and hormones, they respond to stressor factors (Wang et al. 2019a; Zhang et al. 2019; Zhou et al. 2019; Guo et al. 2019).

In stressful situations, hormone coordination regulates plant tolerance and also survival. Phytohormones engage a notable pathway which includes several molecules directing to an arbitrary general pathway (Xiong et al. 2002). Essentially, this begins the reception of signals on the surface of the cell, pursued by the secondary messenger's production like ROS, reactive oxygen species, and NO, nitride oxide, leading to protein phosphorylation cascade in ultimately propelling transcription of stress-responsive genes (Xia et al. 2009a, 2011, 2014; Yin et al. 2016). Nevertheless, for plant acclimatization to stress, the proper spatial coordination among signaling molecules is substantial.

14.2 Brassinosteroids in the Development Regulations of Plant

BRs perform various roles in the plant growing progress (Fang et al. 2019; Peres et al. 2019). In model plants, both signaling pathways and BRs Biosynthetic have been studied widely, which noticeably helped improving the comprehension of BR regulatory mechanisms, especially in various biological practices about development and growth of plants (Zhao and Li 2012; Nolan et al. 2020; Planas-Riverola et al. 2019). At the outset, BRS-induced formation was considered as the result of cell elongation, although a function has been identified in cell division for BR. As Nolan states, in addition to cell elongation and partition, it's perfectly appointed that BRs regulate various aspects of development and growing, for example, photomorphogenesis, xylem differentiation, reproduction of plants, and also response biotic and abiotic stresses. Though biosynthesis of BRs considered to take place just in the endoplasmic reticulum, the realization of BR is mediated by local receptors on the plasma membrane at the surface of cell, BRI1 (Brassinosteroid Insensitive 1), and relevant homologs (Nolan et al. 2020; Northey et al. 2016). After BR is perceived by

BRI1 and the receptors, the BR signals are transmitted through a well-known signaling cascade to BRI1-EMS-Suppressor1 (BES1) and Brassinazole-Resistant1 (BZR1), transcription factors, which ultimately control the transcription of BR-regulated genes (Tong and chu 2018; Nolan et al. 2020; Planas-Riverola et al. 2019).

According to Clouse, low seed germination, delayed flowering and aging, dwarfism, and reduced prolificacy are the reasons of BR deficiency in plants (Clouse 2015). In rice, the expression of a gene as encoding, sterol hydroxylase, enhances interior BR levels and leads in an increase up to 40% in the size of grain (Wu et al. 2008). A new investigation on soy (*Glycine max* L. Merrill) has illustrated BR can lag leaf aging (Yin et al. 2019). The collected evidence also depicts that in the future of agriculture, the inclusion of BR in controlling important agronomic factors has the capacity to deform (Divi and Krishna 2009; Tong and Chu 2018).

14.3 The BR Effects in Abiotic Stress

BRs have a significant part in adapting to the environment, besides improving growth. Except in few cases, BR makes to ameliorate plant adaptation to biotic and abiotic pressures like aridity, salinity, high temperature, low temperature, pesticides, heavy elements, and also stress due to organic pollutants (Kagale et al. 2007; Rajewska et al. 2016; Xia et al. 2018). Nonetheless the process of BR operation in rising the endurance accomplished by plant to abiotic stresses is not yet known. In tomato (*Solanum lycopersicum* L.) while DWF overexpression leads to raised cold tolerance, BRS biosynthesis mutations (dwf) present sensitivity to cold stress (Xia et al. 2018; Fang et al. 2019). Investigations have exposed that through autophagy regulation, the self-destructive process of cell, which is applied with the plant to intercede stress tolerance, BRs are included in plants' tolerance to nitrogen (N) deficiency (Wang et al. 2019d). Exogenous BR involved in autophagosomes formation and autophagy increases the transcript levels of genes. Also tolerance to N starvation and autophagosomes are increased by excessive expression of BZR1, and BR-induced tolerance to N starvation and the formation of autophagosomes are reasons for the BZR1 shutdown. Nevertheless, a duality of BR function in tolerance of plant to nutrient deficiencies and plant susceptibility to iron deficiency is raised by the application of exogenous BR (Wang et al. 2012a). Many studies examining the effects of BR on plants, almost all studies use pharmacological methods that use exogenous to investigate the preventive role of BR (Ahammed et al. 2015). Different modes of applications, such as pre-sowing seed treatment, pre-planting dipping of cuttings, post-emergence root treatment, foliar application, etc. have been used in multiple plant species (Kagale et al. 2007; Sasse 2003; Yu et al. 2004; Amraee et al. 2019; Yue et al. 2019; Yin et al. 2019; Sharma et al. 2016c, 2019). The impacts of BR greatly appertain on several factors such as plant species, dose, types and stress period, and interaction with other criteria like different levels of growth and conditions (with or without tension), signaling molecules, development regulators and hormones, and type and duration of stress (Nolan et al. 2020; Yin et al. 2019).

According to studies on producing crops, the proper brassinolide concentration is between 5 and 50 mg per hectare (Khripach et al. 2000). However, BRs can affect different physiological processes of plant at low doses. In addition, plants' response may be different in the limited doses. For instance, cucumber bushes have high BR concentrations (0.2–1.0 μM EBR). While moderate BR concentrations (0.1–0.15 μM EBR) cause photosynthesis, it eliminates CO_2 uptake capacity (Jiang et al. 2012). On average, a BR low dose (0.1 μM EBR) causes the stomata to open, and a BR high dose (1.0 μM EBR) brings the stomata to close (Xia et al. 2014). Some studies on BR depicts the plants' response to BR concentration widely depends on plant species, growth conditions, the stage of plant growth, and specific application method (Ahammed et al. 2015).

14.3.1 Heat Stress and BRs

Recently, due to climate change, heat tension is identified as a very significant abiotic tension (Nolan et al. 2020), which has a negative impact yield on each part of the world. Damage emerging from heat includes reduced plant growth, leaf burning, abscesses, and aging decreasing productivity of plant, also fruit injuries (Bita and Gerats 2013). Accumulations of BZR1 and BES1 at the molecular stage happen at high temperatures, enhancing PIF4 level (Phytochrome Interacting Factors 4) (Martínez et al. 2018). On gene transcription, the foundation of PIF4-BES1 heterodimers comforts the act of BZR1 and promotes growth of thermogenic. Also it can be stated that reduction of BRI1 cumulation at high temperatures enhances the root growing (Martins et al. 2017). Destructive heat impacts on plants can be reduced by the use of BR exogenously (Sadura and Janeczko 2018). For heat stress, photosynthesis is almost the main delicate physiological procedure (Ahammed et al. 2016). Moreover, high temperature reduces photosynthetic rate, even hindering photosynthesis output (PSII), depending on photochemical activity (Ogweno et al. 2008; Zhang et al. 2013).

According to stress, by rising the antioxidant activity enzymes which minimize lipid peroxidation, EBR pretreatment (0.2 μM) in tomatoes can improve the high-temperature reduction in photosynthesis (Ogweno et al. 2008). Amazingly, in the both genotypes of heat-sensitive and heat-tolerant plant, BR can ameliorate thermotolerance. As an example, EBR pretreatment clearly ameliorate net CO_2 uptake rate, water use efficiency, photosynthetic pigment content, stomatal conductance, and PSI photochemical activity from heat-sensitive and heat-tolerant ecotypes in melons (Zhang et al. 2013).

Treatment of EBR (0.05–0.2 μM) in eggplant by increasing antioxidant potential diminishes heat stress, which ultimately cuts down ROS cumulation at high temperatures (Wu et al. 2014). Thus, the EBR use (0.01 μM) foliage greatly reduces biomass cumulation, growth improvement, photosynthetic regulation, and antioxidant capability at high temperatures in wheat (Hussain et al. 2019). Then, BR mimic 7, 8-Dihydro-8α-20-hydroxyecdysone (αDHECD, 0.0001 μM) usage in reproductive level can enlarge carbohydrate content in rice and also raise photosynthesis and regulation of seed and grain weight below high-temperature stress (Sonjaroon et al. 2018). The accounts demonstrate that BRs own specific stimulative impacts on the antioxidant capacity and photosynthesis of plants, which has an important position in diminishing the damaging outcomes of heat stress. Though major investigations have shown the heat stress preservative role from BR utilizing exogenous applications, just a few ones have concentrated on fully developed procedures with genetic evidence (Ahammed et al. 2015; Yu et al. 2004; Zhou et al. 2014). Also BRs appear to compel heat endurance via complicated mechanisms, of which just a few are declared. Studies have depicted that transient H_2O_2 production in apoplasts is an important signal for stress toleration of BR-induced heat in tomatoes (Zhou et al. 2014). By respiratory burst oxidase homolog 1 (RBOH1), BR-induced H_2O_2 production in NADPH oxidase-dependent apoplasts is encoded. H_2O_2 accumulation is severely repressed, and BR-induced endurance to heat is compromised, when the RBOH1 and Mitogen-activated Kinase 2 (MPK2) or 1 (MPK1) protein genes are extinguished in tomato plants (Nie et al. 2013). According to the researches, extinguishing off MPK1 does not have the required outcome in such effect and suggests MPK2 is more significant in comparison with MPK1 in producing BR2-induced MP2 H_2O_2 in the apoplast and also its heat tolerance; in addition, RBOH1, H_2O_2, and MPK2 may act as affirmative feedback. Moreover BR-induced tolerance to high heats and, also, levels of the stress tolerance and related genes to defense, like APX5, Cu-Zn SOD, CAT1, GR1, HSP90, PR1, NPR1, and WRKY1 which are adjusted by the exogenous BR program in tomatoes and reveal the BR mechanism. Heat tolerance caused by (Zhou et al. 2014). Except antioxidant system, HSPs proteins of high temperature shock do a vital role in BR-induced heatstroke (Dhaubhadel et al. 2002; Kagale et al. 2007). During heat stress, therapy of EBR increases synthesis of HSP by preserving multiple translation machines (Dhaubhadel et al. 2002). However, by rejecting those reports, the increased expression of HSP in the BR2 mutation in dwf4 and det 2–1 under high-temperature stress offers cumulation of HSP which is not essential for BR-induced heatstroke in *Arabidopsis* (Kagale et al. 2007). In 5-day-old seedlings which revealed to salinity stress and high temperatures, the BR biosynthetic gene overexpression in ATDWF4 [in *Arabidopsis*] does not alter stress tolerance. Analysis of BR mutations in atmosphere-deficient BR (mutations in the HvDWARF or HvCPD) or impaired in BR signaling (missense HvBRI1 gene) indicated that all atmospheric mutations are more tolerant of high temperatures than wild-type mutants (Sadura et al. 2019). These studies representing the shapes of physiological impacts and function of endogenous BR and exogenous BR applied in different plant species are very diverse (Ahammed et al. 2015; Kagale et al. 2007; Nie et al. 2019).

14.3.2 Low-Temperature Stress and BRs

In the world's different parts, especially regarding the exothermic plants, low temperature is one of the significant elements in crop producing, which causes cold or frost tolerance (Cui et al. 2011; Zhang et al. 2019). Changes in the actions of macromolecules, membrane fluidity, also mechanical limitations, and reduced osmotic potential in cells are disorders caused by cold stress (Xiong et al. 2002). The mentioned stress type has impacts on the process of photosynthetic in plant, which decrease with lessening CO_2 uptake, Benson-Calvin cycle enzyme activity, and photoinhibition at PSI and PSI (Jiang et al. 2013; Zhang et al. 2019). Cold stress causes ROS to cumulate too and via lipid peroxidation large amounts of ROS can harm biomembranes (Chen et al. 2013). Plants have expanded diverse collections of ROS conservation strategies, in order to prevent overaccumulation of ROS and overdestruction of the photosynthetic systems causing photoinhibition (Fang et al. 2019; Zhang et al. 2019; Ahammed et al. 2020b). The mutant plants by expanding the ability to modify ROS, overexpression of genes which increase ROS inhibitory enzymes or expose better endurance to low-temperature stress (Xiong et al. 2002). Via stimulating protein oxidation and lipid peroxidation, BR deficiency reduces tolerance to low temperatures. In tomato plants, the exogenous EBR usage or DWRF overexpression reduces tolerance to low temperatures by reducing oxidative damage (Xia et al. 2018). Pursuant to studies, ROS against cold stress tolerance can act as a signal in mediating BR-regulated responses (Cui et al. 2011). A new survey found RBOH1, 2-cysteine peroxyrodoxin (2-Cys Prx), and GLUTAREDOXIN (GRX) take part in the signaling flow for BR-induced responses to cold in tomatoes (Xia et al. 2018). RBOH1 encodes NADPH further for signaling oxidase purposes, which is liable for production of ROS in apoplasts (Zhou et al. 2014). With the exception of ROS, NO participates in the cold replication pathway regulated by BR (Cui et al. 2011). Research has shown that there are no downstream functions of H_2O_2 in BR-induced low-temperature endurance. The exogenous EBR (0.1 μM) usage enhances CO_2 uptake and reduces the use of PSII exposure in cold stress. By activating important enzymes involved in the cycle of ascorbate-glutathione (AsA-GSH) also redox homeostasis, photosynthetic recovery after BR therapy is accomplished (Jiang et al. 2013). EBR treatment temporarily moderates AsA-GSH cycle module under cooling stress in grape seedlings (*Vitis vinifera*) and leads to an enhanced cold tolerance (Chen et al. 2019). Furthermore, by enhancing the antioxidant capability that decreases membrane lipid peroxidation under stress, cold tolerance in grapes improves the use of EBR (0.3 μM) leaves (Xi et al. 2013). In maize, EBR pretherapy (1.0 μM) can raise biomass and chlorophyll amounts, plant height, sugar, and also protein in low-temperature stress (Singh et al. 2012). Plants pursue an important strategy called photoprotection in order to prevent exposure to cold stress. BRs have been illustrated to be included in protecting plants which are exposed to cold stress. Plants activate active BR and activate BZR1 after exposure to cold temperatures, which ultimately increases RBOH1 transcript levels and apoplastic H_2O_2 production (Fang et al. 2019). In contrast, a mutation in BZR1 and subduing RBOH1 eliminates BR-induced shielding light, thereby intensifying cooling photoprotection. Notably, BRS-induced apoplastic H_2O_2 is important for cyclic electron flow-dependent protein regulatory gradient protein (PGR5) (CEF) and non-photochemical following induction suppression (NPQ), PSII S subunit protein accumulation (PsbS), D1, Is, and redox signaling, which largely affects the protection of light from the cold (Fang et al. 2019). Transcriptional approach depicts that treatment of EBR increments the transcripts of chlorophyll biosynthesis level and photosynthesis-related genes that secrete PSII oxygen-enhancing proteins, PSI subunits, and complexes of chlorophyll protein I, II, and ferrodoxin at low temperatures (Zhao et al. 2019). Through activation of Cold-Responsive (COR) genes, BRs have been presented to increase cryopreservation tolerance through both dependent and independent C-REPEAT/DEHYDRATION-RESPONSIVE BINDING FACTOR1 (CBF1) genes (Eremina et al. 2016). Medium-enhanced low temperatures tolerance includes the BZR1 and BES1 accumulation in the inactive modes, which transcribe CBF1 and CBF2 to induce cold tolerance (Li et al. 2017). Nevertheless, BR controls the response to low temperatures stress during extended stress by undermining the transcription factor induction of Cbf Expression1 (ICE1) by Brassinosteroidinsensitive2 (BIN2) negatively (Ye et al. 2019). These studies show that BRs can not only enhance response to stress but also can reduce stress tolerances, which is highly dependent on spatial controlling (Nolan et al. 2020).

Increased BRS response to low temperature is not only limited to the entire plants but also to harvest plant products like fruits. Research has presented that BR can enhance the fruits and vegetables quality after harvest by increasing the shelf life affected by cold stress conditions (Aghdam and Mohammadkhani 2014; Wang et al. 2012b). Comparatively high amounts of EBR as compared to amounts used to withstand stress in a healthy plant, for example, cold, significantly reduce quality of fruit in tomatoes (Li et al. 2016a, b); According to the outcomes, 6 μM EBR can reduce cold-induced damage in tomato fruits, which assigned to BR-induced phospholipase D reserved and lipoxygenase activity (Aghdam and Mohammadkhani 2014). By increasing a protein set level such as remorin, proteins which cause ripening such as abscisic acid stress, ripeninglike protein, temperature-induced lipocalin, and type II SK2 dehydroin 10 μM EBR protect fruits from frost damage in the case of mango (Li et al. 2012a). BR enhances the fatty acids rate of unsaturation in the mango fruit plasma membrane fats, reduces the phase increments transition temperature, and also leads to raised fluidity in the low temperatures (Li et al. 2012a). Wang investigated the various amounts of EBR effects on the vegetables quality (*Capsicum annuum* L.) at low temperatures (3 °C) and understood that 15 μM EBR is the most constructive amount that can improve the damage caused by colds in green pepper fruits. The antioxidant enzymes activity, L-ascorbic acid, and the chlorophyll level were higher in EBR-treated pepper fruits, which likely decreased electrolyte leakage and lipid peroxidation under low-temperature pressure (Li et al. 2012a).

14.3.3 BRs and Drought

Drought or aridity attribute to a water deficiency in the soils (Ghassemi et al. 2018; Khoshmanzar et al. 2020; Ahammed et al. 2020a). Failure to precipitation or irrigate results in the dryness, which significantly decreases crop yields. In areas with inadequate or poor rainfall, the issue is more severe. Osmotic stress induced by drought stress causes impacts normal cellular activity by disadjusting homeostasis and ions dispensation, like ion absorption, extrusion and cell individuation (Xiong and Zhu 2002). Drought resistance is related to the abscisic acid (ABA) aggregation. Investigations have shown exogenous use of BR can increase the ABA level and reduces the damaging drought signs on plant (Wang et al. 2019b).

In tomatoes, for example, EBR therapy increases resistance to drought that could be expressed in ameliorate photosynthetic ability, leave water condition and antioxidant defense under stress condition (Yuan et al. 2012). Exogenous BR therapy (0.02 Mm) in pepper leaves will enhance the performance of light usage and the destruction of PSII antennas stimulation energy under drought stress (Hu et al. 2013). Exogenous BR usage (0.1 μM EBR) in *Chorispora bungeana* could increase resistance to drought triggered by polyethylene glycol (PEG) (Li et al. 2012b). BR usage changes genes expression which encodes both constructional and regulating proteins. For example, increased transcript amounts EBR-induced of BnCBF5 and BnDREB (two drought tolerance genes) induced by EBR-induced contribute enough to BR-induced increased drought resistance in *Brassica napus* plants (Kagale et al. 2007).

Experiments as well show that BR therapy could reduce a long range impact of drought stress on different genus species. For example, *Brassica juncea* plants which experience weeklong drought tension in a primary phase of development show growth reduction and photosynthetic performance and after 60 days. Therefore, after 30 days of planting, post-dry therapy with 28-homobrassinolide (HBL, 0.01 μM) may ameliorate growth condition and photosynthesis performance after 60 days of planting (Fariduddin et al. 2009). Although dry tension causes ample aggregation of ROS, BR therapy could greatly decrease ROS rate and lipid peroxidation in drought tension condition (Yuan et al. 2010). However exogenous usage of BRs increases resistance to abiotic stresses, like drought; both insensitive and BR-deficient mutants exhibit improved resistance to stress (Nie et al. 2019; Nolan et al. 2020; Northey et al. 2016). Nevertheless, a study on tomato reveals an exaltation through endogenous BR extent but not BR signaling rate improves resistance to drought (Nie et al. 2019). The research observed a BR I1 negative impact overexpression on tomato drought toleration, indicating defects in the BR mechanism may either enhance or reduction stress endurance and then implicate the complication of the relationship among BRs and stress toleration (Nolan et al. 2020).

14.3.4 BRs and Salinity

Salinity is an important reason in osmotic tension, frequently called physiological drought. This stress has a negative impact on development, extension, and crop yields (Saghafi et al. 2018, 2019a, b, 2020; Khadem Moghadam et al. 2020). BRs were shown to reduce the minus salinity impact in several plants such as arabidopsis (*Arabidopsis thaliana*), rapeseed (*Brassica juncea*), eggplant (*Solanum melongena*), mustard (*Brassica napus*), pepper (*Capsicum annuum*), common bean (*Phaseolus vulgaris*), maize (*Zea mays*), cucumber (*Cucumis sativus*), and black locust (*Robinia pseudoacacia* L.) (Hayat et al. 2010; Yuan et al. 2012; Yue et al. 2019). In eggplants, the enhanced response to salinity stress caused by treatment with EBR is demonstrated by the enhanced acting of antioxidant enzymes, reduced Na^+ and Cl^- cumulation, and enhancement of K^+ and Ca^{+2} cumulation. Comparably, use of EBR could decrease the cumulation of NO_3^- and NH_4^+ in salt-stressed cucumber plants (Yuan et al. 2012). Foliar HBL usage in rapeseed could efficaciously improve the harmful salinity impacts even at 30 days and 45 days after planting (Hayat et al. 2012b). BR-induced increased salinity stress resistance is due to enhanced photosynthesis, productivity in nitrogen usage, and overall polyamines in cucumber plants (Yuan et al. 2012). In black locusts, the exogenous use of EBR decreases leaf Na^+ amount and membrane leaking and ameliorates the photosynthetic rate, stomatal conductivity, transpiration, chlorophyll amount, and highest PSII quantity performance under salinity stress (Yue et al. 2019). BR is as well efficacious in decreasing the impact of composed stress. For example, EBR (1 μM) in *Brassica juncea* could relieve the combined stress caused by NaCl and $NiCl_2$ (Ali et al. 2008), and HBL (0.01 μM) could reduce the combined stress induced by salt and high temperatures among mung beans (Hayat et al. 2010). The wide difference in BR condensation advance underlines the BR dose effect based on the BR forms and plant organisms. BR-induced toleration to salt stress has revealed a task for ubiquitin conjugating enzymes 32 (UBC32) (Cui et al. 2012). UBC32 regulates the BRI1 aggregation receiver in cells as a practical section of the endoplasmic reticulum protein destruction (ERAD) process and also leads the ERAD process to BR-enhanced salt stress in *Arabidopsis*. In fact, BR has been involved in controlling DNA methylation that plays a crucial function in the toleration of salts. Seed priming with EBR, for example, causes complete methylation and ameliorates salt tolerance, indicating BR function in epigenetic change caused by salinity stress (Amraee et al. 2019).

14.3.5 BRs and Heavy Metals Stress

Through widespread anthropogenic activities, including deforestation, urbanization, industrialization, and combustion of fossil fuels, contamination induced by numerous heavy metals has risen tremendously in recent years (Asgari Lajayer et al. 2017a, 2019a). Plants growing in contaminated soils are afflicted by metal-induced stress (Ahammed et al. 2013, 2020c). As compared to different abiotic stress, stress caused by heavy metals has special impacts. At first, the crops cultivated in heavy metals contaminated soils are compromised in terms like efficiency and modality. Furthermore, the use of heavy metal-contaminated plant productions is correlated with major risks due to possible pollution of the food webs (Asgari Lajayer et al. 2017b, 2019b; Hasan et al. 2019; Wang et al. 2019c).

Crops have been grown in such a metal-contaminated soils often contain high concentrations of toxic metals that increase the consumption risks of these contaminated foods (Hasan et al. 2019). The usage of plant growth regularizers, bioactive composed and endogenous hormone adjustment, and signaling pathway indicates tremendous potential for alleviating tension induced by means of heavy metals (Bücker-Neto et al. 2017). Likewise, BRs could alleviate heavy metal tension in a large variety of plant species (Rajewska et al. 2016).

Heavy metals negatively impact CO_2 sorption valence and photosynthetic device in plant (Rajewska et al. 2016). Evidence indicates heavy metals like Cd suppress the photosynthetic procedure by reducing the usage of ATP and NADPH in the Calvin cycle. In tomato, cadmium (cd) tension (100 Mm for 40 days) reduced photosynthetic rate of significantly net, stomatal conduction, highest PSII quantity productivity (Fv/Fm), PSII quantity efficiency, and photochemical quenching coefficient (qP) (Ahammed et al. 2013).

Cd-induced decrease in CO_2 sorption ability is associated positively with the photosynthetic pigment amount and associated with the aggregation of Cd in leafs, negatively. In consequence, aggregation of biomass in plant is significantly inhibited by Cd tension. Nevertheless, foliar use of EBR (0.1 Mm) greatly increments aggregation of biomass by ameliorating potential for sorption of Fv/Fm, CO_2, and photosynthetic pigments amount following Cd tension. In fact, exogenous EBR reduces the absorption of Cd within roots and their transposition to the leaves.

Mesophyll cell transportation electron micrographs in tobacco leaf demonstrated skewed cell membrane and cell wall, and thylakoid dilated under tension of chromium (Cr) (Bukhari et al. 2016). Nevertheless, EBR use preserved the chromium-induced harm to chloroplast and assisted to sustain grana and thylakoids under Cr tension. Like the EBR, HBL as well demonstrates a stress-protective function in reducing heavy metals tension. HBL therapy may mitigate the Cd-induced decrease of growth, photosynthesis, and PSII photochemistry in tomato seedlings (Singh and Prasad 2017).

At the cellular stage, ROS generation is caused when the plants are exposed to heavy metals, which adversely influence the metabolism of plants making oxidative harm to proteins, nucleic acids, and lipids (Song et al. 2013).

Heavy metals like nickel (Ni) in *Brassica juncea* L. enhance the biosynthesis of various BRs (epibrassinolide, typhasterol, dolicholide, and castasterone) (Kanwar et al. 2012). BRs were shown to protect plants from stress caused by induced heavy metal. In tomato plants, e.g., EBR therapy (0.1 Mm) may increase resistance against Cd tension by increasing photosynthesis, photosynthetic pigment amount photochemical performance, and the act of important antioxidant and detoxification-related enzymes at levels of protein and transcript (Ahammed et al. 2013). Likewise, BR (0.01 μM EBR or HBL) foliar use could ameliorate the yield and quality of tomato fruits in polluted soils with ~12 mg kg^{-1} Cd (Hayat et al. 2012a). Within a short time after usage, BRs show a strong protecting impact to Cd stress. For example, a foliar dose of EBR or HBL (0.01 μM) one day before measuring could dramatically ameliorate photosynthesis in tomato leafs within 60-day-long Cd tension (Hasan et al. 2011). BR therapy ameliorates nodule production in leguminous crops under threat from heavy metals. In *Vigna radiata*, the increased nodulation caused by EBR facilitates plant development under Ni tension. Furthermore, HBL therapy mitigates phytotoxicity of Cd by raising the amounts of both enzymatic and nonenzymatic antioxidants in *Cicer arietinum* (Hasan et al. 2008). The completion of EBR (5 nM) in the half strength MS medium improves the resistance seedlings of tomato to the tension caused through ZnO nanoparticles through enhancing antioxidant enzyme function and redox poise (Li et al. 2016b). Exogenous BR-induced response increase to heavy metal is due to essential amelioration in photosynthetic pigment amount, enzymatic and nonenzymatic antioxidant defense, ROS scavenging ability, glutathione contents, phytochelatin quantity, and carbon metabolism affected by heavy metal stress (Choudhary et al. 2012; Rajewska et al. 2016).

14.3.6 BRs and Pesticides

Pesticides are usually organic combinations that are widely used to deter and manage pests, including detrimental pests, phytopathogens (fungi, bacteria, nematodes, weed, etc.) (Sharma et al. 2016a, 2019). The pesticide usage for tolerable crop production worldwide is an important part of modern agriculture (Tiwari et al. 2019). However, pesticides could guarantee crop casualties of up to 80% (Oerke 2006), the levels and quantities of pesticides usage in developed countries are extremely large (Liu et al. 2016). Therefore, non-judicious and unreasonable pesticides usage could raise hytotoxicity and human well-being dangers. Therefore, it is indispensable to ensure food safety by reducing pesticide residues in edible plants (Chen et al. 2019). The plants are capable of detoxifying or degrading toxicant combinations (Zhou et al. 2015). The plants' natural detoxification pathway could be utilized to reduce pesticide remnants (Hou et al. 2018). The detoxification and disintegration of natural contaminant caused by glutathione have a major importance in plant toleration to pesticides and organic polluters. Most findings indicate that BR could increase plant toleration to stress caused by pesticides and heavy metals (Yin et al. 2016; Sharma et al. 2016b, c, 2017; Hou et al. 2018; Xia et al. 2009b). Furthermore, by enhancing the detoxification process, BR can decrease pesticide remnants in plants. BRs have indeed been recognized as safe, environmentally sustainable, natural compounds that are ideal to a broad spectrum of usages to decrease the dangers correlated with pollutant exposure (Hou et al. 2018; Xia et al. 2009b). BR effects on boost plant growth, photosynthesis, and yields are well-known (Nolan et al. 2020). BR-induced increased biomass cumulative is mostly ascribed to the BR-induced betterment in photosynthesis (Yu et al. 2004). BRs individually increment the gene expression of carbon stabilization and RuBisCO's primary action for enhancement CO_2 apperception in plant (Xia et al. 2006; Sharma et al. 2019). Additionally, BRs boost electrons moving related to pathway of photosynthesis and photosystem 1 (PS I) and II overall activity. Nevertheless, using pesticides greatly decreases plant photosynthesis ability (Xia et al. 2006). Xia et al. (2006) studied the effects of nine pesticides (paraquat, cuproxat, imidacloprid, cyazofamid, haloxyfop, xuazifop-p-butyl, chlorpyrifos, xusilazole, and abamectin) on photosynthesis in cucumber plant. They used pragmatic pesticides dosage to evaluate phytotoxicity in presence or absence of EBR pretherapy on photosynthesis (Xia et al. 2006). The use of pesticides resided the ϕPSII, Gs, Qp, Pn, and Fv/Fm. For example, therapy with chlorpyrifos, abamectin, and imidacloprid reduced Pn, respectively, by 36, 81, and 40%. Although the imidacloprid- and chlorpyrifos-induced reduction of CO_2 sorption was linked to pair stomatal and non-stomatal invoices, the decrease in Pn caused by abamectin therapy was primarily due to stomatal parameters. Nevertheless, the pesticide enforced disorders to photosynthetic system are diminished by EBR pretherapy in rare cases. It has proposed which EBR can reduce a terbutryn (*s*-triazine group pesticide that decrease transfer of electron) decreased prevention on PSII by comforting the QB relocation of its binding site in PSII D1 protein. In some other research, once cucumber plant is developed in chlorpyrifos polluted hydroponic solution (20–80 μM), the dose-dependent root length measure was significantly inhibited (Ahammed et al. 2017). Moreover to chlorpyrifos inhibition effect on root length, leaf chlorosis and roots browning were witnessed, and also approved chlorpyrifos toxic impacts on plants development. Nevertheless, foliar use of EBR reduced chlorpyrifos (10 μM)-related deterrence on the length of the shoots and roots. The affirmative impact of EBR on plant grows incremented with the EBR concentrations (0.001–0.1 μM); the 0.1 μM EBR, however, applied the most noteworthy results, with root elongation boosted by ca. 43% caused by therapy with chlorpyrifos (Ahammed et al. 2017). In mustard seed treatment (*Brassica juncea* L.) with EBR pre-seeding improved plant growth (dry weight, seedling length) once developed by imidacloprid (insecticide used for soil management) toxicity (Sharma et al. 2016c). Imidacloprid toxicity EBR-induced reduction was strongly related to EBR-induced improvements in numbers of leaves, shoot length, gas exchange parameters, and photosynthetic pigment contentment (Sharma et al. 2016b).

A numerous researches have shown that BR could defend plant from the oxidative tension caused with ROS by agitating antioxidant (enzymatic and nonenzymatic) advocacy (Rajewska et al. 2016). The Mehler reaction is one of the possible reasons for increased production and toxicity of ROS in plants. EBR-induced increased CO_2 sorption works as an extra electron sink for CO_2 decrease and thereby diverts unused electrons from alternate electron sinks, like reactions the Mehler (Hu et al. 2013). This is well known that EBR could boost plants' ROS cleanse ability in both usual and stressful environments, which is mainly due to enhancing antioxidant enzymes. Therefore, EBR capacity to boost both CO_2 sorption and antioxidant enzyme function contributes significantly to alleviating pesticide reduction phytotoxicity (Xia et al. 2006, 2009a, b). The pesticide remainder in the leaf tissues is primarily extracted via the detoxification process of the plant (Hou et al. 2019). Period-span pesticide remainder study shows that chlorothalonil remainders did not reduction in 6 days of use of pesticides (Xia et al. 2009b). BRs are being implicated to improving the plant detoxification pathway toward reducing pesticide remainders from vegetable and fruit. EBR (0.1 μM) could reduce remainders of different categories of popular pesticide including organochlorine, organophosphorus, and carbamate pesticides in different types of plant species, such as tea, rice, tomato, broccoli, cucumber, and strawberry, by between 30 and 70% EBR (Zhou et al. 2015). Also the minimum EBR condensation (0.02 μM) inside the physiological range substantially reduces the chlorothalonil remainder by 38.9% (Xia et al. 2009b). Study showed exogenous EBR improved the several pesticides metabolism and as a result decreased residues of pesticides in cucumber. They ended up finding that EBR improved the action of the enzymes involved in pesticide metabolism like the glutathione reductase (GR), glutathione S-transfer (GST), and peroxidase (POD). In addition, the transcript amount of the *P450* and *MRP* genes encoding ABC-type transporters and P450 monooxygenase, separately, were inhibited by EBR that led significantly to the increased metabolism of several pesticides like the chlorpyrifos, chlorothalonil, cypermethrin, and carbendazim. Exogenous EBR use often enhances plant secondary metabolism through improving the transcript and acting of enzymes related to secondary metabolism like the phenylalanine ammonia-lyase (PAL) and polyphenol oxidase (PPO) and flavonoid amount and contributes greatly to mitigating organic polluter-induced tension (Ahammed et al. 2013, 2017). A genome-wide microarray study in tomato leaves found that a whole of 301 genes, such as a collection of detoxifying genes that encode cytochrome P450, oxidoreductase, hydrolase, and transferase, were reported to regulated by chlorothalonil fungicide (CHT) and exogenous EBR (Zhou et al. 2015). BRs therefore likely pesticide destruction by enhance the metabolism of glutathione and the acting of GST. BR impact on the metabolism of the pesticides mainly dependent on pharmacological information (Zhou et al. 2015) demonstrated that apoplastic H_2O_2 synthesis through the NADPH oxide-dependent pathway coded with *RBOH1* has a major impact in the BR-induced pesticides metabolism. Extinguishing of *RBOH1* in tomato plant jeopardizes the efficacies of BR on GST acting, glutathione biosynthesis, and redox homeostasis, contributing to improved remainders of pesticides in tomato leafs. It was established that BRs increased the degradation of pesticides by increasing the metabolism of glutathione and GST acting through a pathway based on *RBOH1*. However, *RBOH1*-interceded apoplastic ROS production is necessary for the BR-interceded pesticide in the chemical processes (Zhou et al. 2015). That ROS signs converted downstream to enhance the metabolism remains mostly unclear. A variety of ROS-scavenging enzymes, buffers of redox including glutathione and oxidoreductases like peroxiredoxins (PRXs), thioredoxins, glutaredoxins (GRXs), and peroxidases, are implicated in the estimation of enhance amounts of ROS in plants (Ahammed et al. 2020). It was recently discovered that GRXS16, a CGFS-type GRX, functions streaming of the apoplastic ROS producing and is implicated in the metabolism of BR-induced pesticide in tomatoes (Hou et al. 2018). Placement of GRXS16 both in the nucleus and cytosol was reported and suspected; GRXS16 could activate detoxification genes like GTS by association with presumptive factors of transcription. Many evidence indicates GRX can interfere TGA2, transcription factor, which is involved in process of plant growth. Many evidence indicates that GRX can interfere with TGA2 and transcription factor, which is involved in plant growth, stress response and detoxification processes (Hou et al. 2018; Zander et al. 2014). TGA2 factor could bind straight to the TGACG motif of the detoxification-related *GST3* gene, indicating that the metabolism of BR-induced pesticide is interceded by means of interplay among GRX and transcription factors, which induces the gene expression, implicated in detoxification of pesticide (Hou et al. 2019).

14.4 Conclusion

Plants require those circumstances to grow and develop naturally. Nevertheless, inordinate conditions, also environmental emissions, can therefore have a damaging effect on crop growth. Cellular homeostasis, detoxifying, and growth improvement are three main forms of reactions acted by plant to prevail tension occurrence. Phytohormone BRs have a significant role in the interceding of these reactions through the regulation of specific gene sets. BRs are also used to control the transcription of these genes, which encode proteins protective for stress resistance. However, the BR efficacies on plant under normal conditions are lesser common, but their useful efficacies are well-known in stressful situations. BR-induced increased stress toleration is strongly related to the BR-induced recovery in CO_2 sorption, photoprotection antioxidant potential (enzymatic and nonenzymatic), redox homeostasis, ROS scavenging, defensive reaction, autophagy, secondary metabolism, and the ability for detoxification. Since several stress factors mostly happen to arise under normal circumstances, BRs, in the face of changing climate, have serious consequences for crop production.

References

Aghdam MS, Mohammadkhani N (2014) Enhancement of chilling stress tolerance of tomato fruit by postharvest brassinolide treatment. Food Bioprocess Technol 7:909–914

Ahammed GJ, Choudhary SP, Chen S, Xia X, Shi K, Zhou Y, Yu J (2013) Role of brassinosteroids in alleviation of phenanthrene–cadmium co-contamination-induced photosynthetic inhibition and oxidative stress in tomato. J Exp Bot 64:199–213

Ahammed GJ, Li X, Liu A, Chen S (2020) Brassinosteroids in plant tolerance to abiotic stress. J Plant Growth Regul 39:1451–1464

Ahammed GJ, Xia XJ, Li X, Shi K, Yu JQ, Zhou YH (2015) Role of brassinosteroid in plant adaptation to abiotic stresses and its interplay with other hormones. Curr Prot Peptide Sci 16:462–473

Ahammed GJ, Li X, Zhou H, Zhou YH, Yu JQ (2016) Role of hormones in plant adaptation to heat stress. In: Plant hormones under challenging environmental factors. Springer, pp 1–21

Ahammed GJ, He BB, Qian XJ, Zhou YH, Shi K, Zhou J, Yu JQ, Xia XJ (2017) 24-Epibrassinolide alleviates organic pollutants-retarded root elongation by promoting redox homeostasis and secondary metabolism in *Cucumis sativus* L. Environ Pollut 229:922–931

Ahammed GJ, Li X, Yang Y, Liu C, Zhou G, Wan H, Cheng Y (2020a) Tomato WRKY81 acts as a negative regulator for drought tolerance by modulating guard cell H_2O_2–mediated stomatal closure. Environ Exp Bot 171:103960

Ahammed GJ, Wang Y, Mao Q, Wu M, Yan Y, Ren J, Wang X, Liu A, Chen S (2020b) Dopamine alleviates bisphenol A-induced phytotoxicity by enhancing antioxidant and detoxification potential in cucumber. Environ Pollut 259:113957

Ahammed GJ, Wu M, Wang Y, Yan Y, Mao Q, Ren J, Ma R, Liu A, Chen S (2020c) Melatonin alleviates iron stress by improving iron homeostasis, antioxidant defense and secondary metabolism in cucumber. Sci Hortic 265:109205

Ali B, Hayat S, Fariduddin Q, Ahmad A (2008) 24-Epibrassinolide protects against the stress generated by salinity and nickel in *Brassica juncea*. Chemosphere 72:1387–1392

Amraee L, Rahmani F, Mandoulakani BA (2019) 24-Epibrassinolide alters DNA cytosine methylation of *Linum usitatissimum* L. under salinity stress. Plant Physiol Biochem 139:478–484

Asgari Lajayer B, Ghorbanpour M, Nikabadi S (2017a) Heavy metals in contaminated environment: destiny of secondary metabolite biosynthesis, oxidative status and phytoextraction in medicinal plants. Ecotoxicol Environ Saf 145:377–390

Asgari Lajayer H, Savaghebi GH, Hadian J, Hatami M, Pezhmanmehr M (2017b) Comparison of copper and zinc effects on growth, micro-and macronutrients status and essential oil constituents in pennyroyal (*Mentha pulegium* L.). Rev Bras Bot 40:379–388

Asgari Lajayer B, Khadem Moghadam N, Maghsoodi MR, Ghorbanpour M, Kariman K (2019a) Phytoextraction of heavy metals from contaminated soil, water and atmosphere using ornamental plants: mechanisms and efficiency improvement strategies. Environ Sci Pollut Res 26:8468–8484

Asgari Lajayer B, Najafi N, Moghiseh E, Mosaferi M, Hadian J (2019b) Micronutrient and heavy metal concentrations in basil plant cultivated on irradiated and non-irradiated sewage sludge-treated soil and evaluation of human health risk. Regul Toxicol Pharmacol 104:141–150

Bita C, Gerats T (2013) Plant tolerance to high temperature in a changing environment: scientific fundamentals and production of heat stress-tolerant crops. Front Plant Sci 4:273

Bücker-Neto L, Paiva ALS, Machado RD, Arenhart RA, Margis-Pinheiro M (2017) Interactions between plant hormones and heavy metals responses. Genet Mol Biol 40:373–386

Bukhari SAH, Wang R, Wang W, Ahmed IM, Zheng W, Cao F (2016) Genotype-dependent effect of exogenous 24-epibrassinolide on chromium-induced changes in ultrastructure and physicochemical traits in tobacco seedlings. Environ Sci Pollut Res 23:18229–18238

Chen S, Jin W, Liu A, Zhang S, Liu D, Wang F, Lin X, He C (2013) Arbuscular mycorrhizal fungi (AMF) increase growth and secondary metabolism in cucumber subjected to low temperature stress. Sci Hortic 160:222–229

Chen ZY, Wang YT, Pan XB, Xi ZM (2019) Amelioration of cold-induced oxidative stress by exogenous 24-epibrassinolide treatment in grapevine seedlings: toward regulating the ascorbate–glutathione cycle. Sci Hortic 244:379–387

Choudhary SP, Yu JQ, Yamaguchi-Shinozaki K, Shinozaki K, Tran LSP (2012) Benefits of brassinosteroid crosstalk. Trends Plant Sci 17:594–605

Clouse SD (2015) A history of brassinosteroid research from 1970 through 2005: thirty-five years of phytochemistry, physiology, genes, and mutants. J Plant Growth Regul 34:828–844

Cui JX, Zhou YH, Ding JG, Xia XJ, Shi K, Chen SC, Asami T, Chen Z, Yu JQ (2011) Role of nitric oxide in hydrogen peroxide-dependent induction of abiotic stress tolerance by brassinosteroids in cucumber. Plant Cell Environ 34:347–358

Cui F, Liu L, Zhao Q, Zhang Z, Li Q, Lin B, Wu Y, Tang S, Xie Q (2012) Arabidopsis ubiquitin conjugase UBC32 is an ERAD component that functions in brassinosteroid-mediated salt stress tolerance. Plant Cell 24:233–244

Dhaubhadel S, Browning KS, Gallie DR, Krishna P (2002) Brassinosteroid functions to protect the translational machinery and heat-shock protein synthesis following thermal stress. Plant J 29:681–691

Divi UK, Krishna P (2009) Brassinosteroid: a biotechnological target for enhancing crop yield and stress tolerance. New Biotechnol 26:131–136

Eremina M, Unterholzner SJ, Rathnayake AI, Castellanos M, Khan M, Kugler KG, May ST, Mayer KF, Rozhon W, Poppenberger B (2016) Brassinosteroids participate in the control of basal and acquired freezing tolerance of plants. Proc Natl Acad Sci 113:E5982–E5991

Fang P, Yan M, Chi C, Wang M, Zhou Y, Zhou J, Shi K, Xia X, Foyer CH, Yu J (2019) Brassinosteroids act as a positive regulator of photoprotection in response to chilling stress. Plant Physiol 180:2061–2076

Fariduddin Q, Khanam S, Hasan S, Ali B, Hayat S, Ahmad A (2009) Effect of 28-homobrassinolide on the drought stress-induced changes in photosynthesis and antioxidant system of *Brassica juncea* L. Acta Physiol Plant 31:889–897

Ghassemi S, Farhangi-Abriz S, Faegi-Analou R, Ghorbanpour M, Asgari Lajayer B (2018) Monitoring cell energy, physiological functions and grain yield in field-grown mung bean exposed to exogenously applied polyamines under drought stress. J Soil Sci Plant Nutr 18 (4):1108–1125

Guo D, Wang Z, Li Q, Gu S, Zhang G, Yu Y (2019) Hydrogen peroxide treatment promotes early ripening of Kyoho grape. Aust J Grape Wine Res 25:357–362

Hasan S, Hayat S, Ali B, Ahmad A (2008) 28-Homobrassinolide protects chickpea (*Cicer arietinum*) from cadmium toxicity by stimulating antioxidants. Environ Pollut 151:60–66

Hasan SA, Hayat S, Ahmad A (2011) Brassinosteroids protect photosynthetic machinery against the cadmium induced oxidative stress in two tomato cultivars. Chemosphere 84:1446–1451

Hasan MK, Ahammed GJ, Sun S, Li M, Yin H, Zhou J (2019) Melatonin inhibits cadmium translocation and enhances plant tolerance by regulating sulfur uptake and assimilation in *Solanum lycopersicum* L. J Agric Food Chem 67:10563–10576

Hayat S, Hasan SA, Yusuf M, Hayat Q, Ahmad A (2010) Effect of 28-homobrassinolide on photosynthesis, fluorescence and antioxidant system in the presence or absence of salinity and temperature in *Vigna radiata*. Environ Exp Bot 69:105–112

Hayat S, Alyemeni MN, Hasan SA (2012a) Foliar spray of brassinosteroid enhances yield and quality of *Solanum lycopersicum* under cadmium stress. Saudi J Biol Sci 19:325–335

Hayat S, Maheshwari P, Wani AS, Irfan M, Alyemeni MN, Ahmad A (2012b) Comparative effect of 28 homobrassinolide and salicylic acid in the amelioration of NaCl stress in *Brassica juncea* L. Plant Physiol Biochem 53:61–68

Hou J, Zhang Q, Zhou Y, Ahammed GJ, Zhou Y, Yu J, Fang H, Xia X (2018) Glutredoxin GRXS16 mediates brassinosteroid-induced apoplastic H_2O_2 production to promote pesticide metabolism in tomato. Environ Pollut 240:227–234

Hou J, Sun Q, Li J, Ahammed GJ, Yu J, Fang H, Xia X (2019) Glutaredoxin S25 and its interacting TGACG motif-binding factor TGA2 mediate brassinosteroid-induced chlorothalonil metabolism in tomato plants. Environ Pollut 255:113256

Hu W, Yan XH, Xiao YA, Zeng JJ, Qi HJ, Ogweno JO (2013) 24-Epibrassinosteroid alleviate drought-induced inhibition of photosynthesis in *Capsicum annuum*. Sci Hortic 150:232–237

Hussain M, Khan TA, Yusuf M, Fariduddin Q (2019) Silicon-mediated role of 24-epibrassinolide in wheat under high-temperature stress. Environ Sci Pollut Res 26:17163–17172

Jiang YP, Cheng F, Zhou YH, Xia XJ, Mao WH, Shi K, Chen Z, Yu JQ (2012) Cellular glutathione redox homeostasis plays an important role in the brassinosteroid-induced increase in CO_2 assimilation in *Cucumis sativus*. New Phytol 194:932–943

Jiang YP, Huang LF, Cheng F, Zhou YH, Xia XJ, Mao WH, Shi K, Yu JQ (2013) Brassinosteroids accelerate recovery of photosynthetic apparatus from cold stress by balancing the electron partitioning, carboxylation and redox homeostasis in cucumber. Physiol Plant 148:133–145

Kagale S, Divi UK, Krochko JE, Keller WA, Krishna P (2007) Brassinosteroid confers tolerance in Arabidopsis thaliana and *Brassica napus* to a range of abiotic stresses. Planta 225:353–364

Kanwar MK, Bhardwaj R, Arora P, Chowdhary SP, Sharma P, Kumar S (2012) Plant steroid hormones produced under Ni stress are involved in the regulation of metal uptake and oxidative stress in *Brassica juncea* L. Chemosphere 86:41–49

Khadem Moghadam N, Motesharezadeh B, Maali-Amiri R, Asgari Lajayer B, Astatkie T (2020) Effects of potassium and zinc on physiology and chlorophyll fluorescence of two cultivars of canola grown under salinity stress. Arab J Geosci 13:771. https://doi.org/10.1007/s12517-020-05776-y

Khoshmanzar E, Aliasgharzad N, Neyshabouri MR, Khoshru B, Arzanlou M, Asgari Lajayer B (2020) Effects of Trichoderma isolates on tomato growth and inducing its tolerance to water-deficit stress. Int J Environ Sci Technol 17:869–878

Khripach V, Zhabinskii V, de Groot A (2000) Twenty years of brassinosteroids: steroidal plant hormones warrant better crops for the XXI century. Ann Bot 86:441–447

Li B, Zhang C, Cao B, Qin G, Wang W, Tian S (2012a) Brassinolide enhances cold stress tolerance of fruit by regulating plasma membrane proteins and lipids. Amino Acids 43:2469–2480

Li Y, Liu Y, Xu X, Jin M, An L, Zhang H (2012b) Effect of 24-epibrassinolide on drought stress-induced changes in *Chorispora bungeana*. Biol Plant 56:192–196

Li M, Ahammed GJ, Li C, Bao X, Yu J, Huang C, Yin H, Zhou J (2016a) Brassinosteroid ameliorates zinc oxide nanoparticles-induced oxidative stress by improving antioxidant potential and redox homeostasis in tomato seedling. Front Plant Sci 7:615

Li P, Yin F, Song L, Zheng X (2016b) Alleviation of chilling injury in tomato fruit by exogenous application of oxalic acid. Food Chem 202:125–132

Li H, Ye K, Shi Y, Cheng J, Zhang X, Yang S (2017) BZR1 positively regulates freezing tolerance via CBF-dependent and CBF-independent pathways in Arabidopsis. Mol Plant 10:545–559

Liu S, Che Z, Chen G (2016) Multiple-fungicide resistance to carbendazim, diethofencarb, procymidone, and pyrimethanil in field isolates of Botrytis cinerea from tomato in Henan Province, China. Crop Prot 84:56–61

Martínez C, Espinosa-Ruíz A, de Lucas M, Bernardo-García S, Franco-Zorrilla JM, Prat S (2018) PIF 4-induced BR synthesis is critical to diurnal and thermomorphogenic growth. EMBO J 37: e99552

Martins S, Montiel-Jorda A, Cayrel A, Huguet S, Paysant-Le Roux C, Ljung K, Vert G (2017) Brassinosteroid signaling-dependent root responses to prolonged elevated ambient temperature. Nat Commun 8:1–11

Nie WF, Wang MM, Xia XJ, Zhou YH, Shi K, Chen Z, Yu JQ (2013) Silencing of tomato RBOH1 and MPK2 abolishes brassinosteroid-induced H_2O_2 generation and stress tolerance. Plant Cell Environ 36:789–803

Nie S, Huang S, Wang S, Mao Y, LiuJ MR, Wang X (2019) Enhanced brassinosteroid signaling intensity via SlBRI1 overexpression negatively regulates drought resistance in a manner opposite of that via exogenous BR application in tomato. Plant Physiol Biochem 138:36–47

Nolan TM, Vukašinović N, Liu D, Russinova E, Yin Y (2020) Brassinosteroids: multidimensional regulators of plant growth, development, and stress responses. Plant Cell 32:295–318

Northey JG, Liang S, Jamshed M, Deb S, Foo E, Reid JB, McCourt P, Samuel MA (2016) Farnesylation mediates brassinosteroid biosynthesis to regulate abscisic acid responses. Nat Plants 2:1–7

Oerke E (2006) Crop losses to pests. J Agric Sci 144:31

Ogweno JO, Song XS, Shi K, Hu WH, Mao WH, Zhou YH, Yu JQ, Nogués S (2008) Brassinosteroids alleviate heat-induced inhibition of photosynthesis by increasing carboxylation efficiency and enhancing antioxidant systems in *Lycopersicon esculentum*. J Plant Growth Regul 27:49–57

Peres ALG, Soares JS, Tavares RG, Righetto G, Zullo MA, Mandava NB, Menossi M (2019) Brassinosteroids, the sixth class of phytohormones: a molecular view from the discovery to hormonal interactions in plant development and stress adaptation. Int J Mol Sci 20:331

Planas-Riverola A, Gupta A, Betegón-Putze I, Bosch N, Ibañes M, Caño-Delgado AI (2019) Brassinosteroid signaling in plant development and adaptation to stress. Development 146

Rajewska I, Talarek M, Bajguz A (2016) Brassinosteroids and response of plants to heavy metals action. Front Plant Sci 7:629

Sadura I, Janeczko A (2018) Physiological and molecular mechanisms of brassinosteroid-induced tolerance to high and low temperature in plants. Biol Plant 62:601–616

Sadura I, Pociecha E, Dziurka M, Oklestkova J, Novak O, Gruszka D, Janeczko A (2019) Mutations in the HvDWARF, HvCPD and HvBRI1 genes-involved in brassinosteroid biosynthesis/signalling: altered photosynthetic efficiency, hormonal homeostasis and tolerance to high/low temperatures in barley. J Plant Growth Regul 38:1062–1081

Saghafi D, Ghorbanpour M, Asgari Lajayer B (2018) Efficiency of rhizobium strains as plant growth-promoting rhizobacteria on morphophysiological properties of *Brassica napus* L. under salinity stress. J Soil Sci Plant Nutr 18(1):253–268

Saghafi D, Delangiz N, Asgari Lajayer B, Ghorbanpour M (2019a) An overview of the improvement of crop productivity in saline soils by halotolerant and halophilic PGPRs. 3. Biotech 9:261

Saghafi D, Ghorbanpour M, Shirafkan Ajirloo H, Asgari Lajayer B (2019b) Enhancement of growth and salt tolerance in *Brassica napus* L. seedlings by halotolerant rhizobium strains containing ACC deaminase activity. Plant Physiol Rep 24:225–235
Saghafi D, Asgari Lajayer B, Ghorbanpour M (2020) Engineering bacterial ACC deaminase for improving plant productivity under stressful conditions. In: Sharma V, Salwan R, Al-Ani LKT (eds) Molecular aspects of plant beneficial microbes in agriculture. Elsevier, pp 259–272
Sasse JM (2003) Physiological actions of brassinosteroids: an update. J Plant Growth Regul 22:276–288
Sharma A, Bhardwaj R, Kumar V, Thukral AK (2016a) GC-MS studies reveal stimulated pesticide detoxification by brassinolide application in *Brassica juncea* L. plants. Environ Sci Pollut Res 23:14518–14525
Sharma A, Kumar V, Singh R, Thukral AK, Bhardwaj R (2016b) Effect of seed pre-soaking with 24-epibrassinolide on growth and photosynthetic parameters of *Brassica juncea* L. in imidacloprid soil. Ecotoxicol Environ Saf 133:195–201
Sharma A, Thakur S, Kumar V, Kanwar MK, Kesavan AK, Thukral AK, Bhardwaj R, Alam P, Ahmad P (2016c) Pre-sowing seed treatment with 24-epibrassinolide ameliorates pesticide stress in *Brassica juncea* L. through the modulation of stress markers. Front Plant Sci 7:1569
Sharma A, Thakur S, Kumar V, Kesavan AK, Thukral AK, Bhardwaj R (2017) 24-epibrassinolide stimulates imidacloprid detoxification by modulating the gene expression of *Brassica juncea* L. BMC Plant Biol 17:56
Sharma A, Yuan H, Kumar V, Ramakrishnan M, Kohli SK, Kaur R, Thukral AK, Bhardwaj R, Zheng B (2019) Castasterone attenuates insecticide induced phytotoxicity in mustard. Ecotoxicol Environ Saf 179:50–61
Singh S, Prasad SM (2017) Effects of 28-homobrassinoloid on key physiological attributes of *Solanum lycopersicum* seedlings under cadmium stress: photosynthesis and nitrogen metabolism. Plant Growth Regul 82:161–173
Singh I, Kumar U, Singh S, Gupta C, Singh M, Kushwaha S (2012) Physiological and biochemical effect of 24-epibrassinoslide on cold tolerance in maize seedlings. Physiol Mol Biol Plants 18:229–236
Song Y, Cui J, Zhang H, Wang G, Zhao FJ, Shen Z (2013) Proteomic analysis of copper stress responses in the roots of two rice (*Oryza sativa* L.) varieties differing in Cu tolerance. Plant Soil 366:647–658
Sonjaroon W, Jutamanee K, Khamsuk O, Thussagunpanit J, Kaveeta L, Suksamrarn A (2018) Impact of brassinosteroid mimic on photosynthesis, carbohydrate content and rice seed set at reproductive stage under heat stress. Agric Nat Resour 52:234–240
Tiwari B, Kharwar S, Tiwari D (2019) Pesticides and rice agriculture. In: Cyanobacteria. Elsevier, pp 303–325
Tong H, Chu C (2018) Functional specificities of brassinosteroid and potential utilization for crop improvement. Trends Plant Sci 23:1016–1028
Wang B, Li Y, Zhang WH (2012a) Brassinosteroids are involved in response of cucumber (*Cucumis sativus*) to iron deficiency. Ann Bot 110:681–688
Wang Q, Ding T, Gao L, Pang J, Yang N (2012b) Effect of brassinolide on chilling injury of green bell pepper in storage. Sci Hortic 144:195–200
Wang F, Ahammed GJ, Li G, Bai P, Jiang Y, Wang S, Chen S (2019a) Ethylene is involved in red light-induced anthocyanin biosynthesis in cabbage (*Brassica oleracea*). *Int J Agric Biol* 21:955–963
Wang J, Zhu H, Zhang C, Wang H, Yang Z, Liu Z (2019b) Puerarin protects rat liver and kidney against cadmium-induced oxidative stress. Indian J Anim Sci 89:927–931
Wang YT, Chen ZY, Jiang Y, Duan BB, Xi ZM (2019c) Involvement of ABA and antioxidant system in brassinosteroid-induced water stress tolerance of grapevine (*Vitis vinifera* L.). Sci Horticult 256:108596

Wang Y, Cao JJ, Wang KX, Xia XJ, Shi K, Zhou YH, Yu JQ, Zhou J (2019d) BZR1 mediates brassinosteroid-induced autophagy and nitrogen starvation in tomato. Plant Physiol 179:671–685

Wu C, Trieu A, Radhakrishnan P, Kwok SF, Harris S, Zhang K, Wang J, Wan J, Zhai H, Takatsuto S (2008) Brassinosteroids regulate grain filling in rice. Plant Cell 20:2130–2145

Wu X, Yao X, Chen J, Zhu Z, Zhang H, Zha D (2014) Brassinosteroids protect photosynthesis and antioxidant system of eggplant seedlings from high-temperature stress. Acta Physiol Plant 36:251–261

Xi Z, Wang Z, Fang Y, Hu Z, Hu Y, Deng M, Zhang Z (2013) Effects of 24-epibrassinolide on antioxidation defense and osmoregulation systems of young grapevines (*Vitis vinifera* L.) under chilling stress. Plant Growth Regul 71:57–65

Xia XJ, Huang YY, Wang L, Huang LF, Yu YL, Zhou YH, Yu JQ (2006) Pesticides-induced depression of photosynthesis was alleviated by 24-epibrassinolide pretreatment in *Cucumis sativus* L. Pestic Biochem Physiol 86:42–48

Xia XJ, Wang YJ, Zhou YH, Tao Y, Mao WH, Shi K, Asami T, Chen Z, Yu JQ (2009a) Reactive oxygen species are involved in brassinosteroid-induced stress tolerance in cucumber. Plant Physiol 150:801–814

Xia XJ, Zhang Y, Wu JX, Wang JT, Zhou YH, Shi K, Yu YL, Yu JQ (2009b) Brassinosteroids promote metabolism of pesticides in cucumber. J Agric Food Chem 57:8406–8413

Xia XJ, Zhou YH, Ding J, Shi K, Asami T, Chen Z, Yu JQ (2011) Induction of systemic stress tolerance by brassinosteroid in *Cucumis sativus*. New Phytol 191:706–720

Xia XJ, Gao CJ, Song LX, Zhou YH, Shi K, Yu JQ (2014) Role of H_2O_2 dynamics in brassinosteroid-induced stomatal closure and opening in *Solanum lycopersicum*. Plant Cell Environ 37:2036–2050

Xia XJ, Fang PP, Guo X, Qian XJ, Zhou J, Shi K, Zhou YH, Yu JQ (2018) Brassinosteroid-mediated apoplastic H_2O_2-glutaredoxin 12/14 cascade regulates antioxidant capacity in response to chilling in tomato. Plant Cell Environ 41:1052–1064

Xiong L, Zhu JK (2002) Molecular and genetic aspects of plant responses to osmotic stress. Plant Cell Environ 25:131–139

Xiong L, Schumaker KS, Zhu JK (2002) Cell signaling during cold, drought, and salt stress. Plant Cell 14:S165–S183

Ye K, Li H, Ding Y, Shi Y, Song C, Gong Z, Yang S (2019) Brassinosteroid-Insensitive2 negatively regulates the stability of transcription factor ICE1 in response to cold stress in *Arabidopsis*. Plant Cell 31:2682–2696

Yin YL, Zhou Y, Zhou YH, Shi K, Zhou J, Yu Y, Yu JQ, Xia XJ (2016) Interplay between mitogen-activated protein kinase and nitric oxide in brassinosteroid-induced pesticide metabolism in *Solanum lycopersicum*. J Hazard Mater 316:221–231

Yin W, Dong N, Niu M, Zhang X, Li L, Liu J, Liu B, Tong H (2019) Brassinosteroid-regulated plant growth and development and gene expression in soybean. Crop J 7:411–418

Yu JQ, Huang LF, Hu WH, Zhou YH, Mao WH, Ye SF, Nogués S (2004) A role for brassinosteroids in the regulation of photosynthesis in *Cucumis sativus*. J Exp Bot 55:1135–1143

Yuan GF, Jia CG, Li Z, Sun B, Zhang LP, Liu N, Wang QM (2010) Effect of brassinosteroids on drought resistance and abscisic acid concentration in tomato under water stress. Sci Hortic 126:103–108

Yuan L, Yuan Y, Du J, Sun J, Guo S (2012) Effects of 24-epibrassinolide on nitrogen metabolism in cucumber seedlings under Ca (NO_3) 2 stress. Plant Physiol Biochem 61:29–35

Yue J, You Y, Zhang L, Fu Z, Wang J, Zhang J, Guy RD (2019) Exogenous 24-epibrassinolide alleviates effects of salt stress on chloroplasts and photosynthesis in *Robinia pseudoacacia* L. seedlings. J Plant Growth Regul 38:669–682

Zander M, Thurow C, Gatz C (2014) TGA transcription factors activate the salicylic acid-suppressible branch of the ethylene-induced defense program by regulating ORA59 expression. Plant Physiol 165:1671–1683

Zhang Y, Zhu X, Ding H, Yang S, Chen Y (2013) Foliar application of 24-epibrassinolide alleviates high-temperature-induced inhibition of photosynthesis in seedlings of two melon cultivars. Photosynthetica 51:341–349

Zhang Y, Liang Y, Zhao X, Jin X, Hou L, Shi Y, Ahammed GJ (2019) Silicon compensates phosphorus deficit-induced growth inhibition by improving photosynthetic capacity, antioxidant potential, and nutrient homeostasis in tomato. Agronomy 9:733

Zhao B, Li J (2012) Regulation of brassinosteroid biosynthesis and inactivation F. J Integr Plant Biol 54:746–759

Zhao M, Yuan L, Wang J, Xie S, Zheng Y, Nie L, Zhu S, Hou J, Chen G, Wang C (2019) Transcriptome analysis reveals a positive effect of brassinosteroids on the photosynthetic capacity of wucai under low temperature. BMC Genomics 20:1–19

Zhou J, Wang J, Li X, Xia XJ, Zhou YH, Shi K, Chen Z, Yu JQ (2014) H_2O_2 mediates the crosstalk of brassinosteroid and abscisic acid in tomato responses to heat and oxidative stresses. J Exp Bot 65:4371–4383

Zhou Y, Xia X, Yu G, Wang J, Wu J, Wang M, Yang Y, Shi K, Yu Y, Chen Z (2015) Brassinosteroids play a critical role in the regulation of pesticide metabolism in crop plants. Sci Rep 5:9018

Zhou Y, Guang Y, Li J, Wang F, Ahammed GJ, Yang Y (2019) The CYP74 gene family in watermelon: genome-wide identification and expression profiling under hormonal stress and root-knot nematode infection. Agronomy 9:872

Chapter 15
The Effects of Climate Change on the Alteration of Plant Traits

Nüket Altındal and Demet Altındal

Abstract In the world, the increase in the transportation, industrialization, deforestation, use of fossil fuels, and unconscious use of agricultural land have enhanced the absorption of greenhouse gases in the atmosphere causing global temperatures to rise, and thus the concept of climate change, effects, and consequences have become major issues of concern for all the countries in the world. Stress factors such as drought, water scarcity, greenhouse gases, and salinity, which occur with climate change, have a direct negative effect on plants. The increase in temperature and the change in rainfall frequency also negatively affect food production. Flood disasters caused by excessive rainfall cause pollution and decrease of water resources, and the increase in temperature and change in the frequency of rainfall negatively affect food production. Against stress factors caused by climate change, plants develop some cellular and molecular mechanisms thanks to bioactive compounds, phytohormones, and vitamins.

Keywords Plant · Climate change · Drought · Greenhouse gases · Salinity · Defense

Headings

1. There is a negative relationship between climate change and plants.
2. Climate change has an impact on the ecosystem as well as especially on plants.
3. The world is at risk as a result of climate changes as ecological destruction increases.
4. Against global warming (climate change), plants activate some mechanisms in their metabolism.

N. Altındal
Sivaslı Vocational School, Department of Crop and Animal Production, Programme of Medical and Aromatic Plants, Uşak University, Uşak, Turkey

D. Altındal (✉)
Fethiye Ali Sıtkı Mefharet Koçman Vocational School, Department of Crop and Animal Production, Organic Farming Programme, Muğla Sıtkı Koçman University, Muğla, Turkey

D. K. Choudhary et al. (eds.), *Climate Change and the Microbiome*, Soil Biology 63,
https://doi.org/10.1007/978-3-030-76863-8_15

15.1 Introduction

The climate has tended to change from the formation of the Earth to the present day and still continues. Climate change causes negative consequences, while the relationship between living things and the atmosphere, land, snow, ice, and oceans changes gradually over the short and long periods of time due to changes in internal and external factors. In the short term, climate changes caused by a meteor impact on the Earth or the negative effects of volcano activities resulted in the extinction of many species. The adaptation times of living things to rapidly changing climatic conditions are different, which also directly affects ecological systems. In later times, new species have been trying to adapt to changing conditions, but the plant and animal products, which are important in human life, are facing a reduction.

Agriculture is an activity carried on independently of nature. Therefore, the impact of climate change on agriculture is greater than in any other sector. Since natural resources are constantly used in agricultural activities, agriculture is effective on soil and water resources, and agricultural production is directly affected by changes in natural resources. For all these reasons, the agricultural sector is more affected by climate change, and the extent of the impact is greater. Therefore, changing climatic conditions are becoming a serious threat to agricultural production.

Plants that are sedentary living beings respond to stress conditions such as drought, temperature, salinity, and excessive rainfall with physiological and metabolic changes to minimize their growth and development. However, these changes are still not fully understood due to their complexity.

As a result of global warming, increasing greenhouse gases, carbon dioxide, irregular rainfall, frost, temperature, and drought factors negatively affect plant growth and flowering and cause phenological changes (Parmesan and Hanley 2015). In particular, salinity conditions inhibit seed germination in plants (Fig. 15.1).

Faced with heat, droughts, pests, and diseases, the sustainability of agricultural products in areas where they are adapted and cultivated is adversely affected. Changes in ecology result in a decrease in plant productivity due to abiotic and biotic stresses such as heat, drought, excessive rainfall, nutrition, diseases, and pests. Bioactive compounds, phytohormones, and vitamins found in plants play an important role in defense mechanisms. Therefore, the cellular and molecular mechanisms of plants against biotic and abiotic stresses undergo changes. For example, the amount of proteins, polysaccharides, lipids, minerals, heavy metals, vitamins, and phytochemicals in the grain of wheat, which ranks first among the cultured plants used in human nutrition in terms of cultivation and production in the world, are affected by genotype and environmental factors (Johansson et al. 2020).

Other phytohormones, including abscisic acid, auxins, brassinosteroids, ethylene, gibberellins, and cytokinins, known as the "stress hormone" against stress in plants, have also important contributions. Other phytohormones such as jasmonic acid, salicylic acid, and nitric oxide (NO) enable the plant's adaptation to abiotic stresses and also regulate various physiological processes such as the closure of its stomata,

Fig. 15.1 Seed germination in corn on non-salinity (**a**) and salinity (**b**) conditions

thereby increasing osmotic stress tolerance by reducing water loss (Souri et al. 2020). Again, xylem-induced abscisic acid (ABA) plays a crucial role in the regulation of stomatal opening, which is a priority in defense against drought (Li et al. 2020a).

15.2 Effect of Greenhouse Gases on Plants and Reaction of Plants

Considering that respiration and photosynthesis are the main components of plant carbon balance, the changes in respiration greenhouse gases (carbon dioxide (CO_2), nitrogen oxides (NO_x), ozone (O_3), methane (CH_4)), temperature, and drought factors can affect plant growth, reproduction, and yield. For example, exposure of plants to O_3 causes major changes in transcription and metabolism, ultimately resulting in a reduction in plant yield by up to 14% (Wilkinson et al. 2012) and cell death (Florez-Sarasa et al. 2020).

The high amount of carbon dioxide in the atmosphere can affect plant growth by several factors, increasing temperature, and drought. Therefore, although effects are observed by giving high amounts of carbon dioxide to the environment where the plant is cultivated, correct results may not be obtained for long-term effects. Carbon dioxide, a substrate for photosynthesis, has a direct effect on plants. With the increase in carbon dioxide, nitrogen, one of the building blocks of living things, is used by plants in the form of nitrate, while some of them are converted back into nitrogen by bacteria with denitrification, which means that the N cycle increases and also plant productivity increases in parallel with the increase in temperature. The reactions of plants against carbon dioxide are more complex than expected (Ainsworth et al. 2020).

The presence of carbon dioxide in the environment allows the pores of the leaves to close, and this helps to protect the water in the plant. Due to this mechanism, carbon dioxide has a fertilizing effect on many plants. CO_2 fertilization can improve photosynthesis, nutrient uptake, and plant productivity as well as provide plant resistance to environmental stress. On the contrary, it was found that N uptake decreased in wheat grown in cabins under high atmospheric CO_2; similarly, phosphorus (P) uptake via arbuscular mycorrhizal fungi (AMF) was determined to be affected by atmospheric carbon dioxide (Thirkell et al. 2020).

Increasing atmospheric CO_2 levels in vegetable production under the greenhouse, i.e., making CO_2 fertilization, is a widely used practice to increase productivity. However, this method results in high energy costs and has harmful effects on the environment. Therefore, it suggests that the enrichment of the plant root zone with CO_2 may be a more economical and sustainable alternative to the increase in CO_2 in the atmosphere. Even though carbon dioxide aids the development of plants, it is just one of the many factors that affect the development of plants.

It has been reported that high bicarbonate concentrations (20 mM) in hydroponically grown plants prevent the development of lettuce, pepper, and tomato, but lower concentrations (1 mM) increase biomass accumulation in lettuce and pepper and have no effect on tomato (Estibaliz 2020).

In a study investigating the effect of high carbon dioxide concentration and temperature on the development of cowpea varieties, it was reported that the temperatures of 29 °C (day)/23 °C (night) increase the seed weight and the increase in CO_2 increases the pods and seeds number and seed weight (Angelotti et al. 2020), while in other studies, high CO_2 increased plant productivity in grain crops (Dong et al. 2020) and caused rapid growth in trees (Kim et al. 2019).

Greenhouse gases absorb the rays (infrared) reflected back from the surface of the Earth and cause the formation of radiation. High doses of UV-B radiation can damage Photosystem II and Rubisco enzyme, thus negatively affecting photosynthesis in many plants (Jarma-Orozco et al. 2020).

In a study, it was revealed that in high carbon dioxide conditions in the soil, there was a significant reduction in microelement concentration, especially in iron (Fe) and zinc (Zn) quantities (Chumley and Hewlings 2020).

15.3 Effect of Water and Temperature Stress on Plants and Reaction of Plants

With a significant reduction of rainfall in the atmosphere as a result of climate change, drought occurs as a limiting factor for plant production, which negatively affects soil, water resources, and production systems and leads to severe hydrological imbalances.

One of the most important physiological processes affected by drought stress is carbon partitioning. A study examined some varieties of potatoes grown in drought conditions and found that the most serious effects of drought stress prevented the differentiation of stolons and tuber formation. Some physiological features such as stomata conductivity and chlorophyll content have been reported to affect carbon assimilation, transport, and ultimately tuber yield (Aliche et al. 2020a).

According to the severity of drought conditions in plants, the transport of photo-assimilates in the stem is affected. In water scarcity, the reduction of root growth in plants leads to the reduction of transport distance of assimilates, which is advantageous for plants. Morphological changes in xylem diameter and density under drought in potatoes suggest that they may be associated with xylem flux and dehydration tolerance in potato plants (Aliche et al. 2020b).

Drought stress inhibits plant growth and development by altering cellular and biological activity. However, plants stimulate signaling multiplex mechanisms to be drought-tolerant, and drought stimulates oxidative stress by overproduction of reactive oxygen species (ROS) at the cellular level. A study suggests that drought-induced transcription activation and abscisic acid regulate the stress signaling mechanism in cotton. In the case of drought stress, root development, stomata closure, photosynthesis, hormone production, and ROS take place in the plant. In the genetic studies carried out in cotton, drought-resistance genes are identified and presented as genetic source samples. It has been stated that researching, identifying, and characterizing the genetic basis for superior candidate genes associated with stress physiology will provide information in a better understanding of the complex cellular biology of these plants (Mahmood et al. 2020).

Li et al. (2019) revealed that although long-term drought conditions in cotton caused leaf shedding, water integrity was maintained by early stoma closure and leaf shedding, thus improving drought resistance.

To adapt to the humidity in the soil, *Arabidopsis thaliana* plants reduce root development and alter the flow of cellular signals by closing stoma. These physiological changes lead to early flowering or stunted growth and often reduce yield (Gupta et al. 2020).

Gázquez et al. (2020) studied redox metabolites and enzyme activities in the leaf growth region of paddy plant at the low temperature and found that low-temperature tolerance is achieved by regulation of enzymatic and non-enzymatic antioxidants.

Naraikina et al. (2020) stated that there was a decrease in total saturated fatty acids content in potato leaves, which were kept for 6 days at 5 °C, and that polyunsaturated fatty acids content increased by about 30%.

The seed weight of *Festuca rubra*, a perennial grass species, and nutrient concentrations such as nitrogen and phosphorus in the seed are significantly affected by climatic conditions, while carbohydrate concentrations are not affected. Higher seed weight and temperature lead to more and faster germination (Veselá et al. 2020).

Li et al. (2020b) investigated the effects of climate change on cotton yield and determined that relative air humidity and temperature affect cotton growth during the flowering stage; sunshine duration and average maximum and minimum temperatures affect seed yield; precipitation and minimum temperature affect stalk weight; and also sunshine duration, average temperature, and especially precipitation affect lint percentage.

Pollution of water resources caused by excessive rainfall with flood disasters and drought caused by lack of rainfall can cause heavy metal accumulation in soil. Exposure of plants to heavy metals causes deterioration in various physiological, biochemical, and metabolic processes, including nitrogen (N) uptake and assimilation (Hussain et al. 2020).

Some metals, such as mercury (Hg), zinc (Zn), cadmium (Cd), cobalt (Co), lead (Pb), copper (Cu), and nickel (Ni), inhibit growth of plants. Altındal (2019) studied the effects of some heavy metals (Co, Cu, and Zn) on germination in sunflower (*Helianthus annuus* L.) and stated that increased doses of heavy metals negatively affected germination ratio, mean germination period, shoot and root length, and fresh and dry weight.

15.4 Effect of Salt Stress on Plants and the Reaction of Plants

Salinity is an increasingly common stress factor in the world. Salinity can occur as a result of natural factors and wrong agricultural practices, and stress factors cause average annual loss of 25% crop yield (Altındal and Altındal 2018).

Soil salinity occurs in arid and semiarid (insufficient rainfall) regions as a result of climate changes after accumulating water-soluble salts in the soil at a high rate. Therefore, this situation poses an increasing threat to agriculture in the world.

While salinity affects different physiological and metabolic processes in plants, it causes various symptoms such as decreased trunk length and leaf surface area, increased leaf thickness and fading, absorption of leaves, and necrosis of roots and shoots (Avcı et al. 2020). Plants have developed cellular ion signal reactions, ion transport, gene expression, and physiological mechanisms against salt stress. Plants that grow in salty conditions have to cope with osmotic stress, which leads to a reduction in water intake. Moreover, the increased osmotic pressure in the environment reduces leaf water potential, causes the stomata to close, and continuously limits carbon dioxide fixation, so plants cope with high osmotic pressure (Mahmoud et al. 2020) preventing ion toxicity (Munns et al. 2020). In addition, salinity together with high temperature negatively affects the electron transport chain, reducing photosynthetic capacity (Pérez-Romero et al. 2020).

Yoshida et al. (2020) conducted a research under different conditions of salt concentrations and determined that high levels of Na + were accumulated in roots, stem, and leaves in *Vigna marina* (M1, M4) and that the stomatal conductance, transpiration rate, and photosynthetic rate increased with adaptational regulation of aquaporin gene expression, under salt stress.

15.5 Conclusion

Like all living beings, plants are subject to global warming effects such as drought, greenhouse gases, temperature, and water shortages. Especially temperature changes have a negative effect on plant production. Since the negative effect of temperature changes is greater than the positive effect of rainfall changes, the overall effect of climate change on plant production is negative. Plants create complex physiological and metabolic defense mechanisms to adapt to the environment against adverse effects, which are not still explained by studies. However, these defense mechanisms may be insufficient to sustain the vitality of plants. As it is not possible to fully compensate for the effects of greenhouse gas emissions in the world today, more attention should be paid to mitigate the effects of climate change. Therefore, the fight against plant diseases and pests, organic agriculture and good agricultural practices, ensuring traceability at every stage of production, development of seed varieties resistant to biotic and abiotic stress factors, and the establishment of gene and seed banks are some support applications to adapt to climate change in plant production. It is possible to make effective use of high amounts of new and renewable energy sources, which are inevitable to have significant positive effects on the environment. Increasing the use of renewable energy sources will benefit in terms of maintaining natural balance components such as air, soil, water, and sustainability.

References

Ainsworth EA, Lemonnier P, Wedow JM (2020) The influence of rising tropospheric carbon dioxide and ozone on plant productivity. Plant Biol J 22:5–11. https://doi.org/10.1111/plb.12973

Aliche EB, Theeuwen TPJM, Oortwijn M, Visser RGF, Van der Linden CG (2020a) Carbon partitioning mechanisms in POTATO under drought stres. Plant Physiol Biochem 146:211–219

Aliche EB, Bourke AP, Sanchez MR, Oortwijn M, Gerkema E, As HV, Visser RGF, Van der Linden CG (2020b) Morphological and physiological responses of the potato stem transport tissues to dehydration stress. Planta 251:45

Altındal N (2019) Effect of some heavy metals on seed germination characteristics of sunflower (*Helianthus annuus* L.). 2. International conference on "agriculture, forestry life sciences", 18–20 April 2019, pp 185–191, Prague, Czech Republic

Altındal D, Altındal N (2018) Effect of ethyl methanesulphonate (EMS) applications on in vitro growth of sunflower (*Helianthus annuus* L. cv. Palancı-I) under salinity conditions. J Inst Sci Technol 8(4):351–359

Angelotti F, Barbosa LG, Barros JRA, Dos Santos CAF (2020) Cowpea development under different temperatures and carbon dioxide concentrations. Agric Res Trop 50:e59377. https://doi.org/10.1590/1983-40632020v5059377

Avcı UY, Ahmed HAA, Akdoğan G, Uranbey S (2020) Determination of different cotton genotypes against salt stress tolerance under *in vitro* conditions. Gaziosmanpasa J Sci Res 9(1):13–26

Chumley H, Hewlings S (2020) The effects of elevated atmospheric carbon dioxide [CO_2] on micronutrient concentration, specifically iron (Fe) and zinc (Zn) in rice; a systematic review. J Plant Nutr 43(10):1571–1578. https://doi.org/10.1080/01904167.2020.1739303

Dong J, Gruda N, Li X, Tang Y, Duan Z (2020) Impacts of elevated CO_2 on nitrogen uptake of cucumber plants and nitrogen cycling in a greenhouse soil. Appl Soil Ecol 145:103342. https://doi.org/10.1016/j.apsoil.2019.08.004

Estibaliz LP (2020) The influence of root-zone bicarbonate and carbon dioxide enrichment on lettuce, pepper and tomato growth. PhD thesis, p 203. UNSPECIFIED. https://doi.org/10.17635/lancaster/thesis/832

Florez-Sarasa I, Fernie AR, Gupta KJ (2020) Does the alternative respiratory pathway offer protection against the adverse effects resulting from climate change? J Exp Bot 71(2):465–469. https://doi.org/10.1093/jxb/erz428

Gázquez A, Abdelgawad H, Baggerman G, Raemdonck GV, Asard H, Maiale SJ, Rodríguez AA, Beemster GTS (2020) Redox homeostasis in the growth zone of the rice leaf plays a key role in cold tolerance. J Exp Bot 71(3):1053–1066. https://doi.org/10.1093/jxb/erz455

Gupta A, Rico-Medina A, Caño-Delgado AI (2020) The physiology of plant responses to drought. Science 368(6488):266–269. https://doi.org/10.1126/science.aaz7614

Hussain S, Khaliq A, Noor MA, Tanveer M, Hussain HA, Hussain S, Shah T, Mehmood T (2020) Metal toxicity and nitrogen metabolism in plants: an overview. In: Carbon and nitrogen cycling in soil. Springer, New York, pp 221–248

Jarma-Orozco A, Combatt-Caballero E, Jaraba-Navas J (2020) Growth and development of *Stevia rebaudiana* Bert., in high and low levels of radiation. Current. Plant Biol 22:100144. https://doi.org/10.1016/j.cpb.2020.100144

Johansson E, Branlard G, Cuniberti M, Flagella Z, Hüsken A, Nurit E, Pena RJ, Sissond M, Vazquez D (2020) Genotypic and environmental effects on wheat technological and nutritional quality. In: Igrejas G, Ikeda TM, Guzmán C (eds) Wheat quality for improving processing and human health. Springer, Cham, pp 171–204

Kim D, Medvigy D, Maier CA, Johnsen K, Palmroth S (2019) Biomass increases attributed to both faster tree growth and altered allometric relationships under long-term carbon dioxide enrichment at a temperate forest. Glob Change Biol 26:2519–2533. https://doi.org/10.1111/gcb.14971

Li X, Smith R, Choat B, Tissue DT (2019) Drought resistance of cotton (*Gossypium hirsutum*) is promoted by early stomatal closure and leaf shedding. Funct Plant Biol 47(2):91–98. https://doi.org/10.1071/FP19093

Li S, Li X, Wei Z, Liu F (2020a) ABA-mediated modulation of elevated CO_2 on stomatal response to drought. Curr Opin Plant Biol 56:174–180

Li N, Lin H, Wang T, Li Y, Liu Y, Chen X, Hu X (2020b) Impact of climate change on cotton growth and yields in Xinjiang, China. Field Crops Res 247:107590. https://doi.org/10.1016/j.fcr.2019.107590

Mahmood T, Khalid S, Abdullah M, Ahmed Z, Shah MKN, Ghafoor A, Du X (2020) Insights into drought stress signaling in plants and the molecular genetic basis of cotton drought tolerance. Cell 9:105. https://doi.org/10.3390/cells9010105

Mahmoud OMB, Hidri R, Talbi-Zribi O, Taamalli W, Abdelly C, Djébali N (2020) Auxin and proline producing rhizobacteria mitigate salt-induced growth inhibition of barley plants by enhancing water and nutrient status. S Afr J Bot 128:209–217. https://doi.org/10.1016/j.sajb.2019.10.023

Munns R, Passioura JB, Colmer TD, Byrt CS (2020) Osmotic adjustment and energy limitations to plant growth in saline soil. New Phytol 225:1091–1096. https://doi.org/10.1111/nph.15862

Naraikina NV, Pchelkin VP, Tsydendambaev VD, Trunova TI (2020) Changes in fatty acid composition and lipid content occurring in potato leaves during cold hardening: the role of Δ12-acyl-lipid desaturase. Russ J Plant Physiol 67:267–274

Parmesan C, Hanley ME (2015) Plants and climate change: complexities and surprises. Ann Bot 116(6):849–864. https://doi.org/10.1093/aob/mcv169

Pérez-Romero JA, Barcia-Piedras JM, Redondo-Gómez S, Mateos-Naranjo E (2020) *Sarcocornia fruticosa* photosynthetic response to short-term extreme temperature events in combination with optimal and sub-optimal salinity concentrations. Plant Physiol Biochem 148:45–52. https://doi.org/10.1016/j.plaphy.2019.12.026

Souri Z, Karimi N, Farooq MA, Akhtar J (2020) Phytohormonal signaling under abiotic stress. In: Tripathi DK, Singh VP, Chauhan DK, Sharma S, Prasad SM, Dubey NK, Ramawat N (eds) Plant life under changing environment. Academic Press, Cambridge, pp 397–466. https://doi.org/10.1016/B978-0-12-818204-8.00019-9

Thirkell TJ, Pastok D, Field KJ (2020) Carbon for nutrient exchange between arbuscular mycorrhizal fungi and wheat varies according to cultivar and changes in atmospheric carbon dioxide concentration. Glob Change Biol 26:1725–1738. https://doi.org/10.1111/gcb.14851

Veselá A, Duongová L, Münzbergová Z (2020) Plant origin determines seed mass, seed nutrients and germination behavior of a dominant grass species. bioRxiv. https://doi.org/10.1101/2020.03.02.973552

Wilkinson S, Mills G, Illidge R, Davies WJ (2012) How is ozone pollution reducing our food supply? J Exp Bot 63(2):527–536. https://doi.org/10.1093/jxb/err317

Yoshida J, Tomooka N, Khaing TY, Shantha PGS, Naito H, Matsuda Y, Hiroshi E (2020) Unique responses of three highly salt-tolerant wild *Vigna* species against salt stress. Plant Prod Sci 23 (1):114–128. https://doi.org/10.1080/1343943X.2019.1698968

Chapter 16
Impact of Climate Change on miRNA: Bioinformatics Perspectives

Anshul Tiwari and Prachi Srivastava

Abstract Small, noncoding RNAs are a distinct class of regulatory RNAs that regulate the range of biological functions in animals and plants. In plants, microRNAs (miRNAs) are major groups of small RNAs, which regulate the expression of cognate target genes by binding to reverse complementary sequences, resulting in the target RNAs being cleaved or translationally inhibited. MicroRNAs are also important for adaptation to stress conditions, in addition to their functions in growth and development and retention of genome integrity. In the twenty-first century, climate change is a concerning issue, and the impact of climate change on plant resources, biodiversity, and global food security is a major concern. It is important that we understand at the molecular level how plants react and adapt to higher temperatures, drought, ozone, and UV-B radiation, which are some of the main stresses that will be associated with changing climates, to ensure food production in the expected climates of 2030 and even beyond. In this chapter we discuss the responses and roles of plant miRNAs in climatic changes.

Keywords miRNAs · Climate · Biogenesis · Regulation · Translation · Sequencing · Databases

16.1 Introduction

Climate change in the twenty-first century is a concerning phenomenon. Current studies have suggested that agroecological systems are highly vulnerable to current warming patterns with seasonal variations in precipitation (Hatfield et al. 2018). It is expected that these reforms would have significant consequences for food security.

A. Tiwari
King George's Medical University, Lucknow, UP, India

P. Srivastava (✉)
Amity Institute of Biotechnology, Amity University Uttar Pradesh, Lucknow, UP, India
e-mail: psrivastava@amity.edu

D. K. Choudhary et al. (eds.), *Climate Change and the Microbiome*, Soil Biology 63,
https://doi.org/10.1007/978-3-030-76863-8_16

In accordance with most other environmental stresses, the levels of atmospheric carbon dioxide (CO_2) and ozone (O_3) are rising, including insufficient water sources causing drought or waterlogging situations, rising concentrations of toxic salts and heavy metals in agricultural fields, high temperatures, and UV radiation which are posing a danger to global food supply. Research indicates that global crop production is being limited by the imminent climate change, with no indication of reversing these patterns throughout the coming years. The discovery of intrinsic physiological, metabolic, and gene regulatory mechanisms under expected environmental conditions is of major importance for the execution of breeding or biotechnological approaches. In order to overcome the challenging environmental barriers, breeding methods are accompanied by modern biotechnological techniques, including genomics, proteomics, and transcriptomics, for the development of stress-tolerant agricultural crops with the least harmful impact on yield and even higher yields (Tester and Langridge 2010; Mittler and Blumwald 2010). It is well-known that plants respond to environmental signals in a complex and extremely structured way at physiological, biochemical, and molecular levels (Woodrow et al. 2010, 2016). Stress-mediated interference in normal cell processing requires rapid and broad reprogramming at the molecular level to respond to such adverse circumstances (Megha et al. 2017). The said programming is focused in particular on a strong regulation of stress-responsive genes expression at transcriptional and posttranscriptional levels (Xu et al. 2019; Khare et al. 2018; Shriram et al. 2016; Jeknić et al. 2013).

The small noncoding RNAs comprising 20–24 nucleotides were increasingly examined, either through transcriptional gene silencing (TGS) or posttranscriptional gene silencing (PTGS) allowing these small RNAs to serve as major regulators for protein coding gene expression (Baulcombe 2004). In several cellular and developmental processes, miRNAs are involved and act as master regulators of gene expression. It is well recognized that miRNAs are evolutionarily conserved in eukaryotes, although there are variations between animals and plants, primarily linked to their biogenesis and the target recognition function (Millar and Waterhouse 2005; Moran et al. 2017). In plants, microRNAs are processed in the nucleus and exported to the cytoplasm, while in animals, pri-microRNA and pre-microRNA are produced in the nucleus, while the microRNA/microRNA* is produced in the cytoplasm. Both plant and animal miRNAs are associated in the cytoplasm with the RISC complex, which is important for miRNA activity. The pri-miRNAs are first cleaved by Drosha RNase III in animals, while this is accomplished by Dicer-like (DCL) 1 in plants. There is a 2'-O-methylation of plant miRNAs on the 3′-terminal nucleotide that is not present in animal miRNAs. Considering the target identification mechanisms, in plants it is centered on near-perfect or perfect sequence complementarity (mostly leading to mRNA decay), while in animals the sequence complementarity is imperfect, mostly centered on the "seed law" (base pairing to the 5′ end of miRNAs, especially nucleotides 2–7) (Lewis et al. 2005).

However, understanding and establishing the connection between miRNAs and their coordinated stress signaling regulation is a major challenge (Joshi et al. 2017a, b). Along with the environmental adaptation process, miRNAs are also linked to plant stress tolerance and seem to have considerable potential for crop improvement (Budak and Zhang 2017). In recent years, miRNAs have gained considerable attention from researchers because of their important functions in many aspects of plant molecular life. High-throughput or deep sequencing as well as degradome analyses coupled with advanced analytical techniques and databases have helped to classify a significant number of mRNAs that react to single or multiple biotic/abiotic stresses (Esmaeili et al. 2016; Brant and Budak 2018; Ebrahimi Khaksefidi et al. 2015). Many essential points regarding these noncoding RNA species, however, need to be thoroughly understood.

In this chapter, we focused the responses and roles of plant miRNAs in climatic changes along with recent updates in this field.

16.2 miRNAs in Plants

Posttranscriptionally, miRNAs mainly repress target genes (Bartel 2004; Borges and Martienssen 2015). Plant miRNAs are derived from single-stranded hairpin precursors from unique stem regions, which have distinct characteristics from other forms of small RNAs. Detailed criteria have recently been identified for plant miRNA annotation (Axtell and Meyers 2018). In almost all aspects of normal plant development and growth, but also in response to environmental variations such as light, nutrients, and different abiotic and biotic stresses, miRNAs play crucial roles (Budak et al. 2015; Shriram et al. 2016; Brant and Budak 2018; Wang et al. 2019; Li et al. 2017). To maintain precise regulation of target genes, the temporal-spatial expression of intracellular miRNAs is under multi-tier supervision. In order to cope with stress, environmental stress induces plants to over- or under-express some miRNAs or to synthesize new miRNAs. Under different biotic and abiotic stress conditions, numerous stress-regulated miRNAs have been identified in model plants, including nutritional deficiency (Fujii et al. 2005), drought (Zhao et al. 2007; Zhou et al. 2007; Liu et al. 2008), cold (Zhou et al. 2008), salinity (Liu et al. 2008; Sunkar et al. 2008), bacterial infection (Navarro et al. 2006), UV-B radiation (Zhou et al. 2007), and mechanical stress (Lu et al. 2005). Even though stress regulation may represent a possible feature of the regulated miRNA in stress responses, it is apparent that the fact that miRNA is regulated differently in response to environmental stress does not necessarily imply that miRNA is engaged in responses to stress adaptation. In a latest report, salinity, drought, and low-temperature levels of 117 miRNAs were studied using miRNA chips comprising almost all documented miRNAs found in Arabidopsis (Liu et al. 2008). Seventeen stress-inducible miRNAs were detected, and the results were confirmed by detecting their expression patterns and analyzing the cis-regulatory elements in their promoter sequences (Liu et al. 2008). A library of small RNAs from Arabidopsis seedlings exposed to various abiotic stresses, including cold, dehydration, high salt, and abscisic acid (ABA), was developed by Sunkar and Zhu, and several new miRNAs that are responsive to abiotic stress were identified (Sunkar and Zhu 2004). For reference, miR393 was upregulated by treatments with cold, dehydration, salinity, and ABA; miR397b and miR402 were marginally upregulated by general stress treatments, whereas miR319c was induced by cold, but not by other treatments; miR389a was downregulated by all stress treatments. The results revealed the stress-induced miRNAs target negative stress response regulators or positive stress-inhibited process regulators and that tissue- or developmental stage-specific expression patterns are exhibited by many of the newly recognized miRNAs.

16.3 miRNAs and Temperature

Due to the rapid growth and rapidly inexpensive use of high-throughput sequencing technologies, genome-wide searches for miRNAs involved in temperature stress response have been carried out in many plant species in recent years. Recently, some known miRNAs have been shown to be involved in temperature stress response in some plant species, suggesting a recurrent role in the control of severe changes in temperature. Several temperature stress-associated miRNAs have been exchanged by different plant varieties; their patterns of expression seem to be species-dependent to certain degree during heat or cold stress, e.g., in *Brachypodium* and *Prunus persica*, miR172 was triggered by cold stress (Zhang et al. 2009; Barakat et al. 2012), whereas it was suppressed in grapevine and wheat throughout cold stress (Tang et al. 2012; Sun et al. 2015). In genotypes that exhibit opposing temperature tolerance, miRNA profiles have also been studied to examine the complexities of miRNA expression in plant temperature stress response. A remarkable variation was observed in a latest report between miRNA profiles in heat-tolerant and heat-sensitive rice genotypes (Liu et al. 2017). In total, in the tolerant genotype, 85 heat-responsive miRNAs were detected, while in the sensitive genotype, only 26 heat-responsive miRNAs were recognized. In addition, three miRNAs (miR159a.1, miR159b, and miR528-3p) were upregulated in the sensitive genotype among the nine common responsive miRNAs, while the same set of miRNAs were downregulated in the tolerant genotype. Another research involving comparative analysis between two cultivars of tea plants (*Camellia sinensis*) differing in cold sensitivity ("Yingshuang" cold-tolerant and "Baiye-1" cold-sensitive) also showed differential miRNA expression in response to cold stress (Zhang et al. 2014). These findings indicate that miRNA profiles with opposing temperature sensitivities can be unique in closely related genotypes. It may help to classify the tolerance-associated miRNAs and explore their mediated regulatory mechanisms in plants by means of differential expression comparison of miRNAs between the two genotypes with contrasting temperature stress tolerance.

16.4 Drought

The irregular and decreased precipitation rates, resulting in more frequent and intense drought periods, are one of the key consequences of climate change. Drought, including grain yield, is known to adversely impact plant growth and production. It impacts the physiology of plants and promotes molecular reprogramming by controlling gene expression at transcriptional and posttranscriptional stages, thus adjusting biochemical and physiological mechanisms. In several plants, microRNAs (miRNAs) have been implicated in important roles in stress responses. In several species of plants, such as *A thaliana*, *Oryza sativa*, *Triticum dicoccoides*, *Medicago truncatula*, *Phaseolus vulgaris*, and *P trichocarpa*, drought-responsive miRNAs have been reported (Sunkar and Zhu 2004; Zhou et al. 2007, 2010; Lu et al. 2008; Arenas-Huertero et al. 2009). Upregulation of miR169f and miR169 g, in roots but not in shoots, was detected by microarray-based study of miRNAs in rice seedlings exposed to polyethylene glycol-mediated water deficit (Zhou et al. 2007). Interestingly, 11 downregulated miRNAs (miR170, miR172, miR397, miR408, miR529,miR896, miR1030, miR1035, miR1050, miR1088, and miR1126) and 8 upregulated miRNAs (miR395, miR474, miR845, miR851, miR854, miR901, miR903, and miR1125) were detected by *O sativa* plants exposed to drought (Zhou et al. 2010). Drought-tolerant and drought-sensitive soybean genotypes showed differences in miRNA expression; in sensitive genotypes, miR166-5p; miR169f-3p; miR1513c; miR397a, b; and miR-Seq13 levels were upregulated, while in tolerant genotypes, these miRNAs were downregulated (Kulcheski et al. 2011). Significant upregulation of miR2118, miR159.2, miRS1, miR1514a, and miR2119 and mild upregulation of miR168, miR395, miR397, miR399, miR403, and miR4088 were shown by *P vulgaris* in response to drought stress (Arenas-Huertero et al. 2009). In *T dicoccoides*, *O sativa*, and *P trichocarpa* under drought, miR474 levels were upregulated (Lu et al. 2008; Zhou et al. 2010). MiR474 is known to be involved in the RNA production and control of organelle gene expression (Lu et al. 2005; Zhou et al. 2010). It is established that miR396 targets GRL transcription factors. Upregulated miR396 acts by inhibiting the expression of GRL, which tends to reduce stomatal density and narrow the leaf blade, which might lead to drought tolerance (Liu et al. 2009).

16.5 UV-B Radiation

In plants, UV-B radiation induces physiological and developmental changes. Solar radiation includes characteristics of light that are vital for photosynthesis, but also wavelengths that can cause cell damage. Plants have evolved adaptations to many environmental conditions, especially solar radiation, because of their sessile lifestyle. Plants comprise red/far-red (phytochromes) and blue/UV-A (cryptochromes and phototropins) photoreceptors and at least one UV-B photoreceptor (Rizzini et al.

2011). A wide range of physiological responses are caused by UV-B radiation (280–315 nm) touching the planet. UV-B causes mainly unspecific damage responses in species at elevated levels of intensity. In addition, UV-B-specific photo regulatory responses are also present in plants (Ulm 2003). Pleiotropic effects of elevated UV-B radiation on plant growth, morphology, and physiology were reported (Frohnmeyer and Staiger 2003; Blanding et al. 2007). UV-B stress can cause sudden changes (within 1 h of exposure) in gene expression (Casati and Walbot 2004; Mahalingam et al. 2005). In *A thaliana* and *Populus tremula*, miRNAs regulating gene expression in response to UV-B stress have been identified (Zhou et al. 2007; Jia et al. 2009). *P tremula* showed downregulation of 11 miRNAs in response to UV-B radiation and upregulation of 13 miRNAs as studied using the miRNA array (Jia et al. 2009). In *P tremula*, miR398 was found to be upregulated within 1 h following exposure to UV-B radiation. UV-B exposure typically contributes to oxidative stress; miR398 upregulation is in contrast to the downregulation found under other conditions of oxidative stress such as elevated Cu2þ, Fe3þ, methyl viologen, high sun, fumigation of ozone, and salt stress (Sunkar et al. 2006; Jagadeeswaran et al. 2009).

16.6 Ozone

The most prevalent air pollutant is ozone, which is an integral part of global climate change impacting habitats and human health (Freer-Smith 1990). Ozone is a reactive oxygen species (ROS) abiotic elicitor in plants (Kangasjarvi et al. 1994; Rao and Davis 2001; Kangasjarvi et al. 2005). In response to several different biotic and abiotic stresses, ROS are major signaling molecules that are induced (Mahalingam and Fedoroff 2003; Apel and Hirt 2004; Mittler et al. 2004; Gechev et al. 2006; Miller et al. 2008). Some greenhouse gases (GHG), like CO and NOx, interact toward volatile organics that are photochemically oxidized for the production of tropospheric ozone (Fiore et al. 2002). The ozone of the troposphere, while necessary in the stratosphere, is a GHG and a primary smog component. In the Northern Hemisphere, ozone levels in the pre-industrial period were 1015 ppb, currently rising to 35 ppb (Vingarzan 2004; Fuhrer 2008). Ozone reaches the leaves via stomata; causes oxidative stress; reduces photosynthesis; damages the membranes of DNA, protein, and cells; and thus adversely affects the productivity of plants (Stapleton 1992; Mackerness 2000; Leisner and Ainsworth 2011; Wilkinson et al. 2011; Ainsworth et al. 2012). The response of miRNAs in Arabidopsis to ozone stress was investigated in a recent study which reported an altered response of 22 miRNAs (Iyer et al. 2012). Many of these miRNAs exhibited a similar expression under UV-B stress, suggesting the activation under both types of stressors of common regulating pathways (Iyer et al. 2012). Such similarities may be anticipated in miRNA reactions during ozone and UV-B stress, as both stresses lead to oxidative stress induced by the production of reactive oxygen species (ROS). The upregulation of CSD gene expression, which is known to decrease ROS accumulation, is mediated by miR398 under oxidative stress (Sunkar et al. 2006). Downregulation of miR398 in ozone and UV-B stress (Jagadeeswaran et al. 2009; Jia et al. 2009; Iyer et al. 2012) can therefore induce the expression of the CSD gene and thus prevent the plant from oxidative stress. The significance of miR398-mediated upregulation of CSDs can be evidenced by the fact that miR398 levels escalate to normal levels rapidly after ozone release (Ainsworth et al. 2012). Certain miRNAs such as miR390, miR319, miR159, and miR156 exhibited increase in the levels within 1 h of ozone stress, and their cognate mRNA targets were rapidly reduced (Guan et al. 2013).

16.7 Conclusion

Doubtlessly, in response to temperature, drought, UV-B, and ozone stresses, miRNAs play critical roles. All of the above factors that cause climate change function in conjunction within natural environmental conditions: an increase in ozone can cause warming, which will influence plants more strongly than the impact of the individual stressor that has been investigated in combination with drought associated with oxidative stresses. Consequently, in order to mimic future climate changes, it is important to conduct miRNA analysis in conjunction with these stresses. The small RNA regulatory mechanisms that better support plant adaptability under field conditions will definitely be revealed in this study. Such information has the potential to integrate revolutionary molecular processes in crop plants that, in climate variability, will preserve agricultural production.

References

Ainsworth EA, Yendrek CR, Sitch S, Collins WJ, Emberson LD (2012) The effects of tropospheric ozone on net primary productivity and implications for climate change. Annu Rev Plant Biol 63 (1):637–661. https://doi.org/10.1146/annurev-arplant-042110-103829

Apel K, Hirt H (2004) REACTIVE OXYGEN SPECIES: metabolism, oxidative stress, and signal transduction. Annu Rev Plant Biol 55(1):373–399. https://doi.org/10.1146/annurev.arplant.55.031903.141701

Arenas-Huertero C, Pérez B, Rabanal F, Blanco-Melo D, De la Rosa C, Estrada-Navarrete G, Sanchez F, Covarrubias AA, Reyes JL (2009) Conserved and novel miRNAs in the legume *Phaseolus vulgaris* in response to stress. Plant Mol Biol 70(4):385–401. https://doi.org/10.1007/s11103-009-9480-3

Axtell MJ, Meyers BC (2018) Revisiting criteria for plant MicroRNA annotation in the era of big data. Plant Cell 30(2):272–284. https://doi.org/10.1105/tpc.17.00851

Barakat A, Sriram A, Park J, Zhebentyayeva T, Main D, Abbott A (2012) Genome wide identification of chilling responsive microRNAs in *Prunus persica*. BMC Genomics 13:481–481. https://doi.org/10.1186/1471-2164-13-481

Bartel DP (2004) MicroRNAs. Cell 116(2):281–297. https://doi.org/10.1016/s0092-8674(04)00045-5

Baulcombe D (2004) RNA silencing in plants. Nature 431(7006):356–363. https://doi.org/10.1038/nature02874

Blanding CR, Simmons SJ, Casati P, Walbot V, Stapleton AE (2007) Coordinated regulation of maize genes during increasing exposure to ultraviolet radiation: identification of ultraviolet-responsive genes, functional processes and associated potential promoter motifs. Plant Biotechnol J 5(6):677–695. https://doi.org/10.1111/j.1467-7652.2007.00282.x

Borges F, Martienssen RA (2015) The expanding world of small RNAs in plants. Nat Rev Mol Cell Biol 16(12):727–741. https://doi.org/10.1038/nrm4085

Brant EJ, Budak H (2018) Plant small non-coding RNAs and their roles in biotic stresses. Front Plant Sci 9:1038. https://doi.org/10.3389/fpls.2018.01038

Budak H, Zhang B (2017) MicroRNAs in model and complex organisms. Funct Integr Genomics 17(2–3):121–124. https://doi.org/10.1007/s10142-017-0544-1

Budak H, Kantar M, Bulut R, Akpinar BA (2015) Stress responsive miRNAs and isomiRs in cereals. Plant Sci 235:1–13. https://doi.org/10.1016/j.plantsci.2015.02.008

Casati P, Walbot V (2004) Rapid transcriptome responses of maize (*Zea mays*) to UV-B in irradiated and shielded tissues. Genome Biol 5(3):R16–R16. https://doi.org/10.1186/gb-2004-5-3-r16

Ebrahimi Khaksefidi R, Mirlohi S, Khalaji F, Fakhari Z, Shiran B, Fallahi H, Rafiei F, Budak H, Ebrahimie E (2015) Differential expression of seven conserved microRNAs in response to abiotic stress and their regulatory network in *Helianthus annuus*. Front Plant Sci 6:741. https://doi.org/10.3389/fpls.2015.00741

Esmaeili F, Shiran B, Fallahi H, Mirakhorli N, Budak H, Martínez-Gómez P (2016) In silico search and biological validation of microRNAs related to drought response in peach and almond. Funct Integr Genomics 17(2–3):189–201. https://doi.org/10.1007/s10142-016-0488-x

Fiore AM, Jacob DJ, Field BD, Streets DG, Fernandes SD, Jang C (2002) Linking ozone pollution and climate change: the case for controlling methane. Geophys Res Lett 29(19):25-1–25-4. https://doi.org/10.1029/2002gl015601

Freer-Smith P (1990) Air pollution and forests interaction between air contaminants and forest ecosystems. Environ Pollut 67(1):94–95. https://doi.org/10.1016/0269-7491(90)90176-d

Frohnmeyer H, Staiger D (2003) Ultraviolet-B radiation-mediated responses in plants. Balancing damage and protection. Plant Physiol 133(4):1420–1428. https://doi.org/10.1104/pp.103.030049

Fuhrer J (2008) Ozone risk for crops and pastures in present and future climates. Naturwissenschaften 96(2):173–194. https://doi.org/10.1007/s00114-008-0468-7

Fujii H, Chiou T-J, Lin S-I, Aung K, Zhu J-K (2005) A miRNA involved in phosphate-starvation response in Arabidopsis. Curr Biol 15(22):2038–2043. https://doi.org/10.1016/j.cub.2005.10.016

Gechev TS, Van Breusegem F, Stone JM, Denev I, Laloi C (2006) Reactive oxygen species as signals that modulate plant stress responses and programmed cell death. BioEssays 28 (11):1091–1101. https://doi.org/10.1002/bies.20493

Guan Q, Lu X, Zeng H, Zhang Y, Zhu J (2013) Heat stress induction ofmiR398triggers a regulatory loop that is critical for thermotolerance in Arabidopsis. Plant J 74(5):840–851. https://doi.org/10.1111/tpj.12169

Hatfield JL, Antle J, Garrett KA, Izaurralde RC, Mader T, Marshall E, Nearing M, Philip Robertson G, Ziska L (2018) Indicators of climate change in agricultural systems. Clim Chang. https://doi.org/10.1007/s10584-018-2222-2

Iyer NJ, Jia X, Sunkar R, Tang G, Mahalingam R (2012) microRNAs responsive to ozone-induced oxidative stress in *Arabidopsis thaliana*. Plant Signal Behav 7(4):484–491. https://doi.org/10.4161/psb.19337

Jagadeeswaran G, Saini A, Sunkar R (2009) Biotic and abiotic stress down-regulate miR398 expression in Arabidopsis. Planta 229(4):1009–1014. https://doi.org/10.1007/s00425-009-0889-3

Jeknić Z, Pillman KA, Dhillon T, Skinner JS, Veisz O, Cuesta-Marcos A, Hayes PM, Jacobs AK, Chen THH, Stockinger EJ (2013) Hv-CBF2A overexpression in barley accelerates COR gene transcript accumulation and acquisition of freezing tolerance during cold acclimation. Plant Mol Biol 84(1–2):67–82. https://doi.org/10.1007/s11103-013-0119-z

Jia X, Ren L, Chen Q-J, Li R, Tang G (2009) UV-B-responsive microRNAs in *Populus tremula*. J Plant Physiol 166(18):2046–2057. https://doi.org/10.1016/j.jplph.2009.06.011

Joshi R, Gupta P, Singla-Pareek SL, Pareek A (2017a) Biomass production and salinity response in plants: role of MicroRNAs. Indian J Plant Physiol 22(4):448–457. https://doi.org/10.1007/s40502-017-0327-7

Joshi T, Wang J, Zhang H, Chen S, Zeng S, Xu B, Xu D (2017b) The evolution of soybean knowledge base (SoyKB). Methods Mol Biol 1533:149–159. https://doi.org/10.1007/978-1-4939-6658-5_7

Kangasjarvi J, Talvinen J, Utriainen M, Karjalainen R (1994) Plant defence systems induced by ozone. Plant Cell Environ 17(7):783–794. https://doi.org/10.1111/j.1365-3040.1994.tb00173.x

Kangasjarvi J, Jaspers P, Kollist H (2005) Signalling and cell death in ozone-exposed plants. Plant Cell Environ 28(8):1021–1036. https://doi.org/10.1111/j.1365-3040.2005.01325.x

Khare T, Shriram V, Kumar V (2018) RNAi technology: the role in development of abiotic stress-tolerant crops. In: Biochemical, physiological and molecular avenues for combating abiotic stress tolerance in plants. Elsevier, Amsterdam. https://doi.org/10.1016/b978-0-12-813066-7.00008-5

Kulcheski FR, de Oliveira LF, Molina LG, Almerão MP, Rodrigues FA, Marcolino J, Barbosa JF, Stolf-Moreira R, Nepomuceno AL, Marcelino-Guimarães FC, Abdelnoor RV, Nascimento LC, Carazzolle MF, Pereira GA, Margis R (2011) Identification of novel soybean microRNAs involved in abiotic and biotic stresses. BMC Genomics 12:307–307. https://doi.org/10.1186/1471-2164-12-307

Leisner CP, Ainsworth EA (2011) Quantifying the effects of ozone on plant reproductive growth and development. Glob Chang Biol 18(2):606–616. https://doi.org/10.1111/j.1365-2486.2011.02535.x

Lewis BP, Burge CB, Bartel DP (2005) Conserved seed pairing, often flanked by adenosines, indicates that thousands of human genes are microRNA targets. Cell 120(1):15–20. https://doi.org/10.1016/j.cell.2004.12.035

Li S, Castillo-González C, Yu B, Zhang X (2017) The functions of plant small RNAs in development and in stress responses. Plant J 90(4):654–670. https://doi.org/10.1111/tpj.13444

Liu H-H, Tian X, Li Y-J, Wu C-A, Zheng C-C (2008) Microarray-based analysis of stress-regulated microRNAs in *Arabidopsis thaliana*. RNA 14(5):836–843. https://doi.org/10.1261/rna.895308

Liu D, Song Y, Chen Z, Yu D (2009) Ectopic expression of miR396 suppresses GRF target gene expression and alters leaf growth in Arabidopsis. Physiol Plant 136(2):223–236. https://doi.org/10.1111/j.1399-3054.2009.01229.x

Liu Q, Yang T, Yu T, Zhang S, Mao X, Zhao J, Wang X, Dong J, Liu B (2017) Integrating small RNA sequencing with QTL mapping for identification of miRNAs and their target genes associated with heat tolerance at the flowering stage in Rice. Front Plant Sci 8:43–43. https://doi.org/10.3389/fpls.2017.00043

Lu S, Sun Y-H, Shi R, Clark C, Li L, Chiang VL (2005) Novel and mechanical stress-responsive MicroRNAs in *Populus trichocarpa* that are absent from Arabidopsis. Plant Cell 17 (8):2186–2203. https://doi.org/10.1105/tpc.105.033456

Lu S, Sun Y-H, Chiang VL (2008) Stress-responsive microRNAs in populus. Plant J 55 (1):131–151. https://doi.org/10.1111/j.1365-313x.2008.03497.x

Mackerness SAH (2000) Plant responses to ultraviolet-B (UV-B: 280–320 nm) stress: what are the key regulators? Plant Growth Regul 32(1):27–39. https://doi.org/10.1023/a:1006314001430

Mahalingam R, Fedoroff N (2003) Stress response, cell death and signalling: the many faces of reactive oxygen species. Physiol Plant 119(1):56–68. https://doi.org/10.1034/j.1399-3054.2003.00156.x

Mahalingam R, Shah N, Scrymgeour A, Fedoroff N (2005) Temporal evolution of the Arabidopsis oxidative stress response. Plant Mol Biol 57(5):709–730. https://doi.org/10.1007/s11103-005-2860-4

Megha S, Basu U, Kav NNV (2017) Regulation of low temperature stress in plants by microRNAs. Plant Cell Environ 41(1):1–15. https://doi.org/10.1111/pce.12956

Millar AA, Waterhouse PM (2005) Plant and animal microRNAs: similarities and differences. Funct Integr Genomics 5(3):129–135. https://doi.org/10.1007/s10142-005-0145-2

Miller G, Shulaev V, Mittler R (2008) Reactive oxygen signaling and abiotic stress. Physiol Plant 133(3):481–489. https://doi.org/10.1111/j.1399-3054.2008.01090.x

Mittler R, Blumwald E (2010) Genetic engineering for modern agriculture: challenges and perspectives. Annu Rev Plant Biol 61(1):443–462. https://doi.org/10.1146/annurev-arplant-042809-112116

Mittler R, Vanderauwera S, Gollery M, Van Breusegem F (2004) Reactive oxygen gene network of plants. Trends Plant Sci 9(10):490–498. https://doi.org/10.1016/j.tplants.2004.08.009

Moran Y, Agron M, Praher D, Technau U (2017) The evolutionary origin of plant and animal microRNAs. Nat Ecol Evol 1(3):27. https://doi.org/10.1038/s41559-016-0027

Navarro L, Dunoyer P, Jay F, Arnold B, Dharmasiri N, Estelle M, Voinnet O, Jones JDG (2006) A plant miRNA contributes to antibacterial resistance by repressing auxin signaling. Science 312 (5772):436–439. https://doi.org/10.1126/science.1126088

Rao MV, Davis KR (2001) The physiology of ozone induced cell death. Planta 213(5):682–690. https://doi.org/10.1007/s004250100618

Rizzini L, Favory JJ, Cloix C, Faggionato D, O'Hara A, Kaiserli E, Baumeister R, Schafer E, Nagy F, Jenkins GI, Ulm R (2011) Perception of UV-B by the Arabidopsis UVR8 protein. Science 332(6025):103–106. https://doi.org/10.1126/science.1200660

Shriram V, Kumar V, Devarumath RM, Khare TS, Wani SH (2016) MicroRNAs as potential targets for abiotic stress tolerance in plants. Front Plant Sci 7:817–817. https://doi.org/10.3389/fpls.2016.00817

Stapleton AE (1992) Ultraviolet radiation and plants: burning questions. Plant Cell 4 (11):1353–1358. https://doi.org/10.1105/tpc.4.11.1353

Sun X, Fan G, Su L, Wang W, Liang Z, Li S, Xin H (2015) Identification of cold-inducible microRNAs in grapevine. Front Plant Sci 6:595–595. https://doi.org/10.3389/fpls.2015.00595

Sunkar R, Zhu J-K (2004) Novel and stress-regulated microRNAs and other small RNAs from Arabidopsis. Plant Cell 16(8):2001–2019. https://doi.org/10.1105/tpc.104.022830

Sunkar R, Kapoor A, Zhu J-K (2006) Posttranscriptional induction of two cu/Zn superoxide dismutase genes in Arabidopsis is mediated by downregulation of miR398 and important for oxidative stress tolerance. Plant Cell 18(8):2051–2065. https://doi.org/10.1105/tpc.106.041673

Sunkar R, Zhou X, Zheng Y, Zhang W, Zhu J-K (2008) Identification of novel and candidate miRNAs in rice by high throughput sequencing. BMC Plant Biol 8:25–25. https://doi.org/10.1186/1471-2229-8-25

Tang Z, Zhang L, Xu C, Yuan S, Zhang F, Zheng Y, Zhao C (2012) Uncovering small RNA-mediated responses to cold stress in a wheat thermosensitive genic male-sterile line by deep sequencing. Plant Physiol 159(2):721–738. https://doi.org/10.1104/pp.112.196048

Tester M, Langridge P (2010) Breeding technologies to increase crop production in a changing world. Science 327(5967):818–822. https://doi.org/10.1126/science.1183700

Ulm R (2003) Molecular genetics of genotoxic stress signalling in plants. Topics in current genetics. Springer, Berlin, Heidelberg. https://doi.org/10.1007/978-3-540-39402-0_9

Vingarzan R (2004) A review of surface ozone background levels and trends. Atmos Environ 38 (21):3431–3442. https://doi.org/10.1016/j.atmosenv.2004.03.030

Wang J, Mei J, Ren G (2019) Plant microRNAs: biogenesis, homeostasis, and degradation. Front Plant Sci 10:360–360. https://doi.org/10.3389/fpls.2019.00360

Wilkinson S, Mills G, Illidge R, Davies WJ (2011) How is ozone pollution reducing our food supply? J Exp Bot 63(2):527–536. https://doi.org/10.1093/jxb/err317

Woodrow P, Pontecorvo G, Ciarmiello LF, Fuggi A, Carillo P (2010) Ttd1a promoter is involved in DNA–protein binding by salt and light stresses. Mol Biol Rep 38(6):3787–3794. https://doi.org/10.1007/s11033-010-0494-3

Woodrow P, Ciarmiello LF, Annunziata MG, Pacifico S, Iannuzzi F, Mirto A, D'Amelia L, Dell'Aversana E, Piccolella S, Fuggi A, Carillo P (2016) Durum wheat seedling responses to simultaneous high light and salinity involve a fine reconfiguration of amino acids and carbohydrate metabolism. Physiol Plant 159(3):290–312. https://doi.org/10.1111/ppl.12513

Xu J, Hou QM, Khare T, Verma SK, Kumar V (2019) Exploring miRNAs for developing climate-resilient crops: a perspective review. Sci Total Environ 653:91–104. https://doi.org/10.1016/j.scitotenv.2018.10.340

Zhang J, Xu Y, Huan Q, Chong K (2009) Deep sequencing of Brachypodium small RNAs at the global genome level identifies microRNAs involved in cold stress response. BMC Genomics 10:449–449. https://doi.org/10.1186/1471-2164-10-449

Zhang Y, Zhu X, Chen X, Song C, Zou Z, Wang Y, Wang M, Fang W, Li X (2014) Identification and characterization of cold-responsive microRNAs in tea plant (*Camellia sinensis*) and their targets using high-throughput sequencing and degradome analysis. BMC Plant Biol 14:271–271. https://doi.org/10.1186/s12870-014-0271-x

Zhao B, Liang R, Ge L, Li W, Xiao H, Lin H, Ruan K, Jin Y (2007) Identification of drought-induced microRNAs in rice. Biochem Biophys Res Commun 354(2):585–590. https://doi.org/10.1016/j.bbrc.2007.01.022

Zhou X, Wang G, Zhang W (2007) UV-B responsive microRNA genes in *Arabidopsis thaliana*. Mol Syst Biol 3:103–103. https://doi.org/10.1038/msb4100143

Zhou X, Wang G, Sutoh K, Zhu J, Zhang W (2008) Identification of cold-inducible microRNAs in plants by transcriptome analysis. Biochimica et Biophysica Acta (BBA) - Gene Regulatory Mechanisms 1779(11):780–788. https://doi.org/10.1016/j.bbagrm.2008.04.005

Zhou L, Liu Y, Liu Z, Kong D, Duan M, Luo L (2010) Genome-wide identification and analysis of drought-responsive microRNAs in *Oryza sativa*. J Exp Bot 61(15):4157–4168. https://doi.org/10.1093/jxb/erq237

Chapter 17
Climate Change on Plant Community Structure and Ecosystem Function

Rajeev Singh and Manish Kumar Raj

Abstract Climate change is working in a very unprecedented manner with the land system and mostly affecting the plant communities in either way. The mechanism and action might vary on temporal and spatial scale of plants, but the modifications are inevitable. The modification is driven by many direct and consequential factors involving human-induced land use, land degradation, desertification, biotic interaction of non-native communities and adding to it the role of micro climate of the species. The climate change and the induced alterations in the plant communities bring relatable modification in the working of the ecosystem. Plant cover and composition of biological communities play pivotal role in understanding the dynamic nature of the different ecosystems and their responses towards the climate change. Forecasting their future trajectories requires the in-depth knowledge of driving factors responsible for any shift due to climate-driven change. Recent changes in drought periods and periodically extreme rainfall conditions, carbon sequestration, N deposition and thermal stress due to elevating temperature are the vital signs of a future shift in the composition of the plant community and ecosystem structure.

Keywords Climate change · Thermal stress · Nitrogen deposition · Thermophilization · Biotic invasion

17.1 Introduction

Environment plays a crucial and noteworthy role in defining the structure of a population, and any change in the surrounding tends to shift the population dynamics. The Intergovernmental Panel on Climate Change (IPCC) special report on climate change and land (2019) has discussed the complex interaction of land and climate through multiple bio-geophysical and biogeochemical responses across

R. Singh (✉) · M. K. Raj
Department of Environmental Sciences, Satyawati College, Delhi University, Delhi, India

D. K. Choudhary et al. (eds.), *Climate Change and the Microbiome*, Soil Biology 63,
https://doi.org/10.1007/978-3-030-76863-8_17

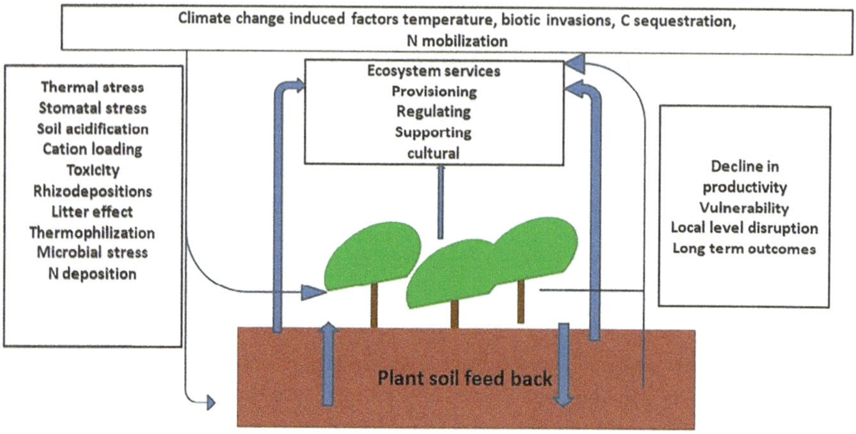

Fig. 17.1 Relation of ecosystem composition structure and ecosystem functioning

different geographical and temporal scales. Species abundance and distribution is expected to be altered, and the modifications consequently influences the ecosystems (Parmesan 2006), as well as gradual or rapid depletion and decline of the services and functioning (Fig. 17.1) is expected that these ecosystems offer (Millennium Ecosystem Assessment 2005). Declining ecological niches due to altered disturbance regimes, desertification and biotic invasions, along with the micro biome interplay, are the multiple drivers for the shifting habitats and plant community structure. Along with the direct human-induced land use patterns, there are many indirect factors like CO_2 fertilization, nitrogen deposition, extended growing season, increase in diffuse radiation, drought, desertification, heat waves, soil acidification, mineralization, etc. Climate-induced precipitation and temperature alterations are more likely to bring the changes in abundance, distribution and plant species interactions and possess the potential to bring structural shifts in plant community composition and consequently can have profound effects on ecosystem functioning, involving plant diversity and productivity (Munson et al. 2011).

17.2 Effect of Temperature

17.2.1 Thermal Stress

Increase in the global temperature is inseparable phenomenon of climate change, and it poses the threats of temperature stress to the biological community (Saxe et al. 2001; Parmesan and Hanley 2015), and in its drastic condition, it carries the potential to limit the individuals and community performance, in the due course leading to the elimination of some species ranges or modification in species ranges if local extinctions are offset by concurrent biotic invasions in case of low temperature areas and local extinctions (Feeley 2012; Lenoir and Svenning 2015). The local extinctions followed by the chain of events including decline in diversity through distribution shift of the species (Colwell et al. 2008; Wiens 2016), changes in community structure and make up (Feeley et al. 2011, 2013; Duque et al. 2015; Fadrique et al. 2018) possible changes in processes such as carbon capture and storage (Clark et al. 2003; Brienen et al. 2015), regulation of the regional climate (Cox et al. 2000; Luo 2007) and yielding of food (Tito et al. 2018) overall indicates the altering nature of climate change-induced thermal stress over ecosystem structure and functions.

The increase in respiration rate and/or the photosynthesis decline causes the thermal stress in plants as high temperatures above a species-specific critical thermal maximum will permanently damage photo system II, leading to reduction in electron transport rates and photosynthetic non-performance, therefore, decreasing effectiveness of leaf physiological processes ensuing leaf and plant death (Baker, 2008). The thermal tolerances of plants can vary remarkably while comparing the communities occupying different latitudinal gradients as it is observed that plant community growing in hot equatorial conditions like rainforests has higher heat tolerances on average compared to communities from colder, higher latitude habitats like temperate or boreal forests (O'Sullivan et al. 2017; Zhu et al. 2018). It is also found that warming effects are context dependent, and multitude of factors works together like precipitation-induced soil moisture (Cowles et al. 2018.) along with other biotic and abiotic interactions. Even an experimental warming study shows the inconsistent productivity response with soil moisture (Wu et al. 2011; Elmendorf et al. 2012). These studies more likely elucidate the not so consistent connection between soil moisture and the escalated temperature response due to biotic factors, such as species composition, and abiotic habitat factors like soil type and infiltration rates (Way and Oren 2010; Dieleman et al. 2012). Another warming-related effect dependency is observed where experimental escalation in warming tends to increase productivity more strongly near the poles (Rustad et al. 2001). Another slope and watering-based study by Ronk et al. (2020), using open-top passive warming chamber (OTC), suggests climate variations have greater consequences for composition of the plant communities under greater soil moisture conditions, at the lower slope location, and at the drier upper slope location with addition of water. Study also suggests more affected community composition at lower slope location with greater soil moisture and the location on the drier upper slope where warming was combined with additional watering. More strong response of the lower slope's community composition to escalating temperature and the associated water deficit of the OTC is consistent with Grime's (1974, 1998) plant strategy theory, which predicts a balanced mechanism and regulation of the plant community in terms of stress tolerance, with potential resistance to environmental disturbances like warming/drought and competitive vigour.

17.2.2 Thermophilization

The process of directional change of the entire plant ecosystems over time to include more and more of the species that prefer warmer climates is called thermophilization. The process suggests the rise in relative abundances of thermophilic or thermal-tolerant species and shifts in precipitation cause change in abundance of tree species which demands more water for the growth and sustenance. Some of the most noteworthy changes is observed in forest plant communities and ecosystem functions due to the changes in climate (Esquivel-Muelbert et al. 2017; Thomas et al. 2004). Various studies relate to the warming climate where the decline in forest biomass and productivity could be related to hydraulic failure in larger trees due to highly increased temperature and intensifying drought (McDowell and Allen 2015; Rowland et al. 2015), and more effectively it is evident in the Amazon tropical forest, where forest diversity is concentrated in the wetter, humid, least seasonal alter forest (Francis and Currie 2003), a continuous lengthening of the hot and precipitation deficient season might threaten a large section of tropical biotic community. The compositional shift is observed to be in consistency with the studied and recognized escalation of the seasonal drought, and also the increase in the abundance of drought-tolerant species shows a slow shift resilience behaviour of Amazon tree communities towards the future drought (Esquivel-Muelbert et al. 2019). Another work by Feeley et al. (2020) on floristic compositions of nearly 200 new world (North, Central and South America) ecoregion collected species over the past 40 years (1970–2011) based on optimal mean annual temperature and optimal total annual precipitation, yearly calculated community temperature index (CTI) and community precipitation index(CPI) suggested positive thermophilization for all biomes which is consistent with the previous studies and expectations that relates the decreasing relative abundance of species in the ecoregions with increasing temperature. Though working at such huge collection over large spatial and temporal scales may have under-represented the urban heat and other land use effects like deforestation, forest recovery, forest fires or biotic interactions, overall, the study represents a scenario of temperature-mediated drought stress effect on plant community composition.

17.3 Effect of Biotic Invasion

Species invasions have been identified as an important aspect of global change, and the invasions influence synergistic effect with other anthropogenic modifications in the environment to shift biodiversity and ecosystem structure and functional outcomes in invaded habitats, and even extinctions are often driven with this synergism (Brook et al. 2008). Due to the continuous changes in climate across the latitudes, more observation-based study is being concentrated on the consequences of these changes on co-existing native and invasive species across various environments (Rahel and Olden 2008) and their relative functions and processes in response to climate change (Stephens et al. 2019). Some striking negative interactions are also observed at different levels of hierarchal organization ranging from native individuals to populations to communities and entire ecosystems (Vilà et al. 2011), but results are well aligned in terms of different physiology and functional aspects of specific species or ecosystem like decline in pollination behaviour which definitely impact the ecosystem structure and functions (Hoover et al. 2014; González-Varo et al. 2013) as it is already predicted that reduction in pollination may lead to effect the plant populations (Bennett et al. 2018). Study of the phonological aspect to climate change at both local and global level (Parmesan and Yohe 2003), species-specific flowering phenological responses and presence of comparatively high phenological plasticity of invasive to native plant species (Gallinat et al. 2018) suggest the invasion role in determining community composition. Further the flowering phenological modification in the native species alters the plant pollinator interaction modifying floral resources in landscapes (Drossart et al. 2017), with sometimes positive (Knight et al. 2018) but generally negative impacts for native plant reproductive process (González-Varo et al. 2013). Change in flower phenology with prolific flower production by many invasive species creates an opportunity for interspecific competition, and it may affect the fitness of the native community (Brown et al. 2002; Mitchell et al. 2009), but again, this would be the prevalent case when the invasive species attract the pollinators with high nutritional value or easily foraged (Brown et al. 2002; González-Varo et al. 2013; Drossart et al. 2017). Few studies have explored that competitive interactions are one of the prevalent ways of reducing the abundance of native species. In support of these interactions, Giejsztowt et al. have studied the experimental warming (+1.7 °C) modulated flowering phenology study by taking flowering overlap between a native plant (*Dracophyllum subulatum*) and an invasive plant (*Calluna vulgaris L.*) as reference and observed only neutral to negative effects of *C. vulgaris* on *D. subulatum* seed mass. The observation suggests some aspects of climate change potential of invasion process impact on the reproductive success over native. Other such example of a negative interaction is observed by Delipetrou (2006) in *Carpobrotus edulis*, a European succulent found in coastal dune and sand habitats that alters the growth, survival, reproduction and germination of native species such as *Gilia millefoliata*, *Scabiosa atropurpurea* and *Malcolmia littorea*, by altering the condition of the soil by way of allelopathy or by terminating it by *C. edulis* (Novoa and Gonzalez 2014) diminishing soil moisture content to native equivalents and by decreasing the pH and improving organic carbon (C) and nitrogen (N) (Delipetrou 2006).

17.4 Nitrogen Deposition

The process of nitrogen deposition, before ending up on the earth's surface, starts with anthropogenic nitrogen emission from various sources and once airborne in the form of atmospheric aerosols is capable of vast coverage in the atmosphere under the

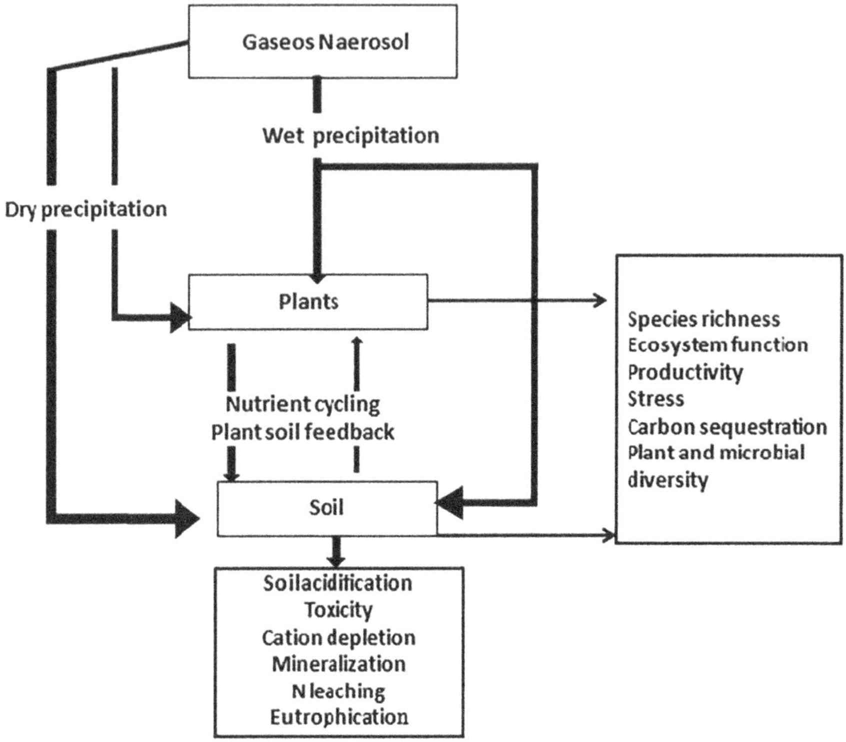

Fig. 17.2 Modified schematic representation process and patterns of the N deposition (Erisman 1990)

dominance of air currents. The range it covers is determined by different factors like emission level, wind speed, turbulence, rate of chemical reactions, etc. Climate change plays a crucial driver plant growth regulation, alterations in species distributions and modifications in biogeochemical cycling. The human-induced creation of reactive N in its various forms emitted to the atmosphere as NOx, NH3 and organic N (Galloway et al. 2004) has effectually increased the losses of reactive N to the environment and emerged in a cascade of environmental consequences (Stevens 2019) and posing a hostile approach to the ecosystem structures and functions (Sala et al. 2000; Bobbink et al. 2010).

Under N-limited conditions, a variety of effects are observed in the form of increased plant growth (Fig. 17.2) as a stimulation of carbon (C) sequestration (de Vries et al. 2014; Schulte-Uebbing and De Vries 2018) and decreased plant biodiversity which follow the principle of competitive exclusion by the expansion of nitrophilous species (Bobbink et al. 2010). The other driving impacts like soil acidification, cation depletion and escalated availability of toxic metals such as Al^{3+} and Fe^{3+} can also downgrade plant health and productivity, modify community composition and cause lowering in species richness (Bowman et al. 2008). Some other aspects have also been observed which includes increased invasive species and nutrient imbalances (Du et al. 2019), elevated level of damages from pests and frost and increased N leaching to water bodies.

A recent study by Cheng et al. on forest ecosystem found remarkably increased net rates of mineralization and nitrification by N deposition. But more research is suggested on gross transformation rates for the proper assessment of the same. Further Lilleskov et al. (2018) reviewed the consequences of N deposition on the structure and function of mycorrhizal communities in forest ecosystems and suggested the withdrawal of key ectomycorrhizal communities, which would bring down the specified role to access organic N and/or phosphorus and slow down the rate of decomposition of organic matter due to strong responsiveness of ectomycorrhizal species to N deposition. The sensitivity of ectomycorrhizal fungi is well understood in boreal and temperate forests, with the help of current approximation of N critical loads which suggest conifers are more sensitive than those of deciduous trees (Lilleskov et al. 2018). However, more intricate studies are required to establish the conclusive effects of N deposition on arbuscular mycorrhizal and tropical ectomycorrhizal communities.

N deposition is also related to the understorey species like bryophytes and lichens based on its sensitivity towards the deposition. It occurs due to the lack of cuticle in understorey species which allow the absorption of pollutants across their surface area. Moreover, N deposition causes increase in productivity in vascular plants which may lead to increased shading and reduction in understorey species. Light and pH conditions can also be the deciding factors for the feedback of the understorey species and community to N deposition (Perring et al. 2018).

17.5 Elevated Atmospheric CO_2

CO_2 has been increasing and piling up in the atmosphere at an accelerating pace, and one of the important aspects of elevated level of CO_2 is knowing and understanding the response towards the plant community. Elevated CO_2 enrichment and their interrelationship with environmental change are responsible for many biological responses at various levels (Ainsworth and Rogers 2007). One of the most studied and interesting research aspects is the response of photosynthesis and stomatal conductance to increased carbon dioxide concentration [CO_2]. The general response of plant community to the elevated CO_2 level is observed in terms of enhancement of the photosynthetic carbon fixation rate by leaves and the plant growth and productivity due to increase in photosynthesis as well as reduction in stomatal conductance (Ainsworth and Rogers 2007). It has also been studied that all other effects of escalating CO_2 on plant communities and interrelated ecosystems are derived from these two fundamental responses of increase in photosynthesis as well as reduction in stomatal conductance (Long et al. 2004; Ainsworth and Rogers 2007). One of the most harmonious and universal results and effects under different studies of rising atmospheric CO_2 on plants is an increase in the photosynthetic carbon fixation rate by leaves. The variation in concentrations of CO_2 is also involved in regulating the stomata opening, gas exchange pores of the plants which helps the plant to function in coordination with the external environment. Opening of stomata not only allows atmospheric CO_2 to diffuse into leaves for the process photosynthesis but also provides a channel for water to diffuse out of leaves. Free-Air CO_2 Enrichment (FACE) experiments allow a noble approach to study the relativeness of elevated CO_2 concentration to the plant communities in open conditions, and the study shows the reduction in efficiency of the whole plant towards water use under elevated CO_2 due to decline in stomatal conductance and eventually lowering of diffusion of water from the leaves to the external environment. This in turn can have related follow-up influence for the hydrological cycle of entire ecosystems, resulting in escalating soil moisture levels as well as runoff under the conditions of rise in atmospheric CO_2 concentration (Leakey et al. 2009). It can be suggested that the degree of stomatal opening is regulated by the plants as a trade-off between high rates of photosynthesis and low rates of water loss, and in case of elevated CO_2 concentrations, plants can support high rates of photosynthesis with relatively low stomatal conductance. Stomatal conductance is a measure of degree of stomatal opening and represents the water deficit status of any plant (Gimenez et al. 2005), and any decline in the stomatal conductance in case of rise in CO_2 concentrations (Long et al. 2004) may narrow down the CO_2 fixation rate in exchange of improving water use efficiency in water-deficit conditions promoting plant growth (Leakey et al. 2009; Sreeharsha et al. 2015). Likewise, while stomatal conductance at escalated CO_2 is usually declined, the consequence is variable and depends on feedback relation of different environmental factors such as deficiency of soil water, inadequate vapour pressure, temperature variability and light either alone or in combination (Šigut et al. 2015; Xu et al. 2016).

To reveal the prime mechanism of the various responses and to test the findings and the proposed hypothesis under open air condition, Free-Air Carbon dioxide Enrichment (FACE) technology is used in many recent studies like comparing the response of C3 and C4 plants towards elevated CO_2. Being more naturalistic in condition and the approach applied is nearer to real world, they should provide the best evidence of the responses of plants to elevated CO_2.

17.5.1 General Response of C3 and C4

Both C3 and C4 plant species show the similar type of response and act in consistent with decline in stomatal conductance at increased concentration of CO_2 (Ainsworth and Rogers 2007). On an average, across all plant species grown at raised CO_2 concentration during the conductance of FACE experiments, stomatal conductance is in reducing trend. But in contrast to general rule, there are evidences of deviation to the rule that stomatal conductance declines at elevated CO_2; even in FACE experiments like in a very specific study of *Pinus taeda*, guard cells appear to be independent to elevated CO_2 (Ellsworth 1999). Also, there are many studies that suggest the variation in stomatal conductance behaviour to elevated CO_2 in presence of various environmental factors.

Studying the photosynthetic responses mediated through Rubisco, it is observed that with the elevation of atmospheric CO_2 levels, the CO_2 affinity towards Rubisco increases in the photosynthetic cells, modifying the ratio of CO_2:O_2 in favour of CO_2 and thereby increasing the rate of carboxylation while decreasing the rate of oxygenase activity of Rubisco (Makino and Mae 1999), but the evidence provided by Ainsworth et al. also demonstrate the downregulation of photosynthesis activity after a long exposure of CO_2 which may sum up the role of degree of plant acclimatization to elevated CO_2, which explains both positive and reducing response of photosynthesis activity. The increase in the rate of photosynthesis caused by rise in CO_2 concentration results in an increase in carbon assimilation, which alters the plant's carbon and nitrogen activity.

17.5.2 A Comparative Response of C3 and C4 Towards Elevated CO_2

C3 plants without any doubt show positive response towards elevated CO_2 (Ainsworth and Long 2005). In case of escalated temperature and drought stress conditions, anticipation is more oriented towards C4 species than C3 species (Ainsworth and Long 2005). However, a few recent studies have reported that increase in CO_2 concentration has increased the growth of several C4 grasses even under hydrated conditions (Ghannoum et al. 2000). Reich et al. (2018) on the basis of 20-year field experiment of 88 grassland plots proposed a complete reversal aspect of biomass enhancement in C4 but not C3 grasses in response to ambient CO_2 and the observation cast some degree of doubt over positive effect on the growth of the plants on a long terms with an increasing atmospheric CO_2, so these conclusions need to be attributed with more research in this field for the consistent results provided in support of more positive photosynthesis growth of C3 over C4 which can be explained on the basis of C3 and C4 primary physiological process of photosynthesis. In conditions of elevated CO_2, enhancement in the photosynthetic rate and efficiency in C3 plants can be advocated on the basis of the general physiology of both plants. Analysing C4 pathway which avoids the photorespiration pathway, fix CO_2 as $HCO3^-$ with the help of phosphoenolpyruvate carboxylase lacks an oxygenase function, and having the high affinity for CO_2 without any competition for oxygenase activity and further decarboxylated to bundle sheath cells for the rubisco active site helps in concentrating CO_2 (Sage 2004). Despite the efforts of eliminating the oxygenase loss, C4 plants required to give away 10–15% of their leaf nitrogen in Rubisco in exchange of its catalytic activity (Ainsworth and Rogers 2007). On the other hand, C3 plants are devoid of CO_2 buildup mechanism, suggesting lower N use efficiency than C4 plants. Due to nature of Rubisco being substrate bound at current CO_2, rate of the carboxylation reaction increases with rising CO_2. Moreover, the absence of photorespiratory pathway reduces the loss of CO_2 in C4 (Long et al. 2004). The increase in the CO_2 concentration is not anticipated to directly effect C4 plants because they avoid photorespiratory pathway and are CO_2-saturated at current level CO_2. Maybe the prediction on the impact of elevated CO_2 at C4 photosynthesis rate is the outcome of indirect effect like water stress (Ghannoum et al. 2000). Results from FACE experiments provide additional support for this conclusion.

17.6 Conclusion

The consistent and expected results of most of the recent studies represent a wider picture of acknowledging the change drivers due to the multitude nature of the responses of the plant communities. The ability of the plant communities to maintain functioning under different circumstances also varies. Observations such as shifting of the communities, phonological plasticity, resilience and adaptability show plant species response in all dimensions. So, it needs a more holistic approach at different levels of biological research. The act of synergism by the induced responses yields greater understanding into current and future possibilities of changes due to climate change. The compositional and functional parameters studied till now also contain some restricted approach towards the experimentation and that inconclusiveness still provides opportunities for a conclusive research on the basis of particular species, communities, functional interactions and many more on local and global basis in different time scales so that it ensures a stable and steadfast supply of ecosystem functions and services in the era of climate change.

References

Ainsworth EA, Long SP (2005) What have we learned from 15 years of free-air CO_2 enrichment (FACE)? A meta-analytic review of the responses of photosynthesis, canopy properties and plant production to rising CO_2. New Phytol 165:351–372. https://doi.org/10.1111/j.1469-8137.2004.01224.x

Ainsworth E, Rogers A (2007) The response of photosynthesis and stomatal conductance to rising [CO_2]: mechanisms and environmental interactions. Plant Cell Environment 30:258–270. https://doi.org/10.1111/j.1365-3040.2007.01641.x

An IPCC Special Report on climate change, desertification, land degradation, sustainable land management, food security, and greenhouse gas fluxes in terrestrial ecosystems, 2019

Baker NR (2008) Chlorophyll fluorescence: a probe of photosynthesis in vivo. Annu Rev Plant Biol 59:89–113. https://doi.org/10.1146/annurev.arplant.59.032607.092759

Bennett JM et al (2018) GloPL, a global data base on pollen limitation of plant reproduction. Scientific Data 5:8

Bobbink R, Hicks K, Galloway J, Spranger T, Alkemade R, Ashmore M et al (2010) Global assessment of nitrogen deposition effects on terrestrial plant diversity: a synthesis. Ecol Appl 20:30e59. https://doi.org/10.1890/08-1140.1

Bowman WD, Cleveland CC, Halada L, Hresko J, Baron JS (2008) Negative impact of nitrogen deposition on soil buffering capacity. Nat Geosci 1:767e770. https://doi.org/10.1038/ngeo339
Brienen RJW, Phillips OL, Feldpausch TR, Gloor E, Baker TR, Lloyd J et al (2015) Long-term decline of the Amazon carbon sink. Nature 519:344–348. https://doi.org/10.1038/nature14283
Brook BW, Sodhi NS, Bradshaw CJA (2008) Synergies among extinction drivers under global change. Trends Ecol Evol 23:453–460. https://doi.org/10.1016/j.tree.2008.03.011
Brown BJ, Mitchell RJ, Graham SA (2002) Competition for pollination between an invasive species (purple loosestrife) and a native congener. Ecology 83:2328–2336. https://doi.org/10.1890/0012-9658(2002)083[2328:CFPBAI]2.0.CO;2
Clark DA, Piper SC, Keeling CD, Clark DB (2003) Tropical rainforest tree growth and atmospheric carbon dynamics linked to interannual temperature variation during 1984–2000. Proc Natl Acad Sci U S A 100:5852–5857. https://doi.org/10.1073/pnas.0935903100
Colwell RK, Brehm G, Cardelus CL, Gilman AC, Longino JT (2008) Global warming, elevational range shifts, and lowland biotic attrition in the wet tropics. Science 322:258–261. https://doi.org/10.1126/science.116254
Cowles J, Boldgiv B, Liancourt P, Petraitis P, Casper B (2018). Effects of increased temperature on plant communities depend on landscape location and precipitation (Ecol Evol 18) sup-1. https://doi.org/10.1002/ece3.3995
Cox PM, Betts RA, Jones CD, Spall SA, Totterdell IJ (2000) Acceleration of global warming due to carbon-cycle feedbacks in a coupledclimate model. Nature 408:184–187. https://doi.org/10.1038/35041539
Delipetrou P (2006) Carpobrotus edulis. ed. D.A.I.S.I.f. Europe
Dieleman WI, Vicca S, Dijkstra FA, Hagedorn F, Hovenden MJ, Larsen KS et al (2012) Simple additive effects are rare: a quantitative review of plant biomass and soil process responses to combined manipulations of CO_2 and temperature. Glob Chang Biol 18(9):2681–2693. https://doi.org/10.1111/j.1365-2486.2012.02745.x
Drossart M, Michez D, Vanderplanck M (2017) Invasive plants as potential food resource for native pollinators: a case study with two invasive species and a generalist bumble bee. Sci Rep 7:12
Du E, Fenn ME, Vries D, Wim, Ok YS (2019) Atmospheric nitrogen deposition to global forests: status, impacts and management options. Environ Pollut 250:1041048. https://doi.org/10.1016/j.envpol.2019.04.014
Duque A, Stevenson P, Feeley KJ (2015) Thermophilization of adult and juvenile tree communities in the northern tropical Andes. Proc Natl Acad Sci USA 112:10744–10749. https://doi.org/10.1073/pnas.1506570112
Ellsworth DS (1999) CO_2 enrichment in a maturing pine forest: are CO_2 exchange and water status in the canopy affected? Plant Cell Environ 22:461–472
Elmendorf SC, Henry GHR, Hollister RD, Björk RG, Boulanger-Lapointe N, Cooper EJ et al (2012) Plot-scale evidence of tundra vegetation change and links to recent summer warming. Nat Clim Chang 2:453–457. https://doi.org/10.1038/nclimate1465
Erisman JW (1990) Estimates of the roughness length at Dutch Air Quality Monitoring stations and on a grid basis over the Netherlands, Report 723001003, National Institute of Public Health and Environmental Protection, the Netherlands
Esquivel-Muelbert A, Baker TR, Dexter KG, Lewis SL, ter Steege H, Lopez-Gonzalez G et al (2017) Seasonal drought limits tree species across the Neotropics. Ecography 40:618–629. https://doi.org/10.1111/ecog.01904
Esquivel-Muelbert A et al (2019) Compositional response of Amazon forests to climate change. Glob Chang Biol 25(1):39–56. https://doi.org/10.1111/gcb.14413
Fadrique B, Báez S, Duque Á, Malizia A, Blundo C, Carilla J et al (2018) Widespread but heterogeneous responses of Andean forests to climate change. Nature 564:207. https://doi.org/10.1038/s41586-018-0715-9
Feeley KJ (2012) Distributional migrations, expansions, and contractions of tropical plant species as revealed in dated herbariumrecords. Glob Change Biol 18:1335–1341. https://doi.org/10.1111/j.1365-2486.2011.02602.x

Feeley KJ, Silman MR, Bush MB, Farfan W, Cabrera KG, Malhi Y et al (2011) Upslope migration of Andean trees. J Biogeogr 38:783–791. https://doi.org/10.1111/j.1365-2699.2010.02444.x

Feeley KJ, Hurtado J, Saatchi S, Silman MR, Clark DB (2013) Compositional shifts in Costa Rican forests due to climate-driven species migrations. Glob Change Biol 19:3472–2480. https://doi.org/10.1111/gcb.12300

Feeley KJ et al (2020) Climate-driven changes in the composition of New World plant communities. Nat Clim Chang 10(10):965–970. https://doi.org/10.1038/s41558-020-0873-2

Francis AP, Currie DJ (2003) A globally consistent richness-climate relationship for angiosperms. Am Nat 161:523–536. https://doi.org/10.1086/368223

Gallinat AS, Russo L, Melaas EK, Willis CG, Primack RB (2018) Herbarium specimens show patterns of fruiting phenology in native and invasive plant species across New England. Am J Bot 105:31–41. https://doi.org/10.1002/ajb2.1189

Galloway JN, Dentener FJ, Capone DG et al (2004) Nitrogen cycles: past, present, and future. Biogeochemistry 70:153–226. https://doi.org/10.1007/s10533-004-0370-0

Ghannoum O, von Caemmerer S, Ziska LH, Conroy JP (2000) The growth response of C4 plants to rising atmospheric CO_2 partial pressure: a reassessment. Plant Cell Environ 23:931–942

Gimenez C, Gallardo M, Thompson RB (2005) Plant–water relations. In: Daniel H (ed) Encyclopedia of soils in the environment. Elsevier, pp 231–238. ISBN 9780123485304. https://doi.org/10.1016/B0-12-348530-4/00459-8

González-Varo JP, Biesmeijer JC, Bommarco R et al (2013) Combined effects of global change pressures on animal-mediated pollination. Trends Ecol Evol 28(9):524–530. https://doi.org/10.1016/j.tree.2013.05.008

Grime JP (1974) Vegetation classification by reference to strategies. Nature 250:26–31. https://doi.org/10.1038/250026a0

Grime JP (1998) Benefits of plant diversity to ecosystems: immediate, filter and founder effects. J Ecol 86:902–910

Hoover DL, Knapp AK, Smith MD (2014) Resistance and resilience of a grassland ecosystem to climate extremes. Ecology 95:2646–2656

Knight TM, Ashman TL, Bennett JM, Burns JH, Passonneau S, Steets JA (2018) Reflections on, and visions for, the changing field of pollination ecology. Ecol Lett 21:1282–1295

Leakey ADB, Ainsworth EA et al (2009) Elevated CO2 effects on plant carbon, nitrogen, and water relations; six important lessons from FACE. J Exp Bot 60:2859–2876. https://doi.org/10.1093/jxb/erp096

Lenoir J, Svenning JC (2015) Climate-related range shifts–a global multidimensional synthesis and new research directions. Ecography 38:15–28. https://doi.org/10.1111/ecog.00967

Lilleskov E, Kuyper T, Bidartondo M, Hobbie E (2018) Atmospheric nitrogen deposition impacts on the structure and function of forest mycorrhizal communities: a review. Environ Pollut 246:148–162. https://doi.org/10.1016/j.envpol.2018.11.074

Long SP, Ainsworth EA, Rogers A, Ort DR (2004) Rising atmospheric carbon dioxide: plants FACE the future. Annu Rev Plant Biol 55:591–628. https://doi.org/10.1146/annurev.arplant.55.031903.141610

Luo Y (2007) Terrestrial carbon–cycle feedback to climate warming. Annu Rev Ecol Evol Sys 38:683–712. https://doi.org/10.1146/annurev.ecolsys.38.091206.095808

Makino A, Mae T (1999) Photosynthesis and plant growth at elevated levels of CO_2. Plant Cell Physiol 40:999–1006. https://doi.org/10.1093/oxfordjournals.pcp.a029493

McDowell NG, Allen CD (2015) Darcy's law predicts widespread forest mortality under climate warming. Nat Clim Chang 5:669–672. https://doi.org/10.1038/nclimate2641

Millennium Ecosystem Assessment (2005) Ecosystems and human well-being: synthesis. Island Press, Washington, DC

Mitchell R, Irwin R, Flanagan R, Karron J (2009) Ecology and evolution of plant-pollinator interactions. Ann Bot 103:1355–1363. https://doi.org/10.1093/aob/mcp122

Munson SM, Belnap J, Okin GS (2011) Responses of wind erosion to climate-induced vegetation changes on the Colorado plateau. Proc Natl Acad Sci 108:3854–3859. https://doi.org/10.1073/pnas.1014947108

Novoa A, Gonzalez L (2014) Impacts of *Carpobrotus edulis* (L.) N.E.Br. Oon the germination, establishment and survival of native plants: a clue for assessing its competitive strength. PLoS One 9:e107557. https://doi.org/10.1371/journal.pone.0107557

O'sullivan OS, Heskel MA, Reich PB, Tjoelker MG, Weerasinghe LK, Penillard A et al (2017) Thermal limits of leaf metabolism across biomes. Glob Change Biol 23:209–223. https://doi.org/10.1111/gcb.13477

Parmesan C (2006) Ecological and evolutionary responses to recent climate change. Ann Rev Ecol Evol Syst 37:637–669. https://doi.org/10.1146/annurev.ecolsys.37.091305.110100

Parmesan C, Hanley ME (2015) Plants and climate change: complexities and surprises. Ann Bot 116:849–864. https://doi.org/10.1093/aob/mcv169

Parmesan C, Yohe G (2003) A globally coherent fingerprint of climate change impacts across natural systems. Nature 421:37–42. https://doi.org/10.1038/nature01286

Perring MP, Diekmann M, Midolo G, Costa DS, Bernhardt-Römermann M, Otto JC et al (2018) Understanding context dependency in the response of forest understorey plant communities to nitrogen deposition. Environ Pollut 242:1787e1799

Rahel FJ, Olden JD (2008) Assessing the effects of climate change on aquatic invasive species. Conserv Biol 22:521–533. https://doi.org/10.1111/j.1523-1739.2008.00950.x

Reich PB, Hobbie SE, Lee TD, Pastore MA (2018) Unexpected reversal of C_3 versus C_4 grass response to elevated CO_2 during a 20-year field experiment. Science 360(6386):317–320. https://doi.org/10.1126/science.aas931

Ronk A, Liancourt P, Boldgiv B, Petraitis PS, Casper BB (2020) Greater effect of warming on community composition with increased precipitation and in moister landscape location. J Veg Sci 31(1):3. https://doi.org/10.1111/jvs.12813

Rowland L, Da Costa ACL, Galbraith DR, Oliveira RS, Binks OJ, Oliveira AA et al (2015) Death from drought in tropical forests is triggered by hydraulics not carbon starvation. Nature 528:119–122. https://doi.org/10.1038/nature15539

Rustad L, Campbell J, Marion G, Norby R, Mitchell M (2001) A meta-analysis of the response of soil respiration, net nitrogen mineralization, and aboveground plant growth to experimental ecosystem warming. Oecologia 126(4):543–562. https://doi.org/10.1007/s004420000544

Sage RF (2004) The evolution of C4 photosynthesis. New Phytol 161:341–370. https://doi.org/10.1111/j.1469-8137.2004.00974.x

Sala OE, Chapin FS 3rd, Armesto JJ et al (2000) Global biodiversity scenarios for the year 2100. Science 287(5459):1770–1774. https://doi.org/10.1126/science.287.5459.1770

Saxe H, Cannell MGR, Johnsen Ø, Ryan MG, Vourlitis G (2001) Tree and forest functioning in response to global warming. New Phytol 149:369–399. https://doi.org/10.1046/j.1469-8137.2001.00057.x

Schulte-Uebbing L, De Vries W (2018) Global-scale impacts of nitrogen deposition on tree carbon sequestration in tropical, temperate, and boreal forests: a meta-analysis. Glob Chang Biol 24:416e431

Šigut L, Holišová P, Klem K, Šprtová M, Calfapietra C, Marek MV et al (2015) Does long-term cultivation of saplings under elevated CO_2 concentration influence their photosynthetic response to temperature? Ann Bot 116:929–939. https://doi.org/10.1093/aob/mcv043

Sreeharsha RV, Sekhar KM, Reddy AR (2015) Delayed flowering is associated with lack of photosynthetic acclimation in pigeon pea (*Cajanus cajan* L.) grown under elevated CO_2. Plant Sci 231:82–93. https://doi.org/10.1016/j.plantsci.2014.11.012

Stephens K, Dantzler-Kyer M, Patten M, Souza L (2019) Differential responses to global change of aquatic and terrestrial invasive species: evidences from a meta-analysis. Ecosphere 10:e02680. https://doi.org/10.1002/ecs2.2680

Stevens CJ (2019) Nitrogen in the environment. Science 363:578–580

Thomas CD, Cameron A, Green RE, Bakkenes M, Beaumont L, Collingham YC, Williams SE (2004) Extinction risk from climate change. Nature 427:145–148. https://doi.org/10.1038/nature02121

Tito R, Vasconcelos HL, Feeley KJ (2018) Global climate change increases risk of crop yield losses and food insecurity in the tropical Andes. Glob Change Biol 24:e592–e602. https://doi.org/10.1111/gcb.13959

Vilà M, Espinar JL, Hejda M et al (2011) Ecological impacts of invasive alien plants: a meta-analysis of their effects on species, communities and ecosystems. Ecol Lett 14(7):702–708. https://doi.org/10.1111/j.1461-0248.2011.01628.x

de Vries W, Du E, Butterbach-Bahl, K. (2014). Short and long-term impacts of nitrogen deposition on carbon sequestration by forest ecosystems. Curr Opin Environ Sustain 9–10:90–104

Way DA, Oren R (2010) Differential responses to changes in growth temperature between trees from different functional groups and biomes: a review and synthesis of data. Tree Physiol 30:669–688. https://doi.org/10.1093/treephys/tpq015

Wiens JJ (2016) Climate-related local extinctions are already widespread among plant and animal species. PLoS Biol 14:e2001104. https://doi.org/10.1371/journal.pbio.2001104

Wu Z, Dijkstra P, Koch GW, Penuelas J, Hungate BA (2011) Responses of terrestrial ecosystems to temperature and precipitation change: a meta-analysis of experimental manipulation. Glob Chang Biol 17:927–942. https://doi.org/10.1111/j.1365-2486.2010.02302.x

Xu Z, Jiang Y, Jia B, Zhou G (2016) Elevated-CO_2 response of stomata and its dependence on environmental factors. Front Plant Sci 7:657. https://doi.org/10.3389/fpls.2016.00657

Zhu L, Bloomfield KJ, Hocart CH, Egerton JJG, O'sullivan OS, Penillard A et al (2018) Plasticity of photosynthetic heat tolerance in plants adapted to thermally contrasting biomes. Plant Cell Environ 41:1251–1262. https://doi.org/10.1111/pce.13133

Chapter 18
Impact of Climate Change on the Importance of Plant Growth-Promoting Microbes in the Rhizosphere

Archi Chaurasia, Chitrakshi Shandilya, Isabell Robert Rupa, Nitin Kumar, Ajit Varma, and Kanchan Vishwakarma

Abstract The global climatic changes have made significant alterations in the environment that can be harmful for microbes, plants, animals, and humans. Climatic parameters include the temperature, atmospheric gases, carbon dioxide levels, etc. that have been known to pose harmful effects to microbes when present in excess limits. Among microbes, plant growth-promoting rhizobacteria and arbuscular mycorrhizal fungi have been shown to be affected by climate fluctuations. Hence, this book chapter will discuss the impact of various climate parameters on microbial communities and the microbial processes going on in the soil with special emphasis on plant growth-promoting rhizospheric bacteria and arbuscular mycorrhizal fungi.

Keywords Climate change · Plant growth-promoting bacteria · Arbuscular mycorrhizal fungi · Temperature · Carbon dioxide

---

Archi Chaurasia and Chitrakshi Shandilya contributed equally with all other contributors

---

A. Chaurasia · C. Shandilya · A. Varma · K. Vishwakarma (✉)
Amity Institute of Microbial Technology, Amity University Uttar Pradesh, Noida, India
e-mail: kvishwakarma@amity.edu

I. R. Rupa
Department of Biotechnology, Periyar Maniammai Institute of Science and Technology, Thanjavur, Tamilnadu, India

N. Kumar (✉)
Department of Biotechnology, Periyar Maniammai Institute of Science and Technology, Thanjavur, Tamilnadu, India

Biotechnology Division, Central Pulp and Paper Research Institute, Saharanpur, Uttar Pradesh, India

D. K. Choudhary et al. (eds.), *Climate Change and the Microbiome*, Soil Biology 63,
https://doi.org/10.1007/978-3-030-76863-8_18

18.1 Introduction

Till the following century, it is said that the global climate will immensely change and will affect various parameters about this alteration in environment (Houghton et al. 2001), leading to continuous increase of atmospheric CO_2 concentration (IPCC Climate Change 2007), along with the prediction of global surface increase in changing temperature between 1.8 and 3.6 °C until 2100, caused due to human and natural activities to increase atmospheric CO_2 level (IPCC Climate Change 2007). It is expected for the soil water content to decrease due to the increase in temperature (Le Houerou 1996), causing drought to various parts of the world. Terrestrial microorganism will be affected by these climate changes. Along with affecting ecosystem and organisms as well as per the recent studies, it has been seen that there is well-established symbiosis of plant taxa with various types of microorganisms (Nicolson 1967). As characterized by De Bary definition, they can be neutral or pathogenic toward their host (de Bary 1879); additionally they increase plant tolerance toward abiotic and biotic stress and support plant (Bent 2006). The rhizosphere is colonized by plant growth-promoting microorganisms. The part of the soil that is attached to the surface of root and affected by exudes of root and via microorganism (Bent 2006). Microorganisms benefit their host by endophytic lifestyle by entering in the root system (Stone et al. 2000). This can be referred to as the plant growth-promoting rhizobacteria (PGPR, Kloepper and Schroth 1978) or plant growth-promoting bacteria (PGPB, Bashan and Holguin 1998). Changes in the plant physiology and root exudation are to be likely seen due to the climate change. The increase in carbon (C) allocation in the root zone with the increase level of CO_2 will lead to root exudates composition change. These changes can be chemoattractant or signaling compound along with various carbon and nitrogen ratio or nutritional presence (Kandeler 2007). The plant-associated microbial activities and diversity will be affected by the climate change (Drigo et al. 2008). Beneficial effect of microorganisms on growth of plant or its overall health might be weakened in reference to possessing their required property and their ability to colonize under specific situations. A definite comprehension on how precisely microbes associated with plants are impacted is either influenced directly by modified plant anatomy, or the way it influences plant performance and the environment functioning is yet missing. Mycorrhizal growth and endophytes just as PGPB are used as biocontrol specialists, biofertilizers, and additionally phytostimulators used in agricultural sector (Vessey 2003; Welbaum et al. 2004) or on the other hand used in phytoremediation applications as degrading microorganisms (Denton 2007). The immense effect of global climate change may alter the performance; thus appropriate study is required to choose the strains that will perform well in changed conditions. Here we will discuss about major environment parameter that are likely to be altered via climate change and impact the microbes and related processes.

18.2 Influence of Climate Change on Soil Microbial Process

Relative abundance and capacity of soil communities is adjusted by growing change in climate since soil network individuals contrast in their physiology, temperature, affectability, and development rates (Castro et al. 2010, 2012; Gray et al. 2011; Lennon et al. 2012; Briones et al. 2014; Delgado-Baquerizo et al. 2014; Whitaker et al. 2014). There has been wide range of exploring due to sudden impacts of climatic changes occurring on composition of microbial composition and its functioning (Blankinship et al. 2011; Manzoni et al. 2012; Chen et al. 2014). In a temperate forest, warming by 5 °C, for instance, expanded the bacterial to fungal proportion of the soil community (DeAngelis et al. 2015).

Changes in ecosystem functioning can be led by the shifts in microbial community when soil microbes contrast in their functional characteristics or controlling a rate-limiting or fate-controlling step (Schimel and Schaeffer 2012). Nitrification and nitrogen fixation are regulated via specific microbial groups (Isobe et al. 2011), denitrification, and methanogenesis (Bodelier et al. 2000). Alterations in the number of organisms who regulate specific process can directly affect the pace of that process. Nonetheless, a few processes are firmly connected to abiotic factors such as nitrogen mineralization, temp, and dampness than microbial community composition in the soil that a variety of life forms are in charge of these processes (Isobe et al. 2011).

The temperature-sensitive process such as microbial soil respiration rates are likely to be altered by warming effect of climate change. The extensive recent consideration is on the function of raised temperature in microbial metabolism (Compant et al. 2010). The natural temperature affectability of microbial action is characterized as the factor by which microbial activity increments with a 10 °C expansion in temperature (known as Q10). Q10 is a frequently used model in climate change to represent sensitive nature of microbes to temperature; in any case, through this a large number of associations that affect the temperature dependence of microbes in processes such as decomposition can be assessed.

Along these lines, temperature sensitivity represented via Q10 in models may give a bad representation of forecasts. During the field study, the process like decaying of soil organic matter, soil respiration, and development of microbial biomass with increase of temperature in reaction to warming are usually short-lived. The short-lived impacts have been speculated to happen as carbon substrates present in soil are expanded by microbial action and due to exchange as the soil microbial diversities display shift in composition or compel their biomass to react to adjusted conditions and substrate availability (Allison and Martiny 2008, Bradford 2013). Alteration of the composition of microbial communities can be initiated by experimental warming and can also affect the population of both gram-positive and gram-negative bacteria; soil dampness is linked to temperature change which may give clarification to contrasting outcomes from tests evaluating microbial community reactions to climate alterations. For instance, microbial activity at hotter temperatures can be restricted by dissemination and its contact with substrate (Zak et al.

1999). While bacterial community may react quickly to moisture pulse, the fungal counterpart may slow in their reaction (Bell et al. 2008; Kardol et al. 2010). Further, drought intensifies the differential temperature affectability of fungal and bacterial gatherings (Briones et al. 2014). With change in soil moisture availability ($<30\%$ decrease in water holding limit), soil fungal communities may shift starting with one dominant member onto the next, while bacterial communities stay consistent. Soil microbial communities acclimated to less water access may inspire to a lesser extent a compositional or functional move to changing water systems. Communications among organisms and background temperature and dampness systems in some random area impact microbial composition and function with evolving climate. Be that as it may, it is yet not satisfactory that (1) how temperature and sogginess and their connection impact unequivocal microbial network, for instance, methanogens inside a community, (2) what impacts microbial network changes that work like rot of new and old soil organic matter, and (3) which system drive the net climate reaction of microbial activities to the ecological change.

18.2.1 Effect of Climate Change on PGPR

Bacteria are also known to colonize the plants other than fungi (Compant et al. 2010) and possess positive effects on their host which can in turn act as beneficiary for them (Welbaum et al. 2004; Lugtenberg and Kamilova 2009). Microscopic organisms improve plant development and health by different components which incorporate plant growth substances, for example, hormones, mitigation of biotic and abiotic stresses, pathogen antagonism, and induction of systemic responses (Lugtenberg and Kamilova 2009). Even in the adverse conditions like high level of heavy metal in plants, bacterial endophytes may grow (Lodewyckx et al. 2002; Idris et al. 2004). Through specific plant genes under stress conditions, they alleviate abiotic and biotic stress (Hallmann 2001).

18.2.2 Effects of Elevated CO_2 Levels

Bacteria-associated plants can be affected by elevated CO_2 level. Drigo et al. (2008) reviewed the repercussions of increased CO_2 on various communities of rhizobacteria. Along with rhizospheric bacteria, endophytic populations may also be affected. The "Free-Air CO_2 Enrichment (FACE) experiment" held in Switzerland showed the effect of increased ambient CO_2 (Hebeisen et al. 1997). This experiment involves interaction of legumes with rhizobia that was triggered by elevated CO_2 (Marilley et al. 1999); they are known for their interaction with legumes for nitrogen fixation in addition to their plant growth-promoting activities (Sessitsch et al. 2002). It also depicted that *leguminosarum* strains were favored for the enrichment of CO_2 over the other plants (Montealegre et al. 2000). 17% more nodulation on the roots were produced by the isolates acclimated to elevated CO_2 than isolates favored by ambient CO_2 levels, telling that under various conditions, plants prefer the selection of specific microorganism (Haase et al. 2007). Nitrogen deficiency symptoms were found under the elevated CO_2 conditions in which N tissue decreased in bean. The reason behind lower N nutritional microbial growth under elevated atmospheric CO_2 concentration is relation between rhizosphere microbial growths is due to increase in competition of plant microbial nitrogen in the rhizosphere (Haase et al. 2007). This experiment provided a lot of information about PGPB such as *Pseudomonas* sp. and even on rhizobia (Marilley et al. 1999), *Actinobacteria*, and *Deltaproteobacteria*. Drigo et al. (2009) gave information on soil-specific effect of elevated CO_2 on *Pseudomonas* and *Burkholderia* sp. and on soil-particular effects on *Pseudomonas* and *Burkholderia* sp. and on the genes producing antibiotics. Elevated CO_2 has different plant responses which led to regulated photosynthate distribution in the rhizosphere. But some strong antibiotic producers such as *Actinomycetes* and *Bacillus* sp. are not influenced. Selective bacterial species responded differently to plants having association with enrichment to atmospheric CO_2 reported by Drigo et al. (2009). The dominance of *Pseudomonas* sp. was increased in presence of elevated CO_2 which includes a variety of plant growth-promoting members, associated with rye and white clover and enhancement of the abundance of *Rhizobium* sp. (Marilley et al. 1999).

Additional data of the relative amount of HCN-delivering *Pseudomonas* strains which are known as likely inhibitors of fungi taken from mass and rhizospheric soil alongside two perpetual perennial grassland root fraction systems (*L. perenne* and *Medicago coerulea*) were diminished in raised CO_2 conditions (Tarnawski and Aragno 2006). There was expanded extent of siderophores and nitrate dissimilating strains. It expresses a model that expanded atmospherically CO_2 levels may affect plant beneficial microbes, which may later have ramifications on their use in agricultural or phytoremediation processes (Fig. 18.1) (Baon et al. 1994; Sanders et al. 1998).

The investigation of practically identical detail on CO_2 restraint of bacterial development is that of King and Nagel. *Pseudomonas aeruginosa* was grown on solitary suboptimum temperature (24 °C) on minimal media and reasoned that inhibition differed directly with CO_2 concentration. A comparative relationship was seen with *P. fluorescens* developing at 30 °C at a restricted range of CO_2 concentration; significant deviation was observed. Low CO_2 fixations created marked hindrance of development in complex medium, yet stimulated growth in least amount of medium, and in complex medium, a most extreme level of restraint was achieved at generally low CO_2 concentration. It is conceivable that a more detailed examination of *P. aeruginosa* would uncover a similar reaction to CO_2.

It has been expected that this perception is relevant to microorganisms; however, plainly with *P. fluorescens*, there is an immediate improvement of CO_2 inhibition with diminishing temperature. The nonlinearity of the connection among temperature and the level of inhibition with minimal medium most likely emerge due to the competition caused due to inhibitory and stimulant impact of CO_2 during development in this medium. Since the inhibitory effect of CO_2 can fluctuate with

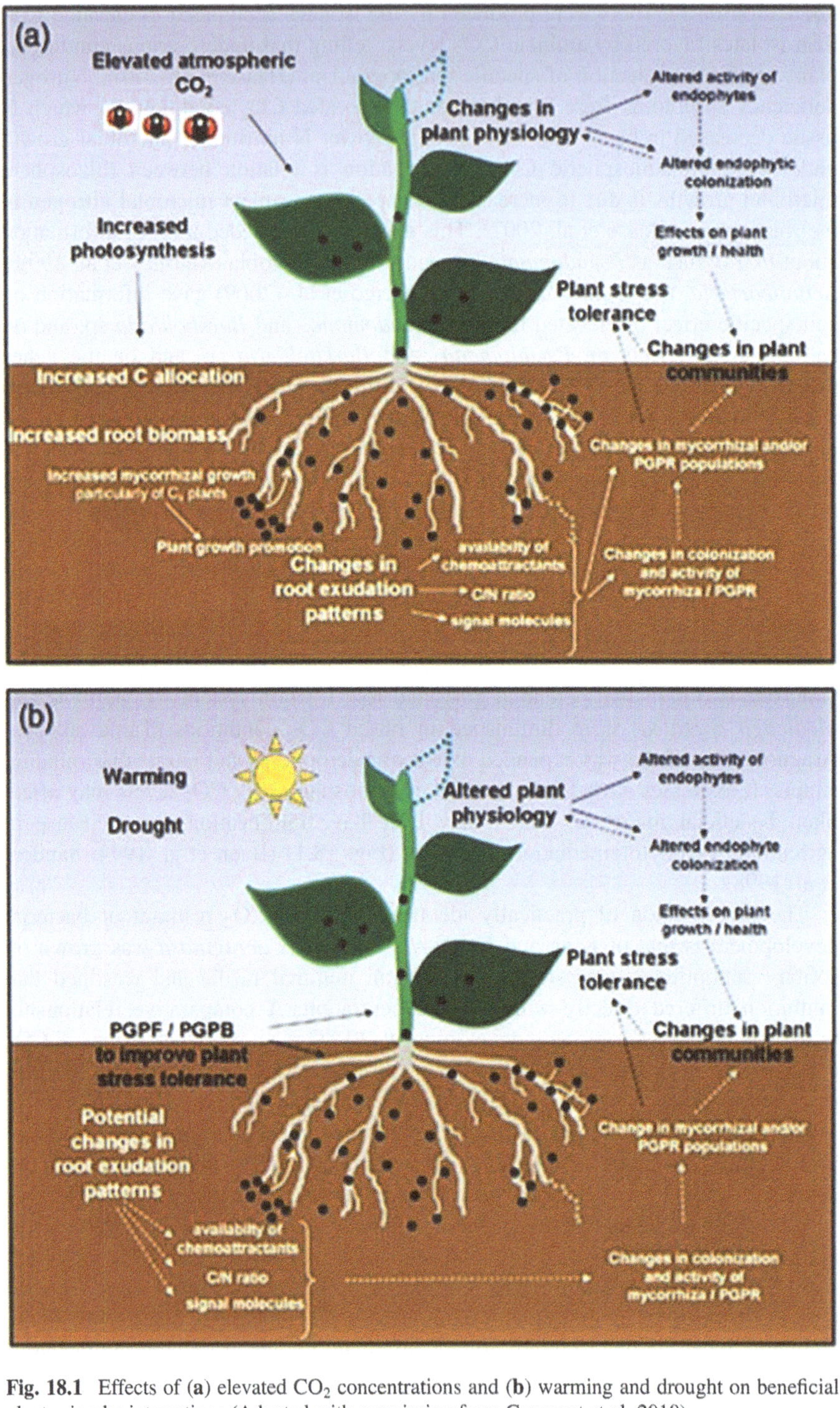

Fig. 18.1 Effects of (**a**) elevated CO_2 concentrations and (**b**) warming and drought on beneficial plant-microbe interactions (Adapted with permission from Compant et al. 2010)

temperature and medium composition, it is clearly not legitimate to look at results from changed sources except if these variables are considered. King and Nagel could not recognize any change in enzyme levels between cells of *P. aeruginosa* cultured in air and inside seeing CO_2 and stated that adaptive enzyme synthesis in the presence of CO_2 does not happen.

18.2.3 Effect of Temperature

Alleviation of temperature stress is depicted by rhizosphere bacteria and endophytes on some plants, and they may cause induction in growth of various crops under altering temperature, climates, and soil conditions (Fig. 18.2) (Bilal et al. 1993;

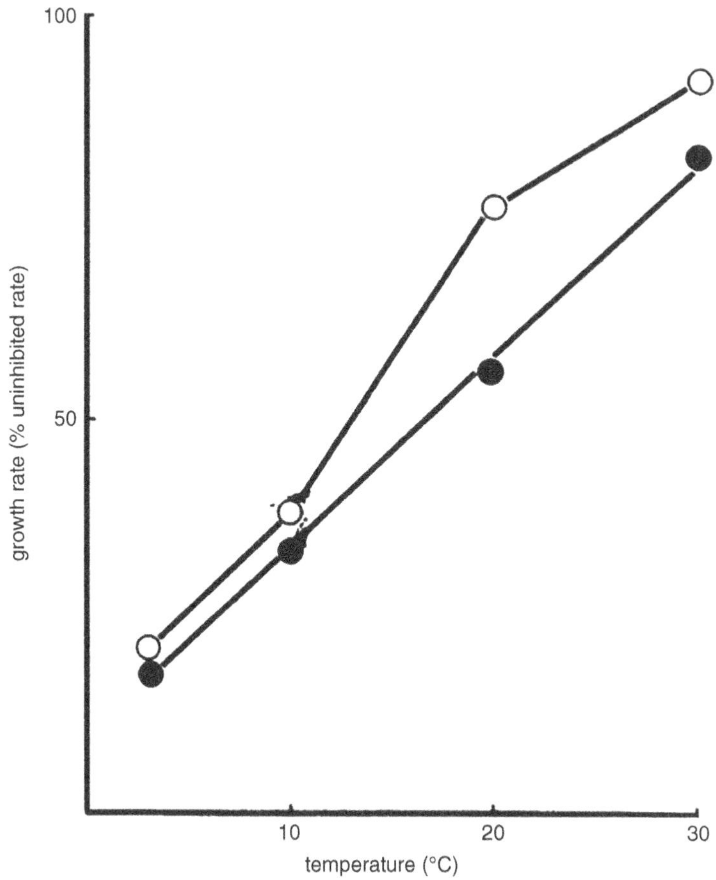

Fig. 18.2 Effect of temperature on the growth rate of *P. fluorescens* growing in simple (O) or complex medium with pCO_2 in solution at 150 mm of Hg (Adapted with permission from Gill and Tan 1979)

Bashan and Holguin 1998). Egamberdiyeva and Höflich (2003) postulated that performance of plant beneficial bacteria may be affected by temperature and soil type. *Mycobacterium* sp. 44, *Pseudomonas fluorescens*, and *Pantoea agglomerans* strain which were isolated from a semi-continental climate showed elevation growth properties of root and shoot at 16 °C in comparison to growth at 26 °C in loamy sand. But in any case, *M. phlei* strain MbP18 delivered good results under the two conditions just like *M. bullata* MpB46, showing genotypic-explicit inclinations for some natural conditions. A few strains of PGPB growing at a hot condition than at cold ones can be used in agriculture. In an assessment, it was evidently shown that rhizobia segregated from the desert wood vegetable *Prosopis glandulosa* nods were grown more noteworthy at 36 °C than at 26 °C. Moreover, microbes colonizing specific destination may react contrastingly to some biological conditions. An assessment with the endophytic microbe *Burkholderia phytofirmans* strain PsJN displayed that increasing the temperature from 10 to 30 °C diminished its establishment in the tomato rhizosphere; however endophytic abundance wasn't impacted (Pillay and Nowak 1997). When successfully colonized in rhizosphere, it may help prevent the temperature or water stress on plants (Aroca and Ruiz-Lozano 2009) by inducing a systemic response (Yang et al. 2009). This shows the expected part of specific strains for their utilization in agriculture; however there is need to consider the impacts of altering conditions on their establishment just as on their plant growth-promoting capacity.

18.3 Influence of Climate Change on Arbuscular Mycorrhizal Fungi

Mycorrhizal fungi are known as one of the crucial factors while considering global changes related to variations in the ecosystem, as these fungi are known to play a major role in the plant-soil interface. Global changes encircle natural as well as human-led changes on the Earth's environment. This section will focus mainly on anthropogenic aspects resulting in global changes related to mycorrhiza. While considering climate change, media heeds about global warming mainly, among other factors. But now people perceive that anthropogenic factors not only influence climate change, but they also steer toward many global changes (Vitousek 1994).

Human-induced changes have caused serious repercussions in the Earth's environment other than climate change. Some important examples are increased concentration of the atmospheric gases (ozone, CO_2, etc.), the ecosystem being exposed to a rising number of nutrients due to atmospheric deposition (Nitrogen), change in climate (altered precipitation as well as temperature regimes), increase in UV radiations, and occurrence of non-native species. These comprise present or upcoming global alterations, displaying good or bad impact on mycorrhizae associations. These aforementioned factors are distinct, based on the fact that mycorrhizal symbiosis involves two classes of organisms. Some factors have direct impact on the host plant solely (example carbon fixation), and some have an indirect influence on the mycorrhizal fungi (mycobionts), like change in carbon distribution by plant (atmospheric causes can affect the presence of nutrients in the soil) in turn affecting the fungi as well. Warming or an altered precipitation is included in factors that directly affect mycobionts. Understanding this difference is substantial in figuring out the mechanism behind the effects of global change factors and also for experimental proposals. An isolated environment can rarely be subjected to factors responsible for global change. We need to figure out the mechanism behind these factors in order to understand the complexity of the regional combinations of such factors, as experimental exploration of all possible scenarios will be nearly impossible in artificial or mimicked environments. For complete assessment of the impact of global changes on mycorrhiza, we need to consider patterns and processes involved in both temporal scale and spatial scale. Mostly experiments mimic short-term responses, but for a complete study, we need to have an insight about long-term responses as well. A possible solution would include in situ experiments, for example, CO_2 springs. Large-scale events such as variations in distribution of plant species globally (it includes regional extinction of a particular plant) as an adverse effect of climate change may in turn affect the mycorrhizal population dependent on that plant as it's host. Therefore, it may alter the mycorrhiza population distribution as well. Some of the mycorrhizal fungi are not dependent on the host (plant) but are still affected by changes in environment. This concept is least understood, but apparently is the most important factor to be considered when analyzing mycorrhiza's response toward global change. Global change is an intricate phenomenon, but here, we focus on how it affects the mycorrhizal fungi. Variations in the symptoms of global changes may be coined important in the future (examples are effect of urbanization, habitat fragmentation, etc.); these eventually may be included in the areas of study concerning mycorrhizal ecology. Factors which are involved in climate change which affect the mycorrhizal fungi are discussed below.

18.3.1 Effect of Warming

Soil is termed as "chief organizing center for ecosystem function", and the part played by soil biota in modifying plant responses as well as ecosystem as a whole, toward global change, is being highly acknowledged. Microbes found in rhizosphere of a plant have a major role in sending strong feedbacks based on the plant's growth (Bever 1994), so any difference in the behavior or functions of these microorganisms as a result of global changes is of significance, as it may alter plant response to certain stimuli. When noting the key role of mycorrhiza at root-soil interface as well as its effect on plant physiology, community, and ecosystem, it is evident after a detailed consideration that mycorrhiza play a major role in studies related to global changes. AMF entirely depend upon their host (plant) for carbon requirement. Therefore, if the plant is affected by global changes, it will indirectly affect the AMF as well. Mycorrhiza will be considered important, if there is a notable interrelation between global changes and presence of symbiosis, i.e., if the symbiont's presence or its function affects the response of plant communities or the ecosystem. Mycorrhizae are present abundantly in natural ecosystems, but in the case of habitat fragmentation or change in land utilization patterns, it may lead to decrease in the population of the fungi. This can contribute as another factor for global change (Vitousek 1994). It is important to know the inter-relation between factors leading to global changes and the presence of mycorrhiza. It will give a better insight on the underlying mechanism of variations occurring in the ecosystem as a response to global changes. The global change factors can affect the ecosystem in complex ways resulting in numerous variations even at regional levels, so it is important to prioritize understanding the mechanism behind it than to document the variations or changes. Based on these circumstances, mycorrhizal fungi (or fungi in general) has crucial role in maintaining and forming the soil structure, thereby benefiting the ecosystem (Tisdall and Oades 1982). If the extraradical mycelium of the mycorrhizal fungi is affected by any factors influenced by global change, it may indirectly impact the soil structure as well (Young et al. 1998; Rillig et al. 1999). Any change in soil structure will affect water infiltration ability, soilborne aspects involved in biogeochemical cycle operations, maintenance of sustainable agriculture, and resistance to soil erosion (Oades 1984; Elliott and Coleman 1988; Hartge and Stewart 1995; Jastrow and Miller 1997). Group of hyphae contribute to forming the mycelia of mycorrhizal fungi, which have a potential effect on the ecosystem. Based on these facts, mycorrhiza should be included in the research conducted to further understand the effects of global change.

18.3.2 Effect of Elevated CO_2 Levels

Approximately 90% of the plants form firm associations with mycorrhiza (Smith and Read 1997). Immediate need has been felt of understanding the impact of carbon (C) due to its increase in atmosphere in the form of carbon dioxide, on symbioses of mycorrhiza. Carbon is the main component for functioning of AM associations which are fixed in photosynthesis and has been seen to move from plant to fungus. Environment rich in carbon dioxide is the main factor for existence of fungus, be it inside the root or in the soil. Thus, if elevated atmospheric carbon increases the fixation of carbon and supply of it to roots then it might help in the promotion of fungal growth. This could lead to the consequence of increased fungal performance, which is best characterized for uptake of phosphate, increased uptake of nutrients rich in minerals, and resistance toward many phytopathogens and drought as well (Newsham et al. 1995). Elevated carbon dioxide can also have a positive aspect in which elevation can lead to enhanced fixation of carbon followed by increased carbon transfer to root system. Improved uptake of phosphate with increased mycorrhizal fungal growth due to elevated fixation of carbon might lead to efficient functioning of plants leading to relief in potential deficiency which could have otherwise caused restriction in ability of responding to increased level of atmospheric carbon (suggestion by photosynthetic downregulation alleviation) (Staddon et al. 1999a). Important and essential processes in the plant metabolism of phosphorus are determined by carbon availability; these metabolisms include uptake, transportation, and utilization. In consideration, photosynthetic rate is determined by status of phosphorus of leaf. The positive feedback which could be possible eventually has constrained by few deficiencies, for example, nitrogen level and water content; thus it can act as a controlling or limiting factor. The main reason for understanding the responses of AM fungi is because they have potential of regulating response of plant toward climatic change; this is quite evident from the interface representation of AM fungi in soil-plant system. Link in transfer chain by which movement of carbon is seen from plant to soil is performed by mycorrhizal fungi, which makes it another major feature of interest (Staddon et al. 1999b). Thus, AM fungi can potentially influence cycling rates of carbon and can even use this structural carbon in construction of extraradical mycelium which is transferred to symbiotic AM fungus. The fine network formed by these mycelia has a faster turnover rate as compared to either root or shoot material, and fungal carbon is therefore a medium for a rapid movement of carbon back to atmosphere, whereas some of the fungal compounds might adapt and become resistant to attack of microbial community, thus entering the slow pool of recalcitrant soil carbon which leads to retardation in carbon cycle (Treseder and Allen 2000). This behavior of some fungal compound can lead to the consequence of either increased accumulated or reduced carbon content in plant soils. The consequence of elevated carbon dioxide in atmosphere for functioning of mycorrhiza could be changes in way of promotion in plant growth or changes in cycling rates of carbon. These effects are considered to be indirect, which is regulated by responses created by plants due to presence of carbon dioxide. Promotion in fungal ability to provide phosphorus to plants and enhance plant growth can be done by elevated carbon fixation and increased availability of carbon to fungus.

With respect to effects of mycorrhiza, elevation in carbon dioxide is considered as the best researched factor of global change. Increase in biomass of root of nitrogen fixing plant by 114%, in non-mycorrhizal individuals, was observed under elevation of carbon dioxide. But reduced or slower rate by 31% of effect was seen when AM species of genus *Glomus* was considered. Mycorrhizal fungi are a means of nutrient accession for plants. Uptake of inorganic phosphorus (Smith and Read 1997) and nitrogen (Ames et al. 1983; Tobar et al. 1994) is enhanced by AM fungi. Access to pool of organic nutrients in soil is usually provided by ecto-mycorrhizal fungi (Smith and Read 1997). Carbon source is required by fungi and uses an important amount of net photosynthate of plant which is usually about 10–20% (Allen 1991). As it is seen that capacity of photosynthesis, efficiency of water use, and growth are enhanced in plants which are exposed to elevation of carbon dioxide (Bazzaz 1990), then increase in investment of mycorrhizal fungi of amount of available carbon can be possible. Along with the rise of carbon demand, demand of nitrogen, phosphorus also rises. Therefore, it is assumed that allocation of resources for plants is done efficiently according to economic theories (Bloom et al. 1985; Read 1991); elevated carbon dioxide might increase mycorrhizal fungi investment. To study the effect of elevated carbon in atmosphere, numerous workers experimented on plants by growing them at ambient (usually 350 μl^{1-}) as well as at elevated levels (variously 500, 600, 610, 700, and 710 μl^{1-}) (eCO_2) of atmospheric carbon dioxide (aCO_2) and measured root colonization degree by AM fungi (Staddon and Fitter 1998). Appropriate evidence has not been found from results acquired from experiment. Nevertheless, rarely seen case was about decreased colonization, and null to increased responses are found equally more frequent. Examination of morphological structures (hyphae, arbuscules, and, where appropriate, vesicles) of fungi was done more efficiently by workers who participated in the experiment. Later, they reported all possible combinations of decrease, increase, and null responses. Argument was made by Staddon and Fitter (1998) on effects which were due to serious flaw made in experiment; according to the researcher as elevated CO_2 was observed to have a positive effect on growth of plants, he suggested that if plants show the same level of growth in both atmospheric and elevated carbon dioxide concentrations, then there must be some flaw in conducting experiment. As we know that plant characteristic does not include nature and degree of root mycorrhizal colonization, instead are dependent on conditions of plants in which they are grown; thus various changes in colonization were expected (Staddon 1998).

18.3.3 Effect of Temperature

Slow or poor spore germination was observed at RT (20–23 °C) when continued experiments were performed specially to obtain axenic growth of various species of AMF. Several temperatures were compared to study its impact on spore germination and incidence of contaminating bacteria and fungi which are in association with surface-sterilized spores. Results were acquired with only those species that gave consistent spore germination rate, although experiments were conducted for evaluating various species of VA mycorrhizal fungi as well. When pot cultures were maintained with VA mycorrhizal fungi, then spores of respective fungi were obtained. As it was possible to maintain few isolates from various samples on pot cultures, it was used to maintain few isolates on pepper and soybean pot cultures. An isolate of *Gigaspora coralloidea* Trappe, Gerdemann and Ho and *G. heterogama* (Nicol. and Gerd.) Gerdemann and Trappe obtained from soybean in Florida were maintained on Bahia grass (*Paspalum notatum* Flugge) pot cultures. An isolate of *Glomus mosseae* (Nicol. and Gerd.) Gerdemann and Trappe from Washington was grown on pepper (*Capsicum annum* L.) pot cultures. Sieving and decanting techniques were used to remove spores from pot culture soil and roots. Surface sterilization was done of spores by using surface sterilizing agent that is 0.5% of sodium hypochlorite for three rounds and was later rinsed two times in a deionized sterile water before transfer to either soil extract agar or Mosse's medium No. 16 (Mosse and Phillips 1971). Preparation of soil extract agar was done by boiling finely sieved sandy loam of amount 200 g in a liter of deionized water. Later filtration was performed under vacuum, and addition of 15 g agar was done to a liter. Small petri dishes were used for transferring spores and incubating them at desired temperatures; petri plates were incubated after transferring it in polyethylene bag to prevent dehydration and were later covered with aluminum foil for excluding light. Contamination from fungi and bacteria along with germination of spores was noted at regular intervals for 2–3 times. Three replicates were used of each treatment per temperature, and six to nine spores were observed at every replication. Studies were carried out with *Gigaspora coralloidea*, *G. heterogama*, and *Glomus mosseae* and were repeated three times, twice, and once, respectively. Maximum germination was obtained at temperature of 34 °C of two isolates taken from Florida (*Gigaspora coralloidea* and *G. heterogama*), of Washington isolate (*Glomus mosseae*) was 20 °C, and no germination was observed at the same temperature or below for *Gigaspora heteroganza* spores. Most surface sterilized spores were found to be contaminated by fungi and bacteria highly at a temperature of 25 °C. *Gigaspora heterogama* spores were highly contaminated with fungi, while no fungal association with spores of *Glomus mosseae* was found. Light was found to be a negative factor for growth of spores, and it was evident that maximum more growth of both *Gigaspora coralloide*a (at temperature of 20 and 25 °C) and *G. heterogama* (at temperature of 34 °C) was observed. Occurrence of contamination through bacteria was not found to differ on the basis of presence of light, whereas fungal contamination was observed to occur on surface-sterilized spores which were exposed to light (× 14.8%). Study conducted by Schenck and Schroder (1974) showed the results of maximum sporulation and vesicle formation of Florida isolate of *Gigaspora calospora* (Nicol. and Gerd.) Gerdemann and Trappe on soybean pot culture [*Glycine max* (L.) Merrill] at temperature of 35 °C. Results of spore germination gave endorse evidence that adaptation at high temperature occurs in Florida isolates *Gigaspora* spp. (Schenck and Schroder 1974). It was also assumed that northern isolates of endomycorrhizal fungi, such as the Washington isolate of *Glomus mosseae*, may be more adapted to lower temperatures.

18.3.4 Effect of Gases on Spore Germination of AM Fungi

Production of aqueous extracts or suspensions was done for testing various effects of beet root HT material on spore germination of AM fungi; the suspension was made by forming a mixture of HTC material with deionized water thereby making adjustments in pH to 6.0, later autoclaved for about 20 min at a temperature of 121 °C. Since pH is found to influence germination of spores, thus it was kept constant (Green et al. 1976). Concentrations used were 0.05, 0.25, 1.25 g HTC 100 mL^{-1} and the parent material (also ground and sieved to 500 m) 1.25 g beet root chips 100 mL^{-1}. Addition of 12 milliliters of suspensions were performed under axenic conditions, to each petri dish which were filled prior with sterilized silica sand (48 g) for total of 50 units of experiment ($n = 10$). Small discs were cut of nitrous cellulose with a help of paper hole punch after sterilizing them with 70% of ethanol. Preparation of spores was done according to the description made by Antunes et al.

(2008). Paraffin tape was used to seal all petri dish, and later petri dishes were kept for incubation at 25 °C for 4 days at random positions. After the completion of incubation duration, spores were stained using 0.05% (w/v); trypan blue examination was done using dissecting microscope (40×); thus, hyphal growth was evident, which was carefully observed, and distinguished germination hyphae from already present hyphae found attached to spores. Previous trails found showed various germination ranges, and it was concluded that a spore is considered non-viable if germination does not take place within duration of 48 h under the provided conditions (Antunes et al. 2008). This experiment finally concluded that range of spore germination is from 50 to 80% across all treatments, besides that significant difference was found in effect of treatments on germination of spores ($p < 0.0001$). Increased germination of spores than parent material was recorded after application of all three HTC treatments. But this experiment had a limitation that addition of only highest rate showed effective high germination of spores than no-addition control. Parent material was found to have negative effect on germination as decrease of spore germination was observed. Thus, evidence was confirmed by concluding deleterious effect of HTC material on growth of plants, but at least symbionts like AM fungi are found to be less sensitive than plants and more stimulated under higher rates of addition.

18.4 Impact of Climate Change in Plant-Microbe Interactions

Elucidation of physical processes occurring in upper soil layers can be done by studying climatic parameters, and atmospheric region at lower level plays a significant role in determining the climate for the local or regional biosphere (Monteith and Unsworth 2007). Migration of plant species to more altitudes and latitudes (Grabherr et al. 1994; Walther et al. 2002; Parmesan and Yohe 2003) has been observed with warming; also these species are leafing out and flowering in the growing season to its earliest (Cleland et al. 2007; Wolkovich et al. 2012) and even seen to be creating alterations in the expression of functional traits (Walker et al. 2006; Hudson et al. 2011; Verheijen et al. 2015). Observations were made when scaling up happened to community level; shrubification of the arctic was seen as a result of warming as woody shrubs have been replaced grasses and forbs in several regions which lead to carbon feedbacks in system and transformation in properties of ecosystem (Sturm et al. 2001; Hinzman et al. 2005; Lawrence and Swenson 2011; Pearson et al. 2013). Plant community transitions may be facilitated or retarded by soil communities, especially the ones which are tightly coupled with plants. For instance, plant survival, phenotype, and expression of functional traits, that are responsible for climatic change, can be strongly influenced by root-associated microbe communities (Van der Heijden et al. 1998; Friesen et al. 2011; Wagner et al. 2014). There is no clear understanding made till date on plant-plant interactions, their association with microbial community, and ecosystem functioning due to climatic changes (Fischer et al. 2014; Mohan et al. 2014). Induction in major shifts in bulk of microbial community by climatic alterations have shown extended effects on establishments and performance of plants as well as balance of carbon in the soil. Responses of ecosystem will be dependent on association between plants and soil microbes, and if alterations are made in soil microbial communities, due to climatic change and plant growth, establishment of plant species is also determined by these changes. Buffering of plants to drought stress can be done by rapid responses of surrounding soil community which is becoming a suggestion by recent studies (Lau and Lennon 2012). Microbial diversity changes are responsible for alteration of functional traits in plants (Lau and Lennon 2011). Significant difference is there in direct and indirect impact of climate change on plants, bulk soil community, and their associated soil communities (Kardol et al. 2010). For instance, Kardol et al. (2010) found out that alterations of soil community and its functions in TN old field occurred with changes in precipitation, but variations were observed with respect to precipitation on soil microbes, its functioning, and composition on basis of location from where the sample soil was collected. Moreover, response of being relatively muted was seen from soils which were collected from different locations, and these were homogenized to access to climatic change impact. Results suggested that offset responses of soil ecosystem to climatic changes could be expected if shift in composition of plant community due to climate change occurs. Therefore, underrepresentation of these soils which are collected across plant species has been seen in most of the studies to cause shift in functional and communal plant species (Kardol et al. 2010).

Temperature Alleviation of negative impact of temperature stress on plants could be done by few organisms such as rhizospheric bacteria and endophytes; these organisms also help in expansion of capability of host plants to show growth at different ranges of temperatures. Example of the symbiosis of the organisms for allowance of both organisms to grow at high temperatures of soil is symbiosis between tropical panic grass *Dichanthelium lanuginosum* and the fungus *Curvularia protuberata*, but neither of them can survive independently at such high temperatures (Marquez et al. 2007). Infection of fungus by thermal tolerance virus, that is *Curvularia*, is required to confer ability of *C. protuberata*, for tolerating heat to the host plant (Marquez et al. 2007). Observation on tomato was made that in addition to panic grass, it could also have *C. protuberata*-mediated heat tolerance (Rodriguez et al. 2008); thus it could suggest that broad application of this underlying mechanism to diversity of plants could cope with increased temperatures. There are some organisms which can even help plants in coping up with different multiple stresses. An intriguing example of this is *Burkholderia phytofirmans* strain PsJN, which helps in improvement in tomato for plant tolerance to heat, cold in grapevine, water stress in wheat, and salinity along with freezing in *Arabidopsis* (Issa et al. 2018; Miotto-Vilanova et al. 2016). This bacterium is not only capable of creating heat tolerance but also has capability of acting as antifungal and can act as a plant defense mechanism primely and formation of better mobilization resource in plants

(Miotto-Vilanova et al. 2016; Timmermann et al. 2017). The mechanism remains elucidated by which PsJN confers multistress tolerance, and this elucidation can act as an interesting topic for crop improvement which will be mediated by microbes.

Drought Water has an essential part in life on Earth, and variation in the amount of water that is too little (lead to osmotic stress) or too much (leads to flooding) can have great impact in many areas of biology of microbes and plants. Regulation of phytohormone level, ABA, occurs during the deficiency of water in plants. Increase of ABA is responsible for triggering a cascade of signals, which results in reprogramming of transpiration at large scale and changes in physiology, which may include stomatal closure for reducing transpiration (Zhu 2016). Studies have been conducted on *Arabidopsis* which showed that pathogens specifically bacterial, such as *P. syringae* or PAMPs, like flg22 (a 22-amino-acid epitope of *Pseudomonas flagellum*), can be perceived by FLS2, which results in closure of stomata, thus helping in reduction of pathogenic entry (Melotto et al. 2006). Thus, stomatal closure, induced by ABA, might help in reduction of bacterial entry through stomata during drought stress. Suppression of SA signaling pathway in mesophyll cell of leaf can occur on elevation of ABA, and thus it can lead to decrease in post-invasion resistance mediated by SA (Jiang et al. 2010). Plant-microbiome is also affected by drought. Increase in intensity of drought leads to increased effect on microbial community composition (bulk soil, rhizosphere, and root endosphere) which resulted in immense intimate association of community with the root, also greater shift of composition in stressed rice plants which was found out by Santos-Medellın et al. (2017). Xu et al. (2018) found out in a study which examined influence of soil moisture on microbiome of sorghum root that there was significant reduction in rhizospheric diversity in root endosphere, while bacterial community diversity in soil surrounding remains mostly unaltered.

Nutritional Status Plant-microbe interaction-driven ultimate force is acquisition of nutrients. There is a significant effect on plant-microbe interactions, and this is possible due to plant nutritional level and availability of nutrients in the environment.

Phosphate Regulation of complex symbiotic association between land plants and phosphate accessing AMF is done by phosphate status in soil and its presence in the plants (Müller and Harrison 2019). Therefore, plant AMF interactions are out of the reach of effect of phosphate on plant-microbe interactions. *Arabidopsis thaliana* is considered to be a non-host for AMF that acquires phosphate (Fernandez et al. 2019). Identification was done by Hiruma et al. (2016) of a natural fungus which is endophytic in nature, known as *Colletotrichum tofieldiae* (Ct), isolated from wild *Arabidopsis* found in central Spain. Ct is responsible for transfer of phosphate to Arabidopsis and for promotion of fertility and growth of plants. Promotion of growth which is mediated by Ct is also observed when study was conducted on plants grown under conditions having deficiency of phosphate.

Nitrogen Symbiotic nodules are formed in root as a result of interactions between legumes and *Rhizobium* spp. Thus, it represents a biological process in which conversion of inert atmospheric N_2 occurs biologically into a useful NH_3 for facilitating growth and development of plants. Although this reaction is expensive energetically, the formation of nodules by legume host may act as cost-effective approach when nitrogen-rich environment is required for growing plants (Morgan et al. 2005). Susceptible state of rhizobium symbiotic interaction is maintained by communication of shoots with roots under conditions having deficiency of nitrogen.

Iron Induced systemic resistance is a form of plant immunity which is triggered by few microbes which live in rhizosphere, forming a mutual relation with prime host, against potential attacks from pathogens. Mutant analysis and microarray were conducted for identifying Arabidopsis transcription factor MYB72 which act as a key regulator of ISR (Van der Ent et al. 2008). Induction of MYB72 factor expression occurs in roots during the deficiency of iron (Buckhout et al. 2009).

18.5 Conclusion

Human-induced changes have led to serious repercussions in the Earth's environment other than climate change. Some important examples are increased concentration of the atmospheric gases (ozone, CO_2, etc.), the ecosystem being exposed to a rising number of nutrients due to atmospheric deposition (nitrogen), change in climate (altered precipitation as well as temperature regimes), increase in UV radiations, and occurrence of non-native species. These components comprise present or upcoming global changes, which can have a positive or negative impact on microbial associations and their existence. Hence, it is necessary to control the climatic alterations and overcome the harmful effects caused to the beneficial microbes such as plant growth-promoting rhizobacteria and arbuscular mycorrhizal fungi.

Acknowledgments Authors would like to acknowledge Amity University for providing necessary facility for writing chapter.

References

Allen MF (1991) The ecology of mycorrhiza. Cambridge Univ Press, Cambridge

Allison SD, Martiny JBH (2008) Resistance, resilience, and redundancy in microbial communities. Proc Nat Acad Sci USA 105:11512–11519

Ames RN, Reid CPP, Porter LK, Cambardella C (1983) Hyphal uptake and transport of nitrogen from two lsN-labeled sources by Glomus mosseae, a vesicular arbuscular mycorrhizal fungus. New Phytol 95:381–396

Antunes PM, Miller J, Carvalho LM, Klironomos JN, Newman JA (2008) Even after death the endophytic fungus of *Schedonorus phoenix* reduces the arbuscular mycorrhizas of other plants. Funct Ecol 22:912–918

Aroca R, Ruiz-Lozano JM (2009) In: Lichtfouse E (ed) Induction of plant tolerance to semi-arid environments by beneficial soil microorganisms –a review. Climate change, intercropping, pest control and beneficial microorganisms, sustainable agriculture reviews, vol 2. Springer, Dordrecht, pp 121–135

Baon JB, Smith SE, Alston AM (1994) Phosphorus uptake and growth of barley as affected by soil temperature and mycorrhizal infection. J Plant Nutr 17:479–492

Bashan Y, Holguin G (1998) Proposal for the division of plant growth-promoting rhizobacteria into two classifications: biocontrol-PGPB (plant growth-promoting bacteria) and PGPB. Soil Biol Biochem 30:1225–1228

Bazzaz FA (1990) The response of natural ecosystems to the rising global CO_2 levels. Annu Rev Ecol Syst 21:167–196

Bell C, McIntyre N, Cox S, Tissue D, Zak J (2008) Soil microbial responses to temporal variations of moisture and temperature in a Chihuahuan Desert grassland. Microb Ecol 56:153–167

Bent E (2006) In: Tuzun S, Bent E (eds) Induced systemic resistance mediated by plant growth-promoting rhizobacteria (PGPR) and fungi (PGPF). Multigenic and induced systemic resistance in plants. Springer, Berlin, pp 225–258

Bever JD (1994) Feedback between plants and their soil communities in an old field community. Ecology 75:1965–1977

Bilal R, Rasul G, Arshad M, Malik KA (1993) Attachment, colonization and proliferation of Azospirillum brasilense and Enterobacter spp. on root surface of grasses. World J Microb Biot 9:63–69

Blankinship JC, Niklaus PA, Hungate BA (2011) A meta-analysis of responses of soil biota to global change. Oecologia 165:553–565

Bloom AJ, Chapin FS III, Mooney HA (1985) Resource limitation in plants – an economic analogy. Annu Rev Ecol Syst 16:363–393

Bodelier PLE, Roslev P, Henckel T, Frenzel P (2000) Stimulation by ammonium-based fertilizers of methane oxidation in soil around rice roots. Nature 403:421–424

Bradford MA (2013) Thermal adaptation of decomposer communities in warming soils. Front Microbiol. https://doi.org/10.3389/fmicb.2013.00333

Briones MJI, McNamara NP, Poskitt J, Crow SE, Ostle NJ (2014) Interactive biotic and abiotic regulators of soil carbon cycling: evidence from controlled climate experiments on peatland and boreal soils. Glob Chang Biol 20:2971–2982

Buckhout TJ, Yang TJW, Schmidt W (2009) Early iron-deficiency- induced transcriptional changes in Arabidopsis roots as revealed by microarray analyses. BMC Genomics 10:147

Castro HF, Classen AT, Austin EE, Norby RJ, Schadt CW (2010) Soil microbial community responses to multiple experimental climate change drivers. Appl Environ Microbiol 76:999–1007

Castro HF, Classen AT, Austin EE, Crawford KM, Schadt CW (2012) Development and validation of a citrate synthase directed quantitative PCR marker for soil bacterial communities. Appl Soil Ecol 61:69–75

Chen S, Zou J, Hu Z, Chen H, Lu Y (2014) Global annual soil respiration in relation to climate, soil properties and vegetation characteristics: summary of available data. Agric For Meteorol 198:335–345

Cleland EE, Chuine I, Menzel A, Mooney HA, Schwartz MD (2007) Shifting plant phenology in response to global change. Trends Ecol Evol 22:357–365

Compant S, Van Der Heijden MG, Sessitsch A (2010) Climate change effects on beneficial plant–microorganism interactions. FEMS Microbiol Ecol 73:197–214

de Bary AH (1879) De la symbiose. Rev Int Sci 3:301–309

DeAngelis KM, Pold G, Topcuoglu BD, van Diepen LTA, Varney RM, Blanchard JL, Melillo J, Frey SD (2015) Long-term forest soil warming alters microbial communities in temperate forest soils. Front Microbiol 6:104

Delgado-Baquerizo M, Maestre FT, Escolar C, Gallardo A, Ochoa V, Gozalo B, Prado Comesana A (2014) Direct and indirect impacts of climate change on microbial and biocrust communities alter the resistance of the N cycle in a semiarid grassland. J Ecol 102:1592–1605

Denton BD (2007) Advances in phytoremediation of heavy metals using plant growth promoting bacteria and fungi. Microbiol Mol Genet 3:1–5

Drigo B, Kowalchuk GA, van Veen JA (2008) Climate change goes underground: effects of elevated atmospheric CO2 on microbial community structure and activities in the rhizosphere. Biol Fert Soils 44:667–679

Drigo B, van Veen JA, Kowalchuk GA (2009) Specific rhizosphere bacterial and fungal groups respond to elevated atmospheric CO2. ISME J 3:1204–1217

Egamberdiyeva D, Höflich G (2003) Influence of growth promoting bacteria on the growth of wheat in different soils and temperatures. Soil Biol Biochem 35:973–978

Elliott ET, Coleman DC (1988) Let the soil work for us. Ecol Bull 39:23–32

Fernandez I, Cosme M, Stringlis IA, Yu K, de Jonge R, van Wees SM, Pozo MJ, Pieterse CMJ, van der Heijden MGA (2019) Molecular dialogue between arbuscular mycorrhizal fungi and the nonhost plant *Arabidopsis thaliana* switches from initial detection to antagonism. New Phytol 223:867–887

Fischer DG, Chapman SK, Classen AT, Gehring CA, Grady KC, Schweitzer JA, Whitham TG (2014) Marschner review: plant genetic effects on soils under climate change. Plant Soil 37:91–19

Friesen ML, Porter SS, Stark SC, von Wettberg EJ, Sachs JL, Martinez-Romero E (2011) Microbially mediated plant functional traits. Annu Rev Ecol Evol Syst 42:23–46

Gill CO, Tan KH (1979) Effect of carbon dioxide on growth of *Pseudomonas fluorescens*. Appl Environ Microbiol 38:237–240

Grabherr G, Gottfried M, Pauli H (1994) Climate effects on mountain plants. Nature 369:448–448

Gray SB, Classen AT, Kardol P, Yermakov Z, Miller RM (2011) Multiple climate change factors interact to alter soil microbial community structure in an old-field ecosystem. Soil Sci Soc Am J 75:2217–2226

Green NE, Graham SO, Schenk NC (1976) The influence of pH on the germination of vesicular–arbuscular mycorrhizal spores. Mycologia 68:929–934

Haase S, Neumann G, Kania A, Kuzyakov Y, Römheld V, Kandeler E (2007) Elevation of atmospheric CO_2 and Nnutritional status modify nodulation, nodule-carbon supply, and root exudation of *Phaseolus vulgaris* L. Soil Biol Biochem 39:2208–2221

Hallmann J (2001) In: Jeger MJ, Spence NJ (eds) Plant interactions with endophytic bacteria. Biotic interactions in plant–pathogen associations. CABI Publishing, Wallingford, UK, pp 87–119

Hartge KH, Stewart BA (1995) Soil structure: its development and function. Advances in soil sciences. CRC/Lewis Publishers, Boca Raton

Hebeisen T, Lüscher A, Nösberger J (1997) Effects of elevated atmospheric CO_2 and nitrogen fertilisation on yield of Trifolium repens and *Lolium perenne*. Acta Oecol 18:277–284

Hinzman LD et al (2005) Evidence and implications of recent climate change in northern Alaska and other arctic regions. Clim Chang 72:251–298

Hiruma K, Gerlach N, Sacristán S, Nakano RT, Hacquard S, Kracher B, Neumann U, Ramírez D, Bucher M, O'Connell RJ, Schulze-Lefert P (2016) Root endophyte *Colletotrichum tofieldiae* confers plant fitness benefits that are phosphate status dependent. Cell 165:464–474

Houghton JT, Ding Y, Griggs DJ, Noguer M, van der Linden PJ, Xiaosu D (2001) In: Houghton JT, Ding Y, Griggs DJ, Noguer M, van der Linder PJ, Dai X, Maskell K, Johnson CA (eds) Climate change 2001: the scientific basis. Cambridge University Press, Cambridge, pp 1–83

Hudson JMG, Henry GHR, Cornwell WK (2011) Taller and larger: shifts in Arctic tundra leaf traits after 16 years of experimental warming. Glob Chang Biol 17:1013–1021

Idris R, Trifonova R, Puschenreiter M, Wenzel WW, Sessitsch A (2004) Bacterial communities associated with flowering plants of the Ni-hyperaccumulator Thlaspi goesingense. Appl Environ Microb 70:2667–2677

IPCC Climate Change (2007) Synthesis report. Summary for policymakers. Available at http://www.ipcc.ch Accessed November 2007

Isobe K, Koba K, Otsuka S, Senoo K (2011) Nitrification and nitrifying microbial communities in forest soils. J For Res 16:351–362

Issa A, Esmaeel Q, Sanchez L, Courteaux B, Guise J-F, Gibon Y, Ballias P, Clément C, Jacquard C, Vaillant-Gaveau N, Aït Barka E (2018) Impacts of *Paraburkholderia phytofirmans* strain PsJN on tomato (*Lycopersicon esculentum* L.) under high temperature. Front Plant Sci 9:1397

Jastrow JD, Miller RM (1997) Soil aggregate stabilization and carbon sequestration: feedbacks through organomineral associations. In: Lal R, Kimble JM, Follett RF, Stewart BA (eds) Soil processes and the carbon cycle. CRC Press, Boca Raton, pp 207–223

Jiang C-J, Shimono M, Sugano S, Kojima M, Yazawa K, Yoshida R, Inoue H, Hayashi N, Sakakibara H, Takatsuji H (2010) Abscisic acid interacts antagonistically with salicylic acid signaling pathway in rice- *Magnaporthe grisea* interaction. Mol Plant-Microbe Interact 23:791–798

Kandeler E (2007) Elevation of atmospheric CO2 and N nutritional status modify nodulation, nodule-carbon supply, and root exudation of *Phaseolus vulgaris* L. Soil Biol Biochem 39:2208–2221

Kardol P, Cregger MA, Campany CE, Classen AT (2010) Soil ecosystem functioning under climate change: plant species and community effects. Ecology 91:767–781

Kloepper JW, Schroth MN (1978) Plant growth-promoting rhizobacteria on radishes. In: station de Pathologie V'eg'etale et Phyto-Bact'eriologie (ed) Proceedings of the 4th international conference on plant pathogenic Bacteria, vol 2. Gilbert-Clarey, Tours, France, pp 879–882

Lau JA, Lennon JT (2011) Evolutionary ecology of plant-microbe interactions: soil microbial structure alters selection on plant traits. New Phytol 192:215–224

Lau JA, Lennon JT (2012) Rapid responses of soil microorganisms improve plant fitness in novel environments. Proc Nat Acad Sci USA 109:14058–14062

Lawrence DM, Swenson SC (2011) Permafrost response to increasing Arctic shrub abundance depends on the relative influence of shrubs on local soil cooling versus large-scale climate warming. Environ Res Lett 6:8

Le Houerou HN (1996) Climate change, drought and desertification. J Arid Environ 34:133–185

Lennon JT, Aanderud ZT, Lehmkuhl BK, Schoolmaster DR Jr (2012) Mapping the niche space of soil microorganisms using taxonomy and traits. Ecology 93:1867–1879

Lodewyckx C, Vangronsveld J, Porteous F, Moore ERB, Taghavi S, Mezgeay M, van der Lelie D (2002) Endophytic bacteria and their potential applications. Crit Rev Plant Sci 21:583–606

Lugtenberg B, Kamilova F (2009) Plant-growth-promoting rhizobacteria. Annu Rev Microbiol 63:541–556

Manzoni S, Schimel JP, Porporato A (2012) Responses of soil microbial communities to water stress: results from a meta-analysis. Ecology 93:930–938

Marilley L, Hartwig UA, Aragno M (1999) Influence of an elevated atmospheric CO2 content on soil and rhizosphere bacterial communities beneath Lolium perenne and *Trifolium repens* under field conditions. Microb Ecol 38:39–49

Marquez LM, Redman RS, Rodriguez RJ, Roossinck MJ (2007) A virus in a fungus in a plant: three-way symbiosis required for thermal tolerance. Science 315:513–515

Melotto M, Underwood W, Koczan J, Nomura K, He SY (2006) Plant stomata function in innate immunity against bacterial invasion. Cell 126:969–980

Miotto-Vilanova L, Jacquard C, Courteaux B, Wortham L, Michel J, Cleément C, Barka EA, Sanchez L (2016) *Burkholderia phytofirmans* PsJN confers grapevine resistance against *Botrytis cinerea* via a direct antimicrobial effect combined with a better resource mobilization. Front Plant Sci 7:1236

Mohan JE et al (2014) Mycorrhizal fungi mediation of terrestrial ecosystem responses to global change: mini-review. Fungal Ecol 10:3–19

Montealegre CM, Van Kessel C, Blumenthal JM, Hur H-G, Hartwig UA, Sadowsky MJ (2000) Elevated atmospheric CO2 alters microbial population structure in a pasture ecosystem. Glob Change Biol 6:475–482

Monteith JL, Unsworth HM (2007) Principles of environmental physics, 3rd edn. Academic Press, New York

Morgan JAW, Bending GD, White PJ (2005) Biological costs and benefits to plant-microbe interactions in the rhizosphere. J Exp Bot 56:1729–1739

Mosse B, Phillips JM (1971) The influence of phosphate and other nutrients on the development of vesicular-arbuscular mycorrhizae in culture. J Gen Microbial 69:157–166

Müller LM, Harrison MJ (2019) Phytohormones, miRNAs, and peptide signals integrate plant phosphorus status with arbuscular mycorrhizal symbiosis. Curr Opin Plant Biol 50:132-9

Newsham KK, Fitter AH, Watkinson AR (1995) Multi-functionality and biodiversity in arbuscular mycorrhizas. Trends Ecol Evol 10:407–411

Nicolson TH (1967) Vesicular–arbuscular mycorrhiza – a universal plant symbiosis. Sci Prog 55:561–581

Oades JM (1984) Soil organic matter and structural stability: mechanisms and implications for management. Plant Soil 76:319–337

Parmesan C, Yohe G (2003) A globally coherent fingerprint of climate change impacts across natural systems. Nature 421:37–42

Pearson RG, Phillips SJ, Loranty MM, Beck PSA, Damoulas T, Knight SJ, Goetz SJ (2013) Shifts in Arctic vegetation and associated feedbacks under climate change. Nat Clim Chang 3:673–677

Pillay VK, Nowak J (1997) Inoculum density, temperature, and genotype effects on in vitro growth promotion and epiphytic and endophytic colonization of tomato (*Lycopersicon esculentum* L.) seedlings inoculated with a pseudomonad bacterium. Can J Microbiol 43:354–361

Read DJ (1991) Mycorrhizas in ecosystems – Nature's response to the “law of the minimum”. In: Hawksworth DL (ed) Frontiers in mycology. CAB International, Regensburg, pp 101–130

Rillig MC, Wright SF, Allen MF, Field CB (1999) Long-term CO2 elevation affects soil structure of natural ecosystems. Nature 400:628

Rodriguez RJ, Henson J, Van Volkenburgh E, Hoy M, Wright L, Beckwith F, Kim Y-O, Redman RS (2008) Stress tolerance in plants via habitat-adapted symbiosis. ISME J 2:404–416

Sanders IR, Streitwolf-Engel R, van der Heijden MGA, Boller T, Wiemken A (1998) Increased allocation to external hyphae of arbuscular mycorrhizal fungi under CO_2 enrichment. Oecologia 117:496–503

Santos-Medellın C, Edwards J, Liechty Z, Nguyen B, Sundaresan V (2017) Drought stress results in a compartment-specific restructuring of the rice root-associated microbiomes. MBio 8:e00764-17. https://doi.org/10.1128/mBio.00764-17

Schenck NC, Schroder VN (1974) Temperature response of Endogone mycorrhiza on soy bean roots. Mycologia 66:600–605

Schimel JP, Schaeffer SM (2012) Microbial control over carbon cycling in soil. Front Microbiol 3:348. https://doi.org/10.3389/fmicb.2012.00348

Sessitsch A, Howieson JG, Perret X, Antoun H, Martinez-Romero E (2002) Advances in rhizobium research. Crit Rev Plant Sci 21:323–378

Smith SE, Read DJ (1997) Mycorrhizal symbiosis. Academic Press, San Diego

Staddon PL (1998) Insights into mycorrhizal colonisation at elevated CO_2: a simple carbon partitioning model. Plant Soil 205:171–180

Staddon PL, Fitter AH (1998) Does elevated atmospheric carbon dioxide affect arbuscular mycorrhizas ? Trends Ecol Evol 13:455–458

Staddon PL, Fitter AH, Robinson D (1999a) Effects of mycorrhizal colonisation and elevated atmospheric carbon dioxide on carbon fixation and below ground carbon partitioning in *Plantago lanceolata*. J Exp Biol 50:853–860

Staddon PL, Robinson D, Graves JD, Fitter AH (1999b) The δ13C signature of the external phase of a Glomus mycorrhizal fungus: determination and implications. Soil Biol Biochem 31:1067–1070

Stone JK, Bacon CW, White JF (2000) In: Bacon CW, White JF (eds) An overview of endophytic microbes: endophytism definded. Microbial endophytes. Marcel Dekker Inc, New York, pp 3–29

Sturm M, Racine C, Tape K (2001) Climate change—increasing shrub abundance in the Arctic. Nature 411:546–547

Tarnawski S, Aragno M (2006) In: Noesberger J, Long SP, Norby RJ, Stitt M, Hendrey GH, Blum H (eds) The influence of elevated pCO_2 on functions and diversity of rhizosphere and soil bacterial communities. Managed ecosystems and CO_2 – case studies, processes and perspectives. Series: ecological studies, vol 187. Springer, Berlin, pp 393–412

Timmermann T, Armijo G, Donoso R, Seguel A, Holuigue L, Gonzá-lez B (2017) *Paraburkholderia phytofirmans* PsJN protects *Arabidopsis thaliana* against a virulent strain of *Pseudomonas syringae* through the activation of induced resistance. Mol Plant-Microbe Interact 30:215–230

Tisdall JM, Oades JM (1982) Organic matter and water-stable aggregates in soils. J Soil Sci 33:141–163

Tobar R, Azcon R, Barea JM (1994) Improved nitrogen uptake and transport from 15N_ labeled nitrate by external hyphae of arbuscular mycorrhiza under water-stressed conditions. New Phytol 126:119–122

Treseder KK, Allen MF (2000) Mycorrhizal fungi have a potential role in soil carbon storage under elevated CO_2 and nitrogen deposition. New Phytol 147:189–200

Van der Ent S, Verhagen BWM, Van Doorn R, Bakker D, Verlaan MG, Pel MJC, Joosten RG, Proveniers MCG, Van Loon LC, Ton J, Pieterse CM (2008) MYB72 is required in early signaling steps of rhizobacteria-induced systemic resistance in Arabidopsis. Plant Physiol 146:1293–1304

van der Heijden MGA, Klironomos JN, Ursic M, Moutoglis P, Streitwolf-Engel R, Boller T, Wiemken A, Sanders IR (1998) Mycorrhizal fungal diversity determines plant biodiversity, ecosystem variability and productivity. Nature 396:69–72

Verheijen L, Aerts R, Brovkin V, Cavender-Bares J, Cornelissen J, Kattge J, Bodegom P (2015) Inclusion of ecologically based trait variation in plant functional types reduces the projected land carbon sink in an earth system model. Global Change Biol 21(8):3074–3086. https://doi.org/10.1111/gcb.12871

Vessey JK (2003) Plant growth promoting rhizobacteria as biofertilizers. Plant Soil 255:571–586

Vitousek PM (1994) Beyond global warming: ecology and global change. Ecology 75:1861–1876

Wagner MR, Lundberg DS, Coleman-Derr D, Tringe SG, Dangl JL, Mitchell-Olds T (2014) Natural soil microbes alter flowering phenology and the intensity of selection on flowering time in a wild Arabidopsis relative. Ecol Lett 17:717–726

Walker MD, Wahren CH, Hollister RD, Henry GH, Ahlquist LE, Alatalo JM, BretHarte MS, Calef MP, Callaghan TV, Carroll AB (2006) Plant community responses to experimental warming across the tundra biome. Proc Nat Acad Sci USA 103:1342–1346

Walther GR, Post E, Convey P, Menzel A, Parmesan C, Beebee TJC, Fromentin JM, Hoegh-Guldberg O, Bairlein F (2002) Ecological responses to recent climate change. Nature 416:389–395

Welbaum G, Sturz AV, Dong Z, Nowak J (2004) Fertilizing soil microorganisms to improve productivity of agroecosystems. Crit Rev Plant Sci 23:175–193

Whitaker J, Ostle N, Nottingham AT, Ccahuana A, Salinas N, Bardgett RD, Meir P, McNamara NP (2014) Microbial community composition explains soil respiration responses to changing carbon inputs along an Andes-to-Amazon elevation gradient. J Ecol 102:1058–1071

Wolkovich EM et al (2012) Warming experiments underpredict plant phenological responses to climate change. Nature 485:494–197

Xu L, Naylor D, Dong Z, Simmons T, Pierroz G, Hixson KK, Kim Y-M, Zink EM, Engbrecht KM, Wang Y et al (2018) Drought delays development of the sorghum root microbiome and enriches for monoderm bacteria. Proc Natl Acad Sci U S A 115:E4284–E4293

Yang J, Kloepper JW, Ryu CM (2009) Rhizosphere bacteria help plants tolerate abiotic stress. Trends Plant Sci 14:1–4

Young M, Blanchart E, Chenu C, Dangerfield M, Fragoso C, Grimaldi M, Ingram J, Monrozier L (1998) The interaction of soil biota and soil structure under global change. Global Change Bioi 4:703–712

Zak DR, Holmes WE, MacDonald NW, Pregitzer KS (1999) Soil temperature, matric potential, and the kinetics of microbial respiration and nitrogen mineralization. Soil Sci Soc Am J 63:575–584

Zhu J-K (2016) Abiotic stress signaling and responses in plants. Cell 167:313–324

Chapter 19
Impact of Climate on Soil Microbes and Plant Health

Swayamsidha Pati, Swati Mohapatra, Kanchan Vishwakarma, Divya Bandekar, Arti Mishra, and Deviprasad Samantaray

Abstract As the global climate continues to change, it becomes more important to understand possible feedbacks from soils to the climate system. This chapter focuses on soil microbial community responses to climate change factors. The omnipresent microbes provide a major contribution in plant, microbes and soil interactions especially during the cycling of soil organic carbon (SOC) and other nutrients. This is the need of the hour to understand and predict the influence of climate change on soil micro flora. In this chapter we compiled the trends on climate change such as influence of greenhouse gas, temperature, pH, salinity and moisture on soil microbiota and its effect on plants.

Keywords Plant microbiota · Endophytes · Plant prebiotic · Mycorrhiza · PGPR · Chemolithotroph

19.1 Introduction

For more than a century, persisted incineration of fossil fuels generates immense amount of carbonic acid gas within our environment that makes our planetary warming and change to a climate crisis (Sheik et al. 2011). These factors have

S. Pati · D. Samantaray
Department of Microbiology, Orissa University of Agriculture and Technology, Odisha, Bhubaneswar, India

S. Mohapatra (✉)
Amity Institute of Microbial Technology, Amity University, Noida, India

Department of Infection Biology, School of Medicine, Wonkang University, Iksan, South Korea
e-mail: smohapatra@amity.edu

K. Vishwakarma · A. Mishra
Amity Institute of Microbial Technology, Amity University, Noida, India

D. Bandekar
Department of Biotechnology, Indian Institute of Technology, Roorkee, Uttarakhand, India

D. K. Choudhary et al. (eds.), *Climate Change and the Microbiome*, Soil Biology 63,
https://doi.org/10.1007/978-3-030-76863-8_19

imprinted a harsh effect on agricultural cultivated field, features of arable land and productiveness. Modification in the ambiance has passive and active implications on up and below ground. Alteration can be change in atmospheric condition in different time intervals which has substantial impact on the selection of the crop for farming, watering, application of compost or manure and protection against different infectious agents needs to be managed. All these activities concomitantly change the microbiota of the ground soil. The climatic response makes the environment more complex especially the environment under the surface soil. As a response, it affects the morphology of plant root and nutrients contain of rhizosphere. Climatic changes are related to moisture, temperature, salinity, charge and compositing of different soils. All these factors modify the microbial communities inside the soil microenvironment that helps the microbes sustain different extreme conditions and can customize nutrient accessibility based on requirement. Due to various nutrient signalling and architecture, the roots are directly controlling the microbial diversity and its load; it also manages the interaction of the plant pathogens. The most profitable contribution of climate change on microbes is it activates genetically for the growth and productivity of plants. Some in-depth monitoring depicted uplifting of carbon dioxide in our ecosystem which leads to the increasing temperature that melts the snow and helps the plant to grow healthy (Bradford et al. 2008). However, the same parameters on the other side become harmful for the plants of terrestrial environment due to increasing temperature. Hence, such controversy needs more scientific attention for sustainable agriculture and the ecosystem. In 2017, an alarming condition was raised in a publication signed by more than 21,000 eminent researchers that it's the need of the hour to shift the focus from economic development to the conservation economy that can help to have a healthy environment and achieve sustainable future. Here this chapter is a compilation of the effect of climate on microbes especially the positive impact on soil and plant (Fig. 19.1).

19.2 Interaction of Soil Microbiota with the Plant

Meta-omics approaches opened the exact pathway of microbes and their metabolism in different environmental conditions. These pathways are related to the primary source of the plant, and it also helps in balancing the C source. This pivotal role of microbes in nutrient (C, N, S, K, S, etc.) cycling and soil fertility highly depends on the soil and the climatic condition, as the actinomycetes, fungi, soil algae, and protozoa help in agricultural productivity of agro-ecosystems (Gehring et al. 2005). Thus uncertainties and complexities of the climate towards these microbes through abiotic and biotic stresses need to be addressed and protected. In a wide range, soil microbes are divided into four variants:

1. The primary variants are decomposers: these variants intake simple carbon source as an example root excreta and litters of plant biomass. The load of decomposer variety increases during the rainy season especially the actinomycetes named as *Micromonosporaceae*, *Nocardiopsis*, *Micromonospora*, *Salinispora* and

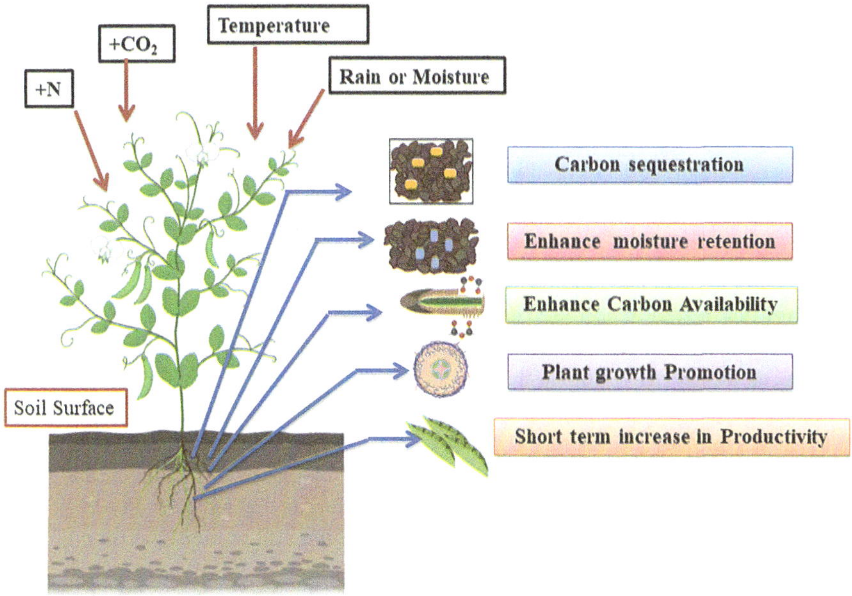

Fig. 19.1 Positive impact of climate on microbe and plants

Pseudonocardia. Due to accumulation of these microbes, the organic matter decomposes easily, and then the energy uptake also speeds up in the soil ecosystem and soil food web. Mostly these decomposers have been proven highly beneficial in farms contaminated with heavy metal, pesticide, pollutants and hydrocarbons. Decomposers facilitating bioremediation keep the nutrients (Pan et al. 2011) (Carbon & nitrogen) intact in the soil.

2. The second variant of microbes makes mutualistic relationship with the plants for their habitat as in return they help for the intake of nitrogen source from the soil. Most of the second variant microbes belong to nitrogen-fixing group (Batterman et al. 2013). In order to keep the environment healthy for the nitrifying and denitrifying bacteria, some parameters such as organic form of dissolved carbon, nitrogen, hydrogen and pH significantly contribute in soil microenvironment (Jana and Roy 1986). It has been observed some plants and microbes (autotroph) that help in nitrogen fixation reduce most common carbon dioxide directly by uptaking CO_2 during photosynthesis sometimes and indirectly building nitrogen-rich compounds like proteins, and its degradation again leads to fertility of the soil that enhances the plant growth. On the other hand, during carbon sequestration due to decomposition on plant litter, emission of NO_2 also produces which is the greenhouse gas; thus a proper balance is essential among the nitrifying and decomposer bacteria for the healthy soil, farming and ecosystem (Kou-Giesbrecht and Menge 2019).
3. The third variant as well-known as chemolithotrophs and chemoautotrophs drives energy and electron source from pollutants (pollutants contain N, Fe, H, etc.). Most of these variants contribute highly in nitrogen cycle (Gehring et al. 2005).

4. The major part of the microbes is fungus that degrades most of the pollutants easily with their secreted enzymes; the growth of fungus needs low temperature for its healthy growth. Diversified fungus and its enzyme promote critical part of the decomposition so that other microbes can easily use for its growth to maintain the soil food web. Apart from the enzyme secreted from fungus, the soil molecules can tie up collectively by fungal hyphae in the ambient temperature (Bond-Lamberty et al. 2018). Due to gathering of the soil compounds by the hyphae, the nutrients of the soil remain intact, and water filtration and holding capacity of the soil increase.

Fungi that reside in soil are of two types: some get energy by decomposing the dead organic material, and some grow in the plant commonly known as mycorrhizae (Torres et al. 2018). Fungus that degrades the organic material like cellulose, hemicellulose, hydrocarbons and lignin generates organic acid, carbon dioxide and other small molecules during its metabolism. This fungus also known as sugar fungus keeps the nutrient in the soil and makes it available for the plant and other bacteria. The second major group of microbes are fungus that requires 18–25 °C with wide range of pH curled the plant root as their habitat and at the same time it.

19.3 Effect of Climate on Endophytic Population

Depending on the climate change, the population of endophytes varies among the same species and plants even in the same region. Reports are available in support of chronological changes in relative frequency of endophytic fungi where matured leaves of teak (*Tectona grandis* L.) and rain tree (*Samanea saman* Merr.) not only had greater number of genera and species but also had higher colonization frequency than in young leaves. Additionally, their occurrence in leaves amplified during rainy season (Rustad et al. 2001). The population and frequency of endophytes differ amid the sampling dates of individual organs studied, like young petiole, leaves and twigs of *Ginkgo biloba* L. Reports also suggested the very first occurrence of *Phyllosticta* sp. in leaves and petioles in August, maximization of numbers in October and declination rate in May. However, *Phomopsis* sp. was detected in twigs throughout the growing season. These results signify that existence and distribution of the two dominant endophytic fungi are organ-specific and differed within seasons.

19.4 Different Factors and Their Impact

19.4.1 Temperature

Temperature is the most important environmental factor that affects microbial growth and metabolism in soil. Recently, temperature dependency of soil microbes comes into limelight; microorganisms are the pivotal group that produces CO_2 during decomposition of organic material in soil. Growth rate and product yield of microbes are determined by temperature, and in response to higher temperature, microbes sometimes experienced reduction in membrane fluidity and expression of heat shock protein (Bond-Lamberty et al. 2018). Although being advanced in functional gene arrangements and gene sequencing revealed functional and community gene shifts of microbes in response to higher temperature, still it is very difficult to assess the in situ response of microbes towards high temperature.

Temperature dependency of soil microbes is typically studied by measurement of respiration rate and various field and laboratory experiments of soil habitats. In case of bacteria, temperature dependency is determined by thymidine and leucine incorporation method, while response of fungi towards temperature is estimated by acetate-in-ergosterol incorporation technique.

The rate of soil respiration increases with increase in temperature up to 40 °C or more, while the growth rate of soil bacteria and fungi is average at temperature below 30 °C and decreases at higher temperatures (Nottingham et al. 2015). Usually, bacteria and fungi of temperate soil grow optimally at temperatures below 30 °C. Additionally, fungi grow easily in dry soil as they are more acclimatize to soil with less moisture than bacteria. However, very less is known about the temperature dependency of fungus. Researchers studied a comparative account of variation in temperature dependency of respiration between forest and agricultural soil, where forest soil has lower minimum dependency for respiration. This difference is due to shift in relative importance of fungi and bacteria as decomposers as fungus are more active in low temperature and in forest soil than bacteria (Wood et al. 2012). Yet, no reports are available with regard to direct comparison of temperature dependency of soil fungal and bacterial communities. Furthermore, climate change favours cyanobacterial blooms both directly and indirectly. High temperature enhances thermal stratification of water bodies resulting in cyanobacteria to float upwards and form dense surface blooms, giving them better access to light. Hence, bloom-forming cyanobacteria grow at high temperature than other non-buoyant phytoplankton organisms.

19.4.2 pH

Neutral pH range of 6–7 favours the beneficial soil microbes and floras; thus, acidic pH of soil can affect the community structure of microbes and their activities. Change in soil pH directly or indirectly affects the structure and function of soil microbial community (Cavicchioli et al. 2019). Interaction with H^+ ions of acidic soil hampers the microbial communities by limiting reproduction, alteration of enzyme production and cell membrane disruption. The interaction of microbes and acidic pH of soil parallelly hinders health and productivity of soil. Acidic pH also shifts the community structure of soil by making it a fungal dominated place, where fungal pathogens can easily invade into the roots (Pugnaire et al. 2019). Besides this,

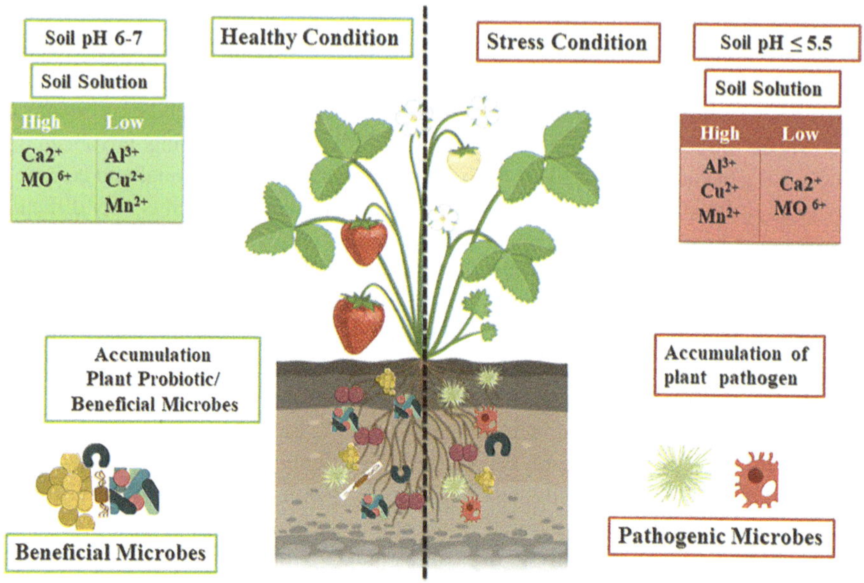

Fig. 19.2 Effect of microbes in different pH of soil and its interaction with plant

both bacteria and fungi play different roles in decomposition of organic residues in soil, so change in soil pH can also modify the processes of organic residue decomposition (Rajkumar et al. 2013). In acidic pH, soil carbon and plant nutrients become immobilized resulting in slow turnover and nutrient release (Fig. 19.2).

19.4.3 Salinity

Maintenance of microbial activity in soil is highly important as they have a pivotal role in mineralization of organic matter into plant available nutrients (Tan et al. 2019). Microbes having ability to tolerate soil salinity synthesize osmolytes for cell metabolism and turgor caused by osmotic stress. Salt content in soil create osmotic potential therefore affected by both salinity and soil water content. Soil salinity and water content vary in time and space. Soil salinity refers to the concentration of salt in soil that occurs naturally in soil and sometimes caused by mineral weathering or slow extraction of an ocean. Studying the impact of soil salinity and water content is extremely essential for crop yield, rehabilitation of saline soils and sustainable land use. High salinity can hamper the microbial activity in soil by osmotic and specific ion effect (Ma et al. 2019).

High salinity raises the osmotic potential of the soil water leading to extraction of water out of cells that may kill microbes and roots through plasmolysis. Also, low osmotic potential makes it tough for the roots and microbes to eradicate water from the soil. Though plants and microbes acclimatize to low osmotic potential by gathering osmolytes, their synthesis requires huge energy leading to reduction in growth and activity of microbes. High salinity not only increases the concentration of Na^+, Cl^- and HCO^{3-} ions that are toxic to plants but also decreases the microbial activity and changes community structure of microorganisms due to cell lysis caused by osmotic stress (Sangiorgio et al. 2020).

Nevertheless, microorganisms have the capacity to become accustomed to salinity by osmolytes accumulation. Organic osmolytes such as proline and glycine betaine and inorganic osmolytes like potassium cations are mostly accumulated by salinity-tolerant microbes. Synthesis of organic osmolytes involves high- mounts of energy, whereas use of inorganic salts as osmolytes has toxic effect; thus, it is confined to halophytic microbes which have salt-tolerant enzymes to survive in high salinity. As fungi are more sensitive to salt stress than bacteria, ratio of bacteria/ fungi is higher than saline soils. Hence, these differences in salinity tolerance can change community structure of microbes as compared to non-saline soils.

19.4.4 Water

The status of soil water refers either to water content of soil or to soil water potential that signifies energy level by which water is held in the soil. Water potential is defined as the pressure required in transferring a solution of known molarity from a referenced elevation to that of pure water, mainly including matric, osmotic and gravitational potential. The processes involving water balance are related to water content of soil, while processes linked to water movement are associated with soil water potential (Classen et al. 2015).

Water is highly essential for transportation of substrates as well as in hydrolysis processes. Besides this, soil water content also controls the mineralization rate and activity of soil microbes. However, excessive soil water content limits O_2 diffusion as diffusion of O_2 in water is much lower than in air. This reduced O_2 diffusion decreases the activity of aerobic microbes and favours the activities of anaerobes. On the other side, lack of water reduces the rate of carbon and nitrogen mineralization and alters the structure of microbial community by limiting the growth and activity of microbes (Cheng et al. 2019). Microbes retain adequate amount of water for cell turgor and metabolism by maintaining high osmotic potential in the cytoplasm than adjacent environment. Soil microbes accumulate organic and inorganic compounds at low water content to increase osmotic potential inside cytoplasm. Thus, osmolyte accumulation is the only principle for tolerance of low water content and high salinity (Aung et al. 2018). Furthermore, substrate supply in dry soil becomes restricted due to drainage of pores and disconnected and thinner water films (Tyagi et al. 2014). Soil microbes like archaea, fungi and Gram-positive bacteria can endure high matric potential than Gram-negative bacteria due to stronger cell walls.

19.4.5 Soil Structure

Soil refers to a multitude of tiny particles known as aggregates, which are separated by pores that allow the movement of vital chemicals, water and nutrients inside the soil and make them available for animals, plants and microbes. Small changes in soil structure not only affect the stability, density and porosity of soil but also strongly affect the function and distribution of soil bacteria (Xue et al. 2018). These bacteria demand some vital conditions for enhancement of crop yield in the plough land. Agricultural machinery, such as tractors, hampers the soil structure by moving over, leading to increasing soil density and demolishing soil porosity. This creates obstruction in distribution of vital chemicals, gases and water in soil and resulting in dismissive impact on plant. Additionally, the compressed soil destroys the soil porosity and restricts the movement of bacteria and other nutrients.

However, microorganisms secrete extracellular polymeric substances (EPS) as a padding approach to improve water retention capacity of soil. Soil microorganisms use carbon as a cellular biomass and transformed to stable metabolites; thus carbon can be sequestered in soil as dead microbial biomass (necromass) (Deltedesco et al. 2020). In addition, plant growth-promoting microbes like symbiotic and associative N_2-fixing bacteria enhance nutrient uptake via mycorrhizal fungi and produce plant growth-promoting hormones, such as indole-3-acetic acid (IAA).

19.5 Conclusion

Soil harbours wide array of microbes regulating the biogeochemical cycling of macro and micro nutrients those are vital for sustenance of life. This is the need of the hour to understand and predict the influence of climate change on soil micro flora. In this chapter we analyse an over view of recent trends on climate change and its influence of soil microbiota in several climate-sensitive soil and its effect on plants.

References

Aung K, Jiang Y, He SY (2018) The role of water in plant-microbe interactions. Plant J 93:771–780. https://doi.org/10.1111/tpj.13795

Batterman SA, Hedin LO, van Breugel M, Ransijn J, Craven DJ, Hall JS (2013) Key role of symbiotic dinitrogen fixation in tropical forest secondary succession. Nature 502:224–227. https://doi.org/10.1038/nature12525

Bond-Lamberty B, Bailey VL, Chen M, Gough CM, Vargas R (2018) Globally rising soil heterotrophic respiration over recent decades. Nature 560:80–83. https://doi.org/10.1038/s41586-018-0358-x

Bradford MA, Davies CA, Frey SD, Maddox TR, Melillo JM, Mohan JE, Reynolds JF, Treseder KK, Wallenstein MD (2008) Thermal adaptation of soil microbial respiration to elevated temperature. Ecol Lett 11:1316–1327. https://doi.org/10.1111/j.1461-0248.2008.01251.x

Cavicchioli R, Ripple WJ, Timmis KN, Azam F, Bakken LR, Baylis M, Behrenfeld MJ, Boetius A, Boyd PW, Classen AT, Crowther TW, Danovaro R, Foreman CM, Huisman J, Hutchins DA, Jansson JK, Karl DM, Koskella B, Mark Welch DB, Martiny JBH, Moran MA, Orphan VJ, Reay DS, Remais JV, Rich VI, Singh BK, Stein LY, Stewart FJ, Sullivan MB, van Oppen MJH, Weaver SC, Webb EA, Webster NS (2019) Scientists' warning to humanity: microorganisms and climate change. Nat Rev Microbiol 17:569–586. https://doi.org/10.1038/s41579-019-0222-5

Cheng YT, Zhang L, He SY (2019) Plant-microbe interactions facing environmental challenge. Cell Host Microbe 26:183–192. https://doi.org/10.1016/j.chom.2019.07.009

Classen AT, Sundqvist MK, Henning JA, Newman GS, Moore JAM, Cregger MA, Moorhead LC, Patterson CM (2015) Direct and indirect effects of climate change on soil microbial and soil microbial-plant interactions: what lies ahead? Ecosphere 6:art130. https://doi.org/10.1890/ES15-00217.1

Deltedesco E, Keiblinger KM, Piepho H-P, Antonielli L, Pötsch EM, Zechmeister-Boltenstern S, Gorfer M (2020) Soil microbial community structure and function mainly respond to indirect effects in a multifactorial climate manipulation experiment. Soil Biol Biochem 142:107704. https://doi.org/10.1016/j.soilbio.2020.107704

Gehring C, Vlek PLG, de Souza LAG, Denich M (2005) Biological nitrogen fixation in secondary regrowth and mature rainforest of Central Amazonia. Agric Ecosyst Environ 111:237–252. https://doi.org/10.1016/j.agee.2005.06.009

Jana BB, Roy SK (1986) Seasonal and spatial distribution pattern of nitrogen fixing bacteria in fish ponds under different management systems. Hydrobiologia 137:45–54. https://doi.org/10.1007/BF00004171

Kou-Giesbrecht S, Menge D (2019) Nitrogen-fixing trees could exacerbate climate change under elevated nitrogen deposition. Nat Commun 10:1493. https://doi.org/10.1038/s41467-019-09424-2

Ma Y, Vosátka M, Freitas H (2019) Editorial: beneficial microbes alleviate climatic stresses in plants. Front Plant Sci 10:595. https://doi.org/10.3389/fpls.2019.00595

Nottingham AT, Whitaker J, Turner BL, Salinas N, Zimmermann M, Malhi Y, Meir P (2015) Climate warming and soil carbon in tropical forests: insights from an elevation gradient in the Peruvian Andes. Bioscience 65:906–921. https://doi.org/10.1093/biosci/biv109

Pan Y, Birdsey RA, Fang J, Houghton R, Kauppi PE, Kurz WA, Phillips OL, Shvidenko A, Lewis SL, Canadell JG, Ciais P, Jackson RB, Pacala SW, McGuire AD, Piao S, Rautiainen A, Sitch S, Hayes D (2011) A large and persistent carbon sink in the World's forests. Science (80-) 333:988–993. https://doi.org/10.1126/science.1201609

Pugnaire FI, Morillo JA, Peñuelas J, Reich PB, Bardgett RD, Gaxiola A, Wardle DA, van der Putten WH (2019) Climate change effects on plant-soil feedbacks and consequences for biodiversity and functioning of terrestrial ecosystems. Sci Adv 5:eaaz1834. https://doi.org/10.1126/sciadv.aaz1834

Rajkumar M, Prasad MNV, Swaminathan S, Freitas H (2013) Climate change driven plant–metal–microbe interactions. Environ Int 53:74–86. https://doi.org/10.1016/j.envint.2012.12.009

Rustad L, Campbell J, Marion G, Norby R, Mitchell M, Hartley A, Cornelissen J, Gurevitch J (2001) A meta-analysis of the response of soil respiration, net nitrogen mineralization, and aboveground plant growth to experimental ecosystem warming. Oecologia 126:543–562. https://doi.org/10.1007/s004420000544

Sangiorgio D, Cellini A, Donati I, Pastore C, Onofrietti C, Spinelli F (2020) Facing climate change: application of microbial biostimulants to mitigate stress in horticultural crops. Agronomy 10:794. https://doi.org/10.3390/agronomy10060794

Sheik CS, Beasley WH, Elshahed MS, Zhou X, Luo Y, Krumholz LR (2011) Effect of warming and drought on grassland microbial communities. ISME J 5:1692–1700. https://doi.org/10.1038/ismej.2011.32

Tan X, Liao H, Shu L, Yao H (2019) Effect of different substrates on soil microbial community structure and the mechanisms of reductive soil disinfestation. Front Microbiol 10:2851. https://doi.org/10.3389/fmicb.2019.02851

Torres N, Antolín MC, Goicoechea N (2018) Arbuscular mycorrhizal Symbiosis as a promising resource for improving berry quality in grapevines under changing environments. Front Plant Sci 9:897. https://doi.org/10.3389/fpls.2018.00897

Tyagi S, Singh R, Javeria S (2014) Effect of climate change on plant-microbe interaction: an overview. Eur J Mol Biotechnol 5:149–156. https://doi.org/10.13187/ejmb.2014.5.149

Wood TE, Cavaleri MA, Reed SC (2012) Tropical forest carbon balance in a warmer world: a critical review spanning microbial- to ecosystem-scale processes. Biol Rev 87:912–927. https://doi.org/10.1111/j.1469-185X.2012.00232.x

Xue P-P, Carrillo Y, Pino V, Minasny B, McBratney AB (2018) Soil properties drive microbial community structure in a large scale transect in south eastern Australia. Sci Rep 8:11725. https://doi.org/10.1038/s41598-018-30005-8

Chapter 20
Climate Change and Plant Diversity: Threats and Opportunities

Usha Mina, Amit Kumar, Ashish K. Chaturvedi, and Pramod Kumar

Abstract Climate change is one of the alarming environmental concerns in the twenty-first century and so on affecting diverse ecosystems at various scales. Diversified plant species provides food, energy, health and other ecosystem services to human livelihood. Severely affected plant diversity due to climate change is a matter of great concern among scientists, policy makers and rising population. Hence, assessment of climate change-associated threats and opportunities to plant diversity become utterly important. Climate change has notable impact on growth, development as well as the reproductive success of plants, majorly due to change in the micro- or macro climate conditions. It also depends on the plant life form or plant groups as per-intrinsic tolerance and adaptation capacity of diverse group of plants. Multiple stresses co-occurring together under climate change vary greatly within plant group or plant types. In this chapter, we highlighted the threat posed by climate change to the plant diversity as a whole categorised under group, namely, algae, bryophyta and pteridophyte extended to gymnosperms and to the advanced or higher groups of plants such as angiosperms. In the ms the adaptive response of plant species distributed among the above group's opportunities available to ensure ecosystems structure, processes and services were also explored and documented.

Keywords Climate change · Plants · Algae · Bryophytes · Gymnosperms · Angiosperms

U. Mina (✉)
School of Environmental Sciences, Jawaharlal Nehru University, New Delhi, India
e-mail: ushamina@mail.jnu.ac.in

A. Kumar
Central Muga Eri Research and Training Institute, Jorhat, Assam, India
e-mail: amitkumar.csb@gov.in

A. K. Chaturvedi
Land and Water Management Research Group, Centre for Water Resources Development and Management, Kozhikode, Kerala, India
e-mail: ashispc@cwrdm.org

P. Kumar
Sri Aurobindo College, Delhi University, New Delhi, India

D. K. Choudhary et al. (eds.), *Climate Change and the Microbiome*, Soil Biology 63,
https://doi.org/10.1007/978-3-030-76863-8_20

20.1 Introduction

Biospheres plant diversity playing a key role in sustaining and supporting the life on it as well as helping humanity to combat emerging environmental challenges. However, human beings consciously or non-consciously through habitat destruction and fragmentation, deforestation, land use changes, introducing species into their non-native areas (i.e. triggering invasion) along with climate change are responsible losses in plant diversity locally, regionally and/or globally (Mina et al. 2018). Main drivers for biodiversity loss are land use change followed by exploitation and climate change (Nic et al. 2020). Global warming has affected and posed stress on natural ecosystems and caused substantial damage to or complete loss of some unique vulnerable ecosystems along with extinction of some species. Loss of biodiversity due to climate changes has been beyond the planetary boundary surpassing even unequal species distribution on Earth (Newbold et al. 2016).

Global climate models have projected that with current GHG emission rates, the likely increase of nearly 5 °C in average temperature has been noted. Even if countries take measures to mitigate or reduce greenhouse gases emission as committed under the 2015 Paris agreement, global warming will exceed 3 °C in the twenty-first century (IPCC 2014). It should also be emphasised that the rise in temperature will not be uniform across the different ecosystems of the planet (IPCC 2014). The Global Land-Ocean Temperature Index will be around 0.98 °C estimated by Goddard Institute for Space Studies (GISS), NASA, and the current atmospheric concentration of carbon dioxide (CO_2) is around 407 ppm. Its level today is maximally increased during the past 800,000 years (NOAA 2019). Besides changes in temperature and rise in CO_2, the world is also facing alteration in rainfall patterns. Hence, for = climate change impact assessment on ecosystems and its components, the measure of speed of change in climate per unit distance representing climate change velocity becomes very important to be addressed.

Estimated global mean velocities which represent the magnitude and range shifts in species = for = temperature and rainfall for a period of 2000–2100 are in the range of 0.42 and 0.22 km year^{-1} (Loarie et al. 2009; Brito et al. 2018). Conversely, temperature change velocities in equatorial region has been reported as >10 km year^{-1}, while at steep slopes it may range in <10 m year^{-1}. The mean temperature change velocity during last glacial maximum and during 2011 was only 5.9 m year^{-1} (Sandel et al. 2011).

Climate fluctuations are not a new phenomenon. Paleo-ecological evidences indicate that the Earth climate has been exhibiting temporal dynamicity since plants originated and evolved. In the past, Earth climate (as extreme cool, warm, dry and wet and high and low CO_2 levels) role in shifting vegetation patterns, speciation and extinction has been deciphered. Plants are immobile or sedentary in nature; each geographic location has a tolerance limit or range to climatic attributes. Biome concept (largest scale at which ecologist classify plant diversity) addresses the broad-scale distribution of world vegetation and associated animals (Kumar and Mina 2020). Climate change has been predicted as the major drivers in the future affecting plant diversity in terms of plant survival, growth and physiological success and along with its effect on ecological interactions involving herbivory, competition and pollination.

Plant response to climate change through acclimation, adaptation and other processes such as migration and/or extinction has been evident at local, regional and global level since origin (Corlett and Westcott 2013; Mina et al. 2018). Evolutionary diversity of plants represents a particularly interesting pattern (Fig. 20.1). Since 400 million years ago from the appearance, plant species number has continuously increased, however, with a dramatic shift through the time in the dominating plant groups. Extinction of the rootless and leafless psilopsids (the early vascular plants) by the Devonian ends with replacement by pteridophytes (ferns) which are flourishing in the Carboniferous period. Further the decreased abundance of pteridophytes by the early Triassic, known for gymnosperm diversification (includes cycad conifers, ginkgos), further declined abundance and diversified as angiosperms (flowering plants) during the past 100 million years (Smith and Smith 2012). Climate change has been acknowledged as a major threat causing loss to biodiversity (Hooper et al. 2012; Nic et al. 2020).Though empirical evidence with respect to climate change driven plant diversity extinctions is limited (Le Roux et al. 2019), evolutionary evidences suggest that plant diversity responds to change in climate spatially and temporally. Response at spatial scale includes shift in distribution range, while temporal scale response includes changes in phenology, diurnal rhythms, physiology and other life history traits/events. Due to increasing temperature and altered precipitation, plant species variably shift their range to track preferred climate, though species differ strongly in the strength and even direction of range shifts (Parmesan and Yohe 2003; Bertrand et al. 2011). Alterations in plant diversity geographical distribution pattern or range are controlled through climatic and non-climatic variables combinations which is idiosyncratic (Bullock 2012). Migration of some plant species to higher latitudes/altitudes under climate change for suitable habitat is well recorded (Beaumont et al. 2007). However, population reduction, loss of habitat and extinction risk for some species unable to endure the climate change are also noted (Root et al. 2003). To persist, plant diversity attribute responses produce adaptive plastic or genetic retorts under changing climate for short- or long-term period, respectively. Plant species exhibited slowly and/or little responses such as slow migration rate probably becoming extinct under this scenario. Climate relict is a term used for those plants which exhibited direct linkage with recent climate change and curtailed distribution (Crawford 2008).

Climate change effects on plant diversity are variable on all geographic region of Earth, as some regions will be benefited and some will be adversely affected. Northern Hemisphere areas—Russia, China and Canada—plant diversity will get more growing days, whereas hot tropical regions plant diversity may reduce nearly 200 growing days per year (Verrall and Pickering 2020). Similarly, in regions under snow cover, freezing days will reduce and support plant growth and may further increase plant diversity on the summits of European mountains, extending growing season of some plants. The actual climate change effect on plant diversity will be modulated by other factors including extreme events of rainfall, rising temperatures,

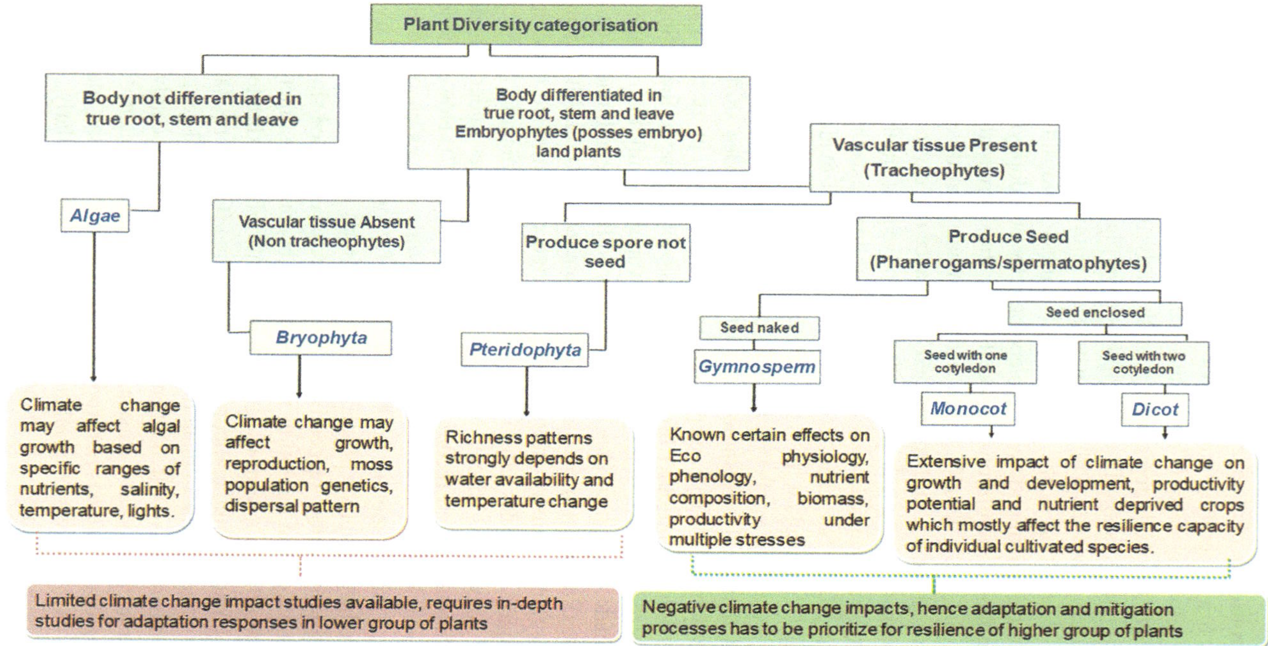

Fig. 20.1 Earth plant diversity categories and climate change impacts. *Orange colour box explains the possible climate change impact on respective plant group. Dotted red lines depict that impact assessment on lower groups of plants are limited, while dotted green lines suggest the negative impacts of climate change on higher group of plants extensively reported*

soil conditions and stress magnitude. Threat on the ability of plants to either adapt/ migrate becomes obvious (Mina et al. 2018). Currently the biggest challenge in plant diversity conservation for sustainability of ecosystem functioning is dependent on the information that how plant diversity attributes respond to climate change. Plant diversity includes multi-cellular, photoautotrophic eukaryotes act as an integral element in climate change impact and adaptations studies.

Present chapter is focused on the impact and adaptations response of plant diversity distributed among the five subgroups of plant kingdom to climate change attributes especially extreme climatic events, CO_2, temperature and precipitation to develop a broad understanding of respective adaptability and response of the plant subgroup. Plant kingdom diversity categorised into five subgroups, namely, algae, bryophyta, pteridophyta, gymnosperms and angiosperms, based on morphology, vascular system and seed formation (Fig. 20.1). According to Brummitt et al. (2015), bryophytes have 34,556 species, pteridophytes have 12,838 species, gymnosperms have 1032 species, and in angiosperms, monocotyledons have 71,445 species, and dicotyledons which are represented by surrogative legume family have 19,874 species. Google scholar repository database on climate change impact on plant diversity for 1990–2020 includes 752,000 papers on the topic. Of this most of literature reported climate change impact on species richness (considered as a resilience indicator for ecosystems to environmental change). Out of the total 752,000 papers, only 23,200 studies have addressed the functional and phylogenetic diversity and 18,100 studies on genetic diversity under climate change research. Moreover, differential response to diversity at specific level with large variability under climate change has been likely to be affected; hence, in a nutshell a broader picture of biodiversity in response to climate change is warranted. We further explored the general climate change impacts on each plant group from algae, bryophyta, pteridophyta and gymnosperms and to the advanced plant groups Angiosperms (Table 20.1, Fig. 20.1) (Table 20.1).

20.2 Algae Diversity and Climate Change

Algae include chlorophyll bearing unicellular, autotrophic, non-vascular, non-embryophytic thalloid organisms forming first trophic level of freshwater, brackish water and marine ecosystems. They are categorised under three main groups, namely, green, red and brown algae. Green algae predominantly occur in freshwater ecosystem, although some species also reported from marine and terrestrial ecosystem. Red and brown algae are predominantly found in marine ecosystems and occasionally in freshwater ecosystems (Kumar and Mina 2019). These three groups are further divided into two groups (macro-algae and micro-algae) based on the morphological size. Macro-algae (seaweeds) are fast-growing multi-cellular thalloid up to 60 m long plants (McHugh 2003), whereas micro-algae are microscopic organisms (Fig. 20.2a and b). The important macro-algae are *Alaria* (arctic

Table 20.1 Response of plant diversity categories to climate change attributes

Plant diversity category and selected species	Elevated levels of climate variables	Response	References
Algae (*Alaria; Sargassum; Laminaria, Corallina; chlorella, Spirulina, Muriellopsis; C. Vulgaris; Zygnema; H. fusiforme*)	CO_2	• Altered growth rate • Increased biomass	Watanabe et al. (1992), Olaizola (2003), Singh and Singh (2014), McHugh (2003)
	Temperature (>35 °C and < 16 °C)	• Algal growth affected	Singh and Singh (2014)
Bryophytes (*Bartramia patens, Hennediella antarctica, Polytrichas trumalpinum* and *Sanionia georgicouncinata*)	Temperature	• Affected reproduction pattern • Altered moss population genetics • Dispersal patterns affected	Casanova-katny et al. (2016), Dorrepaal et al. (2004), Dennis (2001)
	CO_2; temperature	• Moss growth and sexual reproduction	Casanova-Katny et al. (2016), Amesbury et al. (2017)
Pteridophytes (*Trichomanes pinnatum; Cyathea surinamensis; Lindsaea lancea; Adiantum obliquum; Schizaea stricta; Salpichlaena hookeriana*)	CO_2 and temperature	• Limiting habitat due to climatic fluctuations	Stein et al. (2012)
	Other factors	• Plant richness patterns strongly related to environmental conditions	Kreft et al. (2010), Zuquim et al. (2009)
Gymnosperms (*Pinus caribaea* var. *caribaea; Pinus tropicalis; Pinus sylvestris* var. *mongolica; Pinus pinaster; Pinus densiflora; Pinus roxburghii; Pinus koraiensis; Pinus sylvestris*)	CO_2	• *P. caribaea* var. *caribaea* exhibited more photosynthetic efficiency than *P. tropicalis* • Physiological processes such as CO_2assimilation, light compensation point, stomatal conductance and transpiration affected	Lexer and Hönninger (2000)
	Elevated CO_2 and temperature	• Decline in pine tree-ring • Pine utilising relatively greater amounts of isotopically depleted permafrost meltwater • Ecophysiology and tree ring analysis	Jacoby and D'Arrigo (1995)
	CO_2, and water deficit	• Alter the needles metabolism • Elevated [CO2] protects plants from water stress-mediated oxidative damage	de Simón et al. (2020)

(continued)

Table 20.1 (continued)

Plant diversity category and selected species	Elevated levels of climate variables	Response	References
		• Leaves metabolic profile affected	
	Elevated CO_2, temperature and water deficit	• Did not affect litter quantity • Changed lignin/N did not affect soil respiration but affected microbial biomass • Litter quantity and chemistry	de Simón et al. (2020)
	Elevated temperature	• Total seedling biomass was lowest in the 5 °C soil temperature treatment and highest in the 13 °C treatment – Belowground biomass increased • Shoot and root extension growth and biomass and carbohydrate allocation • Effect on phenology needle (leaf) formation • Litter fall, cone formation	Apple et al. (2000)
	Drought and warming	• Reduced the C/N ratio • Rapid decomposition of organic matter • Carbon, nitrogen and mass loss in the fine roots	Sonesson and Eriksson (2000)
Angiosperms Monocot *(Rice, wheat, maize; Sorghum; grasses)*	Elevated CO_2	• **Vegetative growth and metabolic process**: Increased Rubisco activity and concentration in leaf; increased photosynthesis; increase leaf area and biomass; profuse tillering; increased water use efficiency; reduced nutrient especially N due to C fertilisation effect; increased tissue temperature • **Reproductive growth:** Reduced flowering time; decreased spikelet fertility; reduced grain protein and nutrients	Bhatia et al. (2013), Bourgault et al.(2013), Chaturvedi et al. (2017a), Conleyet al. (2001), Craufurd and Wheeler (2009), Högy et al. (2010), Jablonski et al. (2002), Kadam et al. (2014), Li et al. (2004), Mina et al. (2017) Thilakarathne et al. (2013), Matsui et al. (1997), Vu et al. (1997)

(continued)

Table 20.1 (continued)

Plant diversity category and selected species	Elevated levels of climate variables	Response	References
	Elevated temperature	• **Vegetative growth and metabolic process**: Negative impact depending upon the morphological parameter; oxidative stress causing injury to photosynthetic machinery; membrane damage; reduced vegetative phase; decreased tillering; affect species richness in grassland • **Reproductive growth:** Insufficient energy supply at reproductive stage; reduced spikelet production; advanced flowering; hypoplasia in stamen; hyperplasia in pistil; disturbed anther dehiscence; impairment in pollination process; reduced pollen germination and pollen viability; kernel abortion	Barnabás et al. (2008), De Boeck Hans et al. (2007), Dolferuset al. (2011), Jagadish et al. (2010), Rang et al. (2011), Kadam, et al., (2014), Lipiec et al. (2013), Mina et al. (2017), Sadras and Monzon (2006), Sinsawat et al. (2004), Ugarte et al. (2007), Wahid et al. (2007)
	Elevated CO_2 and temperature	• Increase in yield up to a certain threshold value • Poor germinated pollen grains • Reduced spikelet fertility • Decreased seed-set • Reduced pollen production • Reduced grain protein and nutrients • Sink limitation determines CO_2response and high temperature at critical flowering/grain filling stage can limit the yield and quality	Aktar et al. (2006), Chaturvedi et al. (2017b), DaMatta et al. (2010), Myers et al. (2014), Prasad et al. (2011), Taub et al. (2008)
	Nutrients availability	• Higher application rate of both N and P fertiliser: Both (N & P) lead to the increase in	Chaturvedi et al. (2017b)

(continued)

Table 20.1 (continued)

Plant diversity category and selected species	Elevated levels of climate variables	Response	References
		total grain yield and 1000 gram weight	
	Water stress	• **Vegetative growth:** Reduced chlorophyll content and relative water content; decrease in tillering; halted shoot elongation • **Metabolic process:** Reduced photosynthesis rate; increase in canopy temperature; repressed metabolism of sucrose • **Reproductive growth:** Abortion of pollen; anther dehiscence; panicle growth; decrease in peduncle length	Aslam et al. (2013), Lipiec et al. (2013), Liu et al. (2006), Powell et al. (2012), Praba et al. (2009), Rang et al. (2011)
Dicot (*Eucalyptus citriodora; Eucalyptus tereticornis; Eucalyptus* hybrid; *Melia composite Dalbergia sissoo; Populus deltoides*)	CO_2 (800 μmolmol$^{-1)}$. Temperature (2 °C above than ambient), Humidity (5 higher than ambient)	• Increased root weight ratio (RWR) • C/N ratio and net photosynthetic rate • Enhanced biomass allocation in to shoot and root; stem height, stem diameter, number of compound leaves and branches • 2 °C rise of temperature under elevated CO_2 (800 ppm) resulted in decline in plant growth responses thereby resulting in decrease in biomass production • Growth, morphology and biomass; height, number of leaves • Biomass accumulation in terrestrial plants	Singh (2015), Lin et al. (2010)

Fig. 20.2 (**a**) Macro-algae (**b**) Micro-algae

climate), *Sargassum* (tropical climate), *Laminaria*, *Corallina*, etc., and micro-algae are *Chlorella*, *Spirulina*, *Muriellopsis*, etc. Each macro- and micro-algae grow under a specific range of nutrients, salinity, temperature, light, depth and currents. The optimal temperature range for algae phytoplankton is 20–30 °C. Temperature > 35 °C and < 16 °C will affect the algal growth.

Commercial potential of many macro- and micro-algae species has been realised and used in industries. Products of nearly US$5.5–6 billion and US$5–6.5 billion per year from macro-algae and micro-algae, respectively, have been estimated across the world (Pulz and Gross 2004; McHugh 2003). However, few algae species (represents about 2% of total algae species) are causing harm by producing toxin, creating anoxic conditions through large blooms and disrupting ecosystem structure (GEOHAB 2001).

Algae diversity, positive or negative response to climate change, is responsible for either harmful algal bloom (HAB) or coral reef loss, respectively (Fig. 20.3). Variability in algal species in response to CO_2 has been reported. Some of the algae strains negatively affected under its elevated concentrations (Lee and Lee 2003), while few strains optimally grow at 5–10% of its concentration. On the other hand, a drastic decrease in growth rate of some strains above 20% has been well documented (Watanabe et al. 1992). Few algae species grow well at 30–70% saturated CO_2 (Sung et al. 1999). Examples are available with some algal species growing at 100% CO_2 through maintained culture pH and on demand releasing CO_2 (Olaizola 2003).

Phytoplanktons, single-celled algae, freely float at upper layers of the oceans, which served as producers for marine food chain. It has been observed that in some South Pacific regions, 30% decline in phytoplankton biomass have occurred due to climate change and induced drop in thermohaline circulation (Behrenfeld 2006). Although climate change adversely affects most oceanic phytoplankton, however some species respond positively, for example, *Pyrodinium* phytoplankton species known for red tides (Petit and Prudent 2010).

Harmful algal blooms (HAB) is the term to represent proliferative growth of certain photosynthetic organisms that includes phytoplanktons and phytobenthos, macro-algae and cyanobacteria population causing harmful effects on human health, fisheries, aquaculture and other human resources (Behrenfeld 2006; Kudela et al. 2017). Worldwide presence of HABs in aquatic ecosystems due to natural processes and the associated problems has been reported. In recent past concerns have been raised that magnitude of some HABs and their associated impacts may increase in

Fig. 20.2 (continued)

the future due to direct and indirect changes in HAB locations physically (i.e. climate change) and socio-ecological environment (Berdalet et al. 2016; Geohab 2001).

Marine heat waves have been the cause of deterioration of coral populations globally which might be attributed to climate change (Buerger et al. 2020). van der Zande et al. (2020) reported that major coral reefs could disappear in 30 years' time in some of the = World Heritage sites under changing climate. Structural integrity of coral reefs worldwide is maintained by crustose coralline algae (CCA). Climate change adversely impacted CCA, as elevated temperature and CO_2 levels would lead to reduced CCA populations. Controlled experiments with coralline alga (CCA) species—*Porolithononkodes*—exposure to elevated CO_2 (400–1100 ppm) and temperature (26--29 °C) reported that warming caused additive negative effects on algal health. Mortality rate increased from $<1\%$ to 9% for CCA under high CO_2 and exacerbated up to 15% under warming conditions (Diaz-Pulido et al. 2012).

In commercial raceway ponds, the average algal biomass yields at 10–30 t ha^{-1} y^{-1}, and the reproducible productivities stand ranges at 50--60 t ha^{-1} y^{-1} under controlled conditions. The positive growth response characteristic of few algae species is exploited as an opportunity for climate change mitigation option. The enhanced biomass generated by the positively responding algae species will be utilised for the production of bio-fuel (that can replace fossil fuel substantially) and bio-fertilisers. For example, in micro-algae *C. vulgaris* ARC1 strain, an increase of 6% in the biomass under elevated CO_2 when temperature was 30 °C fixed, exhibited fixing of 18.3 mg and 38.4 mg CO_2L/day at ambient (0.036%) and elevated CO_2 (6%), respectively. Under CO_2-enriched conditions, *Zygnema* species commonly occurring in ponds, streams, ditches and similar type of water bodies produce 1.9–38 times more biomass.. In Japan, brown seaweed (*H. fusiforme*) is used as a food, and its relative growth rate under elevated CO_2 was significantly increased (Singh and Singh 2014).

Global investigation is focused towards conservation and replication of the algal-based climate change mitigation solutions for depleting atmospheric CO_2. In this line corals can have the potential to gain heat tolerance by associating with specialised symbiotic partners, and it can prove to be an excellent tool to adapt to climate change. The identification of the promising algal strains which can achieve high growth rate under changing environmental condition will be helpful in designing climate change mitigation strategies.

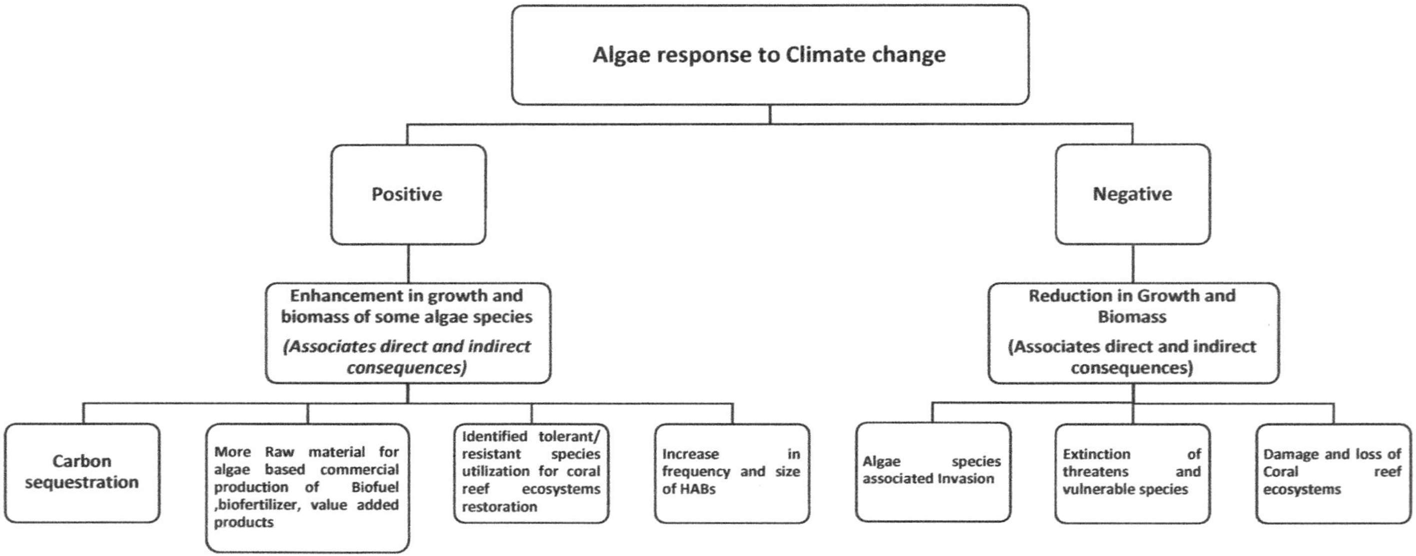

Fig. 20.3 Algae response to climate change and associated impacts

20.3 Bryophytes Diversity and Climate Change

Bryophytes are non-vascular, non-tracheophytes or primitive vascular plants which are having central strand of conducting cell function as xylem and phloem. Bryophytes' have thalloid and leafy form morphology and amphibian's characteristic, requiring lots of moisture and humidity in their habitat. Bryophytes consist of approximately 20,000 species distributed under three groups (hornworts, liverworts, mosses). Maximum and minimum species richness belongs to mosses and hornworts group, respectively. Most bryophyta species exhibited wide variability in physiological tolerance to abiotic stresses such as drought, insolation and temperature extremes due to a poikilohydric strategy (capacity to lose water quickly against the temperature rises and relative humidity drops event) for water and nutrients content. However, some species distributions are mainly to moist and shade-loving habitats and are sensitive to desiccation, whereas some bryophytes can grow on bare rock surfaces and can survive longer in the absence of water .

Moss can colonise on bare ground and rock form and stabilise soil crust through biomass production, nutrient cycling and regulating water flow. Mosses also grow on trees as epiphytes which generally grow on adaxial branch along. In Northern Hemisphere, mosses are more abundant on the north side of the substratum surface (tress and rocks), and opposite is true in Southern Hemisphere due to low moisture loss due to solar radiation. However, in cool, humid, cloudy climates, mosses can grow at any side due to equally moist surface available for them at all sides. Casanova-katny et al. (2016) reported four moss species (*Polytrichas trumalpinum*, *Hennediella antarctica*, *Bartramia patens* and *Sanionia georgicouncinata*) exposure to elevated temperature affected reproduction pattern and suggested that in many Antarctic mosses, elevated temperature exposure may alter moss population genetics and dispersal patterns.

Mosses are utilised for different anthropogenic purposes such as fuel, horticulture, medical and oil absorption activities. The common threats to these economic important bryophytes are habitat fragmentation, selective harvesting, deforestation and developmental activities such as road, building and hotel construction.

Mosses are considered as bioindicator of pollution and other environmental changes including climate change. Paleo-ecological studies used fossilised bryophytes or their parts, i.e. macrofossils or spores, to reconstruct past climate (Gignac 2001). It has been reported that dry state of bryophyte species exhibit extreme temperature tolerance than wet physical state (Longton 1988). Bryophytes distributed in narrow range of moisture and temperature. This characteristic at some locations identify them potential climate change bioindicators. However, some studies suggest bryophytes as weak indicators, as response to microclimate which may be extensively different from the macroclimatic conditions (Longton 1982). On comparing response of bryophytes and vascular plants to climate change, bryophytes are considered to be more responsive (Becker Scarpitta et al. 2017) and have been identified as at risk due to reduction in distribution areas with few exceptions. Research have established that species migrates with changing climatic conditions;

thus bryophyte can be indicative for ecosystem shift due to changing environmental condition and support to understand the climate change in specific geographic location (Gignac 2001). The literature suggests that climate change impact on bryophyte individual species or community can be quantified either through growth attributes monitoring or through changes in the species distribution pattern, assemblages of community and functioning of the ecosystems. Bryophyte response to climate change is not homogeneous, but species-specific. Majority of knowledge produced on the topic of bryophytes and climate change is indirect or was obtained in experiments performed under laboratory conditions or in short-term field studies. Thus, more specific and realistic long-term studies are badly needed. Terrestrial bryophytes from different environments and geographic areas have been studied regarding the effects of climate change. Dorrepaal et al. (2004) reported that in the Arctic tundra, summer warming will lead to a reduction in moss abundance. Replacement of bryophytes species by adjacent shrubs and boreal species in alpine snow beds has been reported (Björk and Molau 2007). In Antarctica, Bramley-Alves et al. (2014) pointed out that the dominance of bryophytes may be threatened by climate change, but increasing temperature improved both moss growth (Amesbury et al. 2017) and sexual reproduction (Casanova-Katny et al. 2016).

The elevated temperature may limit duration of metabolic activities and enhance the desiccation intensity in many bryophytes due to their low range of thermal acclimatisation and susceptibility to loss the moisture under high temperature condition.

Bryophyte species temperature requirement range for photosynthesis varies from region to region (Glime 2014) and observed susceptible to elevated temperatures (He et al. 2016). For example, tundra, temperate and tropical bryophytes can perform photosynthesis in the range of 5–15 °C, 15–25 °C and 25 °C or 26 °C, respectively (Wagner et al. 2013).The temperate species can tolerate elevated temperature up to 2–3 °C, and upper shoots are prone to desiccation above 30 °C, which limits the photosynthesis process in many species of bryophytes.

He et al. (2016) reported that in the boreal forests and temperate and tropical mountains, global warming will shorten the periods of bryophyte metabolic activity (due to an increase in desiccation), which can negatively affect survival and bryophyte diversity. Net CO_2 uptake coincides with increase in temperature up to optimum and then decline due to unbalancing of the photosynthesis and respiration rate. At this point forward, net CO_2 exchange becomes negative due to decline in photosynthetic rate and increased respiration rate. Therefore, respiration affects significantly on photosynthetic accumulation under elevated temperature and affects bryophyte survival.

Currently, bryophyte diversity is under pressure due to global environmental changes; consequently decline in bryophyte diversity is expected. Decline will be more in boreal, alpine biomes (Buytaerta et al. 2010) and peaks of tropical mountains, which leads to alterations in associated ecosystem structures and functioning, for example, peatland ecosystems. Peatlands are the "land with peat" and also known by many names such as mire, marsh, swamp, fen and bog (Rydin and Jeglum 2013). Peatland ecosystems cover >3% of Earth's surface which has been reported from

180 countries (Parish et al. 2008) and stores carbon twice of world's forests and serves as largest terrestrial organic carbon storage. About 80% of all peatlands, mostly in Northern Hemisphere, stores about 15--30% of the global soil carbon and major carbon stocks in frozen mineral soils (Limpens et al. 2008). Bryophytes are key structural and functional species of peatlands ecosystems. The biomass production of the peatland is regulated by *Sphagnum* (drought-sensitive species) and temperature and water variables dominantly regulated the *Sphagnum* productivity (Hájek 2014). *Sphagnum* stores water in capillary to prevent drying and maintained physiological functions and growth. It is highly impossible to resume the physiological function after drought stress in *Sphagnum* (Hájek 2014). Thus, climate change-induced stresses especially drought are having pronounced effects on peatlands. Therefore, these peatlands are critical for terrestrial carbon sinks to minimise the global warming. Therefore, improving understanding on bryophytes response towards elevated temperature is predominantly essential. Overall, bryophytes are ideal plants for climate change bio-indication, due to their sensitivity to ecological changes (Blackstock 2018). Additionally, the responses of desiccation-tolerant bryophytes and non-tolerant higher plants to climate change are not comparable because of their fundamental physiological differences, and thus bryophytes should be separately studied in this context. More observational and manipulative long-term field research at the community level involving the different factors of climate change and their impact on bryophyte species are needed.

20.4 Pteridophytes Diversity and Climate Change

Pteridophytes (non-flowering, vascular and spore-bearing plants) are distributed in different eco-climatic (tundra, tropical forest, humid forests, dry environments, subtropical forest and equatorial) regions (Smith et al. 2006) and divided into two phylogenetically distinct groups, i.e. lycophytes (group of ~1500 species) and ferns (group of ~10,500 species) (PPG I 2016). The pteridophytes comprises over 300 genera and about 13,000 species (Smith et al. 2006) and in India exists ~1000 species, of which 47 are endemic to India (Fraser -Jenkins 2008).

In any ecosystem pteridophytes provide shelter and habitat for many small animals and regulate soil erosion, stream bank stabilisation, pollutants phytoremediation, etc. and serves as ecological indicators (Della and Falkenberg 2019; Paciencia and Prado 2005). Pteridophytes contribute in nutrient cycling, ecological succession and ecosystem dynamics, for example, tree ferns provide microhabitat to various epiphytic plants (de Gasper et al. 2021).The distribution of pteridophytes is highly influence by the microhabitat abiotic characteristics, i.e. temperature, rainfall, humidity, light intensity, water availability, landscape heterogeneity, soil texture, intrinsic soil nutrient properties and fertility (Bergeron and Pellerin 2014). Pteridophytes are resilient in characteristic and are doing better than many flowering species due to their good dispersal abilities (García Criado et al. 2017).

Terrestrial pteridophyte species are under threat due to anthropogenic land use changes and aquatic pteridophytes species are adversely affected by water pollution. About half of pteridophytes from Trinidad and Tobago were deemed possibly at risk (Baksh-Comeau 1996). 1/3 of pteridophytes in the USA are at risk (Stein et al. 2018). Out of 22 rare pteridophytes of Gulf of Guinea, 8 species under critically endangered, 3 under endangered and 11 under vulnerable categories were reported (Figueiredo et al. 2019. In the Philippines, forest destruction and shifting cultivation are serious threat to habitat and diversity of the natural pteridophytes (Amoroso et al. 1996).

The low population size and restricted distribution are the important limitation which imposed extinction pressure on these ferns and lycophytes in various habitats. There are very limited studies conducted on the pteridophytes growth: distribution, species richness and abundance under the changing environmental scenario, i.e. climate change (Bergeron and Pellerin 2014). Thus the prediction of the pteridophytes response against the climate change is highly difficult. The disturbance and fragmentation of habitat and physiological stress along with change climatic condition may worsen the survival and can cause local/regional/global extinctions of these plants.

The spores of pteridophytes are capable for long-distance dispersal; thus they can escape from the stress regions easily compared to seed-bearing plants. This novel characteristic can support these plants to adapt under the changing environmental conditions. However, availability of the new suitable condition also impart to the success of this habitat escape mechanism.

Ferns (*Polystichum munitum*) transfer moisture from fog to the forest floor in California's redwood forests, even when it's not raining. The adaptative feature of these species will support to mitigated the climate change induced drought and population extinctions. However, summertime drought conditions reduce *P. munitum* abundance and vegetative growth. These shows less adaptation to water stress during summer in redwood forest ecosystem. *P. munitum* is not able to response better against the intense effect of climate change and can affect the redwood under story ecosystem functions.

20.5 Gymnosperms Diversity and Climate Change

Gymnosperms (naked seeds) are considered first true terrestrial vascular plants appearing around 245–208 million years ago, i.e. Triassic Period. Living gymnosperms comprise 12 families, 83 genera and ~ 1000 species categorised in four groups, namely, Coniferophyta, Cycadophyta, Ginkgophyta and Gnetophyta (Wang and Ran 2014). The extant gymnosperm species includes cycads (297–331), gnetophytes (80–100), conifers (614) and only one extant Ginkgoales. Of the 1000 species, 342 (33.7%) and 672 (66.3%) are dominantly found in Southern and Northern Hemisphere, respectively. Gymnosperm diversity growth forms are predominantly woody trees and shrubs and rarely climbers. The conifers (pines, cypresses, etc.) represent largest gymnosperm species, while *Ginkgo* represent only one species, i.e. *G. biloba*. Gymnosperms are recorded into each biogeographical region except small and isolated oceanic islands and Antarctica (Fragniere et al. 2015). A total of 506 species (50%) are reported in tropical (i.e. between 23.5° N and 23.5° S) environment.

Gymnosperms are most sensitive to climate change among all spermatophytes (Fragniere et al. 2015; Crisp and Cook 2011). The gymnosperms diversity decreases at equatorial latitudes. About 50% of the extant species was dominantly in tropics and > 43% of gymnosperms at ≤200 m a.s.l., and approximately 60% (mostly cycads) are at the risk of extinction (Fragniere et al. 2015). In a study, about the 41% of the studied gymnosperm (109 species) are at the risk of extinction due to climate change in China (Wu 2020). IUCN has classified *Abies nebrodensis* (Sicilian/Madonian fir) under critically endangered category because of its extremely small population (30 adult trees) in narrow distributed <1.5 km^2 and posed threats, i.e. habitat degradation, hybridisation with non-native firs and increased climate change-associated risk of extreme events (e.g. drought) and wildfire (Pasta and Troia 2017).

Different species of gymnosperms are having variable optimum range of environmental conditions (temperature, humidity, precipitation, etc.) and climate change affecting these conditions variable in different regions.

Gymnosperms are having very ancient in lineage with high cultural and ecological importance at a global scale irrespective of their species diversity lower than the other members of plant kingdoms. Gymnosperms have ability to conserve water, thus allowing them to grow in drier climate. Leaves are just like needle and having a heavy, waxy cuticle. The stomata are sunken in nature and minimise the water loss. The stems have vascular system allowing them to have the ability to grow in the harsh conditions. Gymnosperm seeds develop on scale/cones/stalk which consists embryo within a protective coat. The vascular tissues in a gymnosperm are like a series of pipes or running tubes. These pipes are responsible for transportation of water and nutrients from the roots to the leaves. The main source of pollination and dispersal is wind, ensuring survivability of gymnosperms.

Gymnosperm diversity category has reduced genetic diversity as compared to other categories of plant diversity. Consequences of low genetic diversity will be variable among species and populations, but low climate-related diversity poses a challenge for population to survive under rapid climate change (Burdon and Thrall 2000).

Evolutionary evidence indicates that latitude and elevation gymnosperm diversity shifted in past quaternary climate by migration and adaptations. Rapid climate changes challenge this process exerting strong selection pressure and separating populations from their adapted environments. Along with rapid climate change, rapid land use changes and habitat fragmentation block gene flow, which consequently disrupt adaptation and migration processes and affect growth and survival of many species.

In literature poleward spatial shift has been reported for gymnosperms species with similar drought tolerance, wood density and seed weight traits and evolutionary history and cause changes in forest ecosystem species composition, resilience and sustainability (Fei et al. 2017).

20.6 Angiosperms Diversity and Climate Change

Angiosperm diversity uniqueness is flowering and associated traits. The category comprises more than 220,000 living flowering plants (Seward 2011). Fossil evidence indicates that angiosperms first appeared about 130 million years ago (Seward 2011). The evolution of angiosperm in diverse habitats and restricted gymnosperm evolution to cold or dry habitats make them dominant group of plant diversity. Angiosperms are able to survive in diverse habitats due to following traits—complex and specialised woody tissue system; dense leaf venation; short life cycle; attractive flower; diverse pollination strategies; production of more number of seeds; enclosed seeds with easy dispersal mechanism by wind, water or animals; etc. Angiosperms exist in both woody and herbaceous growth forms (Seward 2011). Angiosperms phenological events occurrence are closely linked or synchronised with abiotic and biotic components of their habitat, climate change attributes will disrupt of affect the synchrony or dynamics of the phenological events of angiosperm diversity. Disruption, deviation or shift of phenological events has variable dramatic effect on the survival, flowering and pollinator interaction, pollination, seed production and fitness of angiosperm species. In this chapter the most dominant plant diversity group, i.e., angiosperm and climate change,, threats and opportunities are discussed with respect to its two main subgroups—monocotyledons and dicotyledons. Angiosperm species placed monocots and dicots group depending upon the single and double cotyledons present in the seeds, respectively.

20.6.1 Monocotyledons

Monocotyledons or monocots seeds have only one cotyledon or embryonic leaf. They contribute one fourth of angiosperm diversity, which is equivalent to approximately 60,000 species of economic and cultural importance such as cereals, grasses, weeds, palms, orchids, etc. Monocots distribution follow latitudinal diversity gradient, maximum at equator. Monocots growth form mostly herbaceous occasionally forming true wood. Monocots consist of parallel venation in long and narrow leaves and a fibrous root (i.e. without tap root) system. Flower parts are arranged in threes or in multiples of three in monocot. Monocots constitutes some of the largest families of angiosperms (e.g. orchids with ca. 20,000 species and grasses with ca. 15,000 species) along with the most economically important plant species which could be further sub-divided as annual and perennial on the basis of average life period.

All important staple food crops (wheat, rice, maize, oat and barley) are generally occurring in the annuals monocot categories (Seppelt et al. 2020). Climate change impacts on the annual monocots (rice and wheat) have been widely quantified with respect to elevated temperature and CO_2 concentration, declining water availability, nutrient conditions, etc. Rice belongs to Poaceae, family that includes other cereals such as wheat and corn. Rice cultivating methods include transplanting, direct seeding, system of rice intensification (SRI) and sub-merged conditions. Rice plant is an annual monocot and C_3 plant, and research across the world have find out that the climate change affected the rice plant in both positive and negative impact. Any negative impact can lead to question of food security. Hence in this chapter, we are focused on the rice as a model example for annual monocots.

Anticipated climate change attributes—elevated CO_2 with temperature—will have adverse effect on rice productivity and quality (Chaturvedi et al. 2017a). Rice is differentially sensitive to high day (major effects on spikelet fertility, anther dehiscence and pollen germination; Jagadish et al. 2010) or night temperature (major effect on night respiration; Bahuguna et al. 2017). Elevated CO_2 exposure resulted in higher biomass and yield in rice due to enhanced leaf photosynthetic rate (Chaturvedi et al. 2017a). Conversely, high temperature at sensitive growth stage associated with elevated CO_2 is documented to negate the beneficial effect of CO_2 (Chaturvedi et al. 2017b). Climate change impact associated with variables including water stress and temperature vary dynamically affecting growth and developmental of crops under field environment (Mittler 2006; Rang et al. 2011). CO_2 exposure to rice and wheat reduces grain protein and overall nutrient content concentration (Taub et al. 2008; DaMatta et al. 2010; Myers et al. 2014). Saseendran et al. (2000) temperature in the range of 26–31 °C leads to positive impact (20–40% increase) on the tiller number. High temperature exposure results in delaying the crop growth duration by as long as 10 days.

Perennial monocotyledons such as banana, ginger, turmeric, cardamom, onion, garlic, horticultural bulbs of lilies, daffodils, orchids, tulips and bluebells provide the diverse essential minerals and vitamins for human health. Different organs of perennial monocots are highly susceptible to various stress caused by the changing environmental scenario such as root systems is more sensitive to droughts than leave. Among them the orchids are having unique places due to their medicinal and ornamental natures. Epiphytic orchids adapted to survive under drought stress by conserving water (Helbsing et al. 2000). Orchid plants which have drought escape mechanism show higher vulnerability in an epiphytic habitat under fluctuating moisture availability attributed to climate change. Similarly, the pseudo-bulb supports many orchidsâ€™ species to conserve water in drought stress condition (Ng and Hew 2000). Under elevated temperature condition, the photosynthesis rate was observed to depress while respiration rate increases. This leads in imbalance in carbon anabolism and catabolism. This imbalance affects vegetative growth (reduced plant size) and reproductive growth (reduced energy supply) (Van Iersel 2003).

Detailed mechanistic physiological understanding of perennial monocots is warranted for their conservation. However, owing to long life-history and slow growth rate, limited studies have been performed in the perennial monocots. Overall it can be said that perennial monocots are sensitive to temperature, drought and other climatic stresses. So, there is a need to conserve the known species of perennial monocots which are economically beneficial.

20.6.2 Dicotyledons

Dicotyledons or dicots or eudicots are the plants possessing seeds have two cotyledons, flowers arranged in whorls, and leaves with vein arrangement that forms networks and stem vascular tissue arranged in ring shape. Dicot growth forms are either herbaceous or woody. Most eudicots produce pollen with distinct morphology. The category includes quarter million flowering plants under more than 300 families with approximately 200,000 species.

Dicot diversity responds to climate change-induced alteration in the native habitat by doing shifting its distribution to new non-native areas as invasive/exotic or non-native species that they can tolerate and compete with other plants of that region and do changes in community structure. USDA also reported that climate change and invasiveness are dominant cause of biodiversity loss.

Climate change-induced changing habitat environments affect dicot diversity natural phenological events and creating asynchrony for species interaction such as symbiotic associations, pest incidence, etc. and related ecosystem processes. Asynchronous phenological events include shift in dicot plants flowering with respect to their pollinators and environmental conditions suitable for pollination. Changing environments have been expected to result in changes in life cycle events for many plant species. Phenological changes in plants under climate change have to be used as a criterion of climate change impact assessment has been advocated. Phenological behaviours of 103 species categorised in ten growth forms in alpine ecosystem of North-West Himalaya were explored. Growth initiation, flowering period and fruiting time of individual species were worked out. Under climate change the phenology of these plants may get altered (Vashistha et al. 2009). Similarly, the biological spectrum of life forms and growth forms in Indian alpine ecosystem due to grazing pressure has been worked (Vashistha et al. 2011) suggesting that native vegetation remained unaffected by grazing. It also explains that whether anthropogenic factors alone or in combination of climatic change can cause changes in biological spectrum, life form and growth form pattern of alpine regions (Vashistha et al. 2011). Impact assessments on some of these alpine plant species/growth forms under elevated CO_2 were performed under in situ CO_2 enrichment facility Chaturvedi et al. 2010). Results revealed growth form or plant species-specific response and further suggested that all the alpine plant species may not respond positively to CO_2 owing to variable sinks (Chaturvedi et al. 2013).

Among dicots, in many agro ecosystems, pulses have been assessed for the climate change attributes—elevated CO_2, temperature and drought impacts and adaptation strategies. Elevated CO_2 exposure delayed leaf senescence and crop maturity of chickpea by 10–12 days (Kumar et al. 2012). Temperature sensitivity and growth dynamics of chickpea was assessed with 14 chickpea genotypes through naturally varying the temperature with different dates of sowing or staggered sowing methodology (Mina and Dubey 2010). It is concluded that under staggered sowing, chickpea which was sown late is vulnerable to both low and high temperature stress, during vegetative and reproductive stages, respectively, under North Indian climatic conditions (Pal et al. 2013). Elevated temperature stress effect could be mitigated by elevated CO_2 exposure on chickpea. The mitigatory role of high CO_2 to adverse impact of temperature was more in *desi* compared to *kabuli* variety (Rai et al. 2016). Pal et al. (2014) reported that rising CO_2 influenced the sunflower yield and seed oil quality. Biomass accumulation and yield increased, but seed quality in terms of mineral and protein content decreased.

20.7 Future Prospects

The experimental evidences suggest plant diversity ranging from algae to advanced angiosperms responds variably to climate change. It is evident that changing climate poses challenge for the plant diversity habitats integrity, distribution range, interaction with other species as well as energy and nutrient demands. Those plant species under different groups have shown higher vulnerability to adverse impact of climate change as compared to others; for them it is prerequisite for conservation. To devise the conservation strategies, the basic information is needed to protect the particular species especially in respect to its distribution, growth and development process and also the reproductive success under the normal as well as the changed climatic conditions. As our knowledge of plant response to CO_2 suggests that CO_2 primarily stimulates photosynthesis, temperature influences all the processes including photosynthesis, cell division, phenology and respiration. Rising temperature implies changing water requirement; hence characterisation of different plant groups based on different single factors or multifactor approach should be done. The required information's can be generated through collaborative network projects. The modelling and bioinformatics strategies for single species traversing species groups (functional types) to communities, ecosystems or biomes for and further for environmental niches, can be utilised to design for the conservation/protection strategies. The blending of the two approaches are producing the effective results than any individual approaches. Environmental variables doesn't act in isolation; other drivers like species invasion, land use changes, habitat degradation and loss, along with pollution, may act synergistically under climate change affecting plant diversity. Knowledge on plant diversity adaptation strategies to climate change may also be of great importance, which is still relatively limited.

References

Aktar J, Haque MM, Haque KMS (2006) Productivity of some selected rice cultivars under elevated CO_2 and high temperature condition. Int J Sustain Crop Prod 1:11–16

Amesbury MJ, Roland TP, Royles J, Hodgson DA, Convey P, Griffiths H, Charman DJ (2017) Widespread biological response to rapid warming on the Antarctic peninsula. Curr Biol 27:1616–1622

Amoroso VB, Acma FM, Pava HP (1996) Diversity status and ecology of Pteridophytes in three forests in Mindanao. In: Canus JM, Johns RJ (eds) Pteridology in perspective royal botanical garden, Kew, pp 53–60

Apple ME, Olszyk DM, Ormrod DP, Lewis J, Southworth D, Tingey DT (2000) Morphology and stomatal function of Douglas fir needles exposed to climate change: elevated CO_2 and temperature. Int J Plant Sci 16:127–132

Aslam M, Zamir MSI, Afzal I, Yaseen M, Mubeen M, Shoai BA (2013) Drought stress, its effect on maize production and development of drought tolerance through potassium application. Cercetări Agron Moldova 46:99–114

Bahuguna RN, Solis CA, Shi W, Jagadish KS (2017) Post-flowering night respiration and altered sink activity account for high night temperature-induced grain yield and quality loss in rice (*Oryza sativa* L.). Physiol Plant 159:59–73

Baksh-Comeau YS (1996) Index rating of threatened ferns in Trinidad and Tobago. Pteridology in perspective. Royal Botanic Gardens, Kew, pp 139–151

Barnabás B, Jäger K, Fehér A (2008) The effect of drought and heat stress on reproductive processes in cereals. Plant Cell Environ 31:11–38

Beaumont LJ, Pitman AJ, Poulsen M, Hughes L (2007) Where will species go? Incorporating new advances in climate modelling into projections of species distributions. Global Change Biol 13:1368–1385

Becker Scarpitta A, Bardat J, Lalanne A, Vellend M (2017) Long-term community change: bryophytes are more responsive than vascular plants to nitrogen deposition and warming. J Veg Sci 28(6):1220–1229

Behrenfeld MJ (2006) Nature phytoplankton absorbs less CO_2. Nature 444:752

Berdalet E, Fleming LE, Gowen R, Davidson K, Hess P, Backer LC, Moore SK, Hoagland P, Enevoldsen H (2016) Marine harmful algal blooms, human health and wellbeing: challenges and opportunities in the 21st century. J Mar Biol Assoc UK 96:61–91

Bergeron A, Pellerin S (2014) Pteridophytes as indicators of urban forest integrity. Ecol Indic 38:40–49

Bertrand R, Lenoir J, Piedallu C, Riofrío-Dillon G, De Ruffray P, Vidal C, Gégout JC (2011) Changes in plant community composition lag behind climate warming in lowland forests. Nature 479:517–520

Bhatia A, Kumar V, Kumar A, Tomer R, Singh B, Singh SD (2013) Effect of elevated ozone and carbon dioxide interaction on growth and yield of maize. Maydica 58:291–298

Björk RG, Molau U (2007) Ecology of alpine snowbeds and the impact of global change. Arct Antarct Alp Res 39:34–42

Blackstock TH (2018) Apparent increase in fertility of Lunulariacruciata (L.) Lind.(Marchantiophyta) in Britain associated with climate change. J Bryol 40:377–383

Bourgault M, Dreccer MF, James AT, Chapman SC (2013) Genotypic variability in the response to elevated CO2 of wheat lines differing in adaptive traits. Funct Plant Biol 40:172–184

Bramley-Alves J, King DH, Robinson SA, Miller RE (2014) Dominating the Antarctic environment: bryophytes in a time of change. In: Photosynthesis in bryophytes and early land plants. Springer, Dordrecht, pp 309–324

Brito-Morales I, García Molino J, Schoema DS, Burrows MT, Poloczanska ES, Brown CJ, Richardson AJ (2018) Climate velocity can inform conservation in a warming world. Trends Ecol Evol 33:441–457. https://doi.org/10.1016/j.tree.2018.03.009

Brummitt NA, Bachman SP, Griffiths-Lee J, Lutz M, Moat JF, Farjon A et al (2015) Green plants in the red: A baseline global assessment for the IUCN sampled red list index for plants. PLoS One 10:e0135152
Buerger P, Alvarez-Roa C, Coppin CW, Pearce SL, Chakravarti LJ, Oakeshott JG, Edwards OR, Van Oppen MJH (2020) Heat-evolved microsymbionts increase coral bleaching tolerance. Sci Adv 6(20):eaba2498
Bullock JM (2012) Plant dispersal and the velocity of climate change. Dispersal Ecol Evol 20:366–377
Burdon JJ, Thrall PH (2000) Coevolution at multiple spatial scales: Linum marginale–Melampsora lini–from the individual to the species. Evol Ecol 14:261–281
Buytaerta W, Cuesta-Camachob F, Tobon C (2010) Potential impacts of climate change on the environmental services of humid tropical alpine regions. Glob Ecol Biogeogr 20:19–33
Casanova-Katny A, Torres-Mellado GA, Eppley SM (2016) Reproductive output of mosses under experimental warming on Fildes peninsula, King George Island, maritime Antarctica. Revistachilena de historia natural 89:1–9
Chaturvedi AK, Prasad P, Nautiyal MC (2010) Lesser chamber effect inside open top chambers provides near-natural microenvironment for CO_2 enrichment studies in an alpine region of India. J Am Sci 6:109–117
Chaturvedi AK, Prasad P, Nautiyal MC (2013) Impact of elevated CO_2 on growth, morphology and dry matter partitioning in alpine growth forms of north western Himalaya. Indian J Plant Physiol 18:118–124
Chaturvedi AK, Bahuguna RN, Pal M, Shah D, Maurya S, Jagadish KSV (2017a) Elevated CO_2 and heat stress interactions affect grain yield, quality and mineral nutrient composition in rice under field conditions. Field Crop Res 206:149–157
Chaturvedi AK, Bahuguna RN, Shah D, Pal M, Jagadish SK (2017b) High temperature stress during flowering and grain filling offsets beneficial impact of elevated CO_2 on assimilate partitioning and sink-strength in rice. Sci Rep 7:1–13
Conley MM, Kimball BA, Brooks TJ, Pinter PJ, Hunsaker DJ, Wall GW, Adam NR, LaMorte RL, Matthias AD, Thompson TL, Leavitt SW, Ottman MJ, Cousins AB, Triggs JM (2001) CO_2 enrichment increases water-use efficiency in sorghum. New Phytol 151:407–412
Corlett RT, Westcott DA (2013) Will plant movements keep up with climate change? Trends Ecol Evol 28:482–488
Craufurd PQ, Wheeler TR (2009) Climate change and the flowering time of annual crops. J. Exp. Botany 60:2529–2539
Crawford RM (2008) Plants at the margin: ecological limits and climate change. Cambridge University Press, Cambridge
Crisp MD, Cook LG (2011) Cenozoic extinctions account for the low diversity of extant gymnosperms compared with angiosperms. N Phytologist 192:997–1009
DaMatta FM, Grandis A, Arenque BC, Buckeridge MS (2010) Impacts of climate change on crop physiology and food quality. Food Res Int 43:1814–1823
De Boeck Hans J, Lemmens CMHM, Vicca S, den Berge JV, Dongen SV, Janssens IA, Ceulemans R, Nijs I (2007) How do climate warming and species richness affect CO2 fluxes in experimental grasslands. New Phytol 175:512–522
de Gasper AL, Grittz GS, Russi CH, Schwartz CE, Rodrigues AV (2021) Expected impacts of climate change on tree ferns distribution and diversity patterns in subtropical Atlantic Forest. Perspect Ecol Conserv. https://doi.org/10.1016/j.pecon.2021.03.007
de Simón FB et al (2020) Rising [CO_2] effect on leaf drought-induced metabolome in *Pinus pinaster* aiton: Ontogenetic-and genotypic-specific response exhibit different metabolic strategies. Plant Physiol Biochem 149:201–216
Della AP, Falkenberg DB (2019) Pteridophytes as ecological indicators: an overview. Hoehnea 46: e522018. https://doi.org/10.1590/2236-8906-52/2018
Dennis GL (2001) Bryophytes as indicators of climate change. The Bryologist 3:410–420

Diaz-Pulido G, Anthony KR, Kline DI, Dove S, Hoegh-Guldberg O (2012) Interactions between ocean acidification and warming on the mortality and dissolution of coralline algae 1. J Phycol 48:32–39

Dorrepaal E, Johannes HC, Cornelissen T, Callaghan V, Richard SP, Logtestijn V (2004) Summer warming and increased winter snow cover affect *Sphagnum fuscum* growth, structure and production in a sub-arctic bog. Glob Chang Biol 10:93–104

Fei S, Desprez JM, Potte KM, Jo I, Knot JA, Oswalt CM (2017) Divergence of species responses to climate change. Sci Adv 3:e1603055

Figueiredo L, Krauss J, Steffan-Dewenter I, Sarmento Cabral J (2019) Understanding extinction debts: Spatio–temporal scales, mechanisms and a roadmap for future research. Ecography 42:1973–1990

Fragniere Y, Bétrisey S, Cardinaux L, Stoffel M, Kozlowski G (2015) Fighting their last stand? A global analysis of the distribution and conservation status of gymnosperms. J Biogeogr 42:809–820

Fraser-Jenkins CR (2008) Endemics and pseudo-endemics in relation to the distribution patterns of Indian pteridophytes. Taiwania 53(3):264–292

García Criado M, Väre H, Nieto A, Bento Elias R, Dyer R, Ivanenko Y, Ivanova D, Lansdown R, Molina JA, Rouhan G, Rumsey F, Troia A, Vrba J, Christenhusz MJM (2017) European red list of Lycopods and ferns. IUCN, Brussels, Belgium. iv + 59 pp

Geohab (2001) In: Glibert P, Pitcher G (eds) Global ecology and oceanography of harmful algal blooms, science plan. SCOR and IOC, Baltimore and Paris, p 87

Gignac LD (2001) Bryophytes as indicators of climate change. The Bryologist 104:410–420

Glime JM (2014) Photosynthesis in aquatic bryophytes. In: Photosynthesis in bryophytes and early land plants. Springer, Dordrecht, pp 201–231

Hájek T (2014) Physiological ecology of peatland bryophytes. In: Photosynthesis in bryophytes and early land plants. Springer, Dordrecht, pp 233–252

He X, He KS, Hyvönen J (2016) Will bryophytes survive in a warming world? Perspect Plant Ecol Evol Syst 19:49–60

Helbsing S, Riederer M, Zotz G (2000) Cuticles of vascular epiphytes: efficient barriers for water loss after stomatal closure? Ann Bot 86:765–769

Högy P, Keck M, Niehaus K, Franzaring J, Fangmeier A (2010) Effects of atmospheric CO_2 enrichment on biomass, yield and low molecular weight metabolites in wheat grain. J Cereal Sci 52:215–220

Hooper DU, Adair EC, Cardinale BJ, Byrnes JE, Hungate BA, Matulich KL, Gonzalez A, Duffy JE, Gamfeldt L, O'Connor MI (2012) A global synthesis reveals biodiversity loss as a major driver of ecosystem change. Nature 486:105–108

IPCC (2014) Summary for policymakers. In: Field CB, Barros VR, Dokken DJ, Mach KJ, Mastrandrea MD, Bilir TE, Chatterjee M, Ebi KL, Estrada YO, Genova RC, Girma B, Kissel ES, Levy AN, Maccracken S, Mastrandrea PR, White LL (eds) Climate change 2014: impacts, adaptation, and vulnerability. Part A: global and sectoral aspects. Contribution of working group II to the fifth assessment report of the intergovernmental panel on climate change. Cambridge University Press, Cambridge, pp 1–32

Jablonski LM, Wang X, Curtis PS (2002) Plant reproduction under elevated CO2 conditions: a meta-analysis of reports on 79 crop and wild species. New Phytol 156:9–26

Jacoby GC, D'Arrigo RD (1995) Tree ring width and density evidence of climatic and potential forest change in Alaska. Glob Biogeochem Cycles 9:227–234

Jagadish SVK, Craufurd PQ, Wheeler TR (2007) High temperature stress and spikelet fertility in rice (*Oryza sativa* L.). J Exp Botany 58:1627–1635

Jagadish SVK, Cairns J, Lafitte R, Wheeler TR, Price AH, Craufurd PQ (2010) Genetic analysis of heat tolerance at anthesis in rice. Crop Sci 50:1633–1641

Kadam NN, Xiao G, Melgar RJ, Bahuguna RN, Quinones C, Tamilselvan A, Prasad PVV, Jagadish KSV (2014) Agronomic and physiological responses to high temperature, drought, and elevated

CO_2 interactions in cereals. In: Sparks D (ed) Advances in agronomy. Academic Press, Elsevier Inc, Amsterdam, pp 111–156

Kreft H, Mutke WJJ, Barthlott W (2010) Contrasting environmental and regional effects on global pteridophyte and seed plant diversity. Ecography 33:408–419

Kudela RM, Berdalet E, Enevoldsen H, Pitcher G, Raine R, Urban E (2017) GEOHAB–the global ecology and oceanography of harmful algal blooms program: motivation, goals, and legacy. Oceanography 30:12–21. https://doi.org/10.5670/oceanog.2017.106

Kumar P, Mina U (2019) Diversity of life. In: Kumar P, Mina U (eds) Lifesciences: fundamentals and practice I. Pathfinder Publication, India, pp 586–632

Kumar P, Mina U (2020) Biomes. In: Kumar P, Mina U (eds) Fundamentals of ecology and environment, 2nd edn. Pathfinder Publication, pp 63–68

Kumar P, Rai P, Chaturvedi AK, Khetarpal S, Pal M (2012) High atmospheric CO_2 delays leaf senescence and crop maturity in chickpea (*Cicer arietinum* L.). Indian J Plant Physiol 17:254–258

Le Roux JJ, Hui C, Castillo ML, Iriondo JM, Keet J-H, Khapugin AA, Hirsch H (2019) Recent anthropogenic plant extinctions differ in biodiversity hotspots and coldspots. Curr Biol 29:2912–2918

Lee JS, Lee JP (2003) Review of advances in biological CO_2 mitigation technology. Biotechnol Bioprocess Eng 8:354–359

Lexer MJ, Hönninger K (2000) The potential for adaptation of Norway spruce forests to climate change: a simulation. FBVA Berichte 111:101–116

Li F, Kang S, Zhang J (2004) Interactive effects of elevated CO_2, nitrogen and drought on leaf area, stomatal conductance, and evapotranspiration of wheat. Agr Water Manage 67:221–233

Limpens J, Berendse F, Blodau C, Canadell JG, Freeman C, Holden J, Roulet N, Rydin H, Schaepman-Strub G (2008) Peatlands and the carbon cycle – a synthesis. Biogeosciences Discuss 5:1379–1419. Available from www.Biogeosciences-discuss.net/5/1379/2008/

Lipiec J, Doussan C, Nosalewicz A, Kondracka K (2013) Effect of drought and heat stresses on plant growth and yield: a review. Int Agrophys 27:463–477

Liu JX, Liao DQ, Oane R, Estenor L, Yang XE, Li ZC, Bennett J (2006) Genetic variation in the sensitivity of anther dehiscence to drought stress in rice. Field Crops Res. 97:87–100

Loarie SR, Duffy PB, Hamilton H, Asner GP, Field CB, Ackerly DD (2009) The velocity of climate change. Nature 462(7276):1052–1055

Longton RE (1982) Physiological ecology of mosses. In: Taylor RJ, Leviton AE (eds) The mosses of North America. Pacific Division of the American Association for the Advancement of Science, San Francisco, CA, pp 77–113

Longton RE (1988) The biology of polar bryophytes and lichens. Cambridge University Press, Cambridge

Matsui T, Namuco OS, Ziska LH, Horie T (1997) Effects of high temperature and CO_2 concentration on spikelet sterility in indica rice. Field Crops Res 51:213–219

McHugh DJ (2003) A guide to the seaweed industry. Rome, FAO. FAO fisheries technical paper no. 441

Mina U, Dubey SC (2010) Effect of environmental variables on development of Fusarium wilt in chickpea (*Cicer arietinum*) cultivars. Indian J Agric Sci 80(3):231

Mina U, Kumar R, Gogoi R, Bhatia A, Harit RC, Singh D, Kumar A, Kumar A (2017) Effect of elevated temperature and carbon dioxide on maize genotypes health index. Ecol Indic. https://doi.org/10.1016/j.ecolind.2017.08.060

Mina U, Singh D, Kumar P (2018) Climate change impacts on plants population and community ecological attributes. Mitigation Strategies and Policy Interventions – A Review Applied Ecology and Environmental Sciences 6:84–92

Mittler R (2006) Abiotic stress, the field environment and stress combination. Trends Plant Sci 11:15–19

Myers SS, Zanobetti A, Kloog I, Huybers P, Leakey AD, Bloom AJ, Carlisle E, Dietterich LH, Fitzgerald G, Hasegawa T, Holbrook NM (2014) Increasing CO_2 threatens human nutrition. Nature 510:139–142

Newbold T, Hudson LN, Arnell AP, Contu S, De Palma A, Ferrier S, Hill SLL, Hoskins AJ, Lysenko I, Phillips HRP, Burton VJ, Chng CWT, Emerson S, Gao D, Pask-Hale G, Hutton J, Jung M, Sanchez-Ortiz K, Simmons BI, Whitmee S, Zhang H, Scharlemann JPW, Purvis A (2016) Hasland use pushed terrestrial biodiversity beyond the planetary boundary? A global assessment. Science 353:288–e291. https://doi.org/10.1126/science.aaf220

Ng CKY, Hew CS (2000) Orchid pseudobulbs–false'bulbs with a genuine importance in orchid growth and survival! Sci Hortic 83:165–172

Nic LE, Bachman SP, CLeão TC et al (2020) Extinction risk and threats to plants and fungi. Plants People Planet 2:389–408. https://doi.org/10.1002/ppp3.10146

NOAA (2019) National centers for environmental information, state of the climate: global climate report for annual 2019, published online January 2020, retrieved on October 2, 2020 from https://www.ncdc.noaa.gov/sotc/global/201913

Olaizola M (2003) Commercial development of microalgal biotechnology: from the test tube to the marketplace. Biomol Eng 20:459–466

Paciencia MLB, Prado J (2005) Effects of forest fragmentation on pteridophyte diversity in a tropical rain forest in Brazil. Plant Ecol 180:87–104

Pal M, Chaturvedi AK, Shah D, Bahuguna RN, Sadhana SS, Mukhopadhyay D, Khetarpal S, Anand A, Deshmukh PS, Bhardwaj C, Kumar J (2013) Growth dynamics and temperature sensitivity of late planting chickpea under Delhi conditions. Indian J Plant Physiol 18:79–82

Pal M, Chaturvedi AK, Pandey SK, Bahuguna RN, Khetarpal S, Anand A (2014) Rising atmospheric CO_2 may affect oil quality and seed yield of sunflower (Helianthus annus L.). Acta Physiol Plant 36(11):2853–2861

Parish F, Sirin A, Charman D, Joosten H, Minayeva T, Silvius M (eds.) (2008) Assessment on peatlands, biodiversity and climate change: main report. Global Environment Centre, Kuala Lumpur and Wetlands International, Wageningen, p 179. Available at http://www.Imcg.net/media/download_gallery/books/assessment_peatland.pdf. Accessed 29 Sep 2017

Parmesan C, Yohe G (2003) A globally coherent fingerprint of climate change impacts across natural systems. Nature 421:37–42

Pasta S, Troia A (2017) Abiesnebrodensis. In Pasta S, Perez-Graber A, Fazan L and De Montmollin B (eds.), The top 50 Mediterranean Island Plants UPDATE 2017. IUCN/SSC/Mediterranean plant specialist group. Neuchâtel (Switzerland). E-book and online, p 141. Available: http://top50.iucn-mpsg.org/book, Accessed 03 Nov 2017

Petit J, Prudent G (eds). (2010) Climate change and biodiversity in the European Union overseas entities. IUCN, Gland, Switzerland and Brussels, Belgium. Reprint, Gland, Switzerland and Brussels, Belgium: IUCN, p 192

Powell N, Ji X, Ravash R, Edlington J, Dolferus R (2012) Yield stability for cereals in a changing climate. Funct Plant Biol 39:539–552

PPG I (2016) A community-derived classification for extant lycophytes and ferns. J Syst Evol 54:563–603

Praba ML, Cairns JE, Babu RC, Lafitte HR (2009) Identification of physiological traits underlying cultivar differences in drought tolerance in rice and wheat. J Agron Crop Sci 195:30–46

Prasad PVV, Boote KJ, Allen LH (2011) Longevity and temperature response of pollen as affected by elevated growth temperature and carbon dioxide in peanut and grain sorghum. Environ Exp Botany 70:51–57

Pulz O, Gross W (2004) Valuable products from biotechnology of microalgae. Appl Microbiol Biotechnol 65:635–648

Rai P, Chaturvedi AK, Shah D, Viswanathan C, Pal M (2016) Impact of elevated CO_2 on high temperature induced effects in grain yield of chickpea (*Cicer arietinum*). Indian J Agric Sci 86:414–417

Rang ZW, Jagadish SVK, Zhou QM, Craufurd PQ, Heuer S (2011) Effect of high temperature and water stress on pollen germination and spikelet fertility in rice. Environ Exp Bot 70:58–65
Root BA, Price JT, Hall K (2003) Fingerprints of global warming on wild animals and plants. Nature 421:47–60
Rydin H, Jeglum JK (2013) The biology of peatlands, 2nd edn. Oxford University Press, Oxford, p 382
Sadras VO, Monzon JP (2006) Modelled wheat phenology captures rising temperature trends: shortened time to flowering and maturity in Australia and Argentina. Field Crops Res. 99:136–146
Sandel B, Arge L, Bo Dalsgaard RG, Davies KJ, Sutherland GWJ, Svenning J-C (2011) The influence of late quaternary climate-change velocity on species endemism. Science 334:660–664
Saseendran SA, Singh KK, Rathore LS, Singh SV, Sinha SK (2000) Effects of climate change on rice production in the tropical humid climate of Kerala, India. Clim Chang 44:495–514
Seppelt R, Arndt C, Beckmann M, Martin EA, Hertel TW (2020) Deciphering the biodiversity–production mutualism in the global food security debate. Trends Ecol Evol. https://doi.org/10.1016/j.tree.2020.06.012
Seward AC (2011) Fossil plants: a text-book for students of botany and geology, vol 4. Cambridge University Press, Cambridge
Singh SP, Singh P (2014) Effect of CO_2 concentration on algal growth: a review. Renew Sust Energ Rev 38:172–179
Sinsawat V, Leipner J, Stamp P, Fracheboud Y (2004) Effect of heat stress on the photosynthetic apparatus in maize (*Zea mays* L.) grown at control or high temperature. Environ Exp Botany 52:123–129
Smith AR, Pryer KM, Schuettpelz E, Korall P, Schneider H, Wolf PG (2006) A classification for extant ferns. Taxon 55:705–731
Smith TM, Smith RL, Waters I (2012) Elements of ecology. Benjamin Cummings, San Francisco
Sonesson J, Eriksson G (2000) Genotypic stability and genetic parameters for growth and biomass traits in a water× temperature factorial experiment with *Pinus sylvestris* L. seedlings. For Sci 46:487–495
Stein RW, Mull CG, Kuhn TS, Aschliman NC, Davidson LNK, Joy JB et al (2018) Global priorities for conserving the evolutionary history of sharks, rays and chimaeras. Nature Ecol Evol 2:288–298
Sung KD, Lee JS, Shin CS, Park SC, Choi MJ (1999) CO_2 fixation by Chlorella sp. KR-1 and its cultural characteristics. Bioresour Technol 68:269–273
Taub DR, Miller B, Allen H (2008) Effects of elevated CO_2 on the protein concentration of food crops: a meta-analysis. Glob Change Biol 14:565–575
Thilakarathne CL, Tausz-Posch S, Cane K, Norton RM, Tausz M, Seneweera S (2013) Intraspecific variation in growth and yield response to elevated CO_2 in wheat depends on the differences of leaf mass per unit area. Funct Plant Biol 40:185–194
Ugarte C, Calderini DF, Slafer GA (2007) Grain weight and grain number responsiveness to pre-anthesis temperature in wheat, barley and triticale. Field Crops Res. 100:240–248
van der Zande RM, Achlatis M, Bender-Champ D, Kubicek A, Dove S, Hoegh-Guldberg O (2020) Paradise lost: end-of-century warming and acidification under business-as-usual emissions have severe consequences for symbiotic corals. Glob Chang Biol Gcb. https://doi.org/10.1111/gcb.14998
Van Iersel MW (2003) Carbon use efficiency depends on growth respiration, maintenance respiration, and relative growth rate. A case study with lettuce. Plant Cell Environ 26:1441–1449
Vashistha RK, Rawat N, Chaturvedi AK, Nautiyal BP, Prasad P, Nautiyal MC (2009) An exploration on the phenology of different growth forms of north – west Himalaya. NY Sci J 2:29–42
Vashistha RK, Rawat N, Chaturvedi AK, Nautiyal BP, Prasad P, Nautiyal MC (2011) Characteristics of life-form and growth-form of plant species in an alpine ecosystem of north-west Himalaya. J For Res 22:501–506

Verrall B, Pickering CM (2020) Alpine vegetation in the context of climate change: a global review of past research and future directions, vol 748. Sci Total Environment, p 141344

Wagner S, Zotz G, Salazar Allen N, Bader MY (2013) Altitudinal changes in temperature responses of net photosynthesis and dark respiration in tropical bryophytes. Ann Bot 111:455–465

Wahid A, Gelani S, Ashraf M, Foolad MR (2007) Heat tolerance in plants: an overview. Environ Exp Botany 61:199–223

Wang X-Q, Ran J-H (2014) Evolution and biogeography of gymnosperms. Mol Phylogenet Evol 75:24–40

Watanabe Y, Ohmura N, Saiki H (1992) Isolation and determination of culturalcharacteristics of microalgae which functions under CO_2 enriched atmosphere. Energy Convers Manag 33:545–552

Wu J (2020) Risk and uncertainty of losing suitable habitat areas under climate change scenarios: a case study for 109 gymnosperm species in China. Environ Manag 65:517–533

Zuquim G, Costa FRC, Prado J, Braga-Neto R (2009) Distribution of pteridophyte communities along environmental gradients in Central Amazonia, Brazil. Biodivers Conserv 18:151–166

Chapter 21
Impact of Climate Change on Functional AM Fungi in Rhizosphere

Manoj Kumar Singh, Sumit Sahni, and Anita Narang

Abstract In the study of global changes and ecosystem impacts, it is very important to consider mycorrhiza, because they hold a critical position at the plant-soil interface. Human-induced environmental changes on earth depend on number of factors such as increasing atmospheric CO_2, nutrient enrichment by atmospheric deposition (N_2), altered precipitation and temperature. All these changes taking place in present and will surely increase in the future can impact the association of fungi with plant roots in a positive or negative direction. These factors are classified on the basis of their impact on colonization of mycorrhiza viz. factors affecting arbuscular mycorrhiza (AM) fungi indirectly by altered allocation of carbon from the host and factors that directly affect AM fungi i.e. altered precipitation, temperature and nitrogen deposition. For the study of global climate change and its impact on AM fungi, this distinction in responses to different factors is very important. These global change factors always occur in association, since experimental examination of a large number of scenarios would not be possible in-situ. Therefore for the study of global changes on AM fungi, large spatial and temporal scale assessments have been considered. The majority of experiments only permit to extract short-term responses, though long-term responses are more appropriate. For example, CO_2 springs, global distribution of plant communities and regional extinction because of climate change. AM fungal community may also be impacted according to host biodiversity at local scales. Further, changes in AM fungal community that are not affected by the changes in plant community should be studied to find precise response of mycorrhizas to global change.

Keywords AM fungal community · Elevated CO_2 · Elevated temperature

M. K. Singh (✉) · S. Sahni · A. Narang
Department of Botany, Acharya Narendra Dev College, University of Delhi, New Delhi, India
e-mail: maneccbhu@gmail.com

D. K. Choudhary et al. (eds.), *Climate Change and the Microbiome*, Soil Biology 63,
https://doi.org/10.1007/978-3-030-76863-8_21

21.1 Introduction

Majority of terrestrial plant species form a symbiotic relationship with arbuscular mycorrhizal fungi (AMF), where the fungi enhance water and nutrient uptake for the host plant and in return obtain photosynthates from the plant (Diagne et al. 2020). The term arbuscular is derived from Latin word 'arbusculum', which means a bush or little tree. Arbuscules are formed within the cortical cells of the host plant roots and function as a major exchange site for plant-assimilated carbon and soil nutrients (Jiang et al. 2013). AMF comes under the phylum *Glomeromycota*, having distinct morphological features such as aseptate hyphae, hyphal coils, vesicles, arbuscules, prominent multinucleate spores, no fruiting body formation and surviving within the roots of the plants (Brundrett 1991). AMF are considered as obligate biotrophs and difficult to culture in the complete absence of host plant (Bago and Bécard 2002).

AMF greatly increase the phosphorus uptake of plant through its hyphal network, which extends more than 20 cm away from the roots and occupy up to 100 m cm^{-3} soil volume, therefore directly increasing the surface area of the plant roots (Miller et al. 1995). AMF extraradical mycelium consists of specialized phosphate transporters which absorb phosphorus from insoluble soil organic compounds (Johri et al. 2015). AMF can also increase the uptake of nitrogen as well as many other micronutrients such as Zn, Cu, Mn, Fe, etc. (Liu et al. 2000).

In addition to nutrient uptake, many other benefits of AMF for plants include carbon cycling (Compant et al. 2010), maintaining plant-water relations (Subramanian et al. 2006), soil stability by glomulin secretion (Wilson et al. 2009), plant biodiversity and increased disease resistance in many economically important plants. Because of their wide range of functional services, AMF can be considered as important microbial symbionts to the majority of terrestrial plant species (Fig. 21.1).

The climate on earth has been changing rapidly since the mid-twentieth century, and this can create a serious threat to all ecosystem components. It is predicted that global temperature will increase in the range of 1–5°C by 2100. Temperature of high latitudes are increasing and predicted to increase in the coming future with greater magnitude, with rapid changes in boreal and arctic ecosystem (Stocker et al. 2013). The effects of these rapid changes are most likely to be significant, and the mean annual temperature isotherms shifted upwards by almost 200 m in temperate mountain zones by 1°C increase in temperature. Number of cold days will decrease and hot days will increase. Other environmental factors like pattern of snowfall and rainfall will also be affected, and extreme disturbances like fire and hurricanes are also expected to increase. In an ecosystem these changes will affect the species adaptability, migration between ecosystems and tolerance within the ecosystem (Bidartondo et al. 2018). In fact the impact of climate change is already visible on AM fungal communities directly or indirectly by altered host physiology. Also the human-induced atmospheric changes affect the AM fungi in a variety of ways (Mohan et al. 2014). Studies have reported that elevated CO_2 in atmosphere enhances the activity and abundance of mycorrhizal fungi; mainly the fungi produce

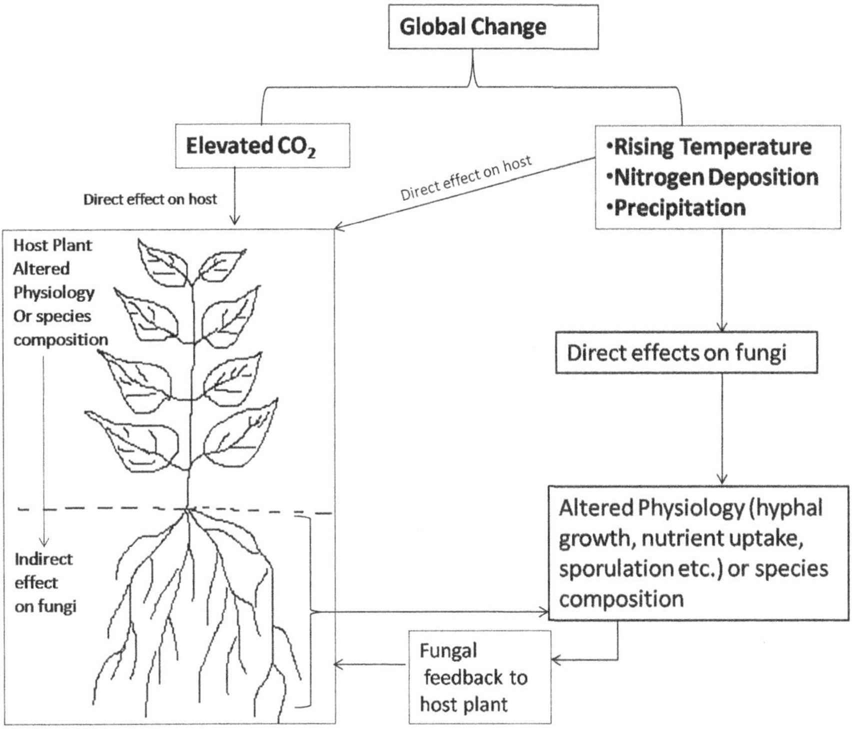

Fig. 21.1 Diagrammatic representation showing the effect of different global changes on mycorrhizal symbiosis

more spore-bearing structures. In warmer temperatures, fungal abundance will increase, while, the nutrient transfer to plant from fungi will decrease. Under drought conditions, AM fungal community provides resistance to plant and increases the productivity; therefore the study in plant community dynamics and fungal shifts is very important. Changes in AM fungal community are directly linked with tolerance of host trees under changed climatic conditions (Bidartondo et al. 2018). Globally the deposition of atmospheric nitrogen is increasing because of increased human activities. AM fungal communities showed strong response to increased nitrogen depositions for longer period of time, and some of the key taxa have disappeared. Ultimately deposition of nitrogen increases N availability and usually acidifies the ecosystems (Tian and Niu 2015; Stevens et al. 2018), both these changes can alter the mycorrhizal fungi interaction with their host and abiotic environment. Considering these complex changes in the environment due to climate change, it is not yet concluded how changes in mycorrhizal abundance, species and activity will affect ecosystem. This chapter summarizes the findings of different studies that how AM fungi are responding to climate change and their ecological interactions and function may affect current and future responses on ecosystem.

21.2 Effect of Elevated CO_2 (eCO_2) on AM Fungi

Since industrial revolution, the concentration of CO_2 in the atmosphere is increasing. Before that era, the average concentration of CO_2 was 280 ppm which has now reached more than 400 ppm, an almost 45% increase (Grover et al. 2015), and will further increase to reach up to 800 ppm by 2100 (Knohl and Veldkamp 2011). AMF is one of the major players in C cycling in ecosystem and symbiotically associated with about 80% of the plants across all the continents. According to an estimate, around 5 billion tonnes of carbon got recycled by AMF alone annually (Bago et al. 2000). Therefore, it has become imperative to study the effects of increasing C concentration in the atmosphere on them as well as on their interaction with plants. Cotton (2018) classified the mechanisms how environmental changes affect AMF communities into direct and indirect. Direct mechanisms involve those changes which directly impacts AM fungi, and indirect mechanism comprises those factors which does not affect AMF community. Effects of eCO_2 and ozone on AMF are classified under indirect mechanisms by Cotton (2018). Plethora of studies has been conducted to understand the effects of eCO_2 on the physiology, nutrient metabolism, growth, transpiration, stomatal conductance of plants and many other parameters (Drake et al. 1997, Ainsworth and Rogers 2007; Augé et al. 2015). But the research elucidating effects of eCO_2 on the AMF and plant-AMF interaction are comparatively less. Moreover, the studies which have covered this concept are sporadic as well as inconsistent in their views. Some are showing positive effects of eCO_2 on colonization of AMF (Sanders et al. 1998; Staddon et al. 2004; Becklin et al. 2016; Jakobsen et al. 2016); contrastingly some are showing detrimental effects of increased carbon dioxide (Goicoechea et al. 2014), while some other studies suggested eCO_2 does not have any effect at all (Gavito et al. 2000; Jifon et al. 2002; Tang et al. 2006; Cotton et al. 2015; Mueller and Bohannan 2015) on AMF. Similar observations were recorded in plants associated with AMF (Dong et al. 2018). The inconsistency observed in effects of eCO_2 on plant-AMF interactions was attributed to difference in distribution patterns of plants, different adaptations and different species of plants as well as AMF (Phillips et al. 2013; Dong et al. 2018). The explanation given for the negative effect of eCO_2 on AMF is that increased growth in plants due to eCO_2 resulted in stress-free resource availability for the plants which resulted in decreased dependency on AMF and hence their limited development. Some researchers are of view that the concentration of CO_2 in the pores of soil is already 10–126 times higher than that of atmospheric CO_2 concentration; hence eCO_2 does not show any effect on AMF communities in soil (Coakley et al. 1999; Grover et al. 2015).

Various parameters have been studied by different researchers in their studies on the effects of eCO_2 on the plant as well as on the AMF. AMF helps in sequestration of CO_2 fixed by the plants by consuming approximately 20% of the plant photosynthate and converting it into slow-degrading organic compounds such as chitin and a recalcitrant glycoprotein-glomalin (Wilson et al. 2009; Giri and Saxena 2017). There are many studies suggesting eCO_2 resulted in increased hyphal growth (internal and external) and root colonization of mycorrhizal fungi, viz. *Prunella vulgaris* showed increased hyphal growth under eCO_2 (600 ppm) (Sanders et al. 1998). Similar results were obtained in various species of *Glomus* in association with *Plantago lanceolata* and *Echinochloa crus-galli* when subjected to 700 ppm of carbon dioxide in microcosm-scale agro-ecosystem (Staddon et al. 2004; Tang et al. 2009). On the contrary there was no significant difference of eCO_2 on the colonization of AMF with *Pisum sativum*, *Citrus aurantium* and *Oryza sativa* (Gavito et al. 2000; Jifon et al. 2002; Monz et al. 1994). There are some reports suggesting reason for such a varied response of AMF colonization depends on the mechanism by which plant fixes carbon, viz. C_3 plants do not show any effect, whereas C_4 plants show better colonization of AMF (Tang et al. 2009; Grover et al. 2015). Poorter and Navas (2003) explained this difference between C_3 and C_4 plant is due to efficient C assimilation cycle in C_4 plants which led to more allocation of C towards mycorrhizal growth.

Different researchers studied different parameters in plants affected with eCO_2 which ranges from plant biomass (shoot, leaves and root individually), rate of photosynthesis, mineral nutrition accumulation, primary metabolites, nitrogen content in different organs, shoot to root ratio, phosphorus content in complete plant and in different organs, etc. The photosynthetic rate got increased under eCO_2 due to higher activity of Rubisco. Further there is synergistic effect of AMF with eCO_2 which further increase the photosynthetic rates (Al jaouni et al. 2018; Saleh et al. 2020).

In accumulation of mineral nutrition under elevated carbon dioxide, plants associated with AMF have varied responses for different minerals. Some plants, viz. peppermint and basil, showed improved accumulation of K, Ca, Zn, Mg and Cu, whereas in others there is decreased mineral accumulation which is attributed to dilution effect of increased biomass due to increased concentration of carbon dioxide (Dong et al. 2018). Saleh et al. (2020) proved higher accumulation of magnesium, phosphorus (P), copper (Cu) and manganese (Mn) in oregano plant in their study. Their study also elucidated that there is synergistic effect of AMF + eCO_2 in accumulation of entire mineral spectrum in oregano but not in phosphorus (P), calcium (Ca) and iron (Fe) accumulation.

Accumulation of primary metabolites such as sugars, starch, soluble sugars, oils, amino acids, etc. also shows an upward trend under elevated CO_2 (Ibrahim and Jaafar 2012; Al Jaouni et al. 2018; Ghasemzadeh and Jaafar 2011). This can be explained as increased photosynthetic rate will lead to increased C assimilation in form of various sugars. Similar results were also shown by Saleh et al. (2020) in oregano where they observed increased levels of organic acids, unsaturated fatty acids and essential and non-essential amino acids. Along with the primary metabolites, there was observation of increased phenolic compounds in oregano.

Due to inconsistent information available in different studies, various meta-analysis were performed; one such meta-analysis was conducted by Dong et al. (2018) with researches available and used 27 different parameters to study the effect of eCO_2 on AMF as well as on various plants where in most of the parameters it was positive effect of eCO_2.

21.3 Effect of Altered Temperature on AM Fungi

A broad range of environmental perturbations affect the community and development of AM fungi and mycorrhiza formation (Liu et al. 2004). There are numerous reports available on diversity of AM fungi under variable environmental conditions due to climate change (Lumini et al. 2011; Camenzind et al. 2014; Botnen et al. 2015). But number of studies are limited to the structure of AM fungal community under variable temperature conditions. Heinemeyer et al. (2004) performed a study on grassland community and reported no significant AM fungal community change in response to the temperature increase by 3°C above ambient. Soil microbial community structure and microbial biomass responses were quantified from 2001 to 2006 under simulated condition of global change at the Jasper Ridge Global Change Experiment by Gutknecht et al. (2012). The authors concluded that only in 2006 AM fungi biomarker biomass was decreased under temperature increase by 1°C, and in rest 5 years no significant difference was reported. Many other studies on the structure of AM fungal community due to climate change including altered temperature were available (Dumbrell et al. 2011; Torrecillas et al. 2013; Bainard et al. 2014), but there is no clear consensus in results that how AM fungal community was altered due to temperature stress. Temperature stress causes many phenological changes in AM fungi like altered growth of mycelia, pattern of growth, extraradical and intraradical AM fungal colonization (Gavito et al. 2005; Compant et al. 2010). Gavito et al. (2005) reported the temperature between 18 and 30 °C as the optimum temperature for the development of AM fungi. But temperature optima are variable for different AM fungal species. In temperate conditions, warming of soil stimulates the growth of extraradical hyphal network and of external hyphal production (Staddon et al. 2004; Hawkes et al. 2008), but lower and higher temperatures reduce the growth of AM fungi. In a study, *Glomus intraradices* showed reduced sporulation at 15 °C, but the metabolic activity of spore was not reduced till 10 °C of temperature (Liu et al. 2004). On the other hand, Schenck et al. (1975) reported spore germination for *Glomus heterogama* and *Glomus coralloidea* was reduced above 34 °C temperature. Haugen and Smith (1992) reported that colonization of *Anacardium occidentale* roots by *Glomus intraradices* was decreased at 38 °C as compared to 22 °C, while Martin and Stutz (2004) studied *Capsicum annuum* root colonization by *Glomus* AZ112 and found increased colonization at 32.1–38 °C compared to 20.7–25.4 °C. Zhu et al. (2011) studied on *Zea mays* roots colonization by *Glomus etunicatum* and found no significant change at 35 and 40 °C. The difference in above findings could be due to species variation in AM fungi and plants and different experimental conditions. In large number of plant species over different vegetation zones, arbuscular mycorrhiza (AM) is the widely distributed mycorrhizas. AM fungi and plants associated with poorer soil organic matter, lower phosphorus, higher nitrogen and higher pH. Soil temperature and humidity have also directly affected the association of AM fungi, which is regulated by climatic factors of the region. In general AM fungi perform better under higher temperature conditions and also provide drought resistant to plant. On the other hand, many studies have reported low AM colonization in cold climatic conditions. Gradient studies showed almost missing colonization of AM fungi in arctic regions as opposite to boreal (Olsson et al. 2004; Kytöviita 2005). Likewise, on the basis of composition of vegetation type on the sites of higher elevated montane, plants harbour more ectomycorrhiza (EM) as compared to arbuscular mycorrhiza (AM), and proration of EM is more in the plants, hosting both EM and AM (Gehring et al. 2006).

The warm seasonal temperature and distribution of AM were positively correlated in a global data analysis, while it is also positively correlated with the incidence of frost; this frost correlation was connected with the AM plants pre-dominance in grasslands of the continental regions in the world, showing considerable seasonal temperature variations (Soudzilovskaia et al. 2015). More experimental setups are required for studying the occurrence of AM in natural field conditions because climate of the regions depends on more factors, such as on high altitude and high latitude the climate is more humid (precipitation minus evapotranspiration). Many research existing on AM by using direct effect of temperature, observed reduced AM formation under lower temperature (Kilpeläinen et al. 2020). The study by Parke et al. (1983) used the EM species *Pinus ponderosa* and *Pseudotsuga menziesii* and an AM species *Trifolium subterraneum* as bait plants. The response of temperature for the formation of mycorrhizal showed a bell-shaped pattern between 7.5 and 35 °C in the three species, through the optimum at 18–25 °C.

21.4 Precipitation Pattern Changes Caused by Climate Change

21.4.1 AM Fungi Provide Tolerance to Plants Under Water Stress

Precipitation is considered as an important factor that contributes in regulating the soil carbon and soil moisture balance in the environment (Fu et al. 2017). Therefore, it has a direct role in determining the soil physical and chemical properties of the region (Ren et al. 2018). Under variable precipitation conditions, AM fungi play an important role to enhance plant robustness by improving nutrients absorption (Smith and Smith 2011), increasing pathogen resistance (Borowicz 2001) and providing drought tolerance (Augé 2001). Water stress causes major physiological changes in plants like reduced photosynthesis, accumulation of reactive oxygen species, reduced aboveground biomass productivity, decreased root proliferation and reduced nutrient and water uptake (Farooq et al. 2009). Under water stress condition, AM fungi might control plant robustness by different mechanisms like altering host physiology and increasing availability of soil water (Santander et al. 2017). The extraradical mycelium of AM fungi extends away from the plant roots and absorbs nutrients that are otherwise inaccessible to plants (Smith and Smith 2011). Additionally mineral nutrients like nitrogen, phosphorus, potassium and calcium in leaves regulates stomatal conductance that improves CO_2 fixation under water stress conditions (Bowles et al. 2016). The extraradical mycelium of AM fungi can change soil structure, but the exact mechanism behind the process is still under investigation. It was hypothesized that the formation of fungal products (glomalin, a glycoprotein) and extraradical mycelium contributes in the formation and stabilization of soil aggregates (Rillig and Mummey 2006). These soil aggregates retain the soil water and nutrients that can be efficiently utilized by the plants (Dıaz-Zorita et al. 2002). Thus even under water stress condition, plant growth is enhanced which is important for their survival, reproduction and competition (Augé 2001). AM fungi also show positive effect on water use efficiency (Kaya et al. 2003) and can maintain root conductivity during water stress period (Sánchez-Blanco et al. 2004). AM fungi regulate aquaporin expression in the roots, which helps the water transport in plant tissue (Aroca et al. 2007). Finally, plant resistance to low water availability is enhanced by AM fungi because they reduce oxidative damage due to decreased rainfall by increasing production of antioxidant in plants (Wu et al. 2006). Response of AM fungi to plants under water stress also depends on the phenotypic characteristics of associated fungal species. Marulanda et al. (2003) reported the effects of many fungal species on the plant *Lactuca sativa.* They reported *Rhizophagus intraradices*, *Funneliformis coronatum*, *Funneliformis mosseae* and *Claroideoglomus claroideum* were the most efficient species in terms of water absorption because these species have developed more arbuscules, external mycelium and other root-associated hyphal structures. On the other hand *Septoglomus constrictum* was less efficient because of less developed water-absorbing structures. Therefore, phenotypic characteristics of AM fungi play important role in water uptake by the plant.

21.4.2 Effects of Altered Rainfall Regimes on AM Fungal Communities

Many studies are available on the effects of rainfall pattern changes on plants and tolerance provided by AM fungi to plants under these circumstances. But there are limited reports available on the effect of these changes on functional AM fungi in plant roots. Many scientists tried to unlock this question by performing pot experiments by using single fungal isolates on a particular plant species (Augé 2001). Few scientists have also performed field experiments for longer duration to study the response of AM fungal communities on altered rainfall, but their main focus is on aggregate measures of fungal abundance.

In a field experiment performed by Martínez-García et al. (2012), rainfall pattern was manipulated for 4 years, to study its effect on AM fungi associated with *Artemisia barrelieri.* The study illustrated detrimental effect of reduced rainfall on vesicles abundance and linked the changes with decrease in rate of photosynthesis in host plant. Therefore, altered precipitation due to climate change influences the interaction between host plant and mycorrhizal fungi, also affecting plant survival and productivity.

A long-term simulating climate change field experiment was performed by Staddon et al. (2003) to observe the temporal changes in grassland AM fungal communities, under changed pattern of rainfall and temperature. The findings showed that soil moisture was an important factor regulating mycorrhizal abundance. Under the applied conditions comparable to summer drought, extraradical mycelium density was decreased, while root length colonization by AM fungi increased. However, in roots and soil, AM fungal density was correlated with both plant diversity and soil moisture. This temporal study approach showed the responses of AM fungal community to climate change. In the study, it is not clear that whether changes in rainfall patterns influenced the species ability for root colonization, extraradical mycelium formation and vesicles production or it changes the composition of species community with different traits.

AM fungal community structure was studied by Deepika and Kothamasi (2015) in a greenhouse experiment. In the experiment *Sorghum vulgare* was inoculated with fungiform semiarid environment, followed by abundance and richness evaluation under variable moisture conditions and changes in community structure which were reported. The flooded treatment condition resulted in less diverse community, dominated by two *Acaulospora* phylotypes. In flooded condition plants' phosphorus content was also reduced; it could be due to changed community structure of AM fungi and the capacity of remaining species to take up phosphorus under those conditions.

Another study on the composition of spores of AM fungi, under increased rainfall conditions at three time points in a year, was conducted by Sun et al. (2013), on a grassland ecosystem situated in Mongolia. They reported changes in spore diversity and relative abundance of common species under increased rainfall condition. On the same site, another study was conducted by Gao et al. (2016) using DNA sequencing method showing altered community composition and reduced richness of AM fungal species, under increased rainfall. The increased rainfall may alter the soil chemical properties like altered soil nitrate, which in succession changes the community structure of AM fungi. On the other hand, Li et al. (2015) reported that increased rainfall changes the composition, but no change in richness of the AM fungal community.

The mechanisms behind the changes in fungal communities by the influence of altered rainfall has been studied to unravel that whether rainfall pattern directly affect fungal community or it happened due to any indirect factor like responses of host plant.

21.4.3 Mechanisms Behind the Changes in Communities of AM Fungi in Altered Rainfall Regimes

Altered rainfall results, changes in soil properties, different physiological responses of host plant and transformed plant community structure, these variations may change the AM fungal communities of the region. Soil moisture content influences the oxygen concentration in the soil; therefore it can affect the AM fungal community (Deepika and Kothamasi 2015). Moreover, under reduced rainfall, the availability of soil nutrients decreased (Farooq et al. 2009), and in response to limited soil nutrients, AM fungal community structure may alter (Liu et al. 2015).

As altered rainfall affects carbon assimilation in plants and AM fungi depend on their host for its carbon requirement, thus it can directly change the AM fungal communities associated with the plant. There are reports available that plant can select different AM fungi by interfering root colonization (Vierheilig 2004) and providing benefit in terms of carbon to other AM fungal species (Bever et al. 2009; Kiers et al. 2011). Therefore altered rainfall regimes could change the AM fungal community composition either by reduced carbon allocation or differentially selecting any other AM fungal species. Another reason of fungal community change due to altered rainfall is differential root traits because of changes in released exudates in the soil, which may affect the fungal interaction and drive the community changes (Larson and Funk 2016; Padilla et al. 2013). A few reports also concluded that morphology of roots influences colonization of AM fungi (Eissenstat et al. 2015), though linking fungal colonization with root traits is very complex because roots show different turnover rates and lifespan. There are different hypotheses, explaining the AM fungal colonization. In driver hypothesis, plant community was shaped by AM fungal community, while in passenger hypothesis, AM fungal community was shaped by host plant (Hart et al. 2001). Alternatively, habitat hypothesis explains that plant and AM fungal communities co-vary according to environmental conditions and not because community is determining the other community (Zobel and Öpik 2014). Few studies also reported that community composition and richness of plant species regulate the AMF community abundance and diversity (Johnson et al. 2004), but AM fungi were not found to be strictly specific to host (van der Heijden et al. 2015). For instance, in a greenhouse experiment performed by Hausmann and Hawkes (2010), order of plant establishment determines the AM fungal community, showing the importance of plant species in regulating the AM fungal community. Therefore due to altered rainfall, changes in plant community may regulate the richness and community composition of AM fungi.

21.4.4 AM Fungal Community Responses Under Altered Rainfall Regimes Through Fungal Traits Study

Traits of an organism are defined as the phenotypic characters, which define the potential of an organism to survive under changing environmental conditions and/or possibly influence the environment, if associated with ecosystem processes (Aguilar-Trigueros et al. 2015). There are some traits which have dual role, where a response trait selection due to any environmental change can lead to a change in community trait composition that ultimately affect functioning of ecosystem (Aguilar-Trigueros et al. 2014). Therefore study of physiological and morphological traits of AM fungi is important to understand the AM fungal community structure in altered rainfall regimes, and the success of a community is also determined by both inter- and intraspecific variability in traits. An extensive group of traits was hypothesized in Table 21.1 to be implicated in capacity of mycorrhizal fungi to survive under altered rainfall. As a result, determinant traits involved in responses under water stress will permit us to predict AM fungal community to changing precipitation regimes.

21.5 Effect of Increasing Nitrogen Deposition on AM Fungi

Nitrogen is the most abundant element in the atmosphere and constitutes approximately 78% of it. Nitrogen is an integral component of all living forms but is unavailable to them in its diatomic form. It enters the ecosystem either in its oxidized form (nitrates, nitrites, N_2O_5, HNO_3, N_2O), reduced form (ammonia, ammonium

Table 21.1 Fungal traits favouring mycorrhizal fungi to survive under altered rainfall

Fungal trait	Function
Spore density	Reproduction
Spore size	Survival and reproduction
Hyphae	Root colonization
Arbuscules	Nutrient exchange between AM fungi and plant
Vesicles	Storage structures
Hyphae spread	Water and nutrient uptake
Hyphal size	Ability to absorb nutrient in different soil conditions
Hyphal density	Nutrient and water uptake
Anastomosis	Maintain homeostasis and restoration of hyphae
Hydrophobin	Host surface adhesion, spores and hyphal hydrophobicity
Aquaporin	Control water lost and uptake
Melanin	Host infection and environmental stress protection
Chitin	Plant-microbe interaction and cell wall structure
Glomalin	Soil structure modification
Starch, sugar alcohol, glycogen	Provide defence with desiccation
Trehalose	Infectivity and sporulation, control on glycolysis

salts) or organic form (urea, amines, uric acid, proteins) also commonly termed as Nr (reactive nitrogen) (Galloway et al. 2004). Diatomic nitrogen can be converted to Nr via three known methods, viz. atmospheric, biological and industrial nitrogen fixation. Atmospheric nitrogen fixation is the most primitive way of fixing nitrogen, but its contribution to the total conversion to Nr is less than other methods. Biologically, few species of *Bacteria* and Archaea can convert nitrogen to Nr. Post industrial revolution, industrial conversion replaced the biological nitrogen fixation (BNF) as the highest contributor of Nr through Haber-Bosch process. Agriculture and energy production are the major anthropogenic activities which greatly increased the Nr deposition (Fowler et al. 2013). According to Behera et al. (2013), the global emission of NH_3 was 48,400 Gg NH3-N annually and 50% of which was contributed by Asia. Food production, animal farming, fertilizers and agricultural residue comprise more than 80% of the total global ammonia emissions. According to an estimate, global Nr deposition level will be doubled by the middle of the twenty-first century from the levels at the end of the twentieth century (Galloway and Cowling 2002).

Of all the global change drivers, impact of increased Nr deposition has received the most attention because it is easier to manipulate. Moreover, availability of historic data on effects of nitrogen fertilization on plants facilitates many long-term studies on effect of nitrogen deposition on plant AMF interactions (Cotton 2018). One of such long-term study was conducted by Egerton-Warburton et al. (2001) where they studied how increase in atmospheric NOx levels over a period of 62 years (1937–1999) impacted the diversity and richness of AMF in the soil. Nitrogen enrichment led to replacement of 29 species AM community with the one composed of only 7 species and average loss of more than one species every year. Increased in atmospheric nitrogen deposition also resulted in decline in AMF abundance with exception of *Glomus aggregatum* and *G. leptotichum*. Treseder et al. (2018) recently reported an observation that with increase in nitrogen content in soil, there is gradual shifting of AMF community from *Gigasporaceae* to *Glomeraceae*.

Although there are many studies on the effects of increased Nr on AMF-plant interactions, the lack of consistency in results is still there as observed in studies on elevated carbon dioxide on such interactions. There are mixed responses recorded on how nitrogen alters the richness and diversity of AM fungi. The parameters considered under these studies include richness, diversity, abundance, root colonization rates and spore density of AM fungi. While many studies illustrated the decrease in diversity, abundance and biomass of AMF with increased Nr (Sigüenza et al. 2006; Blanke et al. 2012; Williams et al. 2017), some reported positive correlation of species richness and abundance (Egerton-Warburton et al. 2007; Zheng et al. 2014; Kim et al. 2015), and few studies found no significant correlation of enhanced nitrogen deposition on AMF (van Diepen et al. 2013; Li et al. 2015). Mohan et al. (2014) did an extensive meta-analysis with 1434 citations to find the effects of various global change drivers including enhancing Nr deposition. They found increase in mycorrhizal abundance and activity on nitrogen enrichment of the environment in 31% and 67% of the studies, respectively, whereas 44% and 33%

studies showed the opposite trends, respectively. Apart from the effects of various global change drivers on AMF, Mohan et al. (2014) also analysed their effects on the productivity of ecosystem where a clear consistency is observed as 94% of the studies showed positive impact of increased Nr on plant growth. Recently, similar experiments were conducted by Lin et al. (2020) to find the effects of nitrogen deposition on 10-year-old Chinese fir forests. The study revealed negative correlation of root colonization rate and spore density with nitrogen deposition.

Cotton (2018) documented the explanation for this disparity in the responses shown by AMF on increased N deposition. According to him, the difference in responses could be attributed to the methodologies adopted in examining the diversity, viz. from microscope-based morphological analysis of spores (Jefwa et al. 2006) to molecular identification (van Diepen et al. 2013). Earlier studies were mostly confined to morphological identification of spores and fungal hyphae, but with the advent of next-gen sequencing (NGS), molecular sequencing has been adopted for identification in recent studies (Peay 2014). Molecular sequencing provides detailed and more precise patterns of AMF community in enhanced Nr environment as it can unearth even those species which might escape morphological identification. With greater details come another problem of classification of newly identified and diverse AMF species (Peay 2014).

Weber et al. (2019) finely dissected the problem of shifting of AMF community and variable responses to Nr deposition. They classify the AMF in to three ecological guilds on the basis of biomass allocation taking cue from Maherali and Klironomos (2007), viz. edaphophilic (showing love towards soil through high allocation to extraradical hyphae), 'rhizophilic' (affinity towards plant rhizosphere through high allocation to intraradical hyphae) and 'ancestral' (lack of allocation preference with lower biomass). These guilds have variable preference for soil nitrogen with edaphophilic AMF (Gigasporaceae) showing negative correlation with soil nitrogen, whereas rhizophilic AMF (Glomeraceae) shows increased abundance with nitrogen enrichment of soil. Weber et al. (2019) also emphasized on the role of host plant in abundance of AMF as shrubs prefer the edaphophilic fungi as it requires greater amount of nutrients, whereas grasses host rhizophilic AMF in abundance to protect them from parasitic fungi.

Although there are many studies explaining the responses of AMF on increasing Nr deposition such as its abundance, richness, etc., very limited information is available on how it affects the ecosystem. There are few sporadic studies indeed compiled by Mohan et al. (2014) but without any conclusive reports.

21.6 Conclusion

AM fungi are widely distributed and exist as important symbionts in terrestrial ecosystem. This review compiles the effects of different changing climatic factors on functioning, composition and diversity of AM fungi. The changing global factors influence the AM fungi in two different ways: they affect either directly by altering their phenology or indirectly via the physiological changes in host plants. Studies available till date are limited in number and generally focused on short-term duration, limited geographical areas, considering only one or two global changes and limited vegetation types. Long-term studies would significantly improve our understanding to AM fungal community responses to global environmental changes. In addition to host-fungus interaction, changes in other soil organisms (which influences AM fungal communities) and soil properties can have potentially large impact on functioning of AM fungi due to global climate changes. But this research direction is not much explored till date. Thus it is clear that further studies are required to explore the mechanism of changes in AM fungal communities by the influence of global climate changes.

References

Aguilar-Trigueros CA, Powell JR, Anderson IC, Antonovics J, Rillig MC (2014) Ecological understanding of root-infecting fungi using trait-based approaches. Trends Plant Sci 19 (7):432–438

Aguilar-Trigueros CA, Hempel S, Powell JR, Anderson IC, Antonovics J, Bergmann J et al (2015) Branching out: towards a trait-based understanding of fungal ecology. Fungal Biol Rev 29 (1):34–41

Ainsworth EA, Rogers A (2007) The response of photosynthesis and stomatal conductance to rising [CO_2]: mechanisms and environmental interactions. Plant Cell Environ 30(3):258–270

Al Jaouni S, Saleh AM, Wadaan MA, Hozzein WN, Selim S, AbdElgawad H (2018) Elevated CO_2 induces a global metabolic change in basil (*Ocimum basilicum* L.) and peppermint (*Mentha piperita* L.) and improves their biological activity. J Plant Physiol 224:121–131

Aroca R, Porcel R, Ruiz-Lozano JM (2007) How does arbuscular mycorrhizal symbiosis regulate root hydraulic properties and plasma membrane aquaporins in *Phaseolus vulgaris* under drought, cold or salinity stresses? New Phytol 173(4):808–816

Augé RM (2001) Water relations, drought and vesicular-arbuscular mycorrhizal symbiosis. Mycorrhiza 11(1):3–42

Augé RM, Toler HD, Saxton AM (2015) Arbuscular mycorrhizal symbiosis alters stomatal conductance of host plants more under drought than under amply watered conditions: a meta-analysis. Mycorrhiza 25(1):13–24

Bago B, Bécard G (2002) Bases of the obligate biotrophy of arbuscular mycorrhizal fungi. In: Mycorrhizal technology in agriculture. Birkhäuser, Basel, pp 33–48

Bago B, Pfeffer PE, Shachar-Hill Y (2000) Carbon metabolism and transport in arbuscular mycorrhizas. Plant Physiol 124(3):949–958

Bainard LD, Bainard JD, Hamel C, Gan Y (2014) Spatial and temporal structuring of arbuscular mycorrhizal communities is differentially influenced by abiotic factors and host crop in a semi-arid prairie agroecosystem. FEMS Microbiol Ecol 88(2):333–344

Becklin KM, Mullinix GW, Ward JK (2016) Host plant physiology and mycorrhizal functioning shift across a glacial through future [CO_2] gradient. Plant Physiol 172(2):789–801

Behera SN, Sharma M, Aneja VP, Balasubramanian R (2013) Ammonia in the atmosphere: a review on emission sources, atmospheric chemistry and deposition on terrestrial bodies. Environ Sci Pollut Res 20(11):8092–8131

Bever JD, Richardson SC, Lawrence BM, Holmes J, Watson M (2009) Preferential allocation to beneficial symbiont with spatial structure maintains mycorrhizal mutualism. Ecol Lett 12 (1):13–21

Bidartondo MI, Ellis C, Kauserud H, Kennedy PG, Lilleskov E, Suz L, Andrew C (2018) Climate change: fungal responses and effects. In: Willis KJ (ed) State of the world's Fungi 2018. Royal Botanical Gardens, Kew, UK, pp 62–69
Blanke V, Bassin S, Volk M, Fuhrer J (2012) Nitrogen deposition effects on subalpine grassland: the role of nutrient limitations and changes in mycorrhizal abundance. Acta Oecologica 45:57–65
Borowicz VA (2001) Do arbuscular mycorrhizal fungi alter plant–pathogen relations? Ecology 82 (11):3057–3068
Botnen S, Kauserud H, Carlsen T, Blaalid R, Høiland K (2015) Mycorrhizal fungal communities in coastal sand dunes and heaths investigated by pyrosequencing analyses. Mycorrhiza 25 (6):447–456
Bowles TM, Barrios-Masias FH, Carlisle EA, Cavagnaro TR, Jackson LE (2016) Effects of arbuscular mycorrhizae on tomato yield, nutrient uptake, water relations, and soil carbon dynamics under deficit irrigation in field conditions. Sci Total Environ 566:1223–1234
Brundrett M (1991) Mycorrhizas in natural ecosystems. In: Advances in ecological research, vol 21. Academic Press, London, pp 171–313
Camenzind T, Hempel S, Homeier J, Horn S, Velescu A, Wilcke W, Rillig MC (2014) Nitrogen and phosphorus additions impact arbuscular mycorrhizal abundance and molecular diversity in a tropical montane forest. Glob Chang Biol 20(12):3646–3659
Coakley SM, Scherm H, Chakraborty S (1999) Climate change and plant disease management. Annu Rev Phytopathol 37(1):399–426
Compant S, Van Der Heijden MG, Sessitsch A (2010) Climate change effects on beneficial plant–microorganism interactions. FEMS Microbiol Ecol 73(2):197–214
Cotton TA (2018) Arbuscular mycorrhizal fungal communities and global change: an uncertain future. FEMS Microb Ecol 94(11):fiy179
Cotton TA, Fitter AH, Miller RM, Dumbrell AJ, Helgason T (2015) Fungi in the future: interannual variation and effects of atmospheric change on arbuscular mycorrhizal fungal communities. New Phytol 205(4):1598–1607
Deepika S, Kothamasi D (2015) Soil moisture—a regulator of arbuscular mycorrhizal fungal community assembly and symbiotic phosphorus uptake. Mycorrhiza 25(1):67–75
Diagne N, Ngom M, Djighaly PI, Fall D, Hocher V, Svistoonoff S (2020) Roles of arbuscular mycorrhizal fungi on plant growth and performance: importance in biotic and abiotic stressed regulation. Diversity 12(10):370
Dıaz-Zorita M, Perfect E, Grove JH (2002) Disruptive methods for assessing soil structure. Soil Tillage Res 64(1–2):3–22
Dong Y, Wang Z, Sun H, Yang W, Xu H (2018) The response patterns of arbuscular mycorrhizal and ectomycorrhizal symbionts under elevated CO_2: a meta-analysis. Front Microbiol 9:1248
Drake BG, Gonzàlez-Meler MA, Long SP (1997) More efficient plants: a consequence of rising atmospheric CO_2? Annu Rev Plant Biol 48(1):609–639
Dumbrell AJ, Ashton PD, Aziz N, Feng G, Nelson M, Dytham C et al (2011) Distinct seasonal assemblages of arbuscular mycorrhizal fungi revealed by massively parallel pyrosequencing. New Phytol 190(3):794–804
Egerton-Warburton LM, Graham RC, Allen EB, Allen MF (2001) Reconstruction of the historical changes in mycorrhizal fungal communities under anthropogenic nitrogen deposition. Proc R Soc Lond Ser B: Biol Sci 268(1484):2479–2484
Egerton-Warburton LM, Johnson NC, Allen EB (2007) Mycorrhizal community dynamics following nitrogen fertilization: a cross-site test in five grasslands. Ecol Monogr 77(4):527–544
Eissenstat DM, Kucharski JM, Zadworny M, Adams TS, Koide RT (2015) Linking root traits to nutrient foraging in arbuscular mycorrhizal trees in a temperate forest. New Phytol 208 (1):114–124
Farooq M, Wahid A, Kobayashi N, Fujita DBSMA, Basra SMA (2009) Plant drought stress: effects, mechanisms and management. In: Sustainable agriculture. Springer, Dordrecht, pp 153–188

Fowler D, Coyle M, Skiba U, Sutton MA, Cape JN, Reis S et al (2013) The global nitrogen cycle in the twenty-first century. Philos Trans R Soc B: Biol Sci 368(1621):20130164
Fu Y, Pan X, Yang Y, Chen F, Liu P (2017) Climatological characteristics of summer precipitation over East Asia measured by TRMM PR: a review. J Meteorol Res 31(1):142–159
Galloway JN, Cowling EB (2002) Reactive nitrogen and the world: 200 years of change. AMBIO J Hum Environ 31(2):64–71
Galloway JN, Dentener FJ, Capone DG, Boyer EW, Howarth RW, Seitzinger SP et al (2004) Nitrogen cycles: past, present, and future. Biogeochemistry 70(2):153–226
Gao C, Kim YC, Zheng Y, Yang W, Chen L, Ji NN et al (2016) Increased precipitation, rather than warming, exerts a strong influence on arbuscular mycorrhizal fungal community in a semiarid steppe ecosystem. Botany 94(6):459–469
Gavito ME, Curtis PS, Mikkelsen TN, Jakobsen I (2000) Atmospheric CO_2 and mycorrhiza effects on biomass allocation and nutrient uptake of nodulated pea (*Pisum sativum* L.) plants. J Exp Bot 51(352):1931–1938
Gavito ME, Olsson PA, Rouhier H, Medina-Peñafiel A, Jakobsen I, Bago A, Azcón-Aguilar C (2005) Temperature constraints on the growth and functioning of root organ cultures with arbuscular mycorrhizal fungi. New Phytol 168(1):179–188
Gehring CA, Mueller RC, Whitham TG (2006) Environmental and genetic effects on the formation of ectomycorrhizal and arbuscular mycorrhizal associations in cottonwoods. Oecologia 149 (1):158–164
Ghasemzadeh A, Jaafar HZ (2011) Effect of CO_2 enrichment on synthesis of some primary and secondary metabolites in ginger (*Zingiber officinale* roscoe). Int J Mol Sci 12(2):1101–1114
Giri B, Saxena B (2017) Response of arbuscular mycorrhizal fungi to global climate change and their role in terrestrial ecosystem C and N cycling. In: Mycorrhiza-function, diversity, state of the art. Springer, Cham, pp 305–327
Goicoechea N, Baslam M, Erice G, Irigoyen JJ (2014) Increased photosynthetic acclimation in alfalfa associated with arbuscular mycorrhizal fungi (AMF) and cultivated in greenhouse under elevated CO_2. J Plant Physiol 171(18):1774–1781
Grover M, Maheswari M, Desai S, Gopinath KA, Venkateswarlu B (2015) Elevated CO_2: plant associated microorganisms and carbon sequestration. Appl Soil Ecol 95:73–85
Gutknecht JL, Field CB, Balser TC (2012) Microbial communities and their responses to simulated global change fluctuate greatly over multiple years. Glob Chang Biol 18(7):2256–2269
Hart MM, Reader RJ, Klironomos JN (2001) Life-history strategies of arbuscularmycorrhizal fungi in relation to their successional dynamics. Mycologia 93(6):1186–1194
Haugen LM, Smith SE (1992) The effect of high temperature and fallow period on infection of mung bean and cashew roots by the vesicular-arbuscular mycorrhizal fungus Glomusintraradices. Plant Soil 145(1):71–80
Hausmann NT, Hawkes CV (2010) Order of plant host establishment alters the composition of arbuscular mycorrhizal communities. Ecology 91(8):2333–2343
Hawkes CV, Hartley IP, Ineson P, Fitter AH (2008) Soil temperature affects carbon allocation within arbuscular mycorrhizal networks and carbon transport from plant to fungus. Glob Chang Biol 14(5):1181–1190
Heinemeyer A, Ridgway KP, Edwards EJ, Benham DG, Young JPW, Fitter AH (2004) Impact of soil warming and shading on colonization and community structure of arbuscular mycorrhizal fungi in roots of a native grassland community. Glob Chang Biol 10(1):52–64
Ibrahim MH, Jaafar HZ (2012) Impact of elevated carbon dioxide on primary, secondary metabolites and antioxidant responses of *Eleais guineensis* Jacq.(oil palm) seedlings. Molecules 17 (5):5195–5211
Jakobsen I, Smith SE, Smith FA, Watts-Williams SJ, Clausen SS, Grønlund M (2016) Plant growth responses to elevated atmospheric CO_2 are increased by phosphorus sufficiency but not by arbuscular mycorrhizas. J Exp Botany, erw383

Jefwa JM, Sinclair R, Maghembe JA (2006) Diversity of glomale mycorrhizal fungi in maize/sesbania intercrops and maize monocrop systems in southern Malawi. Agrofor Syst 67 (2):107–114

Jiang W, Gou G, Ding Y (2013) Influences of arbuscular mycorrhizal fungi on growth and mineral element absorption of chenglu hybrid bamboo seedlings. Pak J Bot 45(1):303–310

Jifon JL, Graham JH, Drouillard DL, Syvertsen JP (2002) Growth depression of mycorrhizal Citrus seedlings grown at high phosphorus supply is mitigated by elevated CO_2. New Phytol 153 (1):133–142

Johnson D, Vandenkoornhuyse PJ, Leake JR, Gilbert L, Booth RE, Grime JP et al (2004) Plant communities affect arbuscular mycorrhizal fungal diversity and community composition in grassland microcosms. New Phytol 161(2):503–515

Johri AK, Oelmüller R, Dua M, Yadav V, Kumar M, Tuteja N et al (2015) Fungal association and utilization of phosphate by plants: success, limitations, and future prospects. Front Microbiol 6:984

Kaya C, Higgs D, Kirnak H, Tas I (2003) Mycorrhizal colonisation improves fruit yield and water use efficiency in watermelon (Citrullus lanatus Thunb.) grown under well-watered and water-stressed conditions. Plant Soil 253(2):287–292

Kiers ET, Duhamel M, Beesetty Y, Mensah JA, Franken O, Verbruggen E et al (2011) Reciprocal rewards stabilize cooperation in the mycorrhizal symbiosis. Science 333(6044):880–882

Kilpeläinen J, Aphalo PJ, Lehto T (2020) Temperature affected the formation of arbuscularmycorrhizas and ectomycorrhizas in Populusangustifolia seedlings more than a mild drought. Soil Biol Biochem 146:107798

Kim YC, Gao C, Zheng Y, He XH, Yang W, Chen L et al (2015) Arbuscularmycorrhizal fungal community response to warming and nitrogen addition in a semiarid steppe ecosystem. Mycorrhiza 25(4):267–276

Knohl A, Veldkamp E (2011) Indirect feedbacks to rising CO 2. Nature 475(7355):177–178

Kytöviita MM (2005) Asymmetric symbiont adaptation to Arctic conditions could explain why high Arctic plants are non-mycorrhizal. FEMS Microbiol Ecol 53(1):27–32

Larson JE, Funk JL (2016) Seedling root responses to soil moisture and the identification of a belowground trait spectrum across three growth forms. New Phytol 210(3):827–838

Li X, Zhu T, Peng F, Chen Q, Lin S, Christie P, Zhang J (2015) Inner Mongolian steppe arbuscular mycorrhizal fungal communities respond more strongly to water availability than to nitrogen fertilization. Environ Microbiol 17(8):3051–3068

Lin C, Wang Y, Liu M, Li Q, Xiao W, Song X (2020) Effects of nitrogen deposition and phosphorus addition on arbuscular mycorrhizal fungi of Chinese fir (*Cunninghamia lanceolata*). Sci Rep 10(1):1–8

Liu A, Hamel C, Hamilton RI, Ma BL, Smith DL (2000) Acquisition of cu, Zn, Mn and Fe by mycorrhizal maize (*Zea mays* L.) grown in soil at different P and micronutrient levels. Mycorrhiza 9(6):331–336

Liu A, Wang B, Hamel C (2004) Arbuscular mycorrhiza colonization and development at suboptimal root zone temperature. Mycorrhiza 14(2):93–101

Liu Y, Mao L, Li J, Shi G, Jiang S, Ma X et al (2015) Resource availability differentially drives community assemblages of plants and their root-associated arbuscular mycorrhizal fungi. Plant Soil 386(1–2):341–355

Lumini E, Vallino M, Alguacil MM, Romani M, Bianciotto V (2011) Different farming and water regimes in Italian rice fields affect arbuscular mycorrhizal fungal soil communities. Ecol Appl 21(5):1696–1707

Maherali H, Klironomos JN (2007) Influence of phylogeny on fungal community assembly and ecosystem functioning. Science 316(5832):1746–1748

Martin CA, Stutz JC (2004) Interactive effects of temperature and arbuscular mycorrhizal fungi on growth, P uptake and root respiration of *Capsicum annuum* L. Mycorrhiza 14(4):241–244

Martínez-García LB, de Dios Miranda J, Pugnaire FI (2012) Impacts of changing rainfall patterns on mycorrhizal status of a shrub from arid environments. Eur J Soil Biol 50:64–67

Marulanda A, Azcon R, Ruiz-Lozano JM (2003) Contribution of six arbuscularmycorrhizal fungal isolates to water uptake by Lactuca sativa plants under drought stress. Physiologia Plantarum 119(4):526–533

Miller RM, Jastrow JD, Reinhardt DR (1995) External hyphal production of vesicular-arbuscular mycorrhizal fungi in pasture and tallgrass prairie communities. Oecologia 103(1):17–23

Mohan JE, Cowden CC, Baas P, Dawadi A, Frankson PT, Helmick K et al (2014) Mycorrhizal fungi mediation of terrestrial ecosystem responses to global change: mini-review. Fungal Ecol 10:3–19

Monz CA, Hunt HW, Reeves FB, Elliott ET (1994) The response of mycorrhizal colonization to elevated CO_2 and climate change in *Pascopyrum smithii* and *Bouteloua gracilis*. Plant Soil 165 (1):75–80

Mueller RC, Bohannan BJ (2015) Shifts in the phylogenetic structure of arbuscularmycorrhizal fungi in response to experimental nitrogen and carbon dioxide additions. Oecologia 179 (1):175–185

Olsson PA, Eriksen B, Dahlberg A (2004) Colonization by arbuscular mycorrhizal and fine endophytic fungi in herbaceous vegetation in the Canadian high Arctic. Can J Bot 82 (11):1547–1556

Padilla FM, Aarts BH, Roijendijk YO, de Caluwe H, Mommer L, Visser EJ, de Kroon H (2013) Root plasticity maintains growth of temperate grassland species under pulsed water supply. Plant Soil 369(1–2):377–386

Parke JL, Linderman RG, Trappe JM (1983) Effect of root zone temperature on ectomycorrhiza and vesicular–arbuscular mycorrhiza formation in disturbed and undisturbed forest soils of Southwest Oregon. Can J For Res 13(4):657–665

Peay KG (2014) Back to the future: natural history and the way forward in modern fungal ecology. Fungal Ecol 12:4–9

Phillips RP, Brzostek E, Midgley MG (2013) The mycorrhizal-associated nutrient economy: a new framework for predicting carbon–nutrient couplings in temperate forests. New Phytol 199 (1):41–51

Poorter H, Navas ML (2003) Plant growth and competition at elevated CO_2: on winners, losers and functional groups. New Phytol 157(2):175–198

Ren H, Gui W, Bai Y, Stein C, Rodrigues JL, Wilson GW et al (2018) Long-term effects of grazing and topography on extra-radical hyphae of arbuscular mycorrhizal fungi in semi-arid grasslands. Mycorrhiza 28(2):117–127

Rillig MC, Mummey DL (2006) Mycorrhizas and soil structure. New Phytol 171(1):41–53

Saleh AM, Abdel-Mawgoud M, Hassan AR, Habeeb TH, Yehia RS, AbdElgawad H (2020) Global metabolic changes induced by arbuscular mycorrhizal fungi in oregano plants grown under ambient and elevated levels of atmospheric CO_2. Plant Physiol Biochem 151:255–263

Sánchez-Blanco MJ, Ferrández T, Morales MA, Morte A, Alarcón JJ (2004) Variations in water status, gas exchange, and growth in Rosmarinus officinalis plants infected with Glomus deserticola under drought conditions. J Plant Physiol 161(6):675–682

Sanders IR, Streitwolf-Engel R, Van der Heijden MGA, Boller T, Wiemken A (1998) Increased allocation to external hyphae of arbuscular mycorrhizal fungi under CO_2 enrichment. Oecologia 117(4):496–503

Santander C, Aroca R, Ruiz-Lozano JM, Olave J, Cartes P, Borie F, Cornejo P (2017) Arbuscular mycorrhiza effects on plant performance under osmotic stress. Mycorrhiza 27(7):639–657

Schenck NC, Graham SO, Green NE (1975) Temperature and light effect on contamination and spore germination of vesicular-arbuscular mycorrhizal fungi. Mycologia 67(6):1189–1192

Sigüenza C, Corkidi L, Allen EB (2006) Feedbacks of soil inoculum of mycorrhizal fungi altered by N deposition on the growth of a native shrub and an invasive annual grass. Plant Soil 286 (1–2):153–165

Smith SE, Smith FA (2011) Roles of arbuscular mycorrhizas in plant nutrition and growth: new paradigms from cellular to ecosystem scales. Annu Rev Plant Biol 62:227–250

Soudzilovskaia NA, Douma JC, Akhmetzhanova AA, van Bodegom PM, Cornwell WK, Moens EJ et al (2015) Global patterns of plant root colonization intensity by mycorrhizal fungi explained by climate and soil chemistry. Glob Ecol Biogeogr 24(3):371–382

Staddon PL, Thompson KEN, Jakobsen I, Grime JP, Askew AP, Fitter AH (2003) Mycorrhizal fungal abundance is affected by long-term climatic manipulations in the field. Glob Chang Biol 9(2):186–194

Staddon PL, Gregersen R, Jakobsen I (2004) The response of two *Glomus* mycorrhizal fungi and a fine endophyte to elevated atmospheric CO_2, soil warming and drought. Glob Chang Biol 10 (11):1909–1921

Stevens BM, Propster J, Wilson GW, Abraham A, Ridenour C, Doughty C, Johnson NC (2018) Mycorrhizal symbioses influence the trophic structure of the Serengeti. J Ecol 106(2):536–546

Stocker TF, Qin D, Plattner GK, Tignor M, Allen SK, Boschung J, …, Alexander LV (2013) Climate Change 2013. The Physical Science Basis. Working Group I Contribution to the Fifth Assessment Report of the Intergovernmental Panel on Climate Change-Abstract for decision-makers; Changementsclimatiques 2013. Les elements scientifiques. Contribution du groupe de travail I au cinquieme rapport d'evaluation du grouped'expertsintergouvernementalsurl'evolution du CLIMAT-Resume al'intention des decideurs

Subramanian KS, Santhanakrishnan P, Balasubramanian P (2006) Responses of field grown tomato plants to arbuscular mycorrhizal fungal colonization under varying intensities of drought stress. Scient Hortic 107(3):245–253

Sun X, Su Y, Zhang Y, Wu M, Zhang Z, Pei K et al (2013) Diversity of arbuscular mycorrhizal fungal spore communities and its relations to plants under increased temperature and precipitation in a natural grassland. Chin Sci Bull 58(33):4109–4119

Tang J, Chen J, Chen X (2006) Response of 12 weedy species to elevated CO_2 in low-phosphorus-availability soil. Ecol Res 21(5):664–670

Tang J, Xu L, Chen X, Hu S (2009) Interaction between C4 barnyard grass and C3 upland rice under elevated CO_2: impact of mycorrhizae. Acta Oecologica 35(2):227–235

Tian D, Niu S (2015) A global analysis of soil acidification caused by nitrogen addition. Environ Res Lett 10(2):024019

Torrecillas E, Torres P, Alguacil MM, Querejeta JI, Roldán A (2013) Influence of habitat and climate variables on arbuscular mycorrhizal fungus community distribution, as revealed by a case study of facultative plant epiphytism under semiarid conditions. Appl Environ Microbiol 79(23):7203–7209

Treseder KK, Allen EB, Egerton-Warburton LM, Hart MM, Klironomos JN, Maherali H, Tedersoo L (2018) Arbuscular mycorrhizal fungi as mediators of ecosystem responses to nitrogen deposition: a trait-based predictive framework. J Ecol 106(2):480–489

van der Heijden MG, Martin FM, Selosse MA, Sanders IR (2015) Mycorrhizal ecology and evolution: the past, the present, and the future. New Phytol 205(4):1406–1423

van Diepen LT, Entwistle EM, Zak DR (2013) Chronic nitrogen deposition and the composition of active arbuscular mycorrhizal fungi. Appl Soil Ecol 72:62–68

Vierheilig H (2004) Regulatory mechanisms during the plant arbuscular mycorrhizal fungus interaction. *canadian.* J Bot 82(8):1166–1176

Weber SE, Diez JM, Andrews LV, Goulden ML, Aronson EL, Allen MF (2019) Responses of arbuscular mycorrhizal fungi to multiple coinciding global change drivers. Fungal Ecol 40:62–71

Williams A, Manoharan L, Rosenstock NP, Olsson PA, Hedlund K (2017) Long-term agricultural fertilization alters arbuscular mycorrhizal fungal community composition and barley (H ordeumvulgare) mycorrhizal carbon and phosphorus exchange. New Phytol 213(2):874–885

Wilson GW, Rice CW, Rillig MC, Springer A, Hartnett DC (2009) Soil aggregation and carbon sequestration are tightly correlated with the abundance of arbuscular mycorrhizal fungi: results from long-term field experiments. Ecol Lett 12(5):452–461

Wu QS, Zou YN, Xia RX (2006) Effects of water stress and arbuscular mycorrhizal fungi on reactive oxygen metabolism and antioxidant production by citrus (Citrus tangerine) roots. Eur J Soil Biol 42(3):166–172

Zheng Y, Kim YC, Tian XF, Chen L, Yang W, Gao C et al (2014) Differential responses of arbuscular mycorrhizal fungi to nitrogen addition in a near pristine Tibetan alpine meadow. FEMS Microbiol Ecol 89(3):594–605

Zhu XC, Song FB, Liu SQ, Liu TD (2011) Effects of arbuscular mycorrhizal fungus on photosynthesis and water status of maize under high temperature stress. Plant Soil 346(1–2):189–199

Zobel M, Öpik M (2014) Plant and arbuscular mycorrhizal fungal (AMF) communities–which drives which? J Veg Sci 25(5):1133–1140

Part III
Impact of Climate Change on Soil-Plant Dynamics and Functionality

Chapter 22
Phytoremediation of Polycyclic Aromatic Hydrocarbons-Contaminated Soils

Nima Dolatabadi, Soheyla Mohammadi Alagoz, Behnam Asgari Lajayer, and Eric D. van Hullebusch

Abstract Polycyclic aromatic hydrocarbons (PAHs), one of the main organic pollutants, are produced mainly during the uncomplete organic material combustion (e.g., wood, petrol, oil, and coal). Catalytic cracking towers, coke and aluminum production, coal-tar pitch and asphalt production, carbon black, coal gasification and liquefying plants, residential heating, and the activities associated with petroleum refineries as well as the exhaust of motor vehicles are the main PAHs anthropogenic sources. PAHs are worldwide concern for soils and water bodies that generate serious threats to human and environmental health. The human exposure pathways are being influenced by the environmental compartments contaminated. Many PAHs possess carcinogenic, mutagenic, and toxic properties. PAHs are mostly hydrophobic compounds. Consequently, they are easily absorbed into the gastrointestinal mammal tracts. To remove these pollutants, various remediation technologies have been examined. One of these technologies is bioremediation, which is cost-effective and safe. As a promising technology, phytoremediation can be applied to remediate soils contaminated with PAHs. Phytoremediation is based on the combination of microbe's actions living in the rhizosphere and plants, which show promises to remediate hydrocarbon-contaminated waterways and land. The appropriate plant combinations and their related endophytes improve organic pollutants biodegradation in the endosphere and/or rhizosphere, contributing to the improved pollutant removal rates and yields.

N. Dolatabadi
Department of Plant Breeding and Biotechnology, Faculty of Agriculture, University of Tabriz, Tabriz, Iran

S. Mohammadi Alagoz
Department of Plant Production and Genetics, Faculty of Agriculture, Urmia University, Urmia, Iran

B. Asgari Lajayer (✉)
Health and Environment Research Center, Tabriz University of Medical Sciences, Tabriz, Iran
e-mail: h-asgari@tabrizu.ac.ir

E. D. van Hullebusch (✉)
Université de Paris, Institut de Physique du Globe de Paris, CNRS, Paris, France
e-mail: vanhullebusch@ipgp.fr

D. K. Choudhary et al. (eds.), *Climate Change and the Microbiome*, Soil Biology 63,
https://doi.org/10.1007/978-3-030-76863-8_22

Keywords PAHs · Pollution · Environment · Soil bioremediation · Biodegradation · Phytoremediation

22.1 Introduction

Polycyclic aromatic hydrocarbons (PAHs) are organic compounds being mainly pale-yellow, white, or colorless solids (Abdel-Shafy and Mansour 2016). PAHs display different environmental toxicity and chemical structures grouped in a family of several chemical compounds that are widely spread in the environment (Veltman et al. 2012). PAHs enter the environment via different routes and mostly exist as a mixture that contains at least two aromatic rings or more (Thorsen et al. 2004). Some PAHs are mostly produced by the industry (Wang et al. 2011). These compounds through several different action modes generate toxic effects on living organisms. The PAHs toxicity is explained by strong interaction with cellular membrane functions as well as in membrane-associated enzyme systems (Arslan et al. 2017). PAHs have been proved to generate mutagenic and carcinogenic effects. Moreover, they are recognized as powerful and effective immune suppressants. It has been documented that they affect humoral immunity, the development of immune system, and resistance to a host (Armstrong et al. 2004). PAHs are also produced during biological conversion of organic materials. In addition, they are a product of unfinished combustion of natural sources (brush fires and forest) as well as that of man-made sources (cigarette smokes and automobile emissions). PAHs are usually encountered in water, soil, and atmospheric compartments (Baklanov et al. 2007). PAHs are a unique class of organic pollutants that contain two or more fused aromatic rings (Fig. 22.1). USEPA has listed 16 PAHs as priority compounds to be analyzed in various environmental matrices.

The physical properties and molecular structure of 16 PAHs are shown in Table 22.1 and in Fig. 22.1, respectively.

Menzie et al. (1992) have well summarized the ubiquitous nature of PAHs in the environment. PAH compounds mainly consist of hydrogen and carbon atoms. PAHs are chemically composed of at least two benzene rings bonded in angular, cluster, or linear arrangements. High boiling and melting points (solid), very low aqueous solubility, and low vapor pressure are the main features of PAHs. The latter two features go down when molecular weight increases, whereas reduction and oxidation resistance go up (Arey and Atinkson 2003). Aqueous PAHs solubility reduces for every extra ring (Masih et al. 2010). Meanwhile, since PAHs are very lipophilic, they are highly soluble in organic solvents. Furthermore, PAHs possess several different functions, including physiological action, resistance to corrosion, ability to emit, conductivity, resistance to heat, and sensitivity to light (Akyuz and Cabuk 2010). PAHs display very characteristic UV absorbance spectra because every aromatic structure of the ring possesses a unique UV spectrum. Consequently, every isomer possesses a different spectrum of UV absorbance, which is particularly useful in identifying PAHs (Mansouri et al. 2020). Furthermore, most PAHs are also

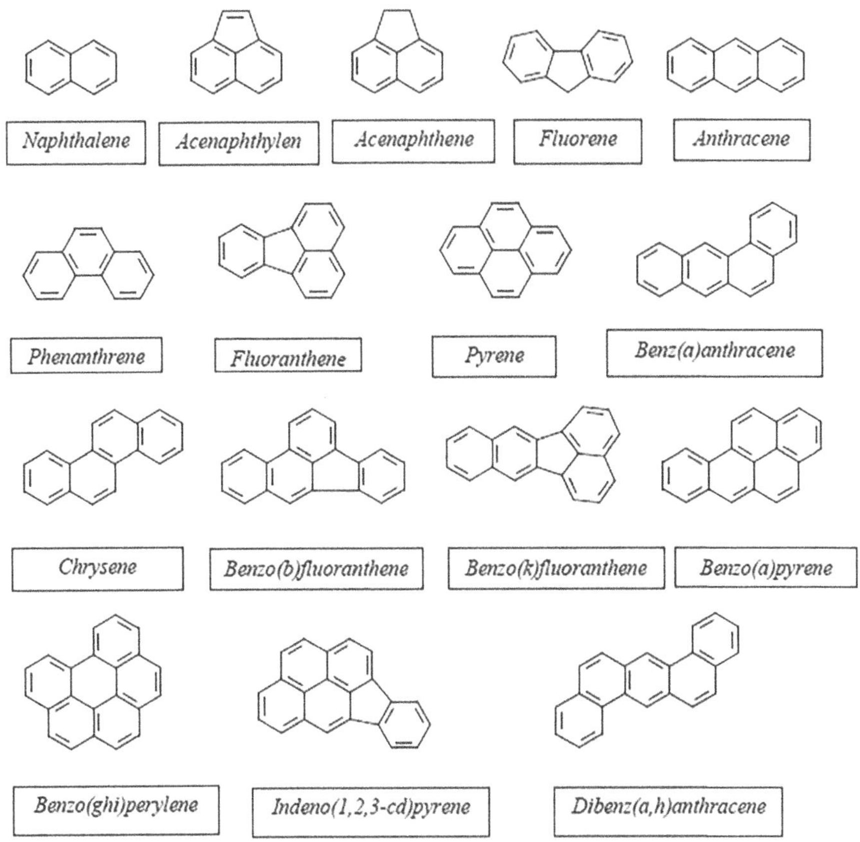

Fig. 22.1 Molecular structures of PAHs (Smol and Włodarczyk-Makuła 2017)

fluorescent, emitting characteristic wavelengths of light when they are excited (Rivera-Figueroa et al. 2004; Abdel-Shafy and Mansour 2016).

22.2 PAHs Sources and Uses

The main PAHs sources are due to the unfinished organic material combustion, including wood, oil, and coal. Industries do not synthesize chemically PAHs for their aims. However, many PAHs are used for a few commercial purposes. PAHs are usually utilized as intermediaries in lubricating materials, thermosetting plastics, photographic products, agricultural products, pharmaceuticals, etc. (Fig. 22.2) (Kaminski et al. 2008). Nevertheless, the following uses are the general ones of some PAHs:

Table 22.1 Physical properties and molecular structure PAHs (WHO 1998)

No.	PAHs	Symbol	Molecular weight (g/mol)	Melting point °C	Boiling point °C	Vapor pressure kPa	Solubility in water (mg/L)	Number of the rings in the molecule	Number of benzene rings in the molecule	Log (K_{ow})	The rate of carcino-genicity
1	Naphthalene	Nap	128.2	80.2	218	1.1×10^{-2}	3.93	2	2	3.37	–
2	Acenaphthylene	Acyl	154.2	92–93	265–280	3.9×10^{-3}	3.93	3	2	4.07	–
3	Acenaphthene	Ac	154.2	90–96	278–279	2.1×10^{-3}	1.93	3	2	4.33	–
4	Fluorene	Fl	166.2	116–118	293–295	8.7×10^{-5}	1.68–1.98	3	2	4.18	–
5	Phenanthrene	Fen	178.2	96–101	339–340	2.3×10^{-5}	1.2	3	3	4.57	–
6	Anthracene	Ant	178.2	216–219	340	36×10^{-6}	0.076	3	3	4.54	0.01
7	Fluoranthene	Flu	202.3	107–111	375–393	6.5×10^{-7}	0.2–2.6	4	3	5.22	–
8	Pyrene	Pir	202.3	150–156	360–404	3.1×10^{-6}	0.077	4	4	5.32	–
9	Benzo[a] anthracene	BaA	228.3	157–167	435	1.5×10^{-8}	0.01	4	4	5.61	0.1
10	Chrysene	Chr	228.3	252–256	441–448	5.7×10^{-10}	0.0028	4	4	5.63	0.01
11	Benzo[b] fluoranthene	BbF	252.3	167–168	481	6.7×10^{-8}	0.0012	5	5	6.06	1
12	Benzo[k] fluoranthene	BkF	252.3	198–217	471–480	2.1×10^{-8}	0.00076	5	4	6.57	0.1
13	Benzo[a]pyrene	BkP	252.3	177–179	493–496	7.3×10^{-10}	0.0023	5	4	6.84	0.1

14	Dibenzo[a,h] anthracene	DahA	278.4	266–270	524	1.3×10^{-11}	0.0005	5	5	6.86	5
15	Benzo[g,h,i] perylene	BPer	276.3	275–278	525	1.3×10^{-11}	0.00026	6	6	7.66	0.01
16	Indeno[1,2,3-cd]pyrene	IP	276.3	162–163	530	Ca.10^{-11}	0.062	6	5	7.23	0.1

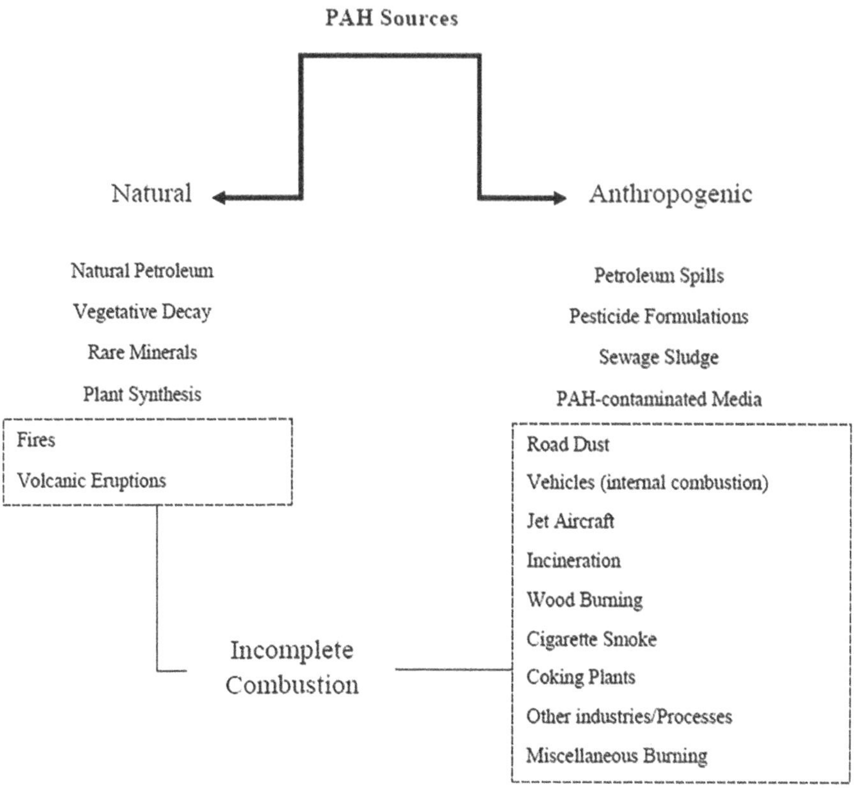

Fig. 22.2 PAHs Sources

- Acenaphthene: Manufacturing pharmaceuticals, pesticides, plastics, dyes, and pigments
- Anthracene: Manufacturing of pigments and dyes and a diluent of wood preservatives
- Fluoranthene: manufacturing pharmaceuticals dyes and agrochemicals
- Fluorene: manufacturing thermoset plastic, pesticides, dyes, pigments, and pharmaceuticals
- Pyrene: manufacturing pigment
- Phenanthrene: manufacturing pesticides and resins

Some of other PAHs contain asphalt, which, in addition to roofing tar, are used for making the surface of roads. Moreover, the precise refined PAHs products are also utilized in liquid crystals, functional plastics, and electronics.

During the pyrolysis process, pyrogenic PAHs are produced when organic matters are under low oxygen or no oxygen and the temperature is high. The pyrolytic processes that deliberately take place consist in the thermal petroleum residual cracking into lighter hydrocarbons or the destructive coal distillation into coal tar and coke. Meanwhile, other accidentally processes take place during the unfinished fuel oil combustion in heating systems, the unfinished wood combustion in forest fireplaces and fires, and the unfinished motor fuel combustion in trucks. The pyrogenic processes take place at temperatures of 350 °C to more than 1200 °C. The pyrogenic PAHs usually exist in particular places close to the main PAHs sources as well as in higher concentrations in urban areas. Moreover, at lower temperatures, PAHs may be created. Crude oils also contain PAHs that have been generated over millions of years at 100–150 °C. In this regard, PAHs that are created over the maturation of crude oil and similar processes are known as petrogenic ones. These petrogenic PAHs are often found because of the wide use, storage, and transportation of crude oil and its products. The accumulation of enormous numbers of small releases of motor oil, gasoline, and substances related to transportation, aboveground and underground storage tank leaks, and freshwater and oceanic oil spills are some of the main petrogenic PAHs sources. It's a well-known fact that PAHs may be created over the unfinished organic substance combustion. In addition, there are PAHs in the petroleum products. Moreover, PAHs can be formed biologically, e.g., these compounds may be synthesized by particular bacteria and plants or created over the vegetative matter degradation. PAHs may be formed by either anthropogenic or natural processes (Tolosa et al. 1996).

PAHs sources in the environment are generally well-known and numerous. The real environmental samples are mainly a basis for identifying the contents of PAHs. Chemical fingerprinting has numerous techniques for distinguishing between the groups of sources of PAHs. These groups include oil-based, wood-based, or coal-based sources. Specific chemical indicators that there are in many samples are utilized to distinguish between these groups. In other words, the techniques of the identification of sources often include the studies related to establishing the main "background" ranges of PAHs in the environment. Furthermore, chemical fingerprinting may be used for identifying and allocating the non-point PAH sources in the environment in both residential and industrial areas (Wang et al. 2011). As already mentioned, the formation temperature is a way to identify the sources of PAHs for the reason that higher formation temperatures compared to lower temperature processes can form the PAHs that possess fewer alkylated chains. For example, PAHs rapidly created at very high temperatures and found in stack effluents at power plants will possess different PAH patterns compared to the distribution of the PAHs that exist at the site of crude oil spill. Recent cases usually occur at low temperatures during millions of years (Parker et al. 2012). Investigating the number of five-member hydrocarbon rings in PAHs is another method for distinguishing pyrogenic PAHs from petrogenic PAHs. Petroleum hydrocarbons possess higher number of five-member rings than pyrogenic substances because the extensive time of forming petroleum hydrocarbons supports marshaling the rings. The material of the source for the substances that are pyrolytically created is quickly converted into the rings with six members that are more stable (Tolosa et al. 1996). Although the PAHs sources in the environment are numerous, advances are continuing to identify their sources and distinguish between pyrogenic PAHs and the non-pyrolytic sources (Wang et al. 2014).

22.3 Ecotoxicological Effects of PAHs

Photooxidation and metabolism affect the PAHs toxicity to aquatic organisms. PAHs are reported to be more toxic in the presence of ultraviolet light, and they have moderate to severe toxicity to aquatic organisms. Toxicity of PAHs on the early development stages of fish, bone metabolism, liver metabolism, and reproduction has been reported in fish such as medaka (*Oryzias latipes*) and zebrafish (*Danio rerio*) (Honda and Suzuki 2020). PAHs are not likely to have toxic effects on the terrestrial invertebrates in soils except for when soils are highly contaminated. Immunity, development, reproduction, and tumors have negative influences on the abovementioned organisms. PAHs may be absorbed by mammals through several different routes, for example, ingestion, dermal contact, and inhalation (Veltman et al. 2012). In addition, PAHs may be absorbed by the root of plant and moved to other parts of plants. The uptake rate is usually controlled by the solubility, concentration, and physicochemical state of water as well as the type of soil. The phytotoxic impacts induced by PAHs are not very often reported. A database of detailed information about this aspect is still limited. Certain types of plants contain the materials that may keep them safe from the effects of PAHs. Some of other plants may synthesize PAHs; consequently, they act as the hormones of growth (Beyer et al. 2010). PAHs exist moderately in the environment and may be accumulated biologically. It is expected that the PAHs concentrations in shellfish and fishes are much higher than in the environment. It has been also shown that there is the biological accumulation in the terrestrial invertebrates. However, PAHs degradative metabolism is enough to prevent the biological magnifications (Tudoran and Putz 2012).

22.4 Effects of PAHs on Human's Health

Seventeen PAHs have been recognized, which given the potential exposure and the negative effects on the human's health are considerable concern, and thus, they are in the same group. Since PAHs are toxic and their diffusion is widespread, biomonitoring studies of human exposure to these compounds are of most important interest. Nevertheless, the effects of PAHs on human health are not precisely very similar. Some PAHs as known carcinogenic to humans have been classified by the International Agency for Research on Cancer (IARC 2010). The most important impact on human health caused by inhalation exposure to PAHs is a serious risk of lung cancer (Kim et al. 2013).

22.5 PAHs Removal

There are various methods for removing PAHs from the environment, including photochemical degradation and biodegradation (Abdel-Shafy et al. 2014). Moreover, PAHs removal from the atmosphere has been reported (Zhong and Zhu 2013).

22.5.1 Degradation of PAHs

PAHs degradation in the environment is mostly taking place via photolysis degradation, chemical degradation, and biodegradation processes (Nadarajah et al. 2002). Each of the abovementioned processes affects the individual fate of PAHs in different way. This is mostly why the structure of every PAH is unique and has a group of biological, chemical, and physical properties. Most studies have investigated PAHs biodegradation either with aerobic or anaerobic microorganisms. Few studies have examined other PAHs degradation processes, including chemical oxidation or photolysis.

22.5.1.1 Photolysis Degradation

The destruction of organic compounds such as PAHs initiated through the light absorption is described as photolysis (Manahan 1999). The abovementioned reactions excite the electrons inside the PAH molecules, thereby creating an unstable structural arrangement and allowing a number of chemical and physical processes to act on the compound (Schwarzenbach et al. 1993). The photolysis reactions of PAHs are similar to the reactions of biodegradation. In other words, PAHs are more effectively degraded whenever they are in the aqueous or vapor phases (Finlayson-Pitts and Pitts 1997). Little photodegradation takes place when PAHs are sorbed onto particles in soil or in the atmospheric compartment. This reaction occurs more effectively when the surface area of particles increases. Furthermore, photolysis reactions on light-colored particles such as alumina or silica gel are greater than darker particles such as carbon black (Abdel-Shafy and Mansour 2013). The photodegradation reactions, like the one occurring via microbial degradation, are directly affected by the PAHs structure. In most situations, the two-ring and linear PAHs and some clustered are degraded under direct light. Degradation of angular PAHs (e.g., dibenzo (a, h), anthracene, and phenanthrene) is slower due to the more stable structure of molecules. Moreover, photolysis is more effective onto PAH compounds with low molecular weight such as naphthalene, because the availability of such compounds is high, thereby possessing longer exposure times to sunlight (Korfmacher et al. 1980). Eventually, the molecules of PAHs can be deposited inside of the pore structure of fly ash whenever they are sorbed to the particles of fly ash.

Consequently, in the pore structure, sorbed PAHs are protected against the sunlight, which stops photolytic reactions (Niu et al. 2007).

22.5.1.2 Chemical Degradation

It seems that chemical oxidation reactions are the main PAHs degradation mechanism occurring under environmental conditions. The chemical oxidation occurs either as part of treatment technologies or during natural attenuation (Abdel-Shafy and Mansour 2013). The PAHs oxidation rate is directly affected by oxidizing agent, temperature, the structure and molecular weight of the compound, and its physical state. Physical treatment such as coagulation (Smol and Włodarczyk-Makuła 2017) and chemical treatment have been reported to remove PAHs from surface water effectively (Moursy and Abdel-Shafy 1983). Fluoranthene is the most stable PAH that was examined for oxidation by ozone (Alebic-Juretic et al. 1990). This can be the reason for the presence of fluoranthene in high concentrations in the studied soil. Hassan et al. (2015) have also shown that the synergistic impact of UV and ZnO or TiO_2 catalysis is impressive in PAH degradation in contaminated soils. Oxalic acid and iron oxides may establish a photo-Fenton-like system lacking extra H_2O_2 in the solid phase to improve the photodegradation of pyrene under ultraviolet (Wang et al. 2009). The use of Fenton's reagent process for remediation of environmental matrices contaminated with PAHs has been also well investigated (Flotron et al. 2005). Nowadays, the Fenton's reagent is used to the remediation of contaminated solid matrices, such as soils, sediments, or sludges (Watts et al. 2002). PAHs degradation with the use of a high ultrasound frequency has also been studied (Manariotis et al. 2011).

22.5.1.3 PAHs Biodegradation

PAHs biodegradation is the most common process of degradation. The first studies regarding PAHs biodegradation have paid special attention to the aerobic degradation. However, anaerobic degradation has been clearly demonstrated under denitrifying conditions (Haritash and Kaushik 2009). For microbial degradation of any particular PAH, this latter should be available for uptake (Fredslund et al. 2008). PAHs begin to be bioavailable whenever they are in either the vapor phase or the dissolved one. Bacteria cannot easily degrade PAHs adsorbed onto soil particles, for the reason that PAHs are separated from the enzymes that bacteria use to break down PAHs (Rappert et al. 2006). The bioavailability of any particular PAH is complex, e.g., it has been shown that aged PAHs soil contamination affects how quickly such compounds desorb from soil (Uyttebroek et al. 2007). Moreover, each PAH compounds desorb at different rates over time. Also, Hatzinger and Alexander (1995) showed that soils freshly spiked with C-14 labeled phenanthrene and chrysene display a rapid desorption of these PAHs compounds. In contrast, chrysene and phenanthrene released from aged soils are desorbed at a much slower rate. At first,

PAHs are quickly desorbed immediately after a period of slower desorption (Cornelissen et al. 1998). Thus, the change in the rate of desorption is mainly induced by the reduction in the concentration gradient as PAHs are desorbed. Since the individual PAHs concentrations reach the limits of aqueous PAHs solubility and a gradient of concentration between the aqueous and sorbed phases decreases, desorption rate is going down. The solubility character of PAHs is a key factor in their bioavailability. The solubility of aqueous PAHs is directly affected by the compounds' molecular weight (Thorsen et al. 2004).

22.6 PAHs Phytoremediation

Phytoremediation means the use of plants for in situ treatments. It has been identified as an alternative to the traditional remediation technologies for the reason that it is socially acceptable and more cost-effective. In addition, it may create economic opportunities, including the production of bioenergy (Fan et al. 2008). Certain indirect attenuation mechanisms are involved in phytoremediation, such as the metabolism of pollutants by plant-related microbes and plant-induced changes in the contaminated environment (Singh and Jain 2003). The phytoremediation mechanisms involved in the removal of PAHs mostly consist in (1) promoted biodegradation: stimulating cometabolic pathways and the microbial activities of soil by the compounds that are released from plant roots, (2) the mobilization or adsorption of PAHs in the rhizosphere because of the surfaces and lipophilic exudates of roots, and (3) the direct metabolism and uptake of plants (Olson et al. 2008). The plant species differ in the exudate compounds composition (as biosurfactant and carbon source) and the characteristics of their roots, affecting the mobilization of sorbed PAHs, the soil physicochemical properties, and the microbial degradation activities (Mueller and Shann 2006). Moreover, plants' ability to metabolize and uptake PAHs from soil is affected by different enzymatic pathways, transpiration, and root adsorption processes (Harvey et al. 2002). Therefore, plant selection must be regarded as an important aspect for optimizing the phytoremediation of soils contaminated with PAHs. Using a multi-process phytoremediation system composed of physical (volatilization), photochemical (photooxidation) and microbial remediation, and phytoremediation (plant-assisted remediation) increase the average efficiency of removal of 16 priority PAHs by 50% more than bioremediation alone, and by 45% more than phytoremediation by itself (Huang et al. 2004). Furthermore, multispecies mixture has been well investigated to improve the potentially toxic metals phytoremediation efficiency compared to monocultures, for the reason that it provides a more favorable condition in the rhizosphere to enhance the bioavailability and biodegradation of contaminants (Wenzel 2009). Nevertheless, not many studies have paid special attention to the effect of multi cropping on the removal of PAHs, and the results of these studies are hard to generalize for the reason that the results sometimes are different. Maila et al. (2005) showed that the combined cultivation of African millet and velvet grass had no positive effect on the removal of pyrene in comparison to monocultures, while Xu et al. (2006) reported that pyrene was significantly removed from the soil with ryegrass and maize. Therefore, more information on the cropping pattern that affects the PAHs removal is required to explain such results.

22.6.1 Phytoremediation by Plants

It is conceivable that plants could be used effectively to contribute to the removal of organic contaminants such as PAHs from soil through two key mechanisms: (1) the accumulation by plant tissues and the metabolization of the contaminants immediately after the root plants uptake and (2) the improved rhizosphere microbial activities caused by the transduction of root enzymes to mineralize and/or transform pollutants (Wild et al. 2005). Therefore, the soil PAHs bioavailability and the uptake capacity of plants would significantly affect the PAHs phytoremediation efficiency. Analysis of the level of nonionic pollutants in plants associated with the external level in soil water (or water) has shown that the abovementioned pollutants go inside plants mainly through the passive processes (Su and Zhu 2006). In the systems of plant-water-soil, the high partition coefficients and low water solubility of PAHs cause the minimal levels of PAHs in pore or external water, especially the soils that have high sorption abilities, which limit the subsequent accessibility of PAHs for plants uptake. Factors such as transpiration stream concentration, shoot concentration, and root concentration (Briggs et al. 1982) have often been utilized to model the absorb and translocation of plants. Meanwhile, other studies have showed that the extent of plants assisted with the removal of PAHs from soil goes down with the contact time of soil (Reid et al. 2000). The aged PAHs contamination displays a lower microbial bioavailability because of a stronger binding to the organic matter of soil. Specifically, it has been shown that an increased association with the insoluble soil humin fraction is a main reason for decrease biodegradation (Nam and Kim 2002). Due to limited mass transport with contamination age, the total concentration of pollutants in soils is not considered as a good indicator of contaminant bioavailability (Alexander 2000). Therefore, an evaluation of the efficiency of phytoremediation without considering the effect of aging may be not completely correct. Many studies into the phytoremediation of PAHs-contaminated soils were carried out with the soils that were freshly contaminated (Xu et al. 2005), and the removal of this contamination was easier due to high PAHs bioavailability. This would be envisaged whenever soils and PAHs have contact for a long time. In other words, the efficiency of phytoremediation observed in previous studies may have been exaggerated. The data that provide many useful facts about the bioavailability and uptake of PAHs by plants in the aged soils are required to accurately evaluate the efficiency of phytoremediation, especially at field sites. Crops used as food are often polluted by organic wastes and several different pesticides (Wennrich et al. 2002). Rice is the food that is normally eaten in Asia. In China, PAHs exist by a large degree as pollutants caused by the agricultural and industrial sources (Tao et al. 2004).

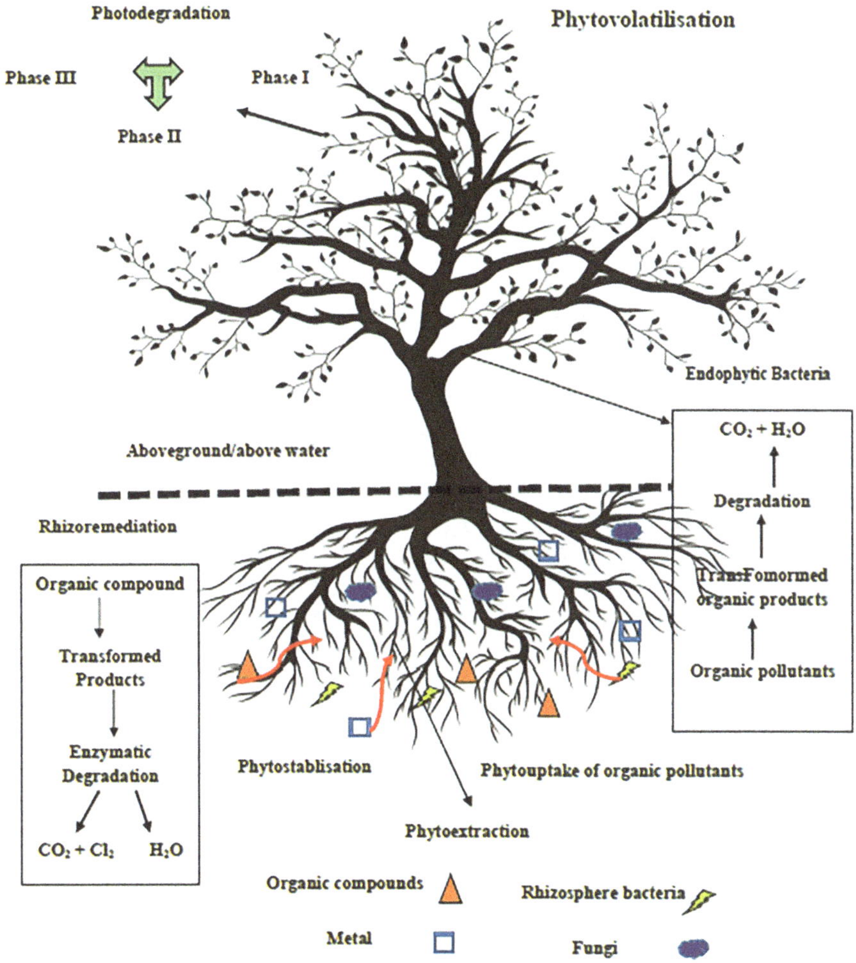

Fig. 22.3 Different plants phytoremediation strategies for remediation of pollutants from the contaminated environment (Modified from Kumar et al. 2018)

Compared to plants that are grown in uplands such as maize, wheat, and ryegrass, the paddy rhizosphere is a key environment for the removal of PAHs. Figure 22.3 refers to various phytoremediation strategies.

22.6.2 *Plant-Endophyte Partnerships*

Phytoremediation technologies are based upon the combined action of microbes living in the rhizosphere and plants, which show promise to remediate hydrocarbon-contaminated waterways and land, but they have not yet been approved formally in

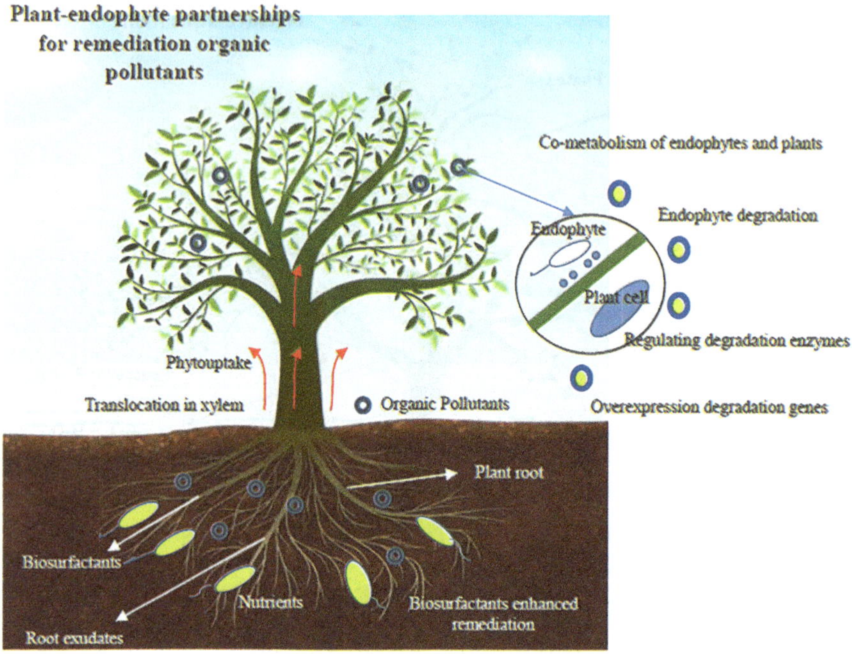

Fig. 22.4 Plant-endophyte partnerships for phytoremediation (Modified from Feng et al. 2017)

the strategies of large-scale remediation. Phytoremediation based upon the synergistic plant actions and the microorganisms associated with them has been identified as an in situ method for remediating soil. Appropriate plant combinations and endophytes associated with them are able to make growth on plants better and improve the organic pollutant biodegradation in the endosphere and/or rhizosphere, making the pollutant removal from soils happen more quickly (Marchut-Mikolajczyk et al. 2018) (Fig. 22.4). Numerous studies have recently been performed to decipher the phytoremediation mechanisms of organic pollutants from assisted by endophyte phytoremediation. Using plants and microorganisms for phytoremediation in order to degrade/remove soil pollutants is a promising, reliable, and cost-effective method (Arslan et al. 2017), especially when the harvested plant biomass may be used for the production of bioenergy (Pandey et al. 2009). Diverse microbial communities inhabiting on plants range from the phyllosphere and rhizosphere to the endosphere (Compant et al. 2010). They play an essential role in the fitness, growth, and development of plants as well as in the removal of pollutants from soils. Endophytes are involved in these friendly interactions with their host plants that are free of infections or other adverse influences, leading to mutualistic associations. Endophytic microbes have a very large number of the genes that degrade pollutants for the process of detoxification and have the right qualities to remove plant from the toxic organic pollutants (Ijaz et al. 2016). Moreover, endophytes (as microbes) may increase plant growth and their tolerance to the phytotoxicity of contaminants because of their activities in promoting the growth of plants. Many investigations have shown that phytoremediation assisted with endophyte phytoremediation has an important role in removing soil contaminants (Santoyo et al. 2016). Since many endophytes have the activities that degrade contaminants and the effects that promote the plant growth, or both of them, a clear understanding of the mechanisms that underlie these beneficial characteristics of endophytes might enhance the application of phytoremediation of pollutants with organic structures.

A group of microbes called "endophytes" lives in the plant's internal tissues for at least part of the life cycle. These microorganisms are present in almost each plant species. A broad range of endophytes have set up a collection of negative, neutral, or positive interactions with host plants, which affect the survival, health, and growth of plants (Compant et al. 2016). Endophytes may keep their close relationships with the host plants. These later provide endophytes with safe and nutrients rich habitats, and thus, they are protected against the abiotic and biotic stress factors (Reinhold-Hurek and Hurek 2011). Endophytes involved in the production of a wide range of natural bio-active materials that facilitate the plant growth and development through several different mechanisms (Santoyo et al. 2016). These characteristics that promote the endophytic plants growth make adaptation to abiotic and biotic stress factors easier and go up the biomass of the plants that are appropriate for phytoremediation. The compatibility of endophytes and host plants and their integral actions are key parameters for remediating the polluted soils (Ijaz et al. 2016). Endophytes may be used effectively to enhance the effectiveness of organic pollutant phytoremediation, since they may deliver the capacities of biodegradation inside/around host plants. This contributes to high activities of metabolism in the endosphere and rhizosphere. When an association between plants and endophytes are set up, endophyte-plant partnerships may develop the new and different strategies of phytoremediation for polluted soils (Afzal et al. 2014).

22.7 Plant-Endophyte Phytoremediation

22.7.1 Degradation and Detoxification of Organic Pollutants

Currently, researches on the phytoremediation of contaminated soils provide an ecological and economic method for the removal of a large array of pollutants. Though plants frequently metabolize or sequester the organic substances because they are photo-autotrophs, they had not evolution for metabolizing the organic pollutants as sources of carbon or energy. However, these organisms can only transform the pollutants into more water-soluble substances and/or prevent their mobilization. When the phytotoxic level of organic contaminants is severe, plant species with phytoremediation activations are usually sensitive to the contaminants. Despite tolerating the contaminants with organic structures by the plants, their growth is damaged, and they cannot play a role in remediation of the polluted soils, probably because of the hydrophobicity and phytotoxicity of organic contaminants. Therefore, this inhibits to obtain the water and the nutrients by plants (Khan et al. 2013). *Burkholderiaceae*, *Pseudomonadaceae*, *Enterobacteriaceae*, and *Bacillaceae* are among the important families of cultivable species of endophytes found at sites with contaminations. *Pseudomonas putida* PD1, as an endophyte, enhance root and shoot growth and protects grass and two different clones of willow against phenanthrene phytotoxicity (Khan et al. 2014). Endophytic bacteria have a greater capacity to enhance phytoremediation of petroleum hydrocarbon than rhizosphere or soil bacteria (Wu et al. 2018). Moreover, endophytes help plants to combat the organic pollutants-induced oxidative stress by antioxidative systems in host plants. These exceptional efficiencies of adaptations and remediation severely show the being suitable of endophytes (as contaminant-degrading) for the degradation and detoxification of organic pollutants (Bacon and White 2016). During soil phytoremediation, endophytes with action of pollutant-degrading have the catabolic genes and appropriately detoxify or mineralize the organic contaminants. *Pseudomonas* sp. (BF1-), as a root endophyte, contain the organophosphorus hydrolase gene ophB, suitable for chlorpyrifos hydrolysis (Barman et al. 2014). *Burkholderia phytofirmans* PsJN has glutathione-S-transferase (GST) genes for detoxifying and degrading the complex organic substances. Furthermore, external microbes in environment can help plants to resist phytotoxic materials from within through colonization. *Burkholderia fungorum* DBT1 (from isolation of oil refinery discharge) can transform PAHs in the hybrid plants of poplar (Mitter et al. 2013).

22.7.2 The Role of Biosurfactants of Endophytes in Phytoremediation

Phytoremediation can partially provide a solution to decontaminate soils contaminated with organic pollutants. Uptake of organic compounds by plants is the first and most important step for phytoremediation and is generally controlled by the hydrophobicity of these organic pollutants. These substances easily enter the xylem for subsequent destruction or accumulation. Thus, it seems that after accumulating these contaminations in plant tissues, endophytes are suitable for plant detoxification (Arslan et al. 2017). Bioavailability is one of the most limited parameters for the phytoremediation of persistent organic contaminants in soil. Microorganisms and plants may release biosurfactants which can desorb and improve organic pollutants degradation in soils by increasing their bioavailability, uptake by plants, or degradation by microbes, and these promote the in situ phytoremediation performance. Based on the researches, endophytes can contribute to the production of biosurfactants, and this increases the bioavailability of organic pollutants and the efficiency of bioremediation (Shekhar et al. 2015). Endophytic bacteria isolated from *Chelidonium majus* L. exhibit potential for hydrocarbons degradation and biosurfactant production. Furthermore, plant growth-promoting ability of the biosurfactant may be used to promote plant growth in hydrocarbon-contaminated sites (Marchut-Mikolajczyk et al. 2018). *Bacillus subtilis* K1, as a root endophyte, produces the heterogeneous mixture of 94 cyclic lipopeptide biosurfactants, and this exhibits suitable activity of emulsification, extremely good surface-active characteristics with stability (Pathak and Keharia 2014). This property makes it usable for bioremediation of oil-polluted soils. Interestingly, a novel actinomycete with endophyte traits, *Nocardiopsis* sp., from leaf tissues of the medicinal plant *Hibiscus rosa-sinensis*, was found to have both biosurfactant production capacities and hydrocarbon biodegradation and showed great potential for bioremediation of diesel-contaminated soils. Many other known endophytic microorganisms *Bacillus*, *Acinetobacter*, *Kocuria*, *Pseudomonas*, *Microbacterium*, and *Rhodococcus genera* have been reported to produce biosurfactants and bio emulsifiers and are used in phytoremediation (Pathak and Keharia 2014). Endophytic microbes with remarkable capacities of biosurfactant production have been isolated from plants. However, most of researches about the production potential of biosurfactant were done in vitro. Based on such studies, it is reasonable to assume that endophytes can produce biosurfactants in vivo. This facilitates the degradation of organic pollutants in plant tissues. Recent indirect evidence suggests that PAH-degrading endophytic bacteria could dramatically reduce the accumulation of hydrophobic PAHs in plant tissues (Sun et al. 2015a). That said, the efficiency of any phytoremediation system is related to the bioavailability of the target contaminants and root-microbial changes for being soluble and their chemical structure in the rhizosphere. Another possibility is that endophyte-derived biosurfactants that colonize root tissues are released into the rhizosphere soils and have a significant effect on rhizosphere processes (Sun et al. 2012), leading to increased degradation of organic contaminants in the soils. It is possible, endophytes that produce biosurfactants react to biophysical factors or cues, which is derived from host, and this promotes the xenobiotic degradation releasing them into soils (Juwarkar et al. 2010). Such processes in rhizosphere are a key strategy to circumvent the limitations of xenobiotic hydrophobicity and may mobilize hydrophobic contaminants from soil particle surfaces (Vergani et al. 2017), allowing their degradation in the rhizosphere or tissues. The use of biosurfactant-producing endophytic inoculants is a promising method to improve the efficiency of the phytoremediation of organic-polluted soils (Mnif et al. 2015). In fact, the application of microorganisms with simultaneous capacities to degrade hydrophobic pollutants and production of biosurfactants can appropriately hasten the bioremediation of organic-polluted soils. In most items, microorganisms naturally co-exist in consortia that make strong and wide capacities of metabolism, and these characteristics are attractive for the bioremediation of organic-polluted soils (Hays et al. 2015).

22.8 Plant Metabolism Effects on Organic Pollutants

The plant metabolism contributing to organic pollutants degradation may offer a cost-effective, environmentally sound approach for the bioremediation of contaminated soils. Plants adapt to organic pollutant-induced stresses via establishment of detoxifying systems of their cells, and organic contaminants are decreased, catabolized, converted, or even removed and because plants have the ability to decrease the detrimental impacts of contaminants known as the "green liver" model. The model was supported in carrot (*Daucus carota* var. *sativus*), which stated the ability for degrading the phthalate esters (PAEs) (Sun et al. 2015b). Higher plant enzymes may detoxify the organic pollutants via processes of degradation and conjugation. Enzymes such as cytochrome P450 monooxygenases, polyphenol peroxidase, peroxidase, catalase, nitro reductase, dehalogenase, hydrolase, laccase, and glutathione-S-transferase play directly the roles in the stabilization, tolerance, detoxification, and deposition of organic pollutants (Schwitzguébel 2017). Oxygenases, such as peroxidases and monooxygenases of cytochrome P450, can play a role in the primary organic pollutant metabolism in plants such as ring-breaking activities that are essential for benzene derivatives such as toluene (Kvesitadze et al. 2009). Furthermore, plants have the reactions, including ester hydrolysis, hydroxylation of aryl and alkyl, and N- or O-dealkylation. The processes change the organic substances structure by the introduction of functional groups and result in more active and polar substances with promoted solubilities (Sun et al. 2015b). The activated substances are conjugated to polar plant molecules such as sulfate, carbohydrates, malonic acid, glutathione, or amino acids. Covalent changes consist in ester or thioether linkages, peptide conjugation, and ether, catalyzed by transferases, including glycosyltransferases and glutathione S-transferases. Soluble conjugates move to vacuoles or apoplast through ATP or Glutathione S-Transferase (GST) binding or can be entirely converted into H_2O and CO_2. Bound conjugates leave from the cell through exocytosis to the apoplast and link to the wall of cells (Kvesitadze et al. 2009).

22.9 Plant-Endophyte Roles in the Removal of Organic Pollutants

22.9.1 Colonization and Survival of Endophyte

Endophytes are useful to study the mechanisms involved in plant growth promotion and are also considered as xenobiotic-degrading microbes for phytoremediation, used by plants and their related endophytes to treat the organically contaminated soils to increase acceptance as a possible technology of cleaning (Santoyo et al. 2016). Nevertheless, inoculants cannot cause the desired effects when re-inoculated in the field. This can be because of undesirable colonization of the plant and/or rhizosphere and the endophyte survival, which has been known as an extremely important stage (Lugtenberg et al. 2001). The "competition-driven" model can determine the survival and colonization of endophytic beneficial inoculants in a polluted rhizosphere or the plant tissues. Plants actively prefer a particular flora of microbes to colonize their rhizosphere and/or internal tissues. This complicated community of microorganisms is often referred to as the second plant genome, increasing the potential of the functional host (Berendsen et al. 2012). Endophytes that may colonize the plant internal tissues may have advantage over microbes in limiting the rhizosphere, since they are in close contact with the plant cells and supply carbon directly, and then affect the plant growth. Rhizosphere microbes may have the potential to enter and colonize plant roots and then spread over the plant, as one of the initial ways of endophytic colonization. After colonization of rhizosphere, endophytes may attack specific tissues of plant and colonize different plant compartments (Compant et al. 2011). The colonization in plants by good endophytes includes rhizoplane and rhizosphere colonization and plant tissue. Before they can have any effect on the plant, endophytic inoculants must be suitable root colonizers, known by strong competition of microbes. Microbe-microbe and host-microorganism interactions differentiate root microbiota through processes that physiologically occur at the root-soil interface. Chemical taxis, both by root secretion and mucilage which plays a key role in successful colorizations in rhizosphere and rhizoplane, increasingly depends on plant phenotype, stress, and farming practices. Additionally, some root secretions are repulsive substances that have negatively effects on inoculant colonization (Kristin and Miranda 2013). In the first stage, endophytes use microorganism-plant interaction by appropriate colonization in the rhizoplane and rhizosphere (Lareen et al. 2016). The changes of ethylene levels in plant 1-Aminocyclopropane-1-carboxylic acid (ACC) deaminase, which is derived from microbes, is an important parameter affecting plant physiology, and the inoculant ability to change the concentrations of plant ethylene is imperative for their competence as endophytes. Appropriate colonization in root makes the dense populations of various heterotrophic microbes preventing the rhizoplane, rhizosphere, and inner tissues of plant. These competent associations of microorganisms enhance the stepwise transformation of organic pollutants through consortia in the rhizosphere and/or within plants, and this provides the opportunities for gene rearrangements due to genetic exchange (Jha et al. 2015). Plant development stage, host plant origin, degree of the inoculum, frequency and timing of inoculation, and the concentrations of pollutants in different substrates of soil have important effects on colonization of endophytic inoculant and survival and following efficiency of phytoremediation (Khan et al. 2013).

22.9.2 Mutualistic Symbiotic Relations Between Endophytes and Host Plants

The application of plant assemblages and related endophytes with synergistic or complementary characteristics instead of the "individual plant" method will likely cause more consistent phytoremediation (Afzal et al. 2014). So-called concerted plant–endophyte synergisms means to gain more and more momentum. The stem biomass and height of *S. triqueter* and the increased rates of removal of diesel oil were promoted by the symbiotic system of *Pseudomonas* sp. J4AJ (as oil-degrading endophyte) and *Scirpus triqueter* compared to individual plants promoted. The inoculation of *Axonopus affinis* with a combination of microbes, two microbes, including *Pantoea* sp. *BTRH79*, *Pseudomonas* sp. ITRH25 (as hydrocarbon-degrading), and one endophyte that enhances the growth of the plant, *Burkholderia* sp. PsJN, caused the promoted growth of the plant and improvement of phytoremediation. The endophytes that reside mutualistically in the inner host plant tissues can maintain the bi- and tri-partite synergistic relations with host plants and related microorganisms (Tara et al. 2014). It must be clarified the molecular mechanisms of interaction with compatible relations between the partners to obtain the synergistic relations. Interestingly, such data are not complete yet. In fact, endophytes have co-evolution with host plants to identify each other by cell-to-cell communication that is similar to quorum sensing (Kusari et al. 2015).

22.9.3 Co-metabolism of Organic Pollutants

Some organic pollutants, particularly recalcitrant substances, including PAHs and polychlorinated biphenyls (PCBs) because of their toxicity and low bioavailability, cannot be as sources of energy and carbon for communities of endophytes during phytoremediation. It can be stated that lacking catabolic enzyme induction and small amounts of the substrates that support growth can decrease the increasing of endophyte populations. In actual fact, the contaminant co-metabolism and inter-species metabolism are the usual situation (Musilova et al. 2016). Biodegradation of recalcitrant substances may proceed via the co-metabolism that have a role in the organic pollutant oxidation with substrates for growth, the important mechanism used by plants and endophytes to degrade of recalcitrant contaminants. The metabolome is a complementary biodegradation linked to a mixed-species community within the plant tissues (Soleimani et al. 2010). For phytoremediation of organic pollutants, plants can beneficially use from endophytes that have suitable pathways of degradation and metabolic capacities and vice versa, both of which can result in the promoted phytoremediation of organic pollutants via co-metabolism. Cooperation to obtain novel catabolism reactions should be considered inside of host plants (Thijs et al. 2016). During phytoremediation, good cooperation occurs when there is a connection between endophytes and their host plants, and this leads to cometabolic. By plant endophytes, an inoculated plant may appropriately enhance its activity in metabolism by supplying the photosynthesis-derived growth substances such as sugars, amino acids, and organic acids. These compounds provide energy and carbon sources for degradation by microbes and/or cause the synthesis of degrading enzymes, and nicotinamide adenine dinucleotide phosphate (NADPH), and nicotinamide adenine dinucleotide (NADH) (as coenzymes), finally leads to the promoted degradation of organic contaminants. Endophytes share a various range of catabolic enzymes with their host plants through preventing plant to obtain genes with contaminant degrading actions. These catabolic enzymes are completely suitable to metabolize and detoxify the xenobiotic substances in plant tissues (Dhir 2013).

22.10 Enzymes Effective on Organic Pollutants Degradation

All plants can host microbes with an extreme variations of endophyte species (Tadych and white 2019). Plants colonized by specific endophytes may stimulate certain levels of transcription for the genes that degrade pollutants and consequently make a remarkable positive effect on the degrading enzyme activities (Compant et al. 2016). The endophytes represent an enormous natural reservoir of degrading enzymes to identify the contaminants and the complete bioremediation of organic-polluted soils (Afzal et al. 2014). Some endophytes whenever they are exposed to organic pollutants possess the possibility to set off and control both inducible and basal enzymes in endophytes or plants (Bacon and White 2016), hastening the changing of organic contaminants in plant or promoting the release of enzymes, which are contaminant-degrading, from roots into the rhizosphere, and this leads to hasten ex planta organic contaminant degradation (Sun et al. 2015a). Most enzymes that degrade contaminants, including nitrilases, nitro reductases, laccases, dehalogenases, dioxygenases, P450 monooxygenases, and peroxidases, are recognized in both plants and endophytes. The abovementioned enzymes directly have role in the bio-transformation of many xenobiotic substances (Bacon and White 2016).

22.11 Conclusion

PAHs are substances that are recognized as polycyclic organic compounds. These substances are mostly originating from the unfinished burning of organic compounds or the organic substance pyrolysis. They are also the result of thermal conversion of gas, coal, wood, and oil to produce energy. PAH substances are categorized as probable carcinogens for human. PAHs belong to the main carcinogen class in the environment. Many of PAHs and their epoxides are very mutagenic, toxic, and/or carcinogenic to microbes as well as higher systems such as human beings. PAHs need to activate the electrophilic metabolites for exerting their carcinogenic or mutagenic impacts. The microbes that degrade PAHs consist in microalgae, cyanobacteria, fungi, and bacteria. They play a role in the breakdown of organic substances via bio-transformation into less complex metabolites and by mineralizing them into inorganic compounds such as CO_2, H_2O, or CH_4. PAHs may be affected by chemical degradation, volatilization, photolysis, and adsorption. Nevertheless, degradation with microbes is the main process of degradation. Bioremediation is a method for transforming these substances to less hazardous/non-hazardous forms with high efficiency. The degradation of PAHs is directly affected by a number of parameters such as the environment conditions (presence of oxygen, pH, etc.), microbes (abundance and type), of the chemical structure of the considered PAH substances.

However, it is widely reported that phenols, anilines, or PAHs are taken up by plants, via the transpiration pathway or through penetrating leaves. However, there are not numerous evidences that they are mineralized by plants. The more hydrophobic substances are hydroxylated and translocated along with the more hydrophilic contaminants to other tissues of plant and either volatilized or excreted into the extracellular cell wall or into the vacuole as conjugates of glucosyl or glutathione.

Even if the plants have especially the capacity to uptake contaminants from surrounding areas through the transpiration path into the rhizosphere, in the rhizosphere, microbes start to degrade the PAH compounds, and this is affected by either physical or chemical properties of the pollutant. Plants may draw pollutants including PAHs into the plant rhizosphere to varying extents via the transpiration stream (Harvey et al. 2002). Therefore, the beneficial effects of ectomycorrhizal fungi and arbuscular to extend the rhizospheric network within soils are valuable. Though most microbes have the limited degradative capacity for high molecular weight contaminants, this is not the situation for white-rot fungi, which have an extracellular system of oxidative enzymes that can degrade polymeric substances with high molecular weight and facilitate their final mineralization. Basic researches of this nature need to make phytoremediation an effective and economic competitive remediation technology.

References

Abdel-Shafy HI, Mansour MS (2013) Removal of selected pharmaceuticals from urine via Fenton reaction for agriculture reuse. J Sustainable Sanitation Practice 17:20–29

Abdel-Shafy HI, Mansour MS (2016) A review on polycyclic aromatic hydrocarbons: source, environmental impact, effect on human health and remediation. Egypt J Pet 25:107–123

Abdel-Shafy HI, Al-Sulaiman AM, Mansour MSM (2014) Greywater treatment via hybrid integrated systems for unrestricted reuse in Egypt. J Water Process Eng 1:101–107

Afzal M, Khan QM, Sessitsch A (2014) Endophytic bacteria: prospects and applications for the phytoremediation of organic pollutants. Chemosphere 117:232–242

Akyuz M, Cabuk H (2010) Gas–particle partitioning and seasonal variation of polycyclic aromatic hydrocarbons in the atmosphere of Zonguldak. Turkey Sci Total Environment 408:5550–5558
Alebic-Juretic A, Cvitas T, Klasinc L (1990) Heterogeneous polycyclic aromatic hydrocarbon degradation with ozone on silica gel carrier. Environ Sci Technol 24:62–66
Alexander M (2000) Aging, bioavailability, and overestimation of risk from environmental pollutants. Environ Sci Technol 34:4259–4265
Arey J, Atkinson R (2003) Photochemical reactions of PAH in the atmosphere. In: Douben PET (ed) PAHs: an eco toxicological perspective. John Wiley and Sons, New York, NY, pp 47–63
Armstrong BG, Hutchinson E, Unwin J, Fletcher T (2004) Lung cancer risk after exposure to polycyclic aromatic hydrocarbons: a review and meta-analysis. Environ Health Perspect 112:970–978
Arslan M, Imran A, Khan QM, Afzal M (2017) Plant–bacteria partnerships for the remediation of persistent organic pollutants. Environ Sci Pollut Res 24:4322–4336
Bacon CW, White JF (2016) Functions, mechanisms and regulation of endophytic and epiphytic microbial communities of plants. Symbiosis 68:87–98
Baklanov A, Hänninen O, Slørdal LH, Kukkonen J, Bjergene N, Fay B, Finardi S, Hoe SC, Jantunen M, Karppinen A, Rasmussen A, Skouloudis A, Sokhi RS, Sørensen JH, Ødegaard V (2007) Integrated systems for forecasting urban meteorology, air pollution and population exposure. Atmospheric Chem Phys 7:855–874
Barman DN, Haque MA, Islam SMA, Yun HD, Kim MK (2014) Cloning and expression of oph B gene encoding organophosphorus hydrolase from endophytic *Pseudomonas* sp. BF1-3 degrades organophosphorus pesticide chlorpyrifos. Ecotoxicol Environ Saf 108:135–141
Berendsen RL, Pieterse CM, Bakker PA (2012) The rhizosphere microbiome and plant health. Trends Plant Sci 17:478–486
Beyer J, Jonsson G, Porte C, Krahn MM, Ariese F (2010) Analytical methods for determining metabolites of polycyclic aromatic hydrocarbon (PAH) pollutants in fish bile: a review. Environ Toxicol Pharmacol 30:224–244
Briggs GG, Bromilow RH, Evans AA (1982) Relationships between lipophilicity and root uptake and translocation of non-ionised chemicals by barley. Pestic Sci 13:495–504
Compant S, Clément C, Sessitsch A (2010) Plant growth-promoting bacteria in the rhizo-and endosphere of plants: their role, colonization, mechanisms involved and prospects for utilization. Soil Biol Biochem 42:669–678
Compant S, Mitter B, Colli-Mull JG, Gangl H, Sessitsch A (2011) Endophytes of grape-vine flowers, berries, and seeds: identification of cultivable bacteria, comparison with other plant parts, and visualization of niches of colonization. Microb Ecol 62:188–197
Compant S, Saikkonen K, Mitter B, Campisano A, Mercado-Blanco J (2016) Editorial special issue: soil, plants and endophytes. Plant Soil 405:1–11
Cornelissen G, Rigterink H, Ferdinandy MM, van Noort PC (1998) Rapidly desorbing fractions of PAHs in contaminated sediments as a predictor of the extent of bioremediation. Environ Sci Technol 32:966–970
Dhir B (2013) Mechanism of removal of contaminants by aquatic plants. Phytoremediation: role of aquatic plants in environmental clean-up. Springer, New Delhi, pp 51–64
Fan SX, Li PJ, Gong ZQ, Ren WX, Na H (2008) Promotion of pyrene degradation in rhizosphere of alfalfa (*Medicago sativa* L.). Chemosphere 71:1593–1598
Feng NX, Yu J, Zhao HM, Cheng YT, Mo CH, Cai QY, Li YW, Li H, Wong MH (2017) Efficient phytoremediation of organic contaminants in soils using plant–endophyte partnerships. Sci Total Environ 583:352–368
Finlayson-Pitts BJ, Pitts JN (1997) Tropospheric air pollution: ozone, airborne toxics, polycyclic aromatic hydrocarbons, and particles. Science 276:1045–1051
Flotron V, Delteil C, Padellec Y, Camel V (2005) Removal of sorbed polycyclic aromatic hydrocarbons from soil, sludge and sediment samples using the Fenton's reagent process. Chemosphere 59:1427–1437

Fredslund L, Sniegowski K, Wick LY, Jacobsen CS, De Mot R, Springael D (2008) Surface motility of polycyclic aromatic hydrocarbon (PAH)-degrading mycobacteria. Res Microbiol 159:255–262

Haritash AK, Kaushik CP (2009) Biodegradation aspects of polycyclic aromatic hydrocarbons (PAHs): a review. J Hazard Mater 169:1–15

Harvey PJ, Campanella BF, Castro PML, Harms H, Lichtfouse E, Schaffner AR, Smrcek S, Werck-Reichhart D (2002) Phytoremediation of polyaromatic hydrocarbons, anilines, and phenols. Environ Sci Pollut Res 9:29–47

Hassan SSM, El Azab WIM, Ali HR, Mansour MSM (2015) The 18th international conference on petroleum, mineral resources and development. EPRI, Cairo, Egypt, pp 8–10

Hatzinger PB, Alexander M (1995) Effect of aging of chemicals in soil on their biodegradability and extractability. Environ Sci Technol 29:537–545

Hays SG, Patrick WG, Ziesack M, Oxman N, Silver PA (2015) Better together: engineering and application of microbial symbioses. Curr Opin Biotechnol 36:40–49

Honda M, Suzuki M (2020) Toxicities of polycyclic aromatic hydrocarbons for aquatic animals. Int J Environ Res Public Health 17:1363

Huang XD, El-Alawi Y, Penrose DM, Glick BR, Greenberg BM (2004) A multi-process phytoremediation system for removal of polycyclic aromatic hydrocarbons from contaminated soils. Environ Pollut 130:465–476

IARC (International Agency for Research on Cancer) (2010) Some non-heterocyclic polycyclic aromatic hydrocarbons and some related exposures. Monogr Eval Carcinog Risks Humans 92:765–771

Ijaz A, Imran A, ul Haq MA, Khan QM, Afzal M (2016) Phytoremediation: recent advances in plant-endophytic synergistic interactions. Plant Soil 405:179–195

Jha P, Panwar J, Jha PN (2015) Secondary plant metabolites and root exudates: guiding tools for polychlorinated biphenyl biodegradation. Int J Environ Sci Technol 12:789–802

Juwarkar AA, Singh SK, Mudhoo A (2010) A comprehensive overview of elements in bioremediation. Rev Environ Sci Biotechnol 9:215–288

Kaminski NE, Faubert Kaplan BL, Holsapple MP (2008) In: Klaassen CD (ed) Casarett and Doull's toxicology, the basic science of poisons, 7th edn. Mc-Graw Hill, New York, p 526

Khan S, Afzal M, Iqbal S, Khan QM (2013) Plant–bacteria partnerships for the remediation of hydrocarbon contaminated soils. Chemosphere 90:1317–1332

Khan Z, Roman D, Kintz T, delas Alas M, Yap R, Doty S (2014) Degradation, phyto-protection and phytoremediation of phenanthrene by endophyte *Pseudomonas putida*, PD1. Environ Sci Technol 48:12221–12228

Kim K, Jahan SA, Kabir E, Brown RJC (2013) A review of airborne polycyclic aromatic hydrocarbons (PAHs) and their human health effects. Environ Int 60:71–80

Korfmacher WA, Wehry EL, Mamantov G, Natusch DFS (1980) Resistance to photochemical decomposition of polycyclic aromatic hydrocarbons vapor-adsorbed on coal fly ash. Environ Sci Technol 14:1094–1099

Kristin A, Miranda H (2013) The root microbiota a fingerprint in the soil? Plant Soil 370:671–686

Kumar V, Shahi SK, Singh S (2018) Bioremediation: an eco-sustainable approach for restoration of contaminated sites. In: Microbial bioprospecting for sustainable development. Springer, Berlin, pp 115–136

Kusari P, Kusari S, Spiteller M, Kayser O (2015) Implications of endophyte-plant crosstalk in light of quorum responses for plant biotechnology. Appl Microbiol Biotechnol 99:5383–5390

Kvesitadze E, Sadunishvili T, Kvesitadze G (2009) Mechanisms of organic contaminants uptake and degradation in plants. World Acad Sci Eng Technol 55:458–468

Lareen A, Burton F, Schäfer P (2016) Plant root-microbe communication in shaping root microbiomes. Plant Mol Biol 90:575–587

Lugtenberg BJJ, Dekkers L, Bloemberg GV (2001) Molecular determinants of rhizosphere colonization by pseudomonas. Annu Rev Phytopathol 39:461–490

Maila MP, Randima P, Cloete TE (2005) Multispecies and monoculture rhizo-remediation of polycyclic aromatic hydrocarbons (PAHs) from the soil. Int J Phytoremediation 7:87–98
Manahan SE (1999) Industrial ecology. Environmental chemistry and hazardous waste. CRC Press, Boca Raton
Manariotis ID, Karapanagioti KH, Chrysikopoulo CY (2011) Degradation of PAHs by high frequency ultrasound. Water Res 45:2587–2594
Mansouri E, Yousefi V, Ebrahimi V, Eyvazi S, Hejazi MS, Mahdavi M, Mesbahi A, Tarhriz V (2020) Overview of ultraviolet-based methods used in polycyclic aromatic hydrocarbons analysis and measurement. Sep Sci Plus 3:112–120
Marchut-Mikolajczyk O, Drożdżyński P, Pietrzyk D, Antczak T (2018) Biosurfactant production and hydrocarbon degradation activity of endophytic bacteria isolated from *Chelidonium majus* L. Microb Cell Factories 17:1–9
Masih J, Masih A, Kulshrestha A, Singhvi R, Taneja A (2010) Characteristics of polycyclic aromatic hydrocarbons in indoor and outdoor atmosphere in the north central part of India. J Hazard Mater 177:190–198
Menzie CA, Potocki BB, Santodonato J (1992) Exposure to carcinogenic PAHs in the environment. Environ Sci Technol 26:1278–1284
Mitter B, Petric A, Shin MW, Chain PSG, Hauberg-Lotte L, Reinhold-Hurek B, Nowak J, Sessitsch A (2013) Comparative genome analysis of *Burkholderia phytofirmans* PsJN reveals a wide spectrum of endophytic lifestyles based on interaction strategies with host plants. Front Plant Sci 4:120
Mnif I, Mnif S, Sahnoun R, Maktouf S, Ayedi Y, Ellouze-Chaabouni S, Ghribi D (2015) Biodegradation of diesel oil by a novel microbial consortium: comparison between co-inoculation with biosurfactant-producing strain and exogenously added biosurfactants. Environ Sci Pollut Res 22:14852–14861
Moursy AS, Abdel-Shafy HI (1983) Removal of hydrocarbons from Nile water. Environ Int 9:165–171
Mueller KE, Shann JR (2006) PAH dissipation in spiked soil: impacts of bioavailability, microbial activity, and trees. Chemosphere 64:1006–1014
Musilova L, Ridl J, Polivkova M, Macek T, Uhlik O (2016) Effects of secondary plant metabolites on microbial populations: changes in community structure and metabolic activity in contaminated environments. Int J Mol Sci 17:1205
Nadarajah N, Van Hamme J, Pannu J, Singh A, Ward O (2002) Enhanced transformation of polycyclic aromatic hydrocarbons using a combined Fenton's reagent, microbial treatment and surfactants. Appl Microbiol Biotechnol 59:540–544
Nam K, Kim JY (2002) Role of loosely bound humic substances and humin in the bioavailability of phenanthrene aged in soil. Environ Pollut 118:427–433
Niu J, Sun P, Schramm KW (2007) Photolysis of polycyclic aromatic hydrocarbons associated with fly ash particles under simulated sunlight irradiation. J Photochem Photobiol A Chem 186:93–98
Olson PE, Castro A, Joern M, DuTeau NM, Pilon-Smits E, Reardon KF (2008) Effects of agronomic practices on phytoremediation of an aged PAH-contaminated soil. J Environ Qual 37:1439–1446
Pandey J, Chauhan A, Jain RK (2009) Integrative approaches for assessing the ecological sustainability of in situ bioremediation. FEMS Microbiol Rev 33:324–375
Parker DSN, Zhang F, Seol Kim Y, Kaiser RI, Landera A, Kislov VV et al (2012) Low temperature formation of naphthalene and its role in the synthesis of PAH in the interstellar medium. Pro Nat Acad Sci 109:53–58
Pathak KV, Keharia H (2014) Application of extracellular lipo-peptide biosurfactant produced by endophytic *Bacillus subtilis* K1 isolated from aerial roots of banyan (*Ficus benghalensis*) in microbially enhanced oil recovery (MEOR). Biotech 4:41–48

Rappert S, Botsch KC, Nagorny S, Francke W, Müller R (2006) Degradation of 2,3-diethyl-5-methylpyrazine by a newly discovered bacterium, *Mycobacterium* sp. strain DM-11. Appl Environ Microbiol 72:1437–1444

Reid BJ, Jones KC, Semple KT (2000) Bioavailability of persistent organic pollutants in soils and sediments e a perspective on mechanisms, consequences, and assessment. Environ Pollut 108:103–112

Reinhold-Hurek B, Hurek T (2011) Living inside plants: bacterial endophytes. Curr Opin Biotechnol 14:435–443

Rivera-Figueroa AM, Ramazan KA, Finlayson-Pitts BJ (2004) Fluorescence, absorption, and excitation spectra of polycyclic aromatic hydrocarbons as a tool for quantitative analysis. J Chem Educ 81:242

Santoyo G, Moreno-Hagelsieb G, Orozco-Mosqueda MC, Glick BR (2016) Plant growth-promoting bacterial endophytes. Microbiol Res 183:92–99

Schwarzenbach RP, Gschwend PM, Imboden DM (1993) Environmental organic chemistry. John Wiley and Sons, New York, p 681

Schwitzguébel JP (2017) Phytoremediation of soils contaminated by organic compounds: hype, hope and facts. J Soils Sediments 17(5):1492–1502

Shekhar S, Sundaramanickam A, Balasubramanian T (2015) Biosurfactant producing microbes and their potential applications: a review. Crit Rev Environ Sci Technol 45:1522–1554

Singh OV, Jain RK (2003) Phytoremediation of toxic aromatic pollutants from soil. Appl Microbiol Biotechnol 63:128–135

Smol M, Włodarczyk-Makuła M (2017) The effectiveness in the removal of PAHs from aqueous solutions in physical and chemical processes: a review. Polycycl Aromat Compd 37:292–313

Soleimani M, Afyuni M, Hajabbasi MA, Nourbakhsh F, Sabzalian MR, Christensen JH (2010) Phytoremediation of an aged petroleum contaminated soil using endophyte infected and non-infected grasses. Chemosphere 8:1084–1090

Su YH, Zhu YG (2006) Bio-concentration of atrazine and chlorophenols into roots and shoots of rice seedlings. Environ Pollut 139:32–39

Sun B, Gao Y, Liu J, Sun Y (2012) The impact of different root exudate components on phenanthrene availability in soil. Soil Sci Soc Am J 76:2041–2050

Sun K, Liu J, Gao Y, Sheng Y, Kang F, Waigi MG (2015a) Inoculating plants with the endophytic bacterium *Pseudomonas* sp. Ph6-gfp to reduce phenanthrene contamination. Environ Sci Pollut Res 22:19529–19537

Sun J, Wu X, Gan J (2015b) Uptake and metabolism of phthalate esters by edible plants. Environ Sci Technol 49:8471–8478

Tadych M, White JF (2019) Endophytic microbes. In: Thomas MS (ed) Encyclopedia of microbiology, 4th edn. Academic Press, New York, pp 123–136

Tao S, Cui YH, Xu FL, Li BG, Cao J, Liu WX, Schmitt G, Wang XJ, Shen WR, Qing BP, Sun R (2004) Polycyclic aromatic hydrocarbons (PAHs) in agricultural soil and vegetables from Tianjin. Sci Total Environ 320:11–24

Tara N, Afzal M, Ansari TM, Tahseen R, Iqbal S, Khan QM (2014) Combined use of alkane-degrading and plant growth-promoting bacteria enhanced phytoremediation of diesel contaminated soil. Int J Phytoremediation 16:1268–1277

Thijs S, Sillen W, Rineau F, Weyens N, Vangronsveld J (2016) Towards an enhanced understanding of plant-microbiome interactions to improve phytoremediation: engineering the metaorganism. Front Microbiol 7:341

Thorsen WA, Cope WG, Shea D (2004) Bioavailability of PAHs: effects of soot carbon and PAH source. Environ Sci Technol 38:2029–2037

Tolosa I, Bayona JM, Albaiges J (1996) Aliphatic and polycyclic aromatic hydrocarbons and sulfurroxygen derivatives in northwestern Mediterranean sediments: spatial and temporal variability, fluxes and budgets. Environ Sci Technol 30:2495–2503

Tudoran MA, Putz MV (2012) Polycyclic aromatic hydrocarbons: from in cerebro to in silico eco-toxicity fate. Ser Chem Environ Eng 57:50–53

Uyttebroek M, Vermeir S, Wattiau P, Ryngaert A, Springael D (2007) Characterization of cultures enriched from acidic polycyclic aromatic hydrocarbon-contaminated soil for growth on pyrene at low pH. Appl Environ Microbiol 73:3159–3164

Veltman K, Huijbregts MAJ, Rye H, Hertwich EG (2012) Including impacts of particulate emissions on marine ecosystems in life cycle assessment: the case of offshore oil and gas production. Integr Environ Assess Manag 7:678–686

Vergani L, Mapelli F, Zanardini E, Terzaghi E, Di Guardo A, Morosini C, Raspa G, Borin S (2017) Phyto-rhizoremediation of polychlorinated biphenyl contaminated soils: an outlook on plant-microbe beneficial interactions. Sci Total Environ 575:1395–1406

Wang Y, Liu CS, Li FB, Liu CP, Liang JB (2009) Photodegradation of polycyclic aromatic hydrocarbon pyrene by iron oxide in solid phase. J Hazard Mater 162:716–723

Wang Z, Yang C, Yang Z, Sun J, Hollebone B, Brown C, Landriault M (2011) Forensic fingerprinting and source identification of the 2009 Sarnia (Ontario) oil spill. J Environ Monit 13:3004–3017

Wang ZD, Yang C, Parrott JL, Frank RA, Yang Z, Brown CE, Hollebone BP, Landriault M, Fieldhouse B, Liu Y, Zhang G, Hewitt LM (2014) Forensic source differentiation of petrogenic, pyrogenic, and biogenic hydrocarbons in Canadian oil sands environmental samples. J Hazard Mater 271:166–177

Watts RJ, Stanton PC, Howsawkeng J, Teel AL (2002) Mineralization of a sorbed polycyclic aromatic hydrocarbon in two soils using catalyzed hydrogen peroxide. Water Res 36:4283–4292

Wennrich L, Popp P, Zeibig M (2002) Polycyclic aromatic hydrocarbon burden in fruit and vegetable species cultivated in allotments in an industrial area. Int J Environ Anal Chem 82:677–690

Wenzel WW (2009) Rhizosphere processes and management in plant-assisted bioremediation (phytoremediation) of soils. Plant Soil 321:385–408

WHO (1998) Guidelines for drinking water quality. World Health Organization, Geneva, p 123

Wild E, Dent J, Tomas GO, Jones KC (2005) Direct observation of organic contaminant uptake, storage, and metabolism with plant roots. Environ Sci Technol 39:3695–3702

Wu T, Xu J, Xie W, Yao Z, Yang H, Sun C, Li X (2018) *Pseudomonas aeruginosa* L10: a hydrocarbon-degrading, biosurfactant-producing, and plant-growth-promoting endophytic bacterium isolated from a reed (*Phragmites australis*). Front Microbiol 9:1087

Xu SY, Chen YX, Lin Q, Wu WX, Xue SG, Shen CF (2005) Uptake and accumulation of phenanthrene and pyrene in spiked soils by ryegrass (*Lolium perenne* L.). J Environ Sci (China) 17:817–822

Xu SY, Chen YX, Wu WX, Wang KX, Lin Q, Liang XQ (2006) Enhanced dissipation of phenanthrene and pyrene in spiked soils by combined plants cultivation. Sci Total Environ 363:206–215

Zhong Y, Zhu L (2013) Distribution, input pathway and soil–air exchange of polycyclic aromatic hydrocarbons in Banshan Industry Park, China. Sci Total Environ 444:177–182

Chapter 23
The Impact of Climate Change on Forest Tree Species Dieback and Changes in Their Distribution

Marcin Pietrzykowski and Bartłomiej Woś

Abstract The paper presents the impact of climate change on the process of range change and tree species dieback in European conditions. In southern Europe, the tree species range is expected to shrink, in particular in the case of Scots pine and Norway spruce. As a result of global warming, the processes of tree stand disturbance and dieback, especially due to drought, occur more and more frequently in European forests. At the same time, due to rising air temperature, carbon dioxide concentration and nitrogen deposition, tree species display a faster growth rate and higher productivity. A faster growth rate translates into earlier culmination of growth and more dynamic tree ageing processes, and furthermore, stress associated with drought weakens stands. For these reasons, in Central Europe of the future, in protected forests of national parks and reserves, a greater rotation of stands should be expected, which will result in a lower felling age in managed forests.

Keywords Climate change · Tree species · Scots pine · Dieback

23.1 Introduction

It is believed that progressive climate change is currently one of the greatest environmental threats in the world. It has been estimated due to human impact, global temperature has increased by an average of 1 ° C (probably from 0.8 to 1.2 ° C) compared to the pre-industrial era (Allen et al. 2018). An increase in global air temperature is often associated with a spatial and temporal change in the distribution of annual precipitation and higher frequency of extreme weather events such as hurricanes or droughts (Räisänen et al. 2004; Stagge et al. 2017).

Rising average air temperature and higher frequency of unfavourable weather conditions may have a significant impact on changes in forest ecosystems and the

M. Pietrzykowski · B. Woś (✉)
Department of Ecology and Silviculture, Faculty of Forestry, University of Agriculture in Krakow, Krakow, Poland
e-mail: b.wos@ur.krakow.pl

D. K. Choudhary et al. (eds.), *Climate Change and the Microbiome*, Soil Biology 63,
https://doi.org/10.1007/978-3-030-76863-8_23

Fig. 23.1 Example of large-scale disturbance of forest ecosystems after hurricane wind—the hurricane that passed over Poland on the night of August 11–12, 2017, destroyed over 100,532 ha of forest in the Lipusz Forest District, Northern Poland. (Photo by M. Pietrzykowski)

distribution of tree species (Cheaib et al. 2012; Li et al. 2020). In the northern hemisphere, these changes may be most pronounced in high mountain locations and at the southern and northern limits of the woody plant range (Brooker et al. 2007; Gatti et al. 2019). Climate change may also have a significant impact on the growth and productivity of forest ecosystems (Ciceu et al. 2020). It has been predicted that, especially in northern and eastern Europe, in the short and medium term, an increase in the concentration of carbon dioxide in the atmosphere and higher temperatures will have a positive impact on the growth and productivity of stands. On the other hand, an increased risk of drought and stand dieback, e.g. due to hurricane winds, will have the opposite effect (Lindner et al. 2010) Fig. 23.1.

However, the intensity and risk of extreme weather events, especially droughts, displays significant regional variation. The frequency of droughts will be greater in southern Europe, while rising air temperature and evapotranspiration will increase rainfall in northern Europe and reduce the frequency of droughts there (Stagge et al. 2017).

23.2 The Impact of Climate Change on Tree Species Distribution

The consequences of the climate change and warming phenomenon have significant influence on the world biome ecosystems. One of the large-scale effects is the melting of permafrost (Figs. 23.2 and 23.3).

As a result of the progressive climate change in the northern hemisphere, it has been predicted that many tree species will extend their ranges north and at the same time lose a significant part of their southern range limit (Hamann and Wang 2006). Widely distributed species with large populations are likely to survive and adapt to climate change. Species with small populations and low fertility or in decline due to introduced pathogenic insects (Fig. 23.4) or fungi (Aitken et al. 2008) will be threatened with extinction. Models predicting the effects of climate change on changes in forest tree ranges in Europe in the period 2061–2080 have divided species into "winners" (mostly late-successional species, *Abies alba*, *Fagus sylvatica*, *Fraxinus excelsior*, *Quercus robur*, *Quercus petraea*), which will increase their current range as the climate warms; "losers" (mostly pioneer and coniferous species, *Betula pendula*, *Larix decidua*, *Picea abies*, *Pinus sylvestris*), which will reduce their range; and alien species (*Pseudotsuga menziesii*, *Quercus rubra*, *Robinia pseudoacacia*), which, similarly to the winners, will increase their range. At the same time, assuming limited migration, most species will limit their current range. Tree range shrinking will most affect the species with the northernmost ranges

Fig. 23.2 Melting Siberia—due to the global warming, the period of snow and ice cover in far Northern Hemisphere (Yakutsk, Russia) is significantly shortened (photo by M. Pietrzykowski)

Fig. 23.3 The consequence of the melting of permafrost and limiting its range are large-scale depressions of the terrain and strong transformations of the surface in the northern forest border zone—(Yakutsk, Russia) (Photo by M. Pietrzykowski, 2015 year)

(Dyderski et al. 2018). Studies on range changes of 80 tree species in the eastern United States indicate that about 30 species may extend their range by at least 10% as a result of climate change and 30 other species may reduce their range by at least 10%. Almost half of the assessed species showed the potential to shift their ecological optimum by at least 100 km, including seven that could shift by >250 km (Iverson and Prasad 1998). In the temperate climate zone of British Columbia (Canada), it was predicted that species with the northern range limit could acquire potential habitats at a rate of 100 km north per decade. Common hardwoods do not seem to be significantly affected by climate change. In turn, most conifers will lose a significant part of their habitat. In British Columbia, the sub-boreal and montane climate region is disappearing (Hamann and Wang 2006). In the Iberian Peninsula (Spain and Portugal), a drastic range reduction has been projected for mountain conifers such as *Pinus sylvestris*, *P. uncinata* and *Abies alba* by 2080 and a range reduction for temperate broadleaf species such as *Fagus sylvatica* and *Quercus robur* and sub-Mediterranean species, especially *Q. pyrenaica* (Garzón et al. 2008). It has been indicated that as a result of progressive climate change in Italy, the local population of *Pinus sylvestris*, *Picea abies* and *Tilia cordata* may become extinct, while Mediterranean species such as *Quercus suber* will increase their range by 2080 (Attorre et al. 2011). The coniferous species *Pinus sylvestris* and *Picea abies* that dominate the northern hardwood zone stands will gradually recede from the south and west of Europe and will be gradually replaced by *Fagus sylvatica* and

Fig. 23.4 Yellowstone National Park—natural processes of bark beetle outbreak causing rhythmic and large-scale dieback of natural forests (Photo by M. Pietrzykowski, 2018 year)

other temperate zone deciduous species that are spreading northwards (Sykes and Prentice 1996). However, more recent studies indicate that beech stands may be damaged by severe droughts in the growing season (Geßler et al. 2007). Research conducted in France indicated that the beech is more sensitive to climatic conditions and high temperatures than *Quercus petraea* and *Pinus sylvestris* (Michelot et al. 2012). The potential for trees to move naturally beyond the current range is also limited. It depends on the ecological properties of tree species, climatic factors, ecological barriers and fragmentation of ecosystems (Malcolm et al. 2002; Iverson et al. 2004). Modelling on five North American tree species found that over 100 years, the probability of colonizing new sites is high at 10–20 km and low at more than 20 km from the range limit. Long-distance colonization does not play a major role in the survival of the species (Iverson et al. 2004). Models developed for the north-eastern United States also do not predict a drastic change in tree ranges by

2100. It is predicted that by 2300, some northern tree species will have become less abundant, while some southern tree species will have become more abundant (Wang et al. 2017). Some models also assume a gradual migration of lowland species to higher altitudes above sea level. However, this process will not be equally intense everywhere. In studies conducted in two national parks in the Carpathian Mountains (eastern Poland), it was found that in one national park *Abies alba* and *Fagus sylvatica* expanded significantly into higher mountainous locations. In the second national park, located only 45 km from the first, the above-mentioned phenomenon was not observed (Gazda et al. 2019).

23.3 The Impact of Climate Changes on Tree Growth and Stand Productivity

According to some predictions, forests in northern Europe may be more productive and grow faster as a result of climate change. On the other hand, southern and western Europe is facing a major challenge due to the projected decline in forest productivity and possible large-scale dieback of forest ecosystems (Schelhaas et al. 2015). According to some researchers, climate change may cause a decline in forest productivity as a result of the replacement of high-productive stands of coniferous species (*Picea abies* and *Pinus sylvestris*) with stands of less productive species, e.g. oaks (*Quercus* spp.) by 2100 (Hanewinkel et al. 2013). So far, however, as a result of an increase in the average air temperature and an increase in CO_2 concentration, tree species stands in the region of Central Europe have usually increased their growth and productivity in recent years (Boisvenue and Running 2006; Pretzsch et al. 2014; Socha et al. 2020). In addition to an increase in temperature and the amount of CO_2, a growth in the deposition of nutrients in recent decades (the so-called fertilization effect), especially nitrogen (Pretzsch et al. 2014; Cienciala et al. 2018; Hess et al. 2018), has also been significant in this process. For example, in Catalonia (Spain), an increase in temperature (an average of 0.19 °C per decade in the twentieth century) had a negative effect on the Scots pine growth rate, especially in arid habitats, but it was big enough to counteract the fertilization effect, and it did not lead to a permanent reduction in water availability in the summer. However, it has been indicated that this situation may change in the future, when the temperature continues to rise or rainfall becomes less frequent (Martínez-Vilalta et al. 2008). It has also been reported that in the Mediterranean, climate warming, including particularly warmer winters, tends to have a positive effect on Scots pine growth rate, and the effect is more pronounced at lower mountain elevations. On the other hand, droughts in the summer significantly reduce the growth rate of trees and stands (Marqués et al. 2017). In case of drought, the Scots pine reacts with reduced annual increments due to the shortening of the growth period compared to the potential period, i.e. the phenological period of the year (Eilmann et al. 2011). The projected most extreme scenarios of carbon dioxide emissions, corresponding to an average air temperature of 1.4–4.8 ° C, indicate that the Scots pine growth rate may decrease by 16.4% after 2050 at the southern limit of the species range. This prediction also assumes a dieback and shrinkage of pine ranges as a result of local extinction of the most vulnerable stands in xerothermic habitats (Sánchez-Salguero et al. 2016). As in the case of the Scots pine, the spruce stands in Central Europe have shown significantly higher growth rate and productivity in recent decades (Pretzsch et al. 2014). In the longer term, however, some researchers suggest a decline in the productivity of this species, especially at lower altitudes above sea level (Altman et al. 2017).

23.4 *Pinus sylvestris* Dieback: A Case Study

One example of forest dieback and a change in natural distribution is the deterioration of the Scots pine (*Pinus sylvestris* L.) stands in Europe. Until recently, Scots pine was considered an undemanding species, with a broad spectrum of ecological requirements (Bigler et al. 2006). However, since the late 1980s, there have been reports in southern Europe of pine dieback, and the aetiology has not been fully investigated (Vertui and Tagliaferro 1998). The reasons for this dieback, which has already been reported in Central Europe, are not fully understood. In the Italian Alps, lack of primary biotic factors and the low incidence of secondary pathogens in dying pines suggest that this process is due to abiotic factors. An analysis of the time series indicated that two consecutive years with a low summer dryness index preceded the dieback phenomenon, while the annual ring width and summer dryness index were strongly correlated with each other (Gonthier et al. 2010). In the Swiss Alps, it has been observed that the highest stand dieback occurs after hot and dry years, and tree defoliation rates show a strong correlation with the amount of precipitation from the previous year (Rebetez and Dobbertin 2004). Not only droughts in the summer (vegetative) period cause a significant weakening of trees and dieback but also winter-droughts (Voltas et al. 2013: Camarero et al. 2016). Research in the Mediterranean Basin (Spain) shows that winter-drought dieback impairs xylem anatomy and phenology, aboveground productivity, xylogenesis and growth in the Mediterranean Scots pine populations (Camarero et al. 2016). Freeze-thaw episodes can unveil such physiological differences by triggering dieback in the trees which are more vulnerable to hydraulic failure (Voltas et al. 2013). It has been indicated that droughts intensify tree susceptibility to insects and parasites, particularly mistletoe (*Viscum album*), which has not been considered a parasite causing significant damage to forest stands so far (Dobbertin and Rigling 2006; Rigling et al. 2010). Mistletoe infection may be a factor in tree dieback by increasing the loss of needles following drought and a factor increasing water stress in trees during droughts (Dobbertin and Rigling 2006). The reduction in needle length due to mistletoe infection is an indication of lower water and nutrient availability in the infested branches. Thus, mistletoe infection might lead to a decrease in the availability of water and carbohydrates, the two most important growth factors, which are already limited due to a chronic drought in the area. Therefore, pine mistletoe increases the risk of drought-induced mortality of the Scots pine when growing in a xeric environment (Rigling et al. 2010). Higher air temperature facilitates the colonization of the Scots pine by nematodes and bark beetles, and increasingly frequent droughts weaken the resistance of the trees to pathogens (Rebetez and Dobbertin 2004; Jaime et al. 2019). In the Alps, it has been indicated that due to more frequent episodes of drought, trees that are even slightly or temporarily weakened will be more vulnerable to attacks by harmful forest insects such as *Ips acuminatus* and *Phaenops cyanea* (Wermelinger et al. 2008). In Central Europe (Slovakia), fungal pathogens such as *Ophiostoma ips* and *Ophiostoma minus* and bark beetles are among the biotic and abiotic factors causing the dieback of pine stands at present (Pastirčáková et al. 2018). However the phenomenon of large-scale outbreaks of bark beetles causing the forest dieback is known in the world and is a part of the natural disturbance of ecosystems, the exacerbation of this phenomenon is clearly related to climatic factors.

The sapwood fungus pathogens may also play a secondary role in the Scots pine dieback (Giordano et al. 2009). It was also observed in dying pine stands that severe drought disturbed the quantitative ratios of nutrients in the Scots pine wood, mainly Mn and Ca imbalances (along with K starvation and imbalance in the Scots pine). It was pointed out that the content of Mn and related ratios (Ca:Mn, Mn:Al or P:Mn) are early-warning signals of forest dieback (Hevia et al. 2019).

In Switzerland, more than half of the Scots pine population in the Alps has died since 1995 (Rebetez and Dobbertin 2004). As a result of a progressive dieback of pine stands, species composition of stands may significantly change, especially on the southern border of the range. In the Pyrenees, it has been observed that in pine stands dying from drought, there is an abundant regeneration of oak species (mainly *Quercus ilex* and *Quercus humilis*). These results indicate that in the Pyrenees, there is a high mountain migration of oaks, mediated by the dieback of the species present there, especially the Scots pine (Galiano et al. 2010). A similar phenomenon was observed in the Swiss Rhone Valley (Valais), one of the driest alpine regions. Pine stands found in lower mountainous locations are gradually being replaced by stands composed of *Quercus pubescens* (Rigling et al. 2012).

Scots pine stands which have been growing rapidly and densely in recent years are particularly exposed to stress related to long-term drought. One of the reasons that may inhibit the observed faster growth trend of stands may also be the depletion of nutrients in the soil as a result of increased tree growth, which may affect the susceptibility of trees to unfavourable abiotic factors. Forest monitoring data from Germany from 1964 to 2019 show that in the twenty-first century, there has been a decrease in sulphur and phosphorus resources in soils, as well as in the concentration of these elements in the Scots pine needles. At the same time, the growth rate of pine stands remained high (Prietzel et al. 2020). Data from Southern Europe show that stands growing in infertile and dry sandy soils are most exposed to the negative impact of climate change and to drought damage (Sánchez-Salguero et al. 2016). In Central Europe, symptoms of species dieback are already visible, and the Scots pine is often the only species that grows well in such habitats, and in the case of mass dieback, it will be difficult to replace it with other native species. The negative consequences will also be associated with the long-term regeneration process of the forest ecosystem after the mass dieback scenario. It has been indicated that countermeasures may include reducing the density of stands in order to increase their resistance to drought (Giuggiola et al. 2013; Bottero et al. 2016) and a gradual introduction of southern species, e.g. the Lebanese cedar (*Cedrus libani*), into managed forests (Messinger et al. 2015). The latter solution is the most controversial. The introduction of alien species brings about adverse changes in ecosystems. It is unknown how these species will acclimatize and adapt in the long term. Moreover, despite global warming, episodes of severe winters are still possible in which case the thermophilic southern European species will not survive.

23.5 *Picea abies* Dieback: A Case Study

Another European species that is currently subject to stand decline and is expected to change its range is the Norway spruce (*Picea abies*). However, in the case of the Norway spruce, the matter is more complicated. The dieback of spruce stands is caused by the combined impact of three main factors: climate change, air pollution and the introduction of spruce monocultures in habitats that are too fertile for the species (Fleischer et al. 2005; Jamroz et al. 2014; Krejza et al. 2020). According to Fleischer et al. (2005), the dieback of the spruce is the synergistic outcome of air pollution and climate change. Research conducted in the Czech Republic shows that in periods of increased acid rain, this factor had a greater impact on the growth and dieback of spruce stands than climate variables (Altman et al. 2017). It was estimated that in the Tatra Mountains, acid rain is responsible for 10–30% of stand dieback (Fleischer et al. 2005). An additional factor causing the dieback of spruce stands in Europe is the introduction of spruce in the form of monocultures into forest habitats which are too fertile for the species. In many of its areas, the spruce was introduced by man in the lower mountain ranges in the early nineteenth century in place of natural mixed forests cut down for industrial purposes. Such unnatural spruce monocultures typically display low resistance to abiotic and biotic stress (Spiecker 2000). Among the climatic factors causing the dieback of the spruce, the most important are extremely high spring and summer temperatures, little rainfall and low air humidity (Fleischer et al. 2005). The results of research carried out in southern Finland indicate that dieback of spruce stands is directly related to drought (Mäkinen et al. 2001). Studies conducted in the Czech Republic from 38 to 995 m a.s.l. indicate that as a result of severe drought, the annual stem radial increment may decrease by up to 78% compared to a wet year (Krejza et al. 2020). The spruce, however, suffers from drought primarily at low and medium altitudes above sea level. At high altitudes, there is a positive impact of higher temperature on the spruce growth rate (Altman et al. 2017; Krejza et al. 2020). Higher summer temperatures at high altitudes favour cell production and xylem hydraulic efficiency. On the other hand, a limited amount of water at lower elevations may adversely affect tracheid enlargement and thus the ability to conduct water (Castagneri et al. 2015). Warm and dry growing seasons weaken the trees and facilitate the development of the bark beetle (*Ips typographus*) outbreaks. It has been predicted that as a result of temperature rise, the stress caused by the spruce bark beetle will increase in the entire range of the Norway spruce due to a positive impact on beetle brood development (Temperli et al. 2013). In the case of the second major biotic factor causing the spruce stand dieback, i.e. fungi of the *Armillaria* sp. genus, climate change may increase its activity and modify the growth of rhizomorphs (Kubiak et al. 2017).

In mountainous conditions, however, two models of disturbance-driven dynamics in the spruce forests are most common: severe stand-replacing disturbances in a large area and gap-phase dynamics composed of distinct age cohorts of trees that originated over extensive areas (Holeksa et al. 2017). The dieback of spruce trees in temperate climate zones may be cyclical and occur approximately every 60–120 years. It has been estimated that in the Polish Tatra Mountains, the frequency of windthrows has not changed significantly for 200 years and disturbances in forest ecosystems were more severe in the nineteenth century than in the twentieth century. The intervals between the disturbances were so long that they allowed for the development of a new generation of stands (Holeksa et al. 2016). In their place, solid spruce stands often appear in spruce stands which are dying due to bark beetle outbreak, it is often found that the decaying spruce stands are replaced again by the Norway spruce with a slight admixture of other species, mainly the rowan (Jonášová and Prach 2004; Kupferschmid et al. 2006) .

23.6 Conclusions

As a result of global warming, the processes of disturbance and dieback of stands, especially from drought, occur more and more frequently in European forests. At the same time, due to rising air temperature, carbon dioxide concentration and nitrogen deposition, tree stands show faster growth rate and higher productivity. Faster growth rate leads to an earlier culmination of growth and more dynamic tree ageing processes, and moreover, the stress associated with droughts weakens stands. For these reasons, in the future in Central Europe, in protected forests in national parks and reserves, a greater rotation of stands should be expected, which will result in a lower felling age in managed forests. In southern Europe, tree ranges are expected to shrink, in particular of the Scots pine and the Norway spruce.

Acknowledgements The study was financed by National Science Centre, Poland, grant No. 2018/31/D/ST10/02137.

References

Aitken SN, Yeaman S, Holliday JA, Wang T, Curtis-McLane S (2008) Adaptation, migration or extirpation: climate change outcomes for tree populations. Evol Appl 1(1):95–111

Allen MR, Dube OP, Solecki W, Aragon-Durand F, Cramer W, Humphreys S, Kainuma M, Kala J, Mahowald N, Mulugetta Y, Perez R, Wairiu M, Zickfeld K (2018) Framing and Context. In: Global warming of 1.5°C. An IPCC Special Report on the impacts of global warming of 1.5°C above pre-industrial levels and related global greenhouse gas emission pathways, in the context of strengthening the global response to the threat of climate change, sustainable development, and efforts to eradicate poverty [V. Masson-Delmotte, P. Zhai, H. O. Portner, D. Roberts, J. Skea, P.R. Shukla, A. Pirani, W. Moufouma-Okia, C. Pean, R. Pidcock, S. Connors, J. B. R. Matthews, Y. Chen, X. Zhou, M. I. Gomis, E. onnoy, T. Maycock, M. Tignor, T. Waterfield (eds.)]. In Press

Altman J, Fibich P, Santruckova H, Dolezal J, Stepanek P, Kopacek J, Hunova I, Oulehle F, Tumajer J, Cienciala E (2017) Environmental factors exert strong control over the climate-growth relationships of *Picea abies* in Central Europe. Sci Total Environ 609:506–516

Attorre F, Alfò M, De Sanctis M, Francesconi F, Valenti R, Vitale M, Bruno F (2011) Evaluating the effects of climate change on tree species abundance and distribution in the Italian peninsula. Appl Veg Sci 14(2):242–255

Bigler C, Bräker OU, Bugmann H, Dobbertin M, Rigling A (2006) Drought as an inciting mortality factor in scots pine stands of the Valais, Switzerland. Ecosystems 9:330–343

Boisvenue C, Running SW (2006) Impacts of climate change on natural forest productivity – evidence since the middle of the 20th century. Glob Chang Biol 12(5):862–882

Bottero A, D'Amato AW, Palik BJ, Bradford JB, Fraver S, Battaglia MA, Asherin LA (2016) Density-dependent vulnerability of forest ecosystems to drought. J Appl Ecol 54(6):1605–1614

Brooker RW, Travis JMJ, Clark EJ, Dytham C (2007) Modelling species' range shifts in a changing climate: the impacts of biotic interactions, dispersal distance and the rate of climate change. J Theor Biol 245(1):59–65

Camarero JJ, Guada G, Sánchez-Salguero R, Cervantes E (2016) Winter drought impairs xylem phenology, anatomy and growth in Mediterranean scots pine forests. Tree Physiol 36 (12):1536–1549

Castagneri D, Petit G, Carrer M (2015) Divergent climate response on hydraulic-related xylem anatomical traits of *Picea abies* along a 900-m altitudinal gradient. Tree Physiol 35 (12):1378–1387

Cheaib A, Badeau V, Boe J, Chuine I, Delire C, Dufrêne E, François C, Gritti ES, Legay M, Pagé C, Thuiller W, Viovy N, Leadley P (2012) Climate change impacts on tree ranges: model intercomparison facilitates understanding and quantification of uncertainty. Ecol Lett 15 (6):533–544

Ciceu A, Popa I, Leca S, Pitar D, Chivulescu S, Badea O (2020) Climate change effects on tree growth from Romanian forest monitoring level II plots. Sci Total Environ 698:134129

Cienciala E, Altman J, Doležal J, Kopáček J, Štěpánek P, Ståhle G, Tumajer J (2018) Increased spruce tree growth in Central Europe since 1960s. Sci Total Environ 619–620:1637–1647

Dobbertin M, Rigling A (2006) Pine mistletoe (*Viscum album* ssp. *austriacum*) contributes to scots pine (*Pinus sylvestris*) mortality in the Rhone valley of Switzerland. For Pathol 36(5):309–322

Dyderski MK, Paź S, Frelich LE, Jagodziński AM (2018) How much does climate change threaten European forest tree species distributions? Glob Chang Biol 24(3):1150–1163

Eilmann B, Zweifel R, Buchmann N, Graf Pannatier E, Rigling A (2011) Drought alters timing, quantity, and quality of wood formation in scots pine. J Exp Bot 62(8):2763–2771

Fleischer P, Godzik B, Bicarova S, Bytnerowicz A (2005) Effects of air pollution and climate change on forests of the Tatra Mountains, Central Europe. In: Omasa K, Nouchi I, De Kok LJ (eds) Plant responses to air pollution and global change. Springer, Tokyo, pp 111–121

Galiano L, Martínez-Vilalta J, Lloret F (2010) Drought-induced multifactor decline of scots pine in the Pyrenees and potential vegetation change by the expansion of co-occurring oak species. Ecosystems 13:978–991

Garzón MB, de Dios RS, Ollero HS (2008) Effects of climate change on the distribution of Iberian tree species. Appl Veg Sci 11(2):169–178

Gatti RC, Callaghan T, Velichevskaya A, Dudko A, Fabbio L, Battipaglia G, Liang J (2019) Accelerating upward treeline shift in the Altai Mountains under last-century climate change. Sci Rep 9:7678

Gazda A, Kościelniak P, Hardy M, Muter E, Kędra K, Bodziarczyk J, Frączek M, Chwistek K, Różański W, Szwagrzyk J (2019) Upward expansion of distribution ranges of tree species: contrasting results from two national parks in Western Carpathians. Sci Total Environ 653:920–929

Geßler A, Keitel C, Kreuzwieser J, Matyssek R, Seiler W, Rennenberg H (2007) Potential risks for European beech (*Fagus sylvatica* L.) in a changing climate. Trees 21:1–11

Giordano L, Gonthier P, Varese GC, Miserere L, Nicolotti G (2009) Mycobiota inhabiting sapwood of healthy and declining scots pine (*Pinus sylvestris* L.) trees in the Alps. Fungal Divers 38:69–83

Giuggiola A, Bugmann H, Zingg A, Dobbertin M, Rigling A (2013) Reduction of stand density increases drought resistance in xeric scots pine forests. For Ecol Manag 310:827–835

Gonthier P, Giordano L, Nicolotti G (2010) Further observations on sudden diebacks of scots pine in the European Alps. For Chron 86(1):110–117

Hamann A, Wang T (2006) Potential effects of climate change on ecosystem and tree species distribution in British Columbia. Ecology 87(11):2773–2786

Hanewinkel M, Cullmann DA, Schelhaas M-J, Nabuurs G-J, Zimmermann NE (2013) Climate change may cause severe loss in the economic value of European forest land. Nat Clim Chang 3:203–207

Hess C, Niemeyer T, Fichtner A, Jansen K, Kunz M, Maneke M, von Wehrden H, Quantec M, Walmsley D, von Oheimb G, Härdtle W (2018) Anthropogenic nitrogen deposition alters growth responses of European beech (*Fagus sylvativa* L.) to climate change. Environ Pollut 233:92–98

Hevia A, Sánchez-Salguero R, Camarero JJ, Querejeta JI, Sangüesa-Barreda G, Gazol A (2019) Long-term nutrient imbalances linked to drought-triggered forest dieback. Sci Total Environ 690:1254–1267

Holeksa J, Zielonka T, Żywiec M, Fleischer P (2016) Identifying the disturbance history over a large area of larch–spruce mountain forest in Central Europe. For Ecol Manag 361:318–327

Holeksa J, Jaloviar P, Kucbel S, Saniga M, Svoboda M, Szewczyk J, Szwagrzyk J, Zielonka T, Żywiec M (2017) Models of disturbance driven dynamics in the west Carpathian spruce forests. For Ecol Manag 388:79–89

Iverson LR, Prasad AM (1998) Predicting abundance of 80 tree species following climate change in the eastern United States. Ecol Monogr 68(4):465–485

Iverson LR, Schwartz MW, Prasad AM (2004) How fast and far might tree species migrate in the eastern United States due to climate change? Glob Ecol Biogeogr 13(3):209–219

Jaime L, Batllori E, Margalef-Marrase J, Navarro MAP, Lloret F (2019) Scots pine (*Pinus sylvestris* L.) mortality is explained by the climatic suitability of both host tree and bark beetle populations. For Ecol Manag 448:119–129

Jamroz E, Kocowicz A, Bekier J, Weber J (2014) Properties of soil organic matter in Podzols under mountain dwarf pine (*Pinus mugo* Turra.) and Norway spruce (*Picea abies* (L.) karst.) in various stages of dieback in the east Sudety Mountains, Poland. For Ecol Manag 330:261–270

Jonášová M, Prach K (2004) Central-European mountain spruce (*Picea abies* (L.) karst.) forests: regeneration of tree species after a bark beetle outbreak. Ecol Eng 23(1):15–27

Krejza J, Cienciala E, Světlík J, Bellan M, Noyer E, Horáček P, Štěpánek P, Marek MV (2020) Evidence of climate-induced stress of Norway spruce along elevation gradient preceding the current dieback in Central Europe. Trees (in press, published online)

Kubiak K, Żółciak A, Damszel M, Lech P, Sierota Z (2017) *Armillaria* pathogenesis under climate changes. Forests 8(4):100

Kupferschmid AD, Brang P, Schönenberger W, Bugmann H (2006) Predicting tree regeneration in *Picea abies* snag stands. Eur J For Res 125:163–179

Li G, Huang J, Guo H, Du S (2020) Projecting species loss and turnover under climate change for 111 Chinese tree species. For Ecol Manag 477:118488

Lindner M, Maroschek M, Netherer S, Kremer A, Barbati A, Garcia-Gonzalo J, Seidl R, Delzon S, Corona P, Kolström M, Lexer MJ, Marchetti M (2010) Climate change impacts, adaptive capacity, and vulnerability of European forest ecosystems. For Ecol Manag 259(4):698–709

Mäkinen H, Nöjd P, Mielikäinen K (2001) Climatic signal in annual growth variation in damaged and healthy stands of Norway spruce [*Picea abies* (L.) karst.] in southern Finland. Trees 15:177–185

Malcolm JR, Markham A, Neilson RP, Garaci M (2002) Estimated migration rates under scenarios of global climate change. J Biogeogr 29(7):835–849

Marqués L, Madrigal-González J, Zavala MA, Camarero JJ, Hartig F (2017) Last-century forest productivity in a managed dry-edge scots pine population: the two sides of climate warming. Ecol Appl 28(1):95–105

Martínez-vilalta J, López BC, Adell N, Badiella L, Ninyerola M (2008) Twentieth century increase of scots pine radial growth in NE Spain shows strong climate interactions. Glob Chang Biol 14 (12):2868–2881

Messinger J, Güney A, Zimmermann R, Ganser B, Bachmann M, Remmele S, Aas G (2015) *Cedrus libani*: a promising tree species for central European forestry facing climate change? Eur J For Res 134:1005–1017

Michelot A, Bréda N, Damesin C, Dufrêne E (2012) Differing growth responses to climatic variations and soil water deficits of *Fagus sylvatica, Quercus petraea* and *Pinus sylvestris* in a temperate forest. For Ecol Manag 265:161–171

Pastirčáková K, Adamčíková K, Pastirčák M, Zach P, Galko J, Kováč M, Laco J (2018) Two blue-stain fungi colonizing scots pine (*Pinus sylvestris*) trees infested by bark beetles in Slovakia, Central Europe. Biologia 73:1053–1066

Pretzsch H, Biber P, Schütze G, Uhl E, Rötzer T (2014) Forest stand growth dynamics in Central Europe have accelerated since 1870. Nat Commun 5:4967

Prietzel J, Falk W, Reger B, Uhl E, Pretzsch H, Zimmermann L (2020) Half a century of scots pine forest ecosystem monitoring reveals long-term effects of atmospheric deposition and climate change. Global change biology, early view (online version of record before inclusion in an issue)

Räisänen J, Hansson U, Ullerstig A, Döscher R, Graham LP, Jones C, Meier HEM, Samuelsson P, Willén U (2004) European climate in the late twenty-first century: regional simulations with two driving global models and two forcing scenarios. Clim Dyn 22:13–31

Rebetez M, Dobbertin M (2004) Climate change may already threaten scots pine stands in the Swiss Alps. Theor Appl Climatol 79:1–9

Rigling A, Eilmann B, Koechli R, Dobbertin M (2010) Mistletoe-induced crown degradation in scots pine in a xeric environment. Tree Physiol 30(7):845–852

Rigling A, Bigler C, Eilmann B, Feldmeyer-Christe E, Gimmi U, Ginzler C, Graf U, Mayer P, Vacchiano G, Weber P, Wohlgemuth T, Zweifel R, Dobbertin M (2012) Driving factors of a vegetation shift from scots pine to pubescent oak in dry alpine forests. Glob Chang Biol 19 (1):229–240

Sánchez-Salguero R, Camarero JJ, Gutiérrez E, Rouco FG, Gazol A, Sangüesa-Barreda G, Andreu-Hayles L, Linares JC, Seftigen K (2016) Assessing forest vulnerability to climate warming using a process-based model of tree growth: bad prospects for rear-edges. Glob Chang Biol 23 (7):2705–2719

Schelhaas M-J, Nabuurs G-J, Hengeveld G, Reyer C, Hanewinkel M, Zimmermann NE, Cullmann D (2015) Alternative forest management strategies to account for climate change-induced productivity and species suitability changes in Europe. Reg Environ Chang 15:1581–1594

Socha J, Tymińska-Czabańska L, Grabska E, Orzeł S (2020) Site index models for Main Forest-forming tree species in Poland. Forests 11(3):301

Spiecker H (2000) Growth of Norway spruce (*Picea abies* [L.] karst.) under changing environmental conditions in Europe. In: Klimo E, Hager H, Kulhavý J (eds.) Spruce monocultures in Central Europe – problems and prospects. EFI proceedings no. 33, pp. 11–26

Stagge JH, Kingston DG, Tallaksen LM, Hannah DM (2017) Observed drought indices show increasing divergence across Europe. Sci Rep 7:14045

Sykes MT, Prentice IC (1996) Climate change, tree species distributions and forest dynamics: a case study in the mixed conifer/northern hardwoods zone of northern Europe. Clim Chang 34:161–177

Temperli C, Bugmann H, Elkin C (2013) Cross-scale interactions among bark beetles, climate change, and wind disturbances: a landscape modeling approach. Ecol Monogr 83(3):383–402

Vertui F, Tagliaferro F (1998) Scots pine (*Pinus sylvestris* L.) die-back by unknown causes in the Aosta Valley, Italy. Chemosphere 36(4–5):1061–1065

Voltas J, Camarero JJ, Carulla D, Aguilera M, Ortiz A, Ferrio JP (2013) A retrospective, dual-isotope approach reveals individual predispositions to winter-drought induced tree dieback in the southernmost distribution limit of scots pine. Plant Cell Environ 36(8):1435–1448

Wang WJ, He HS, Thompson FR, Fraser JS, Dijak WD (2017) Changes in forest biomass and tree species distribution under climate change in the northeastern United States. Landsc Ecol 32:1399–1413

Wermelinger B, Rigling A, Schneider Mathis D, Dobbertin M (2008) Assessing the role of bark- and wood-boring insects in the decline of scots pine (*Pinus sylvestris*) in the Swiss Rhone valley. Ecological Entomology 33(2):239–249

Chapter 24
Climate Change Impacts on Soil Microorganisms that Regulate Nutrient Transformations

Bulbul Gupta, Gurpreet Saggu, Smita Sundaram, Ruby Mishra, and Indu Shekhar Thakur

Abstract Climate change has severe effects on soil microorganisms that regulate the biochemical cycle of nutrients and other components necessary for plants growth. Understanding the nutritional viability and microbial interaction under climate change is vital for future research. Several environmental factors like elevated CO_2, temperature, drought and salinity affect the soil microbial ecosystem, impacting plants nutritional availability. The present known information on the effects of climate change on soil microorganisms that regulate nutrient transformation is analysed in this chapter along with the mitigation mechanisms adopted by the microorganisms to counter the adverse effects of climate change.

Keywords Climate change · Elevated CO_2 · Microbial interaction · Nutrients · Soil · Temperature

24.1 Introduction

The effects of key indicators of climate change on soil microbiomes differ tremendously due to the varied biotic and abiotic properties of the soil. The variations in the biogeochemistry (including pH and salinity) of a particular soil class regulate the types of microorganisms that are present. Furthermore, different compositions of the

B. Gupta (✉) · I. S. Thakur
School of Environmental Sciences, Jawaharlal Nehru University, New Delhi, India

G. Saggu
Thapar Institute of Engineering and Technology, Patiala, Punjab, India

S. Sundaram
Shaheed Rajguru College of Applied Sciences for Women, University of Delhi, New Delhi, India

R. Mishra
Deshbandhu College, University of Delhi, New Delhi, India

D. K. Choudhary et al. (eds.), *Climate Change and the Microbiome*, Soil Biology 63,
https://doi.org/10.1007/978-3-030-76863-8_24

soil influence the development of microbial ecosystems with cascading effects on transformations of nutrients. Therefore, to have a better understanding about the impacts of climate change on species interactions, metabolism and nutrient transformation, it is essential to apprehend the connectivity of soil microbial communities. Although microorganism interactions within soil niches is well-known, the thermodynamics and energy of nutrient transformations that power microbial activity are not well characterized, particularly in the aspect of changing climatic conditions (Dixon and Tilston 2010). Communities and the physiological responses of soil microorganisms are influenced by complex environmental factors across various geographical regions. The diverse climatic conditions make it difficult to generalize mechanisms adopted by the soil microorganisms to cope with the changes in the environment. However, it is possible to provide information about the factors impacting the changes, thus discussing the effect of elevated carbon dioxide (eCO_2), increased temperature, drought and soil moisture on the soil microbial community.

24.1.1 Elevated CO_2

Carbon dioxide is a greenhouse gas (GHG) that acts as the prime indicator of climate change, with methane and nitrous oxide as the secondary ones. Because of the long lifespan of GHGs, they remain in the atmosphere for many decades, playing an essential role in altering the atmospheric temperatures. Even in the absence of any new emissions, there are enough GHGs in the atmosphere to cause temperature rise and destruction of the protective ozone layer. An increased CO_2 level in the atmosphere causes global warming that alters the earth's climate systems in countless ways. With every passing year, an increase in the levels of the three main greenhouse gases was recorded. In 2018, the globally averaged mole fractions of CO_2 were found to be at the highest, i.e. 407.8 ± 0.1 ppm. The concentrations of CO_2 in the atmosphere reveal the difference between its release and uptake by the oceanic systems. Microbial populations play a vital role in terrestrial ecosystems by responding to climate change through the regulation of soil biogeochemical processes. However, little information is available on the effects of eCO_2 and increasing temperatures on soil microorganisms. Recent studies reported the changes in the arrangements of functional genes of the microorganisms under eCO_2 (Haugwitz et al. 2014). These changes stimulate alterations in the key functional genes involved in various activities of soil microorganisms, viz. carbon and nitrogen fixation, nitrification/denitrification, etc.

To evaluate the impact of climate change on forest ecosystems, it becomes essential to have deeper understanding on the responses of soil microbiota to eCO_2. In Australian grasslands, a shift in fungi populations and specific groups of bacteria has been observed with altered levels of CO_2 (Hayden et al. 2012). The microbial genes involved in functional activities of nitrogen and carbon fixation, metabolism and nitrogen mineralization also increase in arid grasslands when exposed to eCO_2 levels. Understanding of changes in functional genes involved in the cycling of organic matter gives knowledge of how eCO_2 affects microorganism metabolisms. Microorganism directly associated with plants also gets affected by varied CO_2 levels. Due to elevated CO_2 levels, the availability of carbon to roots of the plants, plant biomass and soil microbial activities also increases (Yu et al. 2018). However, the differences in the plant species response to eCO_2 resulted in the changed microbial community composition and nutrient consumption patterns. As seen in the nutrient-deficient ecosystems (such as boreal coniferous forest and temperate heathlands), the simultaneous growth of plants and microbial biomass due to eCO_2 might be because of limitation of both nitrogen and phosphorous in the soil.

For long-term impacts of climate change on soil ecosystems, the elevated levels of $CO_2$are also related with plant litter quality, including the carbon to nitrogen ratio of the plant litter. Increased plant growth in the presence of eCO_2 may increase plant demand for nitrogen and plant nitrogen uptake. However, progressive nitrogen constraints eventually restrict plant responses to eCO_2 and affect the microbial community conformation, structure and functional potential when present beneath the ground. Apart from the effects of plant-microbial associations, elevated levels of CO_2 due to global warming also alter the soil microorganism's composition (IPCC 2018). For example, due to the eCO_2 levels, microbial decomposition of soil organic matter increases, but enhanced temperature makes the soil dry. As recorded for the Australian grasslands, under eCO_2, the total microbial abundance increases; however, a decrease was observed with eCO_2 combined with warm weather. Understanding how microbial communities gets affected by other climate change indicators, including extreme weathers, floods and drought, thus, become important for predicting microorganism's responses among different soil ecosystems.

24.1.2 Increased Temperature

Warmer temperatures particularly enhance levels of microbial activity, which in turn boost the metabolic decomposition of organic contaminants (Gouin et al. 2013). For example, warming has been reported to enhance gene mutations responsible for the degradation of soil organic carbon. Reports have shown that, due to a decrease in carbon sinks and biomass production with sustained warming, the elevated levels of soil organic carbon decay decrease (Bardgett et al. 2013). Similarly, soil transpiration, biomass production and organic carbon decay have typically been observed to increase with temperature, but this warming effect is often short termed in the fields. Soil nutrients are presumed to be exhausted by enhanced microbial growth and trade-offs arise as microorganisms (1) acclimatize, (2) change the composition or (3) restrict their adaptation to environmental changes and to the accessibility of substrates (Bradford 2013). With respect to the impact of temperature on microbial populations, the findings of in situ as well as ex situ experiments are mostly inconsistent. These findings suggest that the influences of varied temperatures are extremely complex but can be regulated by phenomena's such as acclimatization, adaptation, mutation, etc. Despite all this, climate change-induced changes in soil temperature should not have a crucial effect on the efficiency of soil bioremediation on a global scale, as we are dealing with a rise of approx. 1 °C–3 °C (Raftery et al. 2017).

The soil microbiome population and their responses to elevated temperature also get affected by the type of habitat under research. For example, boreal forest has shown a differing effect of warming on soil fungi, leading to the enhancement or repression in fungal biomass and their metabolism; these variations are probably attributed to difference in soil humidity and/or biodiversity at different sites. Interestingly, 5 years of comparatively benign 0.3 °C soil warming was found to be adequate to enhance microbial abundance and their productivity in the soil of the boreal heath land. These findings indicate the indirect effect of warming occurring due to the prolonged planting season (Sistla and Schimel 2013). In spite of a slight change in the global temperature (~ 1.5–2 °C), its direct warming impact is still plausible especially in spring season (March to May), which otherwise is considered as a period with soil temperature near to 0 °C.

24.1.3 Drought

Drought is anticipated as the main outcome of potential climate change in habitats of mesonic grasslands. Moreover, in the coming decades, increment in desertification is expected for majority of semi-arid or arid areas. Rising drought is expected to lead to reductions in microbial functions which act as prime key for ecosystem stability. There is not as much water in soil pores as the soil becomes drier, resulting in isolated islands of resources; consequently, less soil organic matter is decayed and exhaled into CO_2 (Kannojia et al. 2019). Together, these variables communicate to yield responses ranging from reduced productivity under drier conditions to turn down the loss of carbon by repressed breathing. Since arid soil habitats are broadly spread, extended from warm to cold deserts with minimal annual rainfall to dry summer and rainy winter Mediterranean grasslands, it is not easy to generalize microbial responses to rising drought as a result of climate change (Clair and Lynch 2010). For example, because of carbon and moisture constraints, life in desert soils is often restricted and is therefore dominated by surface-dwelling photoautotrophs. As a result, biological soil crusts (biocrusts) are formed. For example, more than 40% of drylands have biocrusts. Biocrust species (e.g. cyanobacteria and lichens) undergo carbon and nitrogen fixation and are also the main primary producers where these components are created. Since biocrusts bind and stabilize the surface soil, there are also important consequences for soil depletion (Dhanya and Ramachandran 2016). Drought could cause long-term impact of changes on the soil microbiome in grassland habitats due to changes in vegetation to more drought-tolerant plant species and their subsequent selection for various root-associated microorganisms (Blagodatskaya and Kuzyakov 2013). In mesocosms and multi-year field studies, network observations showed that bacteria in grasslands are more vulnerable to drought than fungi. Therefore, fungi possibly contribute to the preservation of the balance of carbon and nitrogen when water is scarce. Fungal hyphae can also continue to cross spatially distinct resources as soils get drier and microbial dispersion becomes more confined within physically covered soil pores, which can also assist the bulk microbiome. It is therefore necessary to better understand how inter-kingdom experiences produce responses from the population to drought stress. Soil microorganisms developed different physiological pathways for coping with drought stress, such as osmoregulation, dormancy or reactivation and extracellular enzyme synthesis (Berard et al. 2015). For survival at lower water matric potential, solutes (osmolytes) are accumulated by microorganisms to retain cell turgor. Under extreme drying conditions, however, osmolyte aggregation could be too energetically costly. Microorganisms in soil may survive in dehydrated condition and regenerate and regrow when moisture is accessible. The processing of extracellular polymeric compounds for the preservation of water at low matric potential is another physiological technique. Members of certain bacterial, such as actinobacteria and bacilli, may also survive because of their capacity to preserve development and become inactive under dry conditions in drought-impacted soils. How to cope when the soil is re-wet is another biochemical challenge for arid soil microorganisms. The wetting of extracellular enzymes as microbial activity declines implies that a large portion of the wet-up response may be driven by the destruction of dead microbial cells and suggesting a role in this response for viral predation of bacteria. For a higher microbial soil diversity in dry environments, perhaps as there were more disconnected soil niches in dry soil, the modelling simulations and empirical data were contrasted by one analysis. Improved connectivity, further dispersal, further anaerobic niches and a sudden rise of nutrients were found after wetting, resulting in an increase in anaerobic taxa and a decline in diversity. The society was, however, durable and, following re-drying, restored to its former state. In comparison, the microbial biomass rose during the dry season in the California grasslands and then decreased during the rainy season. A comparison of various arid environments, including deserts, showed that aridity increases could lead to a decrease in the stability and genetic capacity of the microbiome of the soil (Cook et al. 2015). The concern, therefore, is how adaptable the soil microbiome can be in the future to growing drought cycles and less stable weather trends.

Drought is among the most major environmental stress factors faced by soil microorganisms, and it is expected that longer drought cycles and exceptionally dry spells would become more widespread in Europe during this century, mostly during the spring end and summer seasons. Drought is commonly predicted to decline both soil microbial activity and soil enzymatic activity, resulting in decreased organic matter turnover and hence the supply of soil nutrients (Schimel 2018). Soil respiration and microbial development have demonstrated the ability to regenerate after soil re-wetting, despite strong signs of drought effects. In addition, various classes of microbial communities react differently to drought, such as drought-favoured acid bacteria compared to proteobacteria. Overall, fungi are expected to cope better with drought than bacteria because of their ability to move water across the hyphal network (Naylor et al. 2017). Therefore, under drier conditions, microbial populations are likely to become more fungal driven, which could affect the decomposition of organic matter, as fungi are able to digest more complex organic molecules. Due to increased evaporation of water from the soil, as longer drought cycles are combined with warming, this can exacerbate drought effects and thereby further decrease microbial activity. In comparison, global warming is anticipated to encourage microbial activity and the decomposition of soil organic matter where water supply is not a limiting factor. This has showed higher levels of mineralization of nitrogen (N) and phosphorus (P), which may theoretically increase microbial biomass and plant growth by releasing more inorganic nutrients into the soil (Upton et al. 2018).

24.1.4 Increased Rainfall and Flooding

Owing to floods and/or extreme and erratic precipitation events, some regions are seeing increases in soil moisture. Increased high rainfall events in wet tropical areas, for example, are expected to occur. Climate change is also expected to move precipitation at the loss of snow to more rain in northern areas, leading to decreasing snowpack and rising freeze-thaw periods (Steven et al. 2015). When soil moisture increases, soil pores are filled with water and anaerobic, providing reliable conditions for methanogenesis and denitrification, with the potential to release CH_4 and N_2O. With respect to microbial community reactions, variations in moisture and vegetation due to shifting rainfall patterns can lead to contrasting effects. For more precise simulations under future climate conditions, predictive metabolic models are also required. Wetlands reflect the intersection between terrestrial and marine ecosystems and are climate-sensitive hotspots. Soil temperature, water-table depth and soil organic carbon composition are the main determinants of CH_4 concentrations in wetlands. Wetland areas, such as peatlands, could shift from carbon sinks to carbon emitters in the coming years, thus exasperating current temperature trends (Belnap et al. 2016). However, oxidative decomposition of organic carbon in soil in peat can be prevented when peatlands are flooded, resulting in net carbon uptake. The development of saltwater soil in association with the increase in sea level is a special case. Due to the rapidly increasing sea level, many coastal regions are witnessing saltwater intrusion, at a rate of 3.2 ± 0.4 mm per annum. Salt and sulphate are added by the introduction of saltwater into fragile coastal soil ecosystems: the latter functions as a terminal electron acceptor and shifts the system's redox cycling mechanisms, resulting in improvements in microbial quantity of soil and improvements in soil organic carbon mineralization, resulting in increased CO_2 output levels. Consequently, the effects of a potential net increase in greenhouse gas production by increasing CO_2 emissions as seawater levels grow. Various coastal soils, however, react to elevated salinity differently (Blazewicz et al. 2014). Microbial activity has been seen to decline with prolonged cycles of floods due to water scarcity, reflecting a "boom and bust" condition. The ultimate climate effect of rising seawater levels would therefore depend on the soil microbial population dynamics and the supply of soil organic carbon and electron acceptors that regulate the equilibrium between carbon and nutrient storage and emission of greenhouse gases.

24.2 Microbial Responses to Climate Change

Climate is described as the conditions of normal or moderate weather conditions of definite areas, including temperature, precipitation and wind system of atmosphere. It is a dynamic, interconnected structure that consists of the atmosphere, the soil, oceans, land, snow and ice. The atmosphere of the planet is most impacted by the latitude, the inclination of the axis of the earth, the motion of the wind belts, the disparity in temperatures of the sea and topography.

Dramatic shifts in the global ecological, chemical and physical climate have been triggered by rising global industrial development, including changes in the structure and distribution of plant species, increased atmospheric CO_2 concentrations, rise in annual average temperature and rainfall patterns. Extreme climate events, including droughts, heat waves and flooding, cause drastic changes in the development and operation of ecosystems (Reichstein et al. 2013; Smith 2011). Microbial population dynamics are changed as the temperature rises, and mechanisms such as respiration, fermentation and methanogenesis are also boosted. Impact of bacteria, fungi, algae and archaea on climate change is driving the global warming by decomposition of organic matter and eventually growing the atmospheric carbon dioxide (CO_2) flux. The microorganisms affect the climate change by increasing global warming through decomposition of organic matter which enhances the carbon dioxide levels in the atmosphere.

24.2.1 *Causes of Climate Change*

Anthropogenic sources have increased the concentration of greenhouse gases into the earth's atmosphere manifold in the recent years. After the Industrial era, the concentration of greenhouse gases started increasing into the atmosphere. The main greenhouse gases are carbon dioxide (CO_2), methane (CH_4), nitrous oxide (N_2O) and the halocarbons. Emissions from manmade sources like industries and vehicles and natural sources like volcanoes and forest fires have outweighed the sinks of these gases, which leads to accumulation of these gases in the air (Castro et al. 2010).

1.1 Carbon dioxide (CO_2) is produced from complete combustion and from fossil fuel sectors like transport, production of cements and other products.
1.2 Methane is also produced from anthropogenic activities like landfills and waste, biomass burning.

1.3 Nitrous oxide gets released naturally from soil and oceans. Human activities which release nitrous oxide include fossil fuel burning and fertilizer use.
1.4 Halocarbons including chlorofluorocarbons (CFC-11 and CFC-12) are used in refrigerants and aerosol and, in industrial processes, were found to cause decrement in the stratospheric ozone hole.
1.5 In the troposphere, the ozone concentration keeps on increasing due to anthropogenic emissions of carbon monoxide, hydrocarbons and nitrogen oxides, which reacts in the presence of sunlight to form ozone.
1.6 Water vapour is the most copious greenhouse gas present in the ambiance. Direct impact is low on water vapours by human activities, but potential for indirect effect by human activities is large by causing warming in the atmosphere.
1.7 Aerosols and particles present in the atmosphere are emitted from both anthropogenic and natural sources. Anthropogenic activities such as fossil fuel and biomass burning have increased concentration of soot (black carbon) in the atmosphere, which leads to warming of the atmosphere.

24.3 Effect of Climate Change on Microorganisms

Earlier research of the impact of climate change on enzyme function, decomposition and heterotrophic respiration have generally focused on the impact of warming. The bacteria, fungi and other microorganisms have huge impact on climate change. They are accelerating warming by organic matter decomposition. Temperature directly affects the microbial activity and physiology. Therefore, determining the role of microorganisms in ever-changing flux of nutrient cycle is important.

24.3.1 Microbial Community and Methane Cycle

Carbon cycling between CO_2 and organic compounds is thought to be ecologically important. Both eukaryotes (such as plants and algae) and autotrophic bacteria (such as cyanobacteria) play a significant role in the fixation of carbon dioxide in organic compounds. Chemical compounds are still used by consumers and carbon dioxide is emitted. Methane (CH_4) is a greenhouse gas that gets emitted into air due to microbial activities much of the time. Microorganisms eating methane are crucial to sustaining a stable atmosphere on earth. It is used by bacteria basically as source of energy for its metabolism (Bousquet et al. 2006; Semrau et al. 2010). Methanotrophic bacteria use methane as their only energy supply and, through their digestion phase, convert it to carbon dioxide. These bacteria can consume huge amounts of methane, which helps to reduce methane emissions from factories and landfills that produce methane. High concentrations of CH_4 compounds that are present everywhere are used by microorganisms. Carbon dioxide is quickly converted into methane in anaerobic environments much as dense compacted mud and is done by methanogenic bacteria. Hydrogen, water and energy for the methanogens are required for the conversion process. To perform the recycling trend, methane-oxidizing bacteria or methanotrophs are another type of bacteria which converts CH_4 to CO_2. This process also produces water and power, which is an aerobic operation. CH_4 is oxidized into CO_2 by methanotrophic bacteria in the presence of oxygen. The oxidation of CH_4 into CO_2 ends the cycle of carbon. Methanotrophs prefer to live on the threshold of aerobic and anaerobic environments. They have exposure to the methane produced by methanogenic anaerobic bacteria, in addition to the oxygen required for the methane conversion (Rajput et al. 2013).

24.3.2 Microbial Community and Carbon Cycle

The annual carbon influx of ~120 Gt into and out of terrestrial environments significantly exceeds the volume of carbon emitted by fossil fuel combustion (Change 2014). Therefore, a lot of scientific research has centred on how to produce more accurate carbon flux estimates in order to determine how much carbon can be captured in terrestrial environments. Soils hold almost 2.5 times more carbon than the atmosphere in conjunction with plant biomass (Singh et al. 2010). Soils have the potential to hold vast volumes of carbon and have helped reduce increasing ambient (CO_2) through their potential to sequester carbon. Several variables govern the amount of carbon soils that can be sequestered, including temperature, organic matter content, soil age and texture, topography, type of vegetation and diversity of the soil population (Jenny 1941). The global carbon cycle relies primarily on microorganisms that fix carbon dioxide from the atmosphere, stimulate plant growth and environmentally degrade or convert organic matter. However, the rate-limiting steps in the decomposition process are essentially governed by microbial decomposers and hence the effect of abiotic factors on decomposition (Gougoulias et al. 2014). Large volume of organic carbon is at present locked in high latitude ice caps, grassland soils, tropical forests and other ecosystems. Microorganisms, on the other hand, play an important role in deciding the durability and persistence of the carbon and whether it is emitted as greenhouse gas into the environment. Photosynthesis and respiration are two processes which regulate the terrestrial carbon (Schulze 2006). Carbon is fixed into the soil by microorganisms as they require it for their metabolic substrate. Figure 24.1 represents the major carbon cycle.

24.3.3 Microbial Community and Nitrogen Cycle

Various nitrogen gaseous molecules, including NH_3, NO and N_2O, also occur in the atmosphere. Nitrogen gas constitutes the highest proportion of around 78% in the

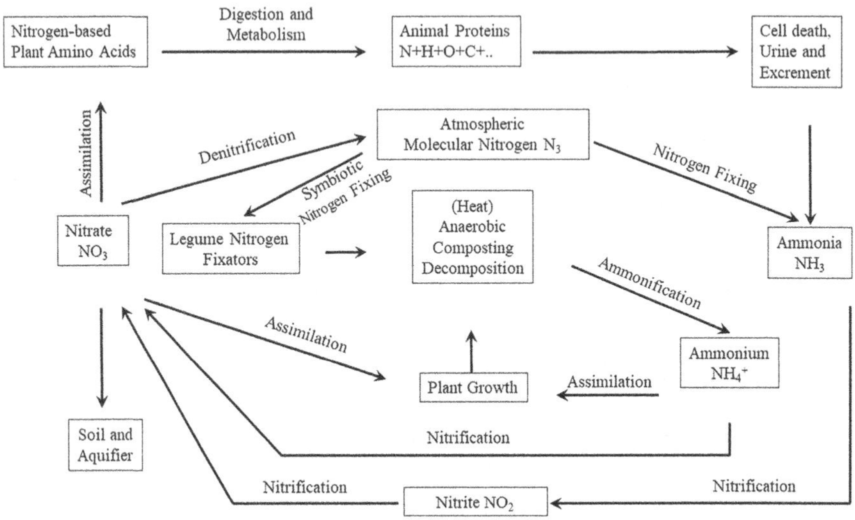

Fig. 24.1 Carbon cycle in the environment

atmosphere (Bange 2008). Nitrogen is present in the shape of a very stable molecule (N_2) and is available to plants and animals with fixation. The method of converting nitrogen into chemical forms that can be used by living organisms is nitrogen fixation. Through biological fixation, N_2 reaches the biosphere. The bacterium attaches to the plant's root hair and responds in response to a hollow thread leading into the root forms the plant. Growing bacteria via this thread of infection ultimately initiate the development on a root nodule. As much as 30 per cent of a nodule's weight can be bacteria. Bacteria get its power and nutrition from plants. Nitrogen from the air is provided by bacteria and fungi in a manner that the plant can utilize by fixation. This signifies the symbiotic type of nitrogen fixation. Particular bacteria possess (*Rhizobium trifolium*) nitrogenase enzymes which can fix a form of nitrogen present in atmosphere into ammonium ion which is utilized in the chemical form by higher species. The plants in a symbiotic relationship change the "fixed" ammonium ion into oxides of nitrogen and amino acids to make proteins and other molecules such as alkaloids. Nitrogen cycle is basically a transformation of nitrogen from one form to another. For majority of the time period, microorganisms work for converting nitrogen into form which they can utilize for their own growth and development. Figure 24.2 represents the nitrogen cycle and the various processes involved at each stage.

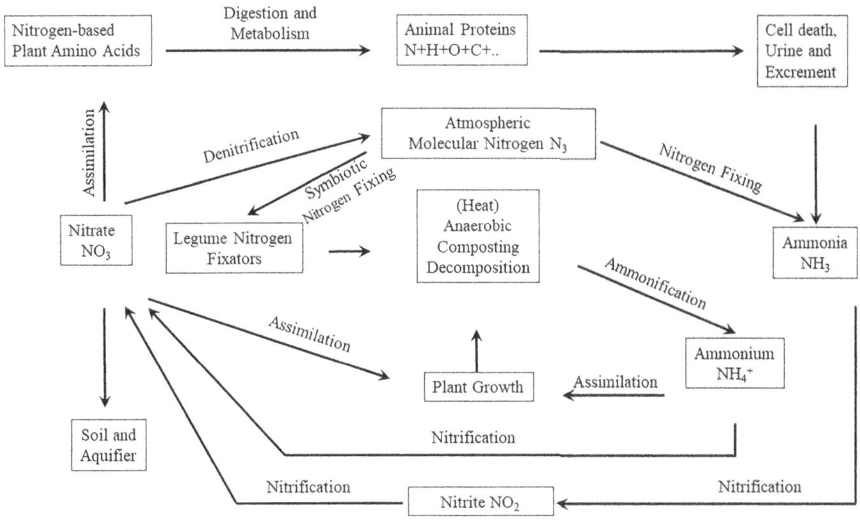

Fig. 24.2 Nitrogen cycle in the environment

24.4 Structure and Role of Microbial Community in Climate Change

Altering environment due to change in climate affects structure and functions of soil microbiome by changing the chemical and physical environment of the soil. Soil desertification, changing of temperature, soil pH, mineralization, transformation and nutrient availability alter the essential function of microbes. These changes significantly affect various remaining functions that the microbial community carries out in agroecosystems, including carbon cycle, decomposition, aggregation and fertility of soil, efficiency of plants, quality of soil water, mineral pools, CO_2 present in air, plant roots and emissions of greenhouse gases. Some changes in the microbial community occurred even after a short duration of changes in the climate. Large amounts of global warming causing gases (GHGs) such as CO_2, N_2O and CH_4 are transformed from one form into another, and they are interlinked by potential positive and negative feedback loops acting in concert in terrestrial ecosystem (Drigo et al. 2008).

For better survival, soil microbes alter some internal mechanism and functional changes like extracellular enzyme development which catalyses organic matter mineralization and alter microbial function, soil quality and ecosystem productivity (Deltedesco et al. 2020). Major enzymes like carbonic anhydrase, Ribulose-1,5-bisphosphate carboxylase/oxygenase, nitrate reductase, nitrite reductase, methane oxygenase and other enzymes involved in storage and transformation processes. Anthropogenic nitrogen (N) enriches the biosphere by fertilizer application and pesticides, and incorporation of manures or compost alters biogeochemical cycling of ecosystems due to presence of recalcitrant organic compounds like lignin and chlorinated compounds in a process that leads to change in flux of biogenic greenhouse gases (GHGs) or inhibiting biosequestration processes. High rise in temperature indirectly affects the process of evapotranspiration, plant physiology and root exudation (Lange et al. 2014).

24.5 Climate Modulators and Native Regimes of the Microbial Community

Microbes are involved in many processes, including the nutrient cycles, food processing, inhabiting gut in many organisms to help digest food, cleaning up pollution in many ways, etc. Performing these versatile processes, microbes either use or produce gases like carbon, nitrogen and sulphur. Carbon and nitrogen are component of the three main gases responsible for global warming, CO_2, CH_4 and N_2O. Thus, by participating large number of indispensable processes, microbes directly or indirectly function as climate modulator. Being a "closed system" earth has to produces everything it needs for the survival and growth of its inhabitants. Various cycles such as the carbon cycle, nitrogen cycle, etc. (a complex series of processes through which the gaseous elements rotate) control and balance the gases that surround the earth (Nauer et al. 2018). Microbes play a key role in cycling and transforming these elements by either generating or consuming these gases in the environment. For example, methanogens are involved in carbon cycle by converting carbon dioxide into methane. Nitrogen-fixing bacteria convert atmospheric nitrogen into biological nitrogen and participate in nitrogen cycle. Photosynthetic algae and cyanobacteria perform photosynthesis and by this means participate in carbon cycle. Fungi and soil bacteria that decompose organic components play a major role in the carbon cycle by breaking down organic matter to release carbon dioxide into the environment. Plants-mycorrhizal fungi also take part in cycling nutrients by plants providing a considerable carbon to the mycorrhizal fungal symbionts, whereas mycorrhizal fungi provide nitrogen and phosphorus to plants (Abatenh et al. 2018).

Various natural and anthropogenic sources are involved in increase in greenhouse gases concentration though the years. In line with these microbes are invisible though chief participants in producing greenhouse gases. Microbe-driven enteric fermentation and manure storage are two important sources of greenhouse gases. Farm animals have hundreds of microbes in their gut of which some help in digesting cellulose. In the process these animals produce considerable quantity of methane gas. Under anaerobic conditions, the organic matter present in manure is partially decomposed by bacteria producing methane and carbon dioxide (Grossi et al. 2019). Along with this the microbial biota present inside wetland termites may participate in 1–3% of global methane gas production (Figs. 24.3 and 24.4) (Nauer et al. 2018; Abatenh et al. 2018).

Impacts of bacteria, fungi and algae on climate change include accelerating global warming by decomposition of organic matter and eventually rising atmospheric CO_2

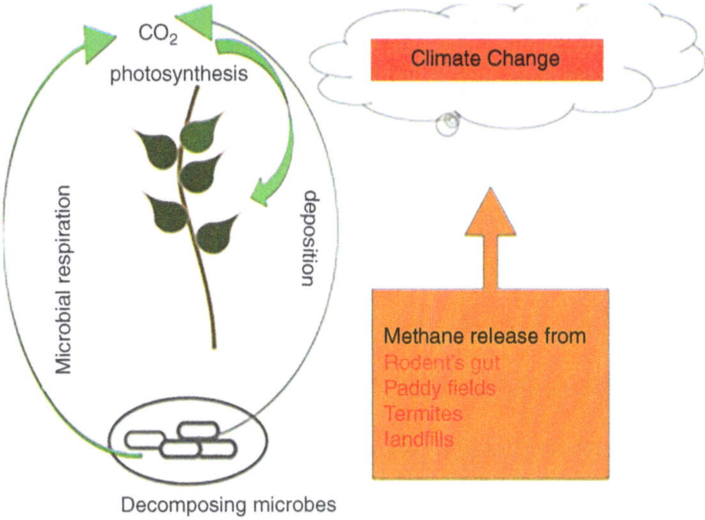

Fig. 24.3 Microbes function as climate modulators

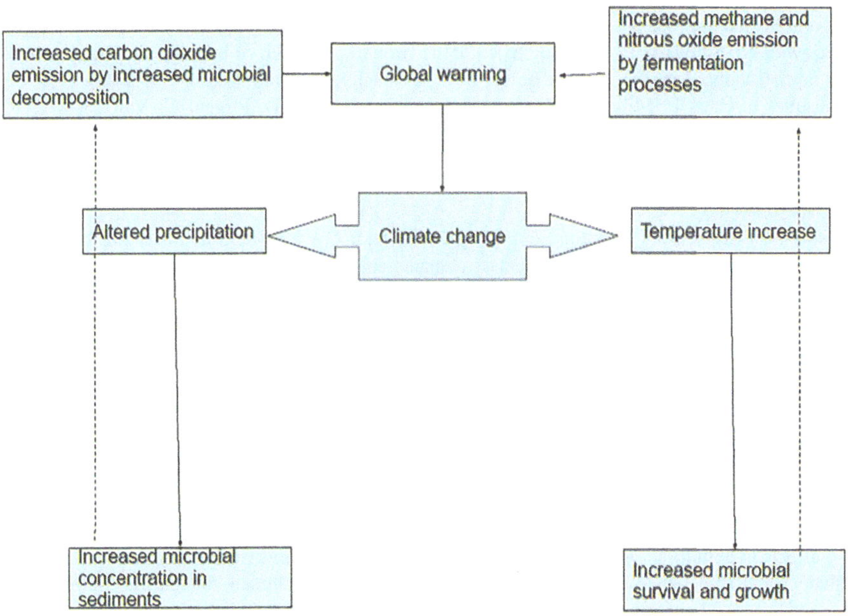

Fig. 24.4 Climate change and microorganisms in aquatic and terrestrial biomes

flow. The adverse effects of climate change on primary productivity of cyanobacteria and phytoplankton are very complex and determined by various factors like vertical mixing of nutrients in seawater. Global warming contributes to increased ocean stratification, which decreases vertical nutrient mixing, leading to a decline in primary productivity, while increasing CO_2 levels increase the production of phytoplankton, unless nutrients are the limiting factor. Microbial population alters as the temperature rises and processes such as respiration, fermentation and methanogenesis also stimulates (Grossi et al. 2019).

References

Abatenh ENDESHAW, Gizaw BIRHANU, Tsegaye ZERIHUN, Tefera G (2018) Microbial function on climate Change-a review. Environ Pollut Climate Change 2:4172

Bange HW (2008) Gaseous nitrogen compounds (NO, N_2O, N_2, NH_3) in the ocean. Elsevier, Amsterdam

Bardgett RD, Manning P, Morriën E, De Vries FT (2013) Hierarchical responses of plant–soil interactions to climate change: consequences for the global carbon cycle. J Ecol 101(2):334–343

Belnap J, Weber B, Büdel B (2016) Biological soil crusts as an organizing principle in dry lands. In: Biological soil crusts: an organizing principle in dry lands. Springer, Cham, pp 3–13

Berard A, Sassi MB, Kaisermann A, Renault P (2015) Soil microbial community responses to heat wave components: drought and high temperature. Clim Res 66(3):243–264

Blagodatskaya E, Kuzyakov Y (2013) Active microorganisms in soil: critical review of estimation criteria and approaches. Soil Biol Biochem 67:192–211

Blazewicz SJ, Schwartz E, Firestone MK (2014) Growth and death of bacteria and fungi underlie rainfall-induced carbon dioxide pulses from seasonally dried soil. Ecology 95:1162–1172

Bousquet P, Ciais P, Miller J, Dlugokencky EJ, Hauglustaine D, Prigent C, Van der Werf G, Peylin P, Brunke E-G, Carouge C (2006) Contribution of anthropogenic and natural sources to atmospheric methane variability. Nature 443:439–443

Bradford MA (2013) Thermal adaptation of decomposer communities in warming soils. Front Microbiol 4:333

Castro HF, Classen AT, Austin EE, Norby RJ, Schadt CW (2010) Soil microbial community responses to multiple experimental climate change drivers. Appl Environ Microbiol 76:999–1007

Change IPOC (2014) Climate change 2013: the physical science basis: working group I contribution to the fifth assessment report of the intergovernmental panel on climate Change. Cambridge University Press, Cambridge

Clair SBS, Lynch JP (2010) The opening of Pandora's box: climate change impacts on soil fertility and crop nutrition in developing countries. Plant Soil 335:101–115

Cook BI, Ault TR, Smerdon JE (2015) Unprecedented 21st century drought risk in the American southwest and Central Plains. Sci Adv 1(1):1400082

Deltedesco E, Keiblinger KM, Piepho HP, Antonielli L, Pötsch EM, Zechmeister-Boltenstern S, Gorfer M (2020) Soil microbial community structure and function mainly respond to indirect effects in a multifactorial climate manipulation experiment. Soil Biol Biochem 142:107704

Dhanya P, Ramachandran A (2016) Farmers' perceptions of climate change and the proposed agriculture adaptation strategies in a semi arid region of South India. J Integr Environ Sci 13 (1):1–18

Dixon GR, Tilston EL (eds) (2010) Soil microbiology and sustainable crop production. Springer Science & Business Media, Dordrecht

Drigo B, Kowalchuk GA, Van Veen JA (2008) Climate change goes underground: effects of elevated atmospheric CO2 on microbial community structure and activities in the rhizosphere. Biol Fertil Soils 44:667–679

Gougoulias C, Clark JM, Shaw LJ (2014) The role of soil microbes in the global carbon cycle: tracking the below ground microbial processing of plant-derived carbon for manipulating carbon dynamics in agricultural systems. J Sci Food Agric 94:2362–2371
Gouin T, Armitage JM, Cousins IT, Muir DC, Ng CA, Reid L, Tao S (2013) Influence of global climate change on chemical fate and bioaccumulation: the role of multimedia models. Environ Toxicol Chem 32(1):20–31
Grossi G, Goglio P, Vitali A, Williams AG (2019) Livestock and climate change: impact of livestock on climate and mitigation strategies. Anim Front 9:69–76
Haugwitz MS, Bergmark L, Priemé A, Christensen S, Beier C, Michelsen A (2014) Soil microorganisms respond to five years of climate change manipulations and elevated atmospheric CO_2 in a temperate heath ecosystem. Plant Soil 374:211–222
Hayden HL, Mele PM, Bougoure DS, Allan CY, Norng S, Piceno YM, Brodie EL, DeSantis TZ, Andersen GL, Williams AL, Hovenden MJ (2012) Changes in the microbial community structure of bacteria, archaea and fungi in response to elevated CO2 and warming in an Australian native grassland soil. Environ Microbiol 14(12):3081–3096
IPCC (2018) IPCC special report on global warming of 1.5°C, chapter 4 – strengthening and implementing the global response. Global warming of 1.5°C an IPCC special report on the impacts of global warming of 1.5°C above pre-industrial levels and related global greenhouse gas emission pathways, in the context of strengthening the global response to the threat of climate change, 313–443
Jenny H (1941) Factors of soil formation. McGraw-Hill, New York
Kannojia P, Sharma PK, Sharma K (2019) Climate change and soil dynamics: effects on soil microbes and fertility of soil, climate change and agricultural ecosystems: current challenges and adaptation. Elsevier, Amsterdam
Lange M, Habekost M, Eisenhauer N, Roscher C, Bessler H, Engels C, Oelmann Y, Scheu S, Wilcke W, Schulze ED, Gleixner G (2014) Biotic and abiotic properties mediating plant diversity effects on soil microbial communities in an experimental grassland. PLoS One 9:96182
Nauer PA, Hutley LB, Arndt SK (2018) Termite mounds mitigate half of termite methane emissions. Proc Nat Acad Sci 115:13306–13311
Naylor D, DeGraaf S, Purdom E, Coleman-Derr D (2017) Drought and host selection influence bacterial community dynamics in the grass root microbiome. ISME J 11(12):2691–2704
Raftery AE, Zimmer A, Frierson DM, Startz R, Liu P (2017) Less than 2°C warming by 2100 unlikely. Nat Clim Chang 7(9):637–641
Rajput P, Saxena R, Mishra G, Mohanty S, Tiwari A (2013) Biogeochemical aspect of atmospheric methane and impact of nanoparticles on Methanotrophs. J Environ Anal Toxicol 3 (195):2161–0525.1000195
Reichstein M, Bahn M, Ciais P, Frank D, Mahecha MD, Seneviratne SI, Zscheischler J, Beer C, Buchmann N, Frank DC (2013) Climate extremes and the carbon cycle. Nature 500:287–295
Schimel JP (2018) Life in dry soils: effects of drought on soil microbial communities and processes. Annu Rev Ecol Evol Syst 49:409–432
Schulze ED (2006) Biological control of the terrestrial carbon sink. Biogeosciences 3(2):147–166
Semrau JD, DiSpirito AA, Yoon S (2010) Methanotrophs and copper. FEMS Microbiol Rev 34:496–531
Singh H, Anderson B, Brune WH, Cai C, Cohen R, Crawford J, Cubison M, Czech E, Emmons L, Fuelberg H (2010) Pollution influences on atmospheric composition and chemistry at high northern latitudes: boreal and California forest fire emissions. Atmos Environ 44:4553–4564
Sistla SA, Schimel JP (2013) Seasonal patterns of microbial extracellular enzyme activities in an arctic tundra soil: identifying direct and indirect effects of long-term summer warming. Soil Biol Biochem 66:119–129

Smith MD (2011) An ecological perspective on extreme climatic events: a synthetic definition and framework to guide future research. J Ecol 99:656–663

Steven B, Kuske CR, Reed SC, Belnap J (2015) Climate change and physical disturbance manipulations result in distinct biological soil crust communities. Appl Environ Microbiol 81 (21):7448–7459

Upton RN, Bach EM, Hofmockel KS (2018) Below ground response of prairie restoration and resiliency to drought. Agric Ecosyst Environ 266:122–132

Yu H, Deng Y, He Z, Van Nostrand JD, Wang S, Jin D, Wang A, Wu L, Wang D, Tai X, Zhou J (2018) Elevated CO_2 and warming altered grassland microbial communities in soil top-layers. Front Microbiol 9:1–10

Chapter 25
Deployment of Benign Bacterial Strains to Improve Soil Productivity Under Drought Stress

Amrita Kasotia, Ajit Varma, and D. K. Choudhary

Abstract Abiotic stresses especially drought affect agriculture lands; thereby they impinge on plant growth and cause in reduction in crop productivity and quality. As reported, drought stress contributes a chief proportion in demolition of cultivated land area and further crop productivity. Plant growth-promoting rhizobacteria (PGPR), a biological agent in agriculture that helps to lessen drought stress, has been considered the most promising approach to enhance production and yield in drought-affected regions. Hence, to mitigate drought stress, adopting an eco-friendly approach to meet future needs is the best way to go in. The present chapter deals with role of important benign microbes and their elicitors in alleviation of drought stress and promotes plant growth promotion.

Keywords Drought stress · Plant growth-promoting bacteria · Drought stress alleviation · Eco-friendly approach · Abiotic stress · Biofertilizer

25.1 Introduction

Drought is the most critical factor restricting crop production in the changing climate scenario, and its severity is expected to rise in the future. Several factors form water deficit condition in soil, viz. low rainfall, salinity and amplified strength of light are among some of them. Contrarily sometimes water is available, but plants are unable to take them up which mediate pseudo-drought or physiological drought (Salehi-Lisar and Bakhshayeshan-Agdam 2016). Drought can persist for even a short interval or several years. Significant losses in soil and crop productivity mediated

A. Kasotia
Mody University of Science & Technology, Lakshmangarh, Rajasthan, India

A. Varma · D. K. Choudhary (✉)
Amity Institute of Microbial Technology (AIMT), Amity University Uttar Pradesh, Noida, UP, India
e-mail: dkchoudhary1@amity.edu

D. K. Choudhary et al. (eds.), *Climate Change and the Microbiome*, Soil Biology 63,
https://doi.org/10.1007/978-3-030-76863-8_25

through an aggravate action of environmental conditions lead to lessen economy and agriculture sustainability. The sustainability of agricultural industries is on the great risk because of severe environmental damage caused by vegetation loss and soil erosion, the after-effects of drought stress (Zingaretti et al. 2013).

Based on IPCC special report, 2012, several countries experienced severe droughts due to some climatic factors wherein global warming has increased the overall temperature of the earth. Because of adverse climatic conditions, the results will be more severe in arid regions like the Middle East and the Sahel in Africa wherein freshwater will persist in some areas after a temperature increase of 1.5 °C but is predicted to dry out if the increase exceeds 2 °C (Masson-Delmotte et al. 2018). Around the world, drought exposure reported higher for US east, Europe south area, India, East China and Nigeria (Kogan et al. 2020; Carrão et al. 2016). Among them, India has been considered most drought-prone country facing at least drought in specific region which accounts for more than 22 major droughts in the past few decades. To mitigate drought risk, water resource management needs an efficient policy structure that includes prospect forecasts of the spatiotemporal distribution of droughts or application of drought managing strategies for crop cultivation using eco-friendly approaches (Gupta et al. 2020).

25.2 Drought Severity and Types

India is a developing nation which is mostly reliant on the agricultural sector especially for economic progress. Based on previous researches, agriculture solely depends on rainfall water wherein 60% of the whole agricultural land area has covered by rainfed land that highly depends on climatic situation in the region, for the most part on precipitation (Sharma et al. 2010; Sharma and Goyal 2020). Bisht et al. (2019) analysed along with other many studies wherein they reported the occurrence of drought in the upcoming years. Likewise, one study reveals an elevation in the incidence of intense droughts in the warmer and humid climate under the climatic change in India (Aadhar and Mishra 2018).

A newly defined event termed flash drought affect the soil and crop sustainability which indirectly pose challenges for about 10–15% *Oryza sativa* and *Zea mays*) grown area in is affected by flash droughts (Mahto and Mishra 2020). Unusually high temperatures, winds and strong incoming solar radiation cause rapid onset and intensification of high rates of evapotranspiration (ET) which forms flash drought (Chen et al. 2019). Low soil moisture, intense heat and elevated evapotranspiration are the indicators of flash drought (Otkin et al. 2018).

In India, 44% of the food production is contributed by 56% of the net cropped land. Due to declined mean summer monsoon rainfall, the droughts were frequent and got increased from 1951 to 2015. Monsoon rainfall is therefore critical for all country's agricultural production and edible safety measures those also providing spin-off effects on other sectors of the economy. The central part of India is most affected, including the parts of Indo-Gangetic Plains, whereas the augmented local weighty rainfall on the sub-daily and daily timescales have increased the risk of flood over India (Mujumdar et al. 2020). South-west monsoon affects drought frequencies of the parts of Indo-Gangetic area, south India coastal land and mid of Maharashtra (Mallya et al. 2016), while north-east monsoon affects central India, viz. Kerala, south peninsula, and north-eastern parts of India (Mujumdar et al. 2020). Anderson et al. (2011) subdivided drought into four key classes: (1) meteorological drought, (2) hydrological drought, (3) agricultural drought and (4) socio-economic drought.

25.3 How Drought Damages Plants?

The most significant abiotic factor restricting plant development is drought; it adversely affects crop productivity and gets increased. Stress leads unwanted physico-biochemical processes which influence cellular phenomenon in the plant. Irregular metabolism, reduced plant growth and plant death are some of the after-effects (Fathi and Tari 2016). Drought stress impacts on the status of water in plants at the level of cells, tissues and organs are causing precise and unspecific reactions and potentially harming tissues and triggering adaptive responses (Cvikrová et al. 2013). The adaptive responses include stomatal closure, the reduced amount of transpiration and photosynthesis, production of new proteins and the growth of osmolytes.

Crops' susceptibility to retain high water potential in tissues under drought is known as lack of moisture evading, and forbearance that determines plant predisposition to survive in water deficiency is called drought resistance (Vadez et al. 2011). The activities of three elements are involved in managing water stress: the soil, the root system and shoot (Lipiec et al. 2013). When the drought stress comes in action, the first effect is shown by modulation of cell membrane permeability and aquaporins (water channels). The second mechanism is demonstrated by an enhanced suberization of root endodermis/exodermis. The third step is xylem embolism, a process by which air is sucked into xylem vessels, hindering the sap flow. Drought resistance shows more excellent resistance to embolism (Li et al. 2009). Recently, plants have also been reported to produce exudates, mucilage and solute accumulation which increase the adaptability to stress. The exchange potential and nutrient uptake of root cations are being substantially reduced in dry environments, and the relative uptake of polyvalent cations (aluminium or heavy metals) can add up more toxicity (Lukowska and Jozefaciuk 2013). During longer time scales, water deficit shows reduced root length. In drought, nodulation activity, duration and formation also get affected (Lipiec et al. 2013). A decreased biomass production is also seen in mild drought stress as compared to photosynthesis. Drought stress is also known to suppress photochemical efficiency of photosystem PS II, inactivation of Rubisco, loss of cell membrane integrity and reduced assimilation of ammonium into organic compounds.

The biochemical and metabolic responses also change significantly during drought stress. The production of reactive oxygen species (ROS) induces oxidative stress. ROS refers to any oxygen derivative that is more reactive than an oxygen molecule (O_2) itself (Mittler 2017). These are known to cause cell membrane damage, protein degradation and enzyme inactivation. The primary hydraulic signalling mechanism during drought stress includes a decreased root dampen absorption and water potential, resulting in reduced turgor pressure in leaves which then leads to stomatal closure, thereby reducing leaf elongation. Several plant hormones have been involved in the directive of physiological processes under different stresses, namely, abscisic acid (ABA), auxin, cytokinin (CK), ethylene, gibberellins, etc. ABA plays a major role in stomatal movement with the help of transcription factors and their target genes. Moreover, ABA, along with cytokinins, implicated in controlling plant senescence and enhanced carbon remobilization.

25.4 Choosing the Crops

The ability of crops to withstand drought differs. Some crops tolerate water deficit while some don't. The genotype of a plant is mainly responsible for drought tolerance. Crops that mature earlier, i.e. early flowering and fruiting, are better suited to drought-prone areas. Moreover, crops with good canopy structure, root structure and root hair number and smaller xylem vessels are more resistant to drought stress. Some of the drought-tolerant crops include millets, sorghum, chickpea and groundnut. Various conventional and modern molecular breeding procedures and transgenic approaches or genetic engineering and application of biofertilizers can increase plant resistance to abiotic stresses. Genetic improvement is the new technique; recently practices since last four decades involve changing the genetic makeup of plant for stress tolerance (Lipiec et al. 2013). In an organism, each gene can influence several different traits. Even a single gene alteration may affect the whole target genome, ensuing in inadvertent effects, all of which may not be decipherable at the very same time. This type of uncertainty is hard to predict (Prakash et al. 2011). For this reason, the use of microbes that maintains balance in agroecosystem without changing the plant's genome can be the best approach to alleviate the stress.

25.5 Plant Growth-Promoting Bacteria

Based on published reports, plant growth-promoting bacteria (PGPB) have been considered as benign agents for plants that improve plant nutrition in plant under abiotic stress such as drought and heavy metal contamination of salinity. PGPB survived as free-living soil, rhizosphere, rhizoplane and phyllosphere bacteria that may contribute to the health, production and growth of plants. Some of the bacteria

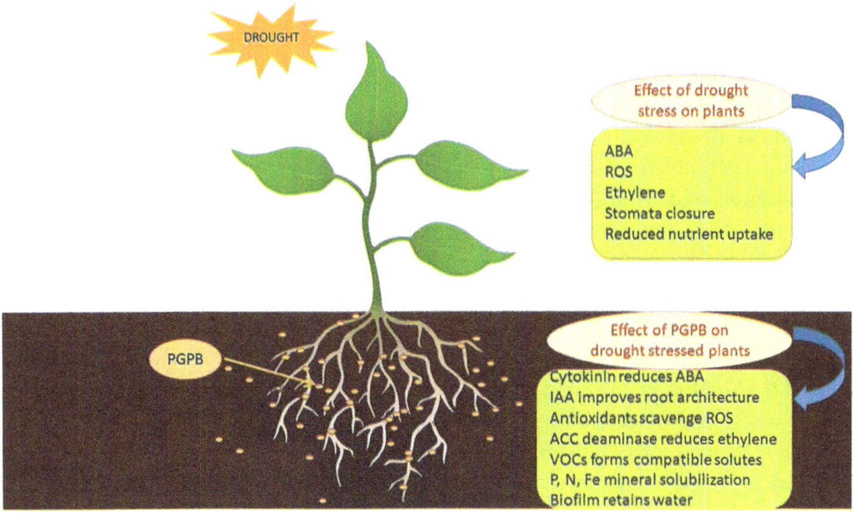

Fig. 25.1 Mitigation of drought stress by PGPB

species which have been found to enhance plant growth include *Pseudomonas* spp., *Enterobacter*, *Bacillus* spp., *Erwinia*, *Azotobacter*, *Burkholderia*, *Azospirillum*, *Arthrobacter*, etc. A variety of mechanisms are mediated by the positive impact of many soil bacteria on plants, including improving mineral nutrition, boosting plant tolerance to biotic and abiotic stress, altering root growth as well as suppressing soilborne diseases (Jacoby et al. 2017) (Fig. 25.1).

Meena et al. (2017) reported that many rhizospheric, viz. *Rhizobium*, *Bacillus*, *Pseudomonas*, *Pantoea*, *Bradyrhizobium*, *Methylobacterium*, *Paeni-bacillus*, *Burkholderia*, *Achromobacter*, *Azotobacter*, *Azospirillum*, *Microbacterium*, *Methylobacterium*, *Variovorax*, *Enterobacter*, *Trichoderma* and *Cyanobacteria*, played important role in plant growth promotion and abiotic stress tolerance to host plant. The development of the plant depends upon abiotic environmental factors such as temperature, humidity, light intensity, the supply of water and minerals and CO_2. An enticing alternative to traditional farming practices is the inoculation of sustainable soil microorganisms with the advancement in plant growth and development (Jacoby et al. 2017). Several microbial inoculants have been successfully formulated, developed, marketed and applied by a rising group of growers nowadays (Ahemad and Kibret 2014; Santos et al. 2019).

25.5.1 Properties of PGPB

25.5.1.1 Phosphate Solubilization

Phosphorus (P) is one of the main macronutrients essential for plant growth and growth. It accounts for 0.2% of the plant's dry weight (Azziz et al. 2012). In soil, the concentration of P is around 0.05% (w/w) soil (Alori et al. 2017). In rhizosphere phosphate-solubilizing bacteria (PSB) hydrolyse organic and inorganic phosphorus from insoluble compounds. Inorganic phosphate solubilization and mineralization are exceptionally dependent on soil pH and form. Various scientists have proposed different mechanisms for phosphate solubilization wherein the most accepted method is the production of organic acids, siderophores, protons, hydroxyl ions and CO_2 (Sharma et al. 2013; Alori et al. 2017). Substitution of H^+ for calcium occurs when PGPB secrete organic acid and drop the pH around the microbial cells around microbial cells causing the release of P ions. Several organic acids are released by PGPB, namely, tartaric acid, fumaric acid, glyoxalic acid, malic acid, citric acid, oxalic acid, acetic acid, isovaleric acid, itaconic acid, glycolic acid, maleic acid, gluconic acid, lactic acid, succinic acid, propionic acid, 2-ketogluconic acid, α-ketobutyric acid, etc. (Alori et al. 2017).

The primary source of organic phosphorus is organic matter in the soil. Organic matter contains phosphates in the form of inositol phosphate (soil phytate), phosphomonoesters, phosphodiesters, phospholipids, nucleic acids and phosphotriesters. Organic phosphorous is mineralized mainly by phosphatases enzymes. These enzymes dephosphorylate the phosphor-ester or phosphoanhydride bond of organic compounds and are also known as non-specific acid phosphatases (NSAPs). Phosphatases work best in acidic to neutral pH which defines their presence in acidic soils. Phytase is another enzyme that is responsible for organic phosphate mineralization. Plant seed and pollen contains stores form of phosphorous known as phytates. Plants cannot take up phosphorus from phytates. The presence of PGPB in the soil can mineralize phytate by secretion of phytase (Richardson and Simpson 2011). The inoculation of over phosphate-solubilizing PGPB mutants alleviated drought stress (Kasotia and Choudhary 2016).

25.5.1.2 Iron Chelation and Siderophores

Iron (Fe) is the second most abundant existed metal in the earth's crust. In cellular biological systems, Fe is a structural component of a large number of proteins that include cytochrome P450, cytochrome oxidases, electron transporters, Fe-sulphur (S) enzymes, dioxygenases, hydroxylases and carboxylases. Moreover, it acts as a catalyst in chlorophyll synthesis and various metabolic processes (Ferreira et al. 2019). Additionally, it is involved in nucleic acid synthesis and repair (Puig et al. 2017).

In aqueous solutions Fe is present in interconvertible form Fe^{2+} and Fe^{3+}. The balance between two forms in soil is dependent on pH, aeration, organic matter content and salinity (Ferreira et al. 2019). The availability of Fe is meagre in well-aerated, neutral pH soils as Fe^{3+} is mostly precipitated as insoluble hydroxides, oxyhydroxides and oxides which makes Fe^{3+} unavailable for plant uptake, whereas in good aeration, acid soil Fe^{3+} transform into more soluble Fe. Both microbes and plants have a relatively high iron requirement. In bacteria, cellular iron deficiency induces the formation of low molecular weight siderophores, molecules with a powerful affinity for Fe^{+3} and membrane receptors capable of binding the complex Fe-siderophore. Bacterial siderophores chelates Fe^{3+} fulfil the iron requirement of both microorganism and plant (Ahmed and Holmström 2014). Recently various bacterial species have been reported to produce siderophore; some of them are *Escherichia coli*, *Salmonella*, *Klebsiella pneumoniae*, *Vibrio cholerae*, *Vibrio anguillarum*, *Aeromonas*, *Aerobacter aerogenes*, *Enterobacter*, *Yersinia* and *Mycobacterium* species (Albelda-Berenguer et al. 2019).

The production of siderophores is also associated with the biocontrol activity induced by PGPB. Plant beneficial bacteria compete with the pathogenic microorganisms for nutrients and space in the plant's vicinity and make the pathogen deficient in nutrients. *Bacillus* and *Pseudomonas* are the most studied genera showing biocontrol activity (Albelda-Berenguer et al. 2019). *Burkholderia*, *Rahnella*, *Curtobacterium* and *Kosakonia* are recently studied biocontrol activity showing bacterial species (Cornelis et al. 2011; Kandel et al. 2017; Lambrese et al. 2018).

25.5.1.3 Modulation of Phytohormone

PGPB promotes the plant growth by production of plant hormone auxin. IAA plays an essential role in increasing the number of root hairs and lateral roots of the plant. Hence, inoculation of PGPB to the plant rhizosphere accelerates nutrient absorption by changing root architecture of the plant. Around 80% of bacteria from the rhizosphere can synthesis IAA, suggesting a possible role in association with the plant (Spaepen and Vanderleyden 2011; Glick 2012). Tryptophan from root exudates or rotting cells improves IAA microbial biosynthesis in the soil. PGPB-synthesizing IAA has been proposed to limit the adverse effects of environmental stresses (Barnawal et al. 2017; Egamberdieva et al. 2017). Cytokinin is another plant hormone which can be synthesized by PGPB (Kapoor and Kaur 2016). Under irrigate deficit CK signalling has evolved as an intercellular communiqué network which is essential to crosstalk (Pavlů et al. 2018) with other types of phytohormones and their regulating pathways in mediating plant stress response (Verma et al. 2016; Huang et al. 2018; Artner and Benkova 2019; Hai et al. 2020). Moreover, CK crosstalk with auxin, SA and brassinosteroids (BR) (Pavlović et al. 2018) and with ABA in mediating drought stress response (Huang et al. 2018). In addition to CK, PGPB exhibited important role in production of gibberellins (GAs) which are reported as prevalent plant hormones which evoke various metabolic functions that include seed germination, stem elongation, sex expression, flowering, fruit formation and senescence, necessary during plant development (Cohen et al. 2009; Kang, et al. 2014).

ABA hormone plays an essential role in plant health, development and responses related to stress (Chen et al. 2020). It is a phytohormone that mediates stomatal closure in stress and, in the absence of stress, regulates aspects of plant growth and development. Drought responses are regulated by ABA-dependent and ABA-independent pathways. ABA can affect the plant either by increasing ABA biosynthesis and (or) by decreasing ABA breakdown. Inoculation of ABA producing *Paenibacillus yonginensis* DCY84T developed changes in *Arabidopsis thaliana* gene expression against aluminium, drought and salt stress (Sukweenadhi et al. 2015). Drought tolerance is mainly achieved via enhancement of ABA levels (Cohen et al. 2015). Another very important stress hormone is ethylene (ET) which is gaseous in nature that enables plant-to-plant communication and regulates varied functions in plant systems, viz. regulation of leaf development, senescence, fruit ripening, stimulation of germination, etc. (Dubois et al. 2018). ET also has been reported to show presence in habitat-imposed stresses. Transgenic alteration of ET biosynthesis has been reported, whereby grain yield took place in maize under field drought-stress conditions (Habben et al. 2014). The ACC-deaminase-producing PGPB converts ACC into ammonia and α-ketobutyrate instead of ethylene (Danish and Zafar-ul-Hye 2019). Regulation of ethylene level in plants mitigates the effects of drought stress wherein application of ACC deaminase-producing PGPB, *Bacillus subtilis* Rhizo SF 48, induced drought tolerance in tomato (Gowtham et al. 2020) which further showed significant growth and yield of wheat under drought stress (Duca et al. 2018; Danish and Zafar-ul-Hye 2019).

PGPB secrete odourless volatile organic compounds (VOCs) with a low molecular weight that, in return, affects the plant growth. These VOCs have also been reported to initiate compatible solute production in plants that maintains cellular osmotic pressure (Li et al. 2019; Bhattacharyya and Jha 2012).

25.5.1.4 Nitrogen Fixation

Nitrogen (N_2) is unavailable to the growing plants despite the abundance in atmosphere which is round 78%. This N_2 is converted to easily reached form to plant by deploying approach biological nitrogen fixation (BNF) that converts N_2 to ammonia by diazotrophs (nitrogen-fixing microbes) using a complex enzyme system known as nitrogenase. Diazotrophs are by and large categorized as symbiotic bacteria that forms a symbiosis with leguminous plants (e.g. rhizobia) and non-leguminous trees (e.g. *Frankia*) and non-symbiotic (free living, associative and endophytes) forms such as cyanobacteria (*Anabaena*, *Nostoc*), *Azospirillum*, *Azotobacter*, etc. (Bhattacharyya and Jha 2012). Nitrogen-fixing bacteria are now used to improve agricultural productivity (Souza et al. 2014), and usually fixation takes place through *nifH* gene expression (Masood et al. 2020). Liu et al. (2011) reported that *Klebsiella pneumoniae* strain NG14 colonized root surface of rice helped in N_2 fixation and exhibited biofilm formation to reduce the detrimental impact of drought.

25.6 Conclusion

It has been envisaged that PGPB exhibited direct effects on root and shoots growth (Saharan and Nehra 2011) and have been characterized as bioinoculant (Vejan et al. 2016) wherein many PGPB strains result in a significant increase in plant height, root length and dry matter production of shoot and root of plants (Glick et al. 2007; Backer et al. 2018). Under habitat-imposed stress especially drought, the ET endogenously regulates plant homeostasis and marks in reduced root and shoot growth. However, this effect is rescued by bacterial ACC deaminase which promotes average plant growth by the degradation of the ACC, the ethylene precursor. ABA hormone level in plants is also increased in water deficit condition in plant leaves which is then regulated by increased bacterial cytokinin. Both cytokinin and ABA share common biosynthetic origin; hence cytokinin-ABA antagonism occurs as a result of metabolic interactions. PGPB helps in scavenging of ROS generated during stress (Kasotia et al. 2015) and rescues the normal plant growth. Phosphate, iron, nitrogen, etc. nutrient acquisition by PGPB adds benefit to the drought-stressed plants for growth promotion.

References

Aadhar S, Mishra V (2018) Impact of climate change on drought frequency over India 14

Ahemad M, Kibret M (2014) Mechanisms and applications of plant growth promoting rhizobacteria: current perspective. J King Saud Univ Sci 26(1):1–20

Ahmed E, Holmström SJM (2014) Siderophores in environmental research: roles and applications. Microb Biotechnol 7(3):196–208. https://doi.org/10.1111/1751-7915.12117

Albelda-Berenguer M, Monachon M, Joseph E (2019) Siderophores: from natural roles to potential applications. In: Advances in applied microbiology, vol 106. Elsevier, pp 193–225. https://doi.org/10.1016/bs.aambs.2018.12.001

Alori ET, Glick BR, Babalola OO (2017) Microbial phosphorus Solubilization and its potential for use in sustainable agriculture. Front Microbiol 8:971. https://doi.org/10.3389/fmicb.2017.00971

Anderson MC, Hain C, Wardlow B, Pimstein A, Mecikalski JR, Kustas WP (2011) Evaluation of drought indices based on thermal remote sensing of evapotranspiration over the continental United States. J Clim 24(8):2025–2044. https://doi.org/10.1175/2010JCLI3812.1

Artner C, Benkova E (2019) Ethylene and Cytokinin: Partners in Root Growth Regulation. Mol Plant 12(10):1312–1314. https://doi.org/10.1016/j.molp.2019.09.003

Azziz G, Bajsa N, Haghjou T, Taulé C, Valverde Á, Igual JM, Arias A (2012) Abundance, diversity and prospecting of culturable phosphate solubilizing bacteria on soils under crop–pasture rotations in a no-tillage regime in Uruguay. Appl Soil Ecol 61:320–326. https://doi.org/10.1016/j.apsoil.2011.10.004

Backer R, Rokem JS, Ilangumaran G, Lamont J, Praslickova D, Ricci E, Subramanian S, Smith DL (2018) Plant growth-promoting Rhizobacteria: context, mechanisms of action, and roadmap to commercialization of biostimulants for sustainable agriculture. Front Plant Sci 9:1473. https://doi.org/10.3389/fpls.2018.01473

Barnawal D, Bharti N, Pandey SS, Pandey A, Chanotiya CS, Kalra A (2017) Plant growth-promoting rhizobacteria enhance wheat salt and drought stress tolerance by altering endogenous phytohormone levels and TaCTR1/TaDREB2 expression. Physiol Plant 161(4):502–514. https://doi.org/10.1111/ppl.12614

Bhattacharyya PN, Jha DK (2012) Plant growth-promoting rhizobacteria (PGPR): emergence in agriculture. World J Microbiol Biotechnol 28(4):1327–1350. https://doi.org/10.1007/s11274-011-0979-9

Bisht DS, Sridhar V, Mishra A, Chatterjee C, Raghuwanshi NS (2019) Drought characterization over India under projected climate scenario. Int J Climatol 39(4):1889–1911. https://doi.org/10.1002/joc.5922

Carrão H, Naumann G, Barbosa P (2016) Mapping global patterns of drought risk: an empirical framework based on sub-national estimates of hazard, exposure and vulnerability. Glob Environ Chang 39:108–124. https://doi.org/10.1016/j.gloenvcha.2016.04.012

Chen LG, Gottschalck J, Hartman A, Miskus D, Tinker R, Artusa A (2019) Flash drought characteristics based on U.S. drought monitor. Atmosphere 10(9):498. https://doi.org/10.3390/atmos10090498

Chen K, Li G-J, Bressan RA, Song C-P, Zhu J-K, Zhao Y (2020) Abscisic acid dynamics, signaling, and functions in plants. J Integr Plant Biol 62(1):25–54. https://doi.org/10.1111/jipb.12899

Cohen AC, Travaglia CN, Bottini R, Piccoli PN (2009) Participation of abscisic acid and gibberellins produced by endophytic Azospirillum in the alleviation of drought effects in maize. Botany 87(5):455–462. https://doi.org/10.1139/B09-023

Cohen AC, Bottini R, Pontin M, Berli FJ, Moreno D, Boccanlandro H, Travaglia CN, Piccoli PN (2015) Azospirillumbrasilense ameliorates the response of *Arabidopsis thaliana* to drought mainly via enhancement of ABA levels. Physiol Plant 153(1):79–90. https://doi.org/10.1111/ppl.12221

Cornelis P, Wei Q, Andrews SC, Vinckx T (2011) Iron homeostasis and management of oxidative stress response in bacteria. Metallomics 3(6):540–549. https://doi.org/10.1039/C1MT00022E

Cvikrová M, Gemperlová L, Martincová O, Vanková R (2013) Effect of drought and combined drought and heat stress on polyamine metabolism in proline-over-producing tobacco plants. Plant Physiol Biochem 73:7–15. https://doi.org/10.1016/j.plaphy.2013.08.005

Danish S, Zafar-ul-Hye M (2019) Co-application of ACC-deaminase producing PGPR and timber-waste biochar improves pigments formation, growth and yield of wheat under drought stress. Sci Rep 9(1):5999. https://doi.org/10.1038/s41598-019-42374-9

Dubois M, Van den Broeck L, Inzé D (2018) The pivotal role of ethylene in plant growth. Trends Plant Sci 23(4):311–323. https://doi.org/10.1016/j.tplants.2018.01.003

Duca DR, Rose DR, Glick BR (2018) Indole acetic acid overproduction transformants of the rhizobacterium Pseudomonas sp. UW4. Antonie Van Leeuwenhoek 111(9):1645–1660. https://doi.org/10.1007/s10482-018-1051-7

Egamberdieva D, Wirth SJ, Alqarawi AA, Abd_Allah EF, Hashem A (2017) Phytohormones and beneficial microbes: essential components for plants to balance stress and fitness. Front Microbiol 8:2104. https://doi.org/10.3389/fmicb.2017.02104

Fathi A, Tari DB (2016) Effect of drought stress and its mechanism in plants. Int J Life Sci 10 (1):1–6. https://doi.org/10.3126/ijls.v10i1.14509

Ferreira MJ, Silva H, Cunha A (2019) Siderophore-producing Rhizobacteria as a promising tool for empowering plants to cope with Iron limitation in saline soils: A review. Pedosphere 29 (4):409–420. https://doi.org/10.1016/S1002-0160(19)60810-6

Glick BR (2012, October 11) Plant growth-promoting Bacteria: mechanisms and applications [review article]. Scientifica; Hindawi 2012:963401. https://doi.org/10.6064/2012/963401

Glick BR, Todorovic B, Czarny J, Cheng Z, Duan J, McConkey B (2007) Promotion of plant growth by bacterial ACC deaminase. Crit Rev Plant Sci 26(5–6):227–242. https://doi.org/10.1080/07352680701572966

Gowtham HG, Brijesh Singh S, Murali M, Shilpa N, Melvin P, Mohammed A, Amruthesh KN, Niranjana SR (2020) Induction of drought tolerance in tomato upon the application of ACC deaminase producing plant growth promoting rhizobacterium Bacillus subtilis Rhizo SF 48. Microbiol Res 234:126422. https://doi.org/10.1016/j.micres.2020.126422

Gupta V, Kumar Jain M, Singh VP (2020) Multivariate modeling of projected drought frequency and Hazard over India. J Hydrol Eng 25(4):04020003. https://doi.org/10.1061/(ASCE)HE.1943-5584.0001893

Habben JE, Bao X, Bate NJ, DeBruin JL, Dolan D, Hasegawa D, Helentjaris TG, Lafitte RH, Lovan N, Mo H, Reimann K, Schussler JR (2014) Transgenic alteration of ethylene biosynthesis increases grain yield in maize under field drought-stress conditions. Plant Biotechnol J 12 (6):685–693. https://doi.org/10.1111/pbi.12172

Hai NN, Chuong NN, Tu NHC, Kisiala A, Hoang XLT, Thao NP (2020) Role and regulation of Cytokinins in plant response to drought stress. Plan Theory 9(4):422. https://doi.org/10.3390/plants9040422

Huang X, Hou L, Meng J, You H, Li Z, Gong Z, Yang S, Shi Y (2018) The antagonistic action of abscisic acid and Cytokinin signaling mediates drought stress response in Arabidopsis. Mol Plant 11(7):970–982. https://doi.org/10.1016/j.molp.2018.05.001

Jacoby R, Peukert M, Succurro A, Koprivova A, Kopriva S (2017) The role of soil microorganisms in plant mineral nutrition—current knowledge and future directions. Front Plant Sci 8:1617. https://doi.org/10.3389/fpls.2017.01617

Kandel SL, Firrincieli A, Joubert PM, Okubara PA, Leston ND, McGeorge KM, Mugnozza GS, Harfouche A, Kim S-H, Doty SL (2017) An in vitro study of bio-control and plant growth promotion potential of Salicaceae endophytes. Front Microbiol 8:386. https://doi.org/10.3389/fmicb.2017.00386

Kang S-M, Waqas M, Khan AL, Lee I-J (2014) Plant-growth-promoting Rhizobacteria: potential candidates for gibberellins production and crop growth promotion. In: Miransari M (ed) Use of microbes for the alleviation of soil stresses, vol 1. Springer, New York, pp 1–19. https://doi.org/10.1007/978-1-4614-9466-9_1

Kapoor R, Kaur M (2016) Cytokinins production by fluorescent Pseudomonas isolated from rhizospheric soils of Malus and Pyrus. Afr J Microbiol Res 6:1274–1279

Kasotia A, Choudhary DK (2016) Induced inorganic phosphate Solubilization through N-methyl-N′-nitro-N-Nitrosoguanidine treated mutants of Pseudomonas koreensis strain AK-1 (MTCC number 12058) under polyethylene glycol. Proc Nat Acad Sci India Sect B: Biol Sci 86 (1):115–123. https://doi.org/10.1007/s40011-014-0416-6

Kasotia A, Varma A, Choudhary DK (2015) Pseudomonas-mediated mitigation of salt stress and growth promotion in *Glycine max*. Agric Res 4(1):31–41. https://doi.org/10.1007/s40003-014-0139-1

Kogan F, Guo W, Yang W (2020) Near 40-year drought trend during 1981–2019 earth warming and food security. Geomat Nat Haz Risk 11(1):469–490. https://doi.org/10.1080/19475705.2020.1730452

Lambrese Y, Guiñez M, Calvente V, Sansone G, Cerutti S, Raba J, Sanz MI (2018) Production of siderophores by the bacterium Kosakoniaradicincitans and its application to control of phytopathogenic fungi. Bioresource Technol Rep 3:82–87. https://doi.org/10.1016/j.biteb.2018.06.003

Li Y, Sperry JS, Shao M (2009) Hydraulic conductance and vulnerability to cavitation in corn (*Zea mays* L.) hybrids of differing drought resistance. Environ Exp Bot 66(2):341–346. https://doi.org/10.1016/j.envexpbot.2009.02.001

Li F, Tang M, Tang X, Sun W, Gong J, Yi Y (2019) Bacillus subtilis – *Arabidopsis thaliana*: A model interaction system for studying the role of volatile organic compounds in the interchange between plants and bacteria. Botany 97(12):661–669. https://doi.org/10.1139/cjb-2019-0093

Lipiec J, Doussan C, Nosalewicz A, Kondracka K (2013) Effect of drought and heat stresses on plant growth and yield: a review. Int Agrophys 27(4):463–477. http://agro.icm.edu.pl/agro/element/bwmeta1.element.agro-c0215af6-ba24-40d6-80ab-1fd1ddb0ba7a

Liu Y, Wang H, Sun X, Yang H, Wang Y, Song W (2011) Study on mechanisms of colonization of nitrogen-fixing PGPB, Klebsiella pneumoniae NG14 on the root surface of Rice and the formation of biofilm. Curr Microbiol 62(4):1113–1122. https://doi.org/10.1007/s00284-010-9835-7

Lukowska M, Jozefaciuk G (2013) Unknown mechanism of plants response to drought: low soil moisture and osmotic stresses induce severe decrease in CEC and increase in acidity of barley roots. J Agric Sci 5(10):p204. https://doi.org/10.5539/jas.v5n10p204

Mahto SS, Mishra V (2020) Dominance of summer monsoon flash droughts in India. Environ Res Lett. https://doi.org/10.1088/1748-9326/abaf1d

Mallya G, Mishra V, Niyogi D, Tripathi S, Govindaraju RS (2016) Trends and variability of droughts over the Indian monsoon region. Weather and Climate Extremes 12:43–68. https://doi.org/10.1016/j.wace.2016.01.002

Masood S, Zhao XQ, Shen RF (2020) *Bacillus pumilus* promotes the growth and nitrogen uptake of tomato plants under nitrogen fertilization. Sci Hortic 272:109581. https://doi.org/10.1016/j.scienta.2020.109581

Masson-Delmotte, V., Zhai, P., Pörtner, H.-O., Roberts, D., Skea, J., Shukla, P. R., Pirani, A., Moufouma-Okia, W., Péan, C., Pidcock, R., Connors, S., Matthews, J. B. R., Chen, Y., Zhou, X., Gomis, M. I., Lonnoy, E., Maycock, T., Tignor, M., & Waterfield, T. (2018). IPCC, 2018: Global Warming of 1.5°C. An IPCC Special Report on the impacts of global warming of 1.5°C above pre-industrial levels and related global greenhouse gas emission pathways, in the context of strengthening the global response to the threat of climate change, sustainable development, and efforts to eradicate poverty

Meena KK, Sorty AM, Bitla UM, Choudhary K, Gupta P, Pareek A, Singh DP, Prabha R, Sahu PK, Gupta VK, Singh HB, Krishanani KK, Minhas PS (2017) Abiotic stress responses and microbe-mediated mitigation in plants: the omics strategies. Front Plant Sci 8:172. https://doi.org/10.3389/fpls.2017.00172

Mittler R (2017) ROS are good. Trends Plant Sci 22(1):11–19. https://doi.org/10.1016/j.tplants.2016.08.002

Mujumdar M, Bhaskar P, Ramarao MVS, Uppara U, Goswami M, Borgaonkar H, Chakraborty S, Ram S, Mishra V, Rajeevan M, Niyogi D (2020) Droughts and floods. In: Krishnan R, Sanjay J, Gnanaseelan C, Mujumdar M, Kulkarni A, Chakraborty S (eds) Assessment of climate change over the Indian region: A Report of the Ministry of Earth Sciences (MoES), government of India. Springer, New Delhi, pp 117–141. https://doi.org/10.1007/978-981-15-4327-2_6

Otkin JA, Svoboda M, Hunt ED, Ford TW, Anderson MC, Hain C, Basara JB (2018) Flash droughts: A review and assessment of the challenges imposed by rapid-onset droughts in the United States. Bull Am Meteorol Soc 99(5):911–919. https://doi.org/10.1175/BAMS-D-17-0149.1

Pavlović I, Petřík I, Tarkowská D, Lepeduš H, Bok VV, Brkanac SR, Novák O, Salopek-Sondi B (2018) Correlations between Phytohormones and drought tolerance in selected Brassica crops: Chinese cabbage, white cabbage and kale. Int J Mol Sci 19(10):2866. https://doi.org/10.3390/ijms19102866

Pavlů J, Novák J, Koukalová V, Luklová M, Brzobohatý B, Černý M (2018) Cytokinin at the crossroads of abiotic stress Signalling pathways. Int J Mol Sci 19(8):2450. https://doi.org/10.3390/ijms19082450

Prakash D, Verma S, Bhatia R, Tiwary BN (2011) Risks and precautions of genetically modified organisms. ISRN Ecol 2011:1–13. https://doi.org/10.5402/2011/369573

Puig S, Ramos-Alonso L, Romero AM, Martínez-Pastor MT (2017) The elemental role of iron in DNA synthesis and repair. Metallomics: Int Biometal Sci 9(11):1483–1500. https://doi.org/10.1039/c7mt00116a

Richardson AE, Simpson RJ (2011) Soil microorganisms mediating phosphorus availability update on microbial phosphorus. Plant Physiol 156(3):989–996. https://doi.org/10.1104/pp.111.175448

Saharan BS, Nehra V (2011) Plant growth promoting Rhizobacteria: A critical review. Life Sci Med Res 21:1–30. https://agris.fao.org/agris-search/search.do?recordID=DJ2012070623

Salehi-Lisar SY, Bakhshayeshan-Agdam H (2016) Drought stress in plants: causes, consequences, and tolerance. In: Hossain MA, Wani SH, Bhattacharjee S, Burritt DJ, Tran L-SP (eds) Drought stress tolerance in plants, vol 1. Springer International Publishing, Manhattan, pp 1–16. https://doi.org/10.1007/978-3-319-28899-4_1

Santos MS, Nogueira MA, Hungria M (2019) Microbial inoculants: reviewing the past, discussing the present and previewing an outstanding future for the use of beneficial bacteria in agriculture. AMB Express 9(1):205. https://doi.org/10.1186/s13568-019-0932-0

Sharma A, Goyal MK (2020) Assessment of drought trend and variability in India using wavelet transform. Hydrol Sci J 65(9):1539–1554. https://doi.org/10.1080/02626667.2020.1754422

Sharma BR, Rao KV, Vittal KPR, Ramakrishna YS, Amarasinghe U (2010) Estimating the potential of rainfed agriculture in India: prospects for water productivity improvements. Agric Water Manag 97(1):23–30. https://doi.org/10.1016/j.agwat.2009.08.002

Sharma SB, Sayyed RZ, Trivedi MH, Gobi TA (2013) Phosphate solubilizing microbes: sustainable approach for managing phosphorus deficiency in agricultural soils. Springerplus 2(1):587. https://doi.org/10.1186/2193-1801-2-587

Souza EM, Chubatsu LS, Huergo LF, Monteiro R, Camilios-Neto D, Wassem R, de Oliveira Pedrosa F (2014) Use of nitrogen-fixing bacteria to improve agricultural productivity. BMC Proc 8(4):O23. https://doi.org/10.1186/1753-6561-8-S4-O23

Spaepen S, Vanderleyden J (2011) Auxin and plant-microbe interactions. Cold Spring Harb Perspect Biol 3(4):a001438. https://doi.org/10.1101/cshperspect.a001438

Special Report IPCC (2012) Managing the risks of extreme events and disasters to advance climate change adaptation: Special Report of the intergovernmental panel on climate change. Cambridge University Press, Cambridge

Sukweenadhi J, Kim Y-J, Choi E-S, Koh S-C, Lee S-W, Kim Y-J, Yang DC (2015) Paenibacillusyonginensis DCY84T induces changes in *Arabidopsis thaliana* gene expression against aluminum, drought, and salt stress. Microbiol Res 172:7–15. https://doi.org/10.1016/j.micres.2015.01.007

Vadez V, Kholova J, Choudhary S, Zindy P, Terrier M, Krishnamurthy L, Kumar PR, Turner NC (2011) Responses to increased moisture stress and extremes: whole plant response to drought under climate change. In: Crop adaptation to climate change. John Wiley & Sons, New York, pp 186–197. https://doi.org/10.1002/9780470960929.ch14

Vejan P, Abdullah R, Khadiran T, Ismail S, Nasrulhaq Boyce A (2016) Role of plant growth promoting Rhizobacteria in agricultural sustainability—A review. Molecules 21(5):573. https://doi.org/10.3390/molecules21050573

Verma V, Ravindran P, Kumar PP (2016) Plant hormone-mediated regulation of stress responses. BMC Plant Biol 16(1):86. https://doi.org/10.1186/s12870-016-0771-y

Zingaretti SM, Inácio MC, Pereira L d M, Paz TA, França S d C (2013) Water stress and agriculture. Responses of Organisms to Water Stress. https://doi.org/10.5772/53877

Chapter 26
Biogeochemical Cycles in Soil Microbiomes in Response to Climate Change

Bita Mohanty, Dipransu Pradhan, Rajashree Das, and Mihir Tanay Das

Abstract Soil microbiomes comprise highly diverse and heterogeneous soil microbes including bacteria, fungi, archaea, viruses, and protozoa which play a major role in biogeochemical cycling of elements. The cycling of elements mediated by soil microbes not only maintains soil health and supports plant growth but also regulates the climate. In the course of elemental cycling, soil microbes can act as a source or sink of major greenhouse gases including CO_2, CH_4 and N_2O due to their functional metabolic diversity. Of late, the contribution and susceptibility of soil microbes to a changing climate is getting recognized, yet our understandings about the microbial ecology, their multitrophic interaction and their interaction with abiotic factors are still limited. In this review, the current state of knowledge about the impacts of climate change on the biogeochemical cycling of major elements such as carbon, nitrogen, and sulfur is highlighted through the survey of existing literature. Further, the challenges, uncertainty factors, and knowledge gaps faced in the soil microbiome research with respect to climate change have been identified to design future research in this direction.

Keywords Carbon cycle · Nitrogen cycle · Sulfur cycle · Ecosystem function · Multitrophic interaction · Microbial evolution

B. Mohanty · D. Pradhan · R. Das
Department of Environmental Science, Fakir Mohan University, Balasore, Odisha, India

M. T. Das (✉)
Department of Environmental Science, Fakir Mohan University, Balasore, Odisha, India

Centre of Excellence for Bioresource Management and Energy Conservation Material Development, Fakir Mohan University, Balasore, Odisha, India

D. K. Choudhary et al. (eds.), *Climate Change and the Microbiome*, Soil Biology 63,
https://doi.org/10.1007/978-3-030-76863-8_26

26.1 Introduction

Throughout the biogeological evolution of earth, microorganisms and climate have been intricately associated with each other. From the Great Oxidation Event around 2.4 billion years ago caused by the marine cyanobacteria (Demoulin et al. 2019) to the evolution of methane during the Carboniferous period by the methanogens (Bartdorff et al. 2008), microbial processes have been the key drivers of, and responders to, climate change (Singh et al. 2010). Pertaining to their unseen majority on our planet in terms of both abundance and diversity, microorganisms play a huge role in maintaining a healthy global ecosystem (Cavicchioli et al. 2019). The backbone of every ecological process is a microbial system which controls the cycling of elements (e.g., carbon, nitrogen, sulfur, etc.) essential for life. The crucial role of microorganisms in maintaining the health and wealth of higher organisms (e.g., humans) via nutrient cycling has been long recognized (Pace 2018); for instance, half of the global production of the oxygen and carbon dioxide fixation is done by the oceanic phytoplanktons (Behrenfeld 2014). The relationship between the biogeochemical cycles and microorganism is very dynamic as they influence each other's characteristics and the net effect of the relation has varied over the evolutionary time scale of our planet. However, in the context of recent ongoing and projected climate change process, the significance and implications of microorganisms have poorly been studied and documented.

In recent times the anthropogenic activities have induced climate change, which in turn is impacting most other life forms of earth. For example, the composition and characteristics of nutrients in biogeochemical cycle like carbon are frequently disturbed by human activities (Griggs et al. 2013), and the resultant environmental and climatic effects cause extinctions of other organisms and ultimately lead to biodiversity loss (Barnosky et al. 2011; Crist et al. 2017; Johnson et al. 2017; Pecl et al. 2017). Over the past century, due to the human activity, the concentration of greenhouse gases including carbon dioxide, methane, and nitrous oxide increased dramatically and global average temperature has risen accordingly (Jansson and Hofmockel 2020). The average temperature of the globe has risen by 1.5 °F during the last century, and it is predicted that additional 0.5–8.6 °F will be added to the average global temperature by 2100 (Dutta and Dutta 2016) leading to the most rapid climatic variations in earth's history (Birch 2014). In this changing scenario, the status of microorganisms needs to be focused in climate change studies as these organisms are responsible for both emitting and sequestering greenhouse gases, thus playing a crucial role in regulating climate change process (Cavicchioli et al. 2019; Dutta and Dutta 2016; Jansson and Hofmockel 2020; Pecl et al. 2017).

Soil ecosystems are highly complex being the interface of all the four environmental components including hydrosphere, atmosphere, lithosphere, and biosphere. The soil microbiome consisting of interacting community of bacteria, archaea, viruses, fungi, and protozoa (Jansson and Hofmockel 2020) contributes to ecosystem health in a variety of ways, including biogeochemical cycling, bioremediation, plant growth, and primary productivity (Amundson et al. 2015; Cavicchioli et al. 2019).

However, being highly exposed to different anthropogenic activities, soil ecosystems are subject to different short- and long-term physicochemical perturbations which affect the diversity and function of the soil microbiome (Wieder et al. 2013). The combination of all these physicochemical and biological fluctuations governs how the nutrients will be exchanged between different environmental components and whether soil will act as a source or sink of greenhouse gases. Although importance of soil microbiome in regulating future climate of earth has been recognized (Amundson et al. 2015; Cavicchioli et al. 2019; Jansson and Hofmockel 2020; Wieder et al. 2013), it still remains a challenge to integrate microbial contribution data in climate models and predict the future scenario.

In the present chapter, the current state of knowledge about how the soil microbiome and climate impact each other with specific reference to the biogeochemical cycling is discussed in detail. Further, the emerging issues in soil microbiome research in the context of climate change have been discussed to highlight data gaps and identify areas of future research.

26.2 Soil Ecosystems and Their Climate Sensitivity

Soil represents one of the most diverse ecosystems on our planet with highly heterogeneous biotic and abiotic components (Kuzyakov and Blagodatskaya 2015). The soil environment is also highly dynamic with the seasonal and diurnal variation in soil moisture, temperature, nutrient composition, pH, and redox states as well as influence of plants and soil fauna (e.g., insects and earthworms). In addition to this, the ongoing and projected climate change is supposed to introduce more uncertainty with unknown consequences on the stability and resilience of the soil microbiome (Jansson and Hofmockel 2020; Norby et al. 2016). The most projected effect of climate change on soil abiotic components is the increased soil temperature (+0.3 °C), extended periods of pre-summer droughts, and raised CO_2 concentration (Haugwitz et al., 2013). However, with the existing soil ecology data, it is difficult to infer how the biotic components will behave under these climatic perturbations. In order to understand better how changing climate may affect soil ecosystem, we need to understand the sensitivity, resistance, and resilience potential of these ecosystems.

26.2.1 Soil Microbiome and the Factors Regulating Its Structure and Function

Soil microbiome comprises millions of microbes including bacteria, fungi, archaea, viruses, and protozoa and their abiotic environment in the soil which altogether form one of the most complex ecological system (Islam et al. 2020; Jansson and Hofmockel 2020). The species richness and diversity of soil microbiome is very high. Microbial species richness in a gram of soil can reach up to the order of 10^4, and the same soil may harbor up to 10^{10} bacterial cells (Roesch et al. 2007). As per an estimate, the total biomass on earth across all taxa is nearly 550 Gt of carbon in which the share of soil microorganisms is nearly 100 Gt of carbon which is next to plants but higher than the animals (Bar-On et al. 2018). This makes soil microbiome as an essential and significant driver of global biogeochemical processes.

Soils environments are complex consisting of a rich mixture of minerals, gases, liquids, organic matter, and living organisms. Though considered as a biome, environmental conditions may differ significantly between the habitats of different microbial colonies in soils (Islam et al. 2020). For instance, an individual soil aggregate of a few millimeters size can have variable oxygen concentrations ranging from 20 to $<1\%$ (Koch 1998). As the abiotic environment varies, the structure of microbial communities also varies considerably within a small region. For example, the microbial community structure in the plant rhizosphere may vary considerably from those found in "bulk" soil environments that are a few centimeters away (Lareen et al. 2016). Furthermore, higher organisms like plants or animals also influence on the composition of the soil microbiome due to their specific association with microorganisms. Several trophic level interactions connect the soil microbes to the species across kingdoms by involving them in a cooperative, inhibitory or syntrophic relationship. The anthropogenic practices like agriculture and animal husbandry also directly affect the soil microbiome characteristics. Thus, the structure and function of soil microbiomes are regulated combined by the biotic and abiotic environmental both spatially and temporally as shown in Fig. 26.1. Apart from the factors discussed above, changes in climate can greatly influence the composition and function of soil microorganisms. Changing climate can affect the soil abiotic environment in many ways including changes in soil temperature, moisture, pH, gaseous and nutrient composition, etc. which in turn will directly affect the microbial processes and ultimately influence the soil-atmosphere nutrient feedback processes. Further, changing climate can cause a number of shifts in aboveground and below-ground communities of higher trophic level which will indirectly influence the microbial processes of the soil (Classen et al. 2015).

26.2.2 The Uncertainty Factors in the Soil Microbiome with Respect to Climate Change

Soil ecosystems are highly complex and diverse, so is the microbiota colonizing them. So it is difficult to infer a generalized conclusion at a global scale with regard to the impacts of climate change on soil microbiome with current level of scientific understanding. Some of the uncertainty factors that limit this global prediction are discussed in this section.

Though the importance of soil microorganisms with respect to global nutrient cycling and climate change is thought to be well established, the fact is that a vast

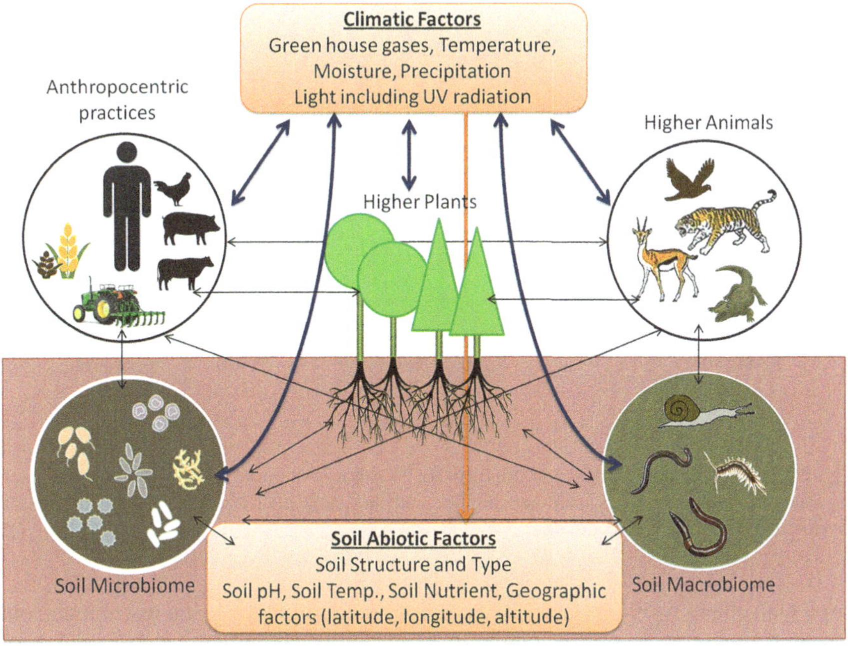

Fig. 26.1 Ecological position of soil microbiome. Bidirectional arrows indicate the ecological interrelationship between the entities. Soil microbiome is directly or indirectly affected by all including aboveground and belowground biotic factors, soil abiotic factors, climatic factors, and anthropogenic practices

majority of the soil microbiome is yet to be studied (Goel et al. 2018). Only a mere fraction of soil microorganisms have been catalogued till date, and with the existing knowledge, the remaining microbiota is uncultivable under laboratory conditions (Naylor et al. 2020) for which their cryptic ecosystem functions are yet to be revealed. This lack of knowledge results in uncertainty for construction of accurate predictive models for global carbon flux under climate change conditions (Friedlingstein et al. 2006; Naylor et al. 2020). Furthermore, depending on their genetic makeup and physiological states, different group of soil microorganisms respond to environmental stress in different ways (Schimel et al. 2007) as it can be seen during drought. Drought is a very common environmental stress factor (Schimel 2004) which believed to reduce both soil microbial diversity (Jensen et al. 2003) and enzyme activity (Sardans et al. 2008) resulting in reduced soil nutrient availability (Allison and Treseder 2008). Different microorganisms react differently, for example, proto-bacteria are less suitable for drought condition than acidobacteria (Castro et al., 2009). Fungi are more resistant to drought condition than bacteria due to presence of their hyphal network through which water can transport (Khalvati et al. 2005). Hence, they are capable to degrade more complex organic material and show quick response after rewetting of the soil (Boer et al. 2005). Most of the soil microorganisms are highly adaptable, and they have evolved strategies to

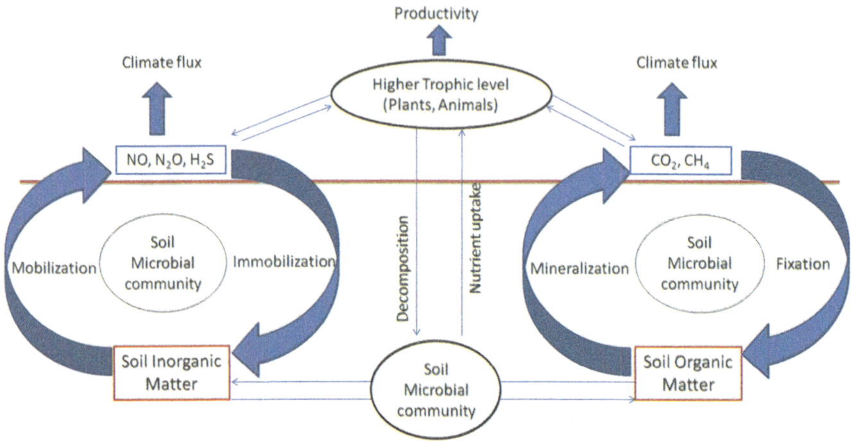

Fig. 26.2 Regulation of climate and productivity by soil microbiome. Soil microbiome delicately regulates the balance between the soil organic and inorganic matter which in turn controls the productivity and affects the climate

cope with changing environmental conditions either by developing novel metabolic mechanisms or by becoming dormant or changing community interaction. Very often these adoptions are induced at genetic level by regulating gene transcription and translation and/or accumulating mutations or new genes through horizontal gene transfer. However, at present, quantifying these genetic and physiological responses of different groups of microorganisms remains a major gap in modeling soil microbiome's response to climate change (Evans and Wallenstein 2014; Jansson and Hofmockel 2020).

Soil microorganisms carry out the decomposition of organic matter and liberate the inorganic constituents, and at the same time, they are also involved in the stabilization of inorganic nutrients into organic forms (Fig. 26.2). The balance between these two processes governs the net flux of CO_2, CH_4 and other greenhouse gases into the atmosphere (Singh et al. 2010). Furthermore, the association of soil microbes with organisms of higher trophic level and effects of climate change on the later too makes the holistic prediction more complex (Bardgett et al. 2013; Pugnaire et al. 2019). For example, increased levels of atmospheric CO_2 result in higher carbon sequestration by plants and subsequent higher litter fall. This increased organic matter input into soil supports increase in copiotrophic-heterotrophic soil microbial activity; however, in the long term, limitation of mineral nutrients such as nitrogen may reverse this, promoting the dominance of oligotrophic-autotrophic and microorganisms (Singh et al. 2010). Moreover, under changing climatic scenario, that plant or animal species that has more competitive advantage will be added to the above soil community which in turn will alter the soil microbiome composition depending on the changed plant-animal-microbe association pattern (Pugnaire et al. 2019). Thus, soil microbial CO_2 and other greenhouse flux into the atmosphere in a given ecosystem are regulated by number of feedback mechanism involving different trophic levels. Therefore, while developing a mathematical predictive model of the microbial ecology with reference to climate change, it is required to consider all possible feedback mechanisms involved.

Another knowledge gap associated with the soil microbiome-climate feedback system is modeling the compounding effects of global climate change on soil microbiome. For example, independent field experiments simulating elevated level of atmospheric CO_2 have been conducted to understand responses of soil microbe (Dunbar et al. 2012; Hayden et al. 2012; Yu et al. 2018). Similarly, responses of soil microbes under experimental warming conditions have also been studied by different research groups (Melillo et al. 2017; Romero-Olivares et al. 2017; Schindlbacher et al. 2011). However, under practical environmental conditions the microbes may behave differently due to the compounding effects of global warming and elevated levels ofatmospheric CO_2. Increasing greenhouse gases or increasing global temperature can lead to cascading environmental changes, and the magnitude of the changes may vary in different geo-climatic regions. Understanding how changes in temperature or atmospheric greenhouse gas concentration interact with other important environmental variables such as precipitation, soil moisture, pH, and nutrients (such as phosphorus, nitrogen, carbon) is therefore essential for predicting the climate change responses of soil microbiome (Jansson and Hofmockel 2020). With lack of these baseline data, it is very difficult to incorporate this variation aspect into the climate change models.

26.2.3 Soil Ecosystems that are Vulnerable to Climate Change

Though all soil ecosystems in general are climate sensitive, some of them, as discussed in this section, are much more vulnerable due to their special ecological characteristics. The Arctic regions of our planet are the most vulnerable terrestrial ecosystems with respect to the climate change (IPCC 2007). The permafrost of the Arctic regions are enormous reservoir of stored carbon which will become susceptible to microbial decomposition due to global warming, and this process will result in substantial emission of CO_2 in the coming years (Schuur et al. 2015). Similarly the soil of forests (Lladó et al. 2017) which cover nearly 30% of the total land surface and the soil of grasslands (Jones et al. 2009) which account for nearly 26% of the total land surface are important carbon sinks and store enormous amount of carbon. With increasing temperature, it is predicted that microbial decomposition of soil organic carbon will be increased leading to CO_2 emission which to some extent will be countered by the increased plant growth. However, with climate change, drought, fire, and flood frequency and severity will also have detrimental effects on these ecosystems (Jansson and Hofmockel 2020). Among other terrestrial ecosystems, the wetlands which are the interface of terrestrial and aquatic ecosystems are vulnerable to climate change given the fact that they are the largest natural source of CH_4 emission and simultaneously act as a carbon sink too (Zhang et al. 2017).

26.3 Ecosystem Functions and Services of Soil Microbiome

Soil microbes are highly diverse and abundant group of living organisms having varied metabolic capabilities that offer vast diversity of ecosystem services and functions (Islam et al. 2020; Saccá et al. 2017). Soil microbial communities, being involved in nutrient cycling process, maintain the soil fertility and affect the primary productivity (Islam et al. 2020; Trivedi et al. 2016), bioremediation of environmental contaminants, and sequester of greenhouse gases (Jacoby et al. 2017). Apart from this, by virtue of their ecosystem functions, they indirectly affect the structure and function of higher trophic level organisms (Bender et al. 2016) and physicochemical nature of their immediate environment (Nielsen and Winding 2002).

The ecosystem functions and services of the soil microbial community are shown in detail in Fig. 26.3.

Soil microbiome diversity is always acknowledged for their role in nutrient management (Six et al. 2006; Jacoby et al. 2017). The soil microorganisms mainly fungi and bacteria serve as chemical engineers by not only carrying out the decomposition of organic matter but also fixing atmospheric nitrogen and thus making the nutrients readily available for the species of higher trophic level especially plants (Saccá et al. 2017). Soil microbiome diversity therefore plays a major role in determining the primary productivity and expression of trait in higher organisms including plants and animals (Islam et al. 2020; Trivedi et al. 2016). The nutrient cycling process mediated by the soil microbes not only helps in maintaining the soil health and fertility, but also by this process they affect the climate by regulating greenhouse gas emission or sequestration (Cavicchioli et al. 2019; Jacoby et al. 2017; Jansson and Hofmockel 2020; Naylor et al. 2020) (Fig. 26.2). They are also responsible for modifying soil physical structure by the formation of pore networks and particle aggregation and affect water infiltration and distribution in soil (Nielsen and Winding 2002). They indirectly regulate the soil biodiversity by regulating resources availability or by producing secondary metabolites like antibiotics. Soil microbes also interact with higher organism like soil invertebrates, plants, etc. and regulate the latter's ecosystem functioning (Thakur and Geisen 2019). Microorganisms possess remarkable potential to originate novel metabolic mechanisms to degrade and decontaminate xenobiotic compounds in the terrestrial ecosystem

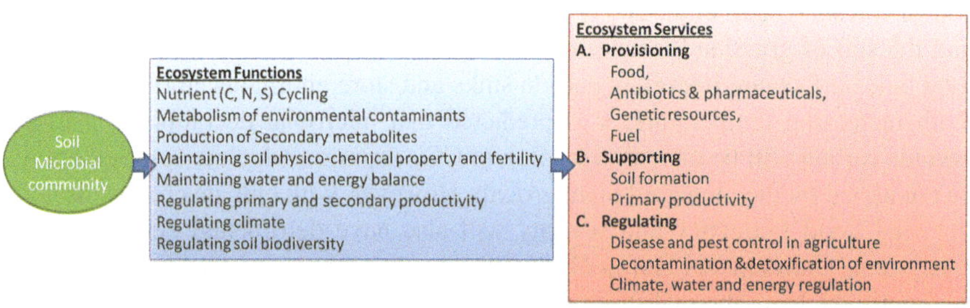

Fig. 26.3 Ecosystem functions and services provided by the soil microbial community

(Arshad et al. 2008) by utilizing these compounds as energy (carbon) and nutrient resources like nitrogen, phosphorus, sulfur, etc. (Hussain et al. 2007a; Hussain et al. 2007b). Furthermore, microbiome-multitrophic interactions like rhizobacteria also facilitate successful phytoremediation strategies.

The current trend of functional aspects of soil microbiome diversity research is limited to microcosm experimentation, explaining the relationship of species diversity with ecosystem properties such as cycling of nutrients, utilization of xenobiotic substrate, and productivity. Thus the soil microbiome has been acknowledged with number of scientific literatures for providing variety of ecosystem functions and services, but the linkage between the structure and function of soil microorganisms under long-term climate change projections is still limited.

26.4 Microorganisms and Biogeochemical Cycling with Reference to Impact on Climate

Microorganisms play a vital role in functioning and transformation of biogeochemical cycles in the ecosystem because of their abundance, adaptation potential, and metabolic capabilities (Madsen 2011). They perform key role as producers, consumers, and decomposers in the ecosystem and are involved in almost all stages of the nutrient cycle. For example, the processes such as nitrogen fixation, denitrification, sulfate reduction, methane production, consumption, etc. are associated with the microbial activity. The continuation of life on earth (plants, animals, and humans) is totally dependent on the microorganisms and their activity, in absence of which, the essential elements of life would trapped in organic matter and wastes. The organic matter content in soil is important not only for biological activities, but it is often used as an indicator of soil health too (De Bona et al. 2008). The soil organic matter and microorganism diversity, abundance, and their biological activity are interdependent with each other which combined together have major influence on the physical and chemical properties of the soil like aerobic conditions, water retention capacity, aggregation, structure stability, etc. These characteristics also affect the bioavailability of nutrients or toxic elements (van Herwijnen et al. 2007; Farenhorst 2006) and most importantly contribute to climate homeostasis processes (Cavicchioli et al. 2019; Jansson and Hofmockel 2020; Wieder et al. 2013).

The microbial activity within the soil varies spatiotemporally depending upon the availability of oxygen, carbon, and other favorable conditions. Intense microbial activity can be seen at locations known as "hot spots" where there is increased availability of organic compounds and other favorable conditions (Ruiz-Rueda et al. 2009). The type of dominant metabolic pathway in soil microbiome is also subjected to variation under different oxygen gradients (oxic, less-oxic to anoxic conditions) due to soil macroporosity level induced by precipitation, flood events, rise in groundwater level, etc. Under aerobic respiration, oxygen is the electron acceptor, whereas under oxygen-limiting conditions, alternate respiratory pathways are possible using electron acceptors like nitrate (by denitrifiers from nitrification or fertilization) or sulfate to recover energy. Under anoxic and nitrate-deficient conditions, the decomposition switches to fermentative pathways where the methanogens or sulfate reducers use the fermentation products like acetate, organic acids, hydrogen to produce sulfides or methane by transferring electrons to sulfate or carbon dioxide. Thus, soil microorganisms play a crucial role in the cycling of carbon, nitrogen, and sulfur, and the cycling of these elements under anthropogenic interference is going to decide the fate of future climate on earth. In the next section, these cycles are discussed in further detail with specific reference to how they are going to respond towards the global climate change.

26.4.1 Carbon Cycle and Climate Change

26.4.1.1 Carbon Cycle and the Role of Soil Microbes

Carbon cycle can be defined as the oxidation-reduction process in which carbon undergoes a series of transformations on earth, starting from its neutral state (zero oxidation state; in the form of graphite or diamond) to the most oxidized (CO_2) and most reduced form (CH_4). Carbon is present in all living organisms; however its concentration is very less in the earth's crust (about 0.27%). The main carbon source required by autotrophic organisms to proliferate and grow is CO_2, formed by association with oxygen. It forms hydrocarbons when combined with hydrogen; with oxygen and hydrogen, it gives rise to lipids and carbohydrates; with sulfur and nitrogen, it forms amino acids (proteins); with phosphorus, it forms nucleotides that stores the genetic information of life. The photosynthetic CO_2 fixation, organic matter transfer in food web, detritus storage, and fossil formation are some of the important aspects of carbon cycle which also enables resource matter cycling in the ecosystem.

A simplified version of carbon cycle highlighting the microbial involvement is shown in Fig. 26.4. Fresh organic matter is transformed by soil microbial activity by any of the three different pathways including (a) decomposition and release of the associated carbon into the atmosphere as CO_2 and CH_4, (b) assimilation and conversion into microbial biomass, and (c) conversion into humic substances. In the microbial decomposition process, major fraction of carbon is released as CO_2 or CH_4, and some fraction of the soil organic matter is assimilated and utilized by microorganisms to build their own biomass. The remaining fraction of organic matter is converted to humus which is a complex heterogeneous material produced as a result of the microbial transformation of organic matter influenced by other abiotic processes. However, humus is chemically not so well defined, and it acts as a rich secondary source of carbon, nitrogen, and energy for further microbial mineralization.

Methane as a greenhouse gas is more potent than CO_2 in bringing about global warming (Shindell et al. 2009), and the soil microbes play the major role in

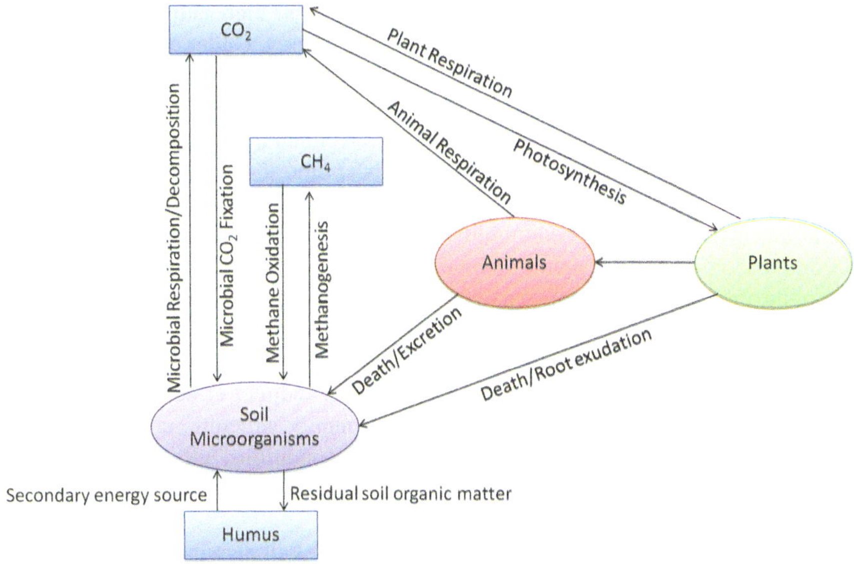

Fig. 26.4 Microbial involvement in terrestrial carbon cycle

producing and consuming methane. There are generally two groups of microbes responsible for balancing the methane concentration on earth, i.e., methanogenic (methane producing) and methanotrophic microbes. Many members of the domain Archaea are responsible for the production of methane, but still most are unidentified which have roles in the methane cycle (Vanwonterghem et al. 2016; Nobu et al., 2016). The methanogens occur in anoxic environments, and they produce methane by degrading organic matter in a fermentative pathway. In contrast, methane is consumed by a unique group of *Proteobacteria* called methanotrophs which are mostly aerobic organisms (Nazaries et al. 2013). They are often found at the anoxic/oxic interface of various habitats such as soils, landfills, peat bogs, wetlands, etc., where they consume the CH_4 produced by the methanogens and thus they are responsible for lowering the overall CH_4 emission from an ecosystem. Methane can also be consumed by anaerobic oxidation process which is carried out by a tight/physical association of anaerobic methanotrophic Archaea and sulfate-reducing bacteria (SRB) where sulfate is used as an electron acceptor to oxidize CH_4 (Nazaries et al. 2013).

26.4.1.2 Possible Impacts of Climate Change on Carbon Cycle

Soil plays a significant role in global carbon cycle as it acts as a sink for long-term storage of carbon in form of soil organic matter. Organic matter inputs into soil are mainly dominated by plant macromolecules like cellulose, hemicelluloses, lignin, and other secondary metabolites in form of root exudates (Lynd et al. 2002). Large amounts of such organic matter are currently locked in high latitude permafrosts, forest, and grassland soils as discussed under Sect. 26.2.3. Under changing climatic conditions, the fate of this stored carbon, whether it is released into the atmosphere as greenhouse gas or not, is dependent on the soil microorganisms as they play vital role in controlling the key steps of carbon cycle (Weiman 2015). The decomposition of organic matter and release of CO_2 or CH_4 back into the atmosphere are mainly carried out by the soil microbes including heterotrophic bacteria and fungi. As environmental conditions like humidity and temperature play significant role in the decomposition of organic matter, climate change would have direct impact on the rate of the process (Carol Adair et al. 2011). Several studies show that soil respiration could be increased by increase in the atmospheric carbon dioxide level (Carol Adair et al. 2011; Wan et al. 2007) or increase in atmospheric temperature (Liu et al. 2016; Wan et al. 2007). Apart from these direct impacts, indirect effects resulting from climate-driven changes such as shifting in community structure, changes in plant productivity, alteration in soil physicochemical conditions, etc. may affect the activity of microbes involved in decomposition processes and carbon release from soil (Bardgett et al. 2008).

Like most of the microbial groups, climate change will also impact the activities of methanogens and methanotrophs, which in turn will affect the net CH_4 flux in to the atmosphere. At high latitudes, warming can also increase methane emission from the peatlands were one-third of the world's soil carbon is stored (Hopple et al. 2020). In addition to this, with increase in carbon dioxide concentration, abundance of methanotrophs may reduce up to 70% (Kolb et al. 2005). Similarly, some other studies have shown that in response to high precipitation and temperature, the population of some methanotrophs may gradually decline (Horz et al. 2005; Mohanty et al. 2007). However, it is difficult to predict the global effect of elevated temperature on methanotrophy, because different ecosystems will respond differently to climate change (Singh et al., 2010). Soil moisture is one of the most important factors that regulate methane flux into the atmosphere as increased moisture content in soil stimulates methanogenesis by creating anaerobic conditions and inhibits oxygen-dependent methanotrophy. But in context of climate change, moisture content of soil will vary depending on the precipitation pattern and geographic location. As a consequence, areas which will become wetter under future climate conditions will experience an increase in CH_4 production, whereas the drier areas may experience a reverse phenomenon (Nazaries et al. 2013).

26.4.2 Nitrogen Cycle and Climate Change

26.4.2.1 Nitrogen Cycle and the Role of Soil Microbes

Nitrogen is a fifth most abundant element on earth which serves as major mineral nutrient for both plants and animals. Complex biomolecules like DNA, RNA, and protein are formed with nitrogen bonded to carbon, and thus nitrogen serves as a major building block for living and nonliving biomass. All living organisms acquire nitrogen from the surrounding environment either in the form of dissolved inorganic chemicals or in the form of organic nitrogen, and they return back the nitrogen to the surrounding environment as organic matter through excretion or through decomposition after death. In food chain, the amount of nitrogen availability to a particular trophic level depends on the efficiency of nitrogen cycle which consists of multiple complex redox reactions for transforming nitrogen compounds mostly mediated by the microbial community.

Molecular nitrogen (N_2) in the gaseous phase is the largest reservoir of nitrogen, and it comprises around 78% of earth's atmosphere. It enters into the biological system through nitrogen fixation, a process where some free living or symbiotic prokaryotes reduce nitrogen gas to ammonium with the help of the enzyme nitrogenase under anaerobic conditions. Nitrogen fixed as ammonium (NH_4^+) is taken up by plants or other microorganisms and is assimilated to form amino acids which are transferred in the food chain among different trophic levels in this organic form. The nitrogen from the organic compounds is returned back to the environment when organisms excrete (animals) or die. Ammonium fixed in the nitrogen fixation process may also get sequentially oxidized to nitrite (NO_2^-) and nitrate (NO_3^-) by a multistep process known as nitrification involving different chemoautotrophic microorganisms under aerobic conditions. Nitrous oxide (N_2O) is generated as a by-product in this process which is an important source of atmospheric N_2O that acts as a greenhouse gas (Canfield et al. 2010; Dore et al. 1998). The "oxidized" nitrogen compounds particularly nitrate (NO_3^-) are taken up by plants where they undergo assimilatory nitrate reduction to form ammonium (NH_4^+) and subsequently into other organic compounds. Nitrate (NO_3^-) can act as a respiratory electron acceptor by many heterotrophic microbes in the absence of oxygen to produce nitrite (NO_2^-), nitric oxide (NO), nitrous oxide (N_2O) and ultimately molecular nitrogen (N_2) gas, and this multistep process is known as denitrification. Diverse soil microbes including bacteria and Archaea as well as some eukaryotes (e.g., fungi, protozoa, and benthic Foraminifera and Gromiida) can act as denitrifiers (Demanèche et al. 2009; Piña-Ochoa et al. 2010). N_2O is again produced as an intermediate in this denitrification process from both marine and terrestrial environments (Canfield et al. 2010; Codispoti 2010; Schlesinger 2009). In recent times an alternate pathway has been reported in some heterotrophic microbes in which nitrate (and some nitrite) is reduced back to ammonium via a process known as dissimilatory nitrate reduction to ammonium (DNRA) (Lam et al. 2009; Zerkle and Mikhail 2017). Further in another alternative route, NO_2^- is used to oxidize NH_4^+ to produce N_2 gas in a process called anaerobic ammonium oxidation (anammox) found among a group of chemoautotrophic bacteria known as a planctomycetes (Strous et al. 2006). This anammox process does not lead to the generation of N_2O unlike classical denitrification process (Strous et al. 2006). Together, the denitrification and anammox processes close the nitrogen cycle by returning N_2 gas back to the atmosphere. A schematic representation of the total process as discussed in the paragraph is shown in Fig. 26.5.

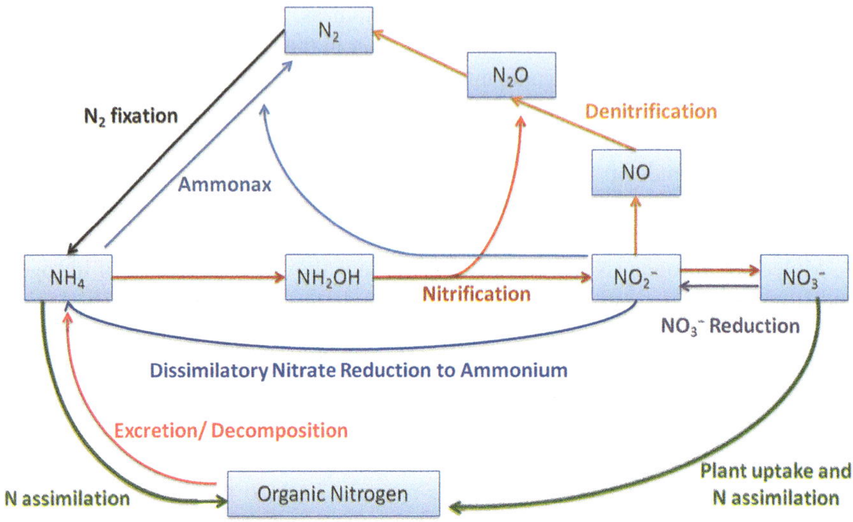

Fig. 26.5 Steps involved in nitrogen cycle

26.4.2.2 Possible Impacts of Climate Change on Nitrogen Cycle

Anthropogenic practices in recent times have caused enormous alterations in the global nitrogen cycle by developing industrial Haber-Bosch processes to produce NH_4^+ from N_2 which account for more than 40% of the total annual terrestrial nitrogen fixation on earth (Canfield et al. 2010). Apart from this, the agricultural practices and burning of fossil fuel have serious impacts on the global nitrogen balance. To these already existing perturbations in global nitrogen cycle, climate change concerns have added more uncertainties with respect to the soil microbiome functioning in nitrogen cycle. Climate change can alter the plant community structure, soil microbial community structure, plant microbe, and microbe-microbe interaction which combined together can affect the functioning of the nitrogen cycle (Classen et al. 2015). For example, rising temperatures can shift the mycorrhizal association from symbiotic to parasitic mode due to the increased biomass addition to mycorrhizal hyphae under changing environmental conditions (Hawkes et al. 2008). In the context of climate change, it is pertinent to mention that microbial processes that regulate specified ecosystem functions such as nitrogen fixation and nitrification will be more severely affected in comparison to the other processes that occur at a coarser scale, such as denitrification, organic nitrogen mineralization, etc. (Classen et al. 2015). Nitrogen fixation and nitrification are carried out by specific group of microorganisms, and a change in the relative abundance of such organisms under changing climatic scenario can have a direct impact on the rate of that process. However, diversity of microorganisms drives the nitrogen mineralization process, and therefore, the ecosystem function in absence of one group may be taken up by another group of microorganisms; but, such functions are affected by the abiotic factors such as temperature and moisture (Hooper et al. 2005).

Apart from the indirect impacts as discussed, few studies have shown that increasing CO_2 or temperature can directly affect the rate of key steps in the nitrogen cycle. A 14-year-long experimental setup with elevated CO_2 condition has revealed that nitrogen (N) fixation and the ammonium-oxidation can be significantly decreased in a semi-arid grassland ecosystem (Yang et al. 2019). Some other studies have also reported similar results that nitrification rates may get decreased in response to elevated CO_2 conditions due to decreased ammonium availability for nitrifiers (Niklaus et al. 2001; Lagomarsino et al. 2008). However, few other reports suggest that the nitrification rate may remain unchanged under similar conditions (Zak et al. 2000; Pinay et al. 2007). On the other hand, elevated CO_2 may stimulate N_2O efflux from soil, either due to enhanced anaerobic denitrification, in response to the increases in soil moisture (Arnone III and Bohlen 1998), or due to stimulated heterotrophic denitrification in presence of high organic carbon inputs to soil (Regan et al. 2011). However, this effect may show spatiotemporal variation in response (Brown et al. 2012; Niboyet et al. 2011).

26.4.3 Sulfur Cycle and Climate Change

26.4.3.1 Sulfur Cycle and the Role of Microbes

Sulfur is an abundant element in nature and is present in various physical states and chemical forms with different oxidation states ranging from −2 to +6. In the atmosphere, sulfur is present both in oxidized (SO_2) and reduced (H_2S, dimethyl sulfide) states, and the oxidized forms like sulfur dioxide are one among the greenhouse gases, and it can give rise to acid rain by formation of sulfuric acid (H_2SO_4). In the lithosphere, the sulfur is found as sulfate (SO_4^{2-}) salts, as a constituent of organic compounds and as metal sulfides like FeS, whereas in the marine environment, it is found mainly as sulfate. In living beings, sulfur like nitrogen and carbon is an essential part and constitutes about 1% of the dry matter of cell. In nature, complex physical, chemical, and biological processes occur under oxic or anoxic conditions to flow sulfur between its different reservoirs. Role of prokaryotes (*Bacteria*, Archaea) are predominant in the redox reactions constituting the biogeochemical cycling of sulfur. This cycle includes the assimilation of sulfate to form organic compounds, decomposition of organic compounds to release H_2S, oxidation of H_2S to elemental sulfur, and again oxidation of elemental and other reduced forms of sulfur to form sulfate. A schematic representation of sulfur cycle is shown in Fig. 26.6.

By assimilatory reduction process, microorganisms and plants convert the sulfate taken up by them from their environment in to sulphydryl compounds (R-SH) which becomes a part of their biomass. Once sulfur from sulfate gets incorporated into a reduced organic S-compound, it moves in the food chain and returns back to the

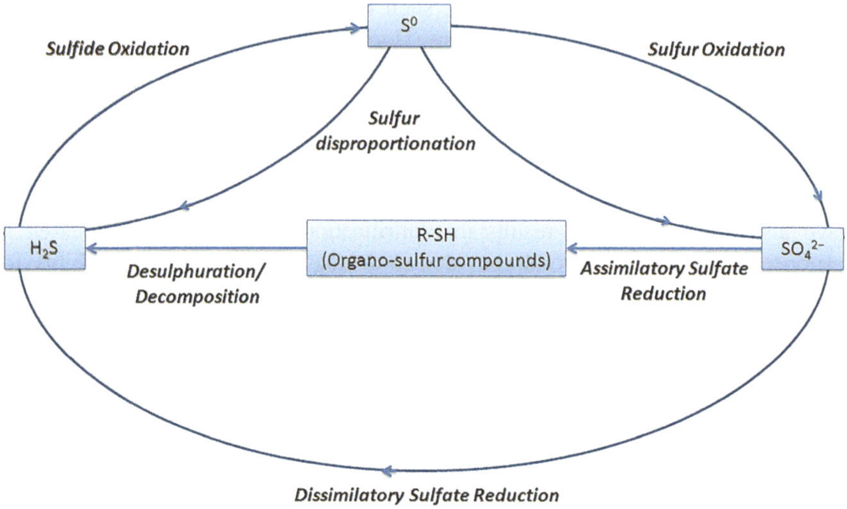

Fig. 26.6 Microbial steps involved in sulfur cycle

environment only with the death of an organism. Organic sulfur is mineralized into inorganic sulfur by the action of aerobic or anaerobic microorganisms (decomposers), and sulfur is released to the environment as sulfide mainly as H_2S and the process is known as desulfuration. Volatile organic compounds having low molecular weight like dimethyl sulfide (DMS), dimethyl disulfide, methanethiol, and mercaptopropionate can also be produced by biodegradation of various organosulfur compounds. One such compound DMS is mainly produced in the marine environment due to biodegradation of the marine biogenic product dimethyl sulfoniopropionate (DMSP) (Stefels et al. 2007). DMS in the atmosphere is oxidized to sulfur dioxide (SO_2) and then sulfate (SO_4^{2-}), which contributes to the formation of non-sea-salt sulfate aerosol, and thus it plays a significant role in regulating cloud albedo and climate (Stefels et al. 2007; Wang et al. 2018). In an alternate pathway, sulfate can be reduced to sulfides under anaerobic conditions by sulfate-reducing bacteria (SRB) and few Archaea by a process known as dissimilatory sulfate reduction (DSR), commonly called sulfate respiration (Hao 2003). Large quantities of sulfides are produced by sulfate-reducers which may reoxidized via chemical or biological processes or react with the metals and precipitate as metallic sulfide precipitates. Apart from the beneficial effects like mineralization of organic matter and precipitation of toxic metallo-sulfides, sulfate reducers can have also adverse effects on the environment like pungent smell, metallo-protein formation leading to intoxication and respiratory obstruction, corrosion phenomena, sulfide accumulation, production of neurotoxins like methylmercury, etc. (Hao 2003).

Hydrogen sulfide undergoes oxidation to produce elemental sulfur by the action of certain anoxygenic photosynthetic sulfur bacteria including purple sulfur bacteria and green sulfur bacteria (Overmann and Garcia-Pichel 2006). Elemental form of sulfur and other forms of reduced sulfur compounds like thiosulfate, sulfide, and sulfite are oxidized to produce sulfate by highly diverse microorganisms known as chemolithotrophic sulfur-oxidizing prokaryotes (CSOP). Robertson and Kuenen (2006) distinguished four groups of CSOP depending upon their energy and carbon sources. The first group known as obligatory chemolithotrophs comprises bacteria such as *Thiobacillus*, *Acidithiobacillus*, *Thiomicrospira*, and archaea like *Sulfolobus*, etc. which can utilize exogenous carbon sources and ferment their cytoplasmic glycogen under unfavorable environmental conditions. The second group of organisms shows the same characteristics as that of first group but can also grow as mixotrophs or heterotrophs and is called as facultative chemolithotrophs. The third group is known as chemolithoheterotrophs (*Thiobacillus*, *Beggiatoa*) who cannot fix mineral carbon and require an organic carbon source for growth but can use reduced sulfur compounds as electron donors. The fourth group is called as chemoorganoheterotrophs (*Beggiatoa* sp., *Macromonas*, *Thiobacterium*, and *Thiothrix*) which can oxidize sulfur compounds. CSOP mainly use dioxygen as primary electron acceptor and lives at the interface between oxic (containing dioxygen) and anoxic (containing sulfides) environments. Further, several microorganisms can also grow by the disproportionation of elemental sulfur, a process in which a sulfur is both oxidized and reduced (Thamdrup et al. 1993). Thus cycling of sulfur in the environment involves diverse metabolic pathways found in diversity of microorganisms.

26.4.3.2 Possible Impacts of Climate Change on Sulfur Cycle

Cycling of sulfur has a key role in future climate homeostasis as sulfate aerosols have the potential to counter the effects of global warming (Levasseur 2011). In this context, patterns of DMS emission from marine ecosystem, which is the largest natural source of atmospheric sulfur, have been studied by several research groups under projected climate change conditions (Grandey and Wang 2015; Six et al. 2013). Six et al. (2013) reported that in the future DMS emissions may decrease as a result of the combined effects of ocean acidification and climate change. In contrast, several other studies have suggested that DMS emissions may enhance with increasing atmospheric greenhouse gas concentrations (Gabric et al. 2001; Grandey and Wang 2015; Levasseur 2011). Though research on sulfur cycling in marine ecosystem with respect to climate change has gained momentum, similar studies on terrestrial sulfur cycle have not yet received due attention by the scientific community. However, the terrestrial microbial community structure and their ecosystem function with respect to sulfur cycle will have to face the impacts of global climate change as dynamics of carbon, nitrogen, and sulfur cycling in terrestrial ecosystems is intimately linked with each other. For example, in wetland ecosystems under anoxic conditions, sulfate-reducing microorganisms (SRM) compete for substrates with methanogenic microorganisms in which the former is thermodynamically favored (Pester et al. 2012). This competition effectively decreases the flux of methane to the atmosphere, thereby naturally countering the greenhouse effect.

Therefore, there is a need to systematically evaluate the role of microbes involved in the terrestrial sulfur cycle in the context of future climate change conditions.

26.5 Emerging Issues and Future Prospects in Soil Microbiome Research in the Context of Nutrient Cycling and Climate Change

26.5.1 Agricultural Practices, Soil Microbes, and Climate Change

Agricultural practices directly and indirectly perturb soil microbial community composition and function, leading to substantial changes in the natural cycle of carbon, nitrogen, sulfur, and other micronutrients. For example, cultivation of rice (van Groenigen et al. 2013) and farming of ruminant animals (Ripple et al. 2014) contribute huge fluxes of CH_4 into the atmosphere. Similarly, nitrogen cycle is also affected due to huge input of nitrogenous fertilizers and monoculture of crops which alter the soil microbial community structure due to their specific plant microbe association (Greaver et al. 2016). Huge amount of nitrous oxides are produced due to the additional input of nitrogen into the agricultural fields by the action of microbial process, i.e., nitrification and denitrification (Itakura et al. 2013). Further, excess nitrogen and phosphorous added to the agricultural fields promote harmful algal bloom in the inland water bodies which disrupts many ecological processes.

To these already existing perturbations with respect to the biogeochemical cycling of elements in the agricultural fields, looming climate change has added more uncertainties (Cavicchioli et al. 2019). Increasing temperature and consequent fluctuation in the weather events such as drought, flood, cyclone, etc. will strongly affect the current agricultural practices by altering the soil microbial structure and function which support plant growth. Therefore, there is an increasing need to understand the plant-microbe and animal-microbe association to mitigate the effects of climate change on food production and increase agricultural sustainability (Godfray et al. 2010). Studies have shown that the fungal dominated soil ecosystems commonly found in grasslands and poorly managed croplands are better able to adapt to drought than the bacterial dominated soil ecosystems which are commonly found in intensive systems (Bahram et al. 2018; de Vries et al. 2012, 2018). While developing agricultural strategies for future under climate change conditions, it is not only important to protect food security, but also mitigating greenhouse gas emissions from agricultural practices should also be emphasized (Lipper et al., 2014). Analyzing the microbiome loop, above- and belowground multitrophic interactions among microbes, plants, and animals may potentially help in addressing the issues like nutrient management, greenhouse gas emissions, and productivity of natural and agro-ecosystem. More research is required to understand how climate, agriculture, and soil microbiome interact with each other so that climate-resilient agricultural practices can be developed to meet the future needs.

26.5.2 Climate Change and Polar Microorganisms

Climate models have predicted that the high latitude regions of the planet are going to be worst affected due to global temperature rise. It has been predicted that Northern Hemisphere is likely to experience an increase in temperature up to 4.8 °C before the end of the twenty-first century (IPCC 2014). Similarly, the Antarctic region has already experienced an average increase of temperature as much as 2.4 °C in the last 50 years (Bromwich et al. 2013). This alteration in polar temperature may lead to a significant change in the microbial community structure and their ecological function (Jansson and Taş 2014; Van Horn et al. 2014). Huge amount of plant, animal, and microbial biomass carbon of the CO_2-rich warm periods in the history of the earth have been preserved in the permafrost since the last glaciation events (Jansson and Taş 2014; Weiman 2015). Studies have shown that the permafrost of both the Arctic and Antarctic region host diverse microbial phyla from different functional guilds including methanogens, sulfate reducers, Fe(iii) reducers, denitrifiers, etc. (Niederberger et al. 2012; Mondav et al. 2014; Rivkina et al. 1998; Steven et al. 2008; Trotsenko and Khmelenina 2005; Yergeau et al. 2010). The methane cycle which plays a key role in climatic processes has been studied extensively in Arctic (Hultman et al. 2015; Tveit et al. 2015) and Antarctic (Yergeau et al. 2007) soils. Increasing temperature and subsequent permafrost thaw have raised the concern of increase in the microbial activity resulting in releasing of considerable amounts of carbon in the form of greenhouse gases (Gilichinsky et al. 2005; Jansson and Taş 2014; McCalley et al. 2014).

Though this aspect has drawn the attention of scientists working in the direction of climate change, comparatively, very little is known about the functional capacity of the microbes of the poles particularly that of the permafrost. As an extreme habitat, microbial communities of permafrost not only have unique climatic adaptations and metabolic capacities, but also they bear the signature of past climatic records. In order to understand the impact of global climate change on the polar edaphic systems, there is a primary need to better understand the polar microbial processes through employing high-throughput modern "omics" approaches.

26.5.3 Species Coexistence and Multitrophic Interactions in Soil Microbiome in the Context of Climate Change

Soil microbiome by virtue of its role in biogeochemical cycling of nutrients plays a significant role in structuring its own community with specific species coexistence and mediating multitrophic interactions which eventually determines the structure and functions of an ecosystem. Understanding the mechanisms of microbiome species coexistence and multitrophic interactions are major challenges confront by ecologists today (Saleem et al. 2015). Though the ecological concepts of food chain, food web, energy, and nutrient flow in the ecosystem, species interaction are not new, establishing a mathematical link between all these functions in the context of climate change is a challenging task. Current pattern of research on "climate change and microbiome" mostly lack multidimensional experimental design, and therefore, it is difficult to link the observed microbiome functional trait under one experimental condition to the ecosystem functioning on local, regional, and global scales. For example, different research groups with their independent set of experiments have reported that soil microbial diversity in response to elevated CO_2 may increase (Lesaulnier et al. 2008), decrease (He et al. 2012), or remain unchanged (Austin et al. 2009). This inconsistency in results could be attributed to high complexity of soil microbial communities (Gans et al. 2005), differences among the techniques used to study microbial communities, differences in the studied ecosystems, and the time scale of the experimental design (Yang et al. 2019). There are a number of unanswered questions in the context of microbiome species coexistence, multitrophic interaction, and their parallel evolution with respect to the changing environmental conditions. Microbiome diversity acts as a foundation of global trophic networks, and by microbial loop concept, the energy transfer to higher trophic levels is acknowledged. Microbiome-induced multitrophic interactions not only regulate diversity but also explain species interactions at different trophic levels. For example, Bonkowski (2004) studied that with increase in bacterial diversity, the micropredator diversity is stabilized resulting in increased primary productivity by increasing bioavailability of limiting nutrients and suppressing plant pathogens (Rosenberg et al. 2009). The higher trophic level organisms such as plants, herbivores, and predators also utilize microbiome diversity to interact with other organisms.

Therefore, it is important to understand the underlying mechanism of microbial species coexistence and multitrophic interaction in an ecosystem while experimenting to study the effects of climate change. An integrated research approach is required to analyze the evolutionary and ecological potential of microorganisms which can help to sustain ecosystem functioning, conservation, and services.

26.5.4 Microbial Evolution with Respect to Climate Change: Past and Future

Microorganisms have the capability of evolving within short duration, which is possible due to their genetic plasticity in the form of their susceptibility to undergo rapid mutations and horizontal gene transfer (Reinold et al., 2019). Therefore, with the ever-changing environment, microbial life has been able to survive uninterruptedly on the planet for nearly 3.85 billion years which suggests that they are one of the most resistant and resilient group of living organisms with respect to climate change and other environmental stress (Cavalier-Smith et al. 2006). They get adapted to extreme environmental changes not only by their genomic changes but also by altered physiology (Collins et al. 2014). For example, during drought conditions for maintaining the volume and fluid balance, some microbes synthesize osmolytes, using carbon, nitrogen, hydrogen, and oxygen (Schimel et al. 2007). Similarly, formation of spores by the microbes in response to limited nutrient supply, development of antibiotic resistance in response to antimicrobial agents, and activation of motility to facilitate movement to more beneficial environments are some other physiological adaptations employed by the microbes to combat environmental stress (Gardiner et al. 2004). However, such physiological adaptations of the microbes during the stress conditions may cause large shifts in fate of C and N cycling in the ecosystem (Schimel et al. 2007). Current research trends mostly focus on the structural or functional dynamics of the community in the soil microbiome in response to climate change; however, the possibility of evolving new microbe under the change is mostly neglected. Most of the climate change studies rely on manipulative experiments under controlled conditions, which lack the multitude of complex environmental factors of a natural setting, limiting the capacity to properly study the resistance, resilience, or evolution pattern of microbial communities. Therefore, future research should focus on this aspect, as the microbial evolution will not affect their own community structure and function, but also, they will be able to give information about the response of more slowly evolving organisms with climate (McCalley et al. 2014).

26.5.5 Mitigating Climate Change by Understanding Microbial Processes

Microbial processes have a central role in the exchange of carbon, nitrogen, and sulfur between the lithosphere, hydrosphere, biosphere, and atmosphere, and thus, they are the prime regulators of the global greenhouse gas flux into the atmosphere. Microbes are likely to respond rapidly to climate change, but whether the changes in microbial processes lead to a net positive or negative feedback for greenhouse gas emissions is unclear (Singh et al. 2010). However, some studies have revealed the potential of microbes to mitigate anthropogenic climate change by manipulating the terrestrial ecosystems (Singh et al. 2010). As soil stores twice the amount of carbon present in the atmosphere and thrice the quantity found in the vegetation (IPCC 2007; Smith 2004), scientific land use management can sequester huge amount of carbon into the soil (Houghton 2007; Smith 2004). However, in order to achieve this, it is important to understand the ecosystem functions and services of diverse microbial groups. Studies have shown that promoting fungal abundance in soil relative to bacteria favors effective carbon sequestration, and this is the reason for which forest soil can store huge amount of carbon for a long term (Bailey et al. 2002; Busse et al. 2009; Castro et al. 2010; De Deyn et al. 2008). In agricultural fields, reduced or no tillage practice can promote soil microbial communities dominated by fungi (Castro et al. 2010). Similarly methane emissions from terrestrial ecosystems can be managed by promoting the activity of methanotrophs or reducing the activity of methanogens which can be done by improving soil aeration (Smith et al. 2000; Tate et al. 2007). Paddy cultivations account for substantial release of methane into the atmosphere, which can be regulated by increasing oxygen availability in soils by improved management of flooding frequency and duration (Yagi et al. 1996). Furthermore, as the sulfate reducers and methanogens compete with each other, for resources under similar environmental conditions, growth of sulfate reducers can be promoted at the expense of methanogens by using ammonium sulfate fertilizers (Neue 1997). Similarly, emission of another anthropogenic greenhouse gas N_2O can be reduced by reducing the availability of nitrogen to microorganisms, which can be achieved by targeted fertilizer applications by understanding the crop requirements (Singh et al. 2010).

In order to realize the real potential of soils microbes to mitigate climate change at a global scale, we need to improve our understanding further about how the soil microbial community respond to as well as interact with other biotic components such as plants and with abiotic components such as climatic variables and soil physiochemical properties in multifactorial experimental conditions which is currently limited.

26.6 Conclusion

The signs of current climatic perturbation and future long-term change are undeniable. Although efforts to develop predictive global climate change models are already in process, these models generally underestimate the role microbial processes in climate change due to which the power of these predictive models are weakened. With increasing number of scientific evidences, it can be safely concluded that microbes act like both the sides of a coin in the context of climate change. They may either help in the mitigation of climate change or may exacerbate effect to the climate change through positive feedback mechanisms. Therefore, the future of climate change and their effect on the ecosystem function is yet uncertain. To overcome this uncertainty, there is an urgent need to gain a holistic understanding about the biogeochemical cycling of elements in the ecosystem mediated by soil microorganisms and to integrate this information into climatic models to make better predictions of climate impacts. Finally, it is ultimately necessary to develop microbial strategies to mitigate further climate change and ecosystem degradation.

References

Allison S, Treseder K (2008) Warming and drying suppress microbial activity and carbon cycling in boreal forest soils. Glob Chang Biol 14:2898–2909

Amundson R, Berhe AA, Hopmans JW et al (2015) Soil and human security in the 21st century. Science 348:1261071

Arnone JA III, Bohlen PJ (1998) Stimulated N_2O flux from intact grassland monoliths after two growing seasons under elevated atmospheric CO_2. Oecologia 116:331–335

Arshad M, Hussain S, Saleem M (2008) Optimization of environmental parameters for biodegradation of alpha and beta endosulfan in soil slurry by *Pseudomonas aeruginosa*. J Appl Microbiol 104:364–370

Austin EE, Castro HF, Sides KE (2009) Assessment of 10 years of CO_2 fumigation on soil microbial communities and function in a sweetgum plantation. Soil Biol Biochem 41:514–520

Bahram M, Hildebrand F, Forslund SK et al (2018) Structure and function of the global topsoil microbiome. Nature 560:233–237

Bailey VL, Smith JL, Bolton H Jr (2002) Fungal-to-bacterial ratios in soils investigated for enhanced C sequestration. Soil Biol Biochem 34:997–1007

Bardgett R, Freeman C, Ostle N (2008) Microbial contributions to climate change through carbon cycle feedbacks. ISME J 2:805–814

Bardgett RD, Manning P, Morriën E et al (2013) Hierarchical responses of plant-soil interactions to climate change: consequences for the global carbon cycle. J Ecol 101:334–343

Barnosky A, Matzke N, Tomiya S et al (2011) Has the Earth's sixth mass extinction already arrived? Nature 471:51–57

Bar-On YM, Phillips R, Milo R (2018) The biomass distribution on earth. Proc Natl Acad Sci U S A 115:6506–6511

Bartdorff O, Wallmann K, Latif M et al (2008) Phanerozoic evolution of atmospheric methane. Global Biogeochem Cycles 22:GB1008

Behrenfeld M (2014) Climate-mediated dance of the plankton. Nature Clim Change 4:880–887

Bender SF, Wagg C, van der Heijden MGA (2016) An underground revolution: biodiversity and soil ecological engineering for agricultural sustainability. Trends Ecol Evol 31:440–452

Birch E (2014) A review of "climate change 2014: impacts, adaptation, and vulnerability" and "climate change 2014: mitigation of climate change". J Am Plan Assoc 80:184–185

Boer W, Folman L, Summerbell R et al (2005) Living in a fungal world: impact of fungi on soil bacterial niche development. FEMS Microbiol Rev 29:795–811

Bonkowski M (2004) Protozoa and plant growth: the microbial loop in soil revisited. New Phytol 162:617–631

Bromwich D, Nicolas J, Monaghan A et al (2013) Central West Antarctica among the most rapidly warming regions on earth. Nat Geosci 6:139–145

Brown JR, Blankinship JC, Niboyet A et al (2012) Effects of multiple global change treatments on soil N2O fluxes. Biogeochemistry 109:85–100

Busse MD, Sanchez FG, Ratcliff AW et al (2009) Soil carbon sequestration and changes in fungal and bacterial biomass following incorporation of forest residues. Soil Biol Biochem 41:220–227

Canfield DE, Glazer AN, Falkowski PG (2010) The evolution and future of Earth's nitrogen cycle. Science 330(6001):192–196

Carol Adair E, Reich PB, Trost JJ et al (2011) Elevated CO_2 stimulates grassland soil respiration by increasing carbon inputs rather than by enhancing soil moisture. Glob Chang Biol 17:3546–3563

Castro H, Classen A, Austin E et al (2009) Soil microbial community responses to multiple experimental climate change drivers. Appl Environ Microbiol 76:999–1007

Castro HF, Classen AT, Austin EE (2010) Soil microbial community responses to multiple experimental climate change drivers. Appl Environ Microbiol 76:999–1007

Cavalier-Smith T, Brasier M, Embley TM (2006) Introduction: how and when did microbes change the world? Philos Trans R Soc Lond Ser B Biol Sci 361:845–850

Cavicchioli R, Ripple WJ, Timmis KN et al (2019) Scientists' warning to humanity: microorganisms and climate change. Nat Rev Microbiol 17:569–586

Classen AT, Sundqvist MK, Henning JA et al (2015) Direct and indirect effects of climate change on soil microbial and soil microbial-plant interactions: what lies ahead? Ecosphere 6(8):130

Codispoti LA (2010) Interesting times for marine N_2O. Science 327:1339–1340

Collins S, Rost B, Rynearson TA (2014) Evolutionary potential of marine phytoplankton under ocean acidification. Evol Appl 7:140–155

Crist E, Mora C, Engelman R (2017) The interaction of human population, food production, and biodiversity protection. Science 356:260–264

De Bona FD, Bayer C, Dieckow J et al (2008) Soil quality assessed by carbon management index in a subtropical acrisol subjected to tillage systems and irrigation. Aust J Soil Res 46:469–475

De Deyn GB, Cornelissen JHC, Bardgett RD (2008) Plant functional traits and soil carbon sequestration in contrasting biomes. Ecol Lett 11:516–531

de Vries FT, Liiri ME, Bjørnlund L et al (2012) Land use alters the resistance and resilience of soil food webs to drought. Nat Clim Chang 2:276–280

de Vries FT, Griffiths RI, Bailey M et al (2018) Soil bacterial networks are less stable under drought than fungal networks. Nat Commun 9:3033

Demanèche S, Philippot L, David MM et al (2009) Characterization of denitrification gene clusters of soil bacteria via a metagenomic approach. Appl Environ Microbiol 75:534–537

Demoulin CF, Lara YJ, Cornet L et al (2019) Cyanobacteria evolution: insight from the fossil record. Free Radic Biol Med 140:206–223

Dore JE, Popp BN, Karl DM et al (1998) A large source of atmospheric nitrous oxide from subtropical North Pacific surface waters. Nature 396:63–66

Dunbar J, Eichorst SA, Gallegos-Graves LV et al (2012) Common bacterial responses in six ecosystems exposed to 10 years of elevated atmospheric carbon dioxide. Environ Microbiol 14:1145–1158

Dutta H, Dutta A (2016) The microbial aspect of climate change. Energ Ecol Environ 1:209–232

Evans SE, Wallenstein MD (2014) Climate change alters ecological strategies of soil bacteria. Ecol Lett 17:155–164

Farenhorst A (2006) Importance of soil organic matter fractions in soil landscape and regional assessments of pesticide sorption and leaching in soil. Soil Sci 70:1005–1012

Friedlingstein P, Cox P, Betts R et al (2006) Climate-carbon cycle feedback analysis: results from the C^4MIP model Intercomparison. J Clim 19(14):3337–3353

Gabric AJ, Whetton PH, Cropp R (2001) Dimethylsulphide production in the subantarctic southern ocean under enhanced greenhouse conditions. Tellus B 53:273–287

Gans J, Wolinsky M, Dunbar J (2005) Computational improvements reveal great bacterial diversity and high metal toxicity in soil. Science 309:1387–1390

Gardiner GE, Casey PG, Casey G et al (2004) Relative ability of orally administered *Lactobacillus murinus* to predominate and persist in the porcine gastrointestinal tract. Appl Environ Microbiol 70:1895–1906

Gilichinsky D, Rivkina E, Bakermans C et al (2005) Biodiversity of cryopegs in permafrost. FEMS Microbiol Ecol 53:117–128

Godfray HC, Beddington JR, Crute IR et al (2010) Food security: the challenge of feeding 9 billion people. Science 327:812–818

Goel R, Kumar V, Suyal DC et al (2018) Towards the unculturable microbes for sustainable agricultural production. In: Meena VS (ed) Role of Rhizospheric microbes in soil. Springer, Singapore, pp 107–123

Grandey B, Wang C (2015) Enhanced marine Sulphur emissions offset global warming and impact rainfall. Sci Rep 5:13055

Greaver TL, Clark CM, Compton JE et al (2016) Key ecological responses to nitrogen are altered by climate change. Nat Clim Chang 6:836–843

Griggs D, Stafford-Smith M, Gaffney O et al (2013) Sustainable development goals for people and planet. Nature 495:305–307

Hao OJ (2003) Sulphate-reducing bacteria. In: Mara D, Horan N (eds) The handbook of water and wastewater microbiology. Academic Press, London, pp 459–469

Haugwitz M, Bergmark L, Prieme A et al (2013) Soil microorganisms respond to five years of climate change manipulations and elevated atmospheric CO_2 in a temperate heath ecosystem. Plant Soil 374:211–222

Hawkes CV, Hartley IP, Ineson P et al (2008) Soil temperature affects carbon allocation within arbuscular mycorrhizal networks and carbon transport from plant to fungus. Glob Chang Biol 14:1181–1190

Hayden HL, Mele PM, Bougoure DS et al (2012) Changes in the microbial community structure of bacteria, archaea and fungi in response to elevated CO_2 and warming in an Australian native grassland soil. Environ Microbiol 14:3081–3096

He Z, Piceno Y, Deng Y et al (2012) The phylogenetic composition and structure of soil microbial communities shifts in response to elevated carbon dioxide. ISME J 6:259–272

Hooper DU, Chapin Iii FS, Ewel JJ et al (2005) Effects of biodiversity on ecosystem functioning: a consensus of current knowledge. Ecol Monogr 75:3–5

Hopple AM, Wilson RM, Kolton M et al (2020) Massive peatland carbon banks vulnerable to rising temperatures. Nat Commun 11:1–7

Horz H, Rich V, Avrahami S et al (2005) Methane-oxidizing Bacteria in a California upland grassland soil: diversity and response to simulated global change. Appl Environ Microbiol 71:2642–2652

Houghton RA (2007) Balancing the global carbon budget. Annu Rev Earth Planet Sci 35:313–347

Hultman J, Waldrop MP, Mackelprang R et al (2015) Multi-omics of permafrost, active layer and thermokarst bog soil microbiomes. Nature 521:208–212

Hussain S, Arshad M, Saleem M, Khalid A (2007a) Biodegradation of alpha- and beta-endosulfan by soil bacteria. Biodegradation 18:731–740

Hussain S, Arshad M, Saleem M, Zahir ZA (2007b) Screening of soil fungi for in vitro degradation of endosulfan. World J Microbiol Biotechnol 23:939–945

IPCC (2007) Intergovernmental panel on climate change. Climate change 2007: synthesis report. Contribution of working groups I, II and III to the fourth assessment report of the intergovernmental panel on climate change. Cambridge University Press, Cambridge, p 104

IPCC (2014) Climate phenomena and their relevance for future regional climate change. In: Climate change 2013 – the physical science basis: working group I contribution to the fifth assessment report of the intergovernmental panel on climate change. Cambridge University Press, Cambridge, pp 1217–1308

Islam W, Noman A, Naveed H et al (2020) Role of environmental factors in shaping the soil microbiome. Environ Sci Pollut Res. https://doi.org/10.1007/s11356-020-10471-2

Itakura M, Uchida Y, Akiyama H et al (2013) Mitigation of nitrous oxide emissions from soils by Bradyrhizobium japonicum inoculation. Nat Clim Chang 3:208–212

Jacoby R, Peukert M, Succurro A et al (2017) The role of soil microorganisms in plant mineral nutrition—current knowledge and future directions. Front Plant Sci 8:1617

Jansson JK, Hofmockel KS (2020) Soil microbiomes and climate change. Nat Rev Microbiol 18:35–46

Jansson JK, Taş N (2014) The microbial ecology of permafrost. Nat Rev Microbiol 12:414–425

Jensen K, Beier C, Michelsen A et al (2003) Effects of experimental drought on microbial processes in two temperate heathlands at contrasting water conditions. Appl Soil Ecol 24:165–176

Johnson C, Balmford A, Brook B et al (2017) Biodiversity losses and conservation responses in the Anthropocene. Science 356:270–275

Jones DL, Nguyen C, Finlay RD (2009) Carbon flow in the rhizosphere: carbon trading at the soil–root interface. Plant Soil 321:5–33

Khalvati M, Hu Y, Mozafar A et al (2005) Quantification of water uptake by arbuscular mycorrhizal hyphae and its significance for leaf growth, water relations, and gas exchange of barley subjected to drought stress. Plant Biol 7:706–712

Koch PL (1998) Isotopic reconstruction of past continental environments. Annu Rev Earth Planet Sci 26:573–613

Kolb S, Carbrera A, Kammann C et al (2005) Quantitative impact of CO_2 enriched atmosphere on abundances of methanotrophic bacteria in a meadow soil. Biol Fertil Soils 41(5):337–342

Kuzyakov Y, Blagodatskaya E (2015) Microbial hotspots and hot moments in soil: concept & review. Soil Biol Biochem 83:184–199

Lagomarsino A, Moscatelli MC, Hoosbeek MR et al (2008) Assessment of soil nitrogen and phosphorus availability under elevated CO_2 and N-fertilization in a short rotation poplar plantation. Plant Soil 308:131

Lam P, Lavik G, Jensen MM (2009) Revising the nitrogen cycle in the Peruvian oxygen minimum zone. Proc Natl Acad Sci U S A 106:4752–4757

Lareen A, Burton F, Schäfer P (2016) Plant root-microbe communication in shaping root microbiomes. Plant Mol Biol 90:575–587

Lesaulnier C, Papamichail D, McCorkle S et al (2008) Elevated atmospheric CO_2 affects soil microbial diversity associated with trembling aspen. Environ Microbiol 10:926–941

Levasseur M (2011) If Gaia could talk. Nat Geosci 4:351–352

Lipper L, Thornton P, Campbell BM et al (2014) Climate-smart agriculture for food security. Nat Clim Chang 4:1068

Liu Y, Liu S, Wan S et al (2016) Differential responses of soil respiration to soil warming and experimental throughfall reduction in a transitional oak forest in Central China. Agric For Meteorol 226:186–198

Lladó S, López-Mondéjar R, Baldrian P (2017) Forest soil bacteria: diversity, involvement in ecosystem processes, and response to global change. Microbiol Mol Biol Rev 81:e00063

Lynd LR, Weimer PJ, van Zyl WH et al (2002) Microbial cellulose utilization: fundamentals and biotechnology. Microbiol Mol Biol Rev 66:506–577

Madsen EL (2011) Microorganisms and their roles in fundamental biogeochemical cycles. Curr Opin Biotechnol 22:456–464

McCalley CK, Woodcroft BJ, Hodgkins SB et al (2014) Methane dynamics regulated by microbial community response to permafrost thaw. Nature 514:478–481

Melillo JM, Frey SD, DeAngelis KM (2017) Long-term pattern and magnitude of soil carbon feedback to the climate system in a warming world. Science 358:101–105

Mohanty S, Bodelier P, Conrad R (2007) Effect of temperature on composition of the methanotrophic community in rice field and forest soil. FEMS Microbiol Ecol 62:24–31

Mondav R, Woodcroft B, Kim E et al (2014) Discovery of a novel methanogen prevalent in thawing permafrost. Nat Commun 5:3212

Naylor D, Sadler N, Bhattacharjee A et al (2020) Soil microbiomes under climate change and implications for carbon cycling. Annu Rev Environ Resour 45:9.1–9.31

Nazaries L, Murrell JC, Millard P et al (2013) Methane, microbes and models. Environ Microbiol 15:2395–2417

Neue HU (1997) Fluxes of methane from rice fields and potential for mitigation. Soil Use Manag 13:258–267

Niboyet A, Le Roux X, Dijkstra P et al (2011) Testing interactive effects of global environmental changes on soil nitrogen cycling. Ecosphere 2:art56

Niederberger TD, Sohm JA, Tirindelli J et al (2012) Diverse and highly active diazotrophic assemblages inhabit ephemerally wetted soils of the Antarctic dry valleys. FEMS Microbiol Ecol 82:376–390

Nielsen MN, Winding A (2002) Microorganisms as indicators of soil health. National Environmental Research Institute, Denmark. Technical Report No. 388

Niklaus PA, Kandeler E, Leadley PW et al (2001) A link between plant diversity, elevated CO2 and soil nitrate. Oecologia 127:540–548

Nobu MK, Narihiro T, Kuroda K et al (2016) Chasing the elusive Euryarchaeota class WSA2: genomes reveal a uniquely fastidious methyl-reducing methanogen. ISME J 10:2478–2487

Norby RJ, De Kauwe MG, Domingues TF et al (2016) Model–data synthesis for the next generation of forest free-air CO2 enrichment (FACE) experiments. New Phytol 209:17–28

Overmann J, Garcia-Pichel F (2006) The phototrophic way of life. In: Dworkin M, Falkow S, Rosenberg E, Schleifer KH, Stackebrandt E (eds) The prokaryotes. Springer, New York, pp 32–85

Pace NR (2018) The small things can matter. PLoS Biol 16(8):e3000009

Pecl GT, Araújo MB, Bell JD et al (2017) Biodiversity redistribution under climate change: impacts on ecosystems and human well-being. Science 355(6332):eaai9214

Pester M, Knorr KH, Friedrich MW et al (2012) Sulfate-reducing microorganisms in wetlands–fameless actors in carbon cycling and climate change. Front Microbiol 3:72

Piña-Ochoa E, Høgslund S, Geslin E et al (2010) Widespread occurrence of nitrate storage and denitrification among foraminifera and Gromiida. Proc Natl Acad Sci U S A 107:1148–1153

Pinay G, Barbera P, Carreras-Palou A et al (2007) Impact of atmospheric CO_2 and plant life forms on soil microbial activities. Soil Biol Biochem 39(1):33–42

Pugnaire FI, Morillo JA, Peñuelas J et al (2019) Climate change effects on plant-soil feedbacks and consequences for biodiversity and functioning of terrestrial ecosystems. Sci Adv 5:eaaz1834

Regan K, Kammann C, Hartung K et al (2011) Can differences in microbial abundances help explain enhanced N_2O emissions in a permanent grassland under elevated atmospheric CO_2? Glob Chang Biol 17:3176–3186

Reinold M, Wong HL, MacLeod FI et al (2019) The vulnerability of microbial ecosystems in a changing climate: potential impact in Shark Bay. Life 9:71

Ripple WJ, Smith P, Haberl H et al (2014) Ruminants, climate change and climate policy. Nat Clim Chang 4:2–5

Rivkina E, Gilichinsky D, Wagener S et al (1998) Biogeochemical activity of anaerobic microorganisms from buried permafrost sediments. Geomicrobiol J 15:187–193

Robertson LA, Kuenen JG (2006) The colorless sulfur bacteria. In: Dworkin M, Falkow S, Rosenberg E, Schleifer KH, Stackebrandt E (eds) The prokaryotes. Springer, New York, pp 985–1011

Roesch L, Fulthorpe R, Riva A et al (2007) Pyrosequencing enumerates and contrasts soil microbial diversity. ISME J 1:283–290

Romero-Olivares AL, Allison SD, Treseder KK (2017) Soil microbes and their response to experimental warming over time: a meta-analysis of field studies. Soil Biol Biochem 107:32–40

Rosenberg K, Bertaux J, Krome K et al (2009) Soil amoebae rapidly change bacterial community composition in the rhizosphere of *Arabidopsis thaliana*. ISME J 3:675–684

Ruiz-Rueda O, Hallin S, Baneras L (2009) Structure and function of denitrifying and nitrifying bacterial communities in relation to the plant species in a constructed wetland. FEMS Microbiol Ecol 67:308–319

Saccá ML, Caracciolo AB, Di Lenola M et al (2017) Ecosystem services provided by soil microorganisms. In: Lukac M, Grenni P, Gamboni M (eds) Soil biological communities and ecosystem resilience, sustainability in plant and crop protection. Springer, Cham, pp 9–24

Saleem M, Pervaiz Z, Traw M (2015) Theories, mechanisms and patterns of microbiome species coexistence in an era of climate change. In: Saleem M (ed) Microbiome community ecology. Springer briefs in ecology. Springer, Cham, pp 13–53

Sardans J, Penuelas J, Estiarte M (2008) Changes in soil enzymes related to C and N cycle and in soil C and N content under prolonged warming and drought in a Mediterranean shrub land. Appl Soil Ecol 39:223–235

Schimel J (2004) Playing scales in the methane cycle: from microbial ecology to the globe. Proc Natl Acad Sci U S A 101(34):12400–12401

Schimel J, Balser TC, Wallenstein M (2007) Microbial stress-response physiology and its implications for ecosystem function. Ecology 88(6):1386–1394

Schindlbacher A, Rodler A, Kuffner M et al (2011) Experimental warming effects on the microbial community of a temperate mountain forest soil. Soil Biol Biochem 43:1417–1425

Schlesinger WH (2009) On the fate of anthropogenic nitrogen. Proc Natl Acad Sci U S A 106:203–208

Schuur EA, McGuire AD, Schädel C et al (2015) Climate change and the permafrost carbon feedback. Nature 520:171–179

Shindell DT, Faluvegi G, Koch DM et al (2009) Improved attribution of climate forcing to emissions. Science 326:716–718

Singh B, Bardgett R, Smith P et al (2010) Microorganisms and climate change: terrestrial feedbacks and mitigation options. Nat Rev Microbiol 8:779–790

Six J, Frey SD, Thiet RK et al (2006) Bacterial and fungal contributions to carbon sequestration in agroecosystems. Soil Sci Soc Am J 70:555

Six KD, Kloster S, Ilyina T (2013) Global warming amplified by reduced Sulphur fluxes as a result of ocean acidification. Nat Clim Chang 3:975–978

Smith P (2004) Soils as carbon sinks: the global context. Soil Use Manag 20:212–218

Smith KA, Dobbie KE, Ball BC et al (2000) Oxidation of atmospheric methane in northern European soils, comparison with other ecosystems, and uncertainties in the global terrestrial sink. Glob Chang Biol 6:791–803

Stefels J, Steinke M, Turner S et al (2007) Environmental constraints on the production and removal of the climatically active gas dimethylsulphide (DMS) and implications for ecosystem modelling. Biogeochemistry 83:245–275

Steven B, Pollard WH, Greer CW et al (2008) Microbial diversity and activity through a permafrost/ ground ice core profile from the Canadian high Arctic. Environ Microbiol 10:3388–3403

Strous M, Pelletier E, Mangenot S et al (2006) Deciphering the evolution and metabolism of an anammox bacterium from a community genome. Nature 440:790–794

Tate KR, Ross DJ, Saggar S et al (2007) Methane uptake in soils from *Pinus radiata* plantations, a reverting shrubland and adjacent pastures: effects of land-use change, and soil texture, water and mineral nitrogen. Soil Biol Biochem 39:1437–1449

Thakur MP, Geisen S (2019) Trophic regulations of the soil microbiome. Trends Microbiol 27:771–780

Thamdrup B, Finster K, Hansen JW et al (1993) Bacterial disproportionation of elemental sulfur coupled to chemical reduction of Iron or manganese. Appl Environ Microbiol 59:101–108

Trivedi P, Delgado-Baquerizo M, Anderson IC et al (2016) Response of soil properties and microbial communities to agriculture: implications for primary productivity and soil health indicators. Front Plant Sci 7:990

Trotsenko YA, Khmelenina VN (2005) Aerobic methanotrophic bacteria of cold ecosystems. FEMS Microbiol Ecol 53:15–26

Tveit AT, Urich T, Frenzel P et al (2015) Metabolic and trophic interactions modulate methane production by Arctic peat microbiota in response to warming. Proc Natl Acad Sci U S A 112: E2507–E2516

van Groenigen KS, van Kessel C, Hungate BA (2013) Increased greenhouse-gas intensity of rice production under future atmospheric conditions. Nat Clim Chang 3:288–291

van Herwijnen R, Laverye T, Poole J et al (2007) The effect of organic materials on the mobility and toxicity of metals in contaminated soils. Appl Geochem 22:2422–2434

Van Horn DJ, Okie JG, Buelow HN et al (2014) Soil microbial responses to increased moisture and organic resources along a salinity gradient in a polar desert. Appl Environ Microbiol 80:3034–3043

Vanwonterghem I, Evans PN, Parks DH et al (2016) Methylotrophic methanogenesis discovered in the archaeal phylum Verstraetearchaeota. Nat Microbiol 1:16170

Wan S, Norby RJ, Ledford J et al (2007) Responses of soil respiration to elevated CO_2, air warming, and changing soil water availability in a model old-field grassland. Glob Chang Biol 13:2411–2424

Wang S, Maltrud M, Elliott S et al (2018) Influence of dimethyl sulfide on the carbon cycle and biological production. Biogeochemistry 138:49–68

Weiman S (2015) Microbes help to drive global carbon cycling and climate change. Microbe Mag 10:233–238

Wieder W, Bonan G, Allison S (2013) Global soil carbon projections are improved by modelling microbial processes. Nature Clim Change 3:909–912

Yagi K, Tsuruta H, Kanda KI et al (1996) Effect of water management on methane emission from a Japanese rice paddy field: automated methane monitoring. Global Biogeochem Cycles 10:255–267

Yang S, Zheng Q, Yuan M et al (2019) Long-term elevated CO_2 shifts composition of soil microbial communities in a Californian annual grassland, reducing growth and N utilization potentials. Sci Total Environ 652:1474–1481

Yergeau E, Kang S, He Z et al (2007) Functional microarray analysis of nitrogen and carbon cycling genes across an Antarctic latitudinal transect. ISME J 1:163–179

Yergeau E, Hogues H, Whyte LG et al (2010) The functional potential of high Arctic permafrost revealed by metagenomic sequencing, qPCR and microarray analyses. ISME J 4:1206–1214

Yu H, Deng Y, He Z (2018) Elevated CO_2 and warming altered grassland microbial communities in soil top-layers. Front Microbiol 9:1790

Zak DR, Pregitzer KS, Curtis PS et al (2000) Atmospheric CO_2 and the composition and function of soil microbial communities. Ecol Appl 10:47–59

Zerkle AL, Mikhail S (2017) The geobiological nitrogen cycle: from microbes to the mantle. Geobiology 15:343–352

Zhang Z, Zimmermann NE, Stenke A et al (2017) Emerging role of wetland methane emissions in driving 21st century climate change. Proc Natl Acad Sci U S A 36:9647–9652

Chapter 27
Climate Change and Nutrients Dynamics of Soil

Ranjana Singh, Kajal Patel, and Meenakshi Chaurasia

Abstract Recent climate change scenario across the globe has evident impact on ecosystem structure and functioning. Pervasive alteration in climatic factors like rise in temperature and change in precipitation regime affects the potential of terrestrial system to act as a source or sink. In this context, soil processes controlled by complex interaction between plant system, soil microbial community, and abiotic factors play a decisive role. Microbial communities are the mainspring for key terrestrial processes such as decomposition, mineralization, and nutrient allocation between different ecosystems and play a critical role in regulating nutrient dynamics. Simultaneously, soil microbiome may contribute immensely to climate change through their feedback responses. Climate change drivers such as increased temperatures, changed precipitation pattern, atmospheric CO_2 deposition, and soil erosion affect soil microbial community composition, physiological and metabolic activities, which can govern the soil nutrients storage, its properties (viz, pH, cation exchange capacity, texture, etc.), and other ecosystem processes. Although the direction and magnitude of the responses of microbes to climate change is uncertain, i.e., it could be both positive and negative as well. These processes and transformations involved in the macronutrient cycling between the atmosphere and terrestrial system are poorly understood and hence require attention to adapt and mitigate the adverse impact of climate changes. This chapter discusses the impact of different climate scenarios on the microbial communities and the process mediated by them; most importantly, the dynamics and transformation of various nutrients like carbon (C), nitrogen (N), and phosphorus (P) at soil interface and their significance in regulation of ecosystem balance and ultimately the future prospects of microbes for mitigating climate changes.

R. Singh (✉)
Government Model Degree College, Arniya, Bulandshahr, Uttar Pradesh, India

K. Patel · M. Chaurasia
Natural Resource Management Laboratory, Department of Botany, University of Delhi, New Delhi, India

D. K. Choudhary et al. (eds.), *Climate Change and the Microbiome*, Soil Biology 63,
https://doi.org/10.1007/978-3-030-76863-8_27

Keywords Climate change · Direct and indirect impacts · Nutrient dynamics · Feedback mechanism · Soil-microbe and mitigation

27.1 Introduction

Climate change and global warming are the two main environmental issues extensively discussed in the present environmental scenario. The continuous rise in greenhouse gases (GHGs) through both anthropogenic activities such as industrialization, urbanization, intensive agricultural activities and several natural processes like volcanic eruptions, forest fires, changes in vegetation cover, snow cover, and so on leads to the changes in global environment at unprecedented rates. Various environmental changes include the changes in gas composition of the atmosphere, spatial and temporal changes in global temperatures, and changes in the precipitation pattern (Várallyay 2010). The Earth's average temperature raised by 1.5 °F in the last century and it has been expected to surge further by 0.5–8.6 °F within 100 years. According to Intergovernmental Panel on Climate Change (IPCC 2001), a slight fluctuation in average global temperature can be a basis of environmental issue as it can cause severe shifts in climate and weather pattern.

The soil system including aboveground plants and belowground microbial communities is considered to be a massive pool of C dynamic and hence decisively shape the trend of climate change. Soil contains around 3.1×10^5 kg of C that is more than two-third of total C present in the terrestrial ecosystem (Davidson and Janssens 2006). The exchange of carbon dioxide (CO_2) between soil and atmosphere is a major part of global carbon cycle (Raich and Potter 1995), where the terrestrial ecosystem plays an important role. In this context, microbial communities hold great importance for maintaining nutrient dynamics and ecosystem functioning. Microbes, being the mainstay of ecological processes, are crucial for C, N, and other biogeochemical cycles and also in structuring plant communities in a particular ecosystem. Soil microbes are found to regulate the emission and removal of GHGs such as CO_2, methane (CH_4), and nitrous oxide (N_2O). Photosynthetic microorganisms remove atmospheric CO_2 while the heterotrophic microbes emit greenhouse gases into the atmosphere by decomposing the organic matter. However, the balance between these two processes is crucial as it is the key determinant of feedback mechanism of microbes in response to climate changes that may vary with the various ecosystems (Singh et al. 2010; Weiman 2015). Nevertheless, it is worth mentioning here that about 1.2×10^{14} kg of C is fixed annually by autotrophic soil microbes while around 1.19×10^{14} kg of C is emitted per year into the atmosphere by heterotrophic soil microbes through the decomposition of organic matters (Singh et al. 2010).

Soil microbial communities' structure and functioning are affected either directly or indirectly through various physiological processes of plant and their community composition (Castro et al. 2010; Bardgett et al. 2013) in response to climate changes. In fact, climate change modifies a number of factors simultaneously, which lead to complex change in soil microbial communities in terrestrial ecosystems and determine the future form and functioning of the ecosystem (Castro et al. 2010). The response of microbial community to individual and interactive climate change factors is very complex and our knowledge about the mechanism through which climate change affects soil microbial communities and its functioning is still very poor. It has been reported that the indirect effect of climate change mediated through plants may be stronger than the direct effect of climate change on soil microbial structure and function. In the recent years, ongoing researches have highlighted the impacts of global climate change and drivers of these changes on microbial community although their results are mostly contradictory (Luo et al. 2009). Plants are important biotic factors that play an important role in shaping structure and function of soil microbial community. As soils are secured from the direct effect of climate conditions, the structure of associated soil microbial communities is indirectly affected by climatic changes through the plants (Kardol et al. 2010) and depends mostly on the effect of climatic conditions on aboveground vegetation (Fierer and Jackson 2006). Plant secretes carbon-rich exudates and induces polyphenols which in turn affects the structure and activity of decomposers in soil. Plant–microbial interaction is also affected by climate change which in turn alters the assemblages of rhizospheric microbial community (Classen et al. 2015).

The C, N, and other biogeochemical cycles are regulated and mediated by microbial communities present in soil. These processes are intricate and very sensitive. A slight fluctuation in climatic conditions, including rise in temperature and CO_2 level, alteration in moisture regime, addition of exogenous organic matter, change in pool of labile and recalcitrant organic matter in soil, may affect greatly the nutrient dynamics of soil by impacting directly or indirectly the structural composition and functional activities of both the soil microbial communities as well as aboveground vegetation (Walker et al. 2006; Hoeppner and Dukes 2012). For example, C and N cycling is affected by elevated CO_2 and high temperature (Dutta and Dutta 2016). Higher temperatures and CO_2 level in soil increase net plant productivity that consequently releases more organic compounds from plant roots to decomposers (Trumbore 1997). Indirect effect of climate change on the microbial community mediated through the plants alters the dominance, diversity, growth, and activities of soil microorganisms that also depend on the availability of nutrients like N and CO_2 level in soil (Bardgett et al. 2009). High temperature (global warming) is expected to increase nutrient content in soil by ameliorating mineralization of N and organic matter by soil microbial communities which in turn increase plant productivity and affects microbial activities again (Hobbie 1996; Ruess et al. 1999).

Hence, the roles and responses of microorganisms with respect to climate change need to be highlighted (Balser et al. 2006; Walsh 2015). However, due to lack of adequate understanding concerning the significance of microbial flora and associated activities, microbes have not gained proper attention in climate change management and models (Walsh 2015). Considering all facts in mind, through this chapter we have tried to unveil the potential roles and responses of soil microbial communities to climate conditions and feedback mechanisms, and overall impact of climate change on nutrient dynamics of soil (Fig. 27.1). Eventually, we highlighted the

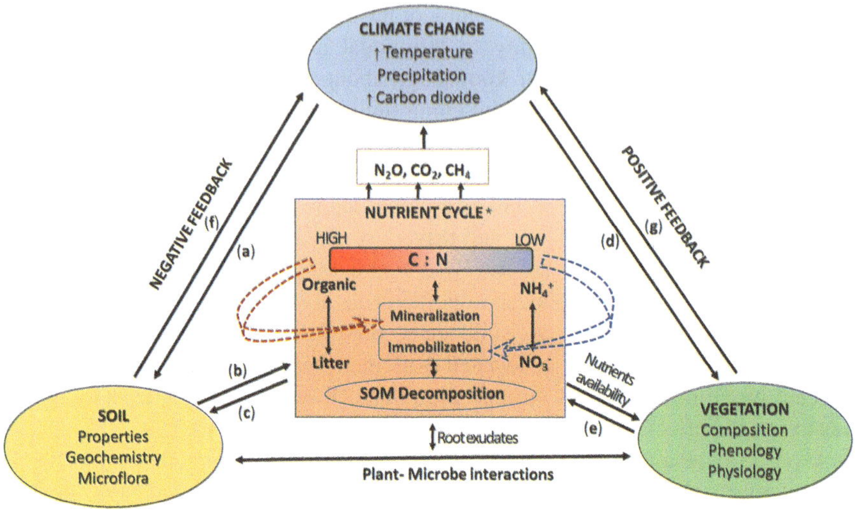

Fig. 27.1 Diagram illustrates the relationship between climate change (rise in temperature, CO_2, and shift in precipitation regime) and terrestrial ecosystem functioning. Global changes induce direct impact on soil microbial flora by influencing the soil properties, geochemistry, and soil microbiome (**a**). Changes in soil microbial activity and microbiome structure affects the soil processes like decomposition, mineralization, and immobilization (**b**) thus, ultimately influencing the net nutrient pool and flux in the ecosystem. Altered nutrient cycling affects the microbial and plant community and the interaction between them by regulating the nutrient availability and also contributes to the soil fertility status (**c**). The indirect impacts of climate change are mainly mediated by changes in diversity and composition of plant community and their physiology and phenology (**d**), which further influences the microbial flora by determining the quantity and quality of substrate input (**e**). Consecutively, shift in microbial flora and associated process may directly cause negative feedback on climatic changes through increased respiration (**f**). While, positive feedback is exerted by vegetation by increased C sequestration (**g**). (Asterisk (*) depicts soil component)

potential application of soil microbial communities and their activities in mitigation of climate changes.

27.2 Impact of Climate Changes

Climate change causes potential direct and indirect impacts on both aboveground and belowground terrestrial ecosystems. The belowground microbial communities comprise the most important component of the terrestrial ecosystem. The significance of soil microbial flora is advocated by their role to sustain ecosystem functioning through biogeochemical cycles, plant growth, and bioremediation which helps in regulating the concentration of various nutrients and gases in the ecosystem (Naylor et al. 2020). Soil microbial flora is the most crucial component and prerequisite for management of C and N fluxes in the ecosystem and plays an important role in balancing the process of emission and storage of GHGs like CO_2, methane

(CH_4), and nitrous oxide (N_2O). For example, about 120 billion tons of atmospheric C is fixed by some specific autotrophic photosynthetic microbes annually while about 119 billion tons of C is emitted to the atmosphere by heterotrophic microbes annually through the degradation of the organic matter (Singh et al. 2010). Their function is regulated by various natural and anthropogenic factors such as temperature, precipitation, litter input, land use management, and others. Prodigious changes in these factors, in the recent time, have imposed serious impact on the microbial community of soil. Numerous studies have showed the negative impact of climatic perturbations on the microbial diversity, activity, and processing (Table 27.1) (Castro et al. 2010). Impact of climate change factors on microbial biomass, enzymatic activity, community profiling has been extensively studied (Janus et al. 2005; Kandeler et al. 1998). Of these changes, shift in the structure and composition of microbial communities is most recognizable, they also depend on various microbial processes such as respiration which in turn rely upon various components of the environment (Balser et al. 2010). Soil microflora play a crucial role with respect to climate change processes; however, their significant roles and responses to climate changes and ecosystem functioning remain neglected.

27.2.1 Impact of Climate Changes on Microbial Structure and Composition

There are diverse studies which focused on ecosystem responses to elevated CO_2, temperature, and precipitation which are major components of climate change. These researches reported a significant contribution of belowground plant root biomass and soil microbial flora associated with them for the complete responses. However, Crowther et al. (2015) observed an interrelationship between climate changes trends and soil microbial flora.

27.2.1.1 Elevated Temperature

It has been reported in various studies that rise in temperature causes alteration in soil microbial abundance, composition, and functioning by affecting microbial enzymatic activities, net primary production, availability of substrate pool, and water flux (Zogg et al. 1997; Pendall et al. 2004; von Lützow and Kögel-Knabner 2009). In a case study, Zogg et al. (1997) observed that gram-positive bacteria respond positively to temperature rise while the negative response was recorded from gram-negative bacteria. It was also observed that at higher temperature, microbes use labile pool for production of energy rather than readily available biomass which gradually results in decline of soil microbial biomass. Besides, elevated temperature is found to induce direct positive effects on arbuscular mycorrhizal (AM) colonization and development and the indirect effects include alteration in

Table 27.1 Studies for diverse impacts of climate change on soil microflora, microbial processes and nutrient dynamics in soil

S. no.	Soil component	Climatic factor	Impacts	References
1	Soil microflora	Elevated temperature	Decline of soil microbial biomass	Zogg et al. (1997)
			Positive effects on arbuscular mycorrhizal (AM) colonization and development	Fitter et al. (2000), Gavito et al. (2003)
			Decrease in fungal and bacterial biomarkers	Waldrop and Firestone (2004)
			Increases fungal abundance but relegates bacterial abundance	Castro et al. (2010)
			Replacement of psychrophiles by non-psychrophilic microbes	Monteux et al. (2018), Ricketts et al. (2020)
		Precipitation	In drought, abundance of Gram-positive while the Gram-negative bacterial population declines	Sylvia et al. (2005), Balser et al. (2010)
			In drought, increases fungal and bacterial diversity	Drenovsky et al. (2004), Jensen et al. (2003)
			In wet season, increase in soil microbial population	Waldrop and Firestone (2004)
			Changes in moistures cause considerable changes in dominance of species in fungal and bacterial communities	Classen et al. (2015)
		Elevated CO_2	Promotes growth of mycorrhiza and nitrogen-fixing microbes	Balser et al. (2010)
			Decline in diversity and population of microflora, resulting in higher fungal diversity and population than that of bacterial	He et al. (2012), Deng et al. (2012)
			Increase in both decomposer and AM fungal hyphae abundance	Klironomos et al. (1996), de Vries and Griffiths (2018)
			Proportion of gram-negative bacteria and other bacterial population increased	Montealegre et al. (2002)
2	Microbial processes	Short-term climate changes	Upgraded nutrients mineralization and altered soil enzyme activities	Zi et al. (2018)
		Low temperature	Favorable factor for soil enzymatic activities	Galloway et al. (2004)
		Elevated temperature	Supports enhanced soil enzyme activities	Baldrian et al. (2013)
			Elevated soil microbial activities and soil respiration	Balser et al. (2006)

(continued)

Table 27.1 (continued)

S. no.	Soil component	Climatic factor	Impacts	References
			Soil respiration shows positive correlation	Luo and Zhou (2006), Aanderud et al. (2013)
			Altered physiology of decomposers and enhance the fungal decomposition	Schindlbacher et al. (2011)
			Growth of belowground fungal hyphae decreases and decline in fungal respiration	Kratz (2014)
			Raise in rates to nitrogen cycling processes	Zhou et al. (2012)
			Accelerated carbon cycling and improved carbon sequestration	Li et al. (2019)
		Moisture content	Affects the soil respiratory activities and soil decomposition dynamics	Luo and Zhou (2006)
			Soil respiration shows negative correlation	Aanderud et al. (2013)
		Declined soil moisture	Induces carbon cycling thus, elevating the CO_2 flux	Singh et al. (2010)
			Alter microbial metabolic activities, decrease hydrolytic and peroxidase enzymes activities	Balser et al. (2010)
		Elevated CO_2	Variations in balance between decomposition and primary productivity	Paterson et al. (1997)
			Decomposition rates were enhanced	Melillo et al. (1982)
			Decline in net gross mineralization rate, greater ammonium immobilization	Hungate et al. (1999)
			Increase in microbial activities	Rice et al. (1994), Dhillion et al. (1995)
			Rise in microbial respiration	De Graaff et al. (2006)
3	Nutrient dynamics	Elevated temperature	Negative impacts on C allocation to soil	Gorissen et al. (2004)
			Increased N mineralization leading to increase in soil N content	Reich et al. (2006), Li et al. (2011)
			Induced a threefold increase in the leaching of inorganic N in soil	Kaste et al. (2006)
		Increased soil moisture	Increase in biological reactions of soil microflora, increased atmospheric N_2O levels	Brevik (2013)

(continued)

Table 27.1 (continued)

S. no.	Soil component	Climatic factor	Impacts	References
		High temperature, less moisture	Soil C got significantly reduced	Brevik (2013)
		Elevated CO_2 levels	Soil organic C decline	Carney et al. (2007)
			Limit others nutrients availability such as of N and P	Niklaus and Körner (2004)

photosynthetic rates, nutrients content, and moisture content (Fitter et al. 2000; Gavito et al. 2003; Pendall et al. 2004). It has been proposed that the fungal and bacterial biomarkers may decrease with projected increase in temperature (Waldrop and Firestone 2004). Contrary to this, Castro et al. (2010) showed that elevated temperature helps to increase fungal abundance but relegates bacterial abundance due to deficiency of labile substrate. In the Arctic tundra region of southern hemisphere which is cold, treeless area with low precipitation, changes in microbial communities in permafrost have been observed due to increasing temperature. Generally, this region is dominated by psychrophiles (adapted to survive in cold) but warming of environment causes their replacement by non-psychrophilic microbes. It is due to thawing of permafrost which results in unavailability of C sources and reduction in microbial growth and efficiency (Monteux et al. 2018; Ricketts et al. 2020).

Researchers across globe are trying to study the impact of climate changes specifically for temperature parameter on the soil microbial communities. Many of them reported that the sources of substrate or C pool utilized by microbes show variations with changes in temperature and thus effect their functions, biomass, and abundance (Ellert and Bettany 1992; Zogg et al. 1997; Andrews et al. 2000; Balser et al. 2010).

27.2.1.2 Precipitation

Another important element of climate change is precipitation, which may result in either flooding or drought condition and thus alter the moisture content in both atmosphere as well as in soil. In various studies, it has been found that the changes in rainfall can significantly affect the composition and functioning of soil microbial flora, like in drought-like conditions (lack of moisture) there is abundance of Firmicutes and Actinobacteria (Gram positive) while the Gram-negative bacterial population declines (Schimel et al. 1994; Sylvia et al. 2005; Drenovsky et al. 2004; Balser et al. 2010). According to Waldrop and Firestone (2004), in the wet season, soil restores its moisture level and causes increase in soil microbial population by improving their functioning. On the other hand, the decline in precipitation causes drying of soil subsequently increasing the fungal diversity, thus resulting in an increased fungi/bacterial community. Also, the soil C:N ratio available for microbial flora is found to be enhanced (Jansen et al. 2003; Drenovsky et al. 2004) which ultimately leads to a shift in soil microbial composition. Modifications in precipitation regime are quite significant as the moisture content is crucial for terrestrial soil microbial community structure. In fact, minute changes in soil moisture content can cause considerable changes in dominance of species in fungal and bacterial communities by improving C pool (Classen et al. 2015).

27.2.1.3 Elevated CO_2

The elevated CO_2 concentration is the third most significant factor of climate change. Some studies purported the positive correlation between CO_2 enrichment and plant N requirement which increases the amount of C allocation to belowground where microbial flora are found (Cardon et al. 2001; Niinistö et al. 2004). Balser et al. (2010) listed some of the common C cycle driver mechanisms for belowground allocation of C and N based on various studies including higher allotment to roots, escalated root turnover, rhizode position, promoting growth of mycorrhiza and nitrogen fixation by microbes. Studies based on PhyloChip hybridization technology suggested that the elevated CO_2 levels significantly affect the soil microbial structure, biomass and population, and diversity. It is reported that the elevated CO_2 causes decline in diversity and population of microflora, resulting in higher fungal diversity and population than that of bacterial (Deng et al. 2012; He et al. 2012). Elevated C level coupled with low concentration of N causes an increase in both decomposer and AM fungal hyphae abundance (Klironomos et al. 1996; de Vries and Griffiths 2018). In a study, Montealegre et al. (2002) investigated the impact of elevated CO_2 levels on soil microflora under different vegetations in a pasture land. The study showed that the proportion of gram-negative bacteria and other bacterial population increased in bulk soil under CO_2 enrichment in comparison to rhizospheric soil. While the proportion of eukaryotes and protozoans was found to increase in rhizospheric soil (when soil is under vegetation) under elevated CO_2.

From the above discussion and studies, it can be concluded that the alterations in temperature, precipitation, and CO_2 level results in changes in growth dynamics and demographical data of various soil microbial species which may lead to changes in primary productivity and ecosystem functioning and balancing.

27.2.2 Impact of Climate Changes on Microbial Processes

According to Schloter et al. (2003), there are several specific processes which help to improve soil fertility and its health to provide ecosystem services, and also play a crucial role in biogeochemical cycles to maintain ecosystem integrity. All these processes are found to be dependent directly or indirectly on various microbial processes which are significantly influenced by several specific factors such as temperature, moisture, and nutrients availability. Soil microbes form an integrated functional part of soil and possessability to perceive and immediately react to environmental changes and subsequently undergoing combination of dynamic seasonal shifts on a shorter time scale compared to the flora and fauna (Schmidt et al. 2007). Soil microbial processes are mediated by microbial enzymatic activities which play a vital role in regulation of nutrient cycles, soil respiration, and mineralization (Burns et al. 2013). Amundson et al. (2015) reported that climatic factors such as temperature, precipitation, soil moisture, and others may stimulate many beneficial functions of soil microbiome.

In case studies of arctic and alpine tundra ecosystems, it has been found that seasonal and spatial changes in typical microbial processes are vastly associated with microbial community composition (Nemergut et al. 2005; Zinger et al. 2009; Luo et al. 2020). In addition, a study in alpine meadow ecosystems concluded that upgraded nutrients mineralization and altered soil enzyme activities can be achieved by short-term climate changes (Zi et al. 2018). There are other contradictory studies about the impact of temperature changes on soil enzymes activities like Galloway et al. (2004) that asserted low temperature as favorable factor for soil enzymatic activities while Baldrian et al. (2013) found that higher temperature supports enhanced soil enzyme activities. Hence, it can be concluded that the relationships between climatic factors and soil enzymatic activities may show variations along with climatic zones and vegetation patterns.

The rising temperature shows a positive impact on soil microbial activities which elevate the soil respiration, which is also influenced by changes in soil microbial community structure, substrate availability, and plant litter quality and quantity (Balser et al. 2006). Soil respiration shows positive correlation with temperature and negative correlation with moisture content and the soil temperature, also moisture shows fluctuations due to changes in precipitation (Luo and Zhou 2006; Aanderud et al. 2013). The impact of changes in temperature and moisture on the rate of enzyme production is due to altered enzyme efficiency, substrate availability, and microbial efficiency (Dutta and Dutta 2016). Schindlbacher et al. (2011) reported that elevated temperature could alter the physiology of decomposers and enhance the fungal decomposition. On instance, results from experiments performed by Kratz (2014) demonstrated that the growth of belowground fungal hyphae decreases in case of elevated temperature for a long period of time due to decline in fungal respiration. Elevated temperature causes raise in rates to N cycling processes such as denitrification, nitrification, and N mineralization (Zhou et al. 2012). It is also reported to accelerate the carbon cycling by improving carbon sequestration mechanisms (Li et al. 2019).

Soil moisture content which depends on the amount and frequency of precipitation affects the soil respiratory activities and soil decomposition dynamics (Aanderud et al. 2011). It has been observed that in peatland and wetland, declined soil moisture causes drying up of soils which enhance oxygen availability and also induce C cycling and thus, elevate the CO_2 flux (Singh et al. 2010). In saturated soils holding high level of moisture content, the moisture creates an effective barrier to aerobic respiration as well as for heterotrophic respiration (Sylvia et al. 2005). According to Balser et al. (2010), increasing soil moisture can alter microbial metabolic activities, and decrease hydrolytic and peroxidase enzymes activities which are interlinked to C sequestration processes (Waldrop and Firestone 2004). Soil with higher moisture content is found to constitute large C sinks.

Paterson et al. (1997) explained that elevated CO_2 concentration seems to affect the terrestrial ecosystem indirectly by variations in balance between decomposition and primary productivity. According to Melillo et al. (1982), decomposition rates were enhanced in litter derived under elevated CO_2 conditions. Hungate et al. (1999) observed a decline in gross mineralization rate and increase in ammonium immobilization in soils under elevated CO_2. Many researchers claimed an increase in microbial activities due to elevated CO_2 levels (Rice et al. 1994; Dhillion et al. 1995). In addition, De Graaff et al. (2006) reported rise in microbial respiration because of accelerated plant productivity under elevated CO_2 which in turn provide more C substrate to soil microorganisms. There has been observation that the C pool in the soil is far more than that of aboveground levels because of higher microbial respiration hence, elevated CO_2 affects the soil microbial diversity and composition. The CO_2 levels in soil are also found to be affected indirectly via changes in quantity and quality of plant-derived inputs into the soil (Hodge 1996; Paterson et al. 1997; Hodge et al. 1998). Plants growing under CO_2 enrichment produce higher amount of available C pool due to enhanced amount of rhizodeposition which are directly utilized by rhizospheric microflora (Van Veen et al. 1991; Rogers et al. 1994). Plant responses vary in various ways such as altered plant growth, nutrients usage and allotment, and altered above and belowground patterns of biomass (Poorter 1993; Hungate et al. 1997; Cotrufo et al. 1998; Balser et al. 2010). A number of studies suggest that elevated CO_2 causes enhanced root exudation and translocation of assimilated C into the soil (Cotrufo and Gorissen 1997; Montealegre et al. 2002).

27.2.3 *Impact of Climate Changes on Nutrient Dynamics*

Production, accumulation, and decomposition of organic matter are three key functions of most of the ecosystems and various nutrient cycles contribute significantly in the maintenance and organization of ecosystems including both inputs and outputs (Han et al. 2012). Among various nutrients, C and N cycle are the most fundamental and are the integral component of the global climate system, and soils are most critical part of these cycles and their regulation (Brevik 2013). In nature, soils sequester C through the soil–plant systems as plant uptake CO_2 during the process of photosynthesis and return back by adding dead tissues in the form of litter to the soil. Carney et al. (2007) reported that even under increased CO_2 levels, soil organic carbon (SOC) declines due to increased microbial activity. The soil microbial respiration is also a way to add C to soil by emitting CO_2 and CH_4 influenced by the oxygen status of the soil system. The water drainage changes the soil environment from anaerobic (lacking oxygen) to aerobic which accelerates the degradation processes and emit GHGs into the atmosphere. Overall C content in soil depends on addition and emission like C increases in soil which is taken from the atmosphere thus, decreased atmospheric CO_2 and vice-versa takes place (Schlesinger 1995; Brevik 2013). In a long-term study, Niklaus and Körner (2004) observed that elevated CO_2 limits other nutrients availability such as of N and P in grassland causing reduced plant growth. It can be concluded that elevated CO_2 levels cause increased C sequestration and C turnover. The rise in atmospheric temperature which is root cause of global warming and climate change was found to have negative impacts on C allocation to soil. This leads to reduced SOC and causes increased CO_2 release from soils to atmosphere which further increase temperature thus, creating a positive-feedback in global C cycle (Gorissen et al. 2004). In a long time period, it can be observed that soil C got significantly reduced due to soil warming and drying (Brevik 2013).

The elevated CO_2 enhances C:N ratio and reduces the N mineralization process and helps to decompose organisms. The N mineralization is a critical process that controls soil N availability to plants and hence, reduced N content in soil negatively impacts the plant productivity and growth. In some studies, it has been found that the rise in temperature increased N mineralization leading to increase in soil N content (Reich et al. 2006; Li et al. 2011). Increased soil moisture causes increase in biological reactions of soil microflora which convert nitrate (NO_3^-) to NO, N_2O, or N_2 ultimately leading to increased atmospheric N_2O levels (Brevik 2013). Another indirect effect of climate change is change in vegetation cover that remarkably affects the nutrients content and turnover in soils thus, affecting the complete nutrient cycle (Melillo et al. 2001). Raise in atmospheric temperature increases soil microbial activities that speed up the decomposition of soil organic matter (SOM) thus, inducing three-fold increase in the leaching of inorganic N in soil (Kaste et al. 2006).

Meehl et al. (2007) and Jobbágy and Jackson (2000) reported that C input to soil by vegetation varies across different biomes because of moisture gradients. It has been also proposed that the balance of nutrients is based on decomposition and productivity. The decomposition of SOM and soil respiration is greatly influenced by temperature and moisture. Rise in temperature increases the fungal decomposition process that in turn increases CO_2 flux from the soil. Moreover, higher temperature also increases soil N content that reduces the rate of fungal decomposition which adversely affects the microbial activity and diversity (American Society for Microbiology 2008). SOC and soil organic nitrogen (SON) stocks are found to have high dependency on temperature and precipitation gradients that may vary in various ecosystem biomes. All these mechanisms together maintain and drive global and regional patterns of SOC in a coordinated way and SOC and SON can be used as an integrative measure of productivity and decomposition (Dutta and Dutta 2016).

Elevated CO_2 enhances the microbial enzymatic activities involved in various nutrients cycling, especially N and P cycle (Xiong et al. 2015; Yang et al. 2019). Various processes have a significant role in depletion of soil mineral N such as plant primary productivity, microbial immobilization, and denitrification rates (Delgado-Baquerizo et al. 2014; Naylor et al. 2020).

In spite of undergoing researches, still there are many uncertainties in the field of influence of climate change factors. There is scope that elevated temperature, altered precipitation, and elevated CO_2 cause warming and thus favor the liberation of CO_2 from soil to atmosphere due to microbial breakdown of SOM. In the final analysis, it can be concluded that nutrients budgets, especially C and N, depend on the balance between photosynthesis and respiration (both autotrophic root respiration and heterotrophic soil microbial respiration).

García-Fayos and Bochet (2009) reported a strong correlation between climate change and soil erosion which is resultant of negative impacts on aggregate stability, bulk density, water holding capacity, pH, organic matter content, total N, and soluble P in the soil of semi-arid Mediterranean ecosystem in Spain. Thus, the conclusion can be withdrawn that climate change negatively affects the soil properties and thus badly influences the soil health and its fertility. Healthy soils are basic requirement for the nutrients supply to plants for their growth, quality, sustainability, and productivity. Unhealthy soils tend to have lesser nutrients availability in soil for plants not only causing reduction in food availability but also the food quality and ultimately raising concern for food and nutrition security for humankind.

27.3 Microbial Feedback to Climate Changes

Microbial flora is the main controlling agent for all the macro- and micro nutrient cycling and therefore main controlling factor for the ecosystem-climate feedbacks, including GHG fluxes in the ecosystem. Small change in microbial community may have profound impact on soil processes like organic matter decomposition and mineralization, inorganic nutrient cycling and respiration leading to efflux of GHGs.

Climate change affects the complex and highly diverse soil microbial communities and the associated biotic interactions, which in turn facilitate their feedback responses like CO_2 and other GHGs concentration in atmosphere having radiating force, which further contributes to climate change, forming a feedback loop (Bardgett et al. 2008). These feedbacks are major determinants of terrestrial ecosystem's capacity to store C (Pendall et al. 2004; Luo 2007; Nie et al. 2013). The microbial feedback responses in terms of GHG emission may amplify or lessen the climate change and hence, play a huge role in determining future climate state (Singh et al. 2010). Alteration in microbial decomposition dynamics can shift the soil C pool to atmosphere, thereby eliciting positive feedback on climate. In contrast, increase in plant productivity driven by plant nutrient status, exceeding the C loss through decomposition processes, contributes to negative feedback to the rate of climate change (Finzi et al. 2011).

Several simulation experiments have showed that these feedback responses could have substantial stimulating or depreciating effect on the rate of climate change (Cox et al. 2000; Friedlingstein et al. 2006; Singh et al. 2010; Ciais et al. 2014). Whether positive or negative or perhaps neutral, feedback responses are decided by several factors like shift in microbial composition and diversity, alteration in microbial efficiency, variation in microbial accessible substrate, and decline in microbial biomass (Melillo et al. 2017). With the projected global climate change and its effect on soil health and fertility, it has become imperative to understand the belowground responses to global climate change.

Microbial community, structure, composition, and metabolic activity that mediate soil nutrient dynamics and GHG emission are basically regulated by direct and indirect feedback mechanisms. Direct microbial feedback to climate change is due to microbial response to temperature, precipitations and other extreme climatic conditions (drought, fires and freezing). Indirect feedbacks are associated with the impact of climatic variation on plant community (diversity and productivity) and coupled biotic interactions (Bardgett et al. 2008; Dutta and Dutta 2016).

27.3.1 Direct Feedback

Climate warming directly accelerates the rate of microbial processes such as decomposition dynamics of SOM and nutrient mineralization and their utilization, thereby affecting the flux between soil and atmosphere. Numerous studies have predicted that enhanced microbial growth efficiency could create positive feedback on climate (Zhou et al. 2012; Classen et al. 2015). It has been anticipated that future climate change is likely to stimulate SOM decay, which may lead to loss of 5% of the currently C stored in soil (approx. 3100 Pg) (Davidson and Janssens 2006). Microbial activity is generalized to be stimulated by warm conditions, thus it is expected that the microbial decomposition and decomposition-derived heterotrophic respiratory C would increase with global warming (Metcalfe 2017). Rest is contributed by autotrophic respiration.

Even though it is extensively known how decomposition rate is regulated by temperature, the correlation between respiration and temperature is poorly understood (Bardgett et al. 2008). Impact of temperature sensitivity on organic matter decomposition is predicted to be mediated through change in the overall quality and inherent decomposability of substrate available for microbes. Greater temperature sensitivity of labile C pools compared with biochemically diverse and complex C source is thought to be the possible reason behind the waning of CO_2 efflux in response to warming after an initial transitory increase in soil warming experiments (Melillo et al. 2002; Davidson and Janssens 2006). However, there is huge uncertainty regarding the dependence of microbial decomposition on temperature, as other studies have showed the recalcitrant C substrate are more (Knorr et al. 2005) or similar (Fang et al. 2005; Conen et al. 2006) in temperature sensitivity than the labile pool. Furthermore, change in temperature can be very imperative in permafrost soils, where permafrost thawing due to rise in temperature could result in decomposition of considerable amount of stored carbon reservoir. According to recent estimation, 5–15% of total C (~1300–1580 Pg of carbon) stored in permafrost is susceptible to be lost as CO_2 incoming decade (Jansson and Hofmockel 2019).

Another decisive factor for microbial feedback responses to climate change is soil moisture which is essential for physiological activity of soil microorganism and also regulates the oxygen availability and gas diffusion (Singh et al. 2010). In consideration of the foregoing, prolonged drier soil condition may exert negative feedback on C flux, through slowing down the microbial activity in some ecosystems. In peatland and wetland systems, drought condition accelerates C cycling by improving oxygen availability for microbial activity, thus posing negative feedback impact on climate (Freeman et al. 2004; Meier and Leuschner 2008). This view has been supported by the increase in extracellular enzyme phenol oxidase, an enzyme responsible for breakdown of complex substrates, leading to depletion of large C stock stored in these systems (Freeman et al. 2004) while drying soil could have opposite effects on methanogenesis. Decreasing water table in such area can greatly lower CH_4 emission by increasing oxygen availability (causing CH_4 oxidation) and so induces negative feedback response to climate (Fetzer et al. 1993). Further, soil drying scenario may prompt positive feedback on climate change by increasing the N_2O efflux from soil (Houghton 1996).

Microbial feedback responses to variation in these climatic factors vary across different ecosystems. For instance, in the ecosystem with water limitation (savanna, deserts, and semiarid grassland), soil moisture can have greater effect on breakdown than temperature. While, in temperate regions where temperature is the limiting factor for microbial growth and activity, change in temperature would have more pronounced effect compared with soil moisture (Singh et al. 2010). Moreover, this picture is further complicated by the effects of climate forcing on composition, diversity, and efficiency of microbial community. Still, there is huge uncertainty regarding the response of different microbial community to the changing environment, differing in their range of action. It has been proposed that the change in organic matter decomposition and subsequent mineralization and respiration are especially susceptible to shift in microbial flora (Hartley et al. 2007; Allison and Treseder 2008). Therefore, it is very difficult to make any presumption about the microbial feedback to climate on the basis of knowledge obtained from warming experiments done so far.

27.3.2 Indirect Feedback

Indirect feedbacks are facilitated by influence of climate change on vegetation composition and plant growth and its consequent effect on microbial community. Plants and microbes are inextricably linked and these plant–microbial interactions play a key role in determining the C and N dynamics and hence, the feedback responses of terrestrial ecosystems to climate warming. The chief element of global warming that regulates these indirect microbial feedback to climate is elevated atmospheric CO_2 level. Rise in CO_2 concentration can enhance plant productivity through fertilization effect and further transfer to associated symbionts and other heterotrophic microorganisms, ultimately contributing to soil C stock (Bardgett et al.

2005; Johnson et al. 2005). Every year, around 120 billion tons of C is fixed by primary production from the terrestrial ecosystem, which may beget negative feedback on changing climate through C sequestration (Singh et al. 2010). However, the fate of this C stored may vary depending upon the plant type, soil fertility status, existing biotic–abiotic interactions, and other factors (Wardle 2013; Bardgett et al. 2008).

Major fraction of this stored C is likely to be lost as respiration fueled by increased microbial diversity and activity. It has been predicted that increased microbial growth efficiency and reduced N immobilization would create N limitation for plant growth, leading to potential negative feedback on climate change (Diaz et al. 1993). Climate-driven shift in plant and microbe community is expected to affect the biotic interaction between them. Warming-induced elevation in the interactions between plant and soil microbial flora coupled with reduced N availability suppress the microbial decomposition process, eventually enhancing the soil function as C sink (Hu et al. 2001).

In addition to the effect of CO_2 enrichment, other climatic factors like temperature and precipitation are also expected to indirectly influence the feedback responses by affecting the plants' growth and phenology. Climate changes affect the growing season and pattern, root and shoot phenology, and subsequent symbiotic interactions. Alteration in root–shoot phenology affects the interaction with rhizosphere microbes, which may further disturb the synchronization between plant, soil, and climate. This commotion can result in shift in composition and productivity of plant community (Classen et al. 2015). The concept of alteration in plant composition and diversity in long course of time, effectuated by change in climatic factors is well evident. The shift in plant community structure and function is of great significance to indirect feedback responses of microbes. Unprecedented changes in temperature and precipitation pattern in recent times have distinctly modified the vegetation of arctic and Canadian tundra, and African savannah (Epstein et al. 2004; Sankaran et al. 2005; Danby and Hik 2007). Such changes in vegetation could have drastic impact on the nutrient balance of terrestrial ecosystem and its feedback to climate. Encroachment of evergreen shrub in arctic tundra is predicted to reduce the respiratory CO_2 emission thus, setting negative feedback on climate change (Cornelissen et al. 2007).

The chief mechanism, through which change in vegetation affects the microbial feedback to climate, is through change in quantity and quality of plant litter, the most important component of SOM, which varies with the plant functional group (Dutta and Dutta 2016). Plant litter as mentioned in the previous section plays a significant role in influencing the rate of decomposition process and microbial respiration (Bardgett et al. 2008). Plant litter alteration might also include change in soil C:N ratio and nutrient availability, crucial for microbial function (Kandeler et al. 1998). For example, evergreen shrubs, which are a slow-growing plant, provide recalcitrant substrate for decomposition, with complex composition and low nutrient content. Plant functional groups like legumes and graminoids produce nutrient-rich and easily decomposable litter under increased microbial activity scenario (Wardle 2013). Dominance of nitrogen fixing legumes in grasslands could increase the atmospheric CO_2 carbon dioxide by accelerating the rate of decomposition and C mineralization (Hanley et al. 2004; Bardgett et al. 2008). These indirect feedbacks are central to biological processes like mineralization, vegetation composition, and other ecological interactions, but remain largely unknown and are more variable than counterpart abiotic factors.

Although, these feedbacks between microbes-mediated nutrient dynamics and climate change have wide implication on climate change mitigation, the underlying mechanism regulating them needs to be further understood. For example, potential role of temperature sensitivity of SOM decomposition on the function of terrestrial system as C sink or storage and how this shift would shape future climate, climate warming driven complex interaction between C and N cycling are poorly recognized (Davidson and Janssens 2006; Melillo et al. 2002). Due to lack of mechanistic insight in this subject, there is a huge gap between the existing knowledge and their practical implementation in climate change mitigation. To better understand these responses and reduce the uncertainty associated with the magnitude and direction of these climate feedbacks, we need to include long-term new biochemical and molecular tools.

27.4 Role of Microbes in Climate Change Mitigation

Climate change is a complex issue with wide and pervasive impacts on ecosystem, socioeconomics, and food security. Despite the growing concern regarding relentless climate warming, atmospheric GHGs concentration is rising at unprecedented rate, urging the global community to take decisive steps to mitigate its negative impacts. Climate mitigation strategies primarily implemented in alleviating the GHGs emission through C sequestration, conservation of C sinks, and adapting the management practices. The recent advancement in the insight in microbial response to climate change in terms of changes in soil processes and related biotic interactions has invoked the idea of potential usage of soil microbes in mitigating the climate change (Singh et al. 2010). These mitigation strategies could be based on direct manipulation of microbes or indirectly through modifications of the microbial processes by adopting changes in land management practices (Fig. 27.2) (Jansson and Hofmockel 2019).

Manipulating soil microbes using biotechnological tool can provide practical and sustainable measures to achieve sustainable development goals including reducing the emission of air pollutants (Cavicchioli et al. 2019). Extensive variability and heterogeneity in the phenotype of microbes helps in adopting to the extremities in surrounding environment (Justice et al. 2008). Incorporating these advantageous mutations in soil microbiome through in situ methods can enhance their C uptake activities, which could be used for management of changing climate (Dastagir 2019). Itakura et al. (2013) proposed deployment of bacterial strain showing higher N_2O reductase activity can help in mitigating N_2O emission from the agriculture system. Similarly, targeting the microbial community of rumens can alter their response and

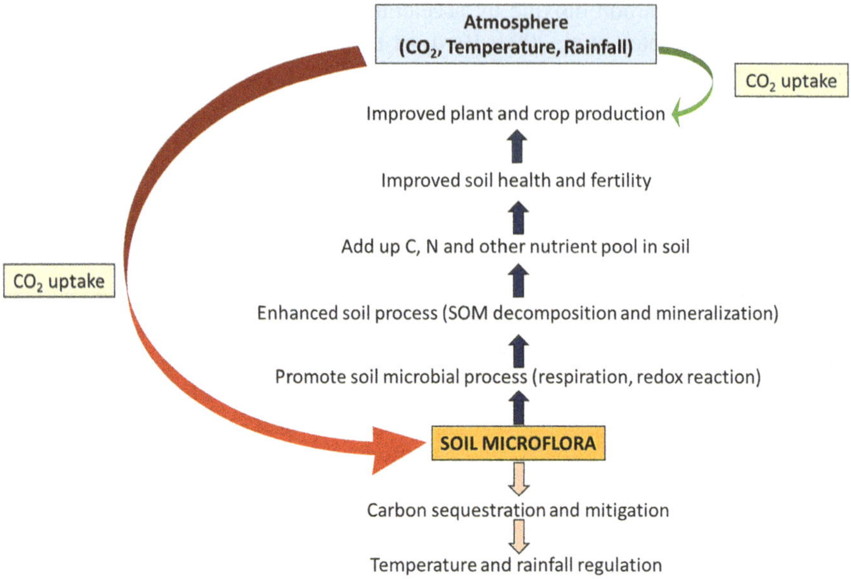

Fig. 27.2 Role of soil microbial flora for mitigating climate change**:** Microbial community can be used as plant growth promoter for enhancing carbon sequestration by plants. Symbiont microbes like arbuscular mycorrhiza, rhizobium spp. can improve plant performance nutrient availability and their productivity in addition to their traditional use as biofertilizer and biopesticide. Microbial metabolites (extracellular polymeric compound) provide resistance to plant against water stress and drought condition. Soil microorganism can directly increase the soil carbon pool by accumulating carbon in the form of living and/or dead microbial biomass. Microbial flora can be used to mitigate drought condition by retaining moisture content in soil. Greenhouse gas emission reduction by manipulating microbes. For example, integrating microbes as biological inhibitor for processes like denitrification can limit N_2O flux, or inoculating nitrous oxide consuming microbes

hence, may help in reducing CH_4 emission from cattle without affecting the host (Henderson et al. 2015; Roehe et al. 2016). Moreover, dead biomass and other recalcitrant polymers in the soil can be off-loaded by modifying microorganism. However, the present research investigations need to incorporate new developing technologies to better understand the alteration in functional efficiency of microbial flora and their consecutive role in the changing environment.

Controlling CO_2 and other GHGs emission is the most fundamental step toward mitigating climate warming. Atmospheric C is naturally fixed and sequestered into much stable forms in vegetation stock through photosynthesis and in soil through autotrophic microbial community (Jansson and Hofmockel 2019). Enhancing plant and the soil microbiome performance by minimizing the prevailing disturbance, promoting biodiversity, can greatly enhance the naturally occurring C assimilation in the terrestrial ecosystem (Lal 2004). Potential of soil microbes to fix the bioavailable forms of C, prone to be lost into more complex and recalcitrant form through their biochemical activity, can be harnessed to restore the soil C pool (Hicks et al. 2017).

In connection to this, manipulation of plant–microbial rhizospheric interactions can contribute in slowing the change in climate change by altering the C sequestration by vegetation (Wallenstein 2017). For instance, inducing input from plant to rhizosphere can protect the C loss to atmosphere by storing in the microbial biomass or by converting into complex metabolites (Jansson et al. 2018). Also, using fungal and bacterial community inhabiting in rhizosphere, as plant growth promoter, can enhance the C sequestration. Using these beneficial microbial communities in nutrient-limited conditions can release the nutrition requirement of plants growing in nutrient-limiting condition and consequently may help in countering the negative consequences of climate forcing (Compant et al. 2010). Not only that, they are capable of imparting tolerance in plants to the extreme climate conditions like drought. Extracellular polymeric compounds and other metabolites secreted by the plant growth-promoting bacteria render resistance against the water stress (Pereyra et al. 2012; Naylor and Coleman-Derr 2018). Using AM can reduce the N_2O emission by consuming available ammonium (Jansson and Hofmockel 2019).

Another important contributor of GHGs in atmosphere that needs to be addressed is the agriculture ecosystem. Around 30% of GHGs emission is contributed by agricultural soil. In addition to this, the extensive use of agrochemicals like pesticide and insecticide to obtain better yield affects the associated symbiont microbial activity and related soil processes ultimately disturbing the C sink-source balance in the ecosystem (Bever 2015). Compounding the relation between agricultural practices and soil C pool suggest the refinement in current agricultural practices for mitigation and adaptation to the climate change. Implementing the agricultural practices involving plant and soil management such as no tillage, intercropping, crop rotation, mulching, and manure application not only improves soil fertility but also enhances soil C storage through slowing down the microbial decomposition process (UNEP 2019). Such agricultural practices preclude C loss into the atmosphere by minimizing the soil disturbance and augmenting organic matter content in soil, preventing soil erosion (Woodward et al. 2009). Application of fertilizer in nutrient-poor soil can greatly improve the aboveground productivity and C return to soil (Neff et al. 2002). However, extensive fertilizer application is a major contributor of N_2O emission, thus the amount of fertilizers and frequency of application should be target-specific to keep the nitrification rate low (Singh et al. 2010). Another recognized strategy to reduce the N_2O efflux from soil to limit the rate of denitrification process is to use biological inhibitor to block the ammonium oxidation pathway (Subbarao et al. 2009) or inoculating N_2O-consuming microorganism (Itakura et al. 2013). Conversion of arable land into grassland increases input from plant and builds up organic matter content in the soil (McLauchlan et al. 2006). Better drainage and other management practices would inhibit the anaerobic condition, thus preventing CH_4 and N_2O emission from soil. These abovementioned strategies for climate change mitigation and adaptation suggest the great potential of microbe to control the rate of climate change and their negative impact.

27.5 Conclusion and Future Perspectives

There is a general agreement among scientists that various anthropogenic activities lead to global climate change. Microbial aspect of climate change is a pertinent issue of the present times that should be properly addressed and need to be given due importance. However, there is uncertainty about the prediction of GHGs emission in future and their response to further changes in atmospheric composition and climate. Global warming is pushing the terrestrial microorganisms into destructive climate feedback loops. Scientists across the globe are raising serious concerns about the impacts of climate changes on soil microorganisms and suggest much more attention is needed.

Although the soil microbial flora is crucial in regulating the climate change, there is a huge gap in our understanding of their response and feedback mechanism to climate change. Hence, it is the need of time to understand the mechanism of microbial feedback responses and their potential to mitigate climate change. This understanding would help to solve uncertainties about the prediction of feedback responses of soil microbes to climate change and will assist to integrate this knowledge in future models of climate change and terrestrial feedbacks. In the present scenario, much attention should be paid to understand how microbial processes that are associated with climate change are altered whether changes in microbial processes result from the effect of climate change on soil microbial communities and on soil physicochemical properties by altering various abiotic factors, or by interactions between these two. Moreover, most of the studies have paid attention inclusively on one GHG that is CO_2, whereas evidences confirm that microbes-mediated emission of different GHGs countered differently to climate change.

It is clear from the available literatures that the nutrients cycling mediated by soil microbes play a crucial role in cycling of C globally under changing climatic conditions by determining the CO_2 efflux from the soil. A better understanding about the microbial communities and associated processes, and impact of climate change on these as well as on soil abiotic factors and its ecological functioning may advance our management for maintaining soil functionality. Therefore, it is mandatory to conduct some experiments to understand the impacts of interaction between multiple factors on soil microbial communities and their functions, on inorganic-N availability and SOC dynamics under changing environmental conditions. It would assist us to comprehend the close relationship between microbial processes and soil nutrient dynamics under changing climatic conditions. Further, this chapter would offer a new set of goals for researchers to identify new drivers, indicators, and theories for soil ecosystem functioning and also to develop some new approaches for mitigation of climate change using soil microorganisms.

Keeping all above facts in mind, here we are proposing some areas that need to be prioritized.

1. There is an urgent need to develop better understanding about the whole mechanism of microbial control of GHGs fluxes and microbial responses to changed

climate conditions like warming, changed precipitation, and elevated CO_2 levels in different ecosystems.
2. Microbes should be properly screened, identified, and classified on the basis of their functional and physiological roles in the field of GHG production and emission.
3. Some research must be conducted in order to observe impacts of climate changes on aboveground and belowground interactions and nutrient cycling as it directly and/or indirectly affects the structure and functions of microorganisms.
4. As soil microbes–plant interaction plays an important role in soil N and C dynamics, research must be undertaken to understand the role of these interactions in modulating the ecosystems response to global climate changes.
5. We need to improve our knowledge to potentially utilize the natural microbial systems in managing climate changes through negative feedback by enhancing C sequestration and/or reducing GHG efflux to atmosphere.
6. Microbes could be proven a very imperative natural resource for managing climate change if we harness their potential properly. Thus, it is high time to study this aspect, understand the acting mechanisms more accurately, and hence, properly utilize it as an emerging solution for climate change mitigation.

Acknowledgments Authors are grateful to Professor K.S. Rao, Natural Resources Management Lab, Department of Botany, University of Delhi, New Delhi, India, for his kind support. Financial support by the UGC, New Delhi, is gratefully acknowledged.

References

Aanderud ZT, Schoolmaster DR, Lennon JT (2011) Plants mediate the sensitivity of soil respiration to rainfall variability. Ecosystems 14:156–167. https://doi.org/10.1007/s10021-010-9401-y

Aanderud ZT, Jones SE, Schoolmaster DR Jr, Fierer N, Lennon JT (2013) Sensitivity of soil respiration and microbial communities to altered snowfall. Soil Biol Biochem 57:217–227. https://doi.org/10.1016/j.soilbio.2012.07.022

Allison SD, Treseder KK (2008) Warming and drying suppress microbial activity and carbon cycling in boreal forest soils. Glob Chang Biol 14:2898–2909. https://doi.org/10.1111/j.1365-2486.2008.01716.x

American Society for Microbiology (2008) Climate change could impact vital functions of microbes. Sci Daily. https://www.sciencedaily.com/releases/2008/06/080603085922.htmbard. Accessed 08 June 2021

Amundson R, Berhe AA, Hopmans JW, Olson C, Sztein AE, Sparks DL (2015) Soil and human security in the 21st century. Science 348:6235. https://doi.org/10.1126/science.1261071

Andrews JA, Matamala R, Westover KM, Schlesinger WH (2000) Temperature effects on the diversity of soil heterotrophs and the ∂13C of soil-respired CO_2. Soil Biol Biochem 32:699–706. https://doi.org/10.1016/s0038-0717(99)00206-0

Baldrian P, Šnajdr J, Merhautová V, Dobiášová P, Cajthaml T, Valášková V (2013) Responses of the extracellular enzyme activities in hardwood forest to soil temperature and seasonality and the potential effects of climate change. Soil Biol Biochem 56:60–68. https://doi.org/10.1016/j.soilbio.2012.01.020

Balser TC, McMahon KD, Bart D, Bronson D, Coyle DR, Craig N, Flores-Mangual ML, Forshay K, Jones SE, Kent AE, Shade AL (2006) Bridging the gap between micro-and

macro-scale perspectives on the role of microbial communities in global change ecology. Plant Soil 289:59–70. https://doi.org/10.1007/s11104-006-9104-5
Balser TC, Gutknecht JL, Liang C (2010) How will climate change impact soil microbial communities? In: Dixon GR, Tilston EL (eds) Soil microbiology and sustainable crop production. Springer, Dordrecht, pp 373–397
Bardgett RD, Bowman WD, Kaufmann R, Schmidt SK (2005) A temporal approach to linking aboveground and belowground ecology. Trends Ecol Evol 20(11):634–641. https://doi.org/10.1016/j.tree.2005.08.005
Bardgett RD, Freeman C, Ostle NJ (2008) Microbial contributions to climate change through carbon cycle feedbacks. ISME J 2:805–814. https://doi.org/10.1038/ismej.2008.58
Bardgett RD, De Deyn GB, Ostle NJ (2009) Plant–soil interactions and the carbon cycle. J Ecol 97:838–839. https://doi.org/10.1111/j.1365-2745.2009.01545.x
Bardgett RD, Manning P, Morriën E, De Vries FT (2013) Hierarchical responses of plant–soil interactions to climate change: consequences for the global carbon cycle. J Ecol 101(2):334–343. https://doi.org/10.1111/1365-2745.12043
Bever JD (2015) Preferential allocation, physio-evolutionary feedbacks, and the stability and environmental patterns of mutualism between plants and their root symbionts. New Phytol 205:1503–1514. https://doi.org/10.1111/nph.13239
Brevik EC (2013) The potential impact of climate change on soil properties and processes and corresponding influence on food security. Agriculture 3:398–417. https://doi.org/10.3390/agriculture3030398
Burns RG, DeForest JL, Marxsen J, Sinsabaugh RL, Stromberger ME, Wallenstein MD, Weintraub MN, Zoppini A (2013) Soil enzymes in a changing environment: current knowledge and future directions. Soil Biol Biochem 58:216–234. https://doi.org/10.1016/j.soilbio.2012.11.009
Cardon ZG, Hungate BA, Cambardella CA, Chapin Iii FS, Field CB, Holland EA, Mooney HA (2001) Contrasting effects of elevated CO2 on old and new soil carbon pools. Soil Biol Biochem 33(3):365–373. https://doi.org/10.1016/S0038-0717(00)00151-6
Carney KM, Hungate BA, Drake BG, Megonigal JP (2007) Altered soil microbial community at elevated CO_2 leads to loss of soil carbon. Proc Natl Acad Sci 104:4990–4995. https://doi.org/10.1073/pnas.0610045104
Castro HF, Classen AT, Austin EE, Norby RJ, Schadt CW (2010) Soil microbial community responses to multiple experimental climate change drivers. Appl Environ Microbiol 76 (40):999–1007. https://doi.org/10.1128/aem.02874-09
Cavicchioli R, Ripple WJ, Timmis KN, Azam F, Bakken LR, Baylis M, Behrenfeld MJ, Boetius A, Boyd PW, Classen AT, Crowther TW (2019) Scientists' warning to humanity: microorganisms and climate change. Nat Rev Microbiol 17:569–586. https://doi.org/10.1038/s41579-019-0222-5
Ciais P, Sabine C, Bala G, Bopp L, Brovkin V, Canadell J, Chhabra A, DeFries R, Galloway J, Heimann M, Jones C (2014) Carbon and other biogeochemical cycles. In: Climate change 2013: the physical science basis. Contribution of Working Group I to the Fifth Assessment Report of the Intergovernmental Panel on Climate Change. Cambridge University Press, Cambridge, pp 465–570. https://doi.org/10.1017/cbo9781107415324.015
Classen AT, Sundqvist MK, Henning JA, Newman GS, Moore JA, Cregger MA, Moorhead LC, Patterson CM (2015) Direct and indirect effects of climate change on soil microbial and soil microbial-plant interactions: what lies ahead? Ecosphere 6:1–21. https://doi.org/10.1890/es15-00217.1
Compant S, Van Der Heijden MG, Sessitsch A (2010) Climate change effects on beneficial plant–microorganism interactions. FEMS Microbiol Ecol 73:197–214. https://doi.org/10.1111/j.1574-6941.2010.00900.x
Conen F, Leifeld J, Seth B, Alewell C (2006) Warming mineralises young and old soil carbon equally. Biogeosciences 3:515–519. https://doi.org/10.5194/bg-3-515-2006
Cornelissen JH, Van Bodegom PM, Aerts R, Callaghan TV, Van Logtestijn RS, Alatalo J, Stuart Chapin F, Gerdol R, Gudmundsson J, Gwynn-Jones D, Hartley AE (2007) Global negative

vegetation feedback to climate warming responses of leaf litter decomposition rates in cold biomes. Ecollett 10:619–627. https://doi.org/10.1111/j.1461-0248.2007.01051.x

Cotrufo MF, Gorissen A (1997) Elevated CO_2 enhances below-ground C allocation in three perennial grass species at different levels of N availability. New Phytol 137:421–431. https://doi.org/10.1046/j.1469-8137.1997.00839.x

Cotrufo MF, Ineson P, Scott A (1998) Elevated CO_2 reduces the nitrogen concentration of plant tissues. Glob Chang Biol 4:43–54. https://doi.org/10.1046/j.1365-2486.1998.00101.x

Cox PM, Betts RA, Jones CD, Spall SA, Totterdell IJ (2000) Acceleration of global warming due to carbon-cycle feedbacks in a coupled climate model. Nature 408:184–187. https://doi.org/10.1038/35041539

Crowther TW, Thomas SM, Maynard DS, Baldrian P, Covey K, Frey SD, van Diepen LTA, Bradford MA (2015) Biotic interactions mediate soil microbial feedbacks to climate change. Proc Natl Acad Sci 112:7033–7038. https://doi.org/10.1073/pnas.1502956112

Danby RK, Hik DS (2007) Variability, contingency and rapid change in recent subarctic alpine tree line dynamics. J Ecol 95:352–363. https://doi.org/10.1111/j.1365-2745.2006.01200.x

Dastagir MR (2019) Role of microorganisms in managing climate change impacts. In: Singh D, Prabha R (eds) Microbial Interventions in Agriculture and Environment. Springer, Singapore. https://doi.org/10.1007/978-981-32-9084-6_1

Davidson EA, Janssens IA (2006) Temperature sensitivity of soil carbon decomposition and feedbacks to climate change. Nature 440:165–173. https://doi.org/10.1038/nature04514

De Graaff MA, Van Groenigen KJ, Six J, Hungate B, van Kessel C (2006) Interactions between plant growth and soil nutrient cycling under elevated CO_2: a meta-analysis. Glob Chang Biol 12:2077–2091. https://doi.org/10.1111/j.1365-2486.2006.01240.x

de Vries FT, Griffiths RI (2018) Impacts of climate change on soil microbial communities and their functioning. In: Horwath W, Kuzyakov Y (eds) Developments in soil science, vol 35. Elsevier, New York, pp 111–129

Delgado-Baquerizo M, Maestre FT, Escolar C, Gallardo A, Ochoa V, Gozalo B, Prado-Comesaña A (2014) Direct and indirect impacts of climate change on microbial and biocrust communities alter the resistance of the N cycle in a semiarid grassland. J Ecol 102:1592–1605. https://doi.org/10.1111/1365-2745.12303

Deng Y, He Z, Xu M, Qin Y, Van Nostrand JD, Wu L, Roe BA, Wiley G, Hobbie SE, Reich PB, Zhou J (2012) Elevated carbon dioxide alters the structure of soil microbial communities. Appl Environ Microbiol 7:82991–82995. https://doi.org/10.1128/aem.06924-11

Dhillion SS, Roy J, Abrams M (1995) Assessing the impact of elevated CO_2 on soil microbial activity in a Mediterranean model ecosystem. Plant Soil 187:333–342. https://doi.org/10.1007/bf00017098

Diaz S, Grime JP, Harris J, McPherson E (1993) Evidence of a feedback mechanism limiting plant response to elevated carbon dioxide. Nature 364:616–617. https://doi.org/10.1038/364616a0

Drenovsky RE, Vo D, Graham KJ, Scow KM (2004) Soil water content and organic carbon availability are major determinants of soil microbial community composition. Microb Ecol 48 (3):424-430. https://doi.org/10.1007/s00248-003-1063-2

Dutta H, Dutta A (2016) The microbial aspect of climate change. Energy Ecol Environ 1:209–232. https://doi.org/10.1007/s40974-016-0034-7

Ellert BH, Bettany JR (1992) Temperature-dependence of net nitrogen and sulfur mineralization. Soil Sci Soc Am J 56:1133–1141. https://doi.org/10.2136/sssaj1992.03615995005600040021x

Epstein HE, Beringer J, Gould WA, Lloyd AH, Thompson CD, Chapin FS III, Michaelson GJ, Ping CL, Rupp TS, Walker DA (2004) The nature of spatial transitions in the Arctic. J Biogeogr 31:1917–1933. https://doi.org/10.1111/j.1365-2699.2004.01140.x

Fang C, Smith P, Moncrieff JB, Smith JU (2005) Similar response of labile and resistant soil organic matter pools to changes in temperature. Nature 433:57–59. https://doi.org/10.1038/nature03138

Fetzer S, Bak F, Conrad R (1993) Sensitivity of methanogenic bacteria from paddy soil to oxygen and desiccation. FEMS Microbiol Ecol 12:107–115. https://doi.org/10.1111/j.1574-6941.1993.tb00022.x

Fierer N, Jackson RB (2006) The diversity and biogeography of soil bacterial communities. Proc Natl Acad Sci 103:626–631. https://doi.org/10.1073/pnas.0507535103

Finzi AC, Austin AT, Cleland EE, Frey SD, Houlton BZ, Wallenstein MD (2011) Responses and feedbacks of coupled biogeochemical cycles to climate change: examples from terrestrial ecosystems. Front Ecol Environ 9:61–67. https://doi.org/10.1890/100001

Fitter AH, Heinemeyer A, Staddon PL (2000) The impact of elevated CO_2 and global climate change on arbuscular mycorrhizas: a mycocentric approach. New Phytol 147:179–187. https://doi.org/10.1046/j.1469-8137.2000.00680.x

Freeman C, Fenner N, Ostle NJ, Kang H, Dowrick DJ, Reynolds B, Lock MA, Sleep D, Hughes S, Hudson J (2004) Export of dissolved organic carbon from peatlands under elevated carbon dioxide levels. Nature 430:195–198. https://doi.org/10.1038/nature02707

Friedlingstein P, Cox P, Betts R, Bopp L, von Bloh W, Brovkin V, Cadule P, Doney S, Eby M, Fung I, Bala G (2006) Climate–carbon cycle feedback analysis: results from the C4MIP model intercomparison. J Clim 19:3337–3353. https://doi.org/10.1175/jcli3800.1

Galloway JN, Dentener FJ, Capone DG, Boyer EW, Howarth RW, Seitzinger SP, Asner GP, Cleveland CC, Green PA, Holland EA, Karl DM (2004) Nitrogen cycles: past, present, and future. Biogeochemistry 70(2):153–226. https://doi.org/10.1007/s10533-004-0370-0

García-Fayos P, Bochet E (2009) Indication of antagonistic interaction between climate change and erosion on plant species richness and soil properties in semiarid Mediterranean ecosystems. Glob Chang Biol 15:306–318. https://doi.org/10.1111/j.1365-2486.2008.01738.x

Gavito ME, Schweiger P, Jakobsen I (2003) P uptake by arbuscular mycorrhizal hyphae: effect of soil temperature and atmospheric CO_2 enrichment. Glob Chang Biol 9:106–116. https://doi.org/10.1046/j.1365-2486.2003.00560.x

Gorissen A, Tietema A, Joosten NN, Estiarte M, Penuelas J, Sowerby A, Emmett BA, Beier C (2004) Climate change affects carbon allocation to the soil in shrublands. Ecosystems 7:650–661. https://doi.org/10.1007/s10021-004-0218-4

Han X, Cheng Z, Meng H (2012) Soil properties, nutrient dynamics, and soil enzyme activities associated with garlic stalk decomposition under various conditions. PLoS One 7:e50868. https://doi.org/10.1371/journal.pone.0050868

Hanley ME, Trofimov S, Taylor G (2004) Species-level effects more important than functional group-level responses to elevated CO_2: evidence from simulated turves. Funct Ecol 18:304–313. https://doi.org/10.1111/j.0269-8463.2004.00845.x

Hartley IP, Heinemeyer A, Ineson P (2007) Effects of three years of soil warming and shading on the rate of soil respiration: substrate availability and not thermal acclimation mediates observed response. Glob Chang Biol 13:1761–1770. https://doi.org/10.1111/j.1365-2486.2007.01373.x

He Z, Piceno Y, Deng Y, Xu M, Lu Z, DeSantis T, Andersen G, Hobbie SE, Reich PB, Zhou J (2012) The phylogenetic composition and structure of soil microbial communities shifts in response to elevated carbon dioxide. ISME J 6:259–272. https://doi.org/10.1038/ismej.2011.99

Henderson G, Cox F, Ganesh S, Jonker A, Young W, Abecia L, Angarita E, Aravena P, Arenas GN, Ariza C, Attwood GT (2015) Rumen microbial community composition varies with diet and host, but a core microbiome is found across a wide geographical range. Sci Rep 5:14567. https://doi.org/10.1038/srep14567

Hicks N, Vik U, Taylor P, Ladoukakis E, Park J, Kolisis F, Jakobsen KS (2017) Using prokaryotes for carbon capture storage. Trends Biotechnol 35:22–32. https://doi.org/10.1016/j.tibtech.2016.06.011

Hobbie SE (1996) Temperature and plant species control over litter decomposition in Alaskan tundra. Ecol Mongr 66:503–522. https://doi.org/10.2307/2963492

Hodge A (1996) Impact of elevated CO_2 on mycorrhizal associations and implications for plant growth. Biol Fert Soils 23:388–398. https://doi.org/10.1007/s003740050188

Hodge A, Paterson E, Grayston SJ, Campbell CD, Ord BG, Killham K (1998) Characterisation and microbial utilisation of exudate material from the rhizosphere of *Lolium perenne* grown under CO_2 enrichment. Soil Biol Biochem 30:1033–1043. https://doi.org/10.1016/s0038-0717(97)00269-1

Hoeppner SS, Dukes JS (2012) Interactive responses of old-field plant growth and composition to warming and precipitation. Glob Chang Biol 18:1754–1768. https://doi.org/10.1111/j.1365-2486.2011.02626.x

Houghton E (1996) Climate change 1995: the science of climate change: contribution of working group I to the second assessment report of the intergovernmental panel on climate change, vol 2. Cambridge University Press, Cambridge. https://doi.org/10.1017/cbo9781107415324.023

Hu S, Chapin FS, Firestone MK, Field CB, Chiariello NR (2001) Nitrogen limitation of microbial decomposition in a grassland under elevated CO_2. Nature 409:188–191. https://doi.org/10.1038/35051576

Hungate BA, Dijkstra P, Johnson DW, Hinkle CR, Drake BG (1999) Elevated CO2 increases nitrogen fixation and decreases soil nitrogen mineralization in Florida scrub oak. Glob Change Biol 5(7):781–789. https://doi.org/10.1046/j.1365-2486.1999.00275.x

Hungate BA, Lund CP, Pearson HL, Chapin FS (1997) Elevated CO_2 and nutrient addition after soil N cycling and N trace gas fluxes with early season wet-up in a California annual grassland. Biogeochemistry 37(2):89–109. https://doi.org/10.1023/A:1005747123463

Intergovernmental Panel on Climate Change (IPCC) (2001) Climate change 2001: synthesis report. Cambridge University Press, New York/Cambridge, p 398. https://doi.org/10.1007/springerreference_28950

Itakura M, Uchida Y, Akiyama H, Hoshino YT, Shimomura Y, Morimoto S, Tago K, Wang Y, Hayakawa C, Uetake Y, Sánchez C (2013) Mitigation of nitrous oxide emissions from soils by *Bradyrhizobium japonicum* inoculation. Nat Clim Chang 3:208–212. https://doi.org/10.1038/nclimate1734

Jansson JK, Hofmockel KS (2019) Soil microbiomes and climate change. Nat Rev Microbiol 18:1–12. https://doi.org/10.1038/s41579-019-0265-7

Jansson C, Vogel J, Hazen S, Brutnell T, Mockler T (2018) Climate-smart crops with enhanced photosynthesis. J Exp Bot 69:3801–3809. https://doi.org/10.1093/jxb/ery213

Janus LR, Angeloni NL, McCormack J, Rier ST, Tuchman NC, Kelly JJ (2005) Elevated atmospheric CO2 alters soil microbial communities associated with trembling aspen (Populus tremuloides) Roots. Microb Ecol 50:102–109. https://doi.org/10.1007/s00248-004-0120-9

Jensen KD, Beier C, Michelsen A, Emmett BA (2003) Effects of experimental drought on microbial processes in two temperate heathlands at contrasting water conditions. Appl Soil Ecol 24 (2):165–176. https://doi.org/10.1016/S0929-1393(03)00091-X

Jobbágy EG, Jackson RB (2000) The vertical distribution of soil organic carbon and its relation to climate and vegetation. Ecol Appl 10:423–436. https://doi.org/10.1890/1051-0761(2000)010[0423:tvdoso]2.0.co;2

Johnson D, Krsek M, Wellington EM, Stott AW, Cole L, Bardgett RD, Read DJ, Leake JR (2005) Soil invertebrates disrupt carbon flow through fungal networks. Science 309:1047–1047. https://doi.org/10.1126/science.1114769

Justice SS, Hunstad DA, Cegelski L, Hultgren SJ (2008) Morphological plasticity as a bacterial survival strategy. Nat Rev Microbiol 6:162–168. https://doi.org/10.1038/nrmicro1820

Kandeler E, Tscherko D, Bardgett RD, Hobbs PJ, Kampichler C, Jones TH (1998) The response of soil microorganisms and roots to elevated CO_2 and temperature in a terrestrial model ecosystem. Plant Soil 20:2251–2262. https://doi.org/10.1023/a:1004309623256

Kardol P, Cregger MA, Campany CE, Classen AT (2010) Soil ecosystem functioning under climate change: plant species and community effects. Ecol 91:767–781. https://doi.org/10.1890/09-0135.1

Kaste Ø, Wright RF, Barkved LJ, Bjerkeng B, Engen-Skaugen T, Magnusson J, Sælthun NR (2006) Linked models to assess the impacts of climate change on nitrogen in a Norwegian river basin

and FJORD system. Sci Total Environ 365:200–222. https://doi.org/10.1016/j.scitotenv.2006.02.035
Klironomos JN, Rillig MC, Allen MF (1996) Below-ground microbial and microfaunal responses to *Artemisia tridentata* grown under elevated atmospheric CO_2. Funct Ecol 10:527–534. https://doi.org/10.2307/2389946
Knorr W, Prentice IC, House JI, Holland EA (2005) Long-term sensitivity of soil carbon turnover to warming. Nature 433:298–301. https://doi.org/10.1038/nature03226
Kratz CJ (2014) Impacts of climate change on soil microorganisms in Northern Hardwood Forests. Dissertation, Michigan Technological University. https://digitalcommons.mtu.edu/etds/755
Lal R (2004) Soil carbon sequestration to mitigate climate change. Geoderma 123:1–22. https://doi.org/10.1016/j.geoderma.2004.01.032
Li Z, Liu WZ, Zhang XC, Zheng FL (2011) Assessing the site-specific impacts of climate change on hydrology, soil erosion and crop yields in the Loess Plateau of China. Clim Chang 105:223–242. https://doi.org/10.1007/s10584-010-9875-9
Li Y, Lv W, Jiang L, Zhang L, Wang S, Wang Q, Xue K, Li B, Liu P, Hong H, Renzen W (2019) Microbial community responses reduce soil carbon loss in Tibetan alpine grasslands under short-term warming. Glob Chang Biol 25:3438–3449. https://doi.org/10.1111/gcb.14734
Luo Y (2007) Terrestrial carbon–cycle feedback to climate warming. Annu Rev Ecol Evol Syst 38:683–712. https://doi.org/10.1146/annurev.ecolsys.38.091206.095808
Luo Y, Zhou X (2006). Soil respiration and the environment. Elsevier. https://doi.org/10.1016/B978-0-12-088782-8.X5000-1
Luo Y, Sherry R, Zhou X, Wan S (2009) Terrestrial carbon-cycle feedback to climate warming: experimental evidence on plant regulation and impacts of biofuel feedstock harvest. GCB Bioenergy:162–174. https://doi.org/10.1111/j.1757-1707.2008.01005.x
Luo SP, He BH, Zeng QP (2020) Effects of seasonal variation on soil microbial community structure and enzyme activity in a Masson pine forest in Southwest China. J Mt Sci 17:1398–1409. https://doi.org/10.1007/s11629-019-5825-9
McLauchlan KK, Hobbie SE, Post WM (2006) Conversion from agriculture to grassland builds soil organic matter on decadal timescales. Ecol Appl 16:143–153. https://doi.org/10.1890/04-1650
Meehl GA, Stocker TF, Collins WD, Friedlingstein P, Gaye AT, Gregory JM, Kitoh A, Knutti R, Murphy JM, Noda A, Raper SCB, Watterson IG, Weaver AJ, Zhao ZC (2007). Global Climate Projections. In: Climate Change 2007: The Physical Science Basis. Contribution of Working Group I to the Fourth Assessment Report of the Intergovernmental Panel on Climate Change [Solomon S, Qin D, Manning M, Chen Z, Marquis M, Averyt KB, Tignor M, Miller HL (eds.)]. Cambridge University Press, Cambridge, United Kingdom and New York, NY, USA.
Meier IC, Leuschner C (2008) Belowground drought response of European beech: fine root biomass and carbon partitioning in 14 mature stands across a precipitation gradient. Glob Chang Biol 14:2081–2095. https://doi.org/10.1111/j.1365-2486.2008.01634.x
Melillo JM, Aber JD, Muratore JF (1982) Nitrogen and lignin control of hardwood leaf litter decomposition dynamics. Ecology 63:621–626. https://doi.org/10.2307/1936780
Melillo JM, Steudler PA, Feigl BJ, Neill C, Garcia D, Piccolo MC, Cerri CC, Tian H (2001) Nitrous oxide emissions from forests and pastures of various ages in the Brazilian Amazon. J Geophys Res Atmos 106:34179–34188. https://doi.org/10.1029/2000jd000036
Melillo JM, Steudler PA, Aber JD, Newkirk K, Lux H, Bowles FP, Catricala C, Magill A, Ahrens T, Morrisseau S (2002) Soil warming and carbon-cycle feedbacks to the climate system. Science 298:2173–2176. https://doi.org/10.1126/science.1074153
Melillo JM, Frey SD, DeAngelis KM, Werner WJ, Bernard MJ, Bowles FP, Pold G, Knorr MA, Grandy AS (2017) Long-term pattern and magnitude of soil carbon feedback to the climate system in a warming world. Science 358:101–105. https://doi.org/10.1126/science.aan2874
Metcalfe DB (2017) Microbial change in warming soils. Science 358:41–42. https://doi.org/10.1126/science.aap7325
Montealegre CM, Van Kessel C, Russelle MP, Sadowsky MJ (2002) Changes in microbial activity and composition in a pasture ecosystem exposed to elevated atmospheric carbon dioxide. Plant Soil 243:197–207. https://doi.org/10.1023/a:1019901828483

Monteux S, Weedon JT, Blume-Werry G, Gavazov K, Jassey VE, Johansson M, Keuper F, Olid C, Dorrepaal E (2018) Long-term in situ permafrost thaw effects on bacterial communities and potential aerobic respiration. ISME J 12:2129–2141. https://doi.org/10.1038/s41396-018-0176-z

Naylor D, Coleman-Derr D (2018) Drought stress and root-associated bacterial communities. Front Plant Sci 8:2223. https://doi.org/10.3389/fpls.2017.02223

Naylor D, Sadler N, Bhattacharjee A, Graham EB, Anderton CR, McClure R, Lipton M, Hofmockel KS, Jansson JK (2020) Soil microbiomes under climate change and implications for carbon cycling. Annu Rev Environ Resour 45. https://doi.org/10.1146/annurev-environ-012320-082720

Neff JC, Townsend AR, Gleixner G, Lehman SJ, Turnbull J, Bowman WD (2002) Variable effects of nitrogen additions on the stability and turnover of soil carbon. Nature 419(6910):915–917. https://doi.org/10.1038/nature01136

Nemergut DR, Costello EK, Meyer AF, Pescador MY, Weintraub MN, Schmidt SK (2005) Structure and function of alpine and arctic soil microbial communities. Res Microbiol 156:775–784. https://doi.org/10.1016/j.resmic.2005.03.004

Nie M, Pendall E, Bell C, Gasch CK, Raut S, Tamang S, Wallenstein MD (2013) Positive climate feedbacks of soil microbial communities in a semi-arid grassland. Ecol Lett 16:234–241. https://doi.org/10.1111/ele.12034

Niinistö SM, Silvola J, Kellomäki S (2004) Soil CO2 efflux in a boreal pine forest under atmospheric CO2 enrichment and air warming. Glob Change Biol 10(8):1363–1376. https://doi.org/10.1111/j.1365-2486.2004.00799.x

Niklaus PA, Körner CH (2004) Synthesis of a six-year study of calcareous grassland responses to in situ CO_2 enrichment. Ecol Monogr 74:491–511. https://doi.org/10.1890/03-4047

Paterson E, Hall JM, Rattray EAS, Griffiths BS, Ritz K, Killham K (1997) Effect of elevated CO_2 on rhizosphere carbon flow and soil microbial processes. Glob Chang Biol 3:363–377. https://doi.org/10.1046/j.1365-2486.1997.t01-1-00088.x

Pendall E, Bridgham S, Hanson PJ, Hungate B, Kicklighter DW, Johnson DW, Law BE, Luo Y, Megonigal JP, Olsrud M, Ryan MG (2004) Below-ground process responses to elevated CO_2 and temperature: a discussion of observations, measurement methods, and models. New Phytol 162:311–322. https://doi.org/10.1111/j.1469-8137.2004.01053.x

Pereyra MA, Garcia P, Colabelli MN, Barassi CA, Creus CM (2012) A better water status in wheat seedlings induced by *Azospirillum* under osmotic stress is related to morphological changes in xylem vessels of the coleoptile. Appl Soil Ecol 53:94–97. https://doi.org/10.1016/j.apsoil.2011.11.007

Poorter H (1993) Interspecific variation in the growth response of plants to an elevated ambient CO_2 concentration. In: Rozema J, Lambers H, Van de Geijn SC, Cambridge ML (eds) CO_2 and biosphere. Springer, Dordrecht, pp 77–98. https://doi.org/10.1007/978-94-011-1797-5_6

Raich JW, Potter CS (1995) Global patterns of carbon dioxide emission from soils. Global Biogeochem Cycles 9:23–36. https://doi.org/10.1029/94gb02723

Reich PB, Hobbie SE, Lee T, Ellsworth DS, West JB, Tilman D, Knops JM, Naeem S, Trost J (2006) Nitrogen limitation constrains sustainability of ecosystem response to CO_2. Nature 440:922–925

Rice CW, Garcia FO, Hampton CO, Owensby CE (1994) Soil microbial response in tallgrass prairie to elevated CO_2. In: Peter SC, Elizabeth GO, James AT, Zak DR, Pregitzer KS (eds) Below-ground responses to rising atmospheric CO_2: implications for plants, soil biota, and ecosystem processes. Springer, Dordrecht, pp 67–74. https://doi.org/10.1007/978-94-017-0851-7_7

Ricketts MP, Matamala R, Jastrow JD, Antonopoulos DA, Koval J, Ping CL, Liang C, Gonzalez-Meler MA (2020) The effects of warming and soil chemistry on bacterial community structure in Arctic tundra soils. Soil Biol Biochem 148:107882. https://doi.org/10.1016/j.soilbio.2020.107882

Roehe R, Dewhurst RJ, Duthie CA, Rooke JA, McKain N, Ross DW, Hyslop JJ, Waterhouse A, Freeman TC, Watson M, Wallace RJ (2016) Bovine host genetic variation influences rumen microbial methane production with best selection criterion for low methane emitting and efficiently feed converting hosts based on metagenomic gene abundance. PLoS Genet 12:1005846. https://doi.org/10.1371/journal.pgen.1005846

Rogers HH, Runion GB, Krupa SV (1994) Plant responses to atmospheric CO_2 enrichment with emphasis on roots and the rhizosphere. Environ Pollut 83:155–189. https://doi.org/10.1016/0269-7491(94)90034-5

Ruess L, Michelsen A, Schmidt IK, Jonasson S (1999) Simulated climate change affecting microorganisms, nematode density and biodiversity in subarctic soils. Plant Soil 212:63–73. https://doi.org/10.1163/156854199508513

Sankaran M, Hanan NP, Scholes RJ, Ratnam J, Augustine DJ, Cade BS, Gignoux J, Higgins SI, Le Roux X, Ludwig F, Ardo J (2005) Determinants of woody cover in African savannas. Nature 438:846–849. https://doi.org/10.1038/nature04070

Schimel DS, Braswell BH, Holland EA, McKeown R, Ojima DS, Painter TH, Parton WJ, Townsend AR (1994) Climatic, edaphic, and biotic controls over storage and turnover of carbon in soils. Global Biogeochem Cycles 8:279–293. https://doi.org/10.1029/94GB00993

Schindlbacher A, Rodler A, Kuffner M, Kitzler B, Sessitsch A, Zechmeister-Boltenstern S (2011) Experimental warming effects on the microbial community of a temperate mountain forest soil. Soil Biol Biochem 43:1417–1425. https://doi.org/10.1016/j.soilbio.2011.03.005

Schlesinger WH (1995) An overview of the carbon cycle. In: Lal R, Kimble H, Levine E, Stewart BA (eds) Soils and global change. CRC Press, Boca Raton, FL, pp 131–142. https://doi.org/10.1201/9780203753187

Schloter M, Dilly O, Munch JC (2003) Indicators for evaluating soil quality. Agric Ecosyst Environ 98:255–262. https://doi.org/10.1016/S0167-8809(03)00085-9

Schmidt SK, Costello EK, Nemergut DR, Cleveland CC, Reed SC, Weintraub MN, Meyer AF, Martin AM (2007) Biogeochemical consequences of rapid microbial turnover and seasonal succession in soil. Ecology 88:1379–1385. https://doi.org/10.1890/06-0164

Singh BK, Bardgett RD, Smith P, Reay DS (2010) Microorganisms and climate change: terrestrial feedbacks and mitigation options. Nat Rev Microbiol 8:779–790. https://doi.org/10.1038/nrmicro2439

Subbarao GV, Nakahara K, Hurtado MDP, Ono H, Moreta DE, Salcedo AF, Yoshihashi AT, Ishikawa T, Ishitani M, Ohnishi-Kameyama M, Yoshida M (2009) Evidence for biological nitrification inhibition in Brachiaria pastures. Proc Natl Acad Sci 106:17302–17307. https://doi.org/10.1073/pnas.0903694106

Sylvia DM, Hartel PG, Fuhrmann JJ, Zuberer DA (2005) Principles and applications of soil microbiology, 2nd edn. Pearson Educational, Upper Saddle River, NJ

Trumbore SE (1997) Potential responses of soil organic carbon to global environmental change. Proc Natl Acad Sci 94:8284–8291. https://doi.org/10.1073/pnas.94.16.8284

UNEP (2019) The huge potential of agriculture to slow climate change. https://www.unenvironment.org/news-and-stories/story/huge-potential-agriculture-slow-climate-change. Accessed 02 August 2020

Van Veen JA, Liljeroth E, Lekkerkerk LJA, Van de Geijn SC (1991) Carbon fluxes in plant-soil systems at elevated atmospheric CO_2 levels. Ecol Appl 1:175–181. https://doi.org/10.2307/1941810

Várallyay G (2010) The impact of climate change on soils and on their water management. Agron Res 8:385–396. https://doi.org/10.1556/agrokem.59.2010.1.2

von Lützow M, Kögel-Knabner I (2009) Temperature sensitivity of soil organic matter decomposition—what do we know?. Biol Fertil Soils 46:1–15. https://doi.org/10.1007/s00374-009-0413-8

Waldrop MP, Firestone MK (2004) Altered utilization patterns of young and old soil C by microorganisms caused by temperature shifts and N additions. Biogeochemistry 67:235–248. https://doi.org/10.1023/B:BIOG.0000015321.51462.41

Walker MD, Wahren CH, Hollister RD, Henry GH, Ahlquist LE, Alatalo JM, Bret-Harte MS, Calef MP, Callaghan TV, Carroll AB, Epstein HE (2006) Plant community responses to experimental warming across the tundra biome. Proc Natl Acad Sci U S A 103:1342–1346. https://doi.org/10.1073/pnas.0503198103

Wallenstein MD (2017) Managing and manipulating the rhizosphere microbiome for plant health: a systems approach. Rhizosphere 3:230–232. https://doi.org/10.1016/j.rhisph.2017.04.004

Walsh DA (2015) Consequences of climate change on microbial life in the ocean. Microbiol Today (Nov 2015 issue). Microbiology Society, England

Wardle DA (2013) Communities and ecosystems: linking the aboveground and belowground components (MPB-34), vol 34. Princeton University Press, Princeton. https://doi.org/10.1515/9781400847297

Weiman S (2015) Microbes help to drive global carbon cycling and climate change. Microbe Mag 10(6):233–238. https://doi.org/10.1128/microbe.10.233.1

Woodward FI, Bardgett RD, Raven JA, Hetherington AM (2009) Biological approaches to global environment change mitigation and remediation. Curr Biol 19:R615–R623. https://doi.org/10.1016/j.cub.2009.06.012

Xiong J, He Z, Shi S, Kent A, Deng Y, Wu L, Van Nostrand JD, Zhou J (2015) Elevated CO_2 shifts the functional structure and metabolic potentials of soil microbial communities in a C4 agroecosystem. Sci Rep 5:1–9. https://doi.org/10.1038/srep09316

Yang S, Zheng Q, Yuan M, Shi Z, Chiariello NR, Docherty KM, Dong S, Field CB, Gu Y, Gutknecht J, Hungate BA (2019) Long-term elevated CO_2 shifts composition of soil microbial communities in a Californian annual grassland, reducing growth and N utilization potentials. Sci Total Environ 652:1474–1481. https://doi.org/10.1016/j.scitotenv.2018.10.353

Zhou J, Xue K, Xie J, Deng Y, Wu L, Cheng X, Fei S, Deng S, He Z, Van Nostrand JD, Luo Y (2012) Microbial mediation of carbon-cycle feedbacks to climate warming. Nat Clim Chang 2:106–110. https://doi.org/10.1038/nclimate1331

Zi HB, Hu L, Wang CT, Wang GX, Wu PF, Lerdau M, Ade LJ (2018) Responses of soil bacterial community and enzyme activity to experimental warming of an alpine meadow. Eur J Soil Sci 69(3):429–438. https://doi.org/10.1111/ejss.12547

Zinger L, Shahnavaz B, Baptist F, Geremia RA, Choler P (2009) Microbial diversity in alpine tundra soils correlates with snow cover dynamics. ISME J 3:850–859. https://doi.org/10.1038/ismej.2009.20

Zogg GP, Zak DR, Ringelberg DB, White DC, MacDonald NW, Pregitzer KS (1997) Compositional and functional shifts in microbial communities due to soil warming. Soil Sci Soc Am J 61:475–481. https://doi.org/10.2136/sssaj1997.03615995006100020015x

Chapter 28
Impact of Climate Change on Soil Fertility

Sudeshna Mondal

Abstract The worldwide mean temperature has increased by nearly about 1.1 °C since the preindustrial era and this increase may reach up to 4 °C by the tip of the twenty-first century due to the rising concentration of greenhouse gases. Since soils are related to climate system in a very complex way through nutrient and hydrologic cycles, global climate change is predicted to have a possible impact on soil fertility through the physical, chemical, and biological properties of soil due to rise in temperature, alternation in precipitation pattern, increase in greenhouse gas concentration in the atmosphere, etc. These detrimental effects of global climate change can be minimized by following both adaptation and mitigation measures. This paper reviews the influence of global change in the climate such as rise in temperature, alteration in precipitation pattern, and increase in atmospheric carbon dioxide on soil properties and processes affecting soil fertility.

Keywords Climate change · Global warming · Greenhouse gas · Carbon sequestration · Soil properties · Soil processes

28.1 Introduction

Climate change addresses any considerable changes in climatic phenomenon persisting over a protracted period, usually decades or longer (IPCC 2007), including changes in average climatic conditions, erratic rainfall events, frequency and magnitude of extreme weather and sea levels, whether due to natural fluctuations or because of human activity. It is a process of global warming, which is generally attributed to the greenhouse gases generated from the burning of fossil fuels like coal, oil, natural gas, etc., caused by human activity (Brett 2009).

S. Mondal (✉)
Department of Agricultural Chemistry and Soil Science, Bidhan Chandra Krishi Viswavidyalaya, Nadia, West Bengal, India

D. K. Choudhary et al. (eds.), *Climate Change and the Microbiome*, Soil Biology 63,
https://doi.org/10.1007/978-3-030-76863-8_28

Climate change affects the environment, including soil (Brevik 2012). It has the potential to threaten global food security through its effects on soil properties and processes (Brevik 2013), since global demands for food and fiber for an increasing population are met by the soil. The Fifth Assessment Report of the Intergovernmental Panel on Climate Change (IPCC) revealed that global warming has increased approximately 1.0 °C above preindustrial levels, with a probable range of 0.8 °C to 1.2 °C, and is likely to be caused by human activities. Global warming may reach 1.5 °C between 2030 and 2052 if it continues to increase at the same rate (IPCC 2018) and is expected to change other regional and global climate-related parameters such as rainfall, soil moisture, and sea level. The predicted global climate change involving increase in temperature and atmospheric carbon dioxide (CO_2) levels, changes in rainfall pattern, and atmospheric nitrogen deposition influences various soil physical, chemical, and biological properties and properties important for restoring soil fertility and productivity. Climate change has potential impacts on soil health including supply of organic matter from biomass, soil temperature regime, soil hydrology, and salinity. Considering all these affairs, the consequences of climate change on soils have been reviewed.

28.2 Environmental Consequences of Climate Change

The rapid rate of climate change and its magnitude is a great concern globally nowadays. Human activities involving higher energy consumption, industrialization, intensive agriculture, and urban and rural development has increased greenhouse gases emission in the atmosphere since the Industrial Revolution, leading to more heat retention and a rise in global temperature and high spatial and temporal variability. This change in temperature regime pattern further leads to appreciable changes in precipitation characteristics like quantity and intensity of rain and snow, their spatial and temporal distribution pattern, etc. Emissions of anthropogenic greenhouse gas (GHG) preceding the preindustrial era largely increased the atmospheric concentrations of carbon dioxide (CO_2), methane (CH_4), and nitrous oxide (N_2O) and it was recorded that about half of the anthropogenic CO_2 emissions since the preindustrial era to 2011 have occurred in the past 40 years. The long-lived greenhouse gases (LLGHGs) increased the total radiative forcing by 43% from 1990 to 2018 (Butler and Montzka 2019) and about 81% of this increase was due to CO_2 concentration. Total annual anthropogenic GHG emissions have increased by about 10 $GtCO_2$ eq. from 2000 to 2010 and it amounts to 52 $GtCO_2$ eq. year^{-1} globally during 2007–2016. The global mean annual temperature at the end of the nineteenth century has increased by 0.4–0.7 °C at the end of the twentieth century, due to the accumulation of GHG (Pathak 2012) and the temperature is predicted to increase by 1.1–6.4 °C by the end of the twenty-first century (IPCC 2007). Increase in the concentration of GHGs (CO_2, CH_4, N_2O) results in the rise in the global mean temperature by 1.1 ± 0.1 °C since the preindustrial era (WMO 2019), increasing at a rate of 0.2 °C per decade (IPCC 2018) and estimated to increase by 4 ± 2 °C at the end of the twenty-first century (IPCC 2013).

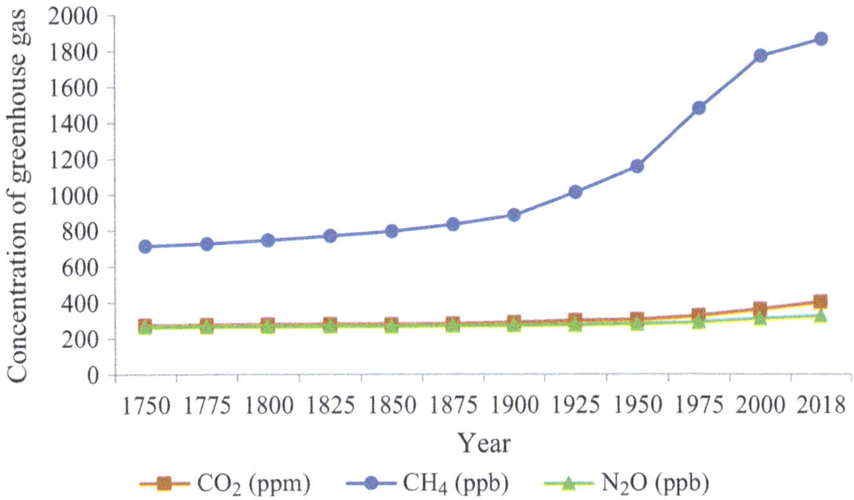

Fig. 28.1 Increase in the global atmospheric concentration of CO_2, CH_4, and N_2O since the preindustrial era to 2018

The increase in the concentration of CO_2 during 2017–2018 was higher than the average growth rate over the last decade. The average rate of increase of CO_2 for three consecutive decades from 1985 to 1995, 1995 to 2005, and 2005 to 2015 increased from 1.42 to 1.86 ppm/year and 2.06 ppm/year, respectively (WMO 2019). The WMO Greenhouse Gas Bulletin (2019) reported that the global atmospheric concentration of GHGs namely CO_2, CH_4, and N_2O has reached a new high in 2018, with 407.8 ppm, 1869 ppb, and 331.1 ppb, respectively, representing about 147%, 259%, and 123%, respectively, of preindustrial era (i.e., before 1750) (Fig. 28.1).

Also, the air temperature at land surface raises nearly twice as much as the global average temperature since the preindustrial period (1750–1850) to 2006–2015. Besides, changing the precipitation characteristics, the rise in temperature also causes a large part of the mountain glaciers, the permafrost soil zone, and the polar ice caps to melt which changes the flow dynamics of water, including flood waves and surface runoff, resulting rise in the eustatic sea level, threatening the low-lying lands, settlements, and agricultural areas. The rise of the sea level increases the chance of sea water intrusion in the major coastal lands resulting in the further extension of salt-affected territories and turning them unsuitable for cultivation (Várallyay 2010).

The climate change results in considerable changes in the natural vegetation and in land-use practices which in turn has a feedback effect on climate by modifying albedo, the heat and energy balance of the near-surface atmosphere, the temperature, and precipitation characteristics. Moreover, the changes in vegetation substantially influence the field water cycle and soil formation processes (Lal et al. 1994).

28.3 Influence of Climate Change on Soil Properties

Global climate change has a major impact on soils as well as the functions performed by the soils. The consequences of climate change are anticipated primarily through rise in soil temperature, alteration in soil moisture content, and increase in CO_2 levels. Soil processes and properties that are responsible for restoring fertility and productivity of soil, expected to be affected by climate mostly through increase in temperature and CO_2 level. The adverse effects of climate change on soil properties are illustrated in Table 28.1.

Table 28.1 Climate change effects on soil

Climatic factors	Effects
Rise in temperature	• Salinization of soil • Soil organic matter decomposition increases • Loss of soil organic matter • Decreases soil porosity • Increases soil compactness • Reduction of soil CEC • Reduction of soil fertility • Deterioration of soil structure • Increases risk of soil erosion • Reduction of water retention capacity • Increases CO_2 release from soil • Reduction of soil organic C • Increases ammonia volatilization • Increases rhizospheric temperature • Stimulation of nutrient acquisition • Enhances soil microbial activity • Increases bioavailability of N and P from organic matter
Heavy and intensive rainfall	• Destruction of soil aggregate • Increases risk of soil erosion • Increases leaching of basic cations • Soil acidification • Reduces soil CEC • Loss of soil nutrients, especially N • Development of hypoxic condition in poorly drained soil • Toxicities of Fe, Mn, Al, and B • Loss of N through denitrification
Decreased rainfall	• Increases salt content • Soil moisture deficit • Decreases diffusion and mass flow of water-soluble nutrients • Possibility of occurring drought • Loss of nutrient from rooting zone through erosion • Reduces nutrient acquisition capacity of root system • Reduces N-fixation in legumes
Increase in atmospheric CO_2	• Increases soil C availability • Increases soil microbial activity • Increase soil fungal population

28.3.1 Soil Physical Properties

Soil physical properties determine the hydrological properties of soils (e.g., water retention, hydraulic conductivity), of which soil texture, structure, bulk density, porosity, and pore-size distribution are the most essential, which ultimately manages the water, air, and temperature of the soil profile (Hillel 1973). The physical properties of the soil have a great impact on soil fertility through the chemical and biological processes of soil, such as adsorption, water, heat and mass transport, nutrient supply, biological activity, etc. (Horel et al. 2015). The physical properties of soils are strongly interrelated to their sensitivity to climate change. Soil hydro-physical properties are highly influenced by the variations in climatic factors such as seasonal temperatures or precipitation intensities, which affect the soil water regime (Horel et al. 2014). The increase in temperature and CO_2 concentration, changes in rainfall pattern and their interactions due to climate change are expected to influence several soil physical processes which will make the soils susceptible to a substantial risk of salinization, reduced water and nutrient availability, altered C and N dynamics, and decrease in soil biodiversity (Benbi and Kaur 2009). Soil moisture stress is reported (Mills et al. 2014) to reduce the soil functions which consequently affect plant productivity. Important soil physical properties that affect soil fertility due to the influence of climate change are as follows.

28.3.1.1 Soil Texture

Although soil texture is a rather constant soil parameter, it can greatly influence the soil characteristics and regulates the sensitivity of soil to change with the climatic factors. The shrinking swelling clay soils are sensitive to climate change if the number of wetting and drying cycles rises, since crack formation is greatly promoted through repeated wetting and drying of the soil. As these cracks deepen rapidly, water moves directly from the surface soil to permeable substrata or drains through bypass flow, which decreases the filtering function of the soil and as a result increases loss of nutrients from soil and pollutes the water bodies (Rounsevell et al. 1999; IPCC 2007). This phenomenon could become intense if longer and more frequent droughts are followed by more intensive precipitation.

The seasonal soil moisture regime is strongly influenced by the regional climate changes and climate-induced changes in capillary water movement from groundwater to root zone. Silt soils are reported (Bormann 2012) to be the most sensitive, while clay soils are least sensitive to climate change, independent of plant cover, impact of groundwater, and regional climate change. Soil texture is the dominant influencing factor determining the response of soil to regional climate change.

28.3.1.2 Soil Structure

Considerable changes in precipitation pattern and increase in temperature due to climate change influences the soil structure (type, spatial distribution, and aggregate stability of soil) in a very complicated process (Várallyay 2010), by involving the processes of slaking, dispersion, mechanical disturbance, and compaction (Reubens et al. 2007). Intensive rainfall, surface runoff and filtrating water during heavy rains, thunderstorms have a direct impact on soil structure (Várallyay 2010) through their aggregate destructing role (Singh et al. 2011). Since, the presence of organic matter in soil, their amount and quality influence the nature and quality of the structure, a decline in its levels in soil results in a decrease in soil aggregate stability, infiltration rates and increase in susceptibility to compaction, run-off as well as to erosion (Karmakar et al. 2016). Climate change may indirectly affect the vegetation pattern and land-use practices (Singh et al. 2011) and also soil biological function (due to the sensitivity of soil macrofauna and microorganisms to climate change), which in turn affect the soil structure (Várallyay 2010).

28.3.1.3 Bulk Density and Porosity

Bulk density is highly influenced by textural properties and organic matter content of soil and is also climate dependent. The loss of soil organic matter due to an increased rate of decomposition owing to elevated temperature (Davidson and Janssens 2006) or soil erosion may increase bulk density, which may further make soil more prone to compaction (Birkás et al. 2009; Singh et al. 2011), resulting in a decrease in porosity and formation of compact layer inhibiting root growth. Variation in soil moisture and temperature regimes due to changes in climate as well as soil hydro-physical properties like bulk density or the development of preferential flow paths alters the root development, as well as microbial activity (Rosenzweig and Hillel 1995). Soil structure along with porosity and pore size distribution primarily influences the moisture and aeration status of the soil. Thus, any changes in porosity causes direct effect on the soil water holding capacity (infiltration, water retention) and emission of CO_2 (in aerobic conditions) and CH_4 (in anaerobic conditions) from soil (Singh et al. 2011; Tóth et al. 2018). Future climate change scenarios like increase in temperature and CO_2 level and variable and extreme rainfall events may affect soil porosity and pore-size distribution and consequently soil functions in unexpected directions which may further alter root development and soil biological activities.

28.3.1.4 Soil Water Retention and Availability

The availability of water for plant growth and important soil processes is determined by some soil properties including porosity, field capacity, lower limit of plant-available water, plant-available water capacity, macropore flow, and texture (Reynolds et al. 2002; Jarvis 2007). Climate change, especially variable and high intensity of rain or drought, greatly influences the soil water availability and its distribution. Soil water availability and water retention are interdependent on the infiltration rate and strongly determine the functioning of the ecosystem. Singh et al. (2011) reported that infiltration is one of the most important soil properties which help to increase soil water retention, soil erosion mitigation, and decrease the risk of flash floods and droughts.

Holsten et al. (2009) evaluated an expected decrease of 4–15% in the average soil water availability in a climate scenario model by the middle of the twenty-first century. The predicted changes in the spatial and temporal distribution of annual rainfall would increase both the duration and severity of water stress in soils. According to Farkas et al. (2014), both an excessive amount of rainfall (e.g., waterlogging) and its scarcity (drought) could cause unfavorable soil conditions for both natural and agroecosystems.

28.3.2 Soil Chemical Properties

Although various chemical reactions occurring in soil accelerated with global warming, chemical composition and/or fertility of soil deteriorated with the increasing soil temperature (Verma and Jayakumar 2012). Rise in sea level and variation in precipitation pattern and flood flows due to climate change may influence acid sulphate soils production. Variation in temperature elevates the freeze and thaw cycle causing an increase in exchangeable ammonium nitrogen and a decrease in exchangeable potassium in soil, inducing water to move along with nutrient and temperature gradient which in turn affects nutrient availability, cation exchange capacity, soil weathering, and also biological activities (Marion 1995). The direct impacts of climate change on some soil chemical properties are discussed below.

28.3.2.1 Soil pH

The pH value of most soil is not expected to vary rapidly due to the direct effects of climate change like elevated temperature, variable precipitation, CO_2 fertilization, and atmospheric N deposition (Brinkman and Sombroek 1999; McCarthy et al. 2001). However, these carriers of climate change influence soil organic matter content, carbon and nutrient cycle, plant-available water, and hence plant productivity, which in turn may affect the soil pH (Reth et al. 2005). Increased precipitation can intensify the leaching of basic cations causing soil acidification. Lowering of soil pH affects soil chemistry by promoting the mobilization of potentially toxic elements (heavy metals), and may result in depletion of basic cations through leaching (Brinkman 1990) in well-drained and structurally stable soil and also heavy and intense rainfall receiving soil. Hence, soil acidification could increase in a wetter climate if buffering pools become exhausted.

Moreover, alterations of rainfall patterns due to diurnal fluctuations as well as seasonal variations as a consequence of climate change may cause an effect on soil pH. During seasons with low to moderate rainfall (evapotranspiration exceeds precipitation), the salt content is increased which forces more exchangeable H^+ ions into the soil solution and tends to reduce pH. Conversely, during wet seasons, salts are diluted or removed from the topsoil by deep percolation and increases the soil pH (Rengel 2002). This seasonal alteration in the total salt content is different from long-term effects occurring over decades and centuries, where basic cations leach out due to increased rainfall, resulting in soil acidification (Tang and Rengel 2003).

28.3.2.2 Cation Exchange Capacity

Cation exchange capacity (CEC) is a key determinant of soil fertility, especially the retention of major cationic nutrients Ca, Mg, and K, and immobilization of potentially toxic cations Al and Mn. Since CEC is associated with the organic matter content in soil (Weil and Magdoff 2004), a higher decomposition rate and decline in SOM due to rise in temperatures (Davidson and Janssens 2006) results in a reduction of CEC of coarse-textured soils and low-activity clay soils. Low CEC of soil may result from increased leaching of basic cations due to high and intense rainfall.

The percentage of CEC occupied by basic cations (Ca^{2+}, Mg^{2+}, K^+, and Na^+) is termed as the percent base saturation, which generally increases with an increase in soil pH. In very acidic soils, H^+ and Al^{3+} are the dominant ions while in neutral and moderately alkaline soils, Ca^{2+} and Mg^{2+} ions predominate (Robertson et al. 1999). Ca^{2+} is effective in maintaining and protecting soil aggregate structure, while a high amount of Na^+ in the case of salt-affected soils (especially in Solonetz) causes poor aggregate structure. Thus, the quality and ratio of adsorbed cations in soil have a significant role on development of the soil aggregate structure, and consequently in soil water management (Amezketa 1999).

28.3.2.3 Soil Salinization

In the last 150–200 years, rapid changes have occurred in the gas content of the atmosphere due to human activities like industrialization, excessive energy consumption, intensive agriculture, urban and rural development, as well as natural events. Evapotranspiration increases with the rise in temperature and salt leaching potential of the rainfall decreases with the decrease in rainfall (Schofield and Kirkby 2003; De Paz et al. 2012). The resultant rapid warming of the atmosphere may also cause glacial melting, a rapid rise in sea levels, excessive rains following floods, ensuing submergence of agricultural areas. As a consequence of the thermal expansion, sea levels may rise by 3 mm $year^{-1}$ due to glacial melting on earth (Vengosh

2005), which may lead to rapid salinization of agricultural lands and groundwater and cannot be controlled. This may completely change the soil formation processes, soil degradation, and water holding capacities, the effect being more rapid in arid and semiarid regions. Salinization and alkalization may naturally make the soil weaker and more susceptible to wind and water erosion. Among the various indicators of climate change, salt dynamics are most sensitive to change, thus aggravating as well as creating soil salinization by climate change (IPCC 1996; Schofield and Kirkby 2003). Climate change is expected to influence salt accumulation and wind deposition, which are the primary causes of salinization (IPCC 1996). This increase in salinization owing to global climate changes has been even faster especially in the last 20 to 30 years (Okur and Örçen 2020).

28.3.2.4 Nutrient Cycle in Soil

The nutrient cycle, especially nitrogen, which is closely associated with the carbon and water cycles, is an important aspect of soil fertility. Thus, factors affecting the carbon cycle and water will ultimately affect the nutrient availability in soil.

Elevated temperatures, variable precipitation, and atmospheric N deposition due to climate change have a major impact on N cycle, which is intimately associated with soil organic carbon cycle (Weil and Magdoff 2004) and possibly on other plant-available nutrient cycles such as phosphorus and sulphur.

Higher temperatures may cause higher decomposition of soil organic matter. Spatial variation in microbial respiration (i.e., soil CO_2 emission) is found to be positively correlated with the soil pH and fine root mass (Reth et al. 2005). Climate change resulting in a hotter and drier climate in many parts of the world may alter C and N cycle in soils and results in soil acidification and consequently alters the community structures of plant, animal, and microbes. However, Smith et al. (2002) analyzed the influence of climate on soil microbial activity and soil properties and concluded that a predicted rise in temperature and reduction in precipitation over the next centennial would subsequently cause an increase in soil pH and decrease in soil electrical conductivity. This results in decreased nitrification potential and an increased ammonium concentration.

Ammonia volatilization, which occurs with the rise in soil pH, leads to reduction in content of manure slurries and may result in soil acidification and eutrophication problems. Rise in temperature due to climate change may lead to an increase in ammonia volatilization (Van der Stelt et al. 2007), which may intensify N pollution.

Soil is considered as an integral part of the C and N cycles of the global climate system. The rise in temperature and increasing levels of ozone owing to climate changes may cause reduction in plant growth under elevated atmospheric CO_2 levels. According to Kirkham (2011), emission of higher levels of CO_2 occur during the decomposition of plant tissues under elevated atmospheric CO_2, and soil organic C levels are found to decline under increased atmospheric CO_2 levels due to increased microbial activity (Carney et al. 2007). Therefore, elevated CO_2 levels instead of increasing soil C sequestration may lead to more C turnover (Eglin et al. 2011).

Rise in global temperature may adversely affect C allocation to the soil, leading to increased release of CO_2 from soils to the atmosphere resulting in reduction in soil organic C and thus creates a positive-feedback in the global C cycle (Gorissen et al. 2004; Wan et al. 2011). Soil warming and drying was found to decrease soil C by 32% over 5 year, the reduction which is much more rapid than due to increased tillage (Link et al. 2003).

28.3.2.5 Soil Fertility and Nutrient Acquisition

The carriers of climate change such as variation in precipitation pattern, elevated temperature, and CO_2 level are expected to influence variably on various soil processes and properties, relevant to soil fertility. Increases in air temperature and variation in precipitation significantly influences root zone temperature and moisture regime, which primarily determines the nutrient availability to plants, root growth and development, carbon allocation to roots, and govern nutrient acquisition. Soil moisture deficit directly influences the availability of soil nutrients and their transport. Moreover, the supply of water to plants, air and temperature regimes, biological activity, and plant nutrient availability in the soil are determined by soil moisture regime.

Drought may increase the susceptibility to nutrient losses from the rooting zone through erosion (Gupta 1993). Nutrient diffusion (for short distances) and mass flow (for longer distances) of water-soluble nutrients such as nitrate, sulphate, calcium, magnesium, and silicon decreased due to soil moisture deficit (Mackay and Barber 1985; Barber 1995). Thus, roots tend to increase their length and surface area and modify their architecture to capture the less mobile nutrients like phosphorus (Lynch and Brown 2001). Under drought conditions, the root growth is reduced and root function is impaired, which causes reduction of the nutrient acquisition capacity of root systems (Marschner 1995). Drought condition also inhibits nitrogen fixation in legumes by reducing both carbon and oxygen fluxes and nitrogen accumulation in root nodules (González et al. 2001; Ladrera et al. 2007; Athar and Ashraf 2009). Furthermore, drought modifies the composition of soil microbial communities and their activity which determines the C and N transformations that regulates soil fertility and nutrient cycling (Schimel et al. 2007).

On the other hand, excessive precipitation causes surface erosion which leads to significant soil nutrient loss (Zougmoré et al. 2009), especially nitrate leaching due to its high mobility in soil resulting in substantial loss of N in agriculture. A poorly drained agricultural soil or that receives frequent and/or intense rainfall becomes waterlogged that develops into hypoxic. The depletion of oxygen leads to change in soil redox status which may cause elemental toxicities of Fe, Al, Mn, and B, reducing the crop yields (Setter et al. 2009) and produces phytotoxic organic solutes by impairing growth and function of the root (Marschner 1995). Hypoxia may also lead to nutrient deficiency since the active transport of ions into root cells is carried out by ATP synthesized in mitochondria through oxidative phosphorylation by electron transport chain (Drew 1988; Atwell and Steer 1990). Under this condition, significant loss of nitrogen through denitrification may also occur as nitrate is used as an alternative electron acceptor by the microorganisms in the absence of oxygen (Marschner 1995).

The rise in temperature increases the transpiration rates, which increases the tendency of plants to acquire water-soluble nutrients more readily. Moreover, the increase in the rhizospheric temperature stimulates nutrient acquisition by increasing nutrient uptake through faster ion diffusion rates and increased root metabolism (Bassirirad 2000). However, nutrient uptake is dependent on adequate soil moisture. Under the dry conditions, higher temperatures decrease nutrient acquisition driven by mass flow (Cramer et al. 2009) since extreme vapor pressure deficits may trigger stomatal closure (Abbate et al. 2004).

28.3.3 Soil Biological Properties

Soil biological properties are also interlinked with other soil physical and chemical properties such as aeration, soil organic matter, or pH, influencing the soil microbial activity which in turn performs relevant activities in carbon and nutrients cycling.

28.3.3.1 Soil Organic Matter

The soil organic matter (SOM), a strong determinant of soil fertility, regulates the majority of soil functions such as cation exchange and water holding capacity and has major control on soil pH. Moreover, it improves soil aggregation and increases water retention for use by plants. Thus, a decrease in SOM may reduce soil fertility and biodiversity and also deteriorates the soil structure, which reduces the water retention capacity, making the soil prone to erosion and also increases bulk density leading to the compactness of soil (Weil and Magdoff 2004).

Generally, the decomposition of organic matter in soil accelerates with the rise in temperature. Thus, an enhancement in soil organic matter resulting from higher plant productivity and its contribution to soil due to an increase in temperature along with precipitation which may increase CO_2 fertilization and deposition of atmospheric N.

In many ecosystems, changes in soil moisture content due to climate change may greatly influence SOM decomposition as compared with rise in temperature. An oxic condition prevails in the unsaturated zone, which may be dominated by efficient aerobic processes, accelerates the rate of decomposition of organic matter (Golovchenko et al. 2007). Maximum microbial activity and SOM decomposition rates are found when the soil moisture content is about 50–60%. In arid or semiarid ecosystems, decomposition process is inhibited by low soil moisture availability and thus increased soil moisture results in enhanced SOM decomposition. In other

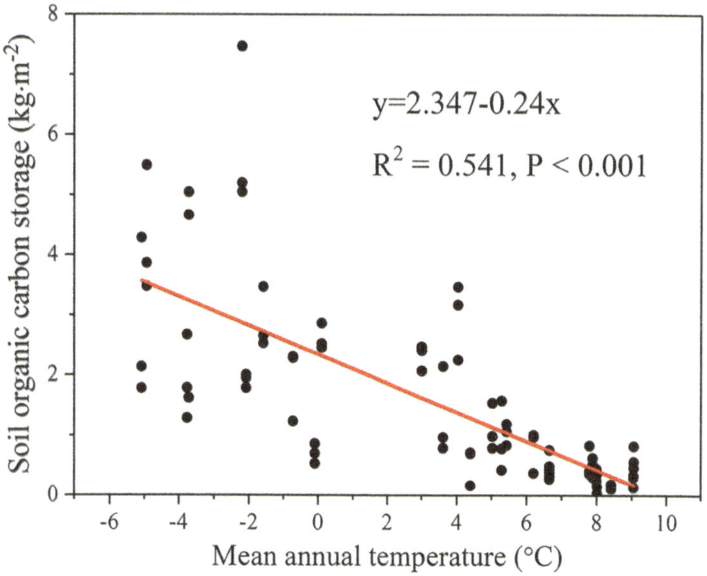

Fig. 28.2 Relationship between SOC storage and mean annual temperature (MAT) (Adapted from Tan et al. 2020)

ecosystems, excess soil moisture may constrain the decomposition process (Plante and Conant 2014).

On the other hand, the content of SOM represents the balance between inputs through surface litter or root exudates and turnover and outputs or removals through decomposition and mineralization to CO_2, leaching of dissolved organic C, or erosion. Thus, the magnitude of potential feedback on soil carbon content and finally on global warming is determined by net gain or loss of SOM by elevated temperatures and altered precipitation patterns (Field et al. 2007). The organic carbon turnover may be increased by drier and warmer soil conditions, as reported by Buytaert et al. (2011).

Global warming has also a profound effect on soil organic carbon (SOC) storage, which was found to decrease with the rise in mean annual temperature (MAT) ($r = -0.735$, $p < 0.001$) (Tan et al. 2020) (Fig. 28.2). The soil organic carbon storage is regulated by two different biological processes. SOC storage is depleted by the decomposition of microbes due to their accelerated activity owing to increased temperature. But, on the contrary, plant production is promoted by increased temperature which results in increased SOC storage. Thus, microbial decomposition to temperature has a greater response relative to plant C input, explaining the decreased SOC storage with MAT.

28.3.3.2 Nutrient Transformation in Soil

Changes in moisture and rise in temperature due to global climate change may strongly influence biological transformation between organic and inorganic pools in soil. Increased temperature results in enhanced soil microbial activity which may increase the release of N and P from organic matter in bioavailable forms (Brown and Braaten 1998; Weintraub and Schimel 2005). Various studies also reported that soil warming may increase the inorganic pools of N and P in soil (Schimel et al. 2004; Natali et al. 2011) through increasing the rate of ammonification, nitrification, and P mineralization (Rustad et al. 2001). Increased temperature may accelerate the rates of adsorption and desorption reactions. Reduction in water table occurring from variation in precipitation pattern causes a rise in the depth of unsaturated zone (Macrae et al. 2013), and the supply of associated additional oxygen in drier soil resulting in an increase in the availability of plant nutrients due to oxidation of organic matter.

28.3.3.3 Soil Microbial Biomass

Soil microbial biomass (SMB), the living part of organic matter and the most labile C pool in soils, reflects microbial size and soil fertility status. The soil microbial biomass carbon (C) and nitrogen (N) consequently tend to reflect the organic matter contents of soils. Soil microbial biomass is responsive to short-term changes of environment (Haynes 2008) and significantly declines with long-term simulated climatic warming experiments (Rinnan et al. 2007). Warming or elevated temperature may stimulate metabolic activity of microorganism, their abundance, and nutrient cycling, particularly in temperate ecosystems (Vinolas et al. 2001), if water and nutrient availability are not limiting the growth (Pilegaard et al. 2006; Castro et al. 2010). Increased temperature may also increase soil respiration (Lin et al. 1999; Niinisto et al. 2004). Elevated carbon dioxide levels over a longer period may have a little direct effect on MBC and community structure (Niklaus et al. 2003). Soil C availability may increase with increasing CO_2 level which stimulates soil microbial activity and the increase of fungal biomass will be more compared with bacterial biomass. An increase in fungal communities may have a strong impact on various soil functions such as degradation of organic matter, nutrient cycling, plant nutrition, and aggregate formation. As fungi are characterized with a higher C/N ratio and lower demand for nitrogen than bacteria, lower availability of N in soil at elevated CO_2 level may increase the fungal population (Hu et al. 2001). The C allocation increases with elevated CO_2 levels which may lead to an increase in bacterial growth leading to higher turnover of the microbial biomass due to grazing by predators, resulting in faster recycling of nutrient from the microbial biomass, increasing the nutrient flux to plant.

28.4 Adaptation and Mitigation of Climate Change

The adverse effects of climate change such as increase or decrease in rainfall and temperatures or other extreme weather conditions can be minimized by following both adaptation and mitigation measures. Adaptation measures attributed to the ability to adjust and increase resilience to a variable and adverse change in the climate. This can be achieved by adopting conservation agriculture practices to maintain sufficient moisture, organic matter, and nutrients in the soil through zero tillage, mulching, crop rotation, avoiding mono-cropping, adjusting the timing of farming operations, application of appropriate quantities of inputs, i.e., irrigation, fertilizers, pesticide, etc., in proper time.

On the other hand, mitigation measures involve attenuation of climate change effects by reducing the greenhouse gas concentration in the atmosphere. Climate change can be mitigated by reducing emissions of GHGs through proper management of C and N in an agricultural system or by absorbing the GHGs by creating C sinks through sequestration and conservation of soil C. Emission of GHGs through soil organic carbon volatilization can be reduced by improving agricultural management practices through minimal soil disturbance and improving grazing management. Emissions can be reduced by following integrated nutrient management through maintaining the appropriate rate and timing of fertilizer application which reduces both leaching and volatilization losses by improving fertilizer-use efficiency. Since the major sources of methane emission in agriculture are from enteric fermentation in ruminants, paddy cultivation, and storing livestock manure, thus, improving the management of livestock waste, covering manure heaps and irrigation in paddy cultivation can reduce its emission. Nitrous oxide emission can be reduced through maintaining appropriate rate, method, and type of N fertilizer application and soil management, using nitrification inhibitors viz., nitrapyrin, dicyandiamide, neem cake, neem oil, karanj cake, etc. GHG emission can also be reduced by maintaining a shallow water table and avoiding deep ploughing unless it may lead to the draining of organic soil. Checking deforestation and maintaining existing forest cover can also help in the reduction of emissions. Emission of carbon dioxide can be mitigated by sequestering carbon in soil through improving the management practices for soil moisture and temperature, reducing biomass burning, and restoring C in soils of degraded land.

Sequestration of soil C, as well as soil functional stability, may be encouraged through following various conservation agricultural practices such as reducing soil disturbance, including cover crops and incorporating green manure or legume crops in crop rotation, practicing contour farming, and mulching, maintaining soil cover to restore soil organic matter and following integrated nutrient management system. Soil C sequestration may be increased by following the management practices that increased the input of C to the soil with a simultaneous reduction in loss of C. The C input in soil may be achieved by selecting high biomass-producing crops, practicing reduced tillage to retain crop residue, regular organic matter application, following crop rotation, etc. Loss of soil C could be reduced by following conservation agriculture and controlling soil erosion by reducing disturbance to soil.

28.5 Conclusion

The influences of global climate change are expected to modify soil physical parameters like texture, structure, bulk density, porosity, nutrient retention, etc., affecting the soil fertility through which it may cause soil salinization, reduce nutrient and water availability, alters C and N dynamics, and decreases soil biodiversity. The adverse effect of climate change mostly affects the chemical properties of soil such as soil pH, salinity, cation exchange capacity, nutrient cycle, and acquisition. Soil physical and chemical properties are highly correlated with soil biological properties which in turn balance the carbon and nutrient cycle of the soil and ultimately the soil fertility. The soil fertility is governed by the soil organic matter content of the soil, since most of the soil functions such as pH, cation exchange capacity, water and nutrient retention, as well as soil structure are dependent on soil organic matter, the variation in decomposition rate of which due to global warming adversely affects the soil fertility. Thus, the detrimental effects of climate change on soil fertility can be minimized by following some of the adaptation and mitigation measures such as conservation agriculture, residue management, integrated nutrient management practices, etc.

References

Abbate PE, Dardanelli JL, Cantarero MG, Maturano M, Melchiori RJM, Suero EE (2004) Climatic and water availability effects on water-use efficiency in wheat. Crop Sci 44:474–483

Amezketa E (1999) Soil aggregate stability: a review. J Sustain Agric 14:83–151

Athar H, Ashraf M (2009) Strategies for crop improvement against salinity and drought stress: an overview. In: Salinity and water stress. Springer, New York, pp 1–16

Atwell BJ, Steer BT (1990) The effect of oxygen deficiency on uptake and distribution of nutrients in maize plants. Plant Soil 122:1–8

Barber SA (1995) Soil nutrient bioavailability: a mechanistic approach, 2nd edn. Wiley, New York, p 414

Bassirirad H (2000) Kinetics of nutrient uptake by roots: responses to global change. New Phytol 147:155–169

Benbi DK, Kaur R (2009) Modeling soil processes in relation to climate change. J Ind Soc Soil Sci 57:433–444

Birkás M, Dexter A, Szemők A (2009) Tillage-induced soil compaction, as a climate threat increasing stressor. Cereal Res Commun 37:379–382

Bormann H (2012) Assessing the soil texture-specific sensitivity of simulated soil moisture to projected climate change by SVAT modelling. Geoderma 185–186:73–83

Brett P (2009) Awareness, opinions about global warming vary Worldwide. The Gallup Organization

Brevik EC (2012) Soils and climate change: gas fluxes and soil processes. Soil Horiz 53. https://doi.org/10.2136/sh12-04-0012

Brevik EC (2013) The potential impact of climate change on soil properties and processes and corresponding influence on food security. Agriculture 3:398–417

Brinkman R (1990) Chapter 5 Resilience against climate change? In: Developments in soil science, pp 51–60

Brinkman R, Sombroek W (1999) The effects of global change on soil conditions in relation to plant growth and food production. In: Bazzaz F, Sombroek W (eds) Global climate change and agricultural production. Food and Agriculture Organization of the United Nations and John Wiley & Sons, Rome, Italy

Brown RD, Braaten RO (1998) Spatial and temporal variability of Canadian monthly snow depths, 1946–1995. Atmos Ocean 36:37–54

Butler JH, Montzka SA (2019) The NOAA Annual Greenhouse Gas Index (AGGI). http://www.esrl.noaa.gov/gmd/aggi/aggi.html

Buytaert W, Cuesta-Camacho F, Tobon C (2011) Potential impacts of climate change on the environmental services of humid tropical alpine regions. Glob Ecol Biogeogr 20:19–33

Carney KM, Hungate BA, Drake BG, Megonigal JP (2007) Altered soil microbial community at elevated CO_2 leads to loss of soil carbon. Proc Natl Acad Sci U S A 104:4990–4995

Castro HF, Classen AT, Austin EE, Norby RJ, Schadt CW (2010) Soil microbial community responses to multiple experimental climate change drivers. Appl Environ Microbiol 76 (40):999–1007

Cramer MD, Hawkins HJ, Verboom GA (2009) The importance of nutritional regulation of plant water flux. Oecologia 161:15–24

Davidson EA, Janssens IA (2006) Temperature sensitivity of soil carbon decomposition and feedbacks to climate change. Nature 440:65–173

De Paz JM, Viscontia F, Molina MJ, Ingelmo F, Martinez D, Sanchezb J (2012) Prediction of the effects of climate change on the soil salinity of an irrigated area under Mediterranean conditions. J Environ Manag 95:53783

Drew M (1988) Effects of flooding and oxygen deficiency on the plant mineral nutrition. In: Lauchli A, Tinker PB (eds) Advances in plant nutrition, vol III. Praeger, New York, pp 115–159

Eglin T, Ciasis P, Piao SL, Barré P, Belassen V, Cadule P, Chenu C, Gasser T, Reichstein M, Smith P (2011) Overview on response of global soil carbon pools to climate and land-use changes. In: Sauer TJ, Norman JM, MVK S (eds) Sustaining soil productivity in response to global climate change: science, policy, and ethics. Wiley, Oxford, UK, pp 183–199

Farkas C, Gelybó G, Bakacsi Z, Horel A, Hagyó A, Dobor L, Kása I, Tóth E (2014) Impact of expected climate change on soil water regime under different vegetation conditions. Biologia 69:1510–1519

Field CB, Lobell DB, Peters HA, Chiariello NR (2007) Feedbacks of terrestrial ecosystems to climate change. Annu Rev Environ Resour 32:1–29

Golovchenko AV, Tikhonova EY, Zvyagintsev DG (2007) Abundance, biomass, structure, and activity of the microbial complexes of minerotrophic and ombrotrophic peatlands. Microbiology 76:630–637

González EM, Gálvez L, Royuela M, Aparicio-Tejo PM, Arrese-Igor C (2001) Insights into the regulation of nitrogen fixation in pea nodules: lessons from drought, abscisic acid and increased photoassimilate availability. Agronomie 21:607–613

Gorissen A, Tietema A, Joosten NN, Estiarte M, Peñuelas J, Sowerby A, Emmett BA, Beier C (2004) Climate change affects carbon allocation to the soil in shrublands. Ecosystems 7:650–661

Gupta JP (1993) Wind erosion of soil in drought-prone areas. In: Sen A, Kar A (eds) Desertification and its control in the Thar, Sahara and Sahel Regions. Jodhpur, Scientific Publishers, pp 91–105

Haynes RJ (2008) Soil organic matter quality and the size and activity of the microbial biomass: their significance to the quality of agricultural soils. In: Soil mineral microbe-organic interactions. Springer, Berlin, pp 201–231

Hillel D (1973) Soil and physical principles and processes, 3rd edn. Academic Press, New York, p 248

Holsten A, Vetter T, Vohland K, Krysanova V (2009) Impact of climate change on soil moisture dynamics in Brandenburg with a focus on nature conservation areas. Ecol Model 220:2076–2087

Horel A, Lichner L, Alaoui A, Czachor H, Nagy V, Tóth E (2014) Transport of iodide in structured clay–loam soil under maize during irrigation experiments analysed using HYDRUS model. Biologia 69:1531–1538

Horel Á, Tóth E, Gelybó GY, Kása I, Bakacsi ZS, Farkas CS (2015) Effect of land use and management on soil hydraulic properties. Open Geosci 1:742–754

Hu S, Chapin FS, Firestone MK, Field CB, Chiariello NR (2001) Nitrogen limitation of microbial decomposition in a grassland under elevated CO_2. Nature 409:188–191

Intergovernmental Panel on Climate Change (1996) Impacts, adaptations and mitigation of climate change: scientific-technical analyses: special report of working group II. Cambridge University Press, New York

Intergovernmental Panel on Climate Change (2018) Summary for policymakers. In: Masson-Delmotte V, Zhai P, Pörtner HO, Roberts D, Skea J, Shukla PR, Pirani A, Moufouma-Okia W, Péan C, Pidcock R (eds) Global warming of 1.5°C; an IPCC special reports on impacts of global warming of 1.5°C above pre industrial levels and related global greenhouse gas emission pathways in context of strengthening the global response to the threat of climate change, sustainable development, and efforts to eradicate poverty. Intergovernmental Panel on Climate Change, Geneva, Switzerland

IPCC (2007) Summary for policymakers. In: Climate change 2007: impacts, Adaptation and Vulnerability. Contribution of Working Group II to the Fourth Assessment Report of the Intergovernmental Panel on Climate Change. Cambridge University Press, Cambridge, UK, p 17

IPCC (2013) Summary for policymakers. In: Stocker TF, Qin D, Plattner GK, Tignor M, Allen SK, Boschung J, Nauels A, Xia Y, Bex V, Midgley PM (eds) Climate change 2013: the physical science basis. Contribution of Working Group I to the Fifth assessment report of the intergovernmental panel on climate change. Cambridge University Press, Cambridge

Jarvis NJ (2007) A review of non-equilibrium water flow and solute transport in soil macropores: principles, controlling factors and consequences for water quality. Eur J Soil Sci 58:523–546

Karmakar R, Das I, Dutta D, Rakshit A (2016) Potential effects of climate change on soil properties: a review. Sci Int 4:51–73

Kirkham MB (2011) Elevated carbon dioxide. CRC Press, Boca Raton, FL

Ladrera R, Marino D, Larrainzar E, Gonzalez EM, Arrese-Igor C (2007) Reduced carbon availability to bacteroids and elevated ureides in nodules, but not in shoots, are involved in the nitrogen fixation response to early drought in soybean. Plant Physiol 145:539–546

Lal R et al (eds) (1994) Soil processes and greenhouse effect. CRC Lewis Publishers, Boca Raton, FL, p 440

Lin GH, Ehleringer JR, Rygiewicz PT, Johnson MG, Tingey DT (1999) Elevated CO_2 and temperature impacts on different components of soil CO_2 efflux in Douglas-fir terracosms. Glob Chang Biol 5:157–168

Link SO, Smith JL, Halverson JJ, Bolton H Jr (2003) A reciprocal transplant experiment within a climatic gradient in a semiarid shrub-steppe ecosystem: effects on bunchgrass growth and reproduction, soil carbon, and soil nitrogen. Glob Change Biol 9:1097–1105

Lynch JP, Brown KM (2001) Topsoil foraging – an architectural adaptation of plants to low phosphorus availability. Plant Soil 237:225–237

Mackay AD, Barber SA (1985) Soil moisture effects on root growth and phosphorus uptake by corn. Agron J 77:519–523

Macrae ML, Devito KJ, Strack M, Waddington JM (2013) Effect of water table drawdown on peatland nutrient dynamics: implications for climate change. Biogeochemistry 112:661–676

Marion GM (1995) Freeze-thaw processes and soil chemistry. CRREL Special Report 95-12, p 29

Marschner H (1995) Mineral nutrition of higher plants, 2nd edn. Academic Press, New York
McCarthy JJ, Canziani OF, Leary NA, Dokken DJ, White KS (eds) (2001) Climate change 2001: impacts, adaption, and vulnerability. Contribution of working Group II to the Third Assessment Report of the Intergovernmental Panel on Climate Change. University Press, Cambridge
Mills RTE, Gavazov KS, Spiegelberger T, Johnson D, Buttler A (2014) Diminished soil functions occur under simulated climate change in a sup-alpine pasture, but heterotrophic temperature sensitivity indicates microbial resilience. Sci Total Environ 473:465–472
Natali SM, Schuur EA, Trucco C, Hicks Pries CE, Crummer KG, Baron Lopez AF (2011) Effects of experimental warming of air, soil and permafrost on carbon balance in Alaskan tundra. Glob Change Biol 17:1394–1407
Niinisto SM, Silvola J, Kellomaki S (2004) Soil CO_2 efflux in a boreal pine forest under atmospheric CO_2 enrichment and air warming. Glob Change Biol 10:1–14
Niklaus PA, Alphei J, Ebersberger D, Kampichler C, Kandeler E, Tscherko D (2003) Six years of in situ CO_2 enrichment evoke changes in soil structure and soil biota of nutrient-poor grasslands. Glob Change Biol 9:585–600
Okur B, Örçen N (2020) Soil salinization and climate change. In: Prasad MNV, Pietrzykowski M (eds) Climate change and soil interactions. Elsevier, New York, pp 331–350
Pathak H (2012) Climate change and agriculture. Climate change impact, adaptation and mitigation in agriculture: methodology for assessment and applications. Indian Agricultural Research Institute, New Delhi, pp 1–3
Pilegaard K, Skiba U, Ambus P, Beier C, Brüggemann N, Butterbach-Bahl K, Dick J, Dorsey J, Duyzer J, Gallagher M, Gasche R, Horvath L, Kitzler B, Leip A, Pihlatie MK, Rosenkranz P, Seufert G, Vesala T, Westrate H, Zechmeister-Boltenstern S (2006) Factors controlling regional differences in forest soil emission of nitrogen oxides (NO and N_2O). Biogeosciences 3:651–661
Plante A, Conant RT (2014) Soil organic matter dynamics, climate change effects. In: Freedman B (ed) Global environmental change. Handbook of global environmental pollution, vol 1. Springer, Dordrecht
Rengel Z (2002) Role of pH in availability of ions in soil. In: Rengel Z (ed) Handbook of plant growth. pH as a master variable in plant growth. Marcel Dekker, New York, pp 323–350
Reth S, Reichstein M, Falge E (2005) The effect of soil water content, soil temperature, soil pH-value and root mass on soil CO_2 efflux - A modified model. Plant Soil 268:21–33
Reubens B, Poesen J, Danjon F, Geudens G, Muys B (2007) The role of fine and coarse roots in shallow slope stability and soil erosion control with a focus on root system architecture: a review. Trees 21:385–402
Reynolds WD, Bowman BT, Drury CF, Tan CS, Lu X (2002) Indicators of good soil physical quality: density and storage parameters. Geoderma 110:131–146
Rinnan R, Michelsen A, Bååth E, Jonasson S (2007) Fifteen years of climate change manipulations alter soil microbial communities in a subarctic heath ecosystem. Glob Chang Biol 13:28–39
Robertson GP, Sollins P, Ellis BG, Lajtha K (1999) Exchangable ions, pH, and cation exchange capacity. In: Robertson GP, Coleman DC, Bledsoe CS, Sollins P (eds) Standard soil methods for long-term ecological research. Oxford University Press, New York, pp 106–111
Rosenzweig C, Hillel D (1995) Potential impacts of climate change on agriculture and food supply. Consequences 1:24–31
Rounsevell M, Evans SP, Bullock P (1999) Climate change and agricultural soils: impacts and adaptation. Clim Chang 43:683–709
Rustad LE, Campbell J, Marion G, Norby R, Mitchell M, Hartley A, Cornelissen J, Gurevitch J (2001) A meta-analysis of the response of soil respiration, net nitrogen mineralization, and aboveground plant growth to experimental ecosystem warming. Oecologia 126:543–562
Schimel JP, Bilbrough C, Welker JM (2004) Increased snow depth affects microbial activity and nitrogen mineralization in two Arctic tundra communities. Soil Biol Biochem 36:217–227
Schimel J, Balser TC, Wallenstein M (2007) Microbial stress-response physiology and its implications for ecosystem function. Ecology 88:1386–1394

Schofield RV, Kirkby MJ (2003) Application of salinization indicators and initial development of potential global soil salinization scenario under climatic change. Glob Biogeochem Cycles 17:1–13
Setter TL, Waters I, Sharma SK, Singh KN, Kulshreshtha N, Yaduvanshi NP, Ram PC, Singh BN, Rane J, McDonald G, Khabaz-Saberi H, Biddulph TB, Wilson R, Barclay I, McLean R, Cakir M (2009) Review of wheat improvement for waterlogging tolerance in Australia and India: the importance of anaerobiosis and element toxicities associated with different soils. Ann Bot 103:221–235
Singh BP, Cowie AL, Chan KY (eds) (2011) Soil health and climate change, soil biology. Springer, Heidelberg, p 414
Smith JL, Halvorson JJ, Bolton H Jr (2002) Soil properties and microbial activity across a 500 m elevation gradient in a semi-arid environment. Soil Biol Biochem 34:1749–1757
Tan Q, Hand W, Lie X, Guoan Wang G (2020) Clarifying the response of soil organic carbon storage to increasing temperature through minimizing the precipitation effect. Geoderma 374:114398
Tang C, Rengel Z (2003) Role of plant cation/anion uptake ratio in soil acidification. In: Rengel Z (ed) Handbook of soil acidity. Marcel Dekker, New York, pp 57–81
Tóth E, Gelybó G, Dencső M, Kása I, Birkás M, Horel Á (2018) Chapter 19, Soil CO_2 emissions in a long-term tillage treatment experiment A2 - Muñoz, María Ángeles. In: Zornoza R (ed) Soil management and climate change. Academic Press, New York, pp 293–307
Van der Stelt B, Temminghoff EJM, Van Vliet PCJ, Van Riemsdijk WH (2007) Volatilization of ammonia from manure as affected by manure additives, temperature and mixing. Bioresour Technol 98:3449–3455
Várallyay G (2010) The impact of climate change on soils and on their water management. Agron Res 8(Special Issue II):385–396
Vengosh V (2005) Salinization and saline environments. Ben Gurion University of the Negev, Beer Sheva, Israel
Verma S, Jayakumar S (2012) Impact of forest fire on physical, chemical and biological properties of soil: a review. Proc Int Acad Ecol Environ Sci 2:168–176
Vinolas LC, Vallejo VR, Jones DL (2001) Control of amino acid mineralization and microbial metabolism by temperature. Soil Biol Biochem 33:1137–1140
Wan Y, Lin E, Xiong W, Li Y, Guo L (2011) Modeling the impact of climate change on soil organic carbon stock in upland soils in the 21st century in China. Agric Ecosyst Environ 141:23–31
Weil RR, Magdoff F (2004) Significance of soil organic matter to soil quality and health. In: Magdoff F, Weil RR (eds) Soil organic matter in sustainable agriculture. CRC Press, Boca Raton, FL, pp 1–43
Weintraub MN, Schimel JP (2005) Nitrogen cycling and the spread of shrubs control changes in the carbon balance of Arctic tundra ecosystems. Bioscience 55:408–415
World Meteorological Organization (WMO) (2019) WMO Greenhouse Gas Bulletin, No.15: The State of Greenhouse Gases in the Atmosphere. Based on Global Observations through 2018
Zougmoré R, Mando A, Leo S (2009) Soil nutrient and sediment loss as affected by Erosion barriers and nutrient source in semi-arid Burkina Faso. Arid Land Res Manag 23:85–101

Chapter 29
Use of Bacterial Strains to Improve Soil Productivity Under Salt Stress

Amrita Kasotia, Ajit Varma, and D. K. Choudhary

Abstract Abiotic stresses such as drought, salinity, metal stress, etc. negatively affects the growth and yield of crops. Therefore, this requires an improved and imperative strategy, i.e., mitigating the adverse environmental conditions, improving per unit productivity, and reducing cost of production of crops. The salinity in the soil may be because of the high concentration of Na^+, K^+, Ca^{2+}, Mg^{2+}, and Cl^- salts, may be coming from weathering minerals, through irrigation water/fertilizers, and sometimes may migrate upward in the soil from shallow groundwater through evaporation. The use of biological products based on plant growth-promoting bacterial inoculants in agriculture offers an environmentally sustainable approach for maintaining the healthy yield of crops in saline habitat. Various traits of benign bacterial pool have been reported on amelioration of salinity stress, e.g., ACC-deaminase, exopolysaccharide, volatile production, Pi solubilization, and IAA production. In the present chapter, we are dealing with deployment of benign microbes in alleviation of salinity to ameliorate soil and plant productivity.

Keywords Abiotic stress · Salinity · Plant growth-promoting bacteria · Environmentally sustainable · Soil amelioration · Rhizosphere

29.1 Introduction

By 2050, the world's human population is estimated to hit 9.7 billion people. Now, it becomes a big challenge for the society to provide a healthy life, food, and shelter to the growing human population. This worldwide increase in human population

A. Kasotia
Mody University of Science & Technology, Lakshmangarh, Rajasthan, India

A. Varma · D. K. Choudhary (✉)
Amity Institute of Microbial Technology (AIMT), Amity University Uttar Pradesh, Noida, Uttar Pradesh, India
e-mail: dkchoudhary1@amity.edu

D. K. Choudhary et al. (eds.), *Climate Change and the Microbiome*, Soil Biology 63,
https://doi.org/10.1007/978-3-030-76863-8_29

Table 29.1 The amount of salt-affected soils in most preponderance region (Wicke et al. 2011)

Region	Area (Mha)
Middle East	189
Oceania (Australia and nearby islands)	169
North Africa	144
Former USSR	126
East Asia	98
South America	84
West Africa	83
USA	77
East Africa	56
South Asia, including India	52
South Africa	22

demands twice the amount of food production as produced in 2012. The demand for meat, dairy, and speciality crops such as fruits, nuts, and vegetables has already raised (FAO 2017). Preserving the environment and natural resources while ensuring healthy, safe, and nutritious food for future generations is also considerably significant (Garcia et al. 2020). Agriculture is purely dependent on freshwater availability. Not only getting desalinized water is enough, but proper drainage of the irrigated water is also critical for keeping it fresh. The climatic change has caused drought and rise in seawater level through which soil salinization has increased. Irrigation and salty groundwater also cause salinization. Irrigation provides most of the world's crops, but about a tenth of the irrigated land available worldwide is affected by salt. Globally, about 1128 m ha area is affected by salinity (Table 29.1) and sodicity stresses (Sharma and Singh 2015). With the rise in population, rapid industrialization, and loss of agriculture productivity, land used in cultivation has become highly salt-affected or can increase with time due to human act or natural process (Table 29.1).

29.2 Salinity

Salinity refers to the presence of dissolved inorganic ions such as Na^+, Cl^-, SO_4^{2-}, HCO_3^-, K^+, Mg^{2+}, NO_3^-, Mg^{2+}, Ca^{2+}, and K^+ in the aqueous solution or soil matrix. Increased salinity of soil has gradual and sometimes abstruse effects on soil composition, water movement, and plant and microbial diversity (Artiola et al. 2019). There are three types of salinities, categorized by their causes: primary salinity (also known as natural salinity), secondary salinity (also known as dryland salinity), and tertiary salinity (also called irrigation salinity).

29.2.1 Types of Salinities

1. Primary Salinity or Natural Salinity: Natural processes such as salt deposition from rainfall over thousands of years or from rock erosion generate natural salinity. In hot arid and semiarid climates where annual rainfall is <27 cm, flushing of salts by rainfall is not possible, and also, evaporation and transpiration from plants lead to accumulation of salts.
2. Secondary Salinity or Dryland Salinity: When groundwater levels increase, it carries the accumulated salt to the surface by 'natural' salinity processes, leading to the secondary salinity. It is the result of nonirrigated landscapes or clearing vegetation and changes in land use.
3. Tertiary or Irrigation Salinity: It happens when reirrigation is done to crops or horticulture in several cycles, either directly or by letting it to filter into the groundwater before being filtered out for rewatering. In every irrigation step, some of it gets evaporated and the salts in the remaining water become more concentrated. Several cycles of watering result in tertiary salinity.

29.2.2 Understanding Salinity

Salt concentrations in the soil solution and the amount of exchangeable Na^+ in the soil characterize the type of salt-affected soil. The concentration of salts in the solution is measured by electroconductometry or by evaporation or spectrometric techniques. Exchangeable Na^+ is calculated by exchanging Na^+ from the soil with another ion such as Ca^{2+} and then measuring Na^+ in solution by flame photometry or spectrometry (e.g., atomic absorption or inductively coupled plasma emission spectrometries) (Sparks 2003).

Electrical conductivity (EC) is the measure of salt concentration in water-saturated soil paste extract. By measuring the soil salinity, the extent of salinity can be ranked (Table 29.2).

An redundant concentration of Na^+ ions in soils produces an unevenness in the ratio of monovalent cations to divalent cations. Exchangeable sodium percent (ESP) measures this ratio. Salt-affected grounds are thus also categorized by their ESP, as shown in Table 29.3. Sodium adsorption ratio (SAR) is also used in the management

Table 29.2 Soil Salinity Ranking (USDA 1954)

Parameter (dS m^{-1})	Nonsaline	Slightly saline	Moderately saline	Saline
Electrical conductivity	<4	4–8	8–16	>16

Table 29.3 Soil exchangeable sodium percentage rankings (USDA 1954)

Parameter	Adequate	Borderline	Inadequate	Comments
ESP	<10	10–15	>15	Sandy soils may tolerate ESP values up to 15

of salinity. It is a ratio of the amount of sodium (Na^+) relative to calcium (Ca^+) and magnesium (Mg^+) in the water extract from saturated soil paste.

29.2.3 *Classification of Different Soil Types*

29.2.3.1 Saline Soil

The calcium and magnesium ions dominate in exchange complex in saline soil. Sodium is rarely available for exchange complex (Table 29.4). The soluble salts help prevent dispersion of soil colloids, which make plant growth uninhibited by low penetration, aggregate stability, or aeration.. Many crops, mainly fruits and ornamentals, are affected by salinity (Sparks 2003).

29.2.3.2 Sodic Soil

These are the most inconvenient of the salt-affected soils. The high amount of sodium is the primary problem. Its low physical characteristics cause dispersion, poor aeration, and reduced water supply, which results in decreased crop productivity. Severe soil erosion also occurs in these soils due to low infiltration rates. Historically, sodic soils were often called black alkali soils. The black color results from the dispersion and dissolution of humic substances (Sparks 2003).

29.2.3.3 Saline-Sodic Soil

These are the intermediate soils between saline and sodic soils. A soil concentrated with neutral salts moderates the dispersing influence of sodium. Cations present in salts move adjacent to the negatively charged colloidal particles, thus reducing their tendency to disperse. Unfavorable levels of both neutral soluble salts and a high proportion of sodium affect plant growth.

Table 29.4 Classification of soil

Type	ECe (dS m^{-1})	ESP (%)	SAR	pH
Saline soil	>4	<15	<13	<8.5
Sodic soil	<4	>15	>13	>8.5
Saline-sodic soil	>4	>15	>13	<8.5

29.2.4 Crop Tolerance to Salt-Affected Soils

Some crops are salt-tolerant, while others do not nurture at all. It relies on variables such as the type and variety of produce, the type of salt in the soil, and its concentration. (Table 29.5).

29.2.5 Effect of Salinity Stress

Soil with ECe (electrical conductivity of saturated paste extract) greater than 4 dS/m or 40 mM NaCl is considered saline (USDA Salinity Laboratory). Hence, rise in these limits leads to mainly three significant stresses for the plant under high salinity, osmotic stress, ionic stress, and secondary stress (Yang and Guo 2018a). The osmotic stress is the first and foremost effect that a plant faces after the increase in salt levels outside the roots, which leads to low water uptake, cell expansion, and lateral bud development (Horie et al. 2012). When the toxic level of Na^+ accumulates in plants especially in leaves above the threshold level, salinity stress develops. This results in leaf mortality with chlorosis and necrosis and a reduced vital cellular metabolic activity together with photosynthesis and reduced enzyme activities (Acosta-Motos et al. 2017). Both osmotic stress and salinity stress cumulatively induce secondary stress, resulting an imbalance in plant nutrients and accumulation of toxic compounds. The after-effects lead to the formation of poisonous reactive oxygen species (ROS) such as hydroxyl radicals, hydrogen peroxide, and superoxide anions. Accumulation of ROS in plant cell damages cell structures and macromolecules like enzymes, DNA, and lipids (Yang and Guo 2018a). Salt-induced changes affect the growth of the plants. At the cellular level, early signaling includes the important role of both calcium and reactive oxygen species (ROS). They are thereby altering gene expression, mRNA stability, and translational regulation. Sodium/potassium homeostasis in the plant is also affected by different ion channels and transporters. Salinity stress affects many aspects of the physiology of a plant, making it difficult to thoroughly study (Negrão et al. 2017). To combat salt stress, plant develops complex mechanisms to tolerate salt in diverse ways, including by ion

Table 29.5 Tolerance of different field crops: (http://www.fao.org/docrep/r4082e/r4082e08.htm)

Highly tolerant		Moderately tolerant		Sensitive	
Date palm	Barley	Wheat	Onion	Red clover	Peas
Sugar beet	Sesbania	Tomato	Cucumber	Almond	Peach
Cotton	Fenugreek	Oats	Pomegranate	Soybean	Pear
Asparagus		Alfalfa	Fig	Sugarcane	Apple
Spinach		Rice	Olive	Orange	Prune
Rice		Maize	Grape	Palm	Apricot
Sinai		Flax	Carrot	Pulses	Gram
Berseem		Potatoes	Castor	Linseed	Sesames

accumulation, tissue-specific growth rates, biomass production, survival, and seed production (van Zelm et al. 2020). To reduce salt phytotoxicity, it is critical to limit Na^+ uptake and shoot Na^+ accumulation (Acosta-Motos et al. 2017).

29.2.6 Plant Signaling During Stress

As soon as the plant faces salinity, various secondary messengers, viz., calcium, ROS, and inositol phosphates (IP), perceive the stress signals and activate the cascade. Simultaneously, calcium-interacting proteins such as Ca^{2+}-dependent protein kinases (CDPKs), calmodulin, and calcineurin B-like proteins (CBLs), the proteins with structural 'EF-hand' calcium-binding motif, get activated (Kim et al. 2011). It leads to activation of transcription factors and stress-responsive elements (Fig. 29.1). In eukaryotes, CDPKs together with Mitogen-Activated Protein Kinases (MAPKs) are two signaling cascades widely stimulated in response to varying environmental abiotic and biotic stress, which are then induced by MPK kinases. They control numerous cellular activities, such as gene expression, mitosis, differentiation, proliferation, and cell survival/cell death (Isayenkov and Maathuis 2019).

As discussed earlier, salinity stress leads to both osmotic stress and ionic stress. The osmotic phase is the result of the accumulation of salt in the shoot. Due to rise in salt concentration in soil water, potential of soil reduces and thereby hinders the uptake of water. It results in water-deficit condition inside plant tissues and the secretion of ABA and activation of ABA-dependent pathway. Within minutes, the signals are transmitted to reduce intracellular turgor pressure and decrease cell expansion (Haswell and Verslues 2015). The lowered stomatal conductance leads to closure of stomata, lower carbon assimilation, biomass production, and decreased yield. ABA-dependent signaling then activates transcriptional activators: AERB/ ABF (ABA-responsive element-binding protein/ABA-binding factor), bZIP (basic leucine zipper), and MYC/MYB (myelocytomatosis oncogene/myeloblastosis oncogene. ABA-independent signaling also gets elicited in response to closed stomata and activates the expression of DREB2 (Dehydration Responsive Element-binding factor); NAC consists of the three described TFs containing NAC domain, namely, NAM (no apical meristem), ATAF (Arabidopsis transcription activator factor), CUC (cup-shaped cotyledons), and ZF-HD (zinc-finger homeodomain) regulon (Ciarmiello et al. 2014). Transcription factors play a crucial role in the regulation of stress-responsive genes in abiotic stress tolerance.

Salt Overly Sensitive (SOS) regulates ionic stress. It is the essential mechanism for cellular-level regulation of Na^+ exclusion and ion homeostasis. Increase in cytosolic Ca^{2+} signals is decoded by the EF-hand Ca^{2+} binding proteins, SOS3, and ScaBP8/CBL10 and translates it to SOS2, a serine/threonine protein kinase (Zhu et al. 2013). Roots express SOS3, and shoots express SCaBP8. SOS3 then activates SOS2 kinase (the sucrose nonfermenting 1-related protein) activity in the plasma membrane and finally to SOS1 (Yang and Guo 2018b). The plasma membrane Na^+/ H^+ antiporter, SOS1, is the main component in Na^+ transport from cytoplasm to the

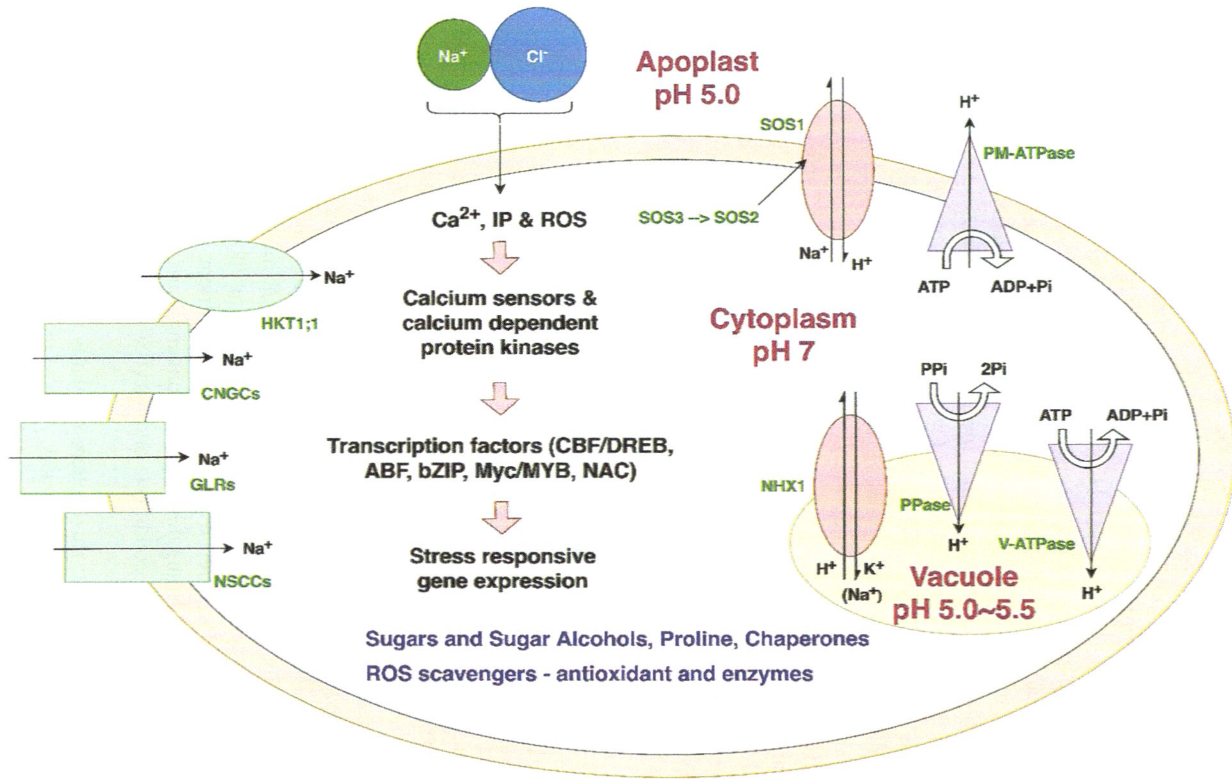

Fig. 29.1 Plant cell signaling during salinity stress. *PPi* pyrophosphatase; *Pi* orthophosphate

apoplast. SOS1 is regulated by SOS2 protein kinase by phosphorylation of its autoinhibitory C-terminal region (Quintero et al. 2011). However, the interaction of the Ca^{2+}-sensing protein SOS3 regulates SOS2 (Seifikalhor et al. 2019).

Under salt stress, the SOS3-SOS2 complex activates a vacuolar Na^{+}/H^{+} exchanger, NHX, and H^{+}-adenosine triphosphatase (ATPase) activity. These are concerned in the transport of Na^{+} from the cytoplasm to the vacuole. The driving force for Na^{+} movement is provided by the H^{+} gradient formed by vacuolar H^{+}-pyrophosphatases (PPases) and H^{+}-ATPases. In addition, HKTs (High-Affinity Potassium Transporters) are concerned in the absorption and exclusion of Na^{+} from xylem sap into root cells, reducing the accumulation of Na^{+} in the shoot (Almeida et al. 2017). In transgenic glycophytes, recirculation of Na^{+} ions by the activation of HKT gene is the central mechanism employed.

In brief, Na^{+} influx is regulated by Cyclic Nucleotide gated Channels (CNGCs), Glutamate Receptor (GLR), and Nonselective Cation Channel (NSCC) in addition to High-Affinity K^{+} Transporter (HKT). The cytoplasmic Na^{+} is extruded by SOS1 (plasma membrane Na^{+}/H^{+} antiporter), and vacuolar osmotic potential is mediated by AtNHX1 through K^{+}/H^{+} exchanges and cytosolic Na^{+}-K^{+} ratio through compartmentalization of K^{+} (Park et al. 2016). Production of osmolytes is also the primary strategy employed by the plant to overcome osmotic imbalance caused by water deficit condition due to continuous exposure to salt. Proline, sugars, and sugar alcohols rise in quantity to maintain the osmotic potential of the cell. The rise in ROS level is limited by activation of ROS scavengers (Fig. 29.1).

29.3 Plant Growth-Promoting Bacteria (PGPB)

A handful of soil contains numerous microorganisms like bacteria, fungi, actinomycetes, protozoa, and algae. Of these, bacteria are the most abundant type of soil organism (~95%). Millions of soil organism exist, but only a fraction of them are identified and are culturable. Bacteria are not uniformly distributed. The types and number of bacteria are also dependent on the soil characteristics, temperature, pH, humidity, etc. The density of bacteria in stressed soil and healthy soil is almost 1: 10^{4} (Glick 2012). The number of bacteria found near plant roots is also high due to the incidence of root exudates consisting of sugars, amino acids, plant hormones, organic acids, etc. Increased availability of the nutrient in the zone around roots helps bacteria flourish. Being unicellular organisms, during evolution, it has learnt to adopt to various climatic conditions and survive on different carbon and nitrogen sources.

Bacteria living near the rhizosphere of plants are attached to the soil particle's surface and are found in soil aggregates. A lot of these soil bacteria associate with the plant roots directly and enable the plant to grow well. These are Plant Growth-Promoting Bacteria (PGPB). In addition to the rhizospheric inhabitant, PGPB also includes several strains of rhizobia that are root nodules forming, e.g., Leguminous plants, and that can live within the tissues of a plant called endophytes (Santoyo et al.

2016). It can be beneficial, harmful, or plant neutral to communicate between bacteria and plant roots. Bacteria that promote the growth of plants represent a massive opportunity for sustainable agriculture production. These are now called as biofertilizers as they improve plant growth and yield by activation of plant growth-promoting substances. PGPB are safe, ecofriendly, and cheaper than chemical fertilizers. Biofertilizers are the better replacement as compared to the conventional use of pesticides, fungicides, and herbicides (Ji et al. 2019). Based on habitat, PGPB are of three categories, the first of which forms a symbiotic association with the plant, eg., Rhizobia spp. and *Frankia spp.*, the other that resides in plant tissue, and the last that is free-living in rhizosphere and on the roots of plants and cyanobacteria.

There are mainly two mechanisms of PGPB's action that promotes plant growth and enables their survival in stress: (1) Direct Mechanism and (2) Indirect Mechanism.

29.3.1 Direct Mechanism

It involves the mechanism in which bacterium enables the acquisition of essential nutrients or controls the level of hormone within a plant. It consists of the nutrient acquisition by nitrogen fixation, phosphate solubilization, and phytohormone production: Auxin, gibberellin, cytokinin, nitic oxide, abscisic acid, and enzyme secretion: ACC deaminase.

29.3.1.1 Nutrient Acquisition

PGPB can directly trigger plant growth in several different ways. They can fulfill plants' need for nutrients by fixing "atmospheric nitrogen", solubilizing inorganic phosphates and uptake iron.

29.3.1.1.1 Nitrogen Fixation

Nitrogen (N) is one of the critical nutrients for every life forms. It is the vital constituent of basic building blocks of an organism, e.g., nucleotides for DNA and RNA and amino acids for proteins. But most living species, including eukaryotes, do not have access to the principal source of N in nature, the atmospheric gaseous nitrogen (N_2). Gaseous N_2 is a stable triple bond in the structure, which makes it unsuitable for plant assimilation (Baas et al. 2014). Diazotrophic microorganisms, specifically archaea and bacteria, are known to perform biological nitrogen fixation (BNF) by fixing N_2 to ammonia (NH_3) (de Souza et al. 2015) (Fig. 29.2).

Diazotrophic microorganisms achieve BNF via nitrogenase, an evolutionarily conserved enzyme consisting of two metalloproteins, FeMo protein and Fe protein (Puri et al. 2017). The nitrogenase enzyme is coded by nif genes that have structural

$$N_2 + 8H^+ + 8e^- + 16\ MgATP \rightarrow 2NH_3 + H_2\uparrow + 16MgADP + 16P_i$$

Fig. 29.2 The overall stoichiometry of the biological nitrogen fixation reaction

genes, genes engaged in Fe protein activation, biosynthesis of the iron-molybdenum cofactor, donation of electrons, and regulatory genes needed for enzyme synthesis and work. Nif genes are usually found in diazotrophic (nitrogen-fixing) bacteria in a cluster of approximately 20–24 kb with seven operons coding 20 distinct proteins (Glick 2012). Nitrogen-fixing bacteria are mostly symbiotically associated with specific plants (primarily legumes). Examples of symbiotic nitrogen fixers are *Rhizobium, Allorhizobium, Sinorhizobium, Azorhizobium, Mesorhizobium, Frankia, Bradyrhizobium, Azoarcus, Achromobacter, Burkholderia, and Herbaspirillum* (Olanrewaju et al. 2017).

29.3.1.1.2 Phosphate Solubilization

Phosphorus (P) is also a basic component of nucleic acid, phospholipids, and adenosine triphosphate (ATP). In soil, phosphorous occurs in two forms, organic and inorganic phosphates. Despite being verily abundance in several soils, it is mostly inaccessible for the plant uptake, and only 5% or less of the total phosphorus is available for plants. Under acidic and alkaline conditions with cations, phosphorous forms nonsoluble complexes. As a result, the input of chemical fertilizer increases to meet agricultural demands (Ali et al. 2019). However, the application of chemical fertilizers increases the phosphate content in the soil, but it is not available to plants. Additionally, chemical fertilizers cause eutrophication of surface water and contamination of groundwater by leaching of ions. Hence, there is need to solubilize phosphorus in the soil to consumable forms.

In nature, soil type and pH of soil determine the extent of solubilization and mineralization of phosphorous. Free oxides and hydroxides of aluminum and iron fix phosphorous in acidic soils, whereas calcium is fixing in alkaline soils. The basic mechanisms for phosphorus solubilization are organic acid production, decrease in medium pH, and/or chelation of cationic partner of phosphate ion (PO_4^{3-}) (Walpola and Yoon 2012).

Phosphate solubilizing bacteria are a collection of bacteria that mobilize poorly available phosphorous (Alori et al. 2017). The organic acid produced by these bacteria helps chelation of metal ion and release of phosphate ion from inorganic phosphates. They synthesize organic acids such as gluconic acid, 2-ketogluconic acid, succinic acid, acetic acid, citric acid, tartaric acid, lactic acid, glycolic acid, oxalic acid, malonic acid, fumaric acid, propionic acid, etc. (Walpola and Yoon 2012). Application of these microorganisms as ecofriendly biofertilizer aids in reducing the use of exclusive phosphatic fertilizers (Mishra and Dash 2014).

Phosphate-solubilizing bacteria solubilize organic phosphates by the production of enzymes. Organic matter in soil consists of mainly inositol phosphate (soil phytate). Other compounds include mono, di, and triesters of phosphates, phospholipids, and nucleic acids. Enzymes involved in mineralization comprise nonspecific acid phosphatases (NSAPs) and phytase (Alori et al. 2017). Examples of phosphorous solubilization and mineralization soil bacteria are *Pseudomonas* spp., *Azotobacter* spp., *Burkholderia* spp., *Agrobacterium* spp., *Enterobacter* spp., *Erwinia* spp., *Bacillus circulans*, *Kushneria* spp., etc.

29.3.1.1.3 Phytohormone Production

Phytohormones regulate biological processes such as growth, differentiation, development, and stomatal movement (Davies 2013). There is a cross talk between hormones. Two or more hormone work together or individually. Plant hormone has a significant effect on the secondary metabolism of plants and plays a critical role in stimulation mechanisms for plant defense against stress. Plant beneficial microorganism can modulate this effect and provide stress tolerance by secretion of phytohormone (Kumar et al. 2019). They alter root hormone concentrations and affect root-to-shoot long-distance signaling to mediate shoot hormone status (Dodd et al. 2010). Inoculation of *B. subtilis* (PGPB) increases the number of constituents in the root exudates, namely, sugars, amino acids, and organic acids (Liu et al. 2013). These components work as a substrate for hormonal synthesis by PGPB. Plant growth-promoting bacteria produce phytohormones: auxin, gibberellin, cytokinin, nitric oxide, and abscisic acid.

29.3.1.1.4 Auxin

Auxin plays a crucial role in inducing cell division; differentiating vascular tissues; elongating stem and root; stimulating lateral and adventitious root; apical dominance; gravitropism and phototropism; and influencing photosynthesis, development of pigments, biosynthesis of numerous metabolites, and tolerance to stress (Spaepen and Vanderleyden 2011). The key synthesis of auxin indole-3-acetic acid (IAA) occurs from tryptophan (Trp) in higher plants via a two-step process that is strongly retained among higher plants. The first step includes the deletion of the amino group from tryptophan, which is catalyzed by tryptophan aminotransferases forming indole-3-pyruvate (IPA). In the second step, YUCCA (YUC) flavin monooxygenase enzyme decarboxylates IPA and forms IAA. Indole-3-acetonitrile and Indole-3-acetamide are also intermediates in auxin biosynthesis with IPA (Zhao 2014).

The precursor of IAA synthesis is tryptophan, whereas anthranilate acts as a precursor for tryptophan synthesis. IAA synthesis is negatively regulated by a negative feedback regulation on the anthranilate synthase as tryptophan inhibits the formation of anthranilate (Spaepen and Vanderleyden 2011). Beneficial microorganism synthesizes IAA and provides it to plant. This IAA now alters the root architecture by accelerating root growth and number. Several reports of IAA synthesis by bacteria that promote plant growth are recently reviewed by Olanrewaju et al. (2017) and Premachandra et al. (2016). *Burkholderia phytofirmans* strain PsJN when inoculated in two potato (*Solanum tuberosum* L.) cultivars with alike root growth, but unlike shoot growth patterns (faster-growing Kennebec and slower-growing Yukon gold), showed an identical and massive increase in root growth after inoculation. Besides, it showed a twofold to threefold rise in IAA and cytokinin (trans-zeatin or tZ) levels, expressed on a per plant basis (Kurepin et al. 2015). Similarly, *Burkholderia phytofirmans* strain PsJN has improved plant growth in maize with a substrate (L-TRP)-derived IAA biosynthesis (Naveed et al. 2015). Foliar application of the leaf-colonizing PGPB improved the growth and nutritional status of maize (Abadi et al. 2020).

29.3.1.1.5 Gibberellin

Gibberellin (GA) was first recognized in the pathogenic fungus *Gibberella fujikuroi*, which causes a disease in rice called 'foolish-seedling'. Due to the production of large quantities of GA, the plants become long and slender, are incompetent of supporting their weight, and are chlorotic and partially infertile (Yabuta 1938). Gibberellins perform different metabolic functions in plants that are important for plant growth and development, including seed germination, stem elongation, flowering dormancy, and fruit formation and senescence (Khan et al. 2015). In addition to enabling stamen elongation, they play an important role in fertility and are required for the growth, release, and germination of pollen and pollen tubes. (Hedden and Thomas 2012). There are 136 GAs identified from higher plants (128 species), 28 GA from fungi (7 species), and only 4 GA (GA_1, GA_3, GA_4, and GA_{20}) from bacteria (7 species), but only a small number of them are bioactive (Hedden 2019; Hedden and Thomas 2012). Bacterial biosynthesis pathway is proposed, based on the previously described GA biosynthesis pathway in plants and fungi. In bacteria, GA is usually biosynthesized from geranylgeranyl-PP, which forms ent-kaurene via ent-copalyl diphosphate, and ent-kaurene is converted into GA_{12} aldehyde via ent-kaurene oxidase and ent-kaurenoic acid oxidase (Kang et al. 2014). GA_{12}-aldehyde is then oxidized into GA_{12} and metabolized into other GA. The involvement of *operan* has also been reported (Morrone et al. 2009).

Several plant growth-promoting bacteria are known to produce gibberellin. These comprise *Acetobacter diazotrophicus*, *Azospirillum lipoferum*, *Azospirillum brasilense*, *Bacillus licheniformis*, *Herbaspirillum seropedicae*, *Rhizobium phaseoli*, *Bacillus pumilus*, *Bacillus pumilus* CJ-69, *Bacillus cereus* MJ-1, *Bacillus macroides* CJ-29, *Acinetobacte*r, *Calcoaceticus*, *Burkholderia cepacia*, *Promicromonospora* sp., etc. (Kang et al. 2014). *Bacillus methylotrophicus* KE2, a gibberellin producer, supports plant growth and boosts nutritional metabolites and food values of lettuce (Radhakrishnan and Lee 2016). Enhanced *Scenedesmus* sp. growth responses to gibberellin secretion by symbiotic bacteria (Dao et al. 2020).

29.3.1.1.6 Cytokinin

Cytokinins (CKs) are another member of phytohormones. They play a noteworthy role in the regulation of the plant cell cycle and many developmental processes. Cytokinins were discovered during the 1950s by Skoog, Miller, and coworkers as factors that facilitate cell division (cytokinesis). Structurally cytokinin has an adenine base and a five-carbon isopentenyl side chain. The first known widespread natural cytokinin was isolated and named 'zeatin' from immature maize kernels. Among these, the most abundant zeatin is *trans*-zeatin. These are present in every cell of plant tissue but predominant in dividing cells, viz., root tip, shoot apex, and juvenile seeds (Schmülling 2013).

Cytokinin biosynthesis occurs when the isopentenyl moiety is transferred from dimethylallyl pyrophosphate (DMAPP) to adenosine monophosphate (AMP), adenosine diphosphate (ADP), or adenosine triphosphate (ATP) by the isopentenyl transferase enzyme (IPT). ADP and ATP are the chosen substrate for plant IPT enzymes, while bacterial enzymes prefer AMP. The reaction then forms cytokinin precursors, isopentenyl-AMP, isopentenyl-ADP, and isopentenyl-ATP. These are now hydroxylated to form cytokinins of the zeatin kind. CKs are degraded by metabolizing through CK oxidases (CKXs) or inactivated through conjugation to sugars (Akhtar et al. 2020). An indirect pathway for CK synthesis includes the turnover of tRNA containing cis-zeatin (Amara et al. 2015).

CKs are produced by plants and some associated microorganism, microalgae, and insects. They can modulate the signaling mechanism in the plant. Cytokinin-producing, plant growth-promoting rhizobacteria that confer resistance to drought stress in *Platycladus orientalis* and interfere with shoot growth suppression have shown a real potential to work as an inhibitor of drought stress in arid environments (Liu et al. 2013). Nodulating rhizobia are prominent producers of CKs (Gamas et al. 2017). In the rhizobial bacteria *Sinorhizobium meliloti* and *Mesorhizobium loti*, CKs were reported as the critical differentiation signal for nodule organogenesis and genes homologous to the *A. tumefaciens ipt* gene (Giron et al. 2013). The development of *Pseudomonas fluorescens* AK1 and *Pseudomonas aeruginosa* AK2 with IPA, dihydroxy zeatin riboside (DHZR) and zeatin riboside (ZR), showed growth promotion in rice seedlings (Karnwal and Kaushik 2011). CKs do cross talk with other phytohormones and enhance plant immunity by underlying mechanism contrary to pathogen infections and can show practical claims in crop production (Akhtar et al. 2020).

29.3.1.1.7 Nitric Oxide

Nitric oxide (NO), a small gaseous, redox-active small molecule, is increasingly becoming a key regulator of plant development, immunity, and environmental interactions (Yu et al. 2014). In plants, there are two possible routes for NO biosynthesis: oxidative and reductive. The oxidative mechanism includes the formation of NO from L-arginine (L-Arg), polyamines, or hydroxylamine. In distinction, the reductive mechanism relies on nitrite as the primary substrate and comprises reduction via nitrate reductase and a plasma membrane-bound nitrate-NO reductase (NiNOR) and mitochondrial nitrite reduction (Gupta et al. 2011; Mur et al. 2013). The era of nitric oxide in plant biology has been recently reviewed by Del Castello et al. (2019).

N0 primary function is to provide immunity to the plant. It involves cell death reaction to hypersensitivity and stimulates the expression of genes and genes involved in salicylic acid signaling of Phenylalanine ammonia-lyase (PAL) and Pathogenesis-related protein 1 (PR1). In the adaptation of plants to abiotic stresses, NO has now become an important endogenous signaling molecule. In involves activation of mitogen-activated protein kinase (MAPK) signaling cascades, which drives closure of stomata during water stress. Role of NO in plant development is also considerable; it is involved in the germination of seed, flower development, flowering time, apical dominance, and root growth and development (Yu et al. 2014). Furthermore, the interaction of roots with microorganisms in the rhizosphere is also modulated (Boscari et al. 2013).

Inoculation of PGPB *Azospirillum brasilense* modifies the configuration of root. Nitric oxide facilitates IAA signaling pathways leading to both adventitious and lateral root creation (Molina-Favero et al. 2008). Denitrification-derived NO modulates biofilm formation in *Azospirillum brasilense* Sp245 (Arruebarrena Di Palma et al. 2013). Occurrence of nitric oxide (NO) metabolism genes, nitrous oxide reductase regulator (*nosR*), nitrous oxide reductase (*nosZ*), and nitric oxide reductase (*norB*) and IAA signaling genes was recognized in the plant-associated bacterium (PAB) *Azospirillum brasilense* SM. Such a robust association suggested the presence of cross talk or common signaling instruments in these two development controllers (Koul et al. 2015).

29.3.1.1.8 Abscisic Acid

Abscisic acid (ABA) is a key phytohormone that controls the development and stress tolerance adaptations of plants. It plays a vital role in various plant developmental processes, including accumulation of cuticular wax, stomatal closure, leaf senescence, germination of seeds, dormancy of buds, osmotic control, and regulation of growth, among many others. Abscisic acid has also been active in downstream reactions by both transcriptional and post-transcriptional pathways to abiotic and biotic fluctuations in the environment (Chen et al. 2020).

Containing 15 atoms of C, Abscisic acid is a sesquiterpenoid. ABA is produced using an indirect mechanism—the carotenoid mechanism. The path is introduced by parting of a C_{40} ancestor like β-carotene (Arc et al. 2013). The clear pathway has been described by Chen et al. (2020). Several PGPB synthesize ABA. Some of the strains include *Azospirillum brasilense* Sp245, *A. lipoferum* USA 59b, *Achromobacter xylosoxidans*, *Bacillus licheniformis*, *Bacillus pumilus*, *Brevibacterium halotolerans* and *Lysinibacillus fusiformis*, etc. (Cohen et al. 2015). It has also been mentioned that a few bacterial strains catabolize ABA and promote plant growth via ABA-dependent mechanism (Belimov et al. 2014). As an innovative measure to build phytoremediation performance in heavy metal polluted soil, ABA metabolism by Aba-catabolizing bacteria could be beneficial (Lu et al. 2020).

29.3.1.2 Enzyme Secretion

29.3.1.2.1 ACC Deaminase

Ethylene, a gaseous hormone, enacts a critical part in plant's development, including control in ripening, abscission, and senescence. Moreover, it restricts vegetative growth by limiting cell elongation, primarily through cross talk with auxins, by being a controller of stress responses (Vaseva et al. 2018). Ethylene-associated changes in plants confer tolerance to stress. It is the first known gaseous biological signaling molecule, and Bakshi et al. (2015) described its various roles.

The amino acid L-methionine helps in the synthesis of ethylene. S-adenosylmethionine (SAM) synthetase that changes methionine to SAM, 1-aminocyclopropane-1-carboxylic acid (ACC) synthase that changes SAM to ACC, and ACC oxidase that changes ACC to ethylene are three enzymes. For ethylene biosynthesis, the reaction catalyzed by ACC synthase has been the rate-limiting response.

When a plant encounters stress, there is a rise in the level of ethylene, and this phenomenon is called stress ethylene. The increased level of ethylene affects the development of roots in the plant. PGPB come in action and decrease the inhibitory level of stress ethylene. PGPB have genes for synthesizing ACC deaminase enzyme. ACC deaminase helps in the hydrolysis of ACC, which is the closest predecessor of ethylene for ammonia and α-ketoglutarate. Bacteria can further metabolize the product of the reaction for their various metabolic processes to produce nitrogen and carbon (Cohen et al. 2015; Singh et al. 2015). PGPB along with ACC deaminase activity improve crop production (Dodd et al. 2010).

Owing to anoxia and increased ethylene generation because of waterlogging in prevailing circumstances under wet weather, *Ocimum sanctum* plants experience severe damage. The application of ACC deaminase-containing PGPB in *O. sanctum* protected the plant from damages due to waterlogging (Barnawal et al. 2012). The studies using *A. thaliana* lines, Columbia (Col) wild-type (WT), ethylene-insensitive mutants *etr1-1* and *ein2-1*, and ethylene overproducing mutant *eto1-1* showed that *V. paradoxus 5C-2* soil inoculation stimulated the development of WT plants and *eto1-1* and also improved the floral initiation. However, despite bacterial colonization, such reactions were not present in the case of ethylene-insensitive mutants (*etr1-1* and *ein2-1*). The study revealed that for V. paradoxus 5C-2 to promote leaf development and flowering of A Thaliana, a fully functioning ethylene signal transduction mechanism is needed (Chen et al. 2013). Likewise, the endophyte *Pseudomonas* spp. is expressed by producing ACC deaminase, which improves NaCl stress tolerance via decreasing stress-related ethylene, which results in improved growth, photosynthetic efficiency, and ionic equilibrium in the plants of tomato (Win et al. 2018).

29.3.2 Indirect Mechanism

It includes the mechanisms by which plants decrease the damage to plants, caused by various phytopathogens (Ma del Orozco-Mosqueda et al. 2020). Antibiotics, cell wall-degrading enzymes, siderophores, and volatile organic compounds are some of them.

29.3.2.1 Antibiotics

PGPB produces many antibiotics for preventing fungal and bacterial phytopathogens. At low concentrations, antibiotics do have property of slowing the development and metabolism of many other microorganisms (Raaijmakers and Mazzola 2011). *Bacillus* yields antibiotic lipopeptides such as bacilysin, fengycin, surfactin, and zwittermicin (Khabbaz et al. 2015; Khabbaz et al. 2019), and PGPB have also been proved as biopesticides (Ruiu 2020).

29.3.2.2 Cell Wall-Degrading Enzymes

The second most crucial mechanism involved in biocontrol activity performed by the PGPB is the activation of fungal cell wall corrupting enzymes, viz., glucanases, protease, chitinases, and lipase. The chitinase degrades chitin, a residue of β-(1, 4)-N-acetyl glucosamine polymer (Husson et al. 2017); β-1,3-glucanase, glucan (cell wall carbohydrate)-degrading enzyme (Vaddepalli et al. 2017); protease, cell wall proteins metabolizing enzyme; and lipase, cell wall-associated lipid degrading enzyme. These enzymes cumulatively act upon the cell wall of the fungal pathogen, thereby disrupting the osmotic strength of the cellular membrane. PGPB that synthesize one or more of these enzymes demonstrate biocontrol action alongside a range of pathogenic fungi comprising *Botrytis cinerea*, *Sclerotium rolfsii*, *Fusarium oxysporum*, *Phytophthora* spp., *Rhizoctonia solani*, and *Pythium ultimum* (Sharma et al. 2017).

29.3.2.3 Siderophores

In nature, although iron is richly found in Earth's crust, and its bioavailability is normally low (Boyd and Ellwood 2010). It is due to the occurrence of mostly insoluble Fe^{3+} ion in most of the natural habitats. At neutral or basic pH levels, iron exists as Fe^{3+} with the formation of low solubility ferric hydroxides

(Albelda-Berenguer et al. 2019). Hence, to overcome iron deficiency, PGPB secrete siderophores. Siderophores are low molecular weight, peptide molecules that have a higher affinity for Fe^{3+} than Fe^{2+} (Hider and Kong 2010). They are commonly referred to as microbial Fe chelators. They chelate Fe^{3+} from the surrounding and transfer iron in the microbial cell via specific receptors (Kramer et al. 2020). Every process right from the synthesis of siderophore to the uptake of iron is tightly regulated. There are around 500 distinct siderophores, out of them 270 are structurally described (Hider and Kong 2010).

Siderophores are also regarded as a critical factor in the biocontrol-induced property by PGPB (Albelda-Berenguer et al. 2019). Siderophores take up the available iron from the plant rhizosphere by chelating and making it unavailable to the pathogen. Iron uptake by PGPB makes the pathogen deprived of essential nutrients, leading to its death. This mechanism is referred to as the siderophore-mediated destruction of pathogens (Zhang et al. 2015). *Bacillus* and *Pseudomonas* have been studied extensively for biocontrol activities.

29.3.2.4 Volatile Organic Compounds (VOCs)

VOCs are of low boiling point, low molecular weight (<300 Da), excessive vapor pressure, odoriferous combinations (<C15), and lipophilic moiety (Li et al. 2019). These are famous to play an essential role in transport of data among creatures. Bacteria converse with host through making VOCs and affect the growth characteristics of the plant positively. VOCs emitted by PGPB, *B. subtilis* GB03, may activate numerous dissimilar hormonal indications in *Arabidopsis thaliana*, which contains auxin, brassinosteroids, cytokinins, gibberellins, and salicylic acid. Biologically active VOCs from *Bacillus* sp. are 2,3-butanediol, acetoin, and tridecane (Ryu et al. 2003). VOCs cause plant tolerance for these abiotic stresses, including salt stress, drought stress, disease resistance, and/or nutrient deficiency, other physiological processes, and insect resistance (Li et al. 2019). Modulation of AtHKT1 has also been reported in *A. thaliana* after the exposure of VOCs in salt stress. VOCs downregulated the manifestation of K^+ transporter (HKT1) in roots and upregulated in shoots, which affects lesser collection of Na^+ in the plant (Zhang et al. 2008). Thereby, promoting induces Systemic Resistance and Tolerance (Bitas et al. 2013; Farag et al. 2013).

29.4 Alleviation of Salinity Stress by Microbes

Salinity stress is an inevitable change in agriculture as it may cause due to several reasons like the poor practice of irrigation or overuse of fertilizers. Ions dissolved in the soil reach every part of the plant and hinder growth of the plant comprising germination, vegetative production, and reproductive growth. Salt stress produces ion toxicity, osmotic stress, lack of nutrients (N, Ca, K, P, Fe, and Zn), and oxidative stress in plants, thereby reducing water absorption (Bano and Fatima 2009). So, adopting ecofriendly methods to minimize the harmful consequences of salinity are the best practice to revive the agroecosystem. The use of PGPB may come as a revolutionary approach to alleviate salinity stress (Fig. 29.3).

The beneficial plant microbes colonize in the locality of roots (rhizosphere) to promote plant development by a direct and indirect mechanism. The nutrient deprivation is beaten by fixing nitrogen, solubilizing phosphate, chelating iron, and uptaking potassium. The higher level of stress ethylene caused due to salinity stress is diminished by ACC deaminase secreted by PGPB and metabolizing it to α-ketoglutarate and ammonia. Moreover, ACC deaminase cross talk with IAA. Plant exudates contain tryptophan, the immediate precursor of IAA. As the PBPB colonizes in root rhizosphere, it takes up the tryptophan and synthesizes IAA. This IAA is now assimilated by plant, initiates cell proliferation and cell elongation, and loosens up the root cell wall. The loosened root tissue directly secretes more tryptophan and more ACC. ACC is then catalyzed through ACC deaminase of bacteria and reducing overall concentration of stress ethylene. ABA is another phytohormone that is involved in providing resistance against salinity by ABA-independent pathway to plants. *Bacillus amyloliquefaciens* RWL-1 (Bacterial endophyte) produces ABA and significantly upregulates the synthesis of precursors or intermediates of stress resistance metabolites in *Oryza sativa*. It includes vital amino acids (like aspartic acid, glutamic acid, proline, and phenylalanine). The stress-sensitive quantities of endogenous ABA have also been greatly diminished. The concentrations of endogenous salicylic acid in RWl-1 inoculated plants, on the other hand, were considerably higher than in control plants subjected to the same amount of salinity tension (Shahzad et al. 2017). Biofilm formation or exopolysaccharide secretion is another beneficial property by PGPB. Inside a self-produced matrix of extracellular polymeric substances (EPS) and biofilms, cultures of microbial cells accumulate. Biofilms are immune to harsh conditions and therefore serve as protective gear, safeguard microorganisms against ultraviolet (UV) radiation, extreme weather conditions, extreme pH, high salinity, high heat, insufficient nutrition, antibiotics, etc. (Yin et al. 2019). EPS retain water layer around the cells and help microbes to survive in salinity stress. EPS may also adsorb nutrients and provide it to plant (Kasim et al. 2016). Primarily, the osmotic balance of cell is changed under stressful condition, and to maintain stabilization of proteins and membranes, cell synthesizes and accumulates osmoprotectants to survive salt-induced osmotic stress (Marwein et al. 2019). Many other research studies (like Li and Jiang 2017; Upadhyay et al. 2012) indicate that proline upregulation and total soluble sugar (TSS) synthesis improve the ability to adapt for the plant in stress. In the salt-tolerant plant, ROS scavenging enzymes play an essential role. Two pathways are used by plants for ROS degeneration: enzymatic and nonenzymatic pathways. An enzymatic pathway is the secretion of antioxidant enzymes, which include catalase (CAT), peroxidase (POD), and superoxide dismutase (SOD). A nonenzymatic pathway for absorbing ROSS is diminished glutathione (GSH) (Mittler and Blumwald 2015). Maize plant showed increased antioxidant enzymes status with PGPB inoculation (Li and Jiang 2017). Moreover, application of

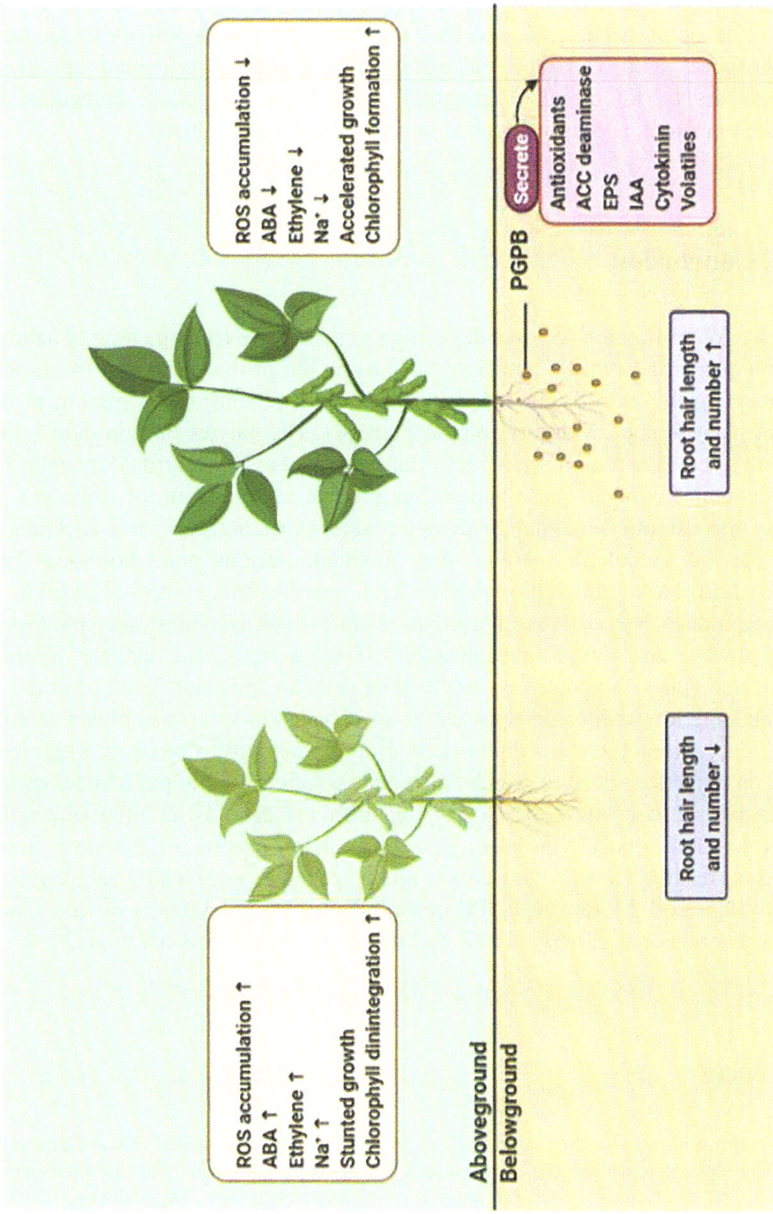

Fig. 29.3 Effect of saline soil on the plant with and without PGPB

halo-tolerant PGPB, *Curtobacterium albidum* SRV4 on paddy under dissimilar salinity settings, enhanced the modulation of osmolyte, photosynthetic efficacy, and antioxidant enzymes status *in planta* (Vimal et al. 2019). Similar effects were shown in *Bacillus* sp. FAB10 isolates treatment plants. It created fit biofilm, higher amount of exopolysaccharides, and IAA and ACC-deaminase activity and solubilized phosphate in vitro (Ansari et al. 2019). Also, PGPB, *Pseudomonas* sp., lower the levels of Na^{+}/K^{+} ratio by activation of HKT gene in plant, upregulate its expression in shoot, and downregulate in root (Kasotia et al. 2016).

29.5 Conclusion

Through a long irrigation, gradual deposition of salts is the leading cause of salinity stress for plants. Moreover, managing salinity is the primary task for the farmer while meeting the world's food demand. Various mechanisms are present in the different plant to cope with physiological stresses. Halophytes that grow in saline soils with salt concentration up to 5 g/l^{-1} have evolved to have various strategies for living in saline settings. These approaches involve accumulation of osmolytes to intensify the osmotic potential of cytoplasm, to expel out Na^{+}, and accumulate sodium in the vacuole. Moreover, they modulate various plant hormones like jasmonic acid, IAA, gibberellin, abscisic acid, and ethylene control absorption of K^{+} ions, produce NO, activate and produce antioxidant enzymes, alter photosynthetic pathways and compartmentalization of various ions, and regulate different stress-related genes. Glycophytes, on the other side, are the plants that grow well in soil containing low sodium and possess no mechanism to survive in higher salinity stress. Hence, when the soil salinity rises, crop productivity is reduced. Therefore, there is an immediate need to search at alternate solutions to minimize losses owing to excessive salt concentrations in the soil. In this respect, as an environmentally friendly solution to enable the plant growth in salinity tension, PGPB have shown tremendous results. They provide almost all the responses as given by halophytes to adapt to saline soil. Moreover, PGPB treated plant show good plant growth characteristics (Kasotia et al. 2015) and reduced sodium in plants (Kasotia et al. 2016).

References

Abadi VAJM, Sepehri M, Rahmani HA, Zarei M, Ronaghi A, Taghavi SM, Shamshiripour M (2020) Role of dominant phyllosphere bacteria with plant growth–promoting characteristics on growth and nutrition of maize (*Zea mays* L.). J Soil Sci Plant Nutr. https://doi.org/10.1007/s42729-020-00302-1

Acosta-Motos J, Ortuño M, Bernal-Vicente A, Diaz-Vivancos P, Sanchez-Blanco M, Hernandez J (2017) Plant responses to salt stress: adaptive mechanisms. Agronomy 7(1):18. https://doi.org/10.3390/agronomy7010018

Akhtar SS, Mekureyaw MF, Pandey C, Roitsch T (2020) Role of cytokinins for interactions of plants with microbial pathogens and pest insects. Front Plant Sci 10. https://doi.org/10.3389/fpls.2019.01777

Albelda-Berenguer M, Monachon M, Joseph E (2019) Siderophores: from natural roles to potential applications. In: Advances in applied microbiology, vol 106. Elsevier, New York, pp 193–225. https://doi.org/10.1016/bs.aambs.2018.12.001

Ali W, Nadeem M, Ashiq W, Zaeem M, Gilani SSM, Rajabi-Khamseh S, Pham TH, Kavanagh V, Thomas R, Cheema M (2019) The effects of organic and inorganic phosphorus amendments on the biochemical attributes and active microbial population of agriculture podzols following silage corn cultivation in boreal climate. Sci Rep 9(1):17297. https://doi.org/10.1038/s41598-019-53906-8

Almeida DM, Oliveira MM, Saibo NJM, Almeida DM, Oliveira MM, Saibo NJM (2017) Regulation of Na^+ and K^+ homeostasis in plants: towards improved salt stress tolerance in crop plants. Genet Mol Biol 40(1):326–345. https://doi.org/10.1590/1678-4685-gmb-2016-0106

Alori ET, Glick BR, Babalola OO (2017) Microbial phosphorus solubilization and its potential for use in sustainable agriculture. Front Microbiol 8:971. https://doi.org/10.3389/fmicb.2017.00971

Amara U, Khalid R, Hayat R (2015) Soil bacteria and phytohormones for sustainable crop production. In: Maheshwari DK (ed) Bacterial metabolites in sustainable agroecosystem. Springer, New York, pp 87–103. https://doi.org/10.1007/978-3-319-24654-3_5

Ansari FA, Ahmad I, Pichtel J (2019) Growth stimulation and alleviation of salinity stress to wheat by the biofilm forming *Bacillus pumilus* strain FAB10. Appl Soil Ecol 143:45–54. https://doi.org/10.1016/j.apsoil.2019.05.023

Arc E, Sechet J, Corbineau F, Rajjou L, Marion-Poll A (2013) ABA crosstalk with ethylene and nitric oxide in seed dormancy and germination. Front Plant Sci:4. https://doi.org/10.3389/fpls.2013.00063

Arruebarrena Di Palma A, Pereyra CM, Moreno Ramirez L, Xiqui Vázquez ML, Baca BE, Pereyra MA, Lamattina L, Creus CM (2013) Denitrification-derived nitric oxide modulates biofilm formation in Azospirillum brasilense. FEMS Microbiol Lett 338(1):77–85. https://doi.org/10.1111/1574-6968.12030

Artiola JF, Walworth JL, Musil SA, Crimmins MA (2019) Soil and land pollution. In: Environmental and pollution science. Elsevier, New York, pp 219–235. https://doi.org/10.1016/B978-0-12-814719-1.00014-8

Baas P, Mohan JE, Markewitz D, Knoepp JD (2014) Assessing heterogeneity in soil nitrogen cycling: a plot-scale approach. Soil Sci Soc Am J 78(S1):S237–S247. https://doi.org/10.2136/sssaj2013.09.0380nafsc

Bakshi A, Shemansky JM, Chang C, Binder BM (2015) History of research on the plant hormone ethylene. J Plant Growth Regul 34(4):809–827. https://doi.org/10.1007/s00344-015-9522-9

Bano A, Fatima M (2009) Salt tolerance in *Zea mays* (L). Following inoculation with rhizobium and Pseudomonas. Biol Fertil Soils 45(4):405–413. https://doi.org/10.1007/s00374-008-0344-9

Barnawal D, Bharti N, Maji D, Chanotiya CS, Kalra A (2012) 1-Aminocyclopropane-1-carboxylic acid (ACC) deaminase-containing rhizobacteria protect *Ocimum sanctum* plants during waterlogging stress via reduced ethylene generation. Plant Physiol Biochem 58:227–235. https://doi.org/10.1016/j.plaphy.2012.07.008

Belimov AA, Dodd IC, Safronova VI, Dumova VA, Shaposhnikov AI, Ladatko AG, Davies WJ (2014) Abscisic acid metabolizing rhizobacteria decrease ABA concentrations in planta and alter plant growth. Plant Physiol Biochem 74:84–91. https://doi.org/10.1016/j.plaphy.2013.10.032

Bitas V, Kim H-S, Bennett JW, Kang S (2013) Sniffing on microbes: diverse roles of microbial volatile organic compounds in plant health. Mol Plant Microbe Interact 26(8):835–843. https://doi.org/10.1094/MPMI-10-12-0249-CR

Boscari A, del Giudice J, Ferrarini A, Venturini L, Zaffini A-L, Delledonne M, Puppo A (2013) Expression dynamics of the *Medicago truncatula* transcriptome during the symbiotic interaction with Sinorhizobiummeliloti: which role for nitric oxide? Plant Physiol 161(1):425–439. https://doi.org/10.1104/pp.112.208538

Boyd PW, Ellwood MJ (2010) The biogeochemical cycle of iron in the ocean. Nat Geosci 3 (10):675–682. https://doi.org/10.1038/ngeo964

Chen L, Dodd IC, Theobald JC, Belimov AA, Davies WJ (2013) The rhizobacterium Variovorax paradoxus 5C-2, containing ACC deaminase, promotes growth and development of *Arabidopsis thaliana* via an ethylene-dependent pathway. J Exp Bot 64(6):1565–1573. https://doi.org/10.1093/jxb/ert031

Chen K, Li G-J, Bressan RA, Song C-P, Zhu J-K, Zhao Y (2020) Abscisic acid dynamics, signaling, and functions in plants. J Integr Plant Biol 62(1):25–54. https://doi.org/10.1111/jipb.12899

Ciarmiello LF, Woodrow P, Piccirillo P, De Luca A, Carillo P (2014) Chapter 3—Transcription factors and environmental stresses in plants. In: Ahmad P, Rasool S (eds) Emerging technologies and management of crop stress tolerance. Academic Press, New York, pp 57–78. https://doi.org/10.1016/B978-0-12-800876-8.00003-5

Cohen AC, Bottini R, Piccoli P (2015) Role of abscisic acid producing PGPR in sustainable agriculture. In: Maheshwari DK (ed) Bacterial metabolites in sustainable agroecosystem. Springer, New York, pp 259–282. https://doi.org/10.1007/978-3-319-24654-3_9

Dao G, Wang S, Wang X, Chen Z, Wu Y, Wu G, Lu Y, Liu S, Hu H (2020) Enhanced Scenedesmus sp. growth in response to gibberellin secretion by symbiotic bacteria. Sci Total Environ 740:140099. https://doi.org/10.1016/j.scitotenv.2020.140099

Davies PJ (2013) Plant hormones: physiology, biochemistry and molecular biology. Springer, New York

de Souza R, Ambrosini A, Passaglia LMP (2015) Plant growth-promoting bacteria as inoculants in agricultural soils. Genet Mol Biol 38(4):401–419. https://doi.org/10.1590/S1415-475738420150053

Del Castello F, Nejamkin A, Cassia R, Correa-Aragunde N, Fernández B, Foresi N, Lombardo C, Ramirez L, Lamattina L (2019) The era of nitric oxide in plant biology: twenty years tying up loose ends. Nitric Oxide 85:17–27. https://doi.org/10.1016/j.niox.2019.01.013

Dodd IC, Zinovkina NY, Safronova VI, Belimov AA (2010) Rhizobacterial mediation of plant hormone status. Ann Appl Biol 157(3):361–379. https://doi.org/10.1111/j.1744-7348.2010.00439.x

FAO (2017) The future of food and agriculture: trends and challenges. Food and Agriculture Organization of the United Nations, Rome

Farag MA, Zhang H, Ryu C-M (2013) Dynamic chemical communication between plants and bacteria through airborne signals: induced resistance by bacterial volatiles. J Chem Ecol 39 (7):1007–1018. https://doi.org/10.1007/s10886-013-0317-9

Gamas P, Brault M, Jardinaud M-F, Frugier F (2017) Cytokinins in symbiotic nodulation: when, where, what for? Trends Plant Sci 22:792–802. https://doi.org/10.1016/j.tplants.2017.06.012

Garcia SN, Osburn BI, Jay-Russell MT (2020) One health for food safety, food security, and sustainable food production. Front Sustain Food Syst 4:1. https://doi.org/10.3389/fsufs.2020.00001

Giron D, Frago E, Glevarec G, Pieterse CMJ, Dicke M (2013) Cytokinins as key regulators in plant–microbe–insect interactions: connecting plant growth and defence. Funct Ecol 27 (3):599–609. https://doi.org/10.1111/1365-2435.12042

Glick BR (2012, October 11) Plant growth-promoting bacteria: mechanisms and applications. Scientifica, Hindawi. https://doi.org/10.6064/2012/963401

Gupta KJ, Fernie AR, Kaiser WM, van Dongen JT (2011) On the origins of nitric oxide. Trends Plant Sci 16(3):160–168. https://doi.org/10.1016/j.tplants.2010.11.007

Haswell ES, Verslues PE (2015) The ongoing search for the molecular basis of plant osmosensing. J Gen Physiol 145(5):389–394. https://doi.org/10.1085/jgp.201411295

Hedden P (2019) A novel gibberellin promotes seedling establishment. Nat Plants 5(5):459–460. https://doi.org/10.1038/s41477-019-0427-7

Hedden P, Thomas SG (2012) Gibberellin biosynthesis and its regulation. Biochem J 444(1):11–25. https://doi.org/10.1042/BJ20120245

Hider RC, Kong X (2010) Chemistry and biology of siderophores. Nat Prod Rep 27(5):637–657. https://doi.org/10.1039/B906679A

Horie T, Karahara I, Katsuhara M (2012) Salinity tolerance mechanisms in glycophytes: an overview with the central focus on rice plants. Rice 5:11. https://doi.org/10.1186/1939-8433-5-11

Husson E, Hadad C, Huet G, Laclef S, Lesur D, Lambertyn V, Jamali A, Gottis S, Sarazin C, Nhien ANV (2017) The effect of room temperature ionic liquids on the selective biocatalytic hydrolysis of chitin via sequential or simultaneous strategies. Green Chem 19(17):4122–4131. https://doi.org/10.1039/C7GC01471F

Isayenkov SV, Maathuis FJM (2019) Plant salinity stress: many unanswered questions remain. Front Plant Sci. https://doi.org/10.3389/fpls.2019.00080

Ji S-H, Kim J-S, Lee C-H, Seo H-S, Chun S-C, Oh J, Choi E-H, Park G (2019) Enhancement of vitality and activity of a plant growth-promoting bacteria (PGPB) by atmospheric pressure non-thermal plasma. Sci Rep 9(1):1044. https://doi.org/10.1038/s41598-018-38026-z

Kang S-M, Waqas M, Khan AL, Lee I-J (2014) Plant-growth-promoting rhizobacteria: potential candidates for gibberellins production and crop growth promotion. In: Miransari M (ed) Use of microbes for the alleviation of soil stresses, vol 1. Springer, New York, pp 1–19. https://doi.org/10.1007/978-1-4614-9466-9_1

Karnwal A, Kaushik P (2011) Cytokinin production by fluorescent Pseudomonas in the presence of rice root exudates. Arch Phytopathol Plant Protect 44(17):1728–1735. https://doi.org/10.1080/03235408.2010.526768

Kasim WA, Gaafar RM, Abou-Ali RM, Omar MN, Hewait HM (2016) Effect of biofilm forming plant growth promoting rhizobacteria on salinity tolerance in barley. Ann Agric Sci 61 (2):217–227. https://doi.org/10.1016/j.aoas.2016.07.003

Kasotia A, Varma A, Choudhary DK (2015) Pseudomonas-mediated mitigation of salt stress and growth promotion in *Glycine max*. Agric Res 4(1):31–41. https://doi.org/10.1007/s40003-014-0139-1

Kasotia A, Varma A, Tuteja N, Choudhary DK (2016) Amelioration of soybean plant from saline-induced condition by exopolysaccharide producing pseudomonas-mediated expression of high affinity K^+-transporter (HKT1) gene. Curr Sci 111:1961–1967

Khabbaz SE, Zhang L, Cáceres LA, Sumarah M, Wang A, Abbasi PA (2015) Characterisation of antagonistic Bacillus and pseudomonas strains for biocontrol potential and suppression of damping-off and root rot diseases. Ann Appl Biol 166(3):456–471. https://doi.org/10.1111/aab.12196

Khabbaz SE, Ladhalakshmi D, Babu M, Kandan A, Ramamoorthy V, Saravanakumar D, Al-Mughrabi T, Kandasamy S (2019) Plant growth promoting bacteria (PGPB)—a versatile tool for plant health management. Can J Pestic Pest Manag 1(1):1–25. https://doi.org/10.34195/can.j.ppm.2019.05.001

Khan AL, Hussain J, Al-Harrasi A, Al-Rawahi A, Lee I-J (2015) Endophytic fungi: resource for gibberellins and crop abiotic stress resistance. Crit Rev Biotechnol 35(1):62–74. https://doi.org/10.3109/07388551.2013.800018

Kim SH, Woo DH, Kim JM, Lee SY, Chung WS, Moon YH (2011) Arabidopsis MKK4 mediates osmotic-stress response via its regulation of MPK3 activity. Biochem Biophys Res Commun 412:150–154

Koul V, Tripathi C, Adholeya A, Kochar M (2015) Nitric oxide metabolism and indole acetic acid biosynthesis cross-talk in Azospirillum brasilense SM. Res Microbiol 166(3):174–185. https://doi.org/10.1016/j.resmic.2015.02.003

Kramer J, Özkaya Ö, Kümmerli R (2020) Bacterial siderophores in community and host interactions. Nat Rev Microbiol 18(3):152–163. https://doi.org/10.1038/s41579-019-0284-4

Kumar A, Patel JS, Meena VS, Srivastava R (2019) Recent advances of PGPR based approaches for stress tolerance in plants for sustainable agriculture. Biocatal Agric Biotechnol 20:101271. https://doi.org/10.1016/j.bcab.2019.101271

Kurepin LV, Park JM, Lazarovits G, Bernards MA (2015) Burkholderiaphytofirmans-induced shoot and root growth promotion is associated with endogenous changes in plant growth hormone levels. Plant Growth Regul 75(1):199–207. https://doi.org/10.1007/s10725-014-9944-6

Li HQ, Jiang XW (2017) Inoculation with plant growth-promoting bacteria (PGPB) improves salt tolerance of maize seedling. Russ J Plant Physiol 64(2):235–241. https://doi.org/10.1134/S1021443717020078

Li F, Tang M, Tang X, Sun W, Gong J, Yi Y (2019) Bacillus subtilis – *Arabidopsis thaliana*: a model interaction system for studying the role of volatile organic compounds in the interchange between plants and bacteria. Botany 97(12):661–669. https://doi.org/10.1139/cjb-2019-0093

Liu F, Xing S, Ma H, Du Z, Ma B (2013) Cytokinin-producing, plant growth-promoting rhizobacteria that confer resistance to drought stress in Platycladusorientalis container seedlings. Appl Microbiol Biotechnol 97(20):9155–9164. https://doi.org/10.1007/s00253-013-5193-2

Lu Q, Weng Y, You Y, Xu Q, Li H, Li Y, Liu H, Du S (2020) Inoculation with abscisic acid (ABA)-catabolizing bacteria can improve phytoextraction of heavy metal in contaminated soil. Environ Pollut 257:113497. https://doi.org/10.1016/j.envpol.2019.113497

Ma del Orozco-Mosqueda C, Glick BR, Santoyo G (2020) ACC deaminase in plant growth-promoting bacteria (PGPB): an efficient mechanism to counter salt stress in crops. Microbiol Res 235:126439. https://doi.org/10.1016/j.micres.2020.126439

Marwein R, Debbarma J, Sarki YN, Baruah I, Saikia B, Boruah HPD, Velmurugan N, Chikkaputtaiah C (2019) Chapter 4—Genetic engineering/genome editing approaches to modulate signaling processes in abiotic stress tolerance. In: Khan MIR, Reddy PS, Ferrante A, Khan NA (eds) Plant signaling molecules. Woodhead Publishing, London, pp 63–82. https://doi.org/10.1016/B978-0-12-816451-8.00004-6

Mishra P, Dash D (2014) Rejuvenation of biofertilizer for sustainable agriculture and economic development. Consilience 11:41–61

Mittler R, Blumwald E (2015) The roles of ROS and ABA in systemic acquired acclimation. Plant Cell 27(1):64–70. https://doi.org/10.1105/tpc.114.133090

Molina-Favero C, Creus CM, Simontacchi M, Puntarulo S, Lamattina L (2008) Aerobic nitric oxide production by *Azospirillumbrasilense* Sp245 and its influence on root architecture in tomato. Mol Plant Microbe Interact 21(7):1001–1009. https://doi.org/10.1094/MPMI-21-7-1001

Morrone D, Chambers J, Lowry L, Kim G, Anterola A, Bender K, Peters RJ (2009) Gibberellin biosynthesis in bacteria: separate ent-copalyl diphosphate and ent-kaurene synthases in Bradyrhizobium japonicum. FEBS Lett 583(2):475–480. https://doi.org/10.1016/j.febslet.2008.12.052

Mur LAJ, Mandon J, Persijn S, Cristescu SM, Moshkov IE, Novikova GV, Hall MA, Harren FJM, Hebelstrup KH, Gupta KJ (2013) Nitric oxide in plants: an assessment of the current state of knowledge. AoB Plants 5. https://doi.org/10.1093/aobpla/pls052

Naveed M, Qureshi MA, Zahir ZA, Hussain MB, Sessitsch A, Mitter B (2015) L-tryptophan-dependent biosynthesis of indole-3-acetic acid (IAA) improves plant growth and colonization of maize by Burkholderia phytofirmans PsJN. Ann Microbiol 65(3):1381–1389. https://doi.org/10.1007/s13213-014-0976-y

Negrão S, Schmöckel SM, Tester M (2017) Evaluating physiological responses of plants to salinity stress. Ann Bot 119(1):1–11. https://doi.org/10.1093/aob/mcw191

Olanrewaju OS, Glick BR, Babalola OO (2017) Mechanisms of action of plant growth promoting bacteria. World J Microbiol Biotechnol 33(11). https://doi.org/10.1007/s11274-017-2364-9

Park HJ, Kim W-Y, Yun D-J (2016) A new insight of salt stress signaling in plant. Mol Cells 39 (6):447–459. https://doi.org/10.14348/molcells.2016.0083

Premachandra D, Hudek L, Brau L (2016) Bacterial modes of action for enhancing of plant growth. J Biotechnol Biomater 6(3):1–8. https://doi.org/10.4172/2155-952X.1000236

Puri A, Padda KP, Chanway CP (2017) Nitrogen-fixation by endophytic bacteria in agricultural crops: recent advances. Nitr Agric Updates. https://doi.org/10.5772/intechopen.71988
Quintero FJ, Martinez-Atienza J, Villalta I, Jiang X, Kim W-Y, Ali Z, Fujii H, Mendoza I, Yun D-J, Zhu J-K, Pardo JM (2011) Activation of the plasma membrane Na/H antiporter salt-overly-sensitive 1 (SOS1) by phosphorylation of an auto-inhibitory C-terminal domain. Proc Natl Acad Sci 108(6):2611–2616. https://doi.org/10.1073/pnas.1018921108
Raaijmakers JM, Mazzola M (2011) Diversity and natural functions of antibiotics produced by beneficial and plant pathogenic bacteria. Annu Rev Phytopathol 50(1):403–424. https://doi.org/10.1146/annurev-phyto-081211-172908
Radhakrishnan R, Lee I-J (2016) Gibberellins producing Bacillus methylotrophicus KE2 supports plant growth and enhances nutritional metabolites and food values of lettuce. Plant Physiol Biochem 109:181–189. https://doi.org/10.1016/j.plaphy.2016.09.018
Ruiu L (2020) Plant-growth-promoting bacteria (PGPB) against insects and other agricultural pests. Agronomy 10(6):861. https://doi.org/10.3390/agronomy10060861
Ryu C-M, Farag MA, Hu C-H, Reddy MS, Wei H-X, Paré PW, Kloepper JW (2003) Bacterial volatiles promote growth in arabidopsis. Proc Natl Acad Sci USA 100(8):4927–4932. https://doi.org/10.1073/pnas.0730845100
Santoyo G, Moreno-Hagelsieb G, Ma del Orozco-Mosqueda C, Glick BR (2016) Plant growth-promoting bacterial endophytes. Microbiol Res 183:92–99. https://doi.org/10.1016/j.micres.2015.11.008
Schmülling T (2013) Cytokinin. In: Lennarz WJ, Lane MD (eds) Encyclopedia of biological chemistry, 2nd edn. Academic Press, New York, pp 627–631. https://doi.org/10.1016/B978-0-12-378630-2.00456-4
Seifikalhor M, Aliniaeifard S, Shomali A, Azad N, Hassani B, Lastochkina O, Li T (2019) Calcium signaling and salt tolerance are diversely entwined in plants. Plant Signal Behav 14 (11):1665455. https://doi.org/10.1080/15592324.2019.1665455
Shahzad R, Khan AL, Bilal S, Waqas M, Kang S-M, Lee I-J (2017) Inoculation of abscisic acid-producing endophytic bacteria enhances salinity stress tolerance in Oryza sativa. Environ Exp Bot 136:68–77. https://doi.org/10.1016/j.envexpbot.2017.01.010
Sharma DK, Singh A (2015) Salinity research in India-achievements, challenges and future prospects. Water Energy Int 58(6):35–45
Sharma M, Tarafdar A, Ghosh R, Gopalakrishanan S (2017) Biological control as a tool for eco-friendly management of plant pathogens. In: Adhya TK, Mishra BB, Annapurna K, Verma DK, Kumar U (eds) Advances in soil microbiology: recent trends and future prospects: volume 2: soil-microbe-plant interaction. Springer, New York, pp 153–188. https://doi.org/10.1007/978-981-10-7380-9_8
Singh RP, Shelke GM, Kumar A, Jha PN (2015) Biochemistry and genetics of ACC deaminase: a weapon to "stress ethylene" produced in plants. Front Microbiol 6. https://doi.org/10.3389/fmicb.2015.00937
Spaepen S, Vanderleyden J (2011) Auxin and plant-microbe interactions. Cold Spring Harb Perspect Biol 3(4):a001438. https://doi.org/10.1101/cshperspect.a001438
Sparks DL (2003) 10—the chemistry of saline and sodic soils. In: Sparks DL (ed) Environmental soil chemistry, 2nd edn. Academic Press, New York, pp 285–300. https://doi.org/10.1016/B978-012656446-4/50010-4
Upadhyay SK, Singh JS, Saxena AK, Singh DP (2012) Impact of PGPR inoculation on growth and antioxidant status of wheat under saline conditions. Plant Biol 14(4):605–611. https://doi.org/10.1111/j.1438-8677.2011.00533.x
USDA (1954) United States salinity laboratory staff
Vaddepalli P, Fulton L, Wieland J, Wassmer K, Schaeffer M, Ranf S, Schneitz K (2017) The cell wall-localized atypical β-1, 3 glucanase ZERZAUST controls tissue morphogenesis in *Arabidopsis thaliana*. Development 144(12):2259–2269. https://doi.org/10.1242/dev.152231
van Zelm E, Zhang Y, Testerink C (2020) Salt tolerance mechanisms of plants. Annu Rev Plant Biol 71(1):403–433. https://doi.org/10.1146/annurev-arplant-050718-100005

Vaseva II, Qudeimat E, Potuschak T, Du Y, Genschik P, Vandenbussche F, Van Der Straeten D (2018) The plant hormone ethylene restricts *Arabidopsis* growth via the epidermis. Proc Natl Acad Sci 115(17):E4130–E4139. https://doi.org/10.1073/pnas.1717649115

Vimal SR, Patel VK, Singh JS (2019) Plant growth promoting Curtobacterium albidum strain SRV4: an agriculturally important microbe to alleviate salinity stress in paddy plants. Ecol Indic 105:553–562. https://doi.org/10.1016/j.ecolind.2018.05.014

Walpola BC, Yoon M-H (2012) Prospectus of phosphate solubilizing microorganisms and phosphorus availability in agricultural soils: a review. Afr J Microbiol Res 6(37):6600–6605. https://doi.org/10.5897/AJMR12.889

Wicke B, Smeets E, Dornburg V, Vashev B, Gaiser T, Turkenburg W, Faaij A (2011) The global technical and economic potential of bioenergy from salt-affected soils. Energy Environ Sci 4 (8):2669–2681. https://doi.org/10.1039/C1EE01029H

Win KT, Tanaka F, Okazaki K, Ohwaki Y (2018) The ACC deaminase expressing endophyte Pseudomonas spp. enhances NaCl stress tolerance by reducing stress-related ethylene production, resulting in improved growth, photosynthetic performance, and ionic balance in tomato plants. Plant Physiol Biochem 127:599–607. https://doi.org/10.1016/j.plaphy.2018.04.038

Yabuta T (1938) On the crystal of gibberellin, a substance to promote plant growth. J Agric Chem Soc Japan 14:1526

Yang Y, Guo Y (2018a) Elucidating the molecular mechanisms mediating plant salt-stress responses. New Phytol 217(2):523–539. https://doi.org/10.1111/nph.14920

Yang Y, Guo Y (2018b) Unraveling salt stress signaling in plants. J Integr Plant Biol 60 (9):796–804. https://doi.org/10.1111/jipb.12689

Yin W, Wang Y, Liu L, He J (2019) Biofilms: the microbial "protective clothing" in extreme environments. Int J Mol Sci 20(14):3423. https://doi.org/10.3390/ijms20143423

Yu M, Lamattina L, Spoel SH, Loake GJ (2014) Nitric oxide function in plant biology: a redox cue in deconvolution. New Phytol 202(4):1142–1156. https://doi.org/10.1111/nph.12739

Zhang H, Kim M-S, Sun Y, Dowd SE, Shi H, Paré PW (2008) Soil Bacteria confer plant salt tolerance by tissue-specific regulation of the sodium transporter HKT1. Mol Plant-Microbe Interact 21(6):737–744. https://doi.org/10.1094/MPMI-21-6-0737

Zhang L, Khabbaz SE, Wang A, Li H, Abbasi PA (2015) Detection and characterization of broad-spectrum antipathogen activity of novel rhizobacterial isolates and suppression of Fusarium crown and root rot disease of tomato. J Appl Microbiol 118(3):685–703. https://doi.org/10.1111/jam.12728

Zhao, Y. (2014). Auxin biosynthesis. The Arabidopsis Book, The American Society of Plant Biologists Rockville 2014(12).

Zhu S, Zhou X, Wu X, Jiang Z (2013) Structure and function of the CBL–CIPK Ca^{2+}–decoding system in plant calcium signaling. Plant Mol Biol Rep 31:1193–1202. https://doi.org/10.1007/s11105-013-0631-y

Chapter 30
Impact of Climate Change on Soil Functionality

Deepika Pandey

Abstract Soil performs the most basic function of providing food to the living world through agriculture, which is the foremost activity to deliver food to the growing population of the world and had been the most primaeval social activity since the beginning of culture in human beings. Forming the basis of wildlife and sustenance of all other living beings on earth, soil health cannot be compromised in the name of development. Soil is not limited to production of food and it also serves several ecological functions, which include water storage and maintenance, nutrient cycling, carbon storage and recycling, regulation of above ground diversity and many more. Climate being the fundamental factor for soil formation is highly effective in changing the soil properties over time. A slow but continuous change in the soil structure and texture is influenced by the climate. However, the changing climate may have adverse effect on soil and reduce its fertility and innumerable vital ecological functions. A comprehensive understanding of soil functions in relation to the climate should be established to maintain the soil functionality in favour of human life.

Keywords Soil functions · Climate change · Water-food-energy nexus · Soil health · Nutrient cycling · Soil ecosystem services

30.1 Introduction

The soil performs numerous functions to balance the ecosystem and its quality and productivity are indispensable for the survival of the man. The interphase of biotic and abiotic components of our environment, carried out well by producers, is possible because of the soil matrix. The most important ecological functions performed by soil are biomass production in the form of forests and agriculture,

D. Pandey (✉)
Amity School of Earth and Environmental Sciences, Amity University Haryana, Gurugram, India

D. K. Choudhary et al. (eds.), *Climate Change and the Microbiome*, Soil Biology 63,
https://doi.org/10.1007/978-3-030-76863-8_30

nutrient and water cycling, carbon pool, biodiversity pool, buffering and transforming potentially harmful elements and compounds and many more to achieve a stable ecological system (Haygarth and Ritz 2009; Balestrini et al. 2015). The formation of soil and its support to the plants and animals started much earlier than the human civilisation. Soil was the basis of terrestrial life with the advent of green plants on the primitive earth. The functions of soil started gaining more importance with the beginning of agriculture and dependence of human civilisation on soil for food. Industrialisation and urbanisation have increased the stress on soil by decreasing the area under plant cover and adding pollutants to the soil ecosystem. However, the complexity of the soil matrix has been able to withstand this stress and continued to serve the humanity by its productivity. The slowly deteriorating soil quality creates a serious problem of decreased primary productivity and increased wasteland, which creates a terrifying situation of scarcity of food for the human population. The stress on agriculture is further enhanced by climate change. A major consequence of the rapid industrialisation and reckless deforestation is the enhanced greenhouse effect, which has brought the climate change globally. Climate is an important factor in soil formation and unpredictable change in climate due to global warming is also bringing in changes in soil functions in the most effected parts of the world. Along with all the living beings, this climate change has brought about the damaging effect on many non-living things and severely affected the natural processes. Extreme events associated with climate change have potential to impact soils to greater extent.

30.2 Soil Functions

The multiple functions performed by soil can be categorised as biophysical and ecological functions. The functions can also be categorised as ecological and non-ecological (Blum 2005). Biophysical functions such as cycling of water and nutrients, exchange of gas with the atmosphere and filtering and buffering of ions through soil profile are the fundamental earth surface processes. The function of soil is to provide physical stability and support to plant systems, terrestrial vegetation in all forms, trees, grasses, etc., which, in succession, support animal life. This is phenomenal and has no other replacement. Non-ecological functions of soil can be summarised as, source of raw materials, physical basis of human activities and basis of geogenic and cultural heritage. Ecological functions include the production of biomass, reservoir of genes and balancing activities for the protection of human and its environment.

Jax (2005) described the use of term 'function' in four ways and this encompasses the complete functionality of soil. The soil function can be used as a synonym for processes and operation/functioning of a system and as a synonym for roles and services. Soil functions are interchangeably described as soil roles and Ecosystem Services provided by soils and thus, it includes a long list. Soil is the centre of water-food-energy nexus (Pandey 2020; Hatfield et al. 2017) and the security of water,

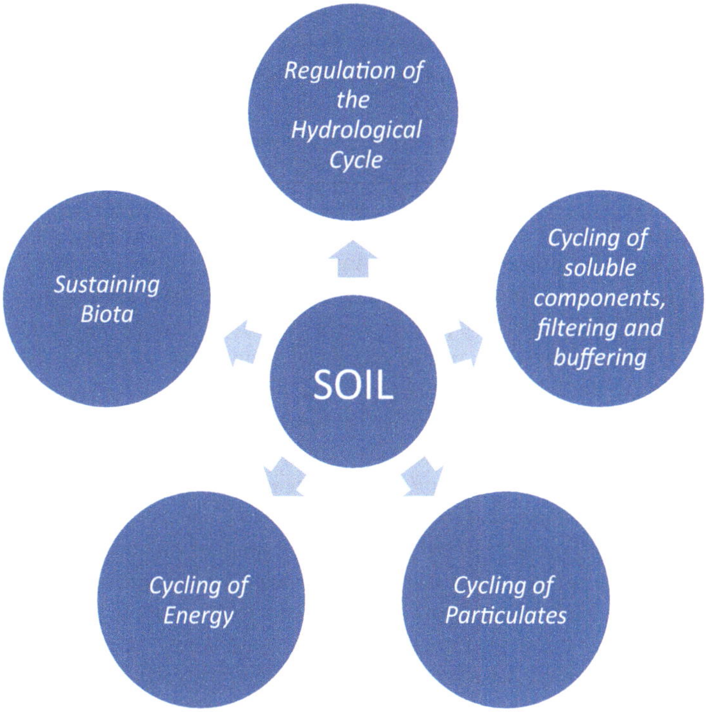

Fig. 30.1 Soil functions summarised in different categories

food and energy symbolises the existing and forthcoming challenge of nurturing life while preserving the environment. European Soil Science community has introduced the concept of soil functions during the early 1970s (Glenk et al. 2012). The seven key soil functions identified for the development of the EU Soil Framework Directive (European Commission 2006) were production of biomass, which include both agriculture and forestry; transformation of nutrients, substances and water by storing and filtering; gene pool and pool for habitats and biodiversity; carbon pool (acting as store and sink) and source of raw materials; for humans and human activities, it gives physical and cultural basis and develops archive of geological and archaeological heritage.

The soil functionality is summarised under five major categories (Fig. 30.1). The complex nature of soil processes is interwoven with the climate and it plays an important role in the water-food-energy nexus. Regulation of the hydrological cycle is one of the basic functions of the ecosystem and essence of life on earth. Sufficient infiltration of water in soil is necessary to control surface-water excess and distribution of overland flow of water. This is crucial to the problem of water logging and fluctuation of water table. Soil plays an interface for uptake of water by the roots and loss of water through transpiration by plants. Root zone storage of water and its disposition and drainage beyond the root zone normalise recharge of deep aquifers and discharge of water through wells, rivers and natural springs. Surface runoff finds its route through soil and its infiltration controls the flow in rills and gullies. The complex functionality of soil is vital for maintaining the water dynamics and is indispensable in cycling of water.

Cycling of soluble components in ecosystem can be carried out only with soil as interface. Hydration, dissolution and re-precipitation of minerals from the rocks undergoing weathering, sorption and exchange of ions in the clay minerals, and minerals formed during weathering, bring the soluble components to the system. Formation and decomposition of organic compounds and reactions such as oxidation-reduction and acidification-alkalization are the fundamental processes of soil profile formation. Leaching process releases solutes to groundwater and streams through eluviation-illuviation in the soil profile and other salination-desalination processes determine the chemistry of soil and of ground water. Volatilisation and outgassing of the soluble compounds to the atmosphere also occur in soil.

Particulates of nutritional importance are available in soil through migration and deposition of particles within soil profile, coating of aggregates on clay particles and accumulation of air-borne particles in the soil accumulates. Water and wind cause soil erosion on the earth surface and water-suspended matter gets transported overland bringing sediments to rivers and faraway lands, which determine their chemistry. This overland transport of suspended matter with water, filtration of suspended particulates in percolation and silting of lakes and water reservoirs complete the cycle of particulates in water bodies and the earth surface.

Energy cycling through soil starts with absorption of incoming shortwave radiation and transmission and exchange of sensible heat through biota and assimilation of chemical energy in biomass. Earth's albedo and emission of terrestrial longwave (thermal) radiation bring energy to the atmosphere in the form of heat. The greenhouse gases such as methane, carbon dioxide and nitrous oxide are affected by various soil processes and enhance or mitigate greenhouse effect.

Sustaining biota is the chief function of soil by providing water, nutrients, and anchorage to roots of all terrestrial plants. The microbial community obtains substrate, water and nutrients through soil, carries out decomposition of plant and animal remains and releases nutrients through aerobiosis and anaerobiosis. Absorbing and neutralising pathogenic and toxic agents is another significant function of soil.

Food and Agriculture Organization of the United Nations (FAO) has shown that the top 30 cm of the world's soil contains carbon, which is about twice as much the entire atmosphere. It is found that soil is the second largest natural sink of carbon, which is greater than forests and other vegetation, and only oceans have more capacity than soil to capture carbon dioxide from air.

30.3 Climate and Soil Interconnection and Impact of Climate Change

At global scale, soils show a strong geographical correlation with climate. Climate affects the type of soil and is determining element in soil formation. Changes in temperature and variability of precipitation strongly influence the parent rocks both physically and chemically (Sadao et al. 2006). Moisture in the atmosphere with changing seasons causes cooling, heating and thawing effects on rocks and break them. Chemical reactions and element mobilisation during weathering of rocks are influenced by climate of the area (Pandey 2007). Precipitation affects translocation of dissolved ions through the soil and favours soil horizon development. Climate indirectly or directly determines vegetation cover of a region, which in turn impacts further soil development. The process of succession continues with both plant community and soil development, directly governed by the climate of the region. With the advancement of time, climate becomes the prime influence on soil and its properties due to the type of vegetation and the factors associated with it, while parent material is left beneath with limited influence on the soil properties. The response to increasing temperatures is often characterised by shift in the dominant species of plants and increase in primary productivity (Lenihan et al. 2003). However, the shifting in seasonal temperature due to climate change may affect the annual cycles of plants and animals. This shift is disastrous for the crop production as blossoming time may not coincide with arrival of pollinators. Food and other plant-based products such as biofuels or raw materials pivot largely on maintaining health of soil and adopting sustainable agriculture practices.

Climate change phenomenon is identified by long-term changes in temperature and precipitation pattern of a region. These factors are directly responsible for the rate of decomposition, leaching of solutes and mobility of soluble molecules. It is predicted that the slow decomposition will be more sensitive to rising temperatures over the centuries (Powlson 2005). The decline of organic matter in peat and mineral soil is among the few identified threats to soil (Stolte et al. 2016). The net primary productivity of the soil is directly affected by change in the climate. The physical properties of soil effected due to precipitation are porosity, soil available water and distribution, infiltration and surface cover (Allen et al. 2011). The soil N, P, K, and S are mobilised and transported by the movement of water within the soil layers and the surface. The mobility of these ions is crucial in bringing these ions to the rivers and alluvial deposits. Macro- and microplant nutrients are distributed in the farmland and as soluble ions in surface and ground water because of their mobility and transport driven by water movement.

Decomposition of organic matter is a significant soil function and affected by increase in soil temperature. The microbial activity and rate of respiration of microfauna present in the soil increase with rise in temperature of the soil (Tóth et al. 2007; Knorr et al. 2005). A positive feedback is possible where extra carbon dioxide released in this manner will accelerate climate change. Soil is the largest store house of terrestrial carbon on earth and small changes in the flow of carbon from the soil to atmosphere by decomposition or respiration activities could bring significant change in atmospheric carbon pool (Davidson and Janssens 2006). Carbon flow models have been developed by many scientists to quantify the soil carbon turnover (Jenkinson et al. 1991; Cox et al. 2000; Jones et al. 2005). Carbon dioxide enrichment in the atmosphere enhances mycorrhizal and N-fixing relationships, but the overall effect of atmospheric temperature increase is variable and difficult to predict. Several studies suggest that fungal activity is enhanced as compared to bacterial because of increase in both atmospheric temperature and carbon dioxide.

The impact of climate change is already visible in soils as soil moisture is significantly decreased globally. Reports have predicted similar effects for the coming decades, increasing the demand of irrigation and exposing the crops to unpredictable precipitation. Yield will be highly effected due to changes in rainfall pattern and may lead to desertification.

Extreme climate events such as heat waves, excessive rainfall, storms or droughts, due to climate change, accelerate the degradation of soil by erosion and loss of land. Rising sea level will bring salt water to the adjoining lands apart from submerging the islands and coastal areas. The contaminants from the sea and loss of productive land will reduce the productivity of agricultural land and further increase the stress.

The biggest and direct consequence of climate change is melting of permafrost region. The major greenhouse gases, carbon dioxide and methane, are stored in surplus quantities in permafrost regions, especially in Siberia, will be released to the atmosphere. The temperature increase will melt the permafrost and thawing will cause decomposition of organic material held in the frozen soil. This massive increase in the greenhouse gases in the atmosphere will be beyond control and have disastrous consequences.

30.4 Role of Soil in Climate Change Crisis

Maintaining and restoring crucial ecosystems such as forests, mangroves, coral reefs, natural seabeds and peatlands and re-establishing them will help to remove the carbon dioxide from the environment in a natural way. The more crucial functions of providing soil ecosystem services such as maintaining biodiversity, air and water purification and supporting spaces for recreation will also be achieved. It is a highly cost-effective measure with multiple impacts. Plants can remove carbon from air through photosynthesis and restoration of degraded soil can retain up to 63 billion tonnes of carbon (FAO report). Natural recreation areas such as mangroves or developed recreation areas may act as effective safeguards against extreme events of climate change. The riparian zones and undisturbed floodplains are highly recommended for protection against floods. The urban green spaces are excellent heat wave absorbers and highly cost effective. A healthy soil can absorb excess water through percolation and avoid floods. Maintaining soil health in cities by keeping them in natural state and using them as parks and other green spaces cools down the atmospheric temperature. Moreover, ecosystem like these slowly release water stored underground to deviate the worst effects of drought during dry season.

Detaining carbon from the air is a crucial soil activity to mitigate climate change. The most rapid method to capture carbon is covering bare land with vegetation or grassland. Arable lands covered by crops residues and a practicing crop rotation have proved to be successful way of increasing carbon stocks in soil. Carbon capturing process by forests is much faster and young forests are more effective in securing the atmospheric carbon. Deforestation affects in two ways as forests store and capture carbon. The release of carbon from forests acts at different scales depending upon the choice of use of forest products, mainly wood. A wrong decision regarding land use may convert reservoir soil into source of emissions. Previously stored carbon in peatlands can add heaps of carbon to the atmosphere due to activities such as burning or draining peatlands.

Careful use of soil and proper land management can be helpful in mitigation of climate change and combating soil degradation. Greenhouse gas emissions from soil can be prevented by practices like conservation tillage or no tillage, crop residue management and multiple cropping. A large part of depleted soil organic carbon pool can be restored by converting more unused and waste lands into beneficial land uses, using crops residue and mulching and promoting natural cycling of nutrients by using compost and manure (Lal 2004). Such mediations will increase organic matter in the soil and prevent enhanced decomposition of soil organic matter.

30.5 Conclusion

Maintaining healthy soil and balanced land ecosystems is imperative to tackle the global crisis of climate change. Soil, if integrated well with other elements, can be a powerful tool in mitigating climate change. As highlighted by Paris Agreement, land use sector plays critical role in climate action. On the other hand, the stress created by climate change on soil may bring deviations in soil functionality. Ecosystem, agriculture, forestry, land use and green infrastructure all have strong links to climate change and soil and their management is the key factor in maintaining soil functions in the time of climate change. However, not completely known, but the dynamics between soil, land and climate is very crucial and we need to have its better understanding for designing sustainable solutions to our problems.

Soils of the world and climate maintain a unique balance, which is responsible for the distribution of world's widespread and unique ecosystems, proving growing medium and essential factors like water and nutrients. The nature, which is all around us, is the result of this unique balance. Soils' ability to support ecosystems will change if the climate changes. Climate change will affect nature, our lives and the places we live and current infrastructures will change, but soil functionality will adjust accordingly and continue in its adapted form, which may or may not be suitable to human life.

References

Allen DE, Singh BP, Dalal RC (2011) Soil health indicators under climate change: review of current knowledge. In: Singh BP et al (eds) Soil health and climate change, soil biology, vol 29. Springer, Berlin, Heidelberg

Balestrini R, Lumini E, Borriello R, Bianciotto V (2015) Plant-soil biota interactions. In: Soil microbiology, ecology and biochemistry. Elsevier, New York, pp 311–338

Blum WEH (2005) Functions of soil for society and the environment. Rev Environ Sci Biotechnol 4:75–79. https://doi.org/10.1007/s11157-005-2236-x

Cox PM, Betts RA, Jones CD, Spall SA, Totterdell IJ (2000) Acceleration of global warming due to carbon-cycle feedbacks in a coupled climate model. Nature 408:184–187

Davidson EA, Janssens IA (2006) Temperature sensitivity of soil carbon decomposition and feedbacks to climate change. Nature 440:165–173

European Commission (2006) Proposal for a directive of the European Parliament and the council establishing a framework for the protection of soil amending directive 2004/35/EC

Glenk K, McVittie A, Moran D (2012) Deliverable D3.1: soil and soil organic carbon within an ecosystem service approach linking biophysical and economic data. Smart SOIL report

Hatfield JL, Sauer TJ, Cruse RM (2017) Soil. Adv Agron 143:1–46

Haygarth PM, Ritz K (2009) The future of soils and land use in the UK: soil systems for the provision of land-based ecosystem services. Land Use Policy 26:S187–S197

Jax K (2005) Function and "functioning" in ecology: what does it mean? Oikos 111:641–648. https://doi.org/10.1111/j.1600-0706.2005.13851.x

Jenkinson DS, Adams DE, Wild A (1991) Model estimates of CO_2 emissions from soil in response to global warming. Nature 351:304–306

Jones C et al (2005) Global climate change and soil carbon stocks: predictions from two contrasting models for the turnover of organic carbon in soil. Glob Change Biol 11:154–166

Knorr W, Prentice IC, House JI, Holland EA (2005) Long-term sensitivity of soil carbon turnover to warming. Nature 433:298–301

Lal R (2004) Soil carbon sequestration to mitigate climate change. Geoderma 123(1–2):1–22

Lenihan JM, Drapek R, Bachelet D, Neilson RP (2003) Climate change effects on vegetation distribution, carbon, and fire in California. Ecol Appl 13:1667–1681

Pandey D (2007) Speciation and mobility of nutrient elements during weathering of Amphibolites in the Cauvery catchment area. PhD Thesis. http://hdl.handle.net/10603/29864

Pandey D (2020) Agricultural sustainability and climate change Nexus. In: Singh P, Singh R, Srivastava V (eds) Contemporary environmental issues and challenges in era of climate change. Springer, Singapore

Powlson D (2005) Will soil amplify climate change? Climatol Nat 433:204–205

Sadao S, Masami N, Tadashi T (2006) Factors of soil formation: climate, As exemplified by volcanic ash soils. In: Certini G, Scalenghe R (eds) Soils: basic concepts and future challenges. Cambridge University Press, Cambridge, pp 131–149

Stolte J, Tesfai M, Øygarden L, Kværnø S, Keizer J, Verheijen F, Panagos P, Ballabio C, Hessel R (eds) (2016) Soil threats in Europe; EUR 27607 EN

Tóth JA, Kate L, Zsolt K, Zsolt K, Bruce C, Richard B, Mária P (2007) The effect of climate change on soil organic matter decomposition. Acta Silv Lign Hung 3:75–85

Part IV
Impact of Climate Change on Ecosystem Functioning

Chapter 31
The Impacts of Climate Change on Soil Fertility in Nigeria

Benjamin Anabaraonye, Joachim Chukwuma Okafor, Beatrice O. Ewa, and Charles C. Anukwonke

Abstract Climate change is one of the global problems, which has dominated the media headlines in recent times. It is a complex global problem because it is intertwined with many other issues such as economic development, poverty reduction, good health, and well-being. Climate change is one of the most important factors affecting the formation of soil with important implications for their development, use, and management perspective with reference to soil structure, stability, topsoil water holding capacity, nutrient availability, and erosion. Scientists have predicted that expected changes in temperature, precipitation, and evaporation as a result of climate change will cause significant change in organic matter turnover and CO_2 dynamics thereby significantly impacting soil fertility. Soil fertility is vital in agricultural processes for farmers in Nigeria. Soils are intricately linked to the climate system through the carbon, nitrogen, and hydrologic cycles. Climate change therefore has a profound effect on soil processes and properties. This study identifies that adaptation to global climate change through improved soil quality by adoption of improved management practices is key to maintaining sustainable agricultural production in Nigeria. Through literature review and participant observation, the researchers identified the impacts of climate change on soil fertility in Nigeria. This research paper discussed how management and protection of soil resources can contribute to sustainable development through sustainable agricultural production while maintaining sustenance of soil fertility. A holistic approach to soil management as the engine for increasing productivity by increasing resource use efficiency and making agriculture more ecofriendly is recommended. This research paper further recommends a deeper research and study of soil-climate interactions in a

B. Anabaraonye (✉) · B. O. Ewa
Institute of Climate Change Studies, University of Nigeria, Nsukka, Nigeria
e-mail: beatrice.ewa@unn.edu.ng

J. C. Okafor
Department of Political Science, University of Nigeria, Nsukka, Nigeria
e-mail: chukwuma.okafor@unn.edu.ng

C. C. Anukwonke
Department of Environmental Management, Chukwuemeka Odumegwu Ojukwu University, Uli, Nigeria

D. K. Choudhary et al. (eds.), *Climate Change and the Microbiome*, Soil Biology 63,
https://doi.org/10.1007/978-3-030-76863-8_31

changing world as critical to addressing future food security concerns to ensure sustainable economic growth in Nigeria.

Keywords Climate change · Soil management · Soil-climate interactions · Food security · Hydrologic cycles

31.1 Introduction

Soil remains the most critical factor in human lives that is essential in food production and is required for the sustenance of human civilizations and is threatened in recent times by the forces of environmental threats. These threats are climate change and global warming escalated by the forces of diverse erosions and floods disasters. In the globalizing world today, Nigeria inclusive, the visible impacts of climate change and global warming on soil fertility and prospects for agricultural productivity occasioned particularly, by the prevailing challenges of flooding, erosion and excessive rainfalls, remain an increasing challenge not just only to governments (state actors) with their various multilateral organizations but to numerous nonstate actors including nongovernmental organizations (NGOs) around the world (Anabaraonye et al. 2018). These environmental threats that include erosion, flooding, drought, and desertification have continued to expose human beings to varieties of humanitarian concerns such as, hunger and starvation, unemployment, poverty, and disease (Anabaraonye et al. 2019; Birsel 2019). In Nigeria also, the impacts of these environmental threats on people livelihoods, displacements of persons, attendants' humanitarian disasters, and conflict it has fostered are numerous (Okafor 2020; Okafor et al. 2019; Okafor and Okafor 2019). In view of the above, the United Nations Food and Agricultural Organizations (FAO) posit that soil erosion remains the greatest challenge to sustainable soil management around the world (Akamigbo and Nnaji 2011; Brevik 2013; Okoroafor et al. 2017; Ahmadu et al. 2019; Pennock 2019). In Nigeria, the major causes of soil erosion include human interference, climatic factors (rainfall, floods), poor geology, undulating topography, and soil nature (Okoroafor et al. 2017).

31.2 Definition of Terms

Adaptation The Intergovernmental Panel on Climate Change (IPCC) defines adaptation as the "adjustment in natural or human systems to a new or changing environment. Adaptation to climate change refers to adjustment in natural or human systems in response to actual or expected climatic stimuli or their effects, which moderates harm or exploits beneficial opportunities. Various types of adaptation can be distinguished, including anticipatory and reactive adaptation, private and public adaptation, and autonomous and planned adaptation" (IPCC 2001).

Climate Change The United Nations Framework Convention on Climate Change (UNFCCC) defines climate change as a change of climate which is attributed directly or indirectly to human activity that alters the composition of the global atmosphere and which is in addition to natural climate variability observed over comparable time periods (UNFCCC 1992). In addition, climate change is identical to global warming.

Climate Finance Climate finance refers to local, national, or transnational financing, which may be drawn from public, private, and alternative sources of financing. Climate finance is critical to addressing climate change because large-scale investments are required to significantly reduce emissions, notably in sectors that emit large quantities of greenhouse gases. Climate finance is equally important for adaptation, for which significant financial resources will be similarly required to allow countries to adapt to the adverse effects and reduce the impacts of climate change (UNEP 2015).

Climate Resilience This can be generally defined as the capacity for a socio-ecological system to absorb stresses and maintain function in the face of external stresses imposed upon it by climate change and adapt, reorganize, and evolve into more desirable configurations that improve the sustainability of the system, leaving it better prepared for future climate change impacts. With the rising awareness of climate change impacts by both national and international bodies, building climate resilience has become a major goal for these institutions (Wikipedia 2018).

Education Education means the impartation of knowledge, skills, and ideas on individuals, people, or organizations for the purpose of equipping them to confront the challenges that life presents. This education could be informal or formal.

Farmers These are people who work on the land (till and cultivate land for planting) and rear and raise livestock either for consumption or commercial purposes.

Vulnerability This is the degree to which a system is susceptible to or unable to cope with adverse effects of climate change, including climate variability and extremes (IPCC 2007). Vulnerability is a function of the character, magnitude, and rate of climate change, and the degree to which a system is exposed, along with its sensitivity and adaptive capacity. It increases as the magnitude of climate change or sensitivity increases and decreases as adaptive capacity increases (OECD 2009).

31.3 Methodology

Data used for this study is derived from published works including academic articles, journals, conference papers, textbooks, and internet materials. The researchers gathered a large number of materials for the research but summarized the characteristics of those that centered more on the impact of climate change on soil fertility in

Nigeria. This enabled the researchers to generate the synthesis of various researchers' views on the impacts of climate change on soil fertility in Nigeria.

31.3.1 Climate Change and Soil Erosion Crisis in Nigeria: The South-East Dilemma

In Nigeria, the impacts of climate change on soil fertility have tremendous effects on agricultural productivity that many farmers in the country in 2019 bemoaned the debilitating calamities brought about by floods and erosion of farmlands, cash crops (Okafor 2020), and the corresponding casualties done by these environmental threats to livelihood displacements and deaths. Though the persistent erosion and flood that affect soil fertility/productivity in Nigeria are prevalence across the federation, the southern parts of the country especially, the South Eastern States such as Abia, Imo, Anambra, Ebonyi, and Enugu are worst hit by soil erosion (Okoroafor et al. 2017). The following are the impacts of soil erosion escalated by climate change on soil productivity in Nigeria.

- Agricultural productivity, sustainability, and management for food security/sustenance in this region have been undermined and greatly limited by the menace posed by soil erosion while the availability of farmlands for agricultural production and construction activities have been greatly reduced by losses caused by the attendant issues of soil erosion (Okoroafor et al. 2017; Okafor 2020).
- Soil erosion is considered to be a major environmental problem since it seriously threatens natural resources and the environment and makes it unfit for agricultural and construction purposes.
- Gully erosion which a product of soil erosion undermines agricultural productivity in the Southeastern Nigeria because, agricultural farmlands are destroyed and become one of the greatest environmental disasters facing many towns and villages in the region (see more in the Table 31.1).

31.3.2 How Management and Protection of Soil Resources Can Contribute to Agricultural Production in Nigeria?

Attempts have been made to characterize the threats to the soil environment as it affects agricultural production. Blum (2013) indicated the soil risks in agricultural production, its general trends, and scenario with respect to the natural and anthropogenic factors. There is a nexus between soil quality degradation and food security and hunger. The thin line can be assessed in the sense that characterizing soil quality degradation such as abrupt changes in soil physical and chemical properties, water quality, food security indices, environmental health, farming and crop yield, trees, and habitats are all attributes of examining soil resources challenges in context.

Table 31.1 Reported Cases of Erosion Crisis in Communities among the Five States in the Southeast Geo-Political Zone that have destroyed Farmlands, Houses, and made Agricultural Activities in the Region Untenable

S/N	Date/month/year	State	Location	Incident	Remark
1.	17th October, 2017	Anambra	Nnewi Ichi, Nnewi Local Government Area	The relocation of many from their ancestral homes for safety and security of their lives	The state government had earlier seek the world bank intervention over the control of erosion sites in this area
2.	16th March, 2016	Anambra	Oko, Ifite Nanka, and Amako Nanka communities in Aguata local government area	The gully erosion threatened to destroy the family house of the former vice president, chief Alex Ekwueme and over 22 other families in the community	The alleged nonchalant attitude of the Federal Government towards addressing the erosion problem in the community and other neighboring areas
3.	9th April, 2013	Anambra	Obuofu Umuike village, in Awka South Local Government Area	Fear of the community being extinct because of the rampaging menace of gully erosion	The gully erosion measures about 60 m deep and 60 m wide, has consumed many houses, farmlands, and property worth millions of naira
4.	7th August, 2012	Anambra	Erosion sites at Uga community in Aguata Local Government	No fewer than 100 houses have been destroyed by the gully erosion in the area. Many villages in the community sleep with one eye open as they do not know when the menace would visit their homes	Ecological fund provided for arresting erosion problems and the essence of the fund was for quick intervention at any slightest occurrence; however, this fund have not been properly utilized to address this menace
5.	10th December, 2012	Anambra	Urumabiam, Egbema, Uamuokpanilo in Ozubulu, and Ibollo Oraifite in Ekwusigo Local Government	Two persons: Mr. Ignatius Igwemadu from Urumabiam, Egbema Ozubulu in Ekwusigo, and his brother allegedly died when their bungalow fell into the gully erosion	Five houses were allegedly swallowed by gully erosion ravaging the four communities. The economy of the inhabitants had adversely been affected as many farmlands had been washed away and roads cut off
6.	20th March, 2016	Anambra	The state is estimated to have 1, 000 erosion sites that are threatening to wash away farmlands and people source of livelihoods	Umuogboo Obiofia Nnewi Ichi and Uruagu Nnewi road in Nnewi North Local Government Area are other communities where erosion had caused destruction	Gully erosion is one of the potent environmental threat in Anambra state

(continued)

Table 31.1 (continued)

S/N	Date/month/year	State	Location	Incident	Remark
7.	25th March, 2020	Abia	Umuogele Umuakwu - Nsulu and Umuezeugwu in Isiala Ngwa North LGA and Umuda Isingwu and Umuagu in Umuahia North LGA	Until recently, these communities were cut off from the neighbors and buildings swept off as a result of gully erosion before Nigeria Erosion and Watershed Management Project (NEWMAP) came to their rescue	The erosion and gully sites were life-threatening that people from these communities through their town union representatives have repeatedly drawn the attention of both the Federal and State Government over their precarious conditions
8.	28th March, 2020	Abia	Amuda Isuochi, Umunneochi LGA, and Amuzukwu Ibeku, Umuahia, North LGA	Devastating erosion sites that are life-threatening are numerous	They have claimed many lives and properties worth millions of naira
9.	8th October, 2019	Abia	Amaugwu, Umuhu-Ezechi. Bende Local Government Area	People in this villages are increasingly been threatened by the prevailing landslide that has completely destroyed buildings and is destroying more	The ravaging landslides are critical that it is far beyond the financial capacity and capability of the indigenes to handle
10.	15th November, 2019	Abia	Umuekwaa, Umuokwasa, and Mgbarakuma, in the Umunwanwa clan of Umuahia South Local Government Area	An estimated 10 houses were swallowed in one fell swoop with many people rendered homeless by erosion in three villages mentioned	People in these areas have lost their streams, no good roads, no good water because erosion has swept the streams away. This has exposed the communities to hardship in looking for water in far away communities.
11.	12th September, 2018	Abia	It is estimated that over 30,000 active erosion sites and gullies exist and are scattered around the major zones in the state that include Umunneochi and Okigwe in the north, Ukwa in the south, Arochukwu in the east, and Isiala Ngwa in the west	These active erosion sites and gullies are life-threatening and have wash away farmlands, houses, and critical infrastructures in the state	Abia state is the second largest state in the south east geo-political zone worst hit by gully erosion sites after Anambra state

12.	17th July, 2019	Imo	Ezemazu Urualla Gully Erosion control site at Urualla in Ideato North Local Government Council	Huge amount of money was mapped out to remedy the menace of erosion in Idea to north but was abandoned	The projects were not carried out and the funds were not accounted for, leaving the residents at the mercy of flooding for the past 5 years
13.	16th August, 2016	Imo	Umueshi Gully Erosion Site in Ideato North Local Council	The pervasiveness of the erosion causes the relocation or resettlement of the indigenes to other places	Government interventions have been on providing palliatives and finding lasting solution to the menace
14.	6th November, 2016	Imo	Alaka Madu gully erosion at Ihim community in Isiala Mbano local council	The gully erosion estimated to be 150 ft. deep and 500 metres in length has claimed many lives with over 12 villages severed from the rest of the community. Both the community primary school and the St Peter's Anglican Church in the area have been destroyed by the surging gully erosion	Efforts to attract the attention of relevant government agencies to save the situation yielded no positive results
15.	3rd October, 2012	Imo	There are over 401 identified erosion sites scattered in all the 27 local government areas of the State namely: Umushievula Avuvu, Ikeduru local council area, Okpala, Aboh Mbaise, Owerri North, Owerri Municipal, Owerri West, Onuimo, Isiala Mbano, Ehime Mbano, Ihitte Uboma, Mbaitoli, Njaba, Ideato North and South local council areas, Orsu, Orlu, among others	Most of the erosion sites are not only life-threatening and fearful. Public buildings, including market stalls, schools, and hospitals, were not spared from the devastations	Several families have been forced to flee from their homes and into safety, sometimes several kilometers away from their ancestral homes
16.	16th March, 2013	Imo	Erosion sites that cut the Orlu-Mgbee-Urualla, Akokwa-Osina Federal Highway into two.	The devastating economic cost of this disaster was numerous especially, inter-state trade from Imo—Anambra state	This particular erosion site was long identified but nothing serious was done to checkmate the advancing gully
17.	17th June, 2020	Enugu	Onuiyi-Nsukka behind Nnamdi Azikiwe building	The Onuiyi-Nsukka gully erosion sites disaster has been a death threat to people during rainy season and has affected communities like Ogbudu, Obukpa, Alor Uno communities with lives and properties worth millions usually lost every year	The Onuiyi-Nsukka road is a federal road linking Benue State to Onitsha in Anambra State

(continued)

Table 31.1 (continued)

S/N	Date/month/year	State	Location	Incident	Remark
18.	16th June, 2020	Enugu	Obinofia Ndiuno in Ezeagu Local Government Area	The erosion was caused by the diversion of flood from neighboring communities of Nachi, Umumba, Obeleagu, and Amokwe	The menace has dragged the community backwards in terms of meaningful development as wealthy indigenes of Obinofia Ndiuno neither invested in the community nor came back home during festive periods. When it rains, the flood will go into people's houses. Our people are now sleeping with their two eyes wide open in the midnight because of erosion
19.	10th September, 2019	Enugu	Umuitodo Obollo Nkwo community in Udenu Local Government Area	The only road leading to the community is threatened by erosion	The lives and properties of the residents of Umuitodo Obollo Nkwo community are equally in a great danger as the erosion is fast encroaching into their buildings
20.	29th August, 2018	Enugu	Four head gully erosion site at Enugu-Ngwo	Immediate threat to the communities surrounding as it threatens them to extinction	The site poses a threat to Enugu-Onitsha express way
21.	22nd May, 2013	Enugu	Odongbu-Amogbeke and Akponge-Amoju in Ezimo, Udenu Local Government Area. Affected areas were: Mkporogwu-Ezimo-Ulo road and Ugwu Odor-Uruneya erosion site in Udunedem ward, Obollo-Afor, four at Amalla community, Ama-Orba and Iyi-Opu Ozalla erosion sites	The scattering of indigenes victims of the communities to neighboring communities	The affected communities lost many arable farmlands as this is their main source of livelihood
22.	18th August, 2016	Ebonyi	The Nguzu-Edda and Abia Iwerre communities in Afikpo South Local Council	The headquarters of Afikpo South council at Nguzu-Edda has been relocated due to the threat by erosion even as a few buildings still sit perilously on the edges of gullies	The menace has affected inter-communities trade and movement of persons thus, the entire Abia-Iwerre community is being threatened by erosion that has also sacked many residents

Compiled by present Authors

31.3.3 Understanding the Soil Resources Problem

Globally, soil depletion and high levels of greenhouse gas emissions are ubiquitous, and can hinder sustainable food and agricultural production. These challenges arise as a result of anthropogenic activities the soils are subjected to (Lal 2009, 2007; Pretty 2008). Authors have identified soil erosion, decline in organic matter, soil contamination and pollution, soil salinization, flooding as challenges of soil environment for agricultural protection (Commission of the European Communities 2002). These challenges are poised on disrupting bumper harvest for agricultural goods and consequences for food insecurity.

FAO (2017) and United Nations, in studying population projections in relation to food supply, posit that the world population is predicted to increase to 9.73 billion by 2050. As the demand for food and other agricultural products is also projected to rise by 50 percent between 2012 and 2050. Thus, it requires an increased agricultural production that will produce 50 percent more food to meet these needs (Alexandratos and Bruinsma 2012). In Sub-Saharan Africa and Nigeria, this requires an improved soil management and protection to contribute to better yield of crops as different streams of factors affect soil quality and its productivity.

Climate change impacts as one of the drivers of soil degradation and loss have been discussed in literature. Qafoku (2015) attempted a succinct discourse on the nitty gritty of climate change effects of soils in areas such as increased disintegration of rock and soil forming minerals and elemental fluxes within the soil ecosystem, methods employed in improving carbon sequestration in soils, soil organic matter conservation, transformation and mineralization, and ambient soil organic matter temperature differentials. In addition, extreme events caused by climate change (Angélil et al. 2014) may have long-term effects on soils with poorly understood consequences.

The potential impact of climate change on crop production has been reviewed by Hatfield et al. (2011) with diverse implications. Thus, improving soil fertility and productivity is required for sustainable soil ecosystem stability.

31.3.4 Improving Soil Fertility and Productivity Through Management

Attention has been focused on aspects of improving the soil organic matter to improve soil productivity generally. The school of thought believed that the organic matter content of soils is a prerequisite to soil fertility and bumper harvest which determines the overall physical, chemical, and biological behavior of soils ecology. This understanding has ties with the submission of Ejike and Osuji (2013).

The capacity of the soil is inherent as it provides goods and services. One of these services according to Blum (2005) includes the production of food through agriculture and forestry. In this background, the land use feature on food production requires maximum yield generated through sustainable soil management practices.

Soil fertility is an incentive for possible gains in the use of soils for agricultural production processes. Finch et al. (2014) identified that the overall productivity of the soil is linked to the net effects of fertility and management. Soil fertility issues have two major factors viz.; natural make-up and variable make-up. Soil management practices according to Finch et al. (2014) affect overall soil conditions by improving the levels of organic matter and biotic events in the soil and plant nutrients; the amount of soil water, prevention of soil loss by erosion and soil losses, correction of soil pH by liming, and soil structure.

Soil management practices refer to various actions which revolve around soil maintenance, conservation, and improvement that help the soil to regain its nutrients, stability, and balanced properties. According to Ullah et al. (2019) soil management practices are helpful for addressing soil quality degradation in different ways that allows for soil recovery and enormous soil productivity and reclamation.

Hatfield et al. (2011) in a study on mitigation opportunities from land management practices in a warming world highlighted the need for an increased soil management as essential in improving soil quality and agricultural production. According to their study, soil management actions can either be defined as those directly affecting tillage, or placement and those revolving around residue and organic matter.

Different soil management practices have peculiar impacts and affect soil water ratio and balance, temperature variations, biotic activities, and gaseous exchange between the soil resources and the atmospheric environment. For example, soil management practices positively influence soil organic carbon viz.; a vi. tillage and soil management systems and major increases in mulching practices (Franzluebbers 2005; Follett 2001).

Additionally, soil management practices, improve soil water ratio in the soil profile, and soil water holding capacity and balance, improve crop yield thereby increase water use efficiency.

Orji (2017) outlined all forms of alley farming methods that improve soil physical and chemical qualities in Nigeria and increase crop yield for different food crops such as maize, rice, growing legumes, cassava, cotton, and cocoyam. Agro-forestry and alley farming as sustainable soil management method offer a mixed approach to soil conservation in forms such as improved tree fallows, plantation crop combination, and shelter belts that helps to achieve soil fertility recovery through improving the chemical and physical conditions of the soil after a phase of usage. According to Orji (2017), alley farming provides added green manure or mulch that maintains soil fertility and reduces fertilizer requirements as agrochemical application is controlled and only used when necessary (Kang et al. 1990). The trees serve as nutrient pumps that recycle soil nutrients. It is pertinent to stress further that the benefit of alley farming to soil management is manifold. The technique provides perfect environmental conditions for soil macro and microorganisms, supplements active soil nitrogen to required crop, improves soil and water conservation practices on a sloping topography while preventing soil erosion disasters. Orji (2001) in alley cropping experiment observed a noticeable increment of basic soil properties in examining the significant impact of alley farming on soil. The author confirmed a noticeable increment on P, K, Na, Mg, and Al likewise noticeable improvement in soil chemical properties. Other results on farm experiments confirmed the efficacy of alley farming in soil protection purposes as it has reduced erosion, improved soil organic matter, and nutrient status while sustaining crop yield under intensive cultivation of the soil (Kang et al. 1995).

Lal (1989) examined that plots tilled and alley cropped with legumes such as Gliricidia and Leucaena were lowered by high values of 73% and 83% in comparison to controlled experiments respectively, while the use of vetiver grass has been proficient in soil erosion too (Orji (2017).

Similarly, Ritchie recommended soil-water conservation efforts by increasing soil albedo possibly by a favorable crop plant spacing and more dense tree population colony.

Other important soil management practices are noteworthy. They are shifting cultivation and bush fallowing. These techniques allow soils to recover its nutrients and prevent nutrient depletion per se as they allow adequate nutrient availability and the maintenance of a stable soil ecosystem Osuji et al. (2012). However, Ejike and Osuji (2013) noted that these soil management techniques are limited due to the fact that land is scarce and the systems of land ownership may not allow a prolonged fallow of such soil.

Osuji et al. (2017) discussed the impacts of sustainable soil management techniques as it improves agricultural production in Imo State. The study identified types of soil management techniques adopted which has assisted in management of soil with wide spread impacts on increasing crop yield while maintaining a balance between all trade-offs. The operation of the techniques helps to alleviate poverty and achieves zero hunger, improves life on land which are all sustainable development goals. These soil management practices are proven to be proficient in soil quality degradation and its restoration in Nigeria, for all ecological zones in the country. Usman (2007) and Osuji et al. (2017) included conservation tillage practices, soil fertility improvement actions, and soil erosion control measures.

In different forms and practices, other examples are organic manuring, crop residue recycling, multiple cropping, and agro-forestry. Usman (2007) submitted that organic manuring is widely used in improving soil fertility and crop yield in different localities in Nigeria. Organic manure is largely practiced by arable crop farmers to improve the fertility of the soil and productivity of the land for different agricultural practices. Multiple cropping as a practice is prevalent and relevant in improving soil organic matter, prevention of soil erosion, excessive runoff, flooding, and harsh soil detachment from raindrops (Emuh 2007).

The implication of exclusive soil management approaches has been rife as well as the nexus between the practices and increased crop yield and soil productivity. Ejike and Osuji (2013) posit that these approaches to soil management increase crop yield per for the local farmers thereby increasing local economy of households. Similarly, a mix of liming, contour stripping, and strip cropping are requisite soil preservation methods. While, a lime is introduced in the soil to improve an efficient acid—base balance in the soil suitable for bumper harvest for some cash crops, taungya system is the system of growing of forest trees on soil to offer a green shade for soil protection from harsh environmental conditions.

Contour and strip cropping according to Enger and Smith (2006), refers to soil management techniques on sloppy terrains. Contour farming is tilling at right angles to the slope of the land and helps to produce a series of small ridges at right angles to the slopes. A combination of contour farming and strip farming is employed when a slope is too steep or too long. Strip farming, however, refers to alternating strips of closely sown crops such as hay, wheat, and grains with row crops such as corn, soya beans, cotton, or sugar cane. These sown crops reduce the forceful flow of water as runoff on soils and prevent soil erosion and eventual soil loss.

According to Akanwa et al. (2019), agroforestry is the system that integrates trees in farms and rangelands. They outlined the potentials of agroforestry and environmental greening for climate change minimization. As a soil management method, it reduces soil erosion, improves soil quality, vegetation cover, and improved land management. Thus, it is a remarkable method in preserving the soil likewise climate change and adaptation.

Improved soil management practices generally provide soft landing for best soil quality performance and eventual profit making potentials that can be accrued. It is a requisite factor of enhancing bumper harvest from crop production operations while offering soft solutions to possible environmental problems arising from anthropogenic tendencies in soil use for agricultural purposes.

31.3.5 Recommendations

Based on the results of the study, the following recommendations were made:

(a) Government, companies, institutions, and even individuals in Nigeria can initiate activities which will lead to the planting of trees in strategic places around and within our primary, secondary schools, and tertiary institutions in Nigeria to help green our environment and release more oxygen into the atmosphere. Strategic Tree planting activities can immensely help to address the impacts of climate change on soil fertility in Nigeria.
(b) The government should construct more dams in the country which could also go a long way as a mitigation strategy.
(c) Relevant government agencies in collaboration with nongovernmental organizations should educate farmers both in rural and urban areas about the impacts of climate change on soil fertility in Nigeria and ways to adapt and mitigate for sustainable development.
(d) Flood Hazard Mapping: To reduce and manage properly the adverse consequences of flood in the country, there should be immediate Flood Hazard Mapping of the affected areas using best practiced modern technologies.

Flooding which is one of the effects of climate change has been discovered to have a great impact on soil fertility in Nigeria.

(e) Disaster risk reduction should be integrated in the syllabus in primary, secondary schools, and tertiary institutions in Nigeria. Principals, teachers, and students in various educational institutions in Nigeria should undergo annual training on disaster risk reduction to mitigate the effects of flooding across various cities and communities in the country.

(f) Government policy on urbanization in Nigeria should be revisited to discontinue the construction of houses, roads, bridges haphazardly. Construction of roads culvert and drainages should be expansive to accommodate the flow of water with special consideration of their proximity to farmlands.

(g) Dumping of biodegradable waste products should be done strategically which can go a long way to increase soil fertility when done properly.

(h) Early warning signs and information should be communicated to Nigerians through radio, television, internet, and numerous social media platforms in order to create awareness about the impacts of climate change on soil fertility in Nigeria.

31.4 Conclusion

The impacts of climate change on soil fertility in Nigeria are very profound. Education is seen as a human right, a key to civilization and enlightenment, and as source of wealth and power. It is the cornerstone of the growth and development of any country's social, economic, and political institutions. Flooding problems should not in any way be allowed to hamper and tamper with Nigeria's agricultural sector. Flooding which leads to soil erosion and adversely affects soil fertility should be properly managed in Nigeria. Disaster risk reduction needs to receive serious attention as a facet of development both to the government and the nongovernmental organizations in Nigeria. There is an urgent need therefore for all stakeholders, researchers, and the farmers who are directly affected to seek strategies to solve the problems of flooding to ensure sustainability in the soil fertility in Nigeria.

References

Ahmadu J, Galadima DH, Chinedu NB, Eze OB (2019) Effect of erosion on agricultural land in Agyana Community in Abaji, Abuja. Int J Appl Agric Sci 5:120–128

Akamigbo FOR, Nnaji GU (2011) Climate change and Nigerian soils: vulnerability, impact and adaptation. J Trop Agric Food Environ Extension 10:80–90

Akanwa AO, Mba HC, Ogbuene EB, Nwachukwu MU, Anukwonke CC (2019) Potential of agroforestry and environmental greening for climate change minimization. In: Abhishek R et al (eds) Climate change impact and agroforestry system. International Standard, CRC-Apple Academic Press and Taylor & Francis, London, p 389

Alexandratos N, Bruinsma J (2012) World agriculture towards 2030/2050: the 2012 revision. ESA Working Paper No. 12–03. Rome, FAO

Angélil O, Stone DA, Tadross M, Tummon F, Wehner M, Knutti R (2014) Attribution of extreme weather to anthropogenic greenhouse gas emissions: sensitivity to spatial and temporal scales. Geophys Res Lett 41:2150–2155

Anabaraonye B, Okafor JC, Hope J (2018) Educating farmers in rural areas on climate change adaptation for sustainability in Nigeria. In: Leal Filho W (ed) Handbook of climate change resilience. Springer, Cham. https://doi.org/10.1007/978-3-319-71025-9_184-1

Anabaraonye B, Okafor JC, Eriobu CM (2019) Green entrepreneurial opportunities in climate change adaptation and mitigation for sustainable development in Nigeria. J Environ Pollut Manag 2:1–6

Birsel R (2019) Flood, fire and plague: climate change blamed for disasters. Retrieved from https://www.theguardian.pe.ca/news/world/flood-fire-and-plague-climate-change-blamed-for-disasters-375852

Blum WEH (2005) Functions of soil for society and the environment. Rev Environ Sci Biotechnol 4:75–79

Blum WEH (2013) Soil and land resources for agricultural production: general trends and future scenario – a worldwide perspective. Int Soil Water Conserv Res 1:1–14

Brevik EC (2013) The potential impact of climate change on soil properties and processes and corresponding influence on food security. Agriculture 3:398–417. https://doi.org/10.3390/Agriculture3030398

Commission of the European Communities (2002) Towards a thematic strategy for soil protection. COM (2002) 179 Final. European Commission, Brussels, Belgium

Ejike RD, Osuji EE (2013) The role of women in sustainable conservation agriculture as a viable alternative to traditional farming practices in Imo State, Nigeria. In: Abstract proceedings of the 27th Annual Conference of Farm Management Association of Nigeria held at University of Ilorin, Kwara State, Nigeria, pp 23–29

Emuh FN (2007) Economic yield and sustainability of maize crop in association with Cowpea and Egwusi-melon in Southwestern, Nigeria. J Agron 6:157–161

Enger ED, Smith BF (2006) Environmental science: a study of inter-relationships, 10th edn. McGraw-Hill, New York

FAO (2017) The future of food and agriculture – trends and challenges. FAO, Rome

Finch HJS, Samuel AM, Lane GPF (2014) Soils and soil management. In: Lockhart & Wiseman's Crop Husbandry including grassland. Elsevier, New York

Follett RF (2001) Soil management concepts and carbon sequestration in cropland soils. Soil Tillage Res 61:77–92

Franzluebbers AJ (2005) Soil organic carbon sequestration and agricultural greenhouse gas emissions in the southeastern USA. Soil Tillage Res 83:120–147

Hatfield JL, Boote KJ, Kimball BA, Ziska LH, Izaurralde RC, Ort D, Thomson AM, Wolfe DW (2011) Climate impacts on agriculture: implications for crop production. Agron J 103:351–370

IPCC (2001) Climate change 2001: the scientific basis. Contribution of working group I to the third assessment report of the intergovernmental panel on climate change. University of Cambridge, Cambridge. First published 2001. https://www.ipcc.ch/ipccreports/tar/wg1/pdf/WGI_TAR_full_report.pdf

IPCC (2007) IPCC fourth assessment report: climate change 2007. https://www.ipcc.ch/publications_and_data/ar4/wg2/en/frontmattersg.html

Kang BT, Wilson GF, Lawson TL (1990) Alley cropping - a sustainable alternative to shifting cultivation. IITA, Ibadan

Kang BT, Akinnifesi FK, Pleysier JL (1995) Effect of agroforestry woody-species on earthworm activity and physio chemical properties of worm casts. Biol Fertil Soils 18:193–199

Lal R (1989) Agroforestry systems and soil surface management of a tropical alfisol. In: Soil moisture and crop yields agroforestry systems, vol 8. Kluwer Academic Publication, Dordrecht, pp 7–29

Lal R (2007) Anthropogenic influences on world soils and implications to global food security. Adv Agron 93:69–93

Lal R (2009) Sequestration of atmospheric CO_2 in global C pools. Energy Environ Sci 1:86–100

OECD (2009) OECD work on climate change. https://www.oecd.org/env/cc/Work-on-Climate-Change-2013-14_web.pdf

Okafor JC (2020) Flood, livelihood displacement, and poverty in Nigeria: plights of flood victims, 2012–2018. In: Leal Filho W et al (eds) African handbook of climate change adaptation. Springer, Cham. https://doi.org/10.1007/978-3-030-42091-8_124-1

Okafor SO, Okafor JC (2019) Government policy approach to environmental conflicts among the Niger Delta communities: implication to 2015 and post 2015 sustainable development goals. J Soc Dev Sci 10:1–14

Okafor JC, Oladejo AO, Ikem PA (2019) Nigeria and the politics of United Nations Climate Financing and Environmental Security: understanding the debacle of Ogoniland. Afr Renaissance 16:189–205

Okoroafor OO, Akinbile CO, Adeyemo AJ (2017) Soil erosion in South Eastern Nigeria: a review. Sci Res J 5:30–37

Orji EC (2001) Prospects and constraints in conserving indigenous fruit trees for sustainable agricultural development in Nigeria. Afr J Environ Stud 2:152–156

Orji EC (2017) Agroforestry and alley farming: a friendly and viable initiative for environmental management. An inaugural lecture presented at the 6th Inaugural Lecture of Chukwuemeka Odumegwu Ojukwu University, Anambra State. Nigeria on 12th December, 2017

Osuji EE, Ohajianya DO, Ehirim NC, Eze EU (2012) Effect of land use patterns on agricultural productivity in Imo State, Nigeria. Int J Food Agric Res 9:81–89

Osuji EE, Anyanwu UG, Ehirim NC, Eze EU, Tim-Ashama A (2017) Economics of processed cassava products in Imo State, Nigeria. J Res Bus Manag, India 5(3):09–19

Pennock D (2019) Soil erosion: the greatest challenge for sustainable soil management. Food and Agricultural Organizations of the United Nations, Rome

Pretty J (2008) Agricultural sustainability: concepts, principles and evidence. Philos Trans R Soc B 363:447–465

Qafoku NP (2015) Climate-change effects on soils: accelerated weathering, soil carbon, and elemental cycling. Pacific Northwest National Laboratory, Richland

Ullah H, Santiago-Arenas R, Ferdous Z, Attiah A, Datta A (2019) Improving water use efficiency, nitrogen use efficiency and radiation use efficiency in field crops under drought stress: a review. Adv Agron 156. https://doi.org/10.1016/bs.agron.2019.02.002

UNEP (2015) Climate finance. https://www.unenvironment.org/explore-topics/climate-change/what-we-do/climate-finance

UNFCCC (1992) United Nations framework convention on climate change. https://unfccc.int/resource/docs/convkp/conveng.pdf

Usman S (2007) Sustainable soil management of the dryland soils of northern Nigeria. GRIN Publishing GmbH, Munich. ISBN (Book): 978-3-640-92122-5, 155 pp

Wikipedia (2018) https://en.wikipedia.org/wiki/Climate_resilience

Chapter 32
Rising Greenhouse Gases in the Atmosphere: The Microbes Can Be a Solution—A Review

Swati, Indu Shekhar Thakur, and Arti Mishra

Abstract In the present world, climate change and global warming are the foremost problems in front of the human race. These problems have arisen due to the increase in Greenhouse gases (GHGs) concentration in the atmosphere. Most of the factors responsible for GHGs emissions are anthropogenic such as fossil fuel combustion, biomass burning, etc. Nature has developed its own way of fighting the GHGs problem by evolving microorganisms responsible for GHGs sequestration. These microorganisms can be utilized for solving the GHGs emission problem. Some of the methods that involve microorganisms for GHGs emission reduction are altering microbial community structure for altering GHGs flux into the atmosphere, microbial pathway engineering for more GHGs sequestration, and utilization of GHGs fixing enzyme derived from microbes to decrease emissions at the point source. Also, the biogeochemical cycling of elements such as C and N cannot be separated from climate change as the cycling of these elements involves the cycling of major GHGs such as carbon dioxide (CO_2), methane (CH_4), and nitrous oxide (N_2O). The present chapter is about microbial role and solution to the GHGs emission problems.

Keywords Greenhouse gases · Carbon dioxide · Methane · Nitrous oxide · Microorganisms and Microbial engineering

32.1 Introduction

The continuous efforts towards a way forward to development are causing great damage to our environment. The various human activities are related to the release of Greenhouse gases such as carbon dioxide (CO_2), Methane, and nitrous oxide that are

---

Swati (✉) · I. S. Thakur
School of Environmental Sciences, Jawaharlal Nehru University, New Delhi, India

A. Mishra
Amity Institute of Microbial Technology, Amity University, Noida, Uttar Pradesh, India
e-mail: amishra9@amity.edu

D. K. Choudhary et al. (eds.), *Climate Change and the Microbiome*, Soil Biology 63,
https://doi.org/10.1007/978-3-030-76863-8_32

the major cause of climate change and increasing global temperature (Eusufzai et al. 2010; IPCC 2007). The industrialization has lead to the exponential release of carbon dioxide in the atmosphere (Ramanathan 1988). The other causes of CO_2 emission include deforestation (9%) and fossil fuel burning (87%). As per Intergovernmental Panel on Climate Change (IPCC), CO_2 is the principal greenhouse gas and constitutes about 76.7% (v/v) of total atmospheric greenhouse gas volume. Among fossil fuels, coal is the major contributor to CO_2 in the atmosphere. Energy and transportation sectors that utilize most of the fossil fuels contributed around two-third of CO_2 emissions globally in 2010 (IEA 2012).

After CO_2, the second most significant greenhouse gas is CH_4. Around 20–30% of the global warming effect is due to the presence of CH_4 in the atmosphere (IPCC 2007). Till the nineteenth century, the concentration of methane was stable or relatively constant in the atmosphere and was about 0.7 ppm (parts per million). With the increase in anthropogenic emission sources of CH_4, there is an increase of about 0.1% per year in CH_4 concentration in the atmosphere (IPCC 2007). Both natural and anthropogenic sources are responsible for CH_4 emission in the atmosphere. Wetlands are responsible for about 62% CH_4 load in the atmosphere and are one of the largest sources of CH_4 emission. CH_4 emission from natural sources is of microbial origin (Conrad 1996).

The third most important greenhouse gas is Nitrous oxide (N_2O). The atmospheric lifetime of N_2O is of 110 years and has a Global warming potential of 298 times that of CO_2. N_2O is also known as laughing gas and is an ozone destroyer (Richardson et al. 2009). It was found that there is an increase in N_2O emission globally by 1.6 (1.4–1.7) TgN y-1 between the year 2010 and 2015 (IIASA 2019). N_2O is emitted from both natural (microbial activity in ocean and soil) and anthropogenic (burning of biomass and fossil fuels, industrial emissions, N-based fertilizers application, the farms, etc.) sources. About 1% of nitrogen-based fertilizers that are used to improve food growth are emitted as N_2O in the atmosphere globally.

The rising level of GHGs is adversely affecting the environment and hence, human health. The presence of high level of CO_2, CH_4, and N_2O along with other GHGs causes global warming and hence climate change. Some other effects that are related to global warming are Ocean warming, increase in heatwave frequency and duration, ice melting, rise in sea level, and atmosphere warming (Dawson et al. 2011; Meinshausen et al. 2009; Rockström et al. 2009). All these phenomena somehow also affect human health such as, increase in heatwave frequency leads to heat-related diseases or illness, temperature, and precipitation variation cause the spread of diseases (Costello et al. 2009). To address the problems related to increasing GHGs in the atmosphere, it is important to develop an efficient technique to sequester these gases from the atmosphere. The different approaches used to date to tackle the problem of GHGs include physical- and chemical-based sequestering technique. But these have some major drawbacks as compared to biological sequestration such as high operational cost, energy-intensive, and produce harmful by-products that adversely affect environmental and human health (Ramanathan 1988). The microbial sequestration not only reduces the number of greenhouse gases from the atmosphere but also converts them into some useful biomolecules such as carbohydrate, proteins, and lipids. Hence, biological sequestration via microbes and other organisms is the most sustainable approach to tackle the GHGs problem (Cole et al. 2007; Sundquist et al. 2008). The present chapter is going to discuss the role of microbes in addressing the rising GHGs problem.

32.2 Sources of GHGs in the Atmosphere

To combat the GHGs-related problem, it is important to understand their sources of emissions. Global GHGs emission is around 50 billion tones each year (in terms of carbon dioxide equivalents) from different natural and anthropogenic sources together. With the advancement of technologies, it is possible to control emissions from some GHGs emission sources but still, we are lacking in controlling the emissions from some of the sources. So, it is important to develop an understanding of priority emission sources so that researchers can progress in developing technologies to get over the problem of GHGs emissions as well as climate change. The division of GHGs emission sources is represented in Fig. 32.1.

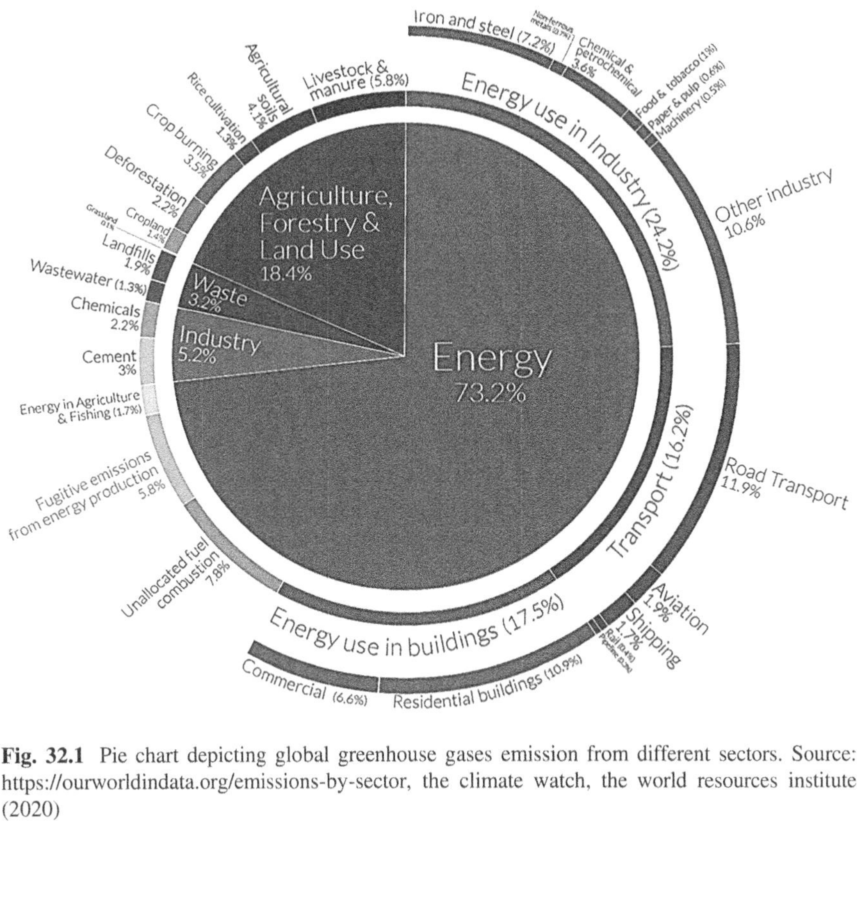

Fig. 32.1 Pie chart depicting global greenhouse gases emission from different sectors. Source: https://ourworldindata.org/emissions-by-sector, the climate watch, the world resources institute (2020)

It has been observed that different sources are responsible for GHGs emissions. It is clear from Fig. 32.1 that the energy sector is contributing more towards the GHGs problem. So to meet the current demand for net-zero emissions, we need to target each and every emission source. Energy sources which are the sum total of emissions from the utilization of different energies, i.e., electricity, heat, and fuels are currently contributing around 73.2% in GHGs global emission out of which 24.2% are industrial emissions, 16.2% are vehicular emissions, and 17.5% are commercial and residential building emissions. Other sources of emission include cement production (3%), chemical and petrochemical manufacturing (2.2%), Organic matter in wastewater (1.3%), landfills (1.9%), Grassland (0.1%), Cropland(1.4%), deforestation (2.2%), crop burning (3.5%), rice cultivation (1.3%), Agricultural soils (4.1%), and Livestock and manure (5.8%). The presence of GHGs in the atmosphere causes various problems if they are present in a large amounts. The problems related to GHGs are explained in the next section.

32.3 Greenhouse Gases and Its Impact on Environment and Human Health

CO_2, CH_4, and N_2O, all are greenhouse gases. They are named so because of their greenhouse effect. These gases have the ability to trap thermal radiations that are radiated from the earth's surface and maintain the Earth's temperature at a suitable level for life development. But various human actions have caused an increase in the concentration of these gases. The incredible increase in GHGs concentration in the atmosphere leads to global warming, i.e., trapping more heat and hence increasing the average earth's temperature. The environmental consequences of rising GHGs and hence global warming are the melting of glacials, rising sea level, flooding of islands and coastal cities, hurricane formation, desertification of fertile land, migration of species, and effect on agriculture and livestock, etc.

The human migration of around 140 million will also occur by 2050 according to the world bank due to flooding or extreme drought problem. The effect on agriculture is due to the proliferation of weeds, crop diseases, and insects. Similarly, the change in global temperature is affecting the reproduction and metabolism of livestock and also responsible for causing various diseases in them. Some of the direct consequences of global warming include Food shortage and the spread of infectious diseases such as malaria, cholera, or dengue. Also, extreme environmental conditions will aggravate respiratory and cardiovascular-related problems in humans.

The above described impacts are common to rise in concentration of all the three gases, i.e., CO_2, CH_4, and N_2O. CH_4 which a powerful greenhouse gas has some other indirect effects on human health as it plays an important role in the formation of tropospheric ozone. The presence of ozone in the tropospheric layer is responsible for around 1 million premature deaths due to respiratory problem globally. Similarly, the traces of N_2O are not harmful but as its concentration is increasing due to various human actions, it is causing some serious environmental issues. The N_2O regulates the ozone concentration in the stratosphere. The high amount of N_2O causes damage to ozone layer and hence reduced the human protection against harmful UV-sunrays that are usually absorbed by ozone layer.

32.4 Microbes and GHGs

The ubiquitous presence of microorganisms and their role in diverse environmental phenomena lead us to think about their role in climate change and GHGs production and regulation. It has been observed that microbes play a central role in regulating the global fluxes of biogenic greenhouse gases (CO_2, CH_4, and N_2O) (Singh et al. 2010; Zimmer 2010) and also respond very rapidly to the changing environmental conditions like climate change. To understand the microbial role in GHGs emissions and regulation, we must consider their interaction with various biotic and abiotic factors so that their interactions can be utilized in a proper way for mitigating a GHGs problem. One way to utilize microbes in tackling the climate change problem is by changing microbial community structure or modifying their genetic material for eliminating GHGs from their vicinity (Zhou et al. 2011). Linking the biogeochemical cycling of GHGs and microbial community structure can prove to be a great mechanism for dealing with the climate change problems.

32.4.1 Microorganisms and Carbon Dioxide

Nutrient elements such as carbon, nitrogen circulate between various biotic and abiotic components via the biogeochemical cycle in the ecosystem. These elements are not destroyed during cycling instead they get accumulated in reservoirs or sinks for a very long period of time. Carbon which is the essential building block for life is mostly present in the form of CO_2 in the atmosphere. Animals cannot utilize this essential component in this form, it must be converted into a usable organic form. The process of absorption of CO_2 from the atmosphere and conversion into organic substances is known as "CO_2 fixation". Photosynthesis is the best-known example of CO_2 fixation. It is either done by plants or cyanobacteria. Even the photosynthetic activity of ancient microorganisms is responsible for the presence of O_2 in today's Earth's environment.

The carbon cycle that includes CO_2 cycling is dependent upon the activity of various microorganisms as the microbial communities present in the environment have the ability to fix atmospheric carbon, can help in promoting plant growth, and can convert organic matter present in the environment (Weiman 2015). That is why the microbes and their enzymatic system responsible for controlling the various processes related to Carbon cycling are considered as the key engines of the CO_2

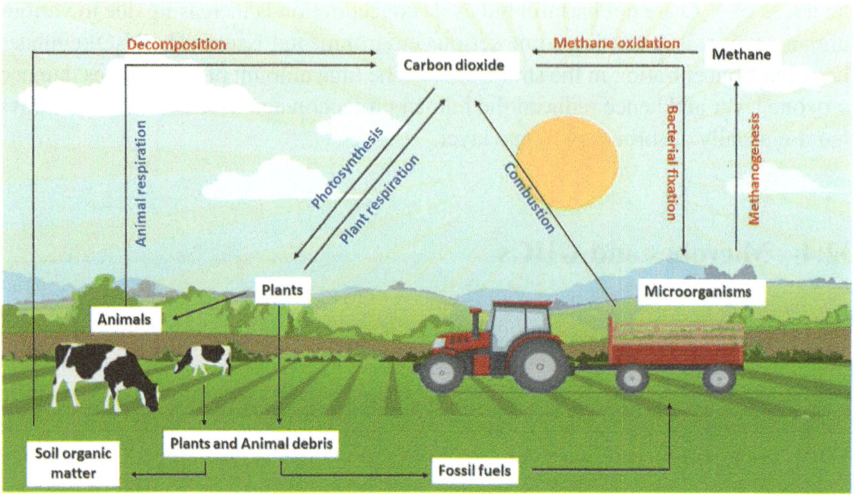

Fig. 32.2 Carbon cycle representing Carbon dioxide and methane-producing microbial process

emission system. The concentration of CO_2 in the atmosphere is regulated mainly by the intricate balance between two processes named photosynthesis and respiration (Falkowski et al. 2008; Gougoulias et al. 2014; Prosser 2007). The photosynthetic activity of plants, chemo- and photoautotrophic microorganisms is responsible for CO_2 fixation from the atmosphere, hence they can help in decreasing CO_2 concentration from the atmosphere if properly utilized in this direction. The carbon cycle along with the processes that involve microbial function has been represented in Fig. 32.2 (Table 32.1).

32.4.2 Microorganisms and Methane

The CH_4 is a carbon compound that is also a part of carbon cycling (Fig. 32.2) which is mainly released into the environment due to microbial action and the microorganisms responsible for its formation and emission are known as methanogens. These microorganisms work in anaerobic conditions and need the presence of hydrogen for CO_2 conversion into CH_4. In this process, they form water and release energy for self-use. Apart from methane-producing bacteria, nature maintains a balance of CH_4 in the atmosphere by the action of methane utilizing bacteria called "methanotrophs". These bacteria utilize CH_4 for energy production and convert it into CO_2 (Bousquet et al. 2006; Nikiema et al. 2005; Semrau et al. 2010). But this process requires the presence of oxygen in the vicinity of the methanotrophs, i.e., this is anaerobic bacteria. Such bacteria are mainly found at the border of aerobic and anaerobic environments so that they access CH_4 produced in the anaerobic environment and convert them into CO_2 in the presence of O_2 in the aerobic environment

Table 32.1 Major microorganism involved in CO_2 sequestration (Mistry et al. 2018)

Microorganism type	Genus
Proteobacteria	*Thiomicrospira denitrificans* *Desulfobacter hydrogenophilus* *Acidithiobacillus thiooxidans* *Ralstonia eutropha* *Xanthobacter flavus* *Rhodobacter capsulatus*
Archaea	*Methanobrevibacter aboriphilus* *Methanothermobacter thermautotrophicus* *Methanospirrillaceae* *Cenarchaeum* spp.
Clostridia	*Clostridium autoethanogenum* *Clostridium thermocellum*
Algae	*Chlorella vulgaris* *Anabaena* sp. *Scenedesmus obliquus* *Nannochloropsis oculata*

(Rajput et al. 2013). Methanotrophs can be useful for reducing emissions from CH_4-producing factories and landfill sites and convert it into comparatively less harmful GHG, i.e., CO_2 (Gupta et al. 2014; Shindell et al. 2012; Zimmerman and Labonte 2015).

32.4.3 Microorganism and Nitrous Oxide

The most potent GHG, i.e., N_2O emission can also be controlled by controlling the activity of microorganisms responsible for its production. N_2O is mainly formed as by-products of the two main processes of N-cycle, i.e., Nitrification and denitrification (Nakagawa et al. 2019; Rodriguez-Caballero et al. 2015). As the name suggests, nitrification is the process of formation of nitrite (NO_2^-) or nitrate(NO_3^-) from ammonium (NH_4^+). This is an aerobic process and requires an aerobic environment for the activity of organisms responsible for the nitrification process to occur. On the other hand, denitrification is the reduction process and occurs in an anaerobic environment. During this process, molecular nitrogen (N_2) or oxides of N are formed from the reduction of NO_2^- or NO_3^- (Nakagawa et al. 2019). The N_2O is formed as a by-product in the processes mentioned above and different microorganisms (e.g., bacteria, archaea, and fungi) follow different metabolic pathways for generating the same (Marusenko et al. 2013).

Some of the widely studied microorganisms responsible for heterotrophic and autotrophic nitrification are following bacteria (*P. stutzeri, A. faecalis, Acinetobacter calcoaceticus, Aspergillus, Nitrosomonas europea, N. eutropha*), Archaea

(*Nitrosopumilus martimus, Nitrososphaera viennensis, Nitrososphaera gargensi)*, and fungi (*Aspergillus, Penicillium, Absidia cylindrospora)*. Similarly some well-known microorganisms responsible for the heterotrophic and autotrophic denitrification process are Bacteria (*Thermomonas, Denitratisoma, Thiobacillus denitrificans, Sulfuritalea hydrogenivorans*), Archaea (*Pyrobaculum aerophilum, Haloferac denitrificans),* and fungi *(F. oxysporum, Gibberella fujikuroi*) (Connan et al. 2018; Cydzik-Kwiatkowska and Zielińska 2016; Hayatsu et al. 2008; Kerou et al. 2016; Nakagawa et al. 2019; Zhang et al. 2017; Zhu et al. 2014). The factors responsible for microbial involvement in N_2O formation are the abundance and composition of nitrifying and denitrifying microorganisms and their association with the various abiotic factors such as O_2 level, temperature, organic content, pH, and nitrogen availability (Hashida et al. 2013). Also, some of the denitrifiers can act as N_2O producer or N_2O reducer which depends upon the activity of nitrous oxide reductase enzyme and hence, influence the N_2O flux in the atmosphere (Nakagawa et al. 2019). The processes responsible for N_2O emission in the atmosphere in N-cycling have been represented in Fig. 32.3.

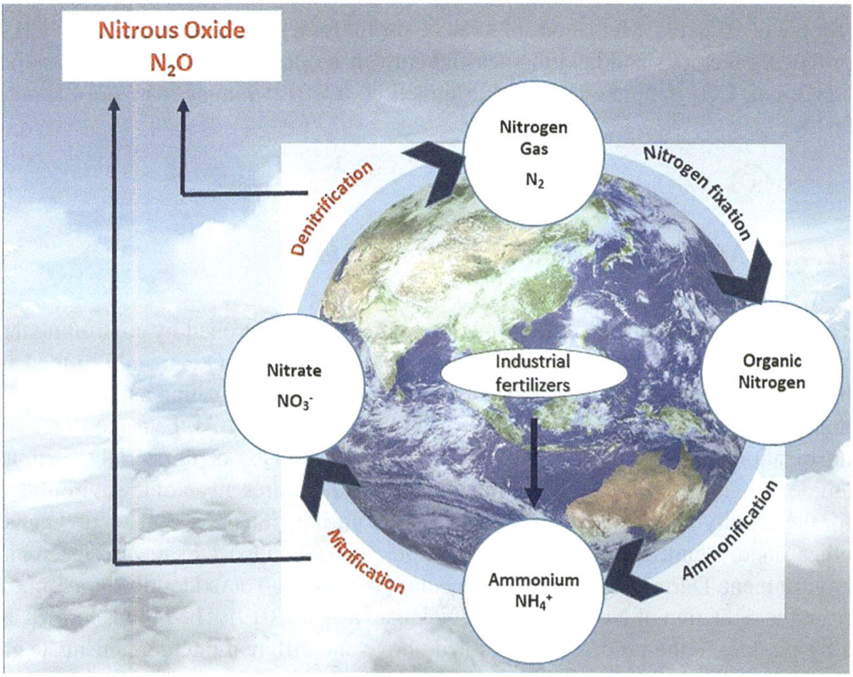

Fig. 32.3 Nitrogen cycle representing nitrous oxide-producing microbial process

32.5 Microbes and Biofuels

Apart from utilizing the biogeochemical cycle alteration method by changing microbial community structure for tacking climate change problems, researchers are going the other way round by generating more environmentally friendly biofuel by using microorganisms. The ability of microorganisms to generate biofuel will help in reducing the global warming problem as it reduces the economy dependence on fossil fuel which in turn would release less GHGs in the atmosphere. The fossil fuels release a very high amount of CO_2 (38.1 billion tonnes in 2019), hence their usage in the energy sector leads to global warming and climate change (Liao et al. 2016). Finding an alternative solution to fossil fuel (nonrenewable) will not only protect our environment from climate change-related problems but also give us a sustainable future by providing us a renewable source of energy.

The most commonly produced biofuel is bioethanol. The bioethanol is generally produced by the activity of *Saccharomyces cerevisiae* on high sugar-containing plants such as sugar beet, sugar cane, and corn. The process of formation of ethanol from sugar is known as fermentation. Although this reduces the harmful emissions such as hydrocarbon emissions (75–90%), carbon monoxide, and smog-producing particulate matter (50%), in a long run using energy crops will not be sustainable. The energy crops usage for biofuel production will lead to the reduction of land for producing food crops, largely in developing countries. Even in developed countries, clearing forests for growing energy crops will lead to an imbalance in the ecosystem. So, there is a need for a much sustainable solution to replace fossil fuel with some renewable sources of energy.

Scientists have found an alternative to using plants for biofuel production but the research is in the native state. The most abundant lignocellulosic plant waste which is dumped into landfills can now be valorized to produce biofuel and other industrial chemicals by microbial treatment. The lignocellulose is composed of cellulose, hemicellulose, and lignin. The enzyme that is responsible for the breakdown of cellulose is known as cellulase and it is actively searched by the researchers and scientists in diverse and unusual environments such as termite's stomach, soil found near volcanoes, etc. Some of the microbial species that are used to produce bioethanol from lignocellulosic waste are Sulfulobus solfatarticus, *Trichoderma reesei,* and algae. Active research is going on genetic modification of these microorganisms for improving their specific enzyme functions for biofuel production. Using microorganisms for biofuel production will solve an array of problems such as GHGs emissions, climate change, land-use changes, and provide a renewable source of energy.

32.6 Other Microbial Techniques Used for Enhancing GHGs Sequestration and Mitigation

32.6.1 Enzymatic Carbon Dioxide Capture and Storage (CCS)

The treatment of GHGs would be much easier and more effective if it will be done at their point sources of emission. It has been observed that the CO_2 concentration in the flue gas that is emitted from steel making plants and thermal power stations is 500 times higher than that is in the atmosphere. Various physical and chemical point treatment methods are in use but a biological technique that utilizes the use of a Carbonic Anhydrase (CA) enzyme is a more viable and environmental benign option (Liu et al. 2005). The CA catalyzes the conversion of CO_2 to bicarbonates (HCO_3^-) and protons.

CA is a zinc metalloenzyme and found in all living organisms such as plants, animals, and microorganisms. Five different classes of CAs (a, b, c, d, and e) exist in nature and are found to be evolved independently from each other (Smith et al. 1999). The a and b class of CA enzyme are widespread in bacteria and archaea. CA is present both, inside the cell, i.e., in cytoplasm and in the cell membrane. CAs are mostly studied in mammals. Recently to combat the GHGs problem, the environmental scientists start focusing on the prokaryotic CAs and have successfully purified an enzyme from a few bacterial species. But the utilization of this technique for CO_2 sequestration is still in native state and needs more research for developing an industrial level process or method.

32.6.2 Engineering Microorganisms for Enhanced CO_2 Sequestration

The undergoing natural processes that utilize or release CO_2 from the microorganisms can be engineered for increasing the efficiency of the process. Different microorganisms are known for CO_2 sequestration. Hu et al. (2018) described the various ways of microbial engineering for enhancing CO_2 sequestration and reducing CO_2 emissions from microorganisms. This is a very promising technique as it fulfills dual functions. One is reducing GHGs emissions and the other is the production of some useful chemical substances. Autotrophs can be strengthened by enhancing the efficiency of the CO_2 fixing pathway, energy harvesting system, and changing the carbon flux distribution. Similarly, heterotrophic ability to fix carbon dioxide can be enhanced by increasing carboxylation ability, more CO_2 fixing bypass creation, and conversion of heterotrophs to autotrophs so that they can utilize CO_2 as a carbon source. Microbial engineering can also be done to reduce CO_2 emission by metabolic pathways rewiring, redox balance improvement, and decrease the production of CO_2 through respiration in microorganisms. Molecular biology technological advancements can partly solve the problem of GHGs emissions by engineering microbes to fix more CO_2 and release less CO_2 during respiration. CO_2 fixing bacteria also produce a large number of other important industrial chemicals which in turn improve resource and energy shortage simultaneously.

32.6.3 *Manipulation of Microbial Community Structure for Mitigation of Nitrous Oxide*

Most of the N_2O load in the atmosphere is released from the microorganisms present in soil (Hu et al. 2017a, b). So these microorganisms can be manipulated or altered for mitigation purposes instead of creating a problem by producing N_2O. It can be done by different microbiological technologies for reducing N_2O emissions.

32.6.3.1 Bioaugmentation with N_2O-Reducing Microbiomes to Roots, Soils, or Fertilizers

The inoculation with N_2O-reducing bacteria has proved to be a successful technique in reducing edaphic N_2O emissions. The key step in this method is the selection of an appropriate strain for achieving the goal. It has been observed that *Bradyhizobium japonicum* has the ability to convert N_2O to N_2 with the action of NOS and mitigated the release of N_2O from soybean fields when inoculated in its roots (Itakura et al. 2012). Nishizawa et al. (2014) have successfully ameliorated N_2O emission from the soil by adding N_2 generating denitrifiers such as *Azoarcus*, *Burkholderia*, and *Niastella*. Also, the inoculation with plant growth-promoting rhizobacteria (*Azospirillum* and *Herbaspirillum)* has performed the dual function, i.e., on one hand, it ameliorated the N_2O emission and on other hand it promoted the plant growth (Chauhan et al. 2015; Gao et al. 2019). Although this technique is useful, the interaction of the inoculated strain with the indigenous microbiome is not studied. The indigenous microorganisms outcompete them and hence, proved as a hurdle in achieving a defined target. This is somehow can be overcome by developing and inoculating a bacterial consortium that helps in reducing N_2O emission from soil (Akiyama et al. 2016; Hu et al. 2017a, b).

32.6.3.2 Altering the Production of Specific Signaling Molecule

This strategy of achieving a goal of reducing GHGs emissions is in the native state and is under research. Some microorganisms are known that release signaling molecules in their vicinity and act as an inhibitor of N_2O conversion. It has been known in *Nitrobacter winogradskyi* which releases acyl-homoserine lactones

(AHLs) in their environment and leads to the inhibition of the *nirK* gene. This gene product is basically involved in the consumption of N_2O by *Nitrobacter winogradskyi.* So, in the presence of AHLs, it will fail to do so (Schuster et al. 2013; Mellbye et al. 2016). The microbial engineering of the genes responsible for AHL production can prove to be a great strategy for N_2O emission control.

32.7 Conclusion

The microorganisms are the key regulator of GHGs emission. They are responsible for both the addition and elimination of GHGs in the atmosphere. The proper management of the microbial community structure in the soil can somehow able to answer the problem of global warming and climate change by altering the emission of GHGs. The phenomenon of inoculation of GHGs sequestering strain and genetic alteration for achieving enhanced GHGs elimination effect are in the native state and need the attention of scientists for achieving the desired result.

References

Akiyama H, Hoshino YT, Itakura M, Shimomura Y, Wang Y, Yamamoto A, Tago K, Nakajima Y, Minamisawa K, Hayatsu M (2016) Mitigation of soil N_2O emission by inoculation with a mixed culture of indigenous *Bradyrhizobium diazoefficiens*. Sci Rep 6:32869

Bousquet P, Ciais P, Miller JB, Dlugokencky EJ, Hauglustaine DA et al (2006) Contribution of anthropogenic and natural sources to atmospheric methane variability. Nature 443:439–443

Chauhan H, Bagyaraj DJ, Selvakumar G, Sundaram SP (2015) Novel plant growth-promoting rhizobacteria-prospects and potential. Appl Soil Ecol 95:38–53

Cole JJ, Prairie YT, Caraco NF, McDowell WH, Tranvik LJ, Striegl RG, Duarte CM, Kortelainen P, Downing JA, Middelburg JJ, Melack J (2007) Plumbing the global carbon cycle: integrating inland waters into the terrestrial carbon budget. Ecosystems 10:172–184

Connan R, Dabert P, Moya-Espinosa M, Bridoux G, Béline F, Magrí A (2018) Coupling of partial nitritation and anammox in two- and one-stage systems: process operation, N_2O emission and microbial community. J Clean Prod 203:559–573

Conrad R (1996) Soil microorganisms as controllers of atmospheric trace gases (H_2, CO, CH_4, OCS, N_2O, and NO). Microbiol Rev 60:609–640

Costello A, Abbas M, Allen A, Ball S, Bell S, Bellamy R, Friel S, Groce N, Johnson A, Kett M, Lee M, Levy C et al (2009) Managing the health effects of climate change. Lancet 373:1693–1733

Cydzik-Kwiatkowska A, Zielińska M (2016) Bacterial communities in full-scale wastewater treatment systems. World J Microb Biot 32:66

Dawson TP, Jackson ST, House JI, Prentice IC, Mace GM (2011) Beyond predictions: biodiversity conservation in a changing climate. Science 332:53–58

Eusufzai MK, Tokida T, Okada M, Sugiyama S, Liu GC, Nakajima M, Sameshima R (2010) Methane emission from rice fields as affected by land use change. Agric Ecosyst Environ 139:742–748

Falkowski PG, Fenchel T, Delong EF (2008) The microbial engines that drive Earth's biogeochemical cycles. Science 320:1034–1039

Gao N, Shen W, Nishizawa T, Isobe K, Guo Y, Ying H, Senoo K (2019) Genome sequences of two *Azospirillum* sp. strains, TSA2S and TSH100, plant growth-promoting rhizobacteria with N_2O mitigation abilities. Microbiol Resour Announc 8:e00459

Gougoulias C, Clark JM, Shaw LJ (2014) The role of soil microbes in the global carbon cycle: tracking the below-ground microbial processing of plant-derived carbon for manipulating carbon dynamics in agricultural systems. J Sci Food Agric 94:2362–2371

Gupta C, Prakash D, Gupta S (2014) Role of microbes in combating global warming. Int J Pharm Sci Lett 4:359–363

Hashida S-N, Johkan M, Kitazaki K, Shoji K, Goto F, Yoshihara T (2013) Management of nitrogen fertilizer application, rather than functional gene abundance, governs nitrous oxide fluxes in hydroponics with rockwool. Plant Soil 374:715–725

Hayatsu M, Tago K, Saito M (2008) Various players in the nitrogen cycle: diversity and functions of the microorganisms involved in nitrification and denitrification. Soil Sci Plant Nutr 54:33–45

Hu HW, Trivedi P, He JZ, Singh BK (2017a) Microbial nitrous oxide emissions in dryland ecosystems: mechanisms, microbiome and mitigation. Environ Microbiol 19:4808–4828

Hu HW, He JZ, Singh BK (2017b) Harnessing microbiome-based biotechnologies for sustainable mitigation of nitrous oxide emissions. Microb Biotechnol 10:1226–1123

Hu G, Li Y, Ye C, Liu L, Chen X (2018) Engineering microorganisms for enhanced CO_2 sequestration. Trends Biotechnol. https://doi.org/10.1016/j.tibtech.2018.10.008

International Energy Agency (2012) CO_2 emissions from fuel combustion 2012. Organisation for Economic Co-operation and Development, Paris

International Institute for Applied Systems Analysis (2019) Nitrous oxide levels are on the rise. ScienceDaily. ScienceDaily. www.sciencedaily.com/releases/2019/11/191118 110816.htm

IPCC (2007) Climate change 2007: the physical science basis. In: Solomon S, Qin D, Manning M, Chen Z, Marquis M, Averyt KB et al (eds) Contribution of working group I to the fourth assessment report of the intergovernmental panel on climate change (IPCC). Cambridge University Press, Cambridge, UK

Itakura M, Uchida Y, Akiyama H, Hoshino YT, Shimomura Y, Morimoto S, Tago K, Wang Y, Hayakawa C, Uetake Y, Sánchez C, Eda S, Hayatsu M, Minamisawa K (2012) Mitigation of nitrous oxide emissions from soils by Bradyrhizobium japonicum inoculation. Nat Clim Chang 3:208–212

Kerou M, Offre P, Valledor L, Abby SS, Melcher M, Nagler M, Weckwerth W, Schleper C (2016) Proteomics and comparative genomics of *Nitrososphaera viennensis* reveal the core genome and adaptations of archaeal ammonia oxidizers. Proc Natl Acad Sci USA 113:E7937–E7946

Liao JC et al (2016) Fuelling the future: microbial engineering for the production of sustainable biofuels. Nat Rev Microbiol 14:288–304

Liu N, Bond GM, Abel TA, McPherson BJ, Stringer J (2005) Biomimetic sequestration of CO_2 in carbonate form: role of produced waters and other brines. Fuel Process Technol 86:1615–1625. https://doi.org/10.1016/J.FUPROC.2005.01.008

Marusenko Y, Huber DP, Hall SJ (2013) Fungi mediate nitrous oxide production but not ammonia oxidation in aridland soils of the southwestern US. Soil Bio Biochem 63:24–36

Meinshausen M, Meinshausen N, Hare W, Raper SCB, Frieler K, Knutti R, Frame DJ, Allen MR (2009) Greenhouse-gas emission targets for limiting global warming to 2°C. Nature 458:1158–1196

Mellbye BL, Giguere AT, Bottomley PJ, Sayavedra-Soto LA (2016) Quorum quenching of *Nitrobacter winogradskyi* suggests that quorum sensing regulates fluxes of nitrogen oxide(s) during nitrification. mBio 7

Mistry AN, Ganta U, Chakrabarty J, Dutta S (2018) A Review on biological systems for CO2 sequestration: organisms and their pathways. Environ Prog Sustain Energy. https://doi.org/10.1002/ep

Nakagawa T, Tsuchiya Y, Ueda S, Fukui M, Takahashi R (2019) Eelgrass sediment microbiome as a nitrous oxide sink in Brackish Lake Akkeshi, Japan. Microbes Environment 34:13–22

Nikiema J, Bibeau L, Lavoie J, Brzezinski R, Vigneux J et al (2005) Biofiltration of methane: an experimental study. Chem Eng J 113:111–117

Nishizawa T, Quan A, Kai A, Tago K, Ishii S, Shen W, Isobe K, Otsuka S, Senoo K (2014) Inoculation with N_2-generating denitrifier strains mitigates N_2O emission from agricultural soil fertilized with poultry manure. Biol Fert Soils 50:1001–1007

Prosser JI (2007) Microorganisms cycling soil nutrients and their diversity. In: Van Elsas JD, Jansson JK, Trevors JT (eds) Modern soil microbiology. CRC Press, New York, NY, pp 237–261

Rajput P, Saxena R, Mishra G, Mohanty SR, Tiwari A (2013) Biogeochemical aspect of atmospheric methane and impact of nanoparticles on methanotrophs. J Environ Anal Toxicol 3:2–10

Ramanathan V (1988) The greenhouse theory of climate change: a test by an inadvertent global experiment. Science 240:293–299

Richardson D, Felgate H, Watmough N, Thomson A, Baggs E (2009) Mitigating release of the potent greenhouse gas N_2O from the nitrogen cycle - could enzymic regulation hold the key? Trends Biotechnol 27:388–397

Rockström J, Steffen W, Noone K, Persson A, Chapin FS, Lambin EF, Lenton TM, Scheffer M, Folke C, Schellnhuber HJ et al (2009) A safe operating space for humanity. Nature 461:472–475

Rodriguez-Caballero A, Aymerich I, Marques R, Poch M, Pijuan M (2015) Minimizing N_2O emissions and carbon footprint on a full-scale activated sludge sequencing batch reactor. Water Res 71:1–10

Schuster M, Sexton DJ, Diggle SP, Greenberg EP (2013) Acyl-homoserine lactone quorum sensing: from evolution to application. Annu Rev Microbiol 67:43–63

Semrau JD, DiSpirito AA, Yoon S (2010) Methanotrophs and copper. FEMS Microbiol Rev 34:496–531

Shindell D, Kuylenstierna JC, Vignati E, Dingenen VR, Amann M et al (2012) Simultaneously mitigating near-term climate change and improving human health and food security. Science 335:183–189

Singh BK, Bardgett RD, Smith P, Dave SR (2010) Microorganisms and climate change: terrestrial feedbacks and mitigation options. Nat Rev Microbiol 8:779–790

Smith KS, Jakubzick C, Whittam TS, Ferry JG (1999) Carbonic anhydrase is an ancient enzyme widespread in prokaryotes. Proc Natl Acad Sci U S A 96:15184–15189

Sundquist E, Burruss R, Faulkner S (2008) Carbon sequestration to mitigate climate change. U.S. Geological Survey, Fact Sheet, 2008–3097

Weiman S (2015) Microbes help to drive global carbon cycling and climate change. Microbe Mag 10:233–238

Zhang X, Zheng S, Sun J, Xiao X (2017) Elucidation of microbial nitrogen-transformation mechanisms in activated sludge by comprehensive evaluation of nitrogen-transformation activity. Bioresour Technol 234:15–22

Zhou J, Xue K, Xie J, Deng Y, Wu L et al (2011) Microbial mediation of carbon-cycle feedbacks to climate warming. Nat Clim Chang:1–5

Zhu T, Meng T, Zhang J, Zhong W, Müller C, Cai Z (2014) Fungi-dominant heterotrophic nitrification in a subtropical forest soil of China. J Soils Sediments 15:705–709

Zimmer C (2010) The microbe factor and its role in our climate future. https://goo.gl/c8wjdu

Zimmerman L, Labonte B (2015) Climate change and the microbial methane banquet. Climate Alert 27

Chapter 33
Battling Climate Change: Improving Crop Productivity and Quality by Increasing Photosynthetic Efficiency, Deploying Microbiome Metagenomics, and Effectively Utilizing Digital Technology

P. S. Chandrashekharaiah, Santosh Kodgire, Debanjan Sanyal, and Santanu Dasgupta

Abstract Change in normal environmental conditions is referred as climate change. The activities such as desertification, emissions of toxic gases from fossil fuel burning, increased livestock farming, use of nitrogenous fertilizers, and fluorinated gases are the main causes for global climate change. These activities release huge quantity of greenhouse gases into atmosphere in addition to those that produce naturally. The increased greenhouse gasses cause greenhouse effect and global warming. The temperature and rainfall are important factors from the point of agriculture and are affected due to climate change. The change in temperature and rainfall pattern affect crop growth and increase the chances of pests and diseases outbreak and finally the productivity. The effect of environmental change on agriculture is unevenly distributed across the world and if not addressed appropriately, it may increase the risk of food security in coming days. Increasing the agriculture productivity under changing climate is the need of the hour. Till date, only marginal efforts were made in enhancing the photosynthetic efficiency of crop plants. Biotechnological tools play significant part in improving the plant photosynthesis and the yield. Besides increasing crop's intrinsic productivity, providing optimum biological environment to the crops is another logical combatting strategy to minimize any adverse effect of changed environment. Plant microbiome is the major component of the biological climate, which denotes the entire genetic makeup of microorganisms live on the soil and plant. The microbiome is unpredictably associated with plant wellbeing and helps in increasing the quality and productivity of crops. The beneficial microbes induce resistance to the plants against pest and diseases, play important role in nutrient recycling, nutrient mobilization. The metagenomic tools

P. S. Chandrashekharaiah · S. Kodgire · D. Sanyal (✉)
Reliance Industries Limited, Jamnagar, Gujarat, India
e-mail: debanjan.sanyal@ril.com

S. Dasgupta
Reliance Industries Limited, Navi Mumbai, India

D. K. Choudhary et al. (eds.), *Climate Change and the Microbiome*, Soil Biology 63,
https://doi.org/10.1007/978-3-030-76863-8_33

help in designing right microbiome to recover the plant and soil wellbeing. The application of digital technology in agriculture will enhance the precision and right prediction. The digital technology can be used to monitor the changing rainfall pattern, temperature, pest and disease outbreak, and databases can be developed for these. The machine learning (ML) algorithms and artificial neural network (ANN) can be used to analyze and process large-scale data. An artificial Intelligence (AI) platform developed by ML and ANN can help further decision making. These biological and digital solutions will bring revolution in the agriculture by increasing the productivity and by minimizing the crop losses. In this book chapter, we are discussing about the role of photosynthesis, microbiome, and digital technology in concern to changed environment and improving agriculture productivity.

Keywords Climate change · Crop · Metagenomics · Microbiome · Photosynthesis · Productivity

33.1 Introduction

Climate change is the rising global concern; it refers to changes in atmospheric gaseous composition by anthropogenic activities, in addition to natural climate variability (Moser and Dilling 2004; Lorenzoni et al. 2007; UNFCC 1994). The atmosphere of earth is majorly composed of nitrogen, oxygen, argon and the concentration is as follows, 78.09%, 20.95%, and 0.93%, respectively. The minor gases, which are present, are methane (CH_4), carbon dioxide (CO_2), chlorofluorocarbons (CFCs), nitrous oxide (N_2O), and ozone (Qiancheng 1998). These minor gases are called as greenhouse gases (GHG), have high absorption for thermal long wave infrared radiations and increase the temperature of atmosphere, seas, and terrestrial planets (http://earthobservatory.nasa.gov/Features/GlobalWarming/page2.php). The heating of earth's atmosphere due to increased concentrations of minor gases is termed as global warming and the process is called as greenhouse effect (Venkata Ramanan and Smitha 2011; Olufemi et al. 2014). The changing global climate will affect various parameters of environment and ecosystem. Intergovernmental Panel on Climatic Change (IPCC 2007) has anticipated the impact of changed environment using computer models and simulation. According to IPCC (2007) by 2100, the worldwide normal surface temperature would rise by 1.1 to 6.4 °C and ocean level would increase between 18 and 59 cm. The other predicted effects include acidification of oceans and occurrence of tropical cyclones (typhoons and hurricanes).

The concentration of CO_2 has increased over period, prior to industrialization it was ~250 ppm and currently the concentration has crossed 380 ppm and would touch 550 ppm by 2050 (Rosegrant and Cline 2003). Altogether, CO_2 significantly affects the environment and stays active in the air for a long time. Approximately 30 years are required for disappearance of 50% of total CO_2 released into the atmosphere, 30% CO_2 remain in atmosphere for many centuries and 20% last for several million years (Solomon et al. 2007). CO_2 comes from burning and use of fossil fuels from different sectors like transportation, manufacturing of cements. It is also released into environment by plant material deterioration, respiration, organic matter breakdown, and deforestation (Davidson and Janssens 2006). Methane (CH_4) is another GHG responsible for climate change. CH_4 is produced by production and combustion of fossil fuels, waste dumping, landfills, and livestock farming (Kusar and Avgustin 2010). Oceanic activities, anaerobic or water logging conditions in rice cultivation also contribute significantly to the methane generation (Singh et al. 2010; Horz et al. 2002). Nitrous oxide is another GHG released into environment due to fossil fuel burning and excess use of fertilizers (Sanford et al. 2012).

The additional utilization of inorganic fertilizers and pesticides in modern agriculture has increased the crop yields but affected soil fertility, soil, and plant microbial community. The effects of global climate changes such as increasing temperature and increased GHG made crop plants more susceptible for pests and diseases and affected the CO_2 fixation, photosynthesis, and crop yield. This has negatively affected the environment and significantly affected the human well-being and food security (Tilman et al. 2002). In this direction plants and microbiome taking a significant part in reducing pollution caused by various greenhouse gases (CO_2, CH_4, and N_2O). Plants absorb and fix CO_2 through photosynthesis and release oxygen, thereby reduce the CO_2 concentration in the atmosphere (Trumbore 2006). However, in the varied climatic conditions, inefficient carbon fixation by photosynthetic enzymes limits the carbon fixation, increases the global warming, and affects the crop productivity (Furuya and Kobayashi 2010).

The growing population is another concern and it is rising at 1.2% per annum. To pace with the increasing population, the agriculture productivity needs to be increased by >50% (Kaneda 2016; Alexandratos and Bruinsma 2012). This can be fulfilled by the increasing plant photosynthetic efficiency through genetic engineering or classical breeding approach that would help in developing plant varieties with higher photosynthetic efficiency for fixing more CO_2 at varied environmental conditions (Long et al. 2006). The other approaches to improve agricultural productivity are the selection of microbiome, which are beneficial for soil and plant health.

Plant's rhizosphere and phyllosphere are associated with various microbes, these microbes induce disease and pest resistance to crops, mobilize nutrients from soil to plants, secrete growth hormones, and fix nitrogen, thereby improve the plant growth (Brundrett 2009). The microbiome of plants and soil also, involves in reducing the climate change caused by GHG through various metabolic activities and plays important role in global nutrient cycle stabilization (Microbiology online 2015). The plant microbiome might be useful or neutral; to understand their involvement in improving plant wellbeing, a high level of research is required. Before the advancement of biotechnological tools, the study on plant-associated microbiome was difficult and time-consuming. The advanced DNA sequencing methods and metagenomics approaches had fastened the identification of single strain or consortia of microbes which involves in promoting plant health and productivity. This approach had helped to design the plant microbiome for different agro ecological regions to improve crop productivity (Mackelprang et al. 2011).

Agriculture plays a promising role in global economy by providing food security (Randy Stringer 2001; FAO 2002). Variations in rainfall patterns, increasing earth's temperature, emission of greenhouse gases, water scarcities, droughts, and depleting resources are some of the negative effects of climate change and these changes have directly affected the plant photosynthesis, growth, and crop productivity (Vitousek et al. 1997; Dengel 2013; Hoffmann 2013; FAO 2013, 2017; Milius 2017). Therefore, the scientific research in agricultural field is focused towards overcoming these challenges. The technological inventions such as Artificial intelligence (AI) and Machine learning (ML) (Karandeep Kaur 2016) were used in agriculture to predict environment conditions (Stone and Hirsh 2005; Kolhe and Gupta 2014). To increase decision making ability of farmers in agriculture, new AI technologies were implemented such as internet of things (IOT), deep learning, wireless sensors, and big data (Alexandros et al. 2012). To take decisions during seed selection, soil preparation, water scheduling, fertilizer, pesticide, and weed management farmers require expert's knowledge (Kolhe et al. 2011; Yelapure and Kulkarni 2012). AI-driven expert systems help farmers to make ecologically strong and economically viable decision at a level or greater than human expert (Robinson et al. 1996; Yelapure and Kulkarni 2012) and can minimize the losses.

To tackle the climate change and to improve the plant productivity, scientists are looking for a new way of farming by increasing plant photosynthetic efficiency, identifying and deploying beneficial microbiome. In this review, the role of plants photosynthesis, microbiome, metagenomics, machine learning, and artificial intelligence to reduce the climate change and to improve agricultural productivities is discussed in detail.

33.2 Role of Plant Photosynthesis in Controlling Climate Change

Due to increased concern on climate change, the pollution caused by CO_2 has to be maximally reduced (Lu and Conrad 2005). Plants are important controllers of CO_2 across worldwide and territorial environment (Zhu et al. 2017). Plant photosynthesis is the main mechanism of CO_2 sequestration (Trumbore 2006). As of now, earthbound ecological system fixes, more environmental CO_2 by photosynthesis than they release through respiration and do not contribute for climate change. Currently plant photosynthesis removes about 25% of pollution caused due to burning of fossil fuels (Le Quere et al. 2009). Annually terrestrial plants fix c. 123 Gt carbon via photosynthesis from the environment (Amthor and Baldocchi 2001; Beer et al. 2010). Out, of the total CO_2 produced by anthropogenic activities, 30% is absorbed by a terrestrial ecosystem through photosynthesis and slows down the climate change by preventing rise in the CO_2 concentration in atmosphere (Le Quere et al. 2016). Thus, vegetation helps in reduction of atmospheric CO_2 and reduces the climate change. Carbon sequestration by plants is considerably higher than CO_2

discharged by anthropogenic actions (Le Quere et al. 2016) (c. 10.3 Gt C/year). Any large-scale alteration in vegetation and carbon transitions could either alleviate or quicken environmental change (Smith and Dukes 2013).

Photosynthesis is the biochemical reaction by which plants fix CO_2 and release O_2, thereby increase the atmospheric O_2 concentration, and simultaneously reduce the CO_2 pollution (Furuya and Kobayashi 2010; Djanaguiraman et al. 2010; Qaderi et al. 2012). Plant photosynthesis is important to mitigate greenhouse effect over long term. Large dominant species of perennial trees that belong to old developing woodlands are considered as world's biggest sinks for environmental carbon. They absorb, capture, and store enormous amount of CO_2. Approximately, 30% of earths land surface is covered by forest and it accounts for 50% of plant productivity. A tree in its prime or old development stage assimilates CO_2 identical to carbon released by a car, which is driven for 42,000 kilometers. The destruction of old forest affects CO_2 sequestration drastically (Geoffrey Craggs 2016). Globally around 19% of the carbon present in the earth's atmosphere is deposited in plants, and 81% in the soil (IPCC 2000).

33.2.1 *Role of C3 and C4 Plants in Reducing Climate Change Contributed by CO_2*

Based on the enzyme involved in CO_2 fixation in photosynthesis, all the terrestrial plant species can be broadly grouped into C3, C4, and Crassulacean acid metabolism (CAM) (Sage 2014). The terms C3 and C4 refer to the 3 or 4 number carbon molecules formed after CO_2 fixation. C3 and C4 plants are widely distributed and play important role in controlling climate change. These two photosynthetic pathways react distinctly to changing atmospheric CO_2 concentration, temperature, and humidity. CAM photosynthesis occurs only in epiphytes and succulents. However, CAM plants are limited in distributions and are not apparent parts of the worldwide carbon cycle.

The vast majority of land plants uses C3 pathway. About 90% of world's energy is supplied by 15 crops out of which, 12 plants have C3 pathway. The commonly found C3 plants are rice, wheat, rye, barley, soybeans, potatoes, cassava, tomatoes, spinach, apple, yams, peach, and eucalyptus. In C3 plants, the enzyme ribulose bisphosphate carboxylase oxygenase (Rubisco) fixes CO_2 and the first product formed after CO_2 fixation is 3-carbon compound, phosphoglyceric acid (PGA) (Benson 2002; Bassham 2003). The enzyme Rubisco has high affinity for CO_2 as well as oxygen. The enzyme activity was limited by the substrate availability at actual atmospheric CO_2 concentration and it has the potential to fix more CO_2 at higher CO_2 availability (Long et al. 2004; Bernacchi et al. 2003). During daytime, C3 plants always keep open their stomata to take CO_2 inside and fix, this leads to a constant loss of water through transpiration and causes dehydration. To minimize the threat of dryness, these plants close their stomata, which causes the reduction in photosynthetic activity and CO_2 absorption (Ainsworth and Rogers 2007). The efficient CO_2 fixation in C3 plants has been related to availability of increased CO_2 (substrate) (Long et al. 2004) at carboxylation site. The Rubisco enzyme has both carboxylase and oxygenase ability. The elevated CO_2 is known to increase the carboxylation efficiency of Rubisco by competitively inhibiting the oxygenase reaction and reducing photorespiration (Ogren 2003), thereby increases the CO_2 sequestration (Spreitzer and Salvucci 2002).

In carboxylation reaction, Rubisco produces one molecule of 3-phosphoglyceric acid (3-PGA) and oxygenation produces one molecule of 2-phosphoglycolate (2-PG). The compound 2-PG is regenerated by photorespiration pathway, which consumes ATP, NADPH, and releases CO_2, thus photorespiration in C3 plants reduces carbon fixation by 30% (Bauwe et al. 2010; Raines 2011). C3 plants are more efficient at elevated CO_2 concentration. The experiments conducted with C3 plant in open-top chambers (OTCs) and in free atmospheric CO_2 enrichment (FACE) showed an improvement in photosynthetic rate of light saturation, with increased CO_2 supply. The increased photosynthesis was due to increased intracellular CO_2 concentration (Mauney et al. 1992). In a classical experimental study by Kimball et al. (2002), C3 plants have shown 10 to 143% increase in biomass at doubled ambient CO_2. Uprety et al. (2010) observed improvement in vegetative growth, flowering phase, and higher grain yield in soybean and peanut at raised CO_2. Farquhar et al. (1980) biochemical model suggests that a rise in climatic CO_2 to 550 ppm from 372 ppm has the ability to enhance Rubisco-mediated photosynthesis by 36% at 25 °C. This suggests that higher CO_2 activates photosynthesis in C3 plants by suppressing photorespiration. Increasing temperature also shown to affect the CO_2 fixation and crop productivity in C3 plants. Increase in temperature above physiological maxima affects crop yield (IPCC 2007). A 2 °C rise in atmospheric temperature found to reduce the grain yield in rice and wheat grain by 15 to 17% (Aggarwal and Mal 2002). Rice yields were found to increase by 0.5 t/ha for every 75 ppm rise in CO_2, but 1 °C rise in temperature reduces the yield by 0.6 t/ha (Sheehy et al. 2005). This suggests that the Rubisco enzyme is inefficient at low CO_2 and higher atmospheric O_2 concentration, and the enzyme is affected by higher temperature. This affects the carbon fixation and releases the fixed carbon into the atmosphere. Due to all these reasons, C3 plants are inefficient to fix carbon at lower atmospheric CO_2 concentration and higher oxygen concentration and temperature.

C4 photosynthesis is not regular among dicots such as shrubs and trees; however, more normal among monocots, likes sedges and grasses. C4 plant leaf has Kranz type of anatomy in which, the interior bundle sheath cells are encircled by mesophyll cells (Edwards et al. 2004). The bundle sheath cells have C3 type of photosynthetic pathway and mesophyll cells have C4 pathway. The enzyme phosphoenolpyruvate (PEP) carboxylase found in mesophyll cells, fixes CO_2 into a 4-carbon compound oxaloacetate (Sage et al. 2012). The four-carbon compound is diffused to mesophyll cells, where it is decarboxylate and re-fixed by Rubisco enzyme. C4 photosynthesis process is a morphological modification of C3 cycle to reduce oxygenase activity of Rubisco, and thereby increases photosynthetic rate in low CO_2 environment and contributes significantly for reducing the climate change. C4 plants have more advantages over C3 plants, as like C3 plants, they keep open their stomata during the day, but these plants have the mechanism to propel CO_2, in any event when the stomata closes surprisingly, hence permitting the process of photosynthesis to proceed (Furbank 2011; Wang et al. 2014). The carbon concentration mechanism in C4 plants reduces the stomatal conductance, thereby improves the water use efficiency and increases the Rubisco carboxylation reactions which bring about more CO_2 being fixed per photon ingested (Skillman 2008). C3 plants need increased CO_2 partial pressure around Rubisco for its maximum carboxylation activity and this is not the case with C4 plants because of the existing carbon concentrating mechanism. PEP carboxylase has strong affinity for bicarbonates instead of CO_2 and operates at low CO_2 partial pressure of (intercellular CO_2 to ambient CO_2) 0.3 in comparison to C3 leaves, which operate at 0.7 in maximum light intensity (Wong et al. 1985). All these attributes make C4 plants to fix additional CO_2 per unit of nitrogen, water, and light than C3 plants under varied environmental conditions (Ghannoum et al. 2011).

33.2.2 Role of Genetic Engineering Approach for Reducing Climate Change

The farming also contributes meaningfully to the discharge of greenhouse gases. The cultivation of low yielding traditional crops requires addition of enormous amount of nitrogenous fertilizers, crop protection chemicals, thereby indirectly contributes for climate change. In this regard, focus should be given for developing high yielding cultivars through classical breeding or by transgenic approach, which needs less nitrogenous fertilizers and requires minimum pesticide application, thereby reduces the GHG emissions (Philippot and Hallin 2011). Developing crops through genetic engineering approach for higher (N) nitrogen use efficiency (NUE) reduces the nitrous oxide (N_2O) and nitrous dioxide pollution caused by excessive use of N fertilizer in intensive agriculture, and thereby improves crop yields in nitrogen-limited environments. This NUE of plant can be achieved by genetic engineering approach (Shrawat et al. 2008). The over expression of the gene from barley crop, the alanine aminotransferase, responsible for the reversible transamination reaction of the N-assimilation pathway has increased the N-uptake at initial stage of growth. The alanine aminotransferase-based gene technology was patented by a private biotech firm and now this company has extended this technology to many other cereals (rice, wheat, sorghum, sugar cane, maize) (Daemrich et al. 2008). Climate change by application of nitrogenous fertilizer could also be reduced by blocking the ammonification and nitrification steps catalyzed by soil bacteria. Certain genes from a tropical grass *Brachiaria humidicola* and wild type wheat *Leymus racemosus* know to release inhibitory compounds that inhibit the growth of *Nitrosomonas* bacteria catalyzing nitrification step (Subbarao et al. 2007).

The genetic engineering approach could be directly used for reducing the pollution caused by halogenated chemical compounds. The over expression of cytochrome P4502E1 in poplar trees played important role in reducing halogenated volatile pollutants (trichloroethylene, hydrocarbons, carbon tetrachloride, vinyl chloride, chloroform, and benzene) from atmosphere through metabolism (Doty et al. 2007). Rice is the major food crop, widely cultivated in Asia and Africa (Muthayya et al. 2014). The flooded paddy fields have been a main source of methane and significantly contribute for global methane emissions. As a part of emission control strategies, cultivation of aerobic paddy crop is promising approach to reduce CH_4 emission. Genetic engineering approach could be used to develop a rice variety suitable for arable cultivation (Wang et al. 2012).

33.2.3 Approaches to Improve Photosynthetic Efficiency and Crop Yield

Improving photosynthesis would improve the crop yield and reduce the climate change caused by CO_2 (Long et al. 2006). The main targets identified by many researchers for improving plant photosynthetic efficiency were, refining Rubisco dynamic properties (Whitney et al. 2011; Parry et al. 2013), introduction of C4 pathway into C3 crops (Gowik and Westhoff 2011), more fast reduction from photoprotection (Murchie and Niyogi 2011), improved action of sedoheptulose-bisphosphatase (Raines 2011), and enriched covering style. Ort et al. (2011) have suggested the reduction of antenna size to improve light infiltration into lower leaves and shelters to decrease the light saturation. Chen and Blankenship (2011) have suggested the genetic transfer of cyanobacterial d and f into higher plant pigment protein complex to improve the light retention in limited light intensity. Chlorophyll a absorbs photons from photosynthetically active radiation in the range of 400–700 nm. Increasing the absorption range of light beyond 700 nm would also enhance the plants photosynthesis and yields. The discovery of novel chlorophyll d and chlorophyll f from cyanobacteria, which has capability to absorb light above photosynthetically active radiation range, would extend absorption up to 710 nm - 750 nm. Engineering of these novel chlorophylls in higher plants would increase photons availability by 19% (Chen and Blankenship 2011) and increase the photosynthesis.

Rubisco enzyme is important in C3 plant photosynthesis and is a prime area of investigation to improve its carboxylation effectiveness (Spreitzer and Salvucci 2002). Rubisco activity is affected with varying concentration of atmospheric CO_2. The Genetic alterations in Rubisco enzyme to enhance the specificity for CO_2 comparative with O_2 (λ) would result in reduction of photorespiration and possibly enhance the photosynthesis and plant yield. Simulation model showed that the increase in Rubisco specificity to the current available CO_2 concentration (380 ppm) would increase carbon assimilation rate by 10% (Sage 2002). The

Rubisco from red algae has 3 times higher specificity than C3 plants (Uemura et al. 1997) and also the crop species which are adapted to dry climate have higher carboxylase activities. The cloning and expression of highly efficient Rubisco from red alga *Griffithsia monilis* in C3 plants were predicted to increase the canopy carbon by 27% (Zhu et al. 2004). Kebeish et al. (2007) introduced glycolate pathway of *Escherichia coli* into *Arabidopsis thaliana* to reduce the energy requirement while recycling the glycolate in C3 pathway. They have predicted at present climatic CO_2 concentration and at 25 °C, the engineered pathway would increase the photosynthetic efficiency by 13% by reducing the respiration rate in C3 plants.

C4 plants generally have reduced photorespiration rate due to carbon concentration mechanism near Rubisco (Sharpe and Offermann 2014). C4 plants photosynthetic efficiency (sorghum, sugar cane, and maize) is 50% greater than C3 species (rice and wheat) at 21 to 23 °C. This is because of photorespiration concealment in the previous. C3 plants grow well in mild temperature regions. Attempts were also made by many researchers in introducing transgenes into nuclear/plastid genomes of C3 plants to increase photosynthetic efficiency (Maurino and Weber 2013; Hibberd et al. 2008). Mathematical models suggest that CO_2 fixation rates would increase by 28% if bicarbonate transporters BicA and SbtA from blue-green algae (*Cyanobacteria*) were introduced into C3 plant chloroplast (Price et al. 2013). The change of C3 plants leaf anatomy to Kranz type by introducing C4 plants photosynthetic enzymes requires introduction of various number of genes (Denton et al. 2013). Modification of the gene Rubisco 19 is the primary target for enhancing the photosynthetic efficiency and plant productivity (Parry et al. 2013). Many models suggest that expression of different types of Rubisco for different environmental conditions may maximize carbon fixation at various CO_2 concentration (Zhu et al. 2004).

33.3 Role of Microbiome in Controlling Climate Change

The term microbiome refers to microbes present in a particular environment or it is the complete genetic material of all microorganisms in a particular niche. Microbes break down organic material and mineralize into usable forms making it available to crops, at the same time produce/consume CO_2, CH_4, nitrogen, and nitrogen oxides. They are considered as key engines of earth's biogeochemical cycle (Falkowski et al. 2008; Microbiology online 2015) and are involved in minimizing the adverse effect of changed environment. Mitigating environmental change by reducing GHG emission through different microbial process is having huge potential and prospects in coming future (Singh et al. 2010; Zimmer 2010). The soil and plant-associated microbes sequester huge amounts of environmental CO_2 and help in alleviating the changed environment (Muller-Stover et al. 2012). The microbes respond to changed environment by altering their community arrangement and structure (Zhou et al. 2011). Microbes take important part in regulating terrestrial global fluctuations of biogenic greenhouse gases (CH_4, N_2O, and CO_2) and were likely to respond rapidly to climate change by utilizing the pollutants (Singh et al. 2010). Microbial communities along with biogeochemical cycles were connected together and act as great combination to address environmental change. This involves complex interactions among microbes and other biotic and abiotic factors. The importance of microbes in reducing the different greenhouse gases and their role in improving agriculture productivities are briefly discussed.

33.3.1 Global Carbon Cycle and CO_2 Mitigation by Microflora

Apart from plants, inorganic carbon in the atmosphere is also fixed to organic material via photo and chemoautotrophic microbes. The photosynthesis in microbes is more frequent than terrestrial plants (Tkemaladze and Makhashvili 2016), contributes significantly for CO_2 fixation (Zimmer 2010), and reduces the impact of climate change. At the same time, microbes degrade dead organic material and release greenhouse gases. The balance between the released GHG and fixed CO_2 is the determinant factor of global carbon flux (Weiman 2015; Prosser 2007; Falkowski et al. 2008). The major portion of atmospheric CO_2 is fixed in pastureland, equatorial forests, and permafrost ecosystems. Microbial processes determine the persistence of carbon fixed in the atmosphere (Weiman 2015). Approximately 93% of the global CO_2 is stored in oceans (Stewart 2003. Micro- and Nano-phytoplanktons, bacteria, and other archaea (Stewart 2003) carry out the oceanic carbon cycle. These oceanic microbes are known to reduce 1/4th of total CO_2 produced by human activities (Weiman 2015). The carbonic acid (H_2CO_3) produced from CO_2 dissolution makes ocean water acidic. The dissolved carbon from CO_2 exists as bicarbonates and consumed by algae present in the ocean, the autotrophic algae play a significant role in fixing dissolved CO_2 into organic compounds. Microorganisms play a very important role in biogeochemical cycle at global scale. Microbes extract carbon from nonliving organic matter and make it available for other organisms, including themselves to grow. Many bacteria and archaea fix CO_2 to sugars and used for cell building and some are escaped as CO_2 through respiration into the soil or atmosphere. Hence, carbon is transferred between organisms from producers to consumers (Zimmer 2010; Crowther et al. 2015).

33.3.2 Methane Mitigation by Methanotrophs

Methane is another powerful GHG, traps and heats up the planet 20 times $>CO_2$ at the same volume (http://www.epa.gov/methane). Methane contributes for 15% of total global warming (Huber-Humer et al. 2008). The atmospheric concentration of methane is ~1.8 ppm and is increasing annually at the rate of 1% (Lelieveld et al.

1993). In anaerobic conditions, the methanogenic bacteria convert CO_2 into CH_4 through Wolfe cycle (Thauer 2012). Removal of CH_4 by capturing or chemical destruction is costly and problematic. Microbial removal by using methanotrophs or methane-oxidizing bacteria is considered as a cheap and an alternative to conventional methods (Brosius 2010; Reeburgh 2007). The CH_4-consuming or -degrading microbes are critical to reduce the greenhouse effect caused by methane. Methanotrophic bacteria use CH_4 as their energy source (Semrau et al. 2010; Bousquet et al. 2006) during their growth and development and reduce the methane concentration (Charu et al. 2014; Shindell et al. 2012). Basically, methanotrophs are microaerophiles and can survive in both aerobic and anaerobic conditions (Rajput et al. 2013). Methane oxidation is only restricted to bacteria such as *Methylococcus*, *Methylocystis*, *Methanomonas*, *Methylosinus*, *Methanobacter*, and *Methylomonas* (https://www.encyclopedia.com). Methanotrophic archaea that exist in marine sediments found to remove ~90% methane emitted by ocean (Zimmerman and Labonte 2015). These microbes can be used to reduce the methane pollution reported in dumping sites and rice fields.

33.3.3 The Nitrogen Cycle and Control of Nitrogen Oxides

Nitrogen constitutes ~78% of the total gases in the atmosphere. Plants and animals directly cannot use atmospheric nitrogen and it has to be fixed before usage. Many bacteria like *Rhizobium, Frankia, Azatobacter,* and *Azospirillum* fix (free form or mutual association) atmospheric N_2 in the most available form to plants (Anne 2010; Vitousek et al. 2013; Orr et al. 2011). The use of these nitrogen-fixing bacteria in agriculture reduces the synthetic chemical fertilizer application and negative effects on the environment. Microorganisms fix an estimated amount of 70–140 million tons of nitrogen annually and decrease the use of synthetic nitrogenous fertilizers. Microbes mineralize the organic nitrogen and catalyze the nitrogen cycle on the earth. The mineralization of organic nitrogen aided by microbes converts all unavailable form of nitrogen to most available form for plant growth (Groffman 2012).

33.4 Microbes and Their Role in Sustainable Agriculture

Conventional agricultural practices are neither economical nor environmentally friendly approaches. For sustainable crop production beneficial plant, associated microbes have to be deployed for enhancement of plant growth and improving quality of food. Microorganisms colonize various parts of plants (roots, stem, and leaf) (Lugtenberg and Kamilova 2009) and some enter through root and live inside the plants as endophytes (Stone et al. 2000; Muller-Stover et al. 2012). They exert various beneficial actions and impart resilience to changing climatic conditions.

Microbes enhance biotic/abiotic stress tolerance level, crop growth and vigor, and disease resistance of plants by various metabolic activities (Brundrett 2009; Bent 2006; Barrow et al. 2008). The bacterial groups, which are beneficial to plants, are called Plant Growth-Promoting Rhizobacteria (PGPR), which include *Pseudomonas, Rhizobium, Azospirillum, Serratia, Bacillus,* and *Stenotrophomonas*. These bacteria fix atmospheric nitrogen, produce growth hormones, and have antagonistic effects against various soil-borne pathogens, and improve the overall plant growth and productivity (Richardson 2001; Stanger and Lauer 2008). The beneficial fungi, which associated with plants and widely studied are vesicular and arbuscular mycorrhizae (Tyagi et al. 2014). These fungi help in solubilization and mobilization of fixed soil phosphorus to plants.

Microbes act as biological control agents and stabilize population levels of plants making them more resilient to changing climate, thereby increases the agricultural productivity (Hoover and Newman 2004). The alkaloids produced by many symbiotic bacteria protect plants from pest attack and reduce the pesticide application (Kuldau and Bacon 2008). Some of these plant- and soil-associated microbiomes aid the nutrient recycling (Barrow et al. 2008; Green et al. 2008) and soil structure stabilization (Gale et al. 2000; Brundrett 2009). The exopolysaccharides produced by microbes increase the soil aggregation and improve soil structure; this in turn improves aeration and water-holding capacity (Gale et al. 2000). There are many microbial-based products, which have various applications in agriculture. Several companies have involved in producing carrier-based microbial inoculants of beneficial microbes such as, nitrogen fixers, biocontrol agents, K and P solubilizers, and mobilizers (Richardson 2001; Stanger and Lauer 2008). The large-scale field trails on economically important crops with beneficial microbes have increased the crop yields by 10–20% (Perez-Jaramillo et al. 2016). The efficacy of beneficial microbes and interaction with plants in the changing microenvironment were studied in large-scale field trials. The recently developed microfluidics based "Microbiome on a Chip" technology had facilitated multifaceted experimental design for studying multiplexed treatments such as environmental stimuli, host response, and microbial colonization in short period of time (Stanley and van der Heijden 2017). This kind of AI-based technique revolutionizes the selection of beneficial microbes for relevant environmental conditions for particular crops and prepares easy to apply formulations with long shelf life. Based on this, Mitter et al. (2017) designed the microbiome and changed the traits of plants by introducing beneficial microbes at flowering stage. The introduced beneficial microbes were seen in next generation through seeds of first generation. This kind of application has several advantages over conventional approach by increasing the colonization percentage of inoculated strain over native and offers better crop protection and improved quality of crop products.

The microflora of plant is influenced by genotypic and phenotypic variations. The identification of plants, which support the beneficial microbiome, is considered as a novel approach (Wallenstein 2017). Cultivation of plants with modern agricultural practices demands inorganic fertilizers, insecticides, and growth hormones (Perez-Jaramillo et al. 2016) which may affect the association of beneficial microflora of plant. Optimization of microbial inoculants as biofertilizer and biopesticides is rapidly advancing for different environmental conditions and crop varieties. In breeding program the plants, which design the rhizosphere and phyllosphere microbiome, should be evaluated. These plants secrete specific metabolites or exudate on root/stem and attract the beneficial microbes through signaling (Abhilash et al. 2012).

33.4.1 Application of Metagenomics for Studying and Developing the Beneficial Microbiome

Plants are associated with diverse groups of microorganisms (Vorholt 2012; Bulgarelli et al. 2013). Microbes live inside plants as endophytes and outside as epiphytes on plants surface, nearby soil, around the roots as epiphytes. Plant-associated microbes may have mutual or nonmutual interactions and might negatively affect the plant growth (Newton et al. 2010). The plant microbe interactions are affected by changing environmental conditions. Hence, it is important to study the type and role of microbiome in varied environmental conditions and to design the appropriate microbiome for particular crop.

Metagenomics is a technological approach by which the entire microbial communities' genome is analyzed. DNA-based genome analysis (individual or metagenomics) provides information on distribution of microbes in ecology and their physiological potential. The RNA studies of these microbes would provide information on regulatory and metabolic pathways. New high-throughput DNA sequencing approaches such as next-generation sequencing (NGS) and 454 sequencing techniques offer opportunities to study and evaluate beneficial interaction of microbe for improving plant growth and productivity (Committee on Metagenomics 2007). Due to these recent technologies, the time and cost of sequencing has been drastically reduced and it is now possible to study microbial diversity, proteomics, transcriptomics, and metabolomics profiles of microbial community. The findings from the molecular analysis could be applied to study the interactions of microbes of plant and soil to understand their role in influencing plant productivity.

The nucleic acid (DNA/RNA) or whole-genome sequencing based on metagenomics and omics study of microbes helps to identify genetic and metabolomics variations in microbial community (Committee on Metagenomics 2007; Lucero et al. 2011). These studies throw more light on microbial interactions with plants, which induce plants to gain resistance against various biotic, abiotic stresses, nutrient mobility and cycling, effect on plants metabolism, and nutrition. These beneficial microbes' genomes are valuable to develop the sequence repository and to study the target genes with respect to plant interaction and beneficial role (Kuldau and Bacon 2008; Schardl et al. 2004). The databases such as Community Cyberinfrastructure for Advanced Microbial Ecology Research and Analysis (CAMERA) and National Center for Biotechnology Information (NCBI/GenBank) maintain the nucleic acid sequences obtained from metagenomics study of various ecological niches for public use. The simulation study using these nucleic acid sequences helps to predict the response of microbes to various conditions and stimuli in larger scale. Metagenomics analysis of uncultivable microbiota helps in reconstruction of individual genomes to obtain useful information. If microbiome community is dominated by one or few phylogenetically distinct microbiota, the sequence read assembly generated for these diverse group of microbial communities are uneven and complex (Scholz et al. 2012) and can be successfully utilized for gene reconstruction (Mackelprang et al. 2011; Albertsen et al. 2013). By metagenomics techniques, specific information of microbe and desired microbes for particular function or dominant microflora across different species of crop plants could be identified. The phyllosphere microbial community analysis by combined metagenomics and metaproteomics data of *Arabidopsis thaliana*, tomato, clover, tamarisk, soybean, and rice revealed the presence of consistent type of microbes at phylum level of all these plants (Vorholt 2012; Atamna-Ismaeel et al. 2012; Knief et al. 2012; Ottesen et al. 2013).

Cretoiu et al. (2012) studied the chiA gene variations in rhizosphere of plants selected from two arctic regions. The *Oxyria digyna* plant rhizosphere shows highest chiA diversity amongst the various rhizosphere soils studied. The Unno and Shinano (2013) analyzed the rhizosphere community of plants which shows superior growth in presence of phytic acid. These results indicate the presence of genes responsible for utilization of phytic acid (such as alkaline phosphatase or citrate synthase). Chhabra et al. (2013) applied the targeted metagenomics approach for mineral phosphate solubilization activity using cosmid library screening and 454 sequencing technology to identify genes and operons which regulate the phosphate solubilization.

33.5 Application of AI and ML for Improving Plant Photosynthesis, Agricultural Productivity, Selection of Beneficial Microbiome, and Control of Climate Change

The negative impact of climate change on agriculture can be minimized by doubling the crop productivity. However, the agricultural yield is depending on several factors like soil quality, rainfall, and environmental conditions, etc. All these factors responsible for agricultural productivity vary with region and time. However, to increase the agricultural productivity, improving photosynthesis efficiency by capturing more carbon was identified as a significant opportunity for optimization (Scoggins 2018). To understand the relation between enhanced photosynthesis efficiency and crop performance, crop simulation technology was the best option (Artificial intelligence helps make the world more food secure 2017). Agriculture production system sIMulator (APSIM) developed in Australia was able to check the variation in plants photosynthetic efficiency. This technology has application to understand how plants efficiently capture carbon dioxide or how well it converted into sugars. The APSIM model was evaluated for performance and findings showed that when 25% increase in photosynthetic enzymes efficiency results in only 5% increase in crop growth. CropOS platform developed by Benson Hill biosystems was used to identify the gene candidates to enhance photosynthesis in many crops (Rakestraw and Acharya 2017). In 2017, Heckmann and team evaluated the potential of leaf reflectance to predict the photosynthesis. They selected C3 and C4 plants such as *Brassica oleracea* and *Zea mays*, for their study and measured the performance of several ML algorithms. The findings of the study showed that recursive feature elimination on untransformed spectra and partial least squares regression has the maximum predictive power. This study showed the importance of leaf reflectance phenotyping for improving photosynthetic capacity of crops.

Based on the available empirical data on plant and environment, many computer-based methods were proposed to obtain optimum conditions for cultivating the crops (Takakura et al. 1974, 1978; Takakura 1975; Takakura and Ohara 1976). The AI-based technologies such as wireless sensor network is used for capturing and recording the physical parameters and the various algorithms are used to predict yield and quality of agricultural crops (Zhou et al. 2012; Aqeel-ur-Rehman Abbasi et al. 2014). The CO_2 emission and photosynthetic rate could be measured by leaf phenotyping and reflectance; this could help in measuring CO_2 emission and global warming. The eXtreme Gradient Boosting (XGBoost) algorithm is used to predict the errors of earlier models. Using these corrections, the final prediction is made, and new models are developed with minimum errors (Chen and Guestrin 2016). XGBoost is one of the very accurate algorithms for more precise crop yield prediction than any other models and it utilizes all the environmental data for prediction. The support vector machine (SVM) algorithm can be used for classification or segregation and regression challenges. By using this algorithm, the crop and weeds differentiation is possible. Also, the photosynthetic rate can be differentiated between the various plant genomes (Drucker et al. 1997). Random forest (RF) algorithm improves accuracy by overfitting the decision, it automates the missing parameter and can be used for both classification and regression analysis (Breiman 2001). RF algorithm can be used for remote estimation of biomass from various crops rapidly without any destruction of the crop. The generalized additive model (GAM) is a hybrid algorithm that can be applied for linear and nonlinear regression problems (Hastie 2017). The effect of abiotic factors on carbon fixation by plants could be effectively evaluated by this model.

To increase agricultural productivity, AI-based solutions were introduced with the investment of more than $500 million during 2012 to 2017 (Brayne et al. 2018). The technology was used to identify the healthy strains of plant microbiome using satellite images. To fight with continuously changing environment, one way is to intensify agriculture with the application of more beneficial microbes. Another way is to manipulate the microbiome for increased agricultural productivity. Combination of robotics, microfluidics, and ML, exploits the microbial community associated with plants for better efficiency and stress resistance (Toju et al. 2018). In 2005, Department of Agriculture (USDA) and National science foundation (NSF) supported joint microbial genome sequencing program (Microbial genome sequencing program 2005). The program provides genetic sequences of agriculturally and ecologically important bacteria and fungi. Recently Monsanto established collaborative partnerships with Atomwise. The output of partnership is AtomNet technology that uses powerful deep learning algorithms and supercomputers for analyzing crop protection products (Brayne et al. 2018). Monsanto has made another collaboration with Second Genome to analyze microbial function for the development of next-generation insect control solution. The analysis was done using several approaches such as predictive analytics, metagenomics, and ML (Brayne et al. 2018). During 2007, National institute of Health (NIH) and Department of Energy (DOE) launched an open source pipeline (Wilke et al. 2016). The Metagenomics rapid annotations using subsystems technology (MG-RAST) provides quantitative information on microbial populations.

Concentric Ag in 2018 developed a phytomicrobiome which is beneficial and found to improve the growth and yield of crops (Semones 2018). Peter Baas (2018) developed a selection platform of microbial consortia to functionally select microbiome for better crop health and yield (Baas 2018). Therefore, AI and ML can be used for identification and application of beneficial plant microbiome for increasing agricultural productivity and mitigating climate change. The development of new technologies AI and ML have several applications in agriculture to fight with changing environment. These technologies were used for forecasting rainfall and precipitation, pest/disease identification, soil management, water management, nutrient deficiency identification, analysis, and crop yield prediction. Rainfall prediction is one of the important and challenging tasks. The management and use of water, require accurate measurement of rainfall data. The prediction of rainfall requires advanced modeling and simulation. Around the world scientific community developed several rainfall prediction models. The soil analysis was important for effective nutrient management. Emamgholizadeh et al. (2017) estimated and compared soil macronutrients by three models. They showed that ANN model was efficient and reliable than geographically weighted regression (GWR) and CoKriging (CK) models. During 2018, the possible defects and nutrient deficiencies in soil were identified by a deep learning platform Plantix, which was developed by PEAT from Berlin (Roy and Kant 2018).

Fahimi and El-Shafie (2015) mapped rainfall data from various stations into one station using Self-organizing maps (SOM). The finding of the study showed that SOM method gives satisfactory performance with reduced dimensionality and computational costs. Mislan and team checked the accuracy of rainfall forecasting using ANN with Backpropagation neural network (BPNN) algorithm (Mislan et al. 2015). The study indicated that BPNN algorithm was performed better for rainfall prediction in Tenggarong, Indonesia. Helen et al. (2016) compared ANN and Fuzzy logic models to understand which one performs better in rainfall prediction. The results from various places such as Ondo and Iju states showed that ANN model performs marginally superior to Fuzzy logic model. Villanueva and Salenga (2018) used ML algorithm with Convolutional neural network (CNN) to predict bitter gourd crop yield through the classification of leaf images. The pest and nutrient deficiency symptoms were diagnosed using the diseased part of leaf images. The same ML tools were also used to get the information on diseases of oilseed crops, such as groundnut, soybean, and rapeseed-mustard. Recently scientists have developed a transfer learning AI-based system for identification of brown leaf spot disease of cassava plant with 98% accuracy (Brayne et al. 2018). The system can be loaded on smartphone and computer for easy identification of disease in timely manner. The image identification algorithm developed by Resson, a Monsanto growth ventures (MGV) company can able to identify and categorize plant pest and disease more accurately than trained human (Rakestraw and Acharya 2017). The Intelligent System for Disease Diagnosis in Crops (ISDDC) was developed to enhance decision-making ability of farmers (Kolhe and Gupta 2014). The objective of AI and ML is to provide recommendation to farmers regarding soil moisture and temperature, soil and crop yield improvement, soil health cards, early warning on pest/disease occurrence, and weather prediction. Thus, in a current scenario AI and ML are important tools in modern agriculture, which can be used accurately for predicting the variations of abiotic factors and their effect on photosynthesis and crop yield under changing environmental conditions.

33.6 Conclusion

The continuously changing climatic conditions affect rainfall and increase the global temperature. The increased GHG emission adversely affects the ecosystem and crop productivity. The plant photosynthesis plays key role in controlling the climate change caused by CO_2, a major greenhouse gas. The lower CO_2 and higher temperature affect the C3 photosynthesis and plant productivity. Distribution and diversity of C3 plants are more common than C4 plants. However, C4 plants have higher photosynthetic efficiency than C3 plants due to their novel photosynthetic machinery arrangement. Synthetic biology approaches would play significant role in improving photosynthetic efficiency of highly diverse C3 plants and to achieve the food security and to mitigate the climate change caused by CO_2. Microbes, which promote plant growth, increase the yield; mobilize the nutrients; and fix atmospheric nitrogen, are alternate options to synthetic fertilizers and pesticides. These unique metabolic activities of microorganisms regulate the biogeochemical cycle and reduce the emission of greenhouse gases and finally the climate change. The advanced biotechnological approaches such as metagenomics and gene sequencing tools have benefitted the scientific community for advanced research. It also helped to develop crop specific microbiome databases, which can be used to design the plant microbiome for improved plant growth yield and quality. The application of AI and ML in agriculture would open new directions of research. The AI and ML play important role in precision farming, prediction of changing climate, and controlling pest and diseases, thereby improve the crop yield and minimize the loss. Overall, the improved plant photosynthetic efficiency, beneficial plant microbiome, metagenomics, AI and ML tools play significant role in achieving sustainable crop productivity and reducing climate change.

References

Abhilash PC, Powell JR, Singh HB, Singh BK (2012) Plant–microbe interactions: novel applications for exploitation in multipurpose remediation technologies. Trends Biotechnol 30:416–420

Aggarwal PK, Mal RK (2002) Climate change and rice yields in diverse agro-environments of India. II. Effects of uncertainties in scenarios and crop models on impact assessment. Clim Chang 52:331–343

Ainsworth EA, Rogers A (2007) The response of photosynthesis and stomatal conductance to rising [CO_2]: mechanisms and environmental interactions. Plant Cell Environ 30:258–270

Albertsen M, Hugenholtz P, Skarshewski A, Nielsen KL, Tyson GW, Nielsen PH (2013) Genome sequences of rare, uncultured bacteria obtained by differential coverage binning of multiple metagenomes. Nat Biotechnol 31:533–538

Alexandratos N, Bruinsma J (2012) World agriculture towards 2030/2050: the 2012 revision. ESA Working paper No. 12-03 Rome, FAO, 154

Alexandros K, Eigenmann R, Teye F, Politopoulou Z, Wolfert S et al (2012) Farm management systems and the future internet era. Comput Electron Agric 89:130–144

Amthor JS, Baldocchi DD (2001) Terrestrial higher plant respiration and net primary production. In: Roy J, Saugier B, Mooney HA (eds) Terrestrial global productivity. Academic Press, San Diego, CA, pp 33–59

Anne B (2010) The nitrogen cycle: processes, players and human impact. Nat Educ Knowl 3:25

Aqeel-ur-Rehman Abbasi AZ, Islam N, Shaikh ZA (2014) A review of wireless sensors and networks' applications in agriculture. Comp Stand Interf 36:263–270

Artificial intelligence helps make the world more food secure (2017) Queensland Alliance for Agriculture and Food Innovation

Atamna-Ismaeel N, Finkel O, Glaser F, Von Mering C, Vorholt JA, Koblizek M et al (2012) Bacterial anoxygenic photosynthesis on plant leaf surfaces. Environ Microbiol Rep 4:209–216

Baas P (2018) Targeted manipulation of the plant microbiome to enhance crop health and yield

Barrow JR, Lucero ME, Reyes-Vera I, Havstad K (2008) Do symbiotic microbes have a role in plant evolution, performance and response to stress? Commun Integr Biol 1:69–93

Bassham JA (2003) Mapping the carbon reduction cycle: a personal retrospective. Photosynth Res 76:35–52

Bauwe H, Haemann M, Fernie AR (2010) Photorespiration: players, partners and origin. Trends Plant sci 15:330–336

Beer C, Reichstein M, Tomelleri E, Ciais P, Jung M, Carvalhais N, Reodenbeck C, Arain MA, Baldocchi D, Bonan GB et al (2010) Terrestrial gross carbon dioxide uptake: global distribution and covariation with climate. Sci 329:834–838

Benson AA (2002) Following the path of carbon in photosynthesis: a personal story. Photosynth Res 73:29–49

Bent E (2006) Induced systemic resistance mediated by plant growth-promoting rhizobacteria (PGPR) and fungi (PGPF). In: Tuzun S, Bent E (eds) Multigenic and induced systemic resistance in plants. Springer, Berlin, pp 225–258

Bernacchi CJ, Pimentel C, Long SP (2003) In vivo temperature response functions of parameters required to model RuBP-limited photosynthesis. Plant Cell Environ 26:1419–1430

Bousquet P, Ciais P, Miller JB, Dlugokencky EJ, Hauglustaine DA et al (2006) Contribution of anthropogenic and natural sources to atmospheric methane variability. Nature 443:439–443

Brayne S, McKellar S, Tzafestas K (2018) Artificial intelligence in the life sciences & patent analytics: market developments and intellectual property landscape. IP Pragmatics Ltd

Breiman L (2001) Random forests. Mach Learn 45(1):5–32

Brosius LS (2010) Investigating controls over methane production and bubbling from interior Alaskan lakes using stable isotopes and radiocarbon ages. University of Alaska, Fairbanks, AK.

Brundrett MC (2009) Mycorrhizal associations and other means of nutrition of vascular plants: understanding the global diversity of host plants by resolving conflicting information and developing reliable means of diagnosis. Plant Soil 320:37–77

Bulgarelli D, Schlaeppi K, Spaepen S, Ver Loren Van Themaat E, Schulze-Lefert P (2013) Structure and functions of the bacterial microbiota of plants. Annu Rev Plant Biol 64:807–838
Charu G, Prakash D, Sneh G (2014) Role of microbes in combating global warming. Int J Pharm Sci Lett 4:359–363
Chen M, Blankenship RE (2011) Expanding the solar spectrum used by photosynthesis. Trends Plant Sci 16:427–431
Chen T, Guestrin C (2016). XGBoost: a scalable tree boosting system. ACM, pp 785–794
Chhabra S, Brazil D, Morrissey J, Burke JI, O'Gara F, Dowling ND (2013) Characterization of mineral phosphate solubilization traits from a barley rhizosphere soil functional metagenome. Microbiol Open 2:717–724
Committee on Metagenomics: Challenges and Functional Applications, National Research Council (2007) The new science of metagenomics: revealing the secrets of our microbial planet. The National Academies Press, Washington, DC, USA. http://www.nap.edu/openbook.php?record_id=11902
Cretoiu MS, Kielak AM, Abu Al-Soud W, Sorensen SJ, Van Elsas JD (2012) Mining of unexplored habitats for novel chitinases - chiA as a helper gene proxy in metagenomics. Appl Microbiol Biotechnol 94:1347–1358
Crowther TW, Thomas SM, Maynard DS, Baldrian P, Covey K, Frey SD, van Diepen LTA, Bradford MA (2015) Biotic interactions mediate soil microbial feedbacks to climate change. Proc Natl Acad Sci 112:7033–7038
Daemrich A, Reinhardt F, Shelman M (2008) Arcadia biosciences: seeds of change. Harvard Business School, Boston, MA
Davidson EA, Janssens IA (2006) Temperature sensitivity of soil carbon decomposition and feedbacks to climate change. Nature 440:165–173
Dengel A (2013) Special issue on artificial intelligence in agriculture. KI 27:309–311
Denton AK, Simon R, Weber APMK (2013) C4 photosynthesis: from evolutionary analyses to strategies for synthetic reconstruction of the trait. Curr Opin Plant Biol 16:315–321
Djanaguiraman M, Prasad PVV, Seppanen M (2010) Selenium protects sorghum leaves from oxidative damage under high temperature high stress by enhancing antioxidant defense system. Plant Physiol Biochem 48:999–1007
Doty SL, Andrew JC, Moore AL, Vajzovic A, Singleton GL, Ma CP, Khan Z, Xin G, Kang JW, Park JY, Meilan R, Strauss SH, Wilkerson J, Farin F, Strand SE (2007) Enhanced phytoremediation of volatile environmental pollutants with transgenic trees. Proc Natl Acad Sci 104(43):16816–16821
Drucker H, Burges CJ, Kaufman L, Smola AJ, Vapnik V (1997) Support vector regression machines, pp 155–161
Edwards GE, Franceschi VR, Voznesenskaya EV (2004) Single-cell C4 photosynthesis versus the dual-cell (Kranz) paradigm. Annu Rev Plant Biol 55:173–196
Emamgholizadeh S, Shahsavani S, Eslami MA (2017) Comparison of artificial neural networks, geographically weighted regression and Cokriging methods for predicting the spatial distribution of soil macronutrients (N, P, and K). Chin Geogr Sci 27(5):747–759
Fahimi F, El-Shafie AH (2015) Application of self-organizing maps (SOMS) method to rainfall stations clustering in multisite regions. Intern J Adv Mech Civ Eng 2:33–37
Falkowski PG, Fenchel T, Delong EF (2008) The microbial engines that drive earth's biogeochemical cycles. Science 320:1034–1039
FAO Agriculture Series (2002) The state of food and agriculture. Food and Agriculture Organization of the United Nations
FAO Agriculture Series (2013) The state of food and agriculture. Food and Agriculture Organization of the United Nations
FAO Agriculture Series (2017) The state of food and agriculture. Food and Agriculture Organization of the United Nations
Farquhar GD, von Caemmerer S, Berry JA (1980) A biochemical model of photosynthetic CO_2 assimilation in leaves of C3 species. Planta 149:78–90
Furbank RT (2011) Evolution of the C4 photosynthetic mechanism: are there really three C4 acid decarboxylation types? J Exp Bot 62:3103–3108

Furuya J, Kobayashi S (2010) Impact of global warming on agriculture product markets, chapter 2. In: Sumi A, Fukushi K, Hiramatsu A (eds) Adaptation and mitigation strategies for climate change. Springer, Tokyo

Gale WJ, Cambardella CA, Bailey TB (2000) Root-derived carbon and the formation and stabilization of aggregates. Soil Sci Soc Am J 64:201–207

Geoffrey Craggs JP (2016) Photosynthesis and its role in climate change and soil regeneration. Strategic Analysis Paper. Future Directions International

Ghannoum O, Evans JR, von Caemmerer S (2011) Nitrogen and water use efficiency of C4 plants. In: Raghavendra AS, Sage RF (eds) C4 photosynthesis and related CO_2 concentrating mechanisms. Springer, Dordrecht, The Netherlands, pp 129–146

Gowik U, Westhoff P (2011) The path from C3 to C4 photosynthesis. Plant Physiol 155:56–63

Green LE, Porras-Alfaro A, Sinsabaugh RL (2008) Translocation of nitrogen and carbon integrates biotic crust and grass production in desert grassland. J Ecol 96:1076–1085

Groffman P (2012) Terrestrial denitrification: challenges and opportunities. Ecol Proc 1:1–11

Hastie TJ (2017) Generalized additive models, pp 249–307

Heckmann D, Schluter U, Weber APM (2017) Machine learning techniques for predicting crop photosynthetic capacity from leaf reflectance spectra. Mol Plant 10:878–890

Helen AA, Bolanle AO, Samuel OF (2016) Comparative analysis of rainfall prediction models using neural network and fuzzy logic. Intern J Soft Comput Eng 5(6)

Hibberd JM, Sheehy JE, Langdale JA (2008) Using C4photosynthesis to increase the yield of rice-rationale and feasibility. Curr Opin Plant Biol 11:228–231

Hoffmann U (2013) Wake up before it is too late: make agriculture truly sustainable now for food security in a changing climate. UNCTAD, UN

Hoover JK, Newman JA (2004) Tritrophic interactions in the context of climate change: a model of grasses, cereal aphids and their parasitoids. Glob Chang Biol 10:1197–1208

Horz HP, Raghubanshi AS, Heyer J, Kammann C, Conrad R, Dunfield PF (2002) Activity and community structure of methane-oxidising bacteria in a wet meadow soil. FEMS Microbiol Ecol 41:247–257

http://earthobservatory.nasa.gov/Features/GlobalWarming/page2.php

http://www.epa.gov/methane

https://www.encyclopedia.com/science/encyclopedias-almanacs-transcripts-and-maps/methane-oxidizing-and-producing-bacteria. Methane oxidizing and producing bacteria. World of Microbiology and Immunology. Encyclopedia.com. 11 Mar 2019

Huber-Humer M, Lechner P, Gortler N (2008) Impact of different biocover designs on methane mitigation. In: Pawlowska M, Pawlowski L (eds) Management of pollutant emission from landfills and sludge. Routledge, New York, pp 21–36

IPCC (2007) Climate change 2007: the physical science basis. In: Solomon SD et al (eds) Contribution of Working Group I to the Fourth Assessment Report of the Intergovernmental Panel on Climate Change. Cambridge University Press, Cambridge

IPCC Climate Change (2007) Synthesis report. Summary for policymakers. http://www.ipcc.ch. Accessed November 2007

IPCC SRES (2000) Nakicenovic N, Swart R (eds) Special report on emissions scenarios: a special report of Working Group III of the Intergovernmental Panel on Climate Change. Cambridge University Press, Cambridge

Kaneda TB (2016) World population data sheet. Population Reference Bureau. www.prb.org, p 22

Kaur K (2016) Machine learning: applications in Indian agriculture. Intern J Adv Res Comput and Commun Eng 5:342–344

Kebeish R, Niessen M, Thiruveedhi K, Bari R, Hirsch H et al (2007) Chloroplastic photorespiratory bypass increases photosynthesis and biomass production in *Arabidopsis thaliana*. Nat Biotechnol 25:593–599

Kimball BA, Kobayashi K, Bindi M (2002) Responses of agricultural crops to free-air CO_2 enrichment. Adv Agron 77:293–368

Knief C, Delmotte N, Chaffron S, Stark M, Innerebner G, Wassmann R et al (2012) Metaproteogenomic analysis of microbial communities in the phyllosphere and rhizosphere of rice. ISME J 6:1378–1390
Kolhe S, Kamal R, Saini HS, Gupta GK (2011) A web-based intelligent disease-diagnosis system using a new fuzzy-logic based approach for drawing the inferences in crops. Comput Electron Agric 76:16–27
Kolhe S, Gupta GK (2014) Intelligent systems for agriculture domain. Intern J Comput Appl:14–18
Kuldau G, Bacon C (2008) Clavicipitaceous endophytes: their ability to enhance resistance of grasses to multiple stresses. Biol Control 46:57–71
Kusar D, Avgustin G (2010) Molecular profiling and identification of methanogenic archaeal species from rabbit caecum. FEMS Microbiol Ecol 74:1–8
Le Quere C, Raupach MR, Canadell JG, Marland G, Bopp L, Ciais P et al (2009) Trends in the sources and sinks of carbon dioxide. Nat Geosci 2:831–836
Le Quere C, Andrew RM, Canadell JG, Sitch S, Korsbakken JI, Peters GP, Manning AC, Boden TA, Tans PP, Houghton RA et al (2016) Global carbon budget. Earth Syst Sci Data 8:605–649
Lelieveld J, Crutzen PJ, Bruhl C (1993) Climate effects of atmospheric methane. Chemosphere 26:739–768
Long SP, Ainsworth EA, Rogers A, Ort DR (2004) Rising atmospheric carbon dioxide: plants FACE the future. Annu Rev Plant Biol 55:591–628
Long SP, Zhu XG, Naidu SL, Ort DR (2006) Can improvement in photosynthesis increase crop yields? Plant Cell Environ 29:315–330
Lorenzoni I, Nicholson-Cole S, Whitmarsh L (2007) Barriers perceived to engaging with Climate Change among the UK public and their policy implications. Glob Environ Change 17:445–459
Lu Y, Conrad R (2005) In situ stable isotope probing of methanogenic archaea in the rice rhizosphere. Science 309:1088–1090
Lucero ME, Unc A, Cooke P, Dowd S, Sun S (2011) Endophyte microbiome diversity in micropropagated *Atriplex canescens* and *Atriplex torreyi* var *griffithsii*. PLoS One 6:e17693
Lugtenberg B, Kamilova F (2009) Plant-growth-promoting rhizobacteria. Annu Rev Microbiol 63:541–556
Mackelprang R, Waldrop MP, Deangelis KM, David MM, Chavarria KL, Blazewicz SJ et al (2011) Metagenomic analysis of a permafrost microbial community reveals a rapid response to thaw. Nature 480:368–371
Mauney JR, Lewin KF, Hendrey GR, Kimball BA (1992) Growth and yield of cotton expose to free-air CO_2 enrichment. Crit Rev Plant Sci 11:213–222
Maurino VG, Weber APM (2013) Engineering photosynthesis in plants and synthetic microorganisms. J Exp Bot 64:743–751
Microbiology Online (2015) Microbes and climate change. http://www.microbiologyonline.org.uk/aboutmicrobiology/microbes and climate change. Accessed 19 June 2020
Milius S (2017) Worries grow that climate change will quietly steal nutrients from major food crops. Sci News
Mislan H, Hardwinarto S, Sumaryono, Aipassa M (2015) Rainfall monthly prediction based on artificial neural network: a case study in Tenggarong Station, East Kalimantan Indonesia. Procedia Comput Sci 59:142–151
Mitter B, Pfaffenbichler N, Flavell R et al (2017) A new approach to modify plant microbiomes and traits by introducing beneficial bacteria at flowering into progeny seeds. Front Microbiol 8:11
Moser SC, Dilling L (2004) Making climate hot: communicating the urgency and challenge of climate change. Environ Sci Policy Sustain Dev 46:32–46
Muller-Stover D, Hauggard-Nielsen H, Eriksen J, Ambus P, Johansen A (2012) Microbial biomass, microbial diversity, soil carbon storage, and stability after incubation of soil from grass clover pastures of different age. Biol Fertil Soils 48:371–383
Murchie EH, Niyogi KK (2011) Manipulation of photoprotection to improve plant photosynthesis. Plant Physiol 155:86–92
Muthayya S, Sugimoto JD, Montgomery S, Maberly GF (2014) An overview of global rice production, supply, trade, and consumption. Ann N Y Acad Sci 1324:7–14

Newton AC, Fitt BDL, Atkins SD, Walters DR, Daniell TJ (2010) Pathogenesis, parasitism and mutualism in the trophic space of microbe-plant interactions. Trends Microbiol 18:365–373
NSF Funding Opportunity Document (2005) Microb Genome Sequencing Program
Ogren WL (2003) Affixing the O to rubisco: discovering the source of photo respiratory glycolate and its regulation. Photosynth Res 76:53–63
Olufemi A, Reuben O, Olufemi O (2014) Global climate change. J Geosci Environ Protect 2:114–122
Orr CH, James A, Leifert C, Cooper JM, Cummings SP (2011) Diversity and activity of free-living nitrogen-fixing bacteria and total bacteria in organic and conventionally managed soils. Appl Environ Microbiol 77:911–919
Ort DR, Zhu X, Melis A (2011) Optimizing antenna size to maximize photosynthetic efficiency. Plant Physiol 155:79–85
Ottesen AR, Pena AG, White JR, Pettengill JB, Li C, Allard S et al (2013) Baseline survey of the anatomical microbial ecology of an important food plant: *Solanum lycopersicum* (tomato). BMC Microbiol 13:114
Parry MAJ, Andralojc PJ, Scales JC, Salvucci ME, Carmo-Silva AE, Alonso H, Whitney SM (2013) Rubisco activity and regulation as targets for crop improvement. J Exp Bot 64:717–730
Perez-Jaramillo JE, Mendes R, Raaijmakers JM (2016) Impact of plant domestication on rhizosphere microbiome assembly and functions. Plant Mol Biol 90:635–644
Philippot L, Hallin S (2011) Towards food, feed and energy crops mitigating climate change. Trends Plant Sci 16:476–480
Price GD, Pengelly JJL, Forster B, Du J, Whitney SM, von Caemmerer S, Badger MR, Howitt SM, Evans JR (2013) The cyanobacterial CCM as a source of genes for improving photosynthetic CO_2 fixation in crop species. J Exp Bot 64:753–768
Prosser JI (2007) Microorganisms cycling soil nutrients and their diversity. In: Modern soil microbiology. CRC Press, Boca Raton, FL, pp 237–261
Qaderi MM, Kurepin LV, Reid DM (2012) Effects of temperature and watering regime on growth, gas exchange and abscisic acid content of canola (*Brassica napus*) seedlings. Environ Exp Bot 75:107–113
Qiancheng M (1998) Greenhouse gases: refining the role of carbon dioxide, NASA Science Briefs
Raines CA (2011) Increasing photosynthetic carbon assimilation in C3 plants to improve crop yield: current and future strategies. Plant Physiol 155:36–42
Rajput P, Saxena R, Mishra G, Mohanty SR, Tiwari A (2013) Biogeochemical aspect of atmospheric methane and impact of nanoparticles on methanotrophs. J Environ Anal Toxicol 3:2–10
Rakestraw R, Acharya A (2017) Can artificial intelligence help feed the world? The mix bowl
Reeburgh WS (2007) Oceanic methane biogeochemistry. Chem Rev 107:486–513
Richardson AE (2001) Prospects for using soil microorganisms to improve the acquisition of phosphorus by plants. Funct Plant Biol 28(9):897–906
Robinson CA, Cruse RM, Ghaffarzadeh M (1996) Cropping system and nitrogen effects on mollisol organic carbon. Soil Sci Soc Am J 60:264–269
Rosegrant MW, Cline SA (2003) Global food security: challenges and policies. Science 1917–1919
Roy A, Kant A (2018) National stratergy for artificial intelligence. Discussion Paper on Nation Strat Artific Intellig
Sage RF (2002) Variation in the k (cat) of rubisco in C-3 and C-4 plants and some implications for photosynthetic performance at high and low temperature. J Exp Bot 53:609–620
Sage RF (2014) Photosynthetic efficiency and carbon concentration in terrestrial plants: the C4 and CAM solutions. J Exp Bot 65:3323–3325
Sage RF, Sage TL, Kocacinar F (2012) Photorespiration and the evolution of C4 photosynthesis. Annu Rev Plant Biol 63:19–47
Sanford RA, Wagner DD, Cu QW, Chee-Sanford J, Thomas SH et al (2012) Unexpected non-denitrifier nitrous oxide reductase gene. Proc Natl Acad Sci U S A 109:19709–19714. https://doi.org/10.1073/pnas.1211238109
Schardl CL, Leuchtmann A, Spiering MJ (2004) Symbioses of grasses with seed-borne fungal endophytes. Annu Rev Plant Biol 55:315–340

Scholz MB, Lo CC, Chain PSG (2012) Next-generation sequencing and bioinformatic bottlenecks: the current state of metagenomic data analysis. Curr Opin Biotechnol 23:9–15
Scoggins G (2018) Benson Hill and Beck's announce the first photosynthetic efficiency trait. NUU Digest 2018
Semones S (2018) Application of microbial consortia to improve plant and soil health
Semrau JD, DiSpirito AA, Yoon S (2010) Methanotrophs and copper. FEMS Microbiol Rev 34:496–531
Sharpe RM, Offermann S (2014) One decade after the discovery of single-cell C4 species in terrestrial plants: what did we learn about the minimal requirement of C4 photosynthesis? Photosyn Res 119:169–180
Sheehy JE, Elmido AE, Centeno HGS, Pablico PP (2005) Searching for new plants for climate change. J Agric Meteorol 60:463–468
Shindell D, Kuylenstierna JCI, Elisabetta V, Rita VD et al (2012) Simultaneously mitigating near-term climate change and improving human health and food security. Science 335:183–189
Shrawat AK, Carroll RT, DePauw M, Taylor GJ, Good AG (2008) Genetic engineering of improved nitrogen use efficiency in rice by the tissue-specific expression of alanine aminotransferase. Plant Biotechnol J 6:722–732
Singh BK, Bardgett RD, Smith P, Dave SR (2010) Microorganisms and climate change: terrestrial feedbacks and mitigation options. Nat Rev Microbiol 8:779–790
Skillman JB (2008) Quantum yield variation across the three pathways of photosynthesis: not yet out of the dark. J Exp Bot 59:1647–1661
Smith NG, Dukes JS (2013) Plant respiration and photosynthesis in global-scale models: incorporating acclimation to temperature and CO_2. Glob Chang Biol 19:45–63
Spreitzer RJ, Salvucci ME (2002) Rubisco: structure, regulatory interactions, and possibilities for a better enzyme. Annu Rev Plant Biol 53:449–475
Stanger TF, Lauer JG (2008) Corn grain yield response to crop rotation and nitrogen over 35 years. Agron J 100:643–650
Stanley CE, van der Heijden MGA (2017) Microbiome-on-a-Chip: new frontiers in plant-microbiota research. Trends Microbiol 25:610–613
Stewart R (2003) Oceanography in the 21st century. Presentation given at the National Marine Educators Association Annual Meeting, Wilmington in North Carolina in July 2003
Stone M, Hirsh H (2005) Artificial intelligence: the next twenty-five years. AI Mag 26(4). https://doi.org/10.1609/aimag.v26i4.1852
Stone JK, Bacon CW, White JF (2000) An overview of endophytic microbes: endophytism defined. In: Bacon CW, White JF (eds) Microbial endophytes. Marcel Dekker, New York, pp 3–29
Stringer R (2001) How important are the 'non-traditional' economic roles of agriculture in development? Centre for Social and Economic Progress
Subbarao GV, Rondon M, Ito O, Ishikawa T, Rao IM, Nakahara K, Lascano C, Berry WL (2007) Biological nitrification inhibition (BNI) – is it a widespread phenomenon? Plant Soil 294:5–18
Takakura T (1975) Maximizing plant growth using a small computer. Acta Hortic 46(147):156
Takakura T, Ohara G (1976) Direct digital control of plant growth (II). J Agric Meteorol 32:107–115
Takakura T, Kozai T, Tachibana K, Jordan KA (1974) Direct digital control of plant growth (I). Trans ASAE 17:1150–1154
Takakura T, Ohara G, Nakamura Y (1978) Direct digital control of plant growth (III). Acta Hortic 78:257–264
Thauer RK (2012) The Wolfe cycle comes full circle. Proc Natl Acad Sci USA 109:15084–15085. https://doi.org/10.1073/pnas.1213193109
Tilman D, Cassman KG, Matson PA, Naylor R, Polasky S (2002) Agricultural sustainability and intensive production practices. Nature 418:671–677
Tkemaladze GS, Makhashvili KA (2016) Climate changes and photosynthesis. Ann Agrarian Sci 14:119–126
Toju H, Peay KG, Yamamichi M, Narisawa K, Hiruma K, Naito K, Fukuda S, Ushio M, Nakaoka S, Onoda Y, Yoshida K, Schlaeppi K, Bai Y, Sugiura R, Ichihashi Y, Minamisawa K, Kiers ET (2018) Core microbiomes for sustainable agroecosystems. Nat Plants 4:247–257

Trumbore S (2006) Carbon respired by terrestrial ecosystems-recent progress and challenges. Glob Chang Biol 12:141–153

Tyagi S, Singh R, Javeria S (2014) Effect of climate change on plant-microbe interaction: an overview. Eur J Mol Biotechnol 5:149–156

Uemura K, Anwaruzzaman M, Miyachi S, Yokota A (1997) Ribulose-1,5-bisphosphate carboxylase/oxygenase from thermophilic red algae with a strong specificity for CO_2 fixation. Biochem Biophys Res Commun 233:568–571

United Nations Framework Convention on Climate Change: Convention (1994) UNFCC, Article 1 - Paragraph 2

Unno Y, Shinano T (2013) Metagenomic analysis of the rhizosphere soil microbiome with respect to phytic acid utilization. Microbes Environ 28:120–127

Uprety DC, Sen S, Dwivedi N (2010) Rising atmospheric carbon dioxide on grain quality in crop plants. Physiol Mol Biol Plants 16:215–227

Venkata Ramanan M, Smitha (2011) Causes and effects of global warming. Indian J Sci Technol 4:226–229

Villanueva MB, Salenga MLM (2018) Bitter melon crop yield prediction using machine learning algorithm. Intern J Adv Comput Sci Appl 9(3). https://doi.org/10.14569/IJACSA.2018.090301

Vitousek PM, Mooney HA, Lubchenco J, Melillo JM (1997) Human domination of Earth's ecosystems. Science 277:494–499

Vitousek PM, Menge DNL, Reed SC, Cleveland CC (2013) Biological nitrogen fixation: rates, patterns and ecological controls in terrestrial ecosystems. Philos Trans R Soc Lond Ser B Biol Sci 368:1–9

Vorholt JA (2012) Microbial life in the phyllosphere. Nat Rev Microbiol 10:828–840

Wallenstein MD (2017) Managing and manipulating the rhizosphere microbiome for plant health: a systems approach. Rhizosphere 3:230–232

Wang J, Zhang X, Xiong Z, Khalil MA, Zhao X, Xie Y, Xing G (2012) Methane emissions from a rice agroecosystems in China: effects of water regime, straw incorporation and nitrogen fertilizer. Nutr Cycl Agroecosyst 93:103–112

Wang Y, Brautigam A, Weber AP, Zhu XG (2014) Three distinct biochemical subtypes of c4 photosynthesis? A modelling analysis. J Exp Bot 65:3567–3578

Weiman S (2015) Microbes help to drive global carbon cycling and climate change. Microbe Mag 10:233–238

Whitney SM, Houtz RL, Alonso H (2011) Advancing our understanding and capacity to engineer nature's CO_2-sequestering enzyme, rubisco. Plant Physiol 155:27–35

Wilke A, Bischof J, Gerlach W, Glass E, Harrison T, Keegan KP, Paczian T, Trimble WL, Bagchi S, Grama A, Chaterji S, Meyer F (2016) The MG-RAST metagenomics database and portal in 2015. Nucleic Acids Res 44

Wong SC, Cowan IR, Farquhar GD (1985) Leaf conductance in relation to rate of CO_2 assimilation. Influence of nitrogen nutrition, phosphorus nutrition, photon flux density, and ambient partial pressure of CO_2 during ontogeny. Plant Physiol 78:821–825

Yelapure SJ, Kulkarni RV (2012) Literature review on expert system in agriculture. Intern J Comput Sci Inf Technol 3:5086–5089

Zhou J, Xue K, Jianping X, Deng Y, Wu L et al (2011) Microbial mediation of carbon-cycle feedbacks to climate warming. Nat Clim Chang 2:106–110

Zhou Y, Song G, Wang M (2012) Wireless sensor network data fusion algorithm based on neural network in the area of agriculture. Sens Transducers J 16:128–136

Zhu XG, Ort DR, Whitmarsh J, Long SP (2004) The slow reversibility of photosystem II thermal energy dissipation on transfer from high to low light may cause large losses in carbon gain by crop canopies: a theoretical analysis. J Exp Bot 55:1167–1175

Zhu P, Zhuang Q, Ciais P, Welp L, Li W, Xin Q (2017) Elevated atmospheric CO_2 negatively impacts photosynthesis through radiative forcing and physiology-mediated climate feedback. Geophys Res Lett 44:1956–1963

Zimmer C (2010) The microbe factor and its role in our climate future. Yale Environ

Zimmerman L, Labonte B (2015) Climate change and the microbial methane banquet. Clim Alert 27(1). Accessed 15 Dec 2015

Chapter 34
Socio-Economic Assessment of Climate Change Impact on Biodiversity and Ecosystem Services

Kartikeya Shukla, Smriti Shukla, Divya Upadhyay, Vartika Singh, Arti Mishra, and Tanu Jindal

Abstract Biodiversity protection and maintenance are important for the elimination of poverty and sustainable development. India is a nation rich in biodiversity and supports 18% of the world's population over just 2.4% of the total land area in the world. It comprises notable sections of four global biodiversity hotspots with high taxa concentrations and some of the highest wild populations of large, widespread mammals. In India, the maintenance of its great biological resources and enhanced human growth and well-being are facing special, challenging challenges. Climate change brings to this problem an overall level. There are many expectations that climate change would have several detrimental effects on wildlife and harmful consequences on human nature. Biodiversity, however, is important both to climate change mitigation and to adaptation by its supportive ecosystem services. Climate change has a major effect on the availability of numerous earthly resources, especially water that supports the life of the earth. Biosphere transitions, habitats and natural resources have harmful consequences for human well-being and life quality. India is predicted to undergo global warming in the twenty-first century. India will also undergo more seasonal temperature fluctuations with more winter warmth than summers. In recent years, the longevity of heatwaves in India has extended with hotter and colder night temperatures and is forecast to continue. It is estimated that

K. Shukla (✉) · D. Upadhyay · V. Singh
Amity Institute of Environmental Sciences, Amity University Uttar Pradesh, Noida, India
e-mail: kshukla1@amity.edu

S. Shukla
Amity Institute of Environmental Toxicology, Safety & Management, Amity University Uttar Pradesh, Noida, India

A. Mishra
Amity Institute of Microbial Technology, Amity University Uttar Pradesh, Noida, India

T. Jindal
Amity Institute of Environmental Sciences, Amity University Uttar Pradesh, Noida, India

Amity Institute of Environmental Toxicology, Safety & Management, Amity University Uttar Pradesh, Noida, India

D. K. Choudhary et al. (eds.), *Climate Change and the Microbiome*, Soil Biology 63,
https://doi.org/10.1007/978-3-030-76863-8_34

the overall temperature rise will be 2.33–4.78 °C with the concentrations doubling. These heatwaves will contribute to increased variability of the summer monsoon vine, which will have significant effects on Indian agriculture. Weather simulations expect a steady increase in concentration and temperature of carbon dioxide (CO_2) worldwide. However, these simulations do not forecast changes in local weather patterns very well. Local climate conditions, such as rain, temperatures and the sun and the wind, along with the locally adaptable plant diversity, cropping systems and soil quality, can optimize food production so long as plants can be regulated by plant conditions. The most successful way of promoting swift improvements required for human populations to respond to future climate change is by maintaining biological diversity on all levels, from genes to biomes. India is well placed to face this challenge due to its vast diversity of human and biological processes.

An appraisal of socio-economic consequences is designed to illustrate the benefits and drawbacks of a policy to society as a whole and to different parties. An significant aspect of the studies that we perform is therefore a socio-economic impact review. Intrinsically dependent on the environment are biodiversity and biodiversity-based ecosystem resources. Climate change faced substantial environmental threats to biodiversity in the twentieth century, with a rise in effects as climate change progresses and could even intensify.

Keywords Biodiversity · Climate change · Socio-economic · Ecosystem services · Cryosphere · Ecotourism · Agricultural production

34.1 Introduction

The socio-economic consequences are jobs, schooling, self-reliance (energy and food safety), community benefits (diversification of rurals, commitment to environmental protection and green tourism potential), recreational sports, fisheries, yachting and boating, other water operations and health risks.

The mixed economy comprises several sectors which are primarily based on natural resource resources, e.g. primary production, electricity and tourism. Its small size, comparatively remote geographical position, and a high reliance on trade and migration are the main thrust of its economy on the global economy (Royal Society of New Zealand 2016). It is impossible to forecast both climatic and socio-economic circumstances. Simply declining global scenarios would not catch the socio-economic trends correctly (Reisinger et al. 2014). The Indian approach to natural resource management and the weight assigned to environmental factors will have an important effect on civilization and the biodiversity as a result of climatic changes in land and water supplies.

Though human socio-economic behaviour contributes to climate change, it impacts climate change, and socio-economic and climate processes connect significantly. While several researches explored the impacts on labour production and gross national product of climate change, the interactions between socio-economic systems and climate systems were not considered. Thus, this research looks at that and the economic effects of changes in labour productivity triggered by climate change. The two scenarios of carbon reduction—2 °C and the representative pathway of concentration 4.5 were adopted as standard. A computer model of general balance and a basic climate model are used for data analysis. The findings suggest that the effects of the changes in labour productivity caused by climate change on the world economy were not important. The relationship between climate change and labour productivity was explored in the economic model and had a negative impact on economic activities. While these results were stronger in the normal market scenario, this was not the case in the 2 °C example. The findings show that higher climate change levels are associated with larger global socio-economic impacts. It has been observed that there is a major effect in particular on high-temperature areas. Interestingly, the economic losses from climate change have not happened in any country. Any of them have a positive economic influence in the low- to medium-temperature areas owing to the competitive benefit due to improvements in productivity of work in the regions. In the end, the relationship measurement was affected by the combined modelling method. Expanded climate change monitoring, mitigation, and transition can lead to an appreciation of climate change and socio-economic interaction.

Over the past 400 years, the exponential rise in human repercussions on the natural climate has prompted scientists to identify a new age in the geological history of the Planet—the Anthropocene phase (Cruzen 2002). The considerably greater human impact on major biogeochemical processes on land, the seas and the atmosphere (Lewis and Maslin 2015) separates this modern era from its ancestor, The Holocene. For example, nitrogen amounts are considerably higher in the earth and carbon in the atmosphere than at any point in the past million years, if not longer (Lewis and Maslin 2015). One of the distinguishing characteristics of the Anthropocene is its continuing shifts in earth conditions which, due to industrial activity, fossil fuel use and deforestation, are caused primarily by accumulation of carbon and other heat-trapping greenhouse gas at a pace unparalleled (IPCC 2014). Climate change is most generally considered to rise internationally and sea temperatures, shifting trends of precipitation and an increasing occurrence of severe events such as heatwaves, heavy rain and drought (IPCC 2018).

In fact, a change in climate will provide environmental input and human–environment relationships. Increased global sea concentrations poses a damaging threat to coastal habitats and human livelihood and well-being in heavily developed coastal regions, for example, due to the shift of ice reservoirs at large latitudes due to global warming and shifts in regional ocean currents (IPCC 2014). There is increasing anxiety about the expected mass movement of people from the coasts and future problems triggered by such large-scale migration (Smith 2007). The human–environment relationship will also be influenced by climate change by modifying the distribution of ecological resources or benefits obtained by humans from environments such as food, water, fuel and raw materials (Mooney et al. 2009). This is possible by mobilizing shifts in biogeochemical ecosystems and processes to modify the distribution of services from those environments, with the way ecological services respond to climate change differing from one another (Mooney et al. 2009).

Understanding the effects of climate change on habitats, ecological processes and environmental resources has been an important focus research field—at least 2100 papers have been published by the Ecological Society of America on climate change and ecology since.

34.2 Impact of Climate Change Globally

It is important to understand that Earth's atmosphere has varied across several time scales from an individual human lifespan up to billions of years to accurately describe the principles of global warming and climate change. In the case of the Pleistocene glacial period (approximately 2,600,000 to 11,700 years ago), for example, major fluctuations in the global scale of glass and ice sheets were noted for their variable climatic history. These developments have taken place over timescales of decades to hundreds of centuries and have been induced by changes in solar radiation propagation on the surface of Earth. The dispersion of sunlight is known as the model of sunlight, and it is highly determined by Earth's geometry around the Sun and by Earth's direction or tilt relative to the Sun's direct rays. About 21,000 years ago, the new world glacial or ice age created a maximum also known as the Last Glacial. Continental ice sheets stretching south as far as London and New York are today and reach deeply into the centre of Europe and North America this century. The global mean annual temperature, it shows, was roughly 4–5 °C cooler (7–9 °F) in contrast with mid-twentieth century. It is worth nothing that these figures are global average. It should be remembered. In fact, the earth's atmosphere in this last ice age was marked by a higher cooling pattern (to the poles) and a very slight cooling of large parts of the tropical ocean (near Ecuador). This ice age ended 11,700 years ago and the increasingly free Holocene period followed. Traditionally, the Holocene is known as the industrial world's history. However, some scientists argue that in the very recent past the Holocene era has ended and today the World is in the atmosphere that is rightly referred to as the Anthropocene era—a time when mankind has dominated the environment.

Over the twentieth century in reaction to human activity, the chemical composition of the environment shifted dramatically. For example, it is well known that since the advent of the industrial revolution, the concentration of greenhouse gases has risen significantly. Human action has modified the Earth's radiative equilibrium on multiple timescales and spatial scales and has affected global surface temperatures. The deepest and most well-known anthropogenic impact was elevated greenhouse gas emissions in the atmosphere. Humans also affect the atmosphere by alteration in aerosol and ozone emissions and by altering the earth's cover. Global warming is correlated with a more general process of climate change that results in changes in all the features of climate change. Apart from air temperature fluctuations, weather varies regularly, including snow, seas, ocean waves and other Earth's climate measures. In general, climate change is seen as a mixture of different environmental influences in different periods. Since the dawn of human civilization, the impact of climate change has been anthropogenic or just human and was more noticeable in the last two centuries' technology era. The word global warming encompasses all indoor warming caused by anthropogenic forces over the last 200 years. The ambient abundance of reactive ozone gases has also been greatly disrupted, especially in industrially producing areas with significant quantities of fossil fuel and in the tropics where biomass burns, mostly due to heavy pollution of combustion products like carbon monoxide, nitrogen oxides and volatile organic compounds. The dramatic rise in the methane content has led to the elevated ozone levels in the troposphere in response to agricultural activities. Depending on the real temperature rise in the next century, the socio-economic effects of global warming may be important. Modelling predicts that, after the world average late in the twentieth century, a net global warming of 1 to 3 °C (1.8 to 5.4 °F) could lead to economic losses, particularly to high tropics, as well as economic benefits in some regions. Warming up above these stages will tend to reduce rewards and raise prices. Models projected the prices will be above normal value, with global average economic losses of between 1 and 5 per cent of the gross domestic product. With warming above 4 °C (39.2 °F). Substantial disturbances, especially in agriculture, food and forestry, water, electricity and human health, can therefore be anticipated under these circumstances.

34.2.1 Impact on Land and Ocean (Terrestrial and Aquatic)

Land is the primary root of people's lives and well-being, including food, freshwater, other eco-related services and biodiversity. The direct human application of ice-free soil has an effect of over 70% (69–76%) worldwide. In climate cycles, land also plays a significant role. People now use 1/4 of the potential net primary of land meat, fibre, wood and electricity generation. Land is the basis for many other ecosystem functions and resources that are important for humanity, including cultural and regulatory services. The terrestrial ecosystems resources of the planet are measured on an annual basis as roughly corresponding to the annual global approach Brute Commodity Domestic. Land is a source and a drain of greenhouse gas (GHG) and places an important role in the trade between the earth and climate of electricity, water and aerosols. Habitats of land and animals are vulnerable to temperature and climate extremes to different degrees. The impacts on ecosystems and populations of numerous stressors, including climate change, can be reduced by sustainable land management.

The effect on live aquatic resources of waste disposal at sea depends to a considerable degree both intentionally and unintentionally on the location and the severity of the procedure. An oil release into open ocean waters may therefore be assumed to have fewer grave consequences than in Arctic Nearshore (though not necessarily negligible). As noted above, moderate waste levels can also increase the efficiency of habitats if disposed of in a responsible manner. The relative uniformity of the bottom topography in many aquatic ecosystems restricts diversity and productivity. Although dumping of materials that satisfy defined requirements in negotiated areas to shape artificial reefs in cases like certain flat bottom areas which infringe upon the London Dumping Convention (IMO) but may be deemed suitable for coastal growth or protection of vulnerable marine environments from trawling. It is also important to address precisely what dumping can be considered hazardous and, in certain conditions to encourage sustainable growth, when and at what rate, and if current treaties require adjustment.

Carbon sequestration, one of the major supportive resources in forestry, is the recycling of carbon from the environment (carbon sequestration) and the long-term preservation for the carbon in wood, dead organic material and carbon reservoirs. Climate change impedes human lives in the regions of degradation or abandonment. The world forest carbon stocks are estimated to be 55 per cent (471 pg C) of the global rain forests and are more than half biomassed. The forests' role in carbon sequestration is apparent in view of the fact, since 57% of global carbon emission of fossil fuels and the land change are being absorbed by soil or ocean reserves halved in the last four decades by the increase in atmospheric CO_2 levels (Le Quéré et al. 2009) In different carbon cycle components including scale, volume and length of the carbon source in soils and plants, species may influence a long-term equilibrium of carbon gains and losses in ecosystems (Diaz et al. 2009; Murphy and Power 2008; Maestre et al. 2012). Studies in the tree plantation of native and introduced plants also have a major and beneficial effects on the different components of the carbon cycle, such as species wealth productivity (Scherer-Lorenzen et al. 2007), soil breath (Murphy and Power 2008) and plant mortality (Scherer-Lorenzen et al. 2007; Healy et al. 2008).

The probability of climate change depends on both the warming degree and the way population, use, agriculture, technical growth and land management trends evolve. Higher demand of food, feed and drink, increased use and development of energy and technical changes to farm yields contribute to increased threats from the lack of water, drylands, earth decay and insecurity of food. The threats associated with dryland water shortages are smaller in low pathways. Even in global warming, the possibility of dryland water scarcity is moderate, even with 3 °C, low population growth, lower demand growth, and higher adaptive potential, such as the socio-economic route 1 (SSP1). Compared to that, there are greater risks to water scarcity in drylands for high populations, higher uncertainties, higher water tension and reduced potential for adaptation such as SSP3. The transition from low to high risks from 1.2 °C to 1.5 °C is protected in the SSP3. Climate-related disruption to lands on roads with increased populations, rapid land use shifts and low ecological capability is the most critical factor in adaptation (e.g. SSP3). This adds to more communities becoming vulnerable to environmental degradation, arson and tidal flooding (medium confidence). For global warming SSP1 at a temperature of 1.8–2.8 °C and for terrestrial degradation SSP3 at 1.4–2 °C occurs the expected transition from low to high risk. The scheduled transition probability from 2.2 °C to 2.8 °C for SSP3 is medium to very high.

The understanding of tolerance limitations and possible maladaptation to combined impacts of climate change and desertification is currently missing. Residual vulnerability and maladaptive outcomes are highly capable in the absence of new or improved adaptation solutions. Social-economic and structural limitations may also raise challenges to their adoption if alternatives are available. Due to their environmental impact such as the drainage of soils or mining, a number of enhancement solutions may become fatal leading to soil water depletion. The complete loss of agricultural productivity can result in severe forms of desertification, restrict adaptation options or exceed the limits of adaptation.

Preparatory reports show that, by the period of 2025–2050, "business as normal" carbon dioxide (a 'greenhouse') in the world is expected to duplicate, triggering a likely rise of the mean global temperature by 1.5–4.5 °C (IPCC, op. cit.). The average sea level of approximately 20 cm by 2030 and 65 cm by 2100 will rise, and temperatures of the sea-surface range from 0.2 °C to 2.5 °C can increase. Bakun (1992) predicted some potential impacts on marine and coastal ecosystems, referring especially to the likely rise by change in coastal air-pressure/wind regime in temperature differences between land and sea. The above-noted global climate changes are not predicted, for example, on terrestrial, lacustrine or river habitats, to have a detrimental impact on fish production, but some populations may suffer and changes in rainfall and river drainage will affect life in semi-close seas and coastal areas in particular. Also affected would be coastal aquaculture. Tropical areas which generate large quantities of fish resources will migrate towards the pole and exacerbate them. However, increased phytoplankton production may decrease the levels of oxygen and create anoxic circumstances locally. The annual variability of services provided by the plant can increase. As in the past, coral reefs will respond to changes in the sea level and mirror them, but only when these changes occur gradually.

34.2.2 Impacts on Cryosphere

Cryosphere is a part of the earth's natural environment interconnecting and communicating with human and biological processes (Qin and Ding 2010; Xiao et al. 2016). Cryosphere is a part of the earth's natural environment interconnecting and communicating with human and biological processes (Derksen and Brown 2012). Mountain cryosphere shifts therefore have major impacts on land habitats, hydrology and human processes (Bales et al. 2006; Derksen and Brown 2012). Therefore, it is important to research the complicated ties between the natural world and individuals, their systems, direct and direct input and consequences on multitude scale, to understanding the impacts of cryosphere transition in human society (Warner et al. 2010; Derksen and Brown 2012; Huggel et al. 2015). Changes in size or some features of the physical local and regional ecosystem and habitat and resources provided by individuals can be affected by the cryosphere (Warner et al. 2010; Fountain et al. 2012; Berman and Schmidt 2019). These developments will influence human society directly by altering hydrology, streamflow, landfill, air temperature, disasters, living standards, transport and infrastructures and indirectly by influencing changing ecosystems (Fountain et al. 2012). There are no consequences of shifts in the cryosphere.

34.2.3 Impact on Fisheries

In less than a century, the global population would double to about 10 billion. UN estimates indicate that 75 per cent would be on a small stretch about 60 km wide along the coasts of continents with the fastest growth rate. About 65% of all big towns and cities in Southeast Asia are now on the shore (UN op. cit.). These estimates highlight the important and drastic effect of human activity on one of the world's most active ecosystems: shallow coastal and brackish waters. The utilization of marine land and ecosystems depends on the population size and socio-economic growth status of the population. The counter crossing of population growth and its accelerated accumulation in coastal areas and coastal urban centres (the so-called littoralization of the populations), around the same time, the unregulated industrialization and discharge of marine by-products are at odds with leisure and resource use and involve fisheries and aquaculture (Pullin 1991). These human effects have an unpredictable effect on the aquatic environment. Bad industrial technologies combined with the effects of high densities of migrant and permanent human populations and industrial agriculture together have contributed to considerably more detrimental effects worldwide on natural coastal environments (UNEP 1990a, b, 1994; Williamson 1992). Regulation of coastal growth and habitat preservation would entail better planning processes, and painful social and political decisions would always arise. As comprehensive Coastal Area Management (ICAM: Clark 1992), the context within which these decisions can be taken is generally referred to. As noted, the Marine Catchment Basin (MCB) (Caddy 1993a) definition, which acknowledges that human growth of inland watersheds can also have a major effect on coastal waters, is seen as an alternate framework for analysis of land-influenced activities in coastal seas.

This management shall consider all uses of the coastal region and human activities, particularly in the river basins (watersheds) which are released to the coast. The unregulated use of waterways and estuary products for the discharge of contaminants and harmful chemicals contributes to major adverse effects on aquatic essential environments, but water diversion and river management in general often have adverse effects on river fish (Welcomme 1994) and diadromous animals such as salmon, eels and shad. Another environmental cause related to human activity in the coastal region and fell on the interface from ecology and economics is the impact of inadequate land use within the waterfront on the siltation of coastal and estuarine environments (Tisdell 1982). Loading of sediment from rivers that modified the seasonal cycle and water supply to coastal waters and the siltation levels of land erosion caused by vegetation covers destruction (e.g. deforestation) (Milliman 1981; Milliman et al. 1987; GESAMP 1994). Wide levels of sediment on waterways such as salmon and stucco are adversely affected by anadromatic species. It is also seriously damageable to fish habitat by impacts on coastal coral reefs (Grigg and Dollar 1988) and water feed.

Exchanges among water interfaces are clearly particularly important for living marine resources, as well as for the combination of sustainable development objectives with nearshore and estuarine ecosystems linked aquatic and terrestrial ecosystems (Martin et al. 1982; Prospero 1981; Liss and Slinn 1983; GESAMP 1986; Martín and Gordeev 1986). As nurseries of shrimp and fish (Boesch and Turner 1984), coastal and shallow nearshore waters play a vital role and function on the one hand in moderating the effect of terrestrial activity on the enclosed aquatic environment (GESAMP 1994), activity generally manifested as nitrogen bursts, suspending sediments and radioactive content entering the coast and the inland seas. Around the same time, the influence of the aquatic climate on earth structures declines with the wetlands and nearshore waters; this effect is manifested by processing such as turbulence, earthquake and consequent coastal erosion (Murty 1977, 1984; Pugh 1987). The world's wide depletion of such wetlands is a disturbing one. New technology and methods, especially remote sensing and spatial information, make cross-border research more workable than ever (FAO 1986a) and warrant widespread development, because they help provide a comprehensive geographic foundation for complex decision-making. Estuary plays a major role in the life cycle of many global economic fishes by the availability of breeding, breastfeeding and feeding grounds currently providing nearly 95% of the world's aquatic supply. The rate of devastation of this critical ecosystem, which is already having disastrous effects on many areas, is under pressure by indiscriminate slurry, restaurants, floods, coasts and mine accumulation and accompanying sediment build-up (ICES 1992; Campbell 1993).

Many populations that are dependent on fishing as a livelihood can be impacted by glacier melting changes (Lehodey et al. 2006; Brander 2007). For instance, fish stocks in Peru's Yanamarey watershed decreased or fully lost as fish habitat in the top watershed was reduced seasonally as a result of the glacier recession. That influenced the economy of the state. Increased ice melts and glacier outbursts in Alaska's Kenai River affected USD 70 million of fisheries a year (Milner et al. 2017). Fish production has become vulnerable in Himalayan rivers too (Allison et al. 2009) threatening income and food protein. Similarly, the warmth of the rivers and lakes has contributed to the creation of introduced species in ecoregions of the Arctic and the Continental Great Lakes, which have expanded trade but have disrupted indigenous food security that rely on subsistence fishing for thousands of years (Warren and Lemmen, 2014). However, the production of Himalayan fish has become sensitive in rivers, threatening income and food protein production (Allison et al. 2009).

The unregulated use of large trees for canoe construction was a direct and unsuccessful product of uncontrolled artisanal fishing, but a very limited explanation for the rapid clearing of coastal forests, in the developing world. Timber demand for houses and boats, the clearance of the cosmic forests and often the need for fuel and trade are among the factors which have led to the deforestation of the coastal belt.

This has been catastrophic in nearshore coral reefs and fish and shellfish nursing areas (UNESCO 1986; UNEP/IUCN 1988), as well as causing the growth of silt or water (Pastorok and Bilyard 1985) and can also allow dangerous species to burst out of it. These kinds of interaction also do not influence coastal populations, although they may be well informed of the solution, which appears to demand that the decision-making resources be clearly passed to the municipal councils. Such practices obviously should be performed in compliance with the general government recommendations set out in the national multisectoral coastal area development strategy (Knox and Miyabara 1984). Coastal fisheries by large ships, particularly trawlers, compete with craft fishing, but, if unchecked, have an adverse impact on the marine ecosystem on the coast, which can lead artisanal fishermen to diminishing shallow water areas, as well as to fish for juvenile fish in coastal kindergarten areas (Caddy 1993b). Under these conditions, some coastal fishermen may respond by using dangerous methods like chlorine, pesticide and dynamites or by destroying the production of coral rock. The strict controls for both of these operations affect essential fish ecosystems (Hoss and Tbayer 1994).

Market demand growth without vessel size controls and technological developments in technology, boats, fish and maritime monitoring equipment, as well as rapid population development, have all helped to increase fishing pressure and to disrupt many conventional fishery management schemes, which have not always been replaced by new management mechanisms. Appropriate land and energy allocation (FAO 1986b) with direct citizen involvement and sufficient resource security from external exploitation are a prerequisite to the fair use of resources and thus to sustainable development. The impact of coastal fishing by big trawlers without local population regulation or equally unchecked results of domestic or industrial contamination can also be seen in this. In maintaining and improving coastal fisheries and aquaculture, the local authorities may themselves play a major role (Barg and Wijkstrom 1994).

Certain variations are now and in future due to global climate change in circumstances under which aquatic habitats live (Francis 1990). All habitats have been exposed to certain modifications to the natural variations. In the case of 'wild' fish, the main effect of environmental change on the class of year intensity has been recognized as a major cause of large shifts in the strength of the hired year groups (Fogarty et al. 1991). In the fields of fisheries, the influence is at best not easy to predict. The Intergovernmental Panel on Climate Change (IPCC) UNEP-WMO has brought together the international research community on the extent of expected climate change global (IPCC 1992), and the ICSU also initiated an International Geosphere Biosphere Programme (IGBP) to determine the speed of all major terrestrial changes (WGBP) measurable (International Science Council). As for the oceans, a Global Ocean Observation Scheme (GOOS) (Kullenberg et al. 1993) has been developed by the IOC.

34.2.4 Impact on Tourism and Ecotourism

Tourism is an important source of income for mountain areas and the ocean, which encourages an outdoor leisure service of a billion-dollar size (Burakowski and Magnusson 2012) but is impacted adversely by the declining number of glaciers in snowfalls, rising sea level and other natural disasters such as hurricane, floods and so on, which have influenced both the state and the national economies. Citizens 'and visitors' entertainment is a significant economic practice in both wealthy and poor countries. In the coast, people who can afford to fly can spend their money everywhere safe beaches, sunshine and all the associated local facilities (hotels, amusement centres, sanitary appliances, banks, post offices, stores, etc.). Some areas such as the Mediterranean coast are many (3–5) occasions during the holiday season for the resident population. For the full number of people, this involves substantial expenditure in sanitation, lodging (hotels) and other services, since such facilities are under-used for perhaps half a year. If authorities do not have installations or facilities to cope with increasing waste in particular, the coastal seas are dependent on the care that is not feasible. Although the ability of the sea to do so may be significant, the rate at which the local sea is altered (flushed) by local currents may be highly dependent. Untreated waste disposal often contributes to increased debris in the sea and in the seabed, and the prevalence of potentially risky microorganisms is greater than normal.

In fact, the notion of dedicating some parts of the coast to tourism activities and others to fishing and/or mariculture (i.e. the implementation of the ICAM concept), although there is de facto geological distinction between fishing and human activity, (usually to the detriment of fishermen) in the course of growth, does not seem to be generally integrated in the public policies of most countries.

Contaminants have also been greatly concerned about their role in impacting aquatic environments and their goods. Thus, the following is stated in GESAMIP (1990):

> The main factors of urgent concern ... are coastal growth and the consequent pollution of maritime materials, plastic debris fouling of the oceans, persistent build-up of chlorinated hydrocarbons, in particular in trope's and subtropics and deposition of tar on beaches.

In this central region of the aquatic climate, the main priority is the protection of present alternatives and, in many cases, the restoration of damaged coastal systems that are important for the early life cycles of many living, more offshore marine species than adults. Fish development of estuarine is also linked to estuarine characteristics, including water run-off patterns (e.g. Houde and Rutherford 1993). The impacts on the coastal west and ecosystems of agricultural pesticides are now more generally recognized, since their effect on coastal low-salinity nurseries (e.g. for shrimp larvae), which can be dimmed by the flash-off from flood water contaminated by pesticides after tropical storms.

Ecotourism and ecorecreation stress the persistence of intact and untapped habitats, but these practices cannot be maintained, as the Australian Great Barrier Reef Marine Park Authority (GBRMPA 1992/1993) was concerned for example. Marine habitat management is increasingly related to protection of the complexity and sustainability of environments, and certain wetlands have a natural role to play in reducing emissions and in mitigating the natural effect of coastal processes on climate. As the public becomes more conscious of the consequences on sea life (e.g. IUCN, UNEP and the WWF 1980), the intangible advantages of restoration are becoming more apparent. An upgraded estimation of nonexploitable uses is planned.

34.2.5 Shifts in Species Distribution

After large conclusions, various species have changed their geographical range towards higher elevations in response to dramatic temperature and precipitation shifts (Chen et al. 2011; Doney et al. 2012). Groffman et al. observed a trend of 0.011 km per decade to higher elevations in terrestrial settings and a rate of 16.9 km/10 at greater latitudes. There have been drastic changes in the geographical distributions of east African animals and habitats owing to climate change. In order for biodiversity to respond to changing environment, current species migration rates would have to be much higher than those during the postglacial period (Malcolm et al. 2002; Hély et al. 2006; Doak and Morris 2010). African species diversity is greatly altered by climate change. Ethiopia's endangered wolf (*Canis simensis*), for example, is trying to adapt to the longer dry times and the decline in water and other resources available. Ecological specialists are especially vulnerable in their comparatively limited dietary and niche widths and respond strongly to changing resources (Altermatt 2010; Lawton et al. 2010; Montoya and Dave 2010; Clavel et al. 2011) and relatively stable climate relationship (Walther et al. 2002).

34.2.6 Disease Regulation

In view of the possible high cost of both new zoonotic viruses and improvements to the spread of current disease vectors, the impacts of changes in human, animal and plant health exacerbated by climate change are a cause for concern. Multiple host species in natural environments retain many pathogens (Woolhouse and Gowtage-Sequeria 2005), where extinction of a host species or biodiversity as a result of climate change could impact the risk of disease and the frequency and severity of potential outbreaks of disease and probably the spread of diseases in Africa are likely to increase. For example, the infection rate of bovine tuberculosis (bTB) was positively related to the loss of mammalian equality (J') organisms. Loss of no qualified or spillage species as a consequence of climate change may increase the migration of cattle herds and the rates of cattle encounter. A "growth of encounter" (Keesing et al. 2010) could lead to an improvement in the probability of bTB threats.

34.2.7 Agricultural Production and Food Security

The average current food intake in India is 550 g per person a day, compared with 980 g in China and 2850 g in the USA. The world is faced with significant challenges in increasing its food supply to 300 m by 2020 to feed its rising population, estimated to hit 1.30 billion by 2020. Farmers in the world must produce 50 per cent more grain by 2020 in order to satisfy demand for food from this rising population. The gross cumulative irrigated area has more than quadrupled since 1950–1951 and 99.1 million ha in 2011–2012, from 22.6 million ha. Though agriculture accounts for 14% of the GDP in India, 64% of the population is subsistence dependent on agriculture. The demand for water has risen over the years because of urbanization, population growth, rapid industrialization and other development efforts. In addition, the hydrologic cycle in many India's climactic regions and river basins has been altered by changes in crops and land use practises, overuse of groundwater and changes in irrigation and drainage. The most significant factor in farm development is the supply of water.

Data since 1961 indicate that global population growth as well as per capita increase in the consumption of food, livestock, fibre, wood and oil have produced unequalled land and freshwater use rates, and agriculture has now compensated for about that. 70 per cent worldwide use of freshwater. The growth in agriculture and forestry, including trade production, has allowed the rising population to feed and have access to food and increased productivity in agriculture and forestry. Those technologies lead to increased net GHG emissions and to a decline in natural ecosystems (such as forests, savannas, natural grasslands and wetlands) and biodiversity because of large regional variations. Approximately one-fourth of the global ice-free region is inductive for humans. Degradation—Agricultural soil depletion currently expected by the soil production rate is 10–20 times higher (no tillage) than 100 times higher (conventional laying). Climate changes intensify soil erosion, particularly within lowlands and coastal areas, river deltas, permafrost and drylands. In the period from 1961–2013, the drylands rose each year, averaging slightly more than 1 per cent each year, with significant interannual variabilities. From the 1980s to the early 21st centuary, about 500 (380–620) million inhabitants lived in decertified areas in 2015. The greatest number of people impacted is South and Eastern Asia and the Sahara Circle, like North Africa, and the Middle East, with the totality of the Arab Peninsula. There has also been desertification in other dryland areas. The air surface on the field has grown since the pre-industrial era, almost twice the world average temperature. In several countries, climate change, including changes in extreme occurrence and severity, adversely affected food protection and terrestrial habitats, as well as desertification and land degradation. Heat events, like warm waves14, in most land regions, have increased frequency, strength and length. In certain regions (including Mediterranean, Western Asia, several areas of South America, most of Africa and Northeast Asia), the extent and severity of the drought have increased and intensity of heavy rainfall on a global scale has increased. Roughly 13% of CO_2, 44% of methane (CH_4), and 82% of the worldwide emissions of nitrous oxide (N_2O) from human activity in 2007–2016 are 23% (12.0 ± 3.0 Gt CO_2/year) of the GHG's overall net anthropogenic emissions. A net drain of around 11.2 Gt CO_2/year in 2007–2016 (equivalent to 29% of the overall CO_2 emissions) has resulted in a natural reaction to human-induced environmental change; the continuity of the sink is unclear due to climate change. If the global food system contains emissions from pre- and post-production activities, it is projected that 21–37% of the global net anthropogenic GHG emissions are estimated.

34.2.8 Impact on Land Use Planning

The most influential socio-economic force is maybe land use transition impulsive improvements and habitat destruction. Urban deforestation the Earth's ecosystem has been dramatically altered by growth, farming and other human activities. Such land disturbance impacts critical ecological processes and resources that can have large and long-term effects. Land usage and policies of land management have a direct effect on natural resources, including water, vegetation, climate, minerals, plants and livestock. Agricultural run-offs are a significant cause of water pollutants of both marine and inland waters. Draining water for the cultivation of crops and drainage of water have had an detrimental effect for many species of wildlife. Irrigated irrigation has altered the water cycle which has contributed to decreases in groundwater in many areas of the world. Continuously planted and deforested soil can erode, salinize, desertify and other soil depletion. The greenhouse impact due to deforestation is increased, ecosystems which foster biodiverse areas have been lost, the hydraulic cycle affects and soil erosion, floods and landslides is strengthened. Urban construction is blamed for air waste, pollution of water and urban rushing and floods. Habitat degradation, separation and urban growth alterations are a significant cause of the reduction in biodiversity and loss of animals. Urban development and industrial agriculture in coastal and inland areas are an important challenge to aquatic ecosystem conservation, productivity and biodiversity around the world.

Land use and land cover change (LULCC) affects directly on the climate and force the human populations to change socio-economically. LULCC is the main source, both local and downstream, of soil erosion and improvements in the functioning and protection of the habitats. LULCC thus primarily decides (or mediates) whether or not the ecosystems and community are vulnerable to and resilient to external threats, such as climate change, national and international policies and other globalization aspects. There is typically a shortage of integrated LULCC study through HKH. The Tibetan Plateau in China, which is particularly based on the vast grasslands, has been the topic of most published case trials. As most available case studies examine temporally and spatially variable satellite data, regional LULCC trends are difficult to clearly discern (Harris 2010).

The transition of farmland and forestry to urban development limits the amount of land for the cultivation of grain and timber. Soil loss, salinization, desertification and other crop-producing loss of soil and deforestation reduce land quality and productivity. Conversions of farms and forestry into urban growth limit local communities' green space and conservation facilities. The 'essential mass' of the land required to sustain local agrarian systems is limited to urban development. Urban trends influence not just people's lives but also the organization of society. Any rural areas have been so impacted by urban development that the identifier of their community has been destroyed. Suburbanization expands the division of income and economic inequalities between neighbourhoods. However, over-controlled land use can impede the operation of the sea and. Regulations aimed at minimizing land usage would increase house prices such that families with a smaller than average revenue will be less available. Regulations on land use can strike a balance between privacy and public interest.

34.2.9 Unsustainable Grazing

Overgrazing of plant and soil capital by domestic livestock is one of the major factors. Most figures suggest that most of the wildlife farms in the HKH are regularly overgrazed (Dong et al. 2007; Harris 2010). Overgrazing does not only change the structure of plants, but there are also other effects, including soil erosion and depletion, soil-nutrient decreases and carbon deposited, and modified flux regimes (Aryal et al. 2015; Sharma 2014). Livestock grazing in HKH rangeland areas may not be enduring (at times loss to land or plant resources is irreversible) only as a result of total cattle numbers, but also because of seasonal patterns of grazing, short-term and long-term pasture intensity and livestock composition (Li et al. 2014).

34.2.10 Mining

Uncontrolled mining may often negatively impact vulnerable mountain ecosystems, contributing often to large-scale landscape destruction and economic growth crises. Several researchers in the field have researched and recorded the mining impact on the socio-economic and natural environment (Pakistan and Sarfraz 2009). Uncontrolled mining methods such as dynamite bombing have resulted in plant destruction, soil degradation and wildlife damage in Pakistan (Donnelly et al. 2004; Wu et al. 2007). The composition and structure in Meghalaya's Nokrek Biosphere Reserve, India, have been adversely affected by coal mining. Large-scale mining, like Gyama Valley close to Lhasa, China, will affect downstream water quality. The main cause or contribution to excessive river-bed mining in the Himalayas for gravel and sand is also unplanned or unchecked disruption to river habitats in HKH.

Boiling and gravel and sand dragging were part of coastal mining. The collapse of coral reefs and the depletion of manganese nodules from the sea floor may also be included (Alcala et al. 1987). It is not so known that minerals such as tin, gold, diamonds, ilmenite, rutile, zircon and monazites are dressed in the seabed. Deep-sea manganese nodules, also made from many other metals including cobalt, nickel and iron, must not be commercially mined unless found in cheaper manganese soil. Such mining is commonly used, and, in fact, it can have significant consequences on fisheries if any industrial treatment takes place at sea (discharge of dust from waste rock into the sea, for instance, to produce singularly greater turbidity, or elevated concentrations of elements that are usually uncommon at sea level but become harmful at higher concentrations). As mentioned earlier, the bulk of marine mining is more costly than land mines.

Oil and gas exploration is currently an important maritime operation in some countries, both for energy and chemical transformation. The fishing can only be impacted if these platforms are widely spaced to allow fishing to close a substantial coral bed area; however, these 'platform parks' will provide the fishing 'refuge' and the potential for sport fishing (Dugas et al. 1979; Reggio 1987) stock. Platform accidents which lead to significant oil discharge in the sea have more negative impacts on other human activities such as tourism than fishing in those places where toxic waste disposal can cause more severe adverse long-term effects. In certain places, gravel, sand and fossil reef beds are primarily harvested to provide materials for civil construction. Coastal gravel and sand dredging are also a significant marine operation (ICES 1992; Campbell 1993) in countries in which construction materials are in scarcity or at a high price. This exploitation has an detrimental impact on benthic species by degradation of reefs and destroying demersal and other fish breeding areas and also affects trawling and other methods of bottom fishing, but the length of the consequences can be comparatively limited in most smooth bottom ecosystems after this form of mining has been terminated. It is important to point out that coral reefs have three main environmental roles that are commonly significantly affected by mining: enclosing a marine lagoon (i.e. making a natural fish pond); protecting the coastal waves and stormy rises (hence a fish habitat); and maintaining a zone of attraction for fish species of interest.

34.2.11 Supporting Services

The ecosystem services facilitate the basic functions of the ecosystems, viz. primary productivity, cycling of nutrients and genetic diversity maintenance. The rise in temperatures decomposes soil organic matter, which can increase soil carbon lost and change C:N balance. This results in indirect improvements in the composition of the microbial soil. Climate change practises, for example, contribute to increased depositions of nitrogen and to major effects, including by eutrophication, on terrestrial and aquatical habitats. The combination of elevated nutrient loads and higher temperatures increases the rate, length and magnitude of cyanobacteria responsible for the dangerous blooms of algal products that can adversely affect human and animal health. Estimates indicate that conditions may enable toxic algae flora of Alexandrium Catenella to be up to 2 months early and to exist for up to an additional month by the end of a century, which may have an effect on aquaculation, leisure and other practises.

34.2.12 Cultural Services

This is the non-material value of habitats and habitats (viz. cultural identity, recreation and mental and physical health). While they are very important for human health, cultural services in relation to other ecosystem services have been understudied. The first voices linking society and environmental change with technology, science and other expressions were local residents.

There is evidence that exposure to natural environments improves human health. Conversely, severe climate-driven consequences such as higher temperatures and storms can reduce mental and physical health. Indirect economic costs (e.g. loss of livelihoods) may also have detrimental social and psychological effects.

34.3 Ecosystem-Based Services

In 1981, the word 'ecosystem services' was coined by most scholars. Sustainable Development was taken back in the 1980s but came back strongly with the introduction of ES into technical literature and an increased emphasis on their economic importance in the 1990s. In the process, with the focus on the ecological basis as the ES, the concept has arisen as the climate and processes in which natural and organic ecosystems preserve and continue to occur in human lives or to a degree of economic value, where ES reflects the benefit that people benefit from the ecosystem's direct and indirect functions. The TEEB Study (2008–2010) identified ES as a balance in terms of the ecosystems' direct and indirect contribution to human health. Despite these differences, the connection between (normal) and human health is stressed in both meanings, as ecosystems and services form a bridge between humans and natural environments, with people basically isolated.

Originally, ecologists used the idea of ecosystem function to define a variety of ecosystem processes in an ecological environment. Some scholars began using the term "functions of nature" in the late 1960s and early 1970s to describe the 'work' of ecological systems, spaces and materials supplied for human communities. The need to separate 'functions' from basic ecological systems and processes has been emphasized in explaining the movement of ES from nature to community and emphasizes that ecosystem functions form the cornerstone of a service. Services are labels for people which give direct and indirect advantages for the 'useful things' ecosystems 'do.' Ecological concern was economically portrayed in the 1970s and 1980s to highlight social dependency on natural ecosystems and increase public interest in the conservation of biodiversity. The definition of 'natural resources' was already used in the 70s and some scholars began to refer to ecosystem (or biological,

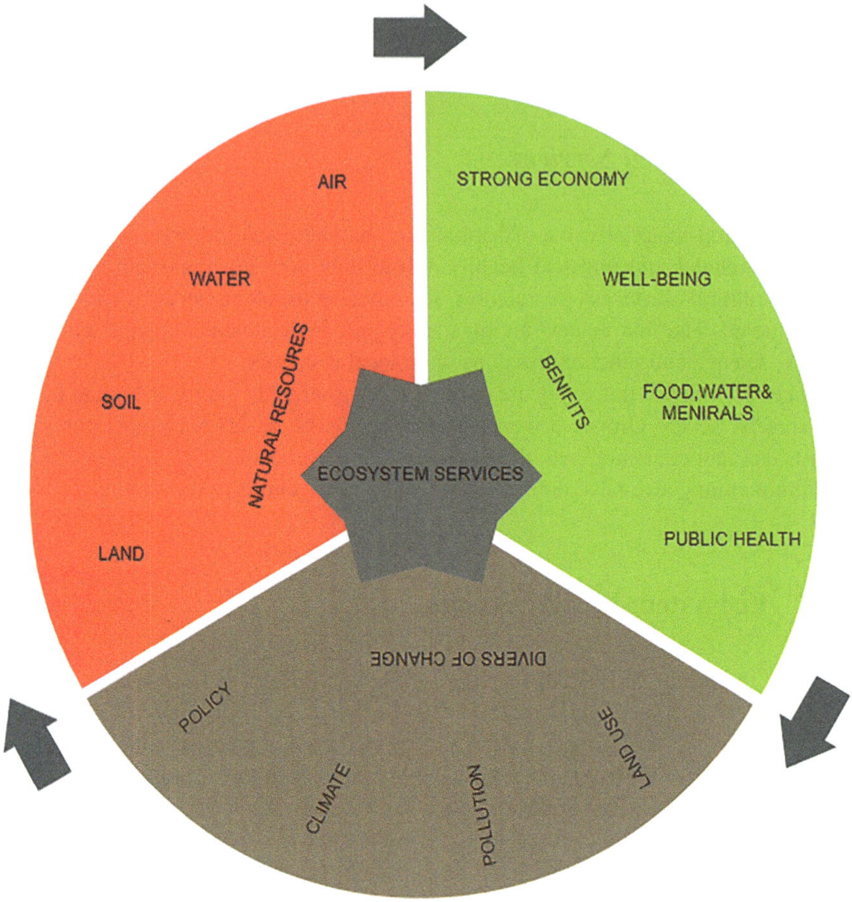

Fig. 34.1 Ecosystem services (environmental benefits based on ecosystem services)

environmental or financial) services shortly afterwards. The purpose behind the definition of ecosystem services was to show how biodiversity disappearance has a significant effect on the ecosystem's functions that underlie vital human well-being services. A landmark in ES 'mainstreaming was the 1997 estimate of the cumulative worth of world natural capital and ES. Another landmark which positions the ES definition firmly on the political agenda is the Millennium Ecosystem Assessment.

As in the above Fig. 34.1, it shows the contributions of the structure of ecosystems and the role (with additional input) to human well-being are ecosystem services (ES). This implies that civilization depends heavily on the well-functioning of natural resources and ecosystems which underlie the continued flow by ES from nature to society. Therefore, ES can become a central instrument for global, national, regional and local policy-making and decision-making. There is a range of possible applications ranging from sustainable natural resources management, land use enhancement, environmental preservation, biodiversity restoration and restructuring, landscape design, natural solutions, climate protection, environmental education and analysis to minimizing the risk of disasters. Ecosystem services (ES) help produce revenue and well-being and deter losses that impact society. The above features such ES, which act for insurance, enforcement and resilience. The awareness and understanding of processes to collect the values and expense of the services rendered (potential losses) are growing. In decision-making, consideration must be taken for all forms of gains and expenses. The range of living organisms in the idea of ecosystem resources has represented the social and economic importance of biodiversity. Safe and biodiversity-rich habitats provide important services including crop pollination, water purification, fertile soils and flood control. The environmental services policy is in an early stage of implementation, while the current policy on biodiversity is coordinated by the various countries.

34.4 Socio-economic Assessment of Biodiversity and Ecosystem Services

Currently, socio-economic and human population size and distribution are the key influences that influence biodiversity and ecological resources. Also important are consumer values, such as increased trade in desirable exotic products which harbour and may be pests. Habitat depletion, deforestation and erosion due to climate, air pollution and improvements in land use have been the main threats to date. Eco and evolutive cycles are active in interactions between the biotic and abiotic elements of ecosystems and gradually contribute to supplies and flows that sustain the ultimate ecosystem services. These are supplied by the regularly controlled habitats for this reason. However, mankind uses 'good(s)' which originate from, rely on and nearly ever require additional input from the end ecosystem services. The primary processing (a process in ecosystems) of the wheat (a final service) is needed, for example, and before it can be eaten, the products that would be flours need even more input (crop, harvest, transport and preparation). Similarly, an ecosystem may grow trees, which are used in a variety of products, including timber and fuelwood, as well as preserving carbon or leisure materials, as a final ecosystem service. The importance of these commodities to people depends on background as well as inputs added. The identification of ecosystem final resources as ecosystem features that lead directly to products and values allows for geographical analysis and the mapping of economic and beneficial alternate ecosystem management regimes.

Provisioning Services Products derived from habitats include genetic material, food and fibre and freshwater, for example, are provided with facilities. Changes in the supply of services, products extracted from habitats, and habitats will affect the economies and well-being of people as a result of climate-induced change. For example, forest watercourse climate impacts like hot weather, rainfall changes and snowfall, as well as threats including wilderness, alter the availability of freshwater to cities, farming and power output for municipalities. Surface water scarcity is predicted in some areas in dry years. Rising flux temperatures also impact the consistency of the water. Likewise, wildfires in rivers, lakes and reservoirs will raise sediment accumulation and waste. These reforms are likely to stress the availability of water and eventually raise water treatment costs.

34.4.1 Valued Socio-Economic Components and Associated Issues

Economic benefit is accessible for both biophysical products and services. When environmental scientists and other scientists coordinate with economists, they can:

1. define biophysically the production of ecosystem products and services.
2. transform biophysical production into economic value projections.

The first major international initiative to investigate the connections between ecological services and human well-being was the Millennial Ecosystem Assessment (MA). The MA structure was developed to recognize the current status of the key ecosystem resources as well as development patterns and flows and significant stresses and challenges. The definition has been broadly embraced in the science and political communities, and new ways to value resources have since been established in order to properly include the concept in Research, Conservation and Creation. Certain water and land management policies and activities that seek to enhance ecological resources and human well-being are focused on untested theories and scarce knowledge (Yang et al. 2020).

Implications for Human Well-Being People rely on harvestable resource habitats as they play their role in management of climate and nutrient cycles, safeguarding against hazards (e.g. conservation of or pollutant transformation, climate modulation, flood reduction) (e.g. sense of place, appreciation of nature, tribal heritage). The Millennium Biodiversity Appraisal suggests a loss or over-exploitation of 60 per cent of ecosystem resources in the global economy. There is also scope for concern about the consequences on human well-being of climate change. The effect of climate change on ecological development and operation by changes in these ecosystem resources is impacting human well-being, livelihoods and overall survival as shown in the Fig. 34.2 that how ecosystem services valuing phases by involves integration of biophysical, economic and social knowledge.

Table 34.1 Valued socio-economic components and its associated issues

Valued socio-economic component	Issues
Health and well-being	• Personal welfare and environmental health • Continuity of the population and ethnic group • Continuity of the family • Preservation of tradition
Sustainable wildlife harvesting, land access and use	• Conventional economies of chassis, catching and collecting. • Proximity to property—recreation and conventional economy • Alternate land use benefit (e.g. tourism vs. hunting vs. industry)
Security of cultural and heritage	• The aesthetic, cultural and/or archaeological importance of the site • Preserving indigenous languages, schooling, regulations and customs
Equal market and prospects for jobs	• Competition of state, provincial and territorial enterprises • State, provincial and territorial residents' job prospects • State, federal, territorial residents' training and job growth • Boom and bust periods avoidance (e.g. via economic diversification)
Sustainability of the population	• Impact of entrance and departure • Change in the socio-economic and cultural composition of the populations affected
Suitable technology and utilities	• Strains on housing, jobs and legal services. • Accessibility, affordability and adequacy of housing stresses • Transport and road safety—physical engineering stresses
Adequate sustainable income and lifestyle	The whole community's capital • Public use of resources—results of improved availability revenue • Regional and local life expenses • Cost/benefit allocation among equity impacted people. • Improvements in adverse lifestyle—gambling, abuse, drug use

Fig. 34.2 Valuing ecosystem services phases (ecosystem valuation involves integration of bio-physical, economic and social knowledge)

34.5 Different Policies for Conservation of Nature and Natural Resources

In the current scenario in which economy prevails over biodiversity, the protection of nature and natural resources was a big challenge. International treaties and agreements are signed and regulations, rules and laws are enforced around the world, yet there remain hostilities by applying protected area policies that significantly compromise the interests of indigenous peoples. It affects your emotions and therefore affects your actions, which has a negative effect on your goal. The principal goals are as follows: to maintain environmental integrity and to restore ecological equilibrium, to spill and to protect water. Natural and genetic capital conservation as shown in Fig.34.3, illustrates substantially growing canopy of forest/tree (33% of land mass and 66% in hills).

(i) Goals of tourism policy; The policy statement on the tourism sector (May 1985)
(ii) Goals of forest, soil and water conservation policy
(iii) Purpose of national park and protected areas system policy
(iv) General policy and principal goals
(v) Specific national park and protected areas system objectives
(vi) Categories of protected areas.
(vii) Forest reserves.
(viii) Managing the national parks and protected areas system.

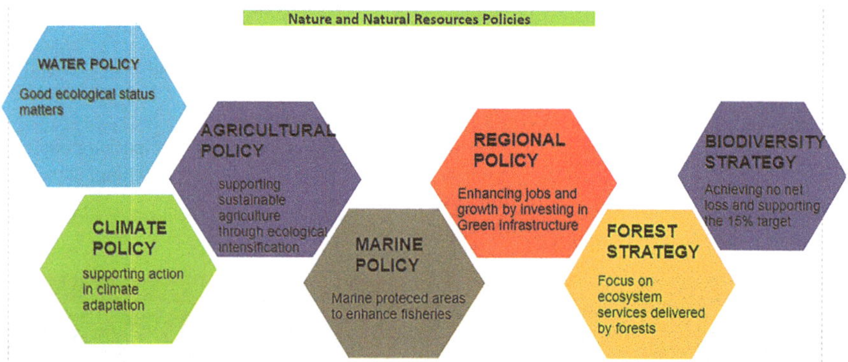

Fig. 34.3 Conservation of nature and natural resources policies

34.6 Sustainability

In temperate areas, global change would lead to early beginnings in the spring and a longer autumn season. The abundance of organisms with ideal conditions for growth and production is inevitable in a longer growing season, thereby raising their productivity (e.g. return on seeds, annual number of plants). It may also make it unfavourable for new animals to be introduced because of low temperatures or because of limited seasons. This would also impact the distribution in the wetlands, pests or pathogens that are also the result of agriculture (Olesen et al. 2011) in the introduction of new crops, for example, grained maize or winter wheat in Northern Europe (Elsgaard et al. 2012).

There are two main elements of sustainability: first—ecological resilience, which is loosely described here as habitats' capability to remain diverse, robust and active overtime, and second—to conserve the flow of critical ecosystem service for people and other organisms. Of course, we must consider the fundamentally dynamic, resilient and rapidly changing structures of ecosystems, including their human components. It is exceedingly unlikely that human action will in any significant way disrupt planetary habitats, but it can lead to abrupt reconfigurations that bring them to swing in various states. A related aspect is economic resilience, which is freely described as an economic system's capacity—to stay diverse, robust and efficient overtime at any scale from individual households to the global economy. Certain ecosystem resources are important for human well-being, and the economic consequences of their depletion may be intolerable. For example, during the Holocene, a geographically stable climate characterized, agriculture and humanity developed. We are now approaching the Anthropocene, a period when natural forces are influenced by human activities on habitats. The danger to environmental and economic security is anthropogenic climatic change.

34.7 Human Well-Being

The environmental and the social and economic developments of the region are influenced by climate changes. It aggravates the already marginalized indigenous peoples' issues, including political and economic isolation, lack of land and wealth, abuses of human rights, discriminations and joblessness. For the vulnerable and disadvantaged people who rely almost entirely on natural resources, the impacts of climate change biodiversity declines are probably greater. Poverty, poor infrastructures (roads, power, water supply, education and healthcare, communication and irrigation), dependence on subsistence farming and forest livelihood goods, inadequate health (high child mortality, low life expectations) and other development issues make HKH more vulnerable to climate change, as HKH is inappropriate to respond.

Through the years, the availability and diffusion of new science and technology, in addition to local impartial awareness of the climate and its management, are among the main motivating forces behind three foundations of mountain survival. Two major implications for conservation as well as development are the generation and spread of traditional and scientific knowledge and the related technologies. Environment and social security will be influenced both positively and adversely by technological implementation. This can be seen, for example, in the increase of food production, greater access to the outside world for mountain populations and improvements in habits of human consumption. In general, limited access in remote mountains of the HKH to suitable technology and science information is a cause (due to their absence) of insecurity and a depletion of natural resources. Any creative technology fields with a significant effect on mountain cultures are illustrated in the following pages. ICT is instrumental in influencing and changing various facets of cultures, states and cultures across the world and is now one of the world's fastest-growing sectors of the twenty-first century in the global economy. Enhanced database availability and knowledge exchange channels are critical to socio-economic growth and raise awareness of environmental problems, associated government legislation and subsidies. The mountain communities in the HKH have no access to fundamental infrastructure, facilities and relevant knowledge. However, several institutions in the area such as governments, NGOs and public and private organizations have shown a weakening emphasis on promoting and using data and technology to enhance the quality and livelihoods of residents, link the HKH with markets outside the region and reduce the marginalization of mountain populations. The number of telephone subscribing businesses using mobile internet networks varies greatly within the HKH. Highly advanced economies such as China now account for over 40 per cent of the economy, and India follows (22 per cent). However, less than 20 per cent of the population has been using mobile internet access in Myanmar, Bangladesh, Pakistan, Nepal and Bhutan. This obviously points to a great opportunity of mobile internet users for developing markets in the area while finding growth opportunities.

In order to solve the problems of food, health and environmental protection and biodiversity restoration, biotechnological applications in agricultural production systems has grown quickly. The latter is used in forestry and agriculture for the production of insect or dry-resistant seeds, high-quality hybrids, biofertilizers, biophytes and biofuel. It is used in medicinal and aromatic plants. This subject has no popular view, nor between countries, and not even among all the writers of this chapter. The topic is very controversial. The moratorium on genetically modified organisms (GM) has, for example, been introduced until 2021 in Switzerland by environmental organizations, as well as small farmers (especially in mountains). On the other hand, many HKH countries regard GM plants as important instruments that are dedicated to developing farm systems, especially those in poor districts, efficiency, viability or sustainability. When applied to farm crops, biotechnology has potential advantages and threats. Its supporters note that the introduction of genetically modified crops has minimized disease transmission, but questions exist about reducing the genetic diversity of native crops due to GM crops. GM interventions have also raised concerns about its potential negative wellness, social and environmental effects. Few studies have also reported negative health effects on farmers due to higher toxin exposure and allergic reactions. Critics further contend that the relationships of genetically modified organisms (GModified organisms) with natural biological diversity are not yet clarified. GMOs are often sterile, and so farmers must continually buy seed from holders of copyrights and make it dependent on both finances and technology. Foreign knowledge also starts substituting native mountain knowledge. In addition, reliance on foreign private corporations undermines national autonomy in food-related matters. The biosafety implications of the use of new organisms that will interfere with the atmosphere to grow sterile seeds and harm locally terrain and marine populations would be important to look at. A healthy outlook also demands that coexistence with biological farming be kept in check—that is to revert to older, non-GM forms in the future. The new CRISPR (Clustered Frequently Interspaced Short Palindromic Repeats) developments should not be impeded.

34.8 Ecological, Economic and Macroallocation Values That Are Incommensurable

The debates between Pareto efficient proponents and EEE proponents are much larger than those in the Pareto efficiency method. The discussions between the EEE supporters are much more substantive. In the debates on productivity and quality addressed in Sects. 34.2 and 34.3, EEE also takes part. EEE advocates support the preservation and rehabilitation of both 'cog and wheel' because natural resources and the services it provides are necessary for the maintenance of human health with no margin replacement option (i.e. high sustainability) and/or because nature has its inherent worth regardless of human desires. While the quality of life can hardly be defined in detail, failure to satisfy basic biophysical needs leads to an inacceptable quality of life and an increasing number of goods and services (ecological or economic), when allocated to those with the lowest standards, is most likely to improve the quality of life. The view of fairness on property rights is therefore excluded by maximizing quality of life. Any system which weights purchase power preferences generally allocates resources to richest people who get the least calculated value of their biological requirements or quality of life. In other words, from the EEE viewpoint, Pareto efficiency is highly inefficient. Economists have been compelled to conclude that we cannot make any substantive comparison of social well-being between people to reconcile the consequentialist approach to justice that led to the emergence of mainstream economics. For example, if an area lacks its water treatment ecosystem operation, it cannot be said that changing health is less for a rich person who has to drink powdered water than for a poor mother forced to watch her infant die of dysentery.

The economic value of goods or services is the goodness that comes from its production and consumption in neoclassical welfare economies. In a perfectly operating market, the demand for and supply of these goods determines the value. The advantage, utility or utility derived from it drives demand for goods. The delivery of goods is driven by the production costs for the producers. The total area A and B shall be labelled 'super plus' and shall be interpreted by means of production and net economic gains or welfare. Consumption at price P with the amount of Q, shall show the value that a product or service gives to its consumer and producer in a market. It is important to realize that we also choose not to allocate these resources to produce alternative goods or services when we decide to allocate the resources. It is frequently the value that drives the loss of ecosystems through the alternative use of resources (e.g. farm, wood production, aquaculture). In the case of ES not traded in a market, it may also be interpreted as a surplus for the welfare deriving from the provision. The biophysical indicators of ES mostly measure the amount provided but not the benefit achieved. The amount of the ecosystem service 'supplied' is determined by other decisions concerning ecosystem protection, use of the land, management, access, etc., not by a market at all. The sum of service rendered is irrespective of its worth. It is depicted as a vertical line.

34.9 An Integrated Climate and Socio-Economic Effects and Its Effects Assessment

Integration is conducted on multiple areas such as different spatial and temporal scales of the ES (regulation, provisioning, cultural). A more complicated methodology does not necessarily have consistent findings or results that are more important. In certain applications, less, as previously mentioned in the fourteenth century, can literally be more (or at least sufficient). "It's pointless to do more than can be achieved for less."

Study on the impacts of adaptive to climate change and alternatives for the agriculture and forestry industries are comprehensive at the industry level in New Zealand. Cross-cutting problems including economic analysis, life-cycle analysis, farming and catchment analysis, social effects, and risk management were also considered. However, climate change can be influenced by changes in financial economies and international relations, along with input on the surrounding world and culture in turn.

If politicians want to adapt to both climate change and change land use concessions and advantages, the ability to consider the entire set of impacts on ecosystem resources is important. In the fifth assessment study (AR5) of the IPCC (IPCC), the latest scenarios established different paths of greenhouse gas emissions from alternative socio-economic routes—their representative concentration pathways (RCPs) (Ebi et al. 2013; IPCC 2014; O'Neill et al. 2014; van Vuuren and Carter 2013). Parallel method is what is referred to. The resulting system gives more freedom to explore methods of prevention and adaptation. The motivation is to give concerned parties an insight into the results of different choices and to transition from a predictive viewpoint into an explorative and optional approach to solutions. Many examples, models and standard techniques to measure indirect and direct factors of transition have been established in Mallampalli et al. and Pichs-Madruga et al. However, multiscale challenges, from global, to regional and territorial, need a participatory approach if substantive participation is to be accomplished in the most possible scenarios and directions to reform and if actions are to be guided and transferable into other regions.

In this relation, we consider three attributes associated with restricting artefacts and connecting research with behaviour to assess the effectiveness of the parallel process:

- integrity relevant to technical facts and claims' scientific adequacy;
- salience that discusses the importance of appraisal of decision-makers' needs
- credibility involving information and technology development that recognizes the divergent principles and opinions of stakeholders.

It is difficult to grasp the input, tip points, side effects, compromise and gains of ecosystem services both climate and socio-economic factors through the dynamics of the relationships and interdependencies of the human–nature environment. The integrated evaluation could help demonstrate these cumulative effects on ecological services (Dunford et al. 2015) when operating across layers in order to resolve multiple challenges, disciplines, processes and scales.

34.10 Challenges to Adaptation

Societal or environmental conditions are classified as socio-economic adaptation challenges which intensify the risks associated with any given climate change prediction by making adaptation more difficult. The danger of climate change comes from a mixture between climate threats, such as the rise of the sea level, change in temperature, precipitation and extreme events, whose risk is exposed to them and their potential for adverse consequences, including geographical, socio-economic and cultural. The risk of climate change is the combination of climate change hazards (to explore how adaptation issues apply to effect definitions, adaptation and risk literature). The climate change risk aspect due to the physical impacts of climate change is expressed in RCP-based climate model forecasts, therefore not included in the SSPs, within scenario matrix architecture. Remaining vulnerability elements are theoretically vulnerable to these risks found in human–environment processes and are thus properly included in the SSPs. Adaptation problems are focused on the socio-economic determinants of climate change exposure, vulnerability to these threats and potential for adaptive coping action. They include the drawbacks of autonomous adaptation and barriers to adaptation policies (e.g. the amount of adapting interventions readily available to individuals and organizations) as well as ineffectual structures and administration that obstruct the application of the policies. Exposure is people's presence; livelihoods; infrastructure; facilities and tools of the ecosystem; physical, social and cultural assets in areas potentially adverse to a climatic danger. For example, a population clustered near a shoreline is likely to be strongly exposed to increasing impacts, while a large population concentrated in urban areas will have high exposure to urban thermal waves. Sensitivity, often related to exposure, suggests the socio-economic environment reaction to a certain level of climate change; the exposure–response relationship may be defined. For instance, if coastal residents reside in low-construction houses, they will be more susceptible than the population living in better built buildings to the rise in storm overflow associated with the sea level. Likewise, urban communities that are more vulnerable to weather temperatures than most other populations will be more sensitive to urban heating waves, as a result of their higher concentrations of the elderly resident.

The resilient capacity demonstrates a company's ability to respond to climate change to mitigate its impacts or to seize its opportunities. Factors that affect this potential include provision of feasible technical adaptation alternatives, the efficacy and dissemination through the population of related organizations (such as agricultural research and development, markets for products under climate change and forest management agencies). A well-working health system, for example, would enhance a society's capacity to boost the heatwave's health effects, while well-functioning food markets and agricultural research and development agencies would increase their potential to reduce the climate change impacts of agriculture, including the ability to benefit from such outcomes as the prolongation.

34.11 Future Perspective

The aim is for climate change to put the main forest regions of the world distinct climate futures. Climate control can have both advantages and global risks (reduced wildfires). Various styles, reward ratios, organizational strategies and adaptation policies need to be introduced, say the authors. Although we research many possibilities and forecasts for the future, they suggest the planet will witness only one version of the future. The authors suggest that reducing climate vulnerability instability and natural fluctuations will boost their forecasts. Additional climate models also enhance the stability of the data.

Countries vary in their opinions as to how to implement international climate policy. In the mid-twenty-first century, long-term targets proposed for Europe and the USA are targeted at reducing greenhouse gas emissions by 80 per cent. In the sense of such acts, the EU has set a goal of limiting temperatures above pre-industrial levels to a maximum of 2 °C (3.6 °F). (Dealing with the global average near-surface air temperature of more than 2 °C (3.6 °F), many climatic scientists believe that major economic and ecosystem risk is likely to occur in the following century.) While there were variations in strategy, it is on a basis of a consensus reached at the UN Climate Change Conference in Bali 2007 that countries initiated a negotiation of a new treaty. A substantive legal environment settlement that would replenish the Kyoto Protocol was developed by the international community. World leaders signed in Paris in 2015 a worldwide, but non-binding pact to restrict the average world temperature rise to no more than 2 °C. Donald J. Trump (President USA) has made signals to withdraw the USA from the climate deal following the structured withdrawal procedure that will take effect on 4 November 2020. By 2019, the number of parties (signatories) to the Treaty was 197 and the agreement had been ratified by 185 nations and in September 2016 by the USA. The Paris Agreement was a historic agreement mandating development on a 5-year basis and the establishment by 2020 of a $100 billion funds to help emerging countries introduce non-greenhouse gas technology. Under the 2005 Climate Protection Pact, which cities have signed, the USA pledged itself to reduce its emissions by 7% below 1990 levels by 2012. The ICLI and its climate security cities define the recommendations and the measures required for action at the local level. Global, international, national and local policy on global warming and climate change continues to evolve. The strategy on mitigation relies on numerous means of reducing greenhouse gas emissions. Adjustment strategy strives to increase the capacity of economies to resolve climate change issues. A model of the private sector initiative to minimize carbon emissions through a swap mechanism is the Chicago Climate Exchange. Long-term cuts in greenhouse gas releases would enable both industrial and big emerging countries to be involved. In 2006, China finally overtook the USA as the world's main greenhouse gas emitter. The writers suggest that in developed nations, many of the most adverse consequences of a changing environment would exist. Societies must identify ways to adjust habits of energy use to combat global warming and climate change. They suggest people will do more to reduce their emissions of greenhouse gases.

The impacts of climate change may have significant cascading effects, with possible devastating repercussions, including negative impacts on the availability of ecosystem service, agricultural production, human health and livelihoods, of million people living in the area (Ariza et al. 2013)—particularly where they are, coupled with social, economical and political pressures (Ariza et al. 2013). Decreased abundance of natural resources and instability caused by climate variability represent a challenge to mountain survival, when a natural resource base is already diminishing. Different methods exist for ES internationally. Biophysical and monetary benefit maps are the most common approaches. Global models can simulate ES patterns over time and space. This makes them important tools for resource management decision-making. Climate change in the world climate is not a recent development. Climate change faces multiple threats; one of the main implications is that water supplies and crop production are altered both in quality and volume. The Indian area is vulnerable to climate change. Present status of global rangeland soil erosion prevention, for the mapping, a further variant of the Universal Soil Loss Equation (USLE) is available with us. Erosion prevention was plotted with a soil erodibility index (0–100). The erosivity of rainfall is both picture-based and a refined land use/cover index derived from fractions of a planetary cover. The decreased erosion prevention of cliffs, vulnerable ground and scanty vegetation cover (e.g. Mediterranean, Central Australia and Chile) is the product of high risk (i.e. high risk of erosion). Cereals, barley, corn, pulse, root and tuber are taken as an indicator for crop development. A changing lifestyle and population pushes demand for crops while innovations, technology Efficiency of production, thus yield, is dictated by environmental factors and management. In the Transparency EU initiative, the two scenarios are modified SSP (e.g. new IPCC scenarios). 'Wealth-being' (WB) stands for growth, and the 'EC' scenario stands for economic growth and promotes worldwide environmental sustainability. The agriculture industry is the most susceptible sector since it directly impacts 1.2 billion people's lives. By 2050, India has set a greenhouse gas emission halving target. Coordinated research activities to determine the effect of climate change on agriculture, forestry, animal husbandry, aquatics and other living creatures are desperately needed.

References

Alcala AC, Gomez ED, Salvat B (1987) Dynamiting coral reefs for fish: A resource-destructive fishing method. Impacts des activités humaines sur les récifs coralliens: connaissances et recommandations, pp 51–60

Allison EH, Perry AL, Badjeck M-C, Neil Adger W, Brown K, Conway D, Halls AS, Pilling GM, Reynolds JD, Andrew NL, Dulvy NK (2009) Vulnerability of national economies to the impacts of climate change on fisheries. Fish Fish 10:173–196. https://doi.org/10.1111/j.1467-2979.2008.00310.x

Altermatt F (2010) Climatic warming increases voltinism in European butterflies and moths. Proc R Soc B Biol Sci 277:1281–1287. https://doi.org/10.1098/rspb.2009.1910

Ariza C, Pérez A, Sánchez-Martínez F, Diéguez M, Espelt A, Pasarín MI, Suelves JM, Torre RDL, Nebot M (2013) Evaluation of the effectiveness of a school-based cannabis prevention program. Drug Alcohol Depend 132(1–2):257–264. https://doi.org/10.1016/j.drugalcdep.2013.02.012

Aryal J, Sapkota T, Jat M, Bishnoi D (2015) On-farm economic and environmental impact of zero-tillage wheat: a case of North-West India. Exp Agric 51(1):1–16. https://doi.org/10.1017/S001447971400012X

Bakun A (1992) Mechanisms of physical-biological interaction in coastal marine ecosystems in relation to projected trends of global change. Paper submitted to the Bremen Marine Training Centre's International Workshop on marine environmental protection and coastal living resources monitoring: training and implementation. 29 September–3 October 1992, Bremerhaven, Germany 10 p

Bales RC, Molotch NP, Painter TH, Dettinger MD, Rice R, Dozier J (2006) Mountain hydrology of the western United States. Water Resour Res 42:W08432. https://doi.org/10.1029/2005WR004387

Barg UC, Wijkstrom UN (1994) Environmental management options for coastal fisheries and aquaculture. Mar Policy 18:127–137

Berman M, Schmidt JI (2019) Economic effects of climate change in Alaska. Weather Clim Soc 11 (2):245–258. https://doi.org/10.1175/WCAS-D-18-0056.1

Boesch DF, Turner RE (1984) Dependence of fishery species on salt marshes: the role of food and refuge. Estuaries 7:460–468

Brander KM (2007) Global fish production and climate change. Proc Natl Acad Sci 104(50):19709–19714. https://doi.org/10.1073/pnas.0702059104

Burakowski EA, Magnusson MF (2012) Climate impacts on the US winter tourism economy

Caddy JF (1993a) Towards a comparative evaluation of human impacts on fishery ecosystems of enclosed and semi-enclosed seas. Rev Fish Sci 1:57–95

Caddy JF (1993b) Some future perspectives for assessment and management of Mediterranean fisheries. Sci Mar 57:121–130

Campbell JA (1993) Guidelines for assessing marine aggregate extraction. Ministry of Agriculture, Fisheries and Food, Directorate of Fisheries Research, Lowestoft. Laboratory Leaflet (73):12

Chen J, Brissette FP, Poulin A, Leconte R (2011) Overall uncertainty study of the hydrological impacts of climate change for a Canadian watershed. Water Resour Res 47:W12509. https://doi.org/10.1029/2011WR010602

Clark JR (1992) Integrated management of coastal zones. FAO Fish Tech Pap (327):167

Clavel J, Julliard R, Devictor V (2011) Worldwide decline of specialist species: toward a global functional homogenization? Front Ecol Environ 9(4):222–228

Cruzen PJ (2002) Geology of mankind. Nature 415:23

Derksen C, Brown R (2012) Spring snow cover extent reductions in the 2008–2012 period exceeding climate model projections. Geophys Res Lett 39:L19504. https://doi.org/10.1029/2012GL053387

Diaz S, Hector A, Wardle D (2009) Biodiversity in forest carbon sequestration initiatives: not just a side benefit. Curr Opin Environ Sustain 1(1):55–60. https://doi.org/10.1016/j.cosust.2009.08.001

Doak D, Morris W (2010) Demographic compensation and tipping points in climate-induced range shifts. Nature 467:959–962. https://doi.org/10.1038/nature09439

Doney SC, Mary R, Emmett Duffy J, Barry JP, Chan F, English CA, Galindo HM, Grebmeier JM, Hollowed AB, Knowlton N, Polovina J, Rabalais NN, Sydeman WJ, Talley LD (2012) Climate change impacts on marine ecosystems. Annu Rev Mar Sci 4(1):11–37

Donnelly A, Jones MB, Sweeney J (2004) A review of indicators of climate change for use in Ireland. Int J Biometeorol 49:1–12. https://doi.org/10.1007/s00484-004-0215-5

Dugas R, Guillory V, Fischer M (1979) Oil rigs and offshore sport fishing in Louisiane. Fisheries (U.S.). 4:2–9

Dunford RW, Smith AC, Harrison PA, Hanganu D (2015) Ecosystem service provision in a changing Europe: adapting to the impacts of combined climate and socio-economic change. Landsc Ecol 30:443–461. https://doi.org/10.1007/s10980-014-0148-2

Ebi KL, Hallegatte S, Kram T, Arnell NW, Carter TR, Edmonds J, Kriegler E, Mathur R, O'Neill BC, Riahi K, Winkler H, van Vuuren DP, Zwickel T (2013) A new scenario framework for climate change research: background, process, and future directions. Clim Change. https://doi.org/10.1007/s10584-013-0905-2

Elsgaard L, Børgesen CD, Olesen JE, Siebert S, Ewert F, PeltonenSainio P, Rötter RP, Skjelvåg AO (2012) Shifts in comparative advantages for maize, oat, and wheat cropping under climate change in Europe. Food Add Contamin 29:1514–1526

FAO (1986a) Study report: Watershed management in Asia and the Pacific: needs and opportunities for action. Tech Rep FO:RAS/85/017:160 p

FAO (1986b) Strategy for fisheries management and development. FAO, Rome, Italy, 26 p

Fogarty MJ, Sissenwine MP, Cohen EB (1991) Recruitment variability and the dynamics of exploited marine populations. TREE 6:241–246

Fountain AG, Campbell JL, Schuur EAG, Stammerjohn SE, Williams MW, Ducklow HW (2012) The disappearing cryosphere: impacts and ecosystem responses to rapid cryosphere loss. Bioscience 62(4):405–415. https://doi.org/10.1525/bio.2012.62.4.11

Francis RC (1990) Climate change and marine fisheries. Fisheries 6(15):7–9

GBRMPA (1992/1993) Annual report of Great Barrier Reef Marine Park Authority. Queensland, Australia, 162 p

GESAMIP (1990) The state of the marine environment. IMO/FAO/UNESCO/WMO/ WHO/IAEA/UN/UNEP Joint Group of Experts on scientific aspects of marine pollution. Reports and studies (39):111. UNEP, Nairobi, Kenya

GESAMP (1986) Environmental capacity: an approach to marine pollution prevention. IMO/FAO/UNESCO-IOC/WHO/WHO/IAEA/UN/UNEP Joint Group of Experts on scientific aspects of marine pollution. Reports and studies (30):49

GESAMP (1994). Anthropogenic influences on sediment discharges to the coastal zone and environmental consequences. IMO/FAO/UNESCO-IOC/WMO/WHO/IAEA/UN/UNEP Joint Group of Experts on the scientific aspects of marine environmental protection. Reports and studies (52):67

Grigg RW, Dollar SJ (1988) Natural and anthropogenic disturbance on coral reefs. In: Coral reef ecosystems. Elsevier Press, Amsterdam, The Netherlands

Harris PG (2010) Misplaced ethics of climate change: political vs. environmental geography. Ethics Place Environ 13(2):215–222. https://doi.org/10.1080/13668791003778875

Healy GN, Dunstan DW, Salmon J, Cerin E, Shaw JE, Zimmet PZ, Owen N (2008) Breaks in sedentary time: beneficial associations with metabolic risk. Diabetes Care 31(4):661–666. https://doi.org/10.2337/dc07-2046

Hély C, Bremond L, Alleaume S, Smith B, Sykes MT, Guiot J (2006) Sensitivity of African biomes to changes in the precipitation regime. Glob Ecol Biogeogr 15:258–270. https://doi.org/10.1111/j.1466-8238.2006.00235.x

Hoss DE, Tbayer GW (1994) The importance of habitat to the early life history of estuarine-dependent fishes. Am Fish Soc Symp (14):3–15

Houde ED, Rutherford ES (1993) Recent trends in estuarine fisheries: predictions of fish production and yield. Estuaries 16:161–176

Huggel C, Stone D, Eicken H et al (2015) Potential and limitations of the attribution of climate change impacts for informing loss and damage discussions and policies. Clim Chang 133:453–467. https://doi.org/10.1007/s10584-015-1441-z

ICES (1992) Effects of extraction of marine sediments on fisheries. International Council for the Exploration of the Sea, Cooperative Research Report (182):78. ICES, Copenhagen, Denmark

IPCC (1992) In: Houghton JT, Callander BA, Varney SK (eds) Climate change 1992: the supplementary report to the IPCC Scientific Assessment-Report. Press Syndicate of the University of Cambridge, Cambridge

IPCC (2014) Climate Change 2014: synthesis report. In: Core Writing Team, Pachauri RK, Meyer LA (eds) Contribution of working groups I, II and III to the fifth assessment report of the intergovernmental panel on climate change. IPCC, Geneva, Switzerland

IPCC (2018) Global warming of 1.5°C. In: Masson-Delmotte V, Zhai P, Pörtner HO, Roberts D, Skea J, Shukla PR, Pirani A, et al. (eds) An IPCC special report on the impacts of global warming of 1.5°C above pre-industrial levels and related global greenhouse gas emission pathways, in the context of strengthening the global response to the threat of climate change, sustainable development, and efforts to eradicate poverty

IUCN, UNEP and WWF (1980) World conservation strategy. World Conservation Organization, Gland, Switzerland

Keesing F, Belden L, Daszak P et al (2010) Impacts of biodiversity on the emergence and transmission of infectious diseases. Nature 468:647–652. https://doi.org/10.1038/nature0957

Knox GA, Miyabara T (1984) Coastal zone resource development and conservation in Southeast Asia. UNESCO East-West Centre, Jakarta Pusat, Indonesia, 182 p

Kullenberg, G., Andersen N, Cole M (1993) Global ocean observing system. In: Coastal Zone '93. Proceedings, 8th Symposium on Coastal and Ocean Management, pp 2986–3000, New Orleans, Louisiana

Lawton JH, Brotherton PNM, Brown VK, Elphick C, Fitter AH, Forshaw J, Haddow RW, Hilborne S, Leafe RN, Mace GM, Southgate MP, Sutherland WA, Tew TE, Varley J, Wynne GR (2010) Making space for nature: a review of England's wildlife sites and ecological network, Report to Defra. Department of Environment, Food and Rural Affairs, London

Le Quéré C, Raupach M, Canadell J et al (2009) Trends in the sources and sinks of carbon dioxide. Nat Geosci 2:831–836. https://doi.org/10.1038/ngeo689

Lehodey P, Alheit J, Barange M, Baumgartner T, Beaugrand G, Drinkwater K, Fromentin J-M, Hare SR, Ottersen G, Perry RI, Roy C, van der Lingen CD, Werner F (2006) Climate variability, fish, and fisheries. J Clim 19(20):5009–5030. https://journals.ametsoc.org/view/journals/clim/19/20/jcli3898.1.xml

Lewis SL, Maslin MA (2015) Defining the anthropocene. Nature 519:171–180

Li X, Zou H, Liu Y et al (2014) The relationships of family socioeconomic status, parent–adolescent conflict, and filial piety to adolescents' family functioning in Mainland China. J Child Fam Stud 23:29–38. https://doi.org/10.1007/s10826-012-9683-0

Liss PS, Slinn WGN (eds) (1983) Air-sea exchange of gases and particles. D. Reidel, Dordrecht, Holland, pp 299–405

Malcolm JR, Markham A, Neilson RP, Garaci M (2002) Estimated migration rates under scenarios of global climate change. J Biogeogr 29:835–849. https://doi.org/10.1046/j.1365-2699.2002.00702.x

Maestre FT, Quero JL, Gotelli NJ, Escudero A, Ochoa V, Delgado-Baquerizo M, García-Gómez M, Bowker MA, Soliveres S, Escolar C, García-Palacios P (2012) Plant species richness and ecosystem multifunctionality in global drylands. Science 335(6065):214–218. https://doi.org/10.1126/science.1215442

Martin J-M, Gordeev VV (1986) River input to the ocean system: a reassessment. In: Proceedings of a Scientific Workshop on Estuarine Processes. UNESCO/IOC/CNA, Lisbon, Portugal, pp 203–240

Martin J-M, Burton JD, Eisma D (eds) (1982) River inputs to ocean systems. Proceedings of a SCOR/ACMRR/ECOR/IAHS/UNESCO/CMG/IABO/IAPSO Review and Workshop, FAO, Rome, Italy, 1979. UNESCO, Paris, France

Milliman JD (1981) Transfer of river-borne particulate material to the oceans. In: Martin J-M, Burton JD, Eisma D (eds) River inputs to ocean systems: proceedings of a SCOR/ACMRR/ECOR/IAHS/UNESCO/CMG/IABO/IAPSO Review and workshop held at FAO, Rome, Italy, 26–30 March 1979, pp 5–12

Milliman JD, Qian YS, Saito Y (1987) Man's influence on the erosion and transport of sediment by Asian rivers: the Yellow River example. J Geol 94:751–762

Milner J, Harpham C, Taylor J, Davies M, Le Quéré C, Haines A, Wilkinson P (2017) The challenge of urban heat exposure under climate change: an analysis of cities in the Sustainable Healthy Urban Environments (SHUE) database. Climate 5:93. https://doi.org/10.3390/cli5040093

Montoya JM, Dave R (2010) Climate change, biotic interactions and ecosystem services. Phil Trans R Soc B 365:2013–2018. https://doi.org/10.1098/rstb.2010.0114

Mooney H, Larigauderie A, Cesario M, Elmquist T, Hoegh-Guldberg O, Lavorel S, Mace GM et al (2009) Biodiversity, climate change, and ecosystem services. Curr Opin Environ Sustain 1:46–54

Murphy JD, Power NM (2008) How can we improve the energy balance of ethanol production from wheat? Fuel 87(10–11):1799–1806. https://doi.org/10.1016/j.fuel.2007.12.011

Murty TS (1977) Seismic sea waves: tsunamis. Bull Fish Res Board Canada 198:337

Murty TS (1984) Storm surges: meteorological ocean tides. Can Bull Fish Aquat Sci (212):897

O'Neill BC, Kriegler E, Riahi K, Ebi KL, Hallegatte S, Carter TR, Mathur R, van Vuuren DP (2014) A new scenario framework for climate change research: the concept of shared socio-economic pathways. Clim Change 122:387–400

Olesen JE, Trnka M, Kersebaum KC, Skjelvåg AO, Seguin B, Peltonen-Saino P, Rossi F, Kozyra J, Micale F (2011) Impacts and adaptation of European crop production systems to climate change. Euro J Agron 34:96–112

Pakistan I, Sarfraz H (2009) Integrated development vision-handbook. IUCN Pakistan, Quetta. portals.iucn.org

Pastorok RA, Bilyard GR (1985) Effects of sewage pollution on coral reef communities. Mar Ecol Prog Ser 21:175–189

Prospero JM (1981) Eolian transport to the world ocean. In: Emiliani C (ed) The sea. (7). The oceanic lithosphere. Wiley, New York, pp 801–874

Pugh DT (1987) Tides, surges and mean sea-level: a handbook for engineers and scientists. Wiley, New York

Pullin RSV (1991) An overview of environmental issues in developing country aquaculture development. In: ICLARM Conference proceedings, p 31

Qin D, Ding Y (2010) Key issues on cryospheric changes, trends and their impacts. Adv Clim Chang Res 1:1–10. https://www.scipedia.com/public/Qin_Ding_2015a

Reggio VC Jr (1987) Rigs-to-reefs. Fisheries (U.S.). 12:2–7

Reisinger A, Kitching RL, Chiew F, Hughes L, Newton PCD, Schuster SS, Tait A, Whetton P (2014) Australasia. In: Barros VR, Field CB, Dokken DJ, Mastrandrea MD, Mach KJ, Bilir TE, Chatterjee M, Ebi KL, Estrada YO, Genova RC, Girma B, Kissel ES, Levy AN, MacCracken S, Mastrandrea PR, White LL (eds) Climate change 2014: impacts, adaptation, and vulnerability. Part B: regional aspects. Contribution of Working Group II to the Fifth Assessment Report of the Intergovernmental Panel on Climate Change. Cambridge University Press, Cambridge, pp 1371–1438

Royal Society of New Zealand (2016) 2016 Annual review. https://www.royalsociety.org.nz/who-we-are/our-reports/our-annual-reviews/2016-annual-review/

Scherer-Lorenzen M, Luis Bonilla J, Potvin C (2007) Tree species richness affects litter production and decomposition rates in a tropical biodiversity experiment. Oikos 116:2108–2124. https://doi.org/10.1111/j.2007.0030-1299.16065.x

Sharma HC (2014) Climate change effects on insects: implications for crop protection and food security. J Crop Improv 28(2):229–259. https://doi.org/10.1080/15427528.2014.881205

Smith PJ (2007) Climate change, mass migration and the military response. Orbis 51:617–633

Tisdell CA (1982) Environmental conservation: economics, ecology, and ethics. Environ Conserv:107–112

UNEP (1990a) Technical annexes to the report on the state of the marine environment. Regional seas reports and studies, (114/2):676

UNEP (1990b) The state of the marine environment. Reports and studies (39):11

UNEP (1994) The impacts of climate on fisheries. UNEP Environment library no. 13:36

UNEP-IUCN (1988) Coral reefs of the world. Vol. 1, 2 and 3. UNEP Regional Seas Directories and Bibliographies. IUCN, Gland, Switzerland and Cambridge, UK; UNEP, Nairobi, Kenya, Vol.1, Atlantic and Eastern Pacific, 373 p, 38 maps. Vol 2, Indian Ocean, Red Sea and Gulf, I + 389 p., 36 maps. Vol 3, Central and Western Pacific, 329 p., 30 maps

UNESCO (1986) Human induced damage to coral reefs. UNESCO reports in marine science 180

van Vuuren DP, Carter TR (2013) Climate and socio-economic scenarios for climate change research and assessment: reconciling the new with the old. Climatic change, special issue. In: Nakicenovic N, Lempert R, Janetos A (eds) A framework for the development of new socio-economic scenarios for climate change research. Springer, New York

Walther GR, Post E, Convey P et al (2002) Ecological responses to recent climate change. Nature 416:389–395. https://doi.org/10.1038/416389a

Warner K, Hamza M, Oliver-Smith A et al (2010) Climate change, environmental degradation and migration. Nat Hazards 55:689–715. https://doi.org/10.1007/s11069-009-9419-7

Warren FJ, Lemmen DS (2014) Synthesis. In: Warren FJ, Lemmen DS (eds) Canada in a changing climate: sector perspectives on impact and adaptation. http://www.nrcan.gc.ca/environment/resources/publications/impactsadaptation/reports/assessments/2014/16309

Welcomme, R.L.K. (1994, July 12–15). Relationships between fisheries and the integrity of river systems. In: Proceedings of international conference on sustaining the ecological integrity of large floodplain rivers. La Crosse, Wisconsin

Williamson P (1992) Global change: reducing uncertainties. In: International Geosphere-Biosphere Programme (of the International Council of Scientific Unions). IGBP Secretariat, Royal Swedish Academy, Stockholm, Sweden, 40 p. Coastal connections, pp 19–21

Woolhouse ME, Gowtage-Sequeria S (2005) Host range and emerging and reemerging pathogens. Emerg Infect Dis 11:1842–1847

Wu Q, Dong X, Liu Y, Jin H (2007) Responses of permafrost on the Qinghai-Tibet Plateau, China, to climate change and engineering construction. Arct Antarct Alp Res 39(4):682–687. https://doi.org/10.1657/1523-0430(07-508)[WU]2.0.CO;2

Xiao X, O'Dwyer JP, White EP (2016) Comparing process-based and constraint-based approaches for modeling macroecological patterns. Ecology 97:1228–1238. https://doi.org/10.1890/15-0962.1

Yang H, Huang J, Liu D (2020) Linking climate change and socioeconomic development to urban land use simulation: analysis of their concurrent effects on carbon storage. Appl Geogr 115:102135

Chapter 35
Impact of Climate Change on Localized Plant–Microbe Signalling and Technology Advancement in Microbial Quorum Sensing

Debanjan Sanyal, G. Venkata Subhash, Vinay Dwivedi, and Santanu Dasgupta

Abstract In the natural habitat, plants have association with abundant microbial population having important direct or indirect roles in plant health and growth and a most of the information is available on the plant microbiota structure. Most of the studies are available on microbial mechanisms on physiological processes with respective to the host plants. In plants, the secreted components could form microbial communities at rhizosphere, endosphere and phyllosphere regions of plants. In each niche, a group of microbial colonies can be established and respond to specific conditions appeared during interaction with plants. Overall, the plant microbiome communities played a major role in control of diseases, nutrient acquisition enhancement and tolerance to stress or aiding in plant growth promotion. In this present chapter, the habitats and features of microbial communities have been discussed in relation to plant growth followed by factors responsible for the plant–microbe interactions, secreting components and signalling mechanisms between plant and microbe communications, and the role quorum sensing in communication and plant protection. The application of synthetic biology tools in deploying plant microbiome in plant protection, plant breeding and plant health for more sustainable agriculture has been also discussed.

Keywords Climate change · Microbe · Plant · Quorum sensing · Signalling · Technology advancement

---

G. Venkata Subhash is also the first author.

D. Sanyal (✉) · V. Dwivedi
Reliance Industries Limited, Jamnagar, Gujarat, India
e-mail: debanjan.sanyal@ril.com

G. V. Subhash · S. Dasgupta
Reliance Industries Limited, Ghansoli, Navi Mumbai, India

D. K. Choudhary et al. (eds.), *Climate Change and the Microbiome*, Soil Biology 63,
https://doi.org/10.1007/978-3-030-76863-8_35

35.1 Introduction

In nature, plants host highly dynamic and diverse microbiota and create a unique microbial ecology. Microbial assemblages from the soil mostly associated with host plants (rhizosphere, phyllosphere and endosphere) and specific plant organs called as plant halobiont and termed as plant microbiota or the plant microbiome show a wide range of functions for supporting the plant health and growth (Philippot et al. 2013). These microbiota contribute nutrition in the host plants, protect from the pathogens and pests and improve the tolerance against stress. The functionality of plant–microbe interactions and factors involved may potentially help researchers to deploy this towards enhanced plant growth and better crop productivity. Most of the earlier surveys focussed on plants and microbes interactions (pathogenic), and after assuming that, some are pathogenic and the others are neutral or beneficial for plant development (Philippot et al. 2013). Beneficial microorganisms involved in the agriculture or plant biomass production by aiding acquisition or availability nutrients and promoting stress tolerance (Kavamura et al. 2013) and these specific microbial groups nitrogen-fixation and fungi belongs to mycorrhiza) involved in promoting the plant growth (Chagnon et al. 2013). The plant–microbe interactions are influenced by a basic principles, such as (1) defence or symbiotic responses which will be activated based on signals from microbes that are perceived by plant immune receptors, (2) modulation of host cell functions which will be mediated by transport effector molecules like microbial DNA and/or protein into the plant cell, and (3) during symbiotic and pathogenic interactions formation of specialized microbial organs (e.g. nodules and galls) which will be developed for nutrient exchanging.

The communication between plant and microflora is known to be initiated by a specific signalling molecule in the environment and they will increase in response to specific microbial population and help to coordinate protective mechanism against adverse conditions called quorum sensing (QS). Under plant and cell communication, microbes can secrete extracellular signal molecules called autoinducers. These plant–microbe interactions can be altered by external conditions, including temperature, moisture and nutrient status. At the plant rhizosphere region, some symbionts (N_2-fixing bacteria or fungi) are known as microbial biofertilizer. For better understanding and practices in agriculture, these plant microbes are important for microbiome-based solutions. The application of knowledge of synthetic biology can significantly aid in understanding of individual or a package of strains in combination with model plants. Under synthetic biology approaches, application of selective microbes for plant developments can create new avenues for the development and use of microbial functions in enhancing crop productivity.

The present chapter discussed the topics of plant and microbe interaction and secretion of various signalling chemical components along with the responsible factors for effective interactions between plant and microbe. In continuation, role of quorum sensing in this communication is also discussed with mechanisms involved and advancements using synthetic biology at molecular level.

35.2 Plant and Microbes

35.2.1 Favourable Zones in Plant for Microbial Growth

The matrix of soil is the diverse microbial reservoir and in interaction with plants and plant parts (Vogel et al. 2009). The microbial population present in the soil is important for processes happening in relation to health of the plant (suppresses plant diseases and infecting plant tissues) (Mendes et al. 2011). These soil microbiomes impart a degree of resistance against "invaders" and show intrinsic and extrinsic activities (van Elsas et al. 2012). Microbial populations at plant region are considered as an active component of the host, being also responsive to changes in environmental (biotic and abiotic) conditions. Segregating the plant–microbe population involves three major compartments: so-called rhizosphere, phyllosphere and endosphere (Hirsch and Mauchline, 2012).

35.2.1.1 Microbiome at Rhizosphere and Secreting Components

In 1904 Hiltner coined the rhizosphere term (Curl and Truelove 1986) and referring to the environment at the root region of plant where the soil at the root of plants and microorganisms (Sugiyama et al. 2014). The microbial population residing at rhizosphere region is organized differentially than the bulk soil and is driven by root exudates and the effect of increased microbial biomass in the rhizosphere.

Rhizodeposits (organic or inorganic compounds) are the components which influence the rhizosphere microbial community at rhizosphere, and this is known as "rhizosphere effect" (Berendsen et al. 2012), which is influenced by the genotype of plant (Bulgarelli et al. 2012), and each plant sp. is specific to microbial populations and coevolution of plants and microbes (Bais et al. 2006). In rhizodeposits, some of them are involved in activation of microbial populations (e.g. glucose) or can activate specific groups of organisms (e.g. flavonoids) (Jones et al. 2004).

These rhizodeposits are directly released to surroundings for microorganisms at roots throughout the plant life (Haichar et al. 2008) and plant growth (Chaparro et al. 2014). Microbial aggregation starts at the germination stage, and microorganisms will be distributed according to root type and zones during root growth (Philippot et al. 2013). Alcohols and sugars will be released in the early stages of plant growth (seedlings), followed by amino acids and phenolic compounds at further plant growth (Chaparro et al. 2014). This selectively will follow the phenomenon like attraction by offering a carbon at the early stage of development, and later selects certain microorganisms by releasing specific compounds (Chaparro et al. 2014). At the root region, microorganisms more number belongs to gram-negative bacteria, gram-positive bacteria, proteobacteria (α, β, γ) etc., usually represent at high level (Philippot et al. 2013).

The rhizosphere microorganisms can have beneficial effects as plant growth-promoting rhizobacteria (PGPR), for plant growth and health (Philippot et al. 2013). PGPR can suppress the diseases from the various sources by a mechanism like creating food (nutrients) competition or antibiosis or parasitism (Philippot et al. 2013).

35.2.1.2 Microbiome at Phyllosphere and Secreting Components

The second component at plant and microbe's interacting location is aerial tissues called phyllosphere (Vorholt 2012a, b). The phyllosphere region is characterized as nutrient poor by comparing with the rhizosphere (Andreote et al. 2014). At phyllosphere, microbial communities play a major role in plant protection from pathogens, nitrogen fixation and phytohormones biosynthesis (Kishore et al. 2005).

The phyllosphere is habituated with bacteria, yeasts, filamentous fungi, viruses, algae and bacteria which are more abundant (10^6 and 10^7 cells cm^2) than fungi and archaea (Vorholt 2012a, b). At the level of community composition, the genus level of bacteria mainly belongs to *Sphingomonas, Pseudomonas, Bacillus, Methylobacterium, Arthrobacter, Pantoea* and *Massilia* present at phyllosphere (Bulgarelli et al. 2013). On phyllosphere region, these bacteria, fungi and other microorganisms occur through the immigration from air, soil, water, seeds or through animal sources and get stabilized (Vorholt 2012a, b). After the stabilization, the microbial communities will depend on nutritive sources (carbon) from leaf surface and leaf veins (Vorholt 2012a, b). The phyllosphere microbiome structure may vary due to a large flux in atmospheric conditions on the day and night (Vorholt 2012a, b; Lindow 1996). These phyllosphere organisms can live under harsh and variable conditions (Andrews and Harris, 2000).

35.2.1.3 Microbiome at Endosphere and Secreting Components

Inner plant tissues can be considered as endosphere region, and the associated microorganisms with the host plant are termed as endosphere microbiome, and these microorganisms reside internally in plant tissues without showing any visible symptom (Hardoim et al. 2008; Mostert et al. 2000; Berg et al. 2014). The endosphere region is composed of the endorhizosphere and endophyllosphere (Truyens et al. 2015).

These endophytic organisms are beneficial or commensal (Malcolm et al. 2013) and, under beneficial role, promote plant development and health (Khan et al. 2012a, b; Waqas et al. 2014), characterized by produced antibiotics and toxicants (Schardl et al. 2013; Gond et al. 2015; Yaish et al. 2015). Endophytes can also involve in altering the expression of plants gene defence and metabolic pathways (Rosenblueth and Martínez-Romero 2006; Mathys et al. 2012; Ownley et al. 2009).

The structure of endosphere is driven by type of soil, soil pH, phylogeny of host and microbes (Baker et al. 2009) and local edaphic conditions (Yandigeri et al. 2012;

Marques et al. 2015). After crossing all these barriers finally, the microbes establish themselves at endosphere. Common microbial population are bacterial phyla, firmicutes (Robinson et al. 2015; Manter et al. 2010) and fungi, including Ascomycota and Basidiomycota (Glynou et al. 2016; Higgins et al. 2014; Toju et al. 2013).

35.2.2 Contribution from Plants to Microbe Interactions by Chemicals and Signals

Plants are multicellular and sessile organisms, which have three well-defined parts: (1) root, (2) stem and (3) shoot (López-Bucio et al. 2005; Ross et al. 2005), and root system is different in its morphology and physiology in the environment. During embryogenesis primary root contrast to be formed, adventitious and lateral roots are formed post-embryonically. Widespread communication occurs between microorganisms (fungal and bacterial species) and plants, and the signalling molecules of both partners have a significant role. Plants can recognize and adjust their defines and growth responses against the microbe-derived compounds (Van Loon et al. 1998). Region of rhizosphere is a complex and associated with increased number of bacterial abundance and other microbial community activities associated with different plant species (Zeng et al. 2017).

At the rhizosphere region, in a sophisticated manner organism will interact with each other and with the plant by chemical communication, and in a response, plants release different metabolites by 'signalomics' at the rhizosphere. Diverse group of compounds from the root system are collectively termed as root exudates. These compounds belong to three main classes: (1) low molecular, (2) high molecular and (3) volatile organic compounds (VOC) (Schulz and Dickschat 2007; Badri and Vivanco 2009; Aulakh et al. 2001).

Microorganisms and their products may respond in various ways (positive or negative or neutral) at the roots (Morgan et al. 2005; Broeckling et al. 2008). Interactions and communication are important in rhizosphere to enhance plant growth. For biotechnology improvement, rhizosphere region is important to biomass production and can be achieved by inoculating or by engineering plants to modify the nature and level of exudate compounds. Plants can produce molecules like elicitors which are involved in defence responses (Mackey and McFall 2006). Exogenous methyl jasmonate, nitric oxide and salicylic acid induce the accumulation of secondary metabolites which play a major role in communication (Noritake et al. 1996).

Usually, the organic carbon forms are present in rhizodeposits to make plant–microbe trophic interactions, and these microbes will respond to various carbon sources (Neumann et al. 2014; Eichorst and Kuske 2012). Under labile chemically recalcitrant substrate conditions, few bacterial communities (*Burkholderiales* and *Pseudomonales*) will show growth (Goldfarb et al. 2011). The types of release sugars by plant influence the microbial diversity; for example, *pseudomonads* (accepts wide range of sugars) can use trehalose (Raaijmakers and Weller 2001) in the tomato rhizosphere (Ghirardi et al. 2012). After sugars, amino acids are abundant class of compounds in the rhizosphere, and they impact rhizosphere microorganisms as signal molecules and nutrient sources which are important in biofilm formation (Moe 2013). Other compounds like phenolics impact the rhizosphere microbiota taxa. For example, the pathogen *Agrobacterium* spp. that shows trophic-mediated communication causes crown gall tumours by the insertion or transfer of T-DNA (bacterial genome) into plant genome and induces tumour cells to produce opines, and these opines are low molecular weight molecules (Chen et al. 2016).

Plant hormones and hormone-like compounds present in small amounts but impact the microbes at rhizosphere. Some of the phytochemicals release signals described for symbiotic or parasitic interactions, for example, flavonoids for symbiotic nitrogen fixation between rhizobia and legumes. These phenolics (acetosyringone) and phytoalexins are involved in parasitic interactions between phytopathogenic bacteria and plants (Subramoni et al. 2014; Kalia 2013).

Hormones involved in plant immunity and defence reactions; for example, the salicylic acid (SA) enriches *A. thaliana* (Lebeis et al. 2015), and the citric acid or malic acid (organic) acts as a signals bacterial chemotaxis towards plants called plant growth-promoting rhizobacteria (Zhang et al. 2014a, b), and they will involve in biofilm formation for the colonization of root bacteria (niche) to protect the bacteria from abiotic and biotic factors.

35.2.3 Contribution from Microbes to Plant Interactions

Rhizosphere-associated microorganisms can influence their host plant by releasing different signalling molecules and are good for plant defences against diseases, etc. Plant beneficial microorganisms are rhizobial bacteria, mycorrhiza, plant growth-promoting fungi (PGPF) and plant growth-promoting rhizobacteria (PGPR) (Cameron et al. 2013).

Plant at initially recognize this as non-self by pattern recognition receptors (PRRs), called microbe-associated molecular patterns (MAMPs) (Zamioudis and Pieterse 2012). These MAMPs trigger the immune defence in the roots (Millet et al. 2010; Vos et al. 2013). PGPR and PGPF are the components involved in plant defence response, which is known as induced systemic resistance (ISR) and acts against pathogens and insects (Pieterse et al. 2014). ISR depends on the jasmonic acid and ethylene signalling pathways (Conrath, 2006).

Rhizosphere microorganisms elicit plant responses not only via MAMPs, Nod and Myc factors released by rhizobia and mycorrhiza (Zamioudis and Pieterse 2012). Small secreted proteins (SSPs) and promoting mycorrhization are produced by mycorrhiza by altering hormonal signalling in the host plant (Plett and Martin 2015), and actually, these molecules function as signal compounds (Hartmann and Schikora 2012).

Other QS molecules from bacteria are less studied and include the Xanthomonas diffusible signal factor (DSF) and cyclodipeptides released by *Pseudomonas aeruginosa*, which affect gene expression for the abscisic acid and salicylic acid in plants (Ortiz-Castro et al. 2011; Xu et al. 2015).

Antimicrobials produced by rhizosphere *pseudomonads* and 2,4-Diacetylphloroglucinol (DAPG) present on roots (Weller et al. 2012) affect the development of root, and these occur via an auxin-dependent signalling pathway (Brazelton et al. 2008). Pyocyanin, a phenazine from *P. aeruginosa*, induces ISR and controls the development of root (Ortiz-Castro et al. 2014; Powers et al. 2015).

VOCs are the signalling molecule from microbes and involved as growth promoters or inhibitors (Bailly and Weisskopf 2012; Zamioudis et al. 2015). 2,3-Butanediol (2,3-BD) from *Bacillus* strains was demonstrated on plants to promote the growth of Arabidopsis (Ryu et al. 2004) and to induce ISR towards phytopathogenic *Erwinia* (Ryu et al. 2004). 2,3-BD produced by a *Pseudomonas chlororaphis* induces ISR against the *Erwinia* in tobacco (*Nicotiana tabacum*) (Han et al. 2006).

Indole is another bacterial VOC produced by PGPRs (Blom et al. 2011) and affects arabidopsis root development via the auxin signalling pathway (Bailly et al. 2014) and also functions as a protectant for plants against attacks from herbivorous insects (Erb et al. 2015).

A further phytohormone like compounds, including gibberellins, auxins, and cytokinins, affect growth, hormonal signalling, organ development and immune responses, in plants (Spaepen et al. 2007).

35.2.4 Climatic Responses on Plant–Microbe Interactions

Seasonal variation on microbial community will change drastically due to the continuously the concentrations levels of atmospheric CO2, and surface temperatures will increase as predicted (IPCC Climate Change 2007). The changes directly show impact on soil water levels and are expected to (Le Houérou 1996) increase drought in the world and affect terrestrial microorganisms such as plants. All land plants have excellent symbiotic relation with microorganisms (Brundrett 2009), and these plant growth-promoting microorganisms (PGPM) colonize at root, and some can also enter into root (endophytic) (Stone et al. 2000). Climate change with altered environmental conditions induces changes in plants and root secretions. These changes fluctuate the availability of chemoattractants or signal compounds (Haase et al. 2007). Similarly, the elevated temperature induces similar changes, and together they influence the plant and microbe associations. Change in climate will influence the diversity and activities and shows direct influence on plant and microbial communities (Drigo et al. 2008).

35.2.4.1 Direct Impact of Climate Change

On planet, most of plants are in association with mycorrhizal (Brundrett 2009), and most of these plants have symbiotic association with AMF (Fitter and Moyersoen 1996). These AMF are enhancing plant nutrient uptake or in exchange for carbohydrates from rhizosphere (Newsham et al. 1995; Sanders et al. 1998; Augé 2001). Climatic change alters the soil communities as the soil community differ plant growth rates have been reviewed extensively. Mostly, the microbial communities respond to warming and other agitations (Allison and Martiny 2008) and shifts in microbial community which leads to changes in ecosystem function (Bodelier et al. 2000).

Warming conditions alter the microbial soil respiration rates, and given no changes in community composition the microbial activity is defined as the factor with Q10. While decomposition of soil organic matter, soil respiration, and growth of microbial biomass generally increase with temperature (Bradford et al. 2008). Initially, warming can alter microbial communities (Zogg et al. 1997), or it may take many years (Rinnan et al. 2007, 2013). This is clear that temperature is coupled with soil and its moisture (Zak et al. 1999), and the bacterial communities mostly will respond to moisture pulses (Cregger et al. 2014). Drought amplifies fungal and bacterial groups (Briones et al. 2014) and leads to shift from one member to another in soil fungal and bacterial communities (Kaisermann et al. 2015).

35.2.4.1.1 Microbial–Plant Interactions

Under warming of soil, plants (Parmesan and Yohe 2003) start flowering earlier and leafing out will occur (Wolkovich et al. 2012). As a result of warming up at arctic region, the woody shrubs have replaced grasses (Pearson et al. 2013). Plant community transitions may be facilitating by the soil communities which are tightly coupled with plants, and these communities show a strong effect on survival of plant, expression and phenology (Wagner et al. 2014).

35.2.4.2 Indirect Effects

35.2.4.2.1 Climate Change on Plant and Microbial Population

Under climate change, the plant species migrations between the soil community and the plant (positive or negative relationship) (van der Putten 2012). Microbial population in soil will respond fast to climate change (van der Putten 2012). At local community level, the climate change can alter plant establishment and plant productivity (Bever et al. 2010). If plants that successfully establish, they induce higher levels of defines compounds (polyphenols) (Engelkes et al. 2008). Geographic disconnects might influence the composition and functioning of the microbial community (van der Putten 2012).

Above ground level, plant structures change due to climate change (Durán et al. 2014), and compared to aboveground level, communities belowground are structured by different environmental conditions (Fierer and Jackson 2006) with different life history characteristics. Due to these, plants are experiencing changes in global climate than soil community (Kardol et al. 2010). By changing distribution levels, the soil communities will respond to climate stress.

35.2.4.2.2 Climate Change Alters Plant Phenology and Microbial Communities

In the growing season, warming may affect the plant species in earlier leafing out and flowering (Wolkovich et al. 2012) and impacts root phenology, interactions of plant-rhizosphere (Iversen et al. 2015). If root growth peaks early, phenologies of below-ground and aboveground synchronous, because they are asynchronous (Abramoff and Finzi 2015). Phenology of root varies by species because it has complex interactions (Abramoff and Finzi 2015). As per IPCC 2013 due to climate change, variation in root–shoot phenology will impact rhizosphere interactions and may influence the soil microbial groups (Zhang et al. 2014a, b).

35.2.5 Microbes in Plant Growth Promotion

To enhancing productivity conventional agricultural practices, have a threat due to the global climate changes and anthropogenic activities in the agroecosystems. To overcome such negative roles of climatic challenges research on plant growth-promoting microbes (PGPM) playing a major role in agro-ecosystems to their original shape. PGPM are the soil and plant growth influencing rhizosphere micro-organisms, colonize plant roots with beneficial activities (Antoun and Prevost 2005). PGPM are of two main groups: plant growth-promoting fungi (PGPF) and plant growth-promoting rhizobacteria (PGPR). PGPR are soil bacteria that colonize the plant roots and enhance the growth in a mutualistic manner (Kapulnik and Okon 2002).

35.2.5.1 Plant Growth-Promoting Rhizobacteria (PGPR)

PGPR stimulate plant growth, form association with roots, leaves and/or in tissues (Glick 2012) and belong to *Rhizobium, Pseudomonas, Azotobacter, Klebsiella, Alcaligenes, Arthrobacter, Bacillus, Serratia,* etc. (Ahemad and Kibret 2014). They provide direct assistance in plant growth by nitrogen (N) fixation, phosphate solubilization (Sharma et al. 2013), iron (Fe) sequestration (Sayyed et al. 2013), phytohormones synthesis (Maheshwari et al. 2015) and phytopathogens control (Mishra and Arora 2012).

Biological Nitrogen Fixation (BNF) BNF is the conversion of atmospheric N to ammonia under the symbiotic process (Dixon and Kahn 2004) and well-developed process, and a vast array of microbes belongs to archaea and bacteria. Bacteria under N-fixing process will form obligate symbiotic association with legumes to form root nodules by colonizing plant's root system (known as rhizobia). Several studies show that at field conditions PGPR increase N content in legume plants (Bruijn 2015).

Phosphate-Solubilizing Bacteria (PSB) PSB involved in uptake of phosphate by plants and species are *Pseudomonas, Bacillus* and *Rhizobium* (Bossis et al. 2000). PGPR play a significant role in secretion of siderophores and are Fe-binding extracellular compounds (Krewulak and Vogel 2008; Boukhalfa et al. 2003),

Other than above-mentioned mechanisms, PGPB can produce broad spectrum of antimicrobial compound called as hydrogen cyanide (HCN) to control root diseases by plant-associated *pseudomonads* (Ramette et al. 2003). PGPR trigger ISR response against plant pathogens (Ramos et al. 2008) and involved in the synthesis of hydrolytic enzymes, which lyse the hyphae of fungi (Maksimov et al. 2011).

In the agriculture sector, members of the genus *Bacillus* sp. produced products which are important and being considered as microbial pesticides, fungicides or fertilizers (Fravel 2005). Another important PGPR organism such as *Pseudomonas* is considered as biocontrol and PGP activities (Tewari and Arora 2015) involved in biocontrol activity against plant pathogenic fungi (Tewari and Arora 2014). *Pseudomonas* produces a variety of components at the rhizospheric region (Fernando et al. 2005).

Actinomycetes in PGP Activities *Micromonospora* spp., *Streptomyces* spp., *Thermobifida* spp. and *Streptosporangium* spp. are involved in control process against root pathogenic fungi by phytohormones production (Solans et al. 2011), enzymes for degrading fungal cell wall (Anitha and Rabeeth 2010) and antibiotics production.

35.2.5.2 Plant Growth-Promoting Fungi (PGPF)

PGPF attributes of rhizospheric fungi, species include *Aspergillus, Penicillium,* and *arbuscular mycorrhizal fungus* (AMF) etc., gained attention due to the involvement in growth of plant and disease control. PGPF produces the plant hormones and involves in decomposition of organic matter and soil solubilization (Khan et al. 2012a, b). In plants, AMF involved in the growth promotion by nutrient uptake, particularly phosphorus (P) and disease suppression (Brundrett 2002; Maherali and Klironomos 2007).

35.3 Quorum Sensing: Microbial Role

Quorum sensing (QS) is coordinal behaviour that occurs between two different organisms by a communication, and this helps the bacteria to respond to scavenging of extracellular nutrients, extracellular hydrolytic enzymes, siderophores, biosynthesis of exopolysaccharides, pigments, antibiotics, aiding motility and for biofilms growth.

35.3.1 Biocommunication and Mechanisms

Rhizosphere around the roots is comprises of the microorganisms and various chemicals secreted from roots are involved in the regulation beneficial and pathogenic microorganisms at rhizosphere (Sharma et al. 2013). Release of chemicals involved through a communication and may export signals to the extracellular environment. The plants as root exudates and generated due to the organic carbon utilization (Bais et al. 2006). Due to its microbial activity it is very high at roots than the non-rhizosphere soil is a hotspot for variety of microorganisms.

During QS, organisms differentiate species-specific signals and interspecies behaviour modulations which enable them to specifically coordinate with species and with other diverse groups. Mycorrhizal fungi come under this category by supporting the growth of bacteria and degrading of complex organic materials to simple for fungi. By extending its hyphae, helps the plant growth by supplying enough nutrients or minerals.

QS in bacteria releases exoenzymes to convert complex food/carbon to simpler molecules and facilitates easy uptake by plants or any other associated cells.

Generally, bacterial QS falls into three classes: (1) AHL-dependent, (2) peptide-mediated QS, and (3) both gram-positive and gram-negative bacteria is luxS-encoded autoinducer 2 (AI-2) QS. In the gram-negative bacteria, QS is mediated by AHL.

In gram-positive bacteria, two types of QS systems are reported—one is autoinducing peptide (AIP), and the other is a two-component signal transduction. Gram-positive bacteria produce a signal peptide precursor which later cleaved at the double-glycine sequence and obtained active AIP, and the two-component signal transduction will regulate AIP which leads to QS.

Autoinducer 2 (AI-2) is observed both in gram-positive and gram-negative bacteria. For interspecies communication, AI-2 helps more (universal language) and characterized *Vibrio harveyi* for the regulation of bioluminescence.

35.3.1.1 Mechanism in Rhizosphere

At rhizosphere region, proteobacteria as a major colonizer produce QS signals and are capable to produce acyl-HSL signals with different specificities by the LuxI family.

In the rhizosphere, actinobacteria is a major organism and QS is dependent on the production of different chemicals class switch includes A-factor of *Streptomyces griseus* and similar compounds found in other *Actinobacteria*.

35.3.1.2 Quorum Sensing in Soil Microorganisms

Soil is a natural media for growth of microorganisms and plant. Several types of microorganisms inhabit the soil and create microenvironment by supporting the biogeochemical cycle, and they show additional benefit for plant growth by developing interactions (Sindhu et al. 2016).

35.3.2 Synthetic Biology in Quorum Sensing

In the present scenario, most of the studies discussed the importance of QC in microbiomes and how it will influence the composition and function of these communities. Synthetic biology is a field where scientists design systems (biological) with predictable design or output to manipulate QS in natural consortia, and it can be used as a tool to construct synthetic co-cultures with desired behaviour. Early studies made the QS circuits for programming cell behaviour. Synthetic biology provides new tools for investigating QS.

35.3.2.1 Synthetic Biology to Manipulate QS Signal and QS-Mediated Cell Phenotypes

Till today, there are a several QS systems available and, for example, AI-1 or acyl-homoserine lactone (AHL) and AI-2 QS systems. In QS systems, well-known systems are AHL (discussed earlier). Synthetic biologists frequently use the AHL QS systems due to the few components, and without specific transporters, they can enter into cell through cell membrane.

For the characterizing responses in AHL and to engineer cells, there are many efforts made by scientists and they did manipulation in the regulator protein LuxR. Wang et al. expressed LuxR to detect different ranges of AHL by a series of varied expression levels using constitutive promoters (Wang et al. 2015; Shong and Collins 2013; Zeng et al. 2017).

35.3.2.2 Synthetic Biology for Plant Microbiome

Plant microbiomes are the main components in plant health and crop yield. Engineered microbiomes can promote plant health by reducing the environmental impacts in agriculture (Busby et al. 2017; Compant et al. 2019). For the growth of plant and productivity, QS plays a major role in plant–microbiome interactions. Some pathogens depend on QS for the virulence production in some of the crops, and the community interactions can inhibit virulence. Virulence of *Pectobacterium carotovorum* can be attenuated by the degradation of AHL signals (Garge and Nerurkar 2016). Similarly, Valente et al. showed a crosstalk mechanism between species (Valente et al. 2017). In synthetic biology concern, engineers have used the QS to engineer commensal bacteria with desired behaviour. Zuniga et al. engineered a Rhizobacterium *Cupriavidus pinatubonensis* to produce indoleacetic acid by an autoinducer-regulated method (Zúñiga et al. 2018) so that the bacteria autonomously produced indoleacetic acid (IAA) to promote plant growth. QS is also an important process to facilitate interactions between species in other ecologically important microbiomes (coral microbiome).

35.3.2.3 Synthetic Biology in Plant Breeding

In agriculture, effective application of microorganisms is important for improved plant response to the environment and plant yield with the pathogen resistance. Under mutualism, plant and microbe will interact with each other and the plant secretion will influence microbiome composition (genotype-dependent) and population and is an evolutionary process. Using modern tools of synthetic biology, pathogen-resistant potato varieties were developed. Mendes et al. (2011) in common bean shown that the microbial taxa selected in breeding for resistance involved in complementing plant protection and finding out of such traits will help breeders to select plant traits with enriched microbial groups.

35.4 Conclusion

For the growth of plant, tolerance to adverse environmental conditions and plant health, a good plant and soil microbiota interactions are essential. Microorganisms adapted at different parts of the plant showed great impacts particularly in secreting of various chemicals, plant–microbe communication, plant growth promotion, etc. Understanding of plant–microbe and microbe–microbe interactions will provide a great help in future to modulating microbes in controlling disease and enhancing plant productivity. Plant-associated microbiota can act against harmful pathogens, thereby resulting in avoiding outbreaks in pathogen attack and increasing plant productivity. Plant-associated microbiomes also aid in stabilization of ecosystem and biodiversity enhancement. In the plant and microbial communication, QS plays a vital role against abiotic factors, environmental challenges and biosurfactant production in rhizosphere, for plant growth promotion, and is proven to be effective in improving the soil quality by soil remediation. In the coming future, these studies will further help in understanding the QS mechanisms and their role in soil fertility and crop productivity. Agriculture plays a major role for economic growth.

Therefore, a major focus on the exploitation of eco-friendly beneficial microorganisms in sustainable crop production in the coming decades would be of paramount importance. It is also critical to explore all potential applications of synthetic biology tools such as genetic engineering for the microbial flora at rhizosphere and generate engineered plants which can alter QS and show multiple functions in agriculture. Screening and identification of quorum quenching compounds for autoinducers or their receptors are another interesting areas for the plant microbial interaction and growth.

References

Abramoff RZ, Finzi AC (2015) Are above- and below-ground phenology in sync? New Phytol 205:1054–1061

Ahemad M, Kibret M (2014) Mechanisms and applications of plant growth promoting rhizobacteria: current perspective. J King Saudi Univ Sci 26:1–20

Allison SD, Martiny JBH (2008) Resistance, resilience, and redundancy in microbial communities. Proc Natl Acad Sci U S A 105:11512–11519

Andreote FD, Gumiere T, Durrer A (2014) Exploring interactions of plant microbiomes. Sci Agric 71(6):528–539. https://doi.org/10.1590/0103-9016-2014-0195

Andrews JH, Harris RF (2000) The ecology and biogeography of microorganisms on plant surfaces. Annu Rev Phytopathol 38:145–180

Anitha A, Rabeeth E (2010) Degradation of fungal cell walls of phytopathogenic fungi by lytic enzyme of *Streptomyces griseus*. African J Plant Sci 4:61–66

Antoun H, Prevost D (2005) Ecology of plant growth promoting rhizobacteria. In: Siddiqui ZA (ed) PGPR: biocontrol and biofertilization. Springer, Dordrecht, pp 1–38

Augé RM (2001) Water relations, drought and vesicular-arbuscular mycorrhizal symbiosis. Mycorrhiza 11:3–42

Aulakh MS, Wassmann R, Bueno C, Kreuzwieser J, Rennenberg H (2001) Characterization of root exudates at different growth stages of ten rice (*Oryza sativa* L.) cultivars. Plant Biol 3:139–148

Badri DV, Vivanco JM (2009) Regulation and function of root exudates. Plant Cell Environ 32:666–681

Bailly A, Weisskopf L (2012) The modulating effect of bacterial volatiles on plant growth: current knowledge and future challenges. Plant Signal Behav 7:79–85

Bailly A, Groenhagen U, Schulz S, Geisler M, Eberl L, Weisskopf L (2014) The inter-kingdom volatile signal indole promotes root development by interfering with auxin signalling. Plant J 80:758–771

Bais HP, Weir TL, Perry LG, Gilroy S, Vivanco JM (2006) The role of root exudates in rhizosphere interactions with plants and other organisms. Annu Rev Plant Biol 57:233–266. https://doi.org/10.1146/annurev.arplant.57.032905.105159

Baker KL, Langenheder S, Nicol GW, Ricketts D, Killham K, Campbell CD, Prosser JI (2009) Environmental and spatial characterisation of bacterial community composition in soil to inform sampling strategies. Soil Biol Biochem 41(11):2292–2298. https://doi.org/10.1016/j.soilbio.2009.08.010

Berendsen RL, Pieterse CM, Bakker PA (2012) The rhizosphere microbiome and plant health. Trends Plant Sci 17(8):478–486. https://doi.org/10.1016/j.tplants.2012.04.001

Berg G, Grube M, Schloter M, Smalla K (2014) Unraveling the plant microbiome: looking back and future perspectives. Front Microbiol 5. https://doi.org/10.3389/fmicb.2014.00148

Bever JD, Dickie IA, Facelli E, Facelli JM, Klironomos J, Moora M, Rillig MC, Stock WD, Tibbett M, Zobel M (2010) Rooting theories of plant community ecology in microbial interactions. Trends Ecol Evol 25:468–478

Blom D, Fabbri C, Connor EC, Schiestl FP, Klauser DR, Boller T, Eberl L, Weisskopf L (2011) Production of plant growth modulating volatiles is widespread among rhizosphere bacteria and strongly depends on culture conditions. Environ Microbiol 13(11):3047–3058. https://doi.org/10.1111/j.1462-2920.2011.02582.x

Bodelier PLE, Roslev P, Henckel T, Frenzel P (2000) Stimulation by ammonium-based fertilizers of methane oxidation in soil around rice roots. Nature 403:421–424

Bossis E, Lemanceau P, Latour X, Gardan L (2000) The taxonomy of *Pseudomonas fluorescens* and *Pseudomonas putida*: current status and need for revision. Agronomie 20:51–63

Boukhalfa H, Lack J, Reilly SD, Hersman L, Neu MP (2003) Siderophore production and facilitated uptake of iron and plutonium in *P. putida*. AIP Conf Proc 673:343–344

Bradford MA, Davies CA, Frey SD, Maddox TR, Melillo JM, Mohan JE, Reynolds JF, Treseder KK, Wallenstein MD (2008) Thermal adaptation of soil microbial respiration to elevated temperature. Ecol Lett 11:1316–1327

Brazelton JN, Pfeufer EE, Sweat TA, Brian B, Gardener M, Coenen C (2008) 2,4-diacetylphloroglucinol alters plant root development. Mol Plant-Microbe Interact 21:1349–1358

Briones MJI, McNamara NP, Poskitt J, Crow SE, Ostle NJ (2014) Interactive biotic and abiotic regulators of soil carbon cycling: evidence from controlled climate experiments on peatland and boreal soils. Glob Chang Biol 20:2971–2982

Broeckling CD, Broz AK, Bergelson J, Manter DK, Vivanco JM (2008) Root exudates regulate soil fungal community composition and diversity. Appl Environ Microbiol 74:738–744

Bruijn FJD (ed) (2015) Biological nitrogen fixation, 2 volume set. Wiley-Blackwell, Hoboken, NJ, p 2060

Brundrett MC (2002) Coevolution of roots and mycorrhizas of land plants. New Phytol 154:275–304

Brundrett MC (2009) Mycorrhizal associations and other means of nutrition of vascular plants: understanding the global diversity of host plants by resolving conflicting information and developing reliable means of diagnosis. Plant Soil 320:37–77

Bulgarelli D, Rott M, Schlaeppi K, Themaat EV, Ahmadinejad N, Assenza F, Rauf P, Huettel B, Reinhardt R, Schmelzer E, Peplies J (2012) Revealing structure and assembly cues for Arabidopsis root-inhabiting bacterial microbiota. Nature 488(7409):91–95. https://doi.org/10.1038/nature11336

Bulgarelli D, Schlaeppi K, Spaepen S, Themaat EV, Schulze-Lefert P (2013) Structure and functions of the bacterial microbiota of plants. Annu Rev Plant Biol 64:807–838. https://doi.org/10.1146/annurev-arplant-050312-120106

Busby PE, Soman C, Wagner MR, Friesen ML, Kremer J, Bennett A, Morsy M, Eisen JA, Leach JE, Dangl JL (2017) Research priorities for harnessing plant microbiomes in sustainable agriculture. PLoS Biol 15:e2001793

Cameron DD, Neal AL, van Wees SC, Ton J (2013) Mycorrhiza-induced resistance: more than the sum of its parts? Trends Plant Sci 18(10):539–545. https://doi.org/10.1016/j.tplants.2013.06.004

Chagnon PL, Bradley RL, Maherali H, Klironomos JN (2013) A trait-based framework to understand life history of mycorrhizal fungi. Trends Plant Sci 18:484–491

Chaparro JM, Bradi DV, Vivanco JM (2014) Rhizosphere microbiome assemblage is affected by plant development. ISME J 8:790–803. https://doi.org/10.1038/ismej.2013.196

Chen K, De Borne FD, Julio E, Obszynski J, Pale P, Otten L (2016) Root-specific expression of opine genes and opine accumulation in some cultivars of the naturally occurring genetically modified organism *Nicotiana tabacum*. Plant J 87(3):258–269. https://doi.org/10.1111/tpj.13196

Compant S, Samad A, Faist H, Sessitsch A (2019) A review on the plant microbiome: ecology, functions, and emerging trends in microbial application. J Adv Res 19:29–37
Conrath U (2006) Systemic acquired resistance. Plant Signal Behav 1:179–184
Cregger MA, Sanders NJ, Dunn RR, Classen AT (2014) Microbial communities respond to experimental warming, but site matters. Peer J 2:e358
Curl EA, Truelove B (1986) The rhizosphere. Springer, New York, pp 1–8
Dixon R, Kahn D (2004) Genetic regulation of biological nitrogen fixation. Nat Rev Microbiol 2:621–631
Drigo B, Kowalchuk GA, Van Veen JA (2008) Climate change goes underground: effects of elevated atmospheric CO_2 on microbial community structure and activities in the rhizosphere. Biol Fert Soils 44:667–679
Durán J, Morse JL, Groffman PM, Campbell JL, Christenson LM, Driscoll CT, Fahey TJ, Fisk MC, Mitchell MJ, Templer PH (2014) Winter climate change affects growing-season soil microbial biomass and activity in northern hardwood forests. Glob Chang Biol 20:3568–3577
Eichorst SA, Kuske CR (2012) Identification of cellulose-responsive bacterial and fungal communities in geographically and edaphically different soils by using stable isotope probing. Appl Environ Microbiol 78:2316e2327
Engelkes T, Morrien E, Verhoeven KJF, Bezemer TM, Biere A, Harvey JA, McIntyre LM, Tamis WLM, van der Putten WH (2008) Successful range-expanding plants experience less above-ground and below-ground enemy impact. Nature 456:946–948
Erb M, Veyrat N, Robert CAM, Xu H, Frey M, Ton J, Turlings TCJ (2015) Indole is an essential herbivore-induced volatile priming signal in maize. Nat Commun 6:6273. https://doi.org/10.1038/ncomms7273
Fernando WGD, Ramarathnam R, Krishnamoorthy AS, Savchuk SC (2005) Identification and use of potential bacterial organic antifungal volatiles in biocontrol. Soil Biol Biochem 37:955–964
Fierer N, Jackson RB (2006) The diversity and biogeography of soil bacterial communities. Proc Natl Acad Sci USA 103:626–631
Fitter AH, Moyersoen B (1996) Evolutionary trends in root–microbe symbioses. Philos Trans R Soc Lond B 351:1367–1375
Fravel D (2005) Commercialization and implementation of biocontrol. Annu Rev Phytopathol 43:337–359
Garge SS, Nerurkar AS (2016) Attenuation of quorum sensing regulated virulence of Pectobacterium carotovorum subsp. carotovorum through an AHL lactonase produced by Lysinibacillus sp. Gs50. PLoS One 11:e0167344
Ghirardi S, Dessaint F, Mazurier S, Corberand T, Raaijmakers JM, Meyer JM, Lemanceau P (2012) Identification of traits shared by rhizosphere-competent strains of fluorescent pseudomonads. Microb Ecol 64:725e737
Glick BR (2012) Plant growth-promoting bacteria: mechanisms and applications. Scientifica 2012:963401
Glynou K, Ali T, Buch AK, Haghi KS, Ploch S, Xia X, Çelik A, Thines M, Maciá-Vicente JG (2016) The local environment determines the assembly of root endophytic fungi at a continental scale. Environ Microbiol 18:2418–2434. https://doi.org/10.1111/1462-2920.13112
Goldfarb KC, Karaoz U, Hanson CA, Santee CA, Bradford MA, Treseder K, Brodie EL (2011) Differential growth responses of soil bacterial taxa to carbon substrates of varying chemical recalcitrance. Front Microbiol 2. https://doi.org/10.3389/fmicb.2011.00094
Gond SK, Bergen MS, Torres MS, White JF Jr (2015) Endophytic Bacillus spp. produce antifungal lipopeptides and induce host defence gene expression in maize. Microbiol Res 172:79–87. https://doi.org/10.1016/j.micres.2014.11.004
Haase S, Neumann G, Kania A, Kuzyakov Y, Römheld V, Kandeler E (2007) Elevation of atmospheric CO_2 and nutritional status modify nodulation, nodule-carbon supply, and root exudation of *Phaseolus vulgaris* L. Soil Biol Biochem 39:2208–2221

Haichar FZ, Marol C, Berge O, Rangel-Castro JI, Prosser JI, Balesdent J, Heulin T, Achouak W (2008) Plant host habitat and root exudates shape soil bacterial community structure. ISME J 2:1221–1230. https://doi.org/10.1038/ismej.2008.80
Han SH, Lee SJ, Moon JH, Park KH, Yang KY, Cho BH, Kim KY, Kim YW, Lee MC, Anderson AJ, Kim YC (2006) GacS-dependent production of 2R, 3R-butanediol by Pseudomonas chlororaphis O6 is a major determinant for eliciting systemic resistance against Erwinia carotovora but not against Pseudomonas syringae pv. tabaci in tobacco. Mol Plant-Microbe Interact 19(8):924–930. https://doi.org/10.1094/MPMI-19-0924
Hardoim PR, Overbeek LS, Elsas JD (2008) Properties of bacterial endophytes and their proposed role in plant growth. Trends Microbiol 16(10):463–471. https://doi.org/10.1016/j.tim.2008.07.008
Hartmann A, Schikora A (2012) Quorum sensing of bacteria and trans-kingdom interactions of N-acyl homoserine lactones with eukaryotes. J Chem Ecol 38:704–713
Higgins KL, Arnold AE, Coley PD, Kursar TA (2014) Communities of fungal endophytes in tropical forest grasses: highly diverse host- and habitat generalists characterized by strong spatial structure. Fungal Ecol 8:1–1. https://doi.org/10.1016/j.funeco.2013.12.005
Hirsch PR, Mauchline TH (2012) Who's who in the plant root microbiome? Nat Biotechnol 30:961–962
IPCC Climate Change (2007) Synthesis report. Summary for policymakers. http://www.ipcc.ch
Iversen CM, Sloan VL, Sullivan PF, Euskirchen ES, McGuire AD, Norby RJ, Walker AP, Warren JM, Wullschleger SD (2015) The unseen iceberg: plant roots in arctic tundra. New Phytol 205:34–58
Jones DL, Hodge A, Kuzyakov Y (2004) Plant and mycorrhizal regulation of rhizodeposition. New Phytol 163:459–480
Kaisermann A, Maron PA, Beaumelle L, Lata JC (2015) Fungal communities are more sensitive indicators to non-extreme soil moisture variations than bacterial communities. Appl Soil Ecol 86:158–164
Kalia VC (2013) Quorum sensing inhibitors: an overview. Biotechnol Adv 31:224e245
Kapulnik Y, Okon Y (2002) Plant growth promotion by rhizosphere bacteria. In: Waisel Y, Eshel A, Kafkafi U (eds) Plant roots: the hidden half. Marcel Dekker, New York, pp 869–885
Kardol P, Cregger MA, Campany CE, Classen TA (2010) Soil ecosystem functioning under climate change: plant species and community effects. Ecology 91:767–781
Kavamura VN, Santos SN, Silva JL, Parma MM, Avila LA, Visconti A, Zucchi TD, Taketani RG, Andreote FD, Melo IS (2013) Screening of Brazilian cacti rhizobacteria for plant growth promotion under drought. Microbiol Res 168:183–191
Khan AL, Hamayun M, Kang SM, Kim YH, Jung HY, Lee JH, Lee IJ (2012b) Endophytic fungal association via gibberellins and indole acetic acid can improve plant growth under abiotic stress: an example of Paecilomyces formosus LHL10. BMC Microbiol 12(1):1. https://doi.org/10.1186/1471-2180-12-3
Khan AL, Hamayun M, Radhakrishnan R et al (2012a) Mutualistic association of *Paecilomyces formosus* LHL10 offers thermotolerance to *Cucumis sativus*. Antonie van Leeuw 101:267–279
Kishore GK, Pande S, Podile AR (2005) Biological control of late leaf spot of peanut (*Arachis hypogaea*) with chitinolytic bacteria. Phytopathology 95:1157–1165
Krewulak KD, Vogel HJ (2008) Structural biology of bacterial iron uptake. Biochim Biophys Acta 1778:1781–1804
Le Houérou HN (1996) Climate change, drought and desertification. J Arid Environ 34:133–185
Lebeis SL, Paredes SH, Lundberg DS, Breakfield N, Gehring J, McDonald M, Dangl JL (2015) Salicylic acid modulates colonization of the root microbiome by specific bacterial taxa. Science 349:860–864
Lindow SE (1996) Role of immigration and other processes in determining epiphytic bacterial populations. Aerial Plant Surface Microbiol. https://doi.org/10.1007/978-0-585-34164-4_10

López-Bucio J, Cruz-Ramírez A, Pérez-Torres A, Ramírez-Pimentel JG, Sánchez-Calderón L, Herrera-Estrella L (2005) Root architecture. In: Turnbull C (ed) Plant architecture and its manipulation. Wiley-Blackwell Annual Review Series, Oxford, pp 181–206
Mackey D, McFall AJ (2006) MAMPs and MIMPs: proposed classifications for inducers of innate immunity. Mol Microbiol 61:1365–1371
Maherali H, Klironomos J (2007) Influence of phylogeny on fungal community assembly and ecosystem functioning. Science 316:1746–1748
Maheshwari DK, Dheeman S, Agarwal M (2015) Phytohormone-producing PGPR for sustainable agriculture. In: Maheshwari DK (ed) Bacterial metabolites in sustainable agroecosystem. Springer, Cham, pp 159–182
Maksimov IV, Abizgil'dina RR, Pusenkova LI (2011) Plant growth promoting rhizobacteria as alternative to chemical crop protectors from pathogens. Appl Biochem Microbiol 47:333–345
Malcolm GM, Kuldau GA, Gugino BK, Jiménez-Gasco MD (2013) Hidden host plant associations of soilborne fungal pathogens: an ecological perspective. Phytopathology 103(6):538–544. https://doi.org/10.1094/PHYTO-08-12-0192-LE
Manter DK, Delgado JA, Holm DG, Stong RA (2010) Pyrosequencing reveals a highly diverse and cultivar-specific bacterial endophyte community in potato roots. Microb Ecol 60(1):157–166. https://doi.org/10.1007/s00248-010-9658-x
Marques JM, Da Silva TF, Vollú RE, De Lacerda JR, Blank AF, Smalla K, Seldin L (2015) Bacterial endophytes of sweet potato tuberous roots affected by the plant genotype and growth stage. Appl Soil Ecol 96:273–281. https://doi.org/10.1016/j.apsoil.2015.08.020
Mathys J, De Cremer K, Timmermans P, Van Kerckhove S, Lievens B, Vanhaecke M, Cammue B, De Coninck B (2012) Genome-wide characterization of ISR induced in *Arabidopsis thaliana* by Trichoderma hamatum T382 against Botrytis cinerea infection. Front Plant Sci 3:1–25. https://doi.org/10.3389/fpls.2012.00108
Mendes R, Kruijt M, Bruijn I, Dekkers E, van der Voort M, Schneider JHM, Piceno YM, Santis TZ, Andersen GL, Bakker PAHM, Raaijmakers JM (2011) Deciphering the rhizosphere microbiome for disease-suppressive bacteria. Science 332:1097–1100
Millet YA, Danna CH, Clay NK, Songnuan W, Simon MD, Werck-Reichhart D, Ausubel FM (2010) Innate immune responses activated in Arabidopsis roots by microbe-associated molecular patterns. Plant Cell 22(3):973–990. https://doi.org/10.1105/tpc.109.069658
Mishra S, Arora NK (2012) Evaluation of rhizospheric *Pseudomonas* and *Bacillus* as biocontrol tool for *Xanthomonas campestris* pv campestris. World J Microbiol Biotechnol 28:693–702
Moe LA (2013) Amino acids in the rhizosphere: from plants to microbes. Am J Bot 100:1692e1705
Morgan JAW, Bending GD, White PJ (2005) Biological costs and benefits to plant-microbe interactions in the rhizosphere. J Exp Bot 56:1729–1739
Mostert L, Crous PW, Petrini O (2000) Endophytic fungi associated with shoots and leaves of *Vitis vinifera*, with specific reference to the Phomopsis viticola complex. Sydowia 52:46–58
Neumann G, Bott S, Ohler M, Mock HP, Lippmann R, Grosch R, Smalla K (2014) Root exudation and root development of lettuce (*Lactuca sativa* L. cv. Tizian) as affected by different soils. Front Microbiol 5:2. https://doi.org/10.3389/fmicb.2014.00002
Newsham KK, Fitter AH, Watkinson AR (1995) Multifunctionality and biodiversity in arbuscular mycorrhizas. Trends Ecol Evol 10:407–411
Noritake T, Kawakita K, Doke N (1996) Nitric oxide induces phytoalexin accumulation in potato tuber tissues. Plant Cell Physiol 37:113–116
Ortiz-Castro R, Díaz-Pérez C, Martínez-Trujillo M, del Río RE, Campos-García J, López-Bucio J (2011) Transkingdom signaling based on bacterial cyclodipeptides with auxin activity in plants. Proc Natl Acad Sci U S A 108:7253–7258
Ortiz-Castro R, Pelagio-Flores R, Méndez-Bravo A, Ruiz-Herrera LF, Campos-García J, López-Bucio J (2014) Pyocyanin, a virulence factor produced by *Pseudomonas aeruginosa*, alters root development through reactive oxygen species and ethylene signaling in Arabidopsis. Mol Plant-Microbe Interact 27(4):364–378. https://doi.org/10.1094/MPMI-08-13-0219-R

Ownley BH, Gwinn KD, Vega FE (2009) Endophytic fungal entomopathogens with activity against plant pathogens: ecology and evolution. In: The ecology of fungal entomopathogens. Springer, Dordrecht, pp 113–128. https://doi.org/10.1007/978-90-481-3966-8_9

Parmesan C, Yohe G (2003) A globally coherent fingerprint of climate change impacts across natural systems. Nature 421:37–42

Pearson RG, Phillips SJ, Loranty MM, Beck PSA, Damoulas T, Knight SJ, Goetz SJ (2013) Shifts in Arctic vegetation and associated feedbacks under climate change. Nat Clim Chang 3:673–677

Philippot L, Raaijmakers JM, Lemanceau P, Putten WH, Van D (2013) Going back to the roots: the microbial ecology of the rhizosphere. Nat Rev Microbiol 11:789–799

Pieterse CMJ, Zamioudis C, Berendsen RL, Weller DM, Van Wees SCM, Bakker PAHM (2014) Induced systemic resistance by beneficial microbes. Annu Rev Phytopathol 52:347–375

Plett JM, Martin F (2015) Reconsidering mutualistic plant–fungal interactions through the lens of effector biology. Curr Opin Plant Biol 26:45–50

Powers MJ, Sanabria-Valentín E, Bowers AA, Shank EA (2015) Inhibition of cell differentiation in Bacillus subtilis by *Pseudomonas protegens*. J Bacteriol 197(13):2129–2138

Raaijmakers JM, Weller DM (2001) Exploiting genotypic diversity of 2,4- diacetylphloroglucinol-producing Pseudomonas spp.: characterization of superior root-colonizing *P. fluorescens* strain Q8r1-96. Appl Environ Microbiol 67:2545–2554

Ramette A, Frapolli M, Défago G, Moënne-Loccoz Y (2003) Phylogeny of HCN synthase-encoding hcnBC genes in biocontrol fluorescent *Pseudomonads* and its relationship with host plant species and HCN synthesis ability. Mol Plant-Microbe Interact 16:525–535

Ramos SB, Barriuso MJ, de la Iglesia Pereyra MT, Domenech J, Gutierrez MFJ (2008) Systemic disease protection elicited by plant growth promoting rhizobacteria strains: relationship between metabolic responses, systemic disease protection, and biotic elicitors. Phytopathology 98:451–457

Rinnan R, Michelsen A, Baath E, Jonasson S (2007) Fifteen years of climate change manipulations alter soil microbial communities in a subarctic heath ecosystem. Glob Chang Biol 13:28–39

Rinnan R, Michelsen A, Baath E (2013) Fungi benefit from two decades of increased nutrient availability in tundra heath soil. PLoS One 8:10

Robinson RJ, Fraaije BA, Clark IM, Jackson RW, Hirsch PR, Mauchline TH (2015) Endophytic bacterial community composition in wheat (*Triticum aestivum*) is determined by plant tissue type, developmental stage and soil nutrient availability. Plant Soil 4:1–6. https://doi.org/10.1007/s11104-015-2495-4

Rosenblueth M, Martínez-Romero E (2006) Bacterial endophytes and their interactions with hosts. Mol Plant-Microbe Interact 19(8):827–837. https://doi.org/10.1094/MPMI-19-0827

Ross JJ, Reid JB, Weller JL, Simmons GM (2005) Shoot architecture I: regulation of stem length. In: Turnbull C (ed) Plant architecture and its manipulation. Wiley- Blackwell Annual Review Series, Oxford, pp 57–91

Ryu CM, Farag MA, Hu CH, Reddy MS, Kloepper JW, Paré PW, P.W. (2004) Bacterial volatiles induce systemic resistance in Arabidopsis. Plant Physiol 134:1017–1026

Sanders IR, Streitwolf-Engel R, Van Der Heijden MGA, Boller T, Wiemken A (1998) Increased allocation to external hyphae of arbuscular mycorrhizal fungi under CO2 enrichment. Oecologia 117:496–503

Sayyed RZ, Chincholkar SB, Reddy MS, Gangurde NS, Patel PR (2013) Siderophore producing PGPR for crop nutrition and phytopathogen suppression. In: Maheshwari KD (ed) Bacteria in agrobiology: disease management. Springer, Berlin, pp 449–471

Schardl CL, Florea S, Pan J, Nagabhyru P, Bec S, Calie PJ (2013) The epichloae: alkaloid diversity and roles in symbiosis with grasses. Curr Opin Plant Biol 16(4):480–488. https://doi.org/10.1016/j.pbi.2013.06.012

Schulz S, Dickschat JS (2007) Bacterial volatiles: the smell of small organisms. Nat Prod Rep 24:814–842

Sharma SB, Sayyed RZ, Trivedi MH, Gobi TA (2013) Phosphate solubilizing microbes: sustainable approach for managing phosphorus deficiency in agricultural soils. SpringerPlus 2:1–14

Shong J, Collins CH (2013) Engineering the esaR promoter for tunable quorum sensing- dependent gene expression. ACS Synth Biol 2:568–575
Sindhu SS, Parmar P, Phour M, Sehrawat A (2016) Potassium-solubilizing microorganisms (KSMs) and its effect on plant growth improvement. In: Meena VS, Maurya BR, Verma JP, Meena RS (eds) Potassium solubilizing microorganisms for sustainable agriculture. Springer, New Delhi, pp 171–185. https://doi.org/10.1007/978-81-322-2776-2_13
Solans M, Vobis G, Cassán F, Luna V, Wall LG (2011) Production of phytohormones by root-associated saprophytic actinomycetes isolated from the actinorhizal plant *Ochetophila trinervis*. World J Microbiol Biotechnol 27:2195–2202
Spaepen S, Vanderleyden J, Remans R (2007) Indole-3-acetic acid in microbial and microorganism-plant signaling. FEMS Microbiol Rev 31:425–448
Stone JK, Bacon CW, White JF (2000) An overview of endophytic microbes: endophytism defined. In: Bacon CW, White JF (eds) Microbial endophytes. Marcel Dekker, New York, pp 3–29
Subramoni S, Nathoo N, Klimov E, Yuan ZC (2014) Agrobacterium tumefaciens responses to plant-derived signaling molecules. Front Plant Sci 5(322)
Sugiyama A, Ueda Y, Zushi T, Takase H, Yazaki K (2014) Changes in the bacterial community of soybean rhizospheres during growth in the field. PLoS One. https://doi.org/10.1371/journal.pone.0100709
Tewari S, Arora N (2014) Multifunctional exopolysaccharides from *Pseudomonas aeruginosa* PF23 involved in plant growth stimulation, biocontrol and stress amelioration in sunflower under saline conditions. Curr Microbiol 4:484–494
Tewari S, Arora NK (2015) Plant growth promoting fluorescent *Pseudomonas* enhancing growth of sunflower crop. Int J Sci Technol Soi 1(1):1
Toju H, Yamamoto S, Sato H, Tanabe AS, Gilbert GS, Kadowaki K (2013) Community composition of root-associated fungi in a Quercus-dominated temperate forest: "codominance" of mycorrhizal and root-endophytic fungi. Ecol Evol 3:1281–1293. https://doi.org/10.1002/ece3.546
Truyens S, Weyens N, Cuypers A, Vangronsveld J (2015) Bacterial seed endophytes: genera, vertical transmission and interaction with plants. Environ Microbiol Rep 7:40–50
Valente RS, Nadal-Jimenez P, Carvalho AFP, Vieira FJD, Xavier KB (2017) Signal integration in quorum sensing enables cross-species induction of virulence in Pectobacterium wasabiae. MBio 8:e00398–e00317
van der Putten WH (2012) Climate change, aboveground-belowground interactions and species range shifts. Annu Rev Ecol Evol Syst 43:365–383
van Elsas JD, Chiurazzi M, Mallon CA, Elhottova D, Krištufek V, Salles JF (2012) Microbial diversity determines the invasion of soil by a bacterial pathogen. Proc Natl Acad Sci USA 109:1159–1164
Van Loon LC, Bakker PAHM, Pieterse CMJ (1998) Systemic resistance induced by rhizosphere bacteria. Annu Rev Phytopathol 36:453–483
Vogel TM, Simonet P, Jansson JK, Hirsch PR, Tiedje JM, Elsas JD, Bailey MJ, Nalin R, Philippot L (2009) Terra genome: a consortium for the sequencing of a soil metagenome. Nat Rev Microbiol 7:252
Vorholt JA (2012a) Microbial life in the phyllosphere. Nat Rev Microbiol 10(12):828–840. https://doi.org/10.1038/nrmicro2910
Vorholt JA (2012b) Microbial life in the phyllosphere. Nat Rev 10:828–840
Vos IA, Pieterse CMJ, Van Wees SCM (2013) Costs and benefits of hormone regulated plant defences. Plant Pathol 62:43–55
Wagner MR, Lundberg DS, Coleman-Derr D, Tringe SG, Dangl JL, Mitchell-Olds T (2014) Natural soil microbes alter flowering phenology and the intensity of selection on flowering time in a wild Arabidopsis relative. Ecol Lett 17:717–726
Wang B, Barahona M, Buck M (2015) Amplification of small molecule-inducible gene expression via tuning of intracellular receptor densities. Nucleic Acids Res 43:1955–1964

Waqas M, Khan AL, Lee IJ (2014) Bioactive chemical constituents produced by endophytes and effects on rice plant growth. J Plant Interact 9(1):478–487. https://doi.org/10.1080/17429145.2013.860562

Weller DM, Mavrodi DV, van Pelt JA, Pieterse CM, van Loon LC, Bakker PA (2012) Induced systemic resistance in Arabidopsis thaliana against Pseudomonas syringae pv. tomato by 2,4-diacetylphloroglucinol-producing Pseudomonas fluorescens. Phytopathology 102:403–412

Wolkovich EM, Cook BI, Allen JM, Crimmins TM, Betancourt JL, Travers SE, Pau S, Regetz J, Davies TJ, Kraft NJ, Ault TR, Bolmgren K, Mazer SJ, McCabe GJ, McGill BJ, Parmesan C, Salamin N, Schwartz MD, Cleland EE (2012) Warming experiments underpredict plant phenological responses to climate change. Nature 485:494–497. https://doi.org/10.1038/nature11014

Xu J, Zhou L, Venturi V et al (2015) Phytohormone-mediated interkingdom signaling shapes the outcome of rice–Xanthomonas oryzae pv. Oryzae interactions. BMC Plant Biol 15:10. https://doi.org/10.1186/s12870-014-0411-3

Yaish MW, Antony I, Glick BR (2015) Isolation and characterization of endophytic plant growth-promoting bacteria from date palm tree (*Phoenix dactylifera* L.) and their potential role in salinity tolerance. Antonie Van Leeuwenhoek 107(6):1519–1532. https://doi.org/10.1007/s10482-015-0445-z

Yandigeri MS, Meena KK, Singh D, Malviya N, Singh DP, Solanki MK, Yadav AK, Arora DK (2012) Drought-tolerant endophytic actinobacteria promote growth of wheat (*Triticum aestivum*) under water stress conditions. Plant Growth Regul 68(3):411–420. https://doi.org/10.1007/s10725-012-9730-2

Zak DR, Holmes WE, MacDonald NW, Pregitzer KS (1999) Soil temperature, matric potential, and the kinetics of microbial respiration and nitrogen mineralization. Soil Sci Soc Am J 63:575–584

Zamioudis C, Pieterse CM (2012) Modulation of host immunity by beneficial microbes. Mol Plant-Microbe Interact 25:139–150

Zamioudis C, Korteland J, Van Pelt JA, Van Hamersveld M, Dombrowski N, Bai Y, Hanson J, Van Verk MC, Ling HQ, Schulze-Lefert P, Pieterse CM (2015) Rhizobacterial volatiles and photosynthesis-related signals coordinate MYB72 expression in Arabidopsis roots during onset of induced systemic resistance and iron-deficiency responses. Plant J 84(2):309–322. https://doi.org/10.1111/tpj.12995

Zeng W, Du P, Lou Q, Wu L, Zhang HM, Lou C, Wang H, Quyang Q (2017) Rational design of an ultrasensitive quorum-sensing switch. ACS Synth Biol 6:1445–1452

Zhang HS, Li G, Qin FF, Zhou MX, Qin P, Pan SM (2014b) Castor bean growth and rhizosphere soil property response to different proportions of arbuscular mycorrhizal and phosphate-solubilizing fungi. Ecol Res 29:181–190

Zhang N, Wang D, Liu Y, Li S, Shen Q, Zhang R (2014a) Effects of different plant root exudates and their organic acid components on chemotaxis, biofilm formation and colonization by beneficial rhizosphere-associated bacterial strains. Plant Soil 374:689e700

Zogg GP, Zak DR, Ringelberg DB, MacDonald NW, Pregitzer KS, White DC (1997) Compositional and functional shifts in microbial communities due to soil warming. Soil Sci Soc Am J 61:475–481

Zúñiga A, Fuente F, Federici F, Lionne C, Bônnet J, de Lorenzo V, González B (2018) An engineered device for indoleacetic acid production under quorum sensing signals enables Cupriavidus pinatubonensis JMP134 to stimulate plant growth. ACS Synth Biol 7:1519–1527

Chapter 36
Molecular and Cellular Mechanisms Underlying the Microbial Survival Strategies: Insights into Temperature and Nitrogen Adaptations

Khaled A. Selim, Erik Zimmer, Heba Yehia, and Sofía Doello

Abstract Bacteria inhabit almost all ecological niches, including harsh environments of desert, oceans, hypersaline, volcanic and thermal biospheres, representing therefore one of the quantitatively most abundant organisms on earth. To survive under such a variety of ecological habitats, bacteria developed a number of strategies to rapidly adapt and respond to environmental changes by tuning down their metabolic activities, thus overcoming periods of unfavorable growth conditions. Generally, the processes of entering into and exiting from the metabolic stand-by mode are tightly regulated and characterized by a series of signaling events involving various secondary messenger molecules, signaling proteins, and regulatory RNAs. For example, the availability of nitrogen is highly variable in nature and, hence, considered as the limiting factor of microbial growth and development. Therefore, the nitrogen assimilation reactions require a tight regulation and a constant sensing of the quantity and quality of the available nitrogen. Temperature sensing is also essential for microbial survival. Consistently, microbes have developed diverse molecular strategies to sense temperature fluctuations and readjust their metabolism to survive and resume growth at a different temperature. In this chapter,

K. A. Selim (✉)
Organismic Interactions Department, Interfaculty Institute for Microbiology and Infection Medicine, Cluster of Excellence 'Controlling Microbes to Fight Infections', Tübingen University, Tübingen, Germany

Chemistry of Natural and Microbial Products Department, Pharmaceutical and Drug Industries Research Division, National Research Centre, Cairo, Egypt
e-mail: Khaled.selim@uni-tuebingen.de

E. Zimmer · S. Doello
Organismic Interactions Department, Interfaculty Institute for Microbiology and Infection Medicine, Cluster of Excellence 'Controlling Microbes to Fight Infections', Tübingen University, Tübingen, Germany

H. Yehia
Chemistry of Natural and Microbial Products Department, Pharmaceutical and Drug Industries Research Division, National Research Centre, Cairo, Egypt

D. K. Choudhary et al. (eds.), *Climate Change and the Microbiome*, Soil Biology 63,
https://doi.org/10.1007/978-3-030-76863-8_36

we summarize the recent advances in our understanding of the microbial adaptation strategies toward environmental changes, specifically those related to temperature fluctuations and changes in nitrogen availability.

Keywords Chlorosis and resuscitation · Glutamine synthase · Nitrogenase · PII signaling protein · Stress response · RNA thermometers · Cyanobacteria

36.1 Introduction

Climate change, global warming, and greenhouse effects are terms that have been pressingly discussed since the middle of the twentieth century in both scientific and political contexts. They describe the phenomenon of increasing average temperature on Earth, which is posited to reach 4 °C increment by 2100, and the corresponding dramatic and versatile consequences (Yang et al. 2017). Changes in average temperatures are reflected in changes in atmospheric gas composition, water surface and polar regions, the frequency of the freeze–thaw cycles in the alpine region, vegetation, and many other phenomena that affect and/or endanger many forms of life as well as the global food security situation.

Microorganisms are ubiquitous: the microbial pool in any definite ecosystem, whether terrestrial or aquatic, consists of a community of different members each playing a different role and interacting uniquely with their habitat and with their "neighbors" via various metabolic processes (Docherty and Gutknecht 2012; Abatenh et al. 2018; Cronan 2018). Unquestionably, biogeochemistry or the dynamics of any ecosystem, regardless of its size, cannot be studied without considering the microbial community diversity, structure, and contribution to the niche. Classically, microbes are regarded as enzymes' bags that carry out different metabolic activities and control the nutrient load for plants and animals or influence the suitability of the environment for these higher organisms, i.e., controlling the elemental cycles of different key elements in the biosphere (e.g., C and N cycling, redox cycling of different elements, organic compounds decomposition, molecule fixation, and oxygenic photosynthesis). Thus, the diversity of the microbial communities not only is important from the environmental microbiology or taxonomy perspective, but also directly affects the connected environment and living organisms (i.e., mineral resources, agriculture, crop yield, livestock) and hence the global biodiversity, food security, strategic industries, and energy resources (Rodriguez and Durán 2020; Voolstra and Ziegler 2020).

It is only natural then to infer that the microbial pools' structure and function are impacted by environmental changes such as temperature, pH, humidity, and emitted gases as the organisms' bioprocesses respond to the new conditions (Bradford 2013). However, until recently, the microbial factor was underrepresented in most of the modeling studies that describe the influence of varying environmental conditions on the ecosystem cycles and the elemental source-sink flux. Furthermore, studies that attempted to include microbial-related parameters into ecosystem change models lacked experimental validation due to: (1) the complexity of the microbial communities, (2) the limitation of the necessary techniques, (3) the difficulty to culture some strains under laboratory conditions, or (4) the absence of long-term in situ datasets to monitor the changes in microbiota structure and function (Singh et al. 2010; Dutta and Dutta 2016; Cavicchioli et al. 2019).

Due to the progress in metaomics, it is becoming more feasible to study the mechanisms and dynamics of microorganisms' growth and function (Rodriguez and Durán 2020; Voolstra and Ziegler 2020). Some of the different concepts, strategies, and methods have been summarized by Zak et al. (2006). However, some of these models did not thoroughly consider the ability of microorganisms to tolerate and adapt to changes both on the short and long terms (Docherty and Gutknecht 2012; Bradford 2013; Rousk and Bengtson 2014; Hallin and Bodelier 2020). Usually, short-term adaptation entails a temporary stress response through metabolic regulation and feedback mechanisms, while a long-lasting adaptation is the result of a permanent genetic acclimatization taking place over several generations. Currently, with the average earth temperature getting warmer and the concentration of greenhouse gases, namely carbon dioxide, ammonia, methane, nitrous oxide, and chlorofluorocarbons increasing, many research groups are trying to elucidate how the different microorganism communities react to these disturbances. Yet, it is very challenging to carry out such investigations due to the complexity and heterogeneity of the different microbiomes and the lack of thorough descriptions of their structure, dynamics, function, and relevant food webs at their specific locations (Singh et al. 2010; Dutta and Dutta 2016).

Nevertheless, all prokaryotes, especially the photoautotrophic cyanobacteria (Forchhammer and Selim 2020; Selim et al. 2020c), developed sophisticated strategies on both the molecular and cellular levels to overcome environmental stresses (Selim and Maldener 2021). In contrast, in comparison to prokaryotes, fungi existence in a certain location is minimally affected by ecological perturbations. This is attributed to their ability to form dormant spores, and to their thicker cell wall, especially in response to stress, mycelial growth, etc. (Gionchetta et al. 2019; Perez-Mon et al. 2020). In this chapter, we will discuss in detail the bacterial survival strategies in response to sudden changes in temperature and in nitrogen availability, the latter as an example of nutrient adaptation.

36.2 Various Microbial Responses to Changes in Environmental Conditions

Microorganisms are found all around us, even in the most extreme environments, which are inhabited by so-called extremophiles endowed with special adaptation mechanisms to suit harsh habitats (temperature, salinity, pH, carbon and nitrogen availability) (Rampelotto 2013; Merino et al. 2019). Each microbial genus exhibits optimal features of growth rate, enzyme activities, generation time, etc., in its native habitat conditions (Yang et al. 2017; Kosaka et al. 2019). Thus, mesophiles, which can only survive at a subglobal warming temperature, are the ones that face the greatest damage threats, in addition to the consequent perturbations in the metabolic cycles in which they are involved (Cavicchioli et al. 2019; Kosaka et al. 2019).

Changes or disturbances can occur at different magnitudes and frequencies resulting in quorum quenching, i.e., unsettling both the normal gene expression and physiologic homeostatic functions of individual microorganisms and the ongoing crosstalk between the microbial communities and their hosts and metaorganisms (Grandclement et al. 2016). These disturbances may either be short-term pulses or long-term pressures that change the nature of local environments (Shade et al. 2012). The organisms' reaction depends on several factors that include, but are not limited to: (1) microorganisms' generation time or doubling time and the disturbance duration relative to this time (discrete or continuous), (2) the microbial robustness against the disturbance (e.g., temperature range, greenhouse gas concentrations, and light intensity in deep waters), (3) the redundancy of the physiologic function(s) of the microorganism in the community (also referred to as the metaorganism), i.e., the ability of the neighboring species to serve the same roles, and (4) whether the microorganism is associated with a host (higher organism) that adds more pressure to the adaptation capacity and speed (Shade et al. 2012; Cavicchioli et al. 2019; Voolstra and Ziegler 2020).

Due to their higher turnover rates, microbial populations show the fastest responses and/or adaptations in comparison to plants and animals. With only the environmental factors in mind, it is generally accepted that the different microbial species respond either (1) **directly**, by adapting to the warming as allowed by their own features, e.g., critical high temperature (CHT), specific thermal optimum, functional enzymes, and thermal performance curve or spectrum; consequently, the ones that maintain vital functions at warmer temperatures are the ones that survive; or (2) **indirectly**, as a result of the response of the whole ecosystem to the changes, e.g., altered vegetation, plants and animals migration, and/or dying out (Drigo et al. 2007; Bradford 2013; Dutta and Dutta 2016; Yang et al. 2017; Kosaka et al. 2019; Voolstra and Ziegler 2020).

Different organisms show different survival responses when subjected to changes of the surroundings. They could either take advantage of the changes and undergo a developmental transition (e.g., develop special types of cells like akinetes, heterocyst, and hormogonia of cyanobacteria) or simply alter their food rhythms by consuming different sources of food (Selim and Maldener 2021). However, they could also be vulnerable to the changes and lose their fitness and ability to function on account of the different setup or selection pressure. Hence, what we can describe as adaptation is in fact either resilience or functional plasticity, exploitation or tolerance to the new conditions or disturbances. Several studies interpret the ability to accommodate changing conditions or disturbances as the result of microbial genetic variation (e.g., increasing the frequency of a favorable operon and effective DNA mutations) and, therefore, consider it as evolutionary adaptation (Franks and Hoffmann 2012; Shade et al. 2012; Gionchetta et al. 2019; Kosaka et al. 2019; Voolstra and Ziegler 2020; Wooliver et al. 2020). Nevertheless, it is still a matter of debate whether the adaptation mechanisms of an organism could truly serve to salvage it against the selection parameter of climate change, or it is inevitable to declare that the world is undergoing a widespread loss of biodiversity and mass extinction.

36.3 Microbial Responses to Warming with Underlying Genetic Disposition

Amid the very few studies that meant to study the microbial genetic adaptation to warming is the one conducted by Xue et al. (2016). These authors screened the active functional genes pool responsible for carbon (C), nitrogen (N), phosphorous (P), and sulfur (S) metabolism in terrestrial microbial communities subjected to temperature increase of 2 °C over 9 years. However, they did not investigate the behavior of individual organisms as certain species are activated/deactivated more than others. They found that the expression of the genes encoding enzymes responsible for decomposing recalcitrant C sources (e.g., vanillate *O*-demethylase oxygenase, glyoxal oxidase, and manganese peroxidase) increased significantly in comparison to the decrease of those that metabolize labile C (e.g., mannanase, xylanase, and acetylglucosaminidase) (Xue et al. 2016). This is comparable to the results from organic soil reported by Yang and his group. Both the microbial diversity and the availability of functional genes involved in labile C metabolism decreased in the soil incubated at 8 °C more than that incubated at near-freezing temperature −2 °C. Yet, genes responsible for recalcitrant carbon digestion did not increase in this report due to the incubation in the absence of oxygen (Yang et al. 2017). With respect to N degradation, Xue and his group documented that four out of 13 relevant genes increased, while three decreased. In a way, this could be attributed to the altered vegetation nature, total organic soil carbon, and C:N ratio after the temperature increase. All the genes playing roles in P and S cycles showed higher abundance after warming, reflecting the increased need for plant growth and carbon fixation as a feedback mechanism to the increased carbon dioxide concentration (Drigo et al. 2007; Xue et al. 2016).

A sound postulation by Kosaka and his group states that in a native habitat, organisms are usually subjected to scarcity of certain elements and, thus, more prone to mutations if compared to those grown in vitro in rich media. Using a thermotolerant mutant of the mesophile *Zymomonas mobilis*, they were able to show that thermal adaptation included reduced activity of NADH dehydrogenase (respiration activity) and, consequently, reduced harmful effects of oxidative stress (accumulation of reactive oxygen species) (Kosaka et al. 2019). Interestingly, different thermo-adapted *Z. mobilis* and *Escherichia coli* strains conserved the wild type-like cell size and shape at their CHT (i.e., not related to σ^s responses), and they all showed ratios of mutations in the same order (10^{-2}) to achieve thermal fitness, mostly in genes responsible for membrane stabilization, transporters

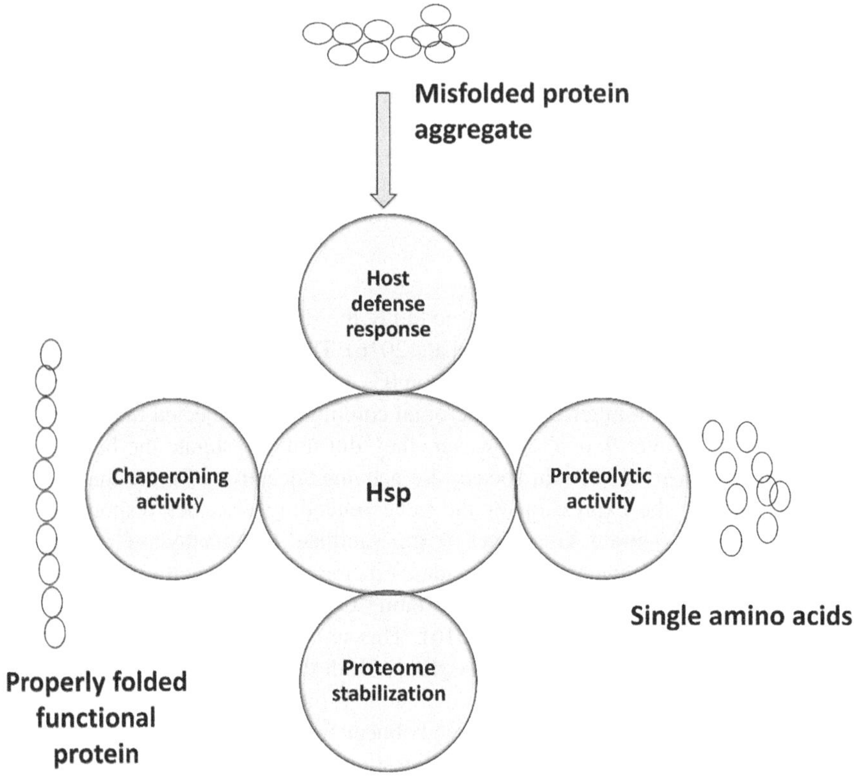

Fig. 36.1 Schematic illustration of heat shock proteins (Hsp) role in controlling proteostasis, modified from Maleki et al. (2016)

synthesis, transcription regulation, and protein proofing (Rudolph et al. 2010; Sandberg et al. 2014; Kosaka et al. 2019).

Furthermore, other mechanisms exist that regulate how bacteria respond to thermal changes, albeit not previously discussed in the context of global warming. However, they definitely fit the narrative when the whole evolutionary picture is evoked. Among these mechanisms are the universally conserved (a) RNA thermometers (RNA-Ts), which modulate the expression of the encoding downstream cistron, and (b) the enhanced or differential expression of heat shock proteins (Hsps), whose function is either to solubilize misfolded proteins that aggregated at high temperature (chaperone activity) or to facilitate their breakdown (protease activity) (discussed below in details; Fig. 36.1).

RNA thermometers are regulatory elements located in the intergenic regions of open reading frames (ORF). Being in the noncoding 5′-untranslated region of some mRNA molecules, they are able to coordinate the ribosomal binding, i.e., controlling the gene expression at the translational level (Narberhaus et al. 2006). At lower temperatures, the mRNA exists in a hairpin-like conformation that masks the

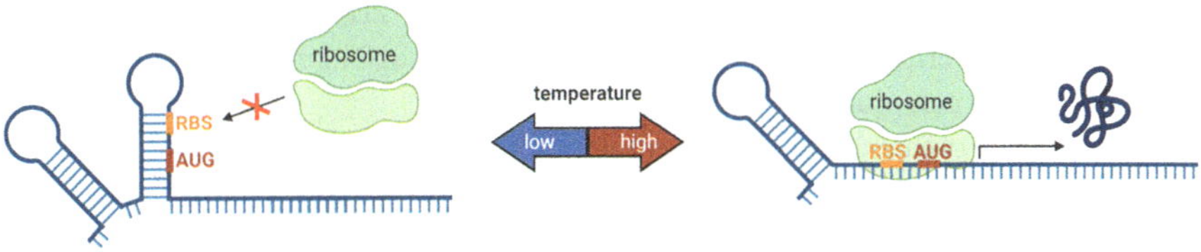

Fig. 36.2 Model for the zipper-like RNA thermometer (RNA-T). RNA-T occluding the ribosomal binding site (RBS) and/or the translation start codon (AUG) of the mRNA by base pairing until temperature is increased, which causes a reversible disruption of the zipper-like RNA structure, to initiate the protein translation, modified from Kortmann and Narberhaus (2012), Righetti and Narberhaus (2014), and Loh et al. (2018)

ribosome binding site (RBS) from ribosomal unit attachment. As the nucleic acid melts and unwinds at higher than ambient temperature, the RNA-T functions as a zipper that switches on/off the expression of the downstream genes that are under its control (Fig. 36.2). These are mainly cold and heat shock proteins and virulence-controlling factors in some pathogenic bacteria. RNA-Ts are extremely sensitive to temperature changes and start reacting to an increase of 1 °C, with the intensity of the response augmenting as the temperature further increases (Narberhaus et al. 2006; Kortmann and Narberhaus 2012). RNA-Ts contain very short conserved motifs and sometimes none, making their identification in the genome not possible via bioinformatics tools. They can be divided into three families with varying nucleotide length, the common feature among them being the mismatched noncanonical base pairing. The most common family is the repressor of heat shock gene expression family (ROSE), the members of which consist of up to 100 nucleotides that form up to four stem-loops with different heat stabilities and, thus, variable levels of control of RBS occlusion. The double-strand binding strength decreases in the 3′ direction where the first to unfold is the one exposing the Shine–Dalgarno sequence (SD) and the AUG start codon (Chowdhury et al. 2006; Kortmann and Narberhaus 2012). The second family is known as the fourU thermometer and was first reported in *Salmonella enterica* to control the production of the Hsp aggregation suppression A (agsA). As the name implies, members of this family consist of four uridine nucleotides that form hydrogen bonds with AGGA in the SD sequence (Kortmann and Narberhaus 2012; Abduljalil 2018). The third and simplest group is the cyanobacterial thermometer detected at the 5′-end of the heat shock protein 17 (Hsp17) gene. This thermometer comprises one hairpin-like secondary structure that normally blocks the expression of Hsp17, which is responsible for both the solubility and structural integrity of cellular proteins (Kortmann and Narberhaus 2012; Cimdins et al. 2014).

The first discovered RNA-T was reported in connection to the alternative sigma factor rpoH gene product (σ^{32}, sigma H) in *E. coli*. Under stress conditions, σ^{32} is liberated from a complex with the chaperones DnaK and DnaJ and binds to the RNA polymerase core enzyme (E) shifting it from the promoters of housekeeping genes to those of stress-related genes. Thereby, the transcription of more than 30 different Hsps transcription is initiated (e.g., ClpB, DnaK, GroEL, GroES, HtpG, YedU)

(Narberhaus and Balsiger 2003; Wang and DeHaseth 2003; Calloni et al. 2012; Righetti and Narberhaus 2014), enabling the cell to salvage aggregated proteins and, hence, secure protein quality control and proteostasis.

Heat shock proteins are ubiquitous molecules, produced in response to biotic and abiotic insults (e.g., temperature, osmotic or oxidative stress, starvation, and infectious agents). They are subdivided into groups depending on their size and function and are located in different parts of the prokaryotic or eukaryotic living cell (cytosol, mitochondria, nucleus, etc). Their sizes range from 8 to 28 kDa in the case of small ATP-independent Hsps to 40–105 kDa for the larger ATP-dependent ones. They either function in solubilizing the misfolded proteins that aggregated at high temperature (chaperone activity) or in facilitating their breakdown (protease activity). They were also reported to have other purposes unrelated to stress protection such as immune system modulation in higher eukaryotes (Tiwari et al. 2015; Maleki et al. 2016; Miller and Fort 2018). In addition to their protein trafficking and chaperoning activity and cell homeostasis/stabilization functions, Hsps were also implicated in rescuing cells from death via an antiapoptotic mechanism. This involves the interaction with several caspase proteins to disrupt the formation of an effective apoptosome, thus inhibiting caspase proteolytic cascade-dependent cell death (e.g., Hsp70 to Apaf-1 and Hsp27 to pro-caspase-3) (Beere 2004).

When it comes to the underlying conceptualization of the microbial response to global warming, many theories are discussed. However, none is yet thoroughly understood and deemed as a correct fact, and many are contradictory and require further investigation. On the one hand, some reports support the idea of "hotter-is-better; HiB." This means that, up to a maximum temperature, all microbial performance and metabolic vital signs (e.g., respiration rate, and growth rate) will increase with increasing temperature, and hence, the microbial population should thrive (Deutsch et al. 2008; Angilletta et al. 2010; Dell et al. 2011; Huey et al. 2012; Smith et al. 2019). The positive thermal response was, for instance, proven by Smith and his group for many mesophilic prokaryotic strains after both short- and long-term exposure to a higher culturing temperature (Smith et al. 2019). Moreover, the increased greenhouse gas concentrations, another feature of global warming, trigger a feedback cycle to control the flux of these gases as modeled in many climate studies. Of special concern is the CO_2 levels as it overlaps with the global carbon cycle and affects the aquatic and terrestrial environments. Elevated temperatures are assumed to result in higher rates of microbial respiration which, in addition to accelerated organic matter decomposition, leads to a positive feedback with a resulting increase in temperature (Bardgett et al. 2008; Singh et al. 2010; Dutta and Dutta 2016; Cavicchioli et al. 2019). On the other hand, other scientists posit reduced metabolic activities and proliferation of microbes in response to environmental warming. This stress response may be mediated by the hyperphosphorylated alarmone tetra- or pentaphosporylated guanosine (p)ppGpp. Albeit usually studied in relation to starvation, (p)ppGpp can also be released in response to changes in temperature, pH, and osmotic pressure. It acts as a sink for activated guanine (GTP) and directly affects some enzymes like GTPases, primases, and RNA polymerases modulating all their relevant activities. Among the negatively influenced bioprocesses are DNA synthesis (initiation and elongation) as well as RNA and protein synthesis (especially the ribosomal RNA) and ribosomal unit assembly (Steinchen and Bange 2016; Kosaka et al. 2019; Ronneau and Hallez 2019).

36.4 Microbial Adaptation Strategies to Changes in Nutrient Availability (Nitrogen Starvation)

Nitrogen is the most abundant gaseous element in the Earth's atmosphere, representing about 78% of the air. Nitrogen is the simplest building block of life and is essential for all living organisms, since it is a major component of amino acids and nucleic acids. In nature, the availability of nitrogen is highly variable, as most microorganisms are unable to fix atmospheric N_2 (discussed below). Hence, nitrogen is considered one of the limiting factors for microbial growth and development due to its presence in a limited amount of useful forms (i.e., combined nitrogen sources such as ammonia, urea, nitrite, and nitrate) in microbial habitats, especially in the ocean. In this section, we will discuss in detail the microbial adaptation strategies for efficient utilization of nitrogen, with special emphasis on nitrogen assimilation reactions and the mechanisms for overcoming limitation of combined nitrogen.

36.4.1 Adaptation to Variable Nitrogen Availability Via Regulation of the Glutamine Synthetase

Bacterial metabolism requires a tight regulation and a constant sensing of the quantity and quality of the nitrogen and carbon availability. The nitrogen assimilation reactions lead to a consumption of 2-oxoglutarate (2-OG), which represents an indicator of the intracellular carbon/nitrogen balance due to its intermediate positioning between the TCA cycle and nitrogen assimilation cycle (Muro-Pastor et al. 2001; Commichau et al. 2006; Luque and Forchhammer 2008; Forchhammer 2010; Huergo and Dixon 2015; Forchhammer and Selim 2020; Selim et al. 2020c). Generally, the nitrogen and carbon metabolisms are coordinated by a complex crosstalk between different input signals (Fig. 36.3) (Commichau et al. 2006; Luque and Forchhammer 2008). The sensing and regulation of the nitrogen/carbon metabolisms in several bacteria mainly depend on the signal transduction PII protein, which senses the energy/carbon/nitrogen status of the cell through binding ATP/ADP and 2-OG in presence of ATP (see below) (Fig. 36.3; Fokina et al. 2010; Lapina et al. 2018; recently reviewed in Forchhammer and Selim 2020 and Selim et al. 2020c).

For efficient nitrogen assimilation, bacteria possess two pathways to integrate inorganic nitrogen in the form of ammonium (NH_4^+) into organic molecules: the glutamate dehydrogenase (GDH) reaction (Fig. 36.3) and the glutamine synthase/

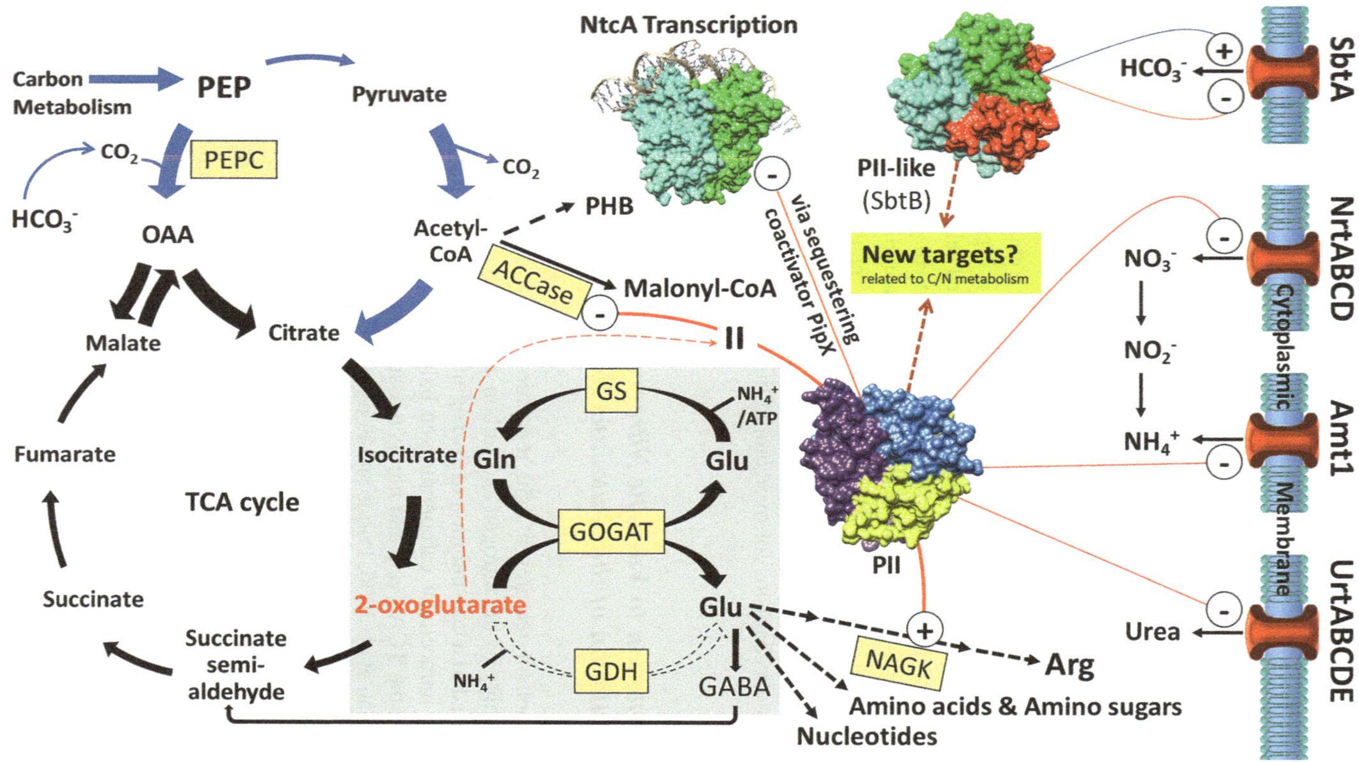

Fig. 36.3 Crosstalk between nitrogen and carbon metabolism is modulated by cell signaling protein of PII superfamily, modified from Forchhammer and Selim (2020). Metabolic basis of C:N balance sensing at the intersection of carbon and nitrogen metabolisms and the role of PII signaling protein in regulating key physiological processes (indicated by red lines). Changes in the 2-oxoglutarate level, which report changes in the C/N balance, are sensed by PII protein. The

2-OG levels are refilled through TCA cycle via carbon anabolic reactions and consumed by the GS-GOGAT cycle (nitrogen assimilation reactions). Glutamate is the primary amino acid and acts as nitrogen donors for several anabolic reactions (indicated by dotted lines), with arginine biosynthesis being of particular interest. PII signaling protein regulates: (1) enzymes through controlling the key reactions of arginine biosynthesis via N-acetyl glutamate kinase (NAGK) (Selim et al. 2020c, 2020b) and of fatty acid synthesis via Acetyl-CoA carboxylase, (ACCase) (Selim et al. 2020c); (2) transcription factors like NtcA (discussed below) (Forchhammer and Selim 2020); (3) as well as the uptake of the nitrogen sources nitrate, ammonium, and urea through their respective uptake systems as indicated (Forchhammer and Selim 2020). Recently, new class of PII-like protein called SbtB was found to control carbon metabolism (Selim et al. 2018; recently reviewed in Forchhammer and Selim 2020). Notably, another class of PII-like protein called CutA was found to be highly abundant in the cells, implying a key role in the intracellular signaling processes. However, at present, the importance of CutA proteins in cell signaling processes waits for further biochemical and physiological investigations (Selim et al. 2020a) glutamine oxoglutarate aminotransferase (GS-GOGAT) cycle (Figs. 36.3 and 36.4). GDH can aminate 2-oxoglutarate (2-OG) to glutamate (Glu), using NH_4^+ and NAD (P)H/H$^+$. Since GDH has a relatively low affinity for NH_4^+ and requires no ATP, this reaction mostly takes place under NH_4^+-excess and energy-limiting conditions. The most widely used pathway for nitrogen incorporation in bacteria is the GS-GOGAT cycle, in which NH_4^+ and Glu are condensed to glutamine (Gln) in a reaction catalyzed by the glutamine synthetase (GS; encoded by *glnA*) with the use of one ATP molecule (Forchhammer and Selim 2020). Following this reaction, Gln and 2-OG are transaminated into two molecules of Glu, thereby recovering the Glu molecule initially used by the GS and providing an additional Glu molecule, which can be used as a building block to synthesize other amino sugars and amino acids, such as arginine. This second reaction is catalyzed by the Gln:2-OG aminotransferase (GOGAT) and requires oxidation of one NAD(P)H/H$^+$ or of one ferredoxin (Figs. 36.3 and 36.4). In summary, the GS-GOGAT cycle uses one 2-OG, one NH_4^+, one ATP, and one reduction equivalent to yield one Glu. Given that the GS-GOGAT cycle is the main metabolic pathway of NH_4^+ assimilation in bacteria, this makes GS one of the central enzymes in nitrogen metabolism and its regulation is of key importance for the optimization of nitrogen utilization (Bolay et al. 2018).

GS is found in all three domains of organisms and can be categorized into three types: GSI, GSII, and GSIII, which differ in their protein structure and regulatory mechanisms. Due to the omnipresence of GS in all clades of life, it is believed that these different types emerged before the evolutionary divergence into eukaryotes and prokaryotes. GSI and GSIII occur in bacteria and archaea, and both are dodecamers consisting of two parallel hexameric rings (Fig. 36.4a). Furthermore, GSIII occurs in a few eukaryotic species. Some bacterial species possess one or multiple GSI genes, while others possess genes for both GSI and GSIII or only for GSIII. It is suggested that GSI and GSIII genes occur in both bacteria and archaea due to multiple lateral gene transfer events. GSI is further divided into GSI-α and GSI-β, whereby GSI-β contains an additional 25-amino-acid insertion and is usually posttranslationally regulated by adenylylation at conserved tyrosine (Tyr) residues that are missing in GSI-α. However, there are some exceptions to this classification. Finally, the decameric GSII is mainly present in eukaryotes, with the exception of few bacterial species (Brown et al. 1994).

Moreover, new types of bacterial GS-like enzymes have been recently discovered as unique adaptation strategies to fulfill new metabolic needs in these bacteria and to utilize a variety of nitrogen sources, other than NH_4^+. The evolution pressure on the *glnA* gene encoding GS led to the emergence of various *glnA*-like genes, which have thus far been less studied (Krysenko et al. 2017, 2019). For example, *Streptomyces coelicolor* possesses two classical *glnA* genes, encoding for GSI and GSII, whose functions are well characterized, as well as three other genes, *glnA2-glnA4*, annotated as GS-like enzymes (Rexer et al. 2006). The GS-like enzyme GlnA3 was found to encode for a gamma-glutamylpolyamine synthetase, which is required for polyamine metabolism and detoxification, allowing *S. coelicolor* to utilize and grow on the toxic polyamines like spermine, spermidine, putrescine, or cadaverine, as a sole nitrogen source (Krysenko et al. 2017). Furthermore, the GS-like enzyme GlnA4

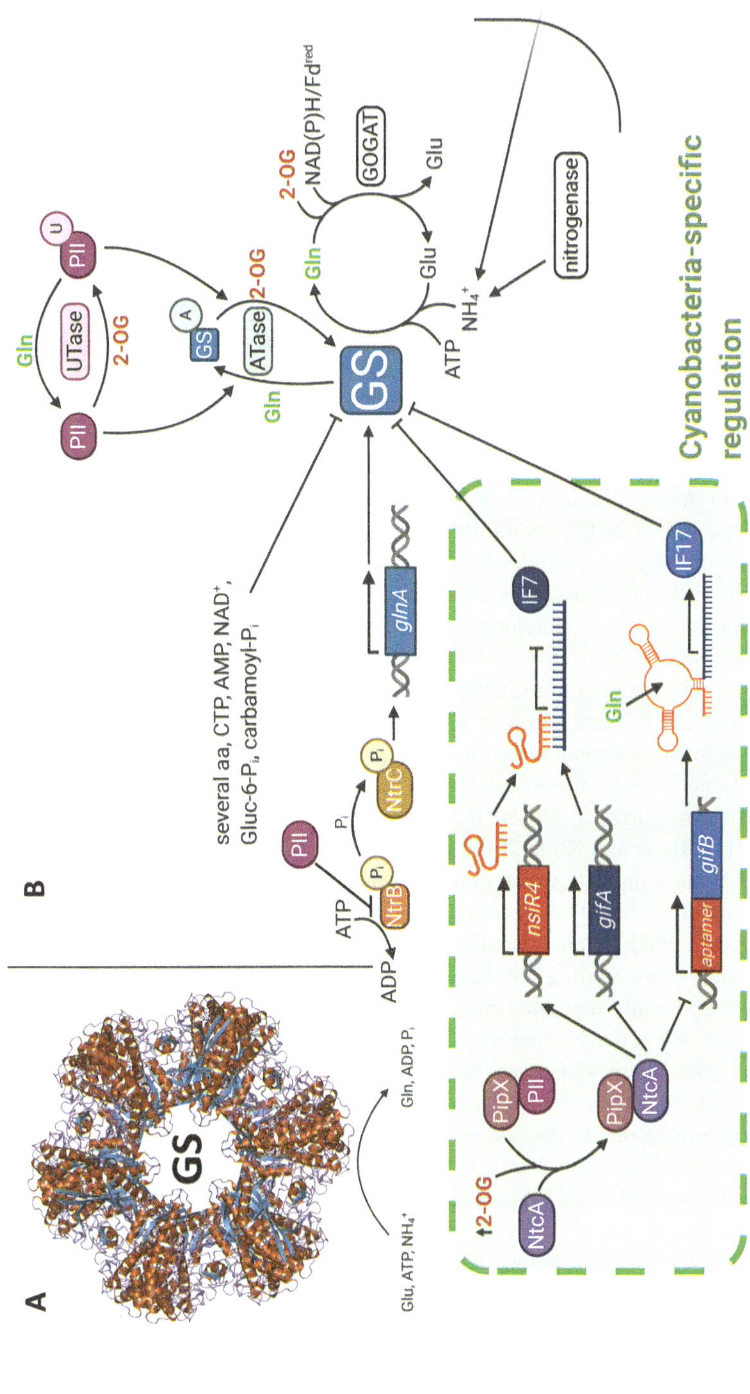

Fig. 36.4 Model for the general modes of GS regulations, with indirect role for PII on GS

was found to encode for a gamma-glutamyl ethanolamine synthetase, which is required for ethanolamine utilization. Through the activity of GS-like GlnA-4, *S. coelicolor* is able to utilize ethanolamine as a sole carbon or nitrogen source, although it grows poorly on ethanolamine as a nitrogen source as it lacks a specific transporter for ethanolamine (Krysenko et al. 2019). The emergence of the evolutionary GS-like enzymes like GlnA3 and GlnA4 clearly reflects survival adaptation strategies of bacteria to compete with other microbes for resources and to occupy a wide variety of ecological niches.

The regulation of GS is complex and takes place at multiple levels (e.g., transcriptional, posttranscriptional, and posttranslational levels) within different bacterial phyla, as each phylum or even bacterium possesses unique regulatory mechanisms for its own GS enzyme(s) (Fig. 36.4b). For instance, in many proteobacteria, GSI is transcriptionally regulated by the NtrB/NtrC two-component histidine kinase system, which has been well studied in *E. coli* (Huergo et al. 2013). Additionally, GS is posttranslationally regulated by feedback inhibition of different metabolites, such as amino acids and adenine nucleotides, and by modification such as (de-)adenylylation of the enzyme. One of the central status reporter metabolites involved in this regulation is 2-OG. Since 2-OG is consumed by the GS-GOGAT cycle to assimilate NH_4^+, its intracellular concentration decreases under N-excess and, vice versa, increases under N-limitation. In addition, 2-OG reports on the C:N balance in bacteria because it is a metabolite in the tricarboxylic acid (TCA) cycle and the origin of many anaplerotic reactions within this metabolic pathway. The signal transduction from 2-OG to GS is indirect, and it is mediated by the PII protein, as well as further signal transduction proteins specific for different bacterial phyla (Bolay et al. 2018; Forchhammer and Selim 2020). Remarkably, the NtrB component of the two-component histidine kinase (NtrB/NtrC) system, which regulates GS transcriptionally, is regulated as well by the PII protein via direct protein–protein interaction to form a PII-NtrB complex under nitrogen-replete conditions (Fig. 36.4b) (Jiang and Ninfa 2009; Huergo et al. 2013).

This complex regulation is required because the availability of nitrogen and the occurrence of different inorganic nitrogen species, like nitrate (NO_3^-), nitrite (NO_2^-), and NH_4^+, are usually highly variable in bacterial environments (Forchhammer and Selim 2020; Selim and Haffner 2020). Therefore, bacteria need to be able to flexibly tune their nitrogen metabolism in accordance with sudden as well as to season-related changes in the environmental conditions. This is especially true for cyanobacteria as primary producers: They possess a particular flexibility in their ability to tune their nitrogen metabolism, since they inhabit all types of aqueous habitats and must often cope with nitrogen limitation (Forchhammer and Schwarz 2019; Selim and Maldener 2021). In the following paragraphs, we will take a closer look into the regulation of GSI in different cyanobacterial species.

Cyanobacteria have evolved unique mechanisms to regulate their GSI (Fig. 36.4b). They do not possess the NtrBC two-component system. The transcription of *glnA* gene is instead regulated by the global nitrogen regulator NtcA, a dimeric transcription factor unique to cyanobacteria (Fig. 36.5). The transcriptional regulation of GSI (*glnA*) by NtcA responds to the levels of 2-OG. Under nitrogen

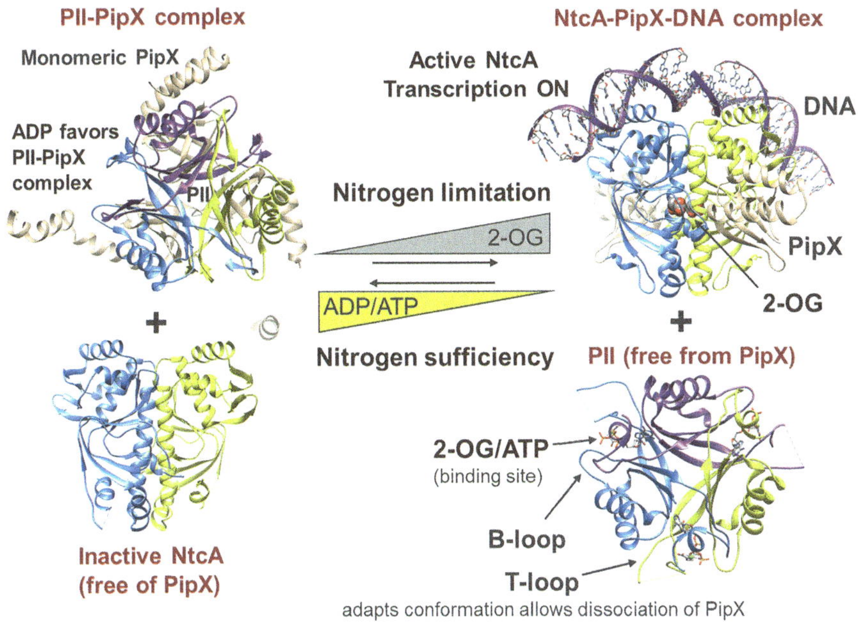

Fig. 36.5 Regulation of NtcA via PipX and PII protein. PipX is an activator of nitrogen transcription factor regulator, NtcA. Under nitrogen excess condition indicated by low 2-OG, high levels of ADP favor the PII-PipX complex formation (PDB: 2XG8), leaving NtcA in an inactive form. When 2-OG levels raise under nitrogen limiting conditions, 2-OG binds to PII and causes dissociation of PII-PipX complex. NtcA binds 2-OG as well and successfully competes for PipX binding. The NtcA-PipX complex (PDB: 2XKO; Llácer et al. 2010) has high affinity to the NtcA-DNA binding sites and switches on the transcription of NtcA-dependent genes, adapted from Forchhammer and Selim (2020)

limiting conditions, the intracellular concentration of 2-OG increases, and the increased 2-OG concentration mediates the interaction of one NtcA dimer with two PipX monomers (summarized in Forchhammer and Selim 2020). PipX is an activator of NtcA and increases the affinity of NtcA for its target promoters characterized by the conserved consensus sequence GTA-N_8-TAC (Espinosa et al. 2007). Thereby, NtcA can either act as an activator or as a repressor of transcription depending on the relative position of the NtcA consensus motif GTA-N_8-TAC to the transcriptional start site (TSS). Expression of *glnA* and other nitrogen assimilation genes is activated by NtcA. Again, the NtcA-PipX interaction is under the control of the master nitrogen regulator PII protein, which also senses the intracellular 2-OG concentrations (Lapina et al. 2018; Selim et al. 2019; Forchhammer and Selim 2020). Upon increasing N availability, lower 2-OG levels lead to the interaction of one PII trimer with three PipX monomers, which prevents PipX (the NtcA activator) from binding to NtcA (Fig. 36.5). Under these conditions, NtcA has low affinity for its target promoters, resulting in a decreased *glnA* expression and, in turn, a reduction in the intracellular 2-OG concentrations due to GOGAT activity

(Forcada-Nadal et al. 2017). The increase in 2-OG levels under N-limitation conditions disrupts the PII-PipX interaction and leaves PipX free again to interact with and activate NtcA (Fig. 36.5).

On the posttranslational regulatory level, GSI of cyanobacteria like GS of other bacteria is regulated via feedback inhibition by several amino acids as well as ATP and AMP. GSI is not regulated by adenylylation. Instead, cyanobacteria possess two regulatory proteins, namely the inactivating factor (IF)7 and IF17 functioning as protein inhibitors of GSI (García-Domínguez et al. 1999; Pantoja-Uceda et al. 2016). In this function, IF7-like proteins exist in most cyanobacterial species. The IF7 and IF17 are small intrinsically disordered proteins, i.e., proteins missing a proper folded structure, which seem to fold upon binding to GSI, although a recent NMR study revealed that IF7 remains disordered even upon binding to GS (Saelices et al. 2011; Neira et al. 2020). Biochemical and physiological analyses revealed that both IF7 and IF17 are required for full inhibition of GS (García-Domínguez et al. 1999). The C-terminus of the 17 kDa IF17 protein displays sequence similarity to the smaller 7 kDa IF7 protein (Saelices et al. 2011). By an unknown mechanism, the additional N-terminus of IF17 seems to enhance the stability of IF17 protein, which is proteolytically degraded in the absence of NH_4^+. In contrast, the degradation of IF7 is performed by the Prp1/Prp2 metalloprotease and is not affected by the availability of nitrogen sources. Both inactivating factors are proposed to bind GSI via electrostatic interactions. Three arginine (Arg) residues in IF7 and IF17 were identified to be important for the interaction with the negatively charged face of GSI. Compared to IF7, IF17 contains an additional lysine (Lys) residue participating in the binding of GSI, which leads to a higher binding affinity of IF17 to GSI than that of IF7. This enhanced binding affinity together with the N-terminus of IF17 is hypothesized to be the reasons for the stronger inhibitory effect of IF17 on GSI compared to IF7. However, the C-terminal part of IF7 appears to be involved as well in the binding of GSI, thus contributing to form the GSI-IF7 complex (Neira et al. 2020). The exact mechanism by which how binding of IF7 inhibits GS activity is so far unknown. However, it is known that the two inactivating factors inhibit GSI independently from each other in a concentration-dependent manner and are able to completely inactivate GS enzymatic activity together. Like *glnA*, also *gifA* and *gifB*, the genes encoding for IF7 and IF17, respectively, are subjected to transcriptional regulation by the master nitrogen transcription factor NtcA (Fig. 36.4b). In contrast to *glnA*, expression of the inactivating factors is repressed by NtcA leading to decreased IF7 and IF17 accumulation under nitrogen-limited conditions. Apart from this transcriptional control by regulatory proteins, both *gifA* and *gifB* genes were recently shown to be regulated by noncoding RNA (ncRNA) molecules as well (Bolay et al. 2018).

In the cyanobacterial genus *Synechocystis*, translation of the *gifA* mRNA into IF7 is regulated by the small regulatory RNA (sRNA) NsiR4 (nitrogen stress-induced RNA 4). *Synechocystis* NsiR4 has a length of 63 nucleotides (nt) and displays a secondary structure with two stem-loops. Furthermore, a shorter form of NsiR4 occurs in some cyanobacterial species inhabiting a *nsiR4* gene. This form is missing the 20 nt at the 5′-end forming one of the stem-loops described for *Synechocystis* NsiR4. The 16-nt-long unpaired region between the two stem-loops is hypothesized to bind to the 5′-untranslated region (UTR) of its target *gifA* mRNA by complementary base pairing. Consequently, translation of this mRNA is inhibited, for example, by blocking the ribosomal binding site (RBS) of the mRNA. Furthermore, binding of a sRNA to its target mRNA usually decreases the stability of the mRNA. The transcription of NsiR4 and, therefore, the repression of *gifA* expression are induced under N-limiting conditions by NtcA, which functions as an activating transcription factor for NsiR4. This leads to increased GSI activity, thereby decreasing the concentration of 2-OG via the GS-GOGAT cycle. The decreased 2-OG concentration then decreases the activity of NtcA and, consequently, enhances the translation of IF7 to inhibit GSI activity. The parallel regulation of IF7 expression by NtcA directly as a transcriptional repressor of *gifA* and indirectly as a transcriptional activator of NsiR4 is known as a feed-forward loop. It is hypothesized that the translational control via NsiR4 is important for the cells to compensate the delayed response of the transcriptional repression alone (Klahn et al. 2015). This is necessary to respond to sudden and strong fluctuations in the concentration of the available nitrogen sources, as it is the case in the aquatic habitats of cyanobacteria (Selim and Maldener 2021). The gene *nsiR4* is conserved in all species of the β-cyanobacterial subsections but is absent in α-cyanobacteria (Klahn et al. 2015). These α-cyanobacteria inhabit oceans, while β-cyanobacteria are found in freshwater and coastal areas. Therefore, it is reasonable to speculate that NsiR4 only evolved in β-cyanobacteria or that it was evolutionary lost in α-cyanobacteria as an adaptation to the often nutrient-deficient saltwater environment. This would be consistent with the hypothesis of marine bacteria's environmental adaptation to N-cost minimizing measures, which will be discussed later on.

In *Synechocystis,* the translation of the *gifB* mRNA into IF17 is controlled by a riboswitch, which is another type of ncRNA. The riboswitch consists of a type 1 Gln aptamer, formerly known as glnA aptamer, located upstream of the *gifB* gene in the 5'-UTR and, therefore, cotranscribed with *gifB* in one mRNA. When transcribed, the aptamer mRNA forms an RNA ring with two stem-loops as a secondary structure. This secondary structure binds Gln specifically promoting a conformational change in the mRNA secondary structure which then leads to translation of the downstream *gifB* mRNA (Fig. 36.4b). The existence of a Gln-binding aptamer is the first proof that Gln functions as a status reporter metabolite in cyanobacteria in parallel to 2-OG. Mechanistically, the binding of Gln to the aptamer is thought to promote a long-range Watson–Crick interaction that subsequently melts a short mRNA double-strand. As this double-strand blocks the RBS, binding of Gln and the ensuing conformational changes free the RBS and allow translation initiation. The activated expression of IF17 then leads to a strong inhibition of GSI activity, thereby decreasing the Gln concentration, which in turn frees the type 1 Gln aptamer in a feedback loop. The regulation of IF17 via the riboswitch is independent from NtcA. However, like for IF7, the expression of IF17 is regulated on the level of transcription (by NtcA) as well as on the level of translation. Like described above, the translational regulation is particularly important for the cells to react to sudden changes of N availability in the cyanobacterial habitats. The binding affinity of the type 1 Gln aptamer to Gln is in the range of the dissociation constant *(K_D)* values 0.5–5 mM.

This corresponds to the highly dynamic intracellular concentrations of Gln. With respect to the binding kinetics, this perfectly makes sense since binding affinity differs the most with ligand concentrations in the range of the K_D of the target. The binding affinity of the type 1 Gln aptamer to Gln is therefore perfectly tuned to the physiological relevant concentrations of ligand. Type 1 Gln aptamers occur in virtually all cyanobacterial species possessing an IF17 homolog. The wide distribution again undermines the importance of the translational regulation of *gifB* in the control of the cellular nitrogen metabolism in cyanobacteria. Another type of Gln-regulated riboswitch, namely the type 2 Gln aptamer, is solely found in marine picocyanobacteria like *Prochlorococcus*. This type 2 Gln aptamer is located in the 5'-UTR of *gifB*-like genes suggesting a comparable regulatory function on IF17-like proteins as the type 1 Gln aptamer on IF17 in other cyanobacteria. Even though the identified IF17-like proteins lack the C-terminal part of IF17, which is crucial for binding to GSI, another binding mechanism to the target enzyme could have coevolved in *Prochlorococcus*. Marine picocyanobacteria inhabit extremely nitrogen-deficient environments, and they are able to survive along a wide range of the vertical water column. Furthermore, *Prochlorococcus* is believed to be the most widespread and most abundant living organism on earth, and it is responsible for the major stake of oceanic primary production. Therefore, *Prochlorococcus* is a very successful bacterium, and this is in accordance with a perfect adaptation to its nutrient-poor habitat. Like mentioned before, multiple N-cost minimizing mechanisms are believed to be responsible for this adaptation, for instance, genome streamlining. One of these mechanisms is the substitution of regulatory proteins with ncRNAs because less nitrogen is consumed in the transcription of ncRNAs than in the expression of proteins. This is a possible explanation for the occurrence of a unique type 2 Gln aptamer in picocyanobacteria (Klahn et al. 2018).

36.4.2 Metabolic Adaptation to Nitrogen Deprivation

Limitation of a combined nitrogen source (like NO_3 or NH_4^+) is one of the most common hurdles bacteria face in natural environments. Depending on their ability to utilize atmospheric N_2, bacteria can be grouped into two different categories: diazotrophic and nondiazotrophic. Diazotrophic bacteria are able to fix atmospheric N_2 into a more biologically usable form, such as ammonia, via the activity of the nitrogenase enzyme (e.g., *Azospirillum* sp., *Cyanothece* sp., *Nostoc* sp., and *Anabaena variabilis*). On the contrary, nondiazotrophic bacteria cannot fix atmospheric N_2 and require a source of combined nitrogen for growth and for anabolic and catabolic cellular processes (e.g., *E. coli*, *Synechococcus elongatus* PCC 7942 and *Synechocystis* sp. PCC 6803) (Selim and Haffner 2020). In the following sections, we will review the different metabolic strategies adopted by diazotrophic and nondiazotrophic bacteria in response to nitrogen starvation.

36.4.2.1 Nitrogen Fixation by Diazotrophic Bacteria

Nitrogen fixation is a process in which atmospheric N_2 is converted to ammonia or other biologically usable nitrogen compounds. As the nitrogenase is the only enzyme known to catalyze the fixation of gaseous N_2 (Bothe et al. 2010), we therefore would like to summarize in this section the evolutionary and adaptation aspects of the nitrogenase in different bacteria. Nitrogenase is a protein complex formed by the two different proteins dinitrogenase reductase and dinitrogenase. The dinitrogenase reductase, which is also called iron (Fe) protein, is a homodimer and contains a [4Fe-4S] cluster responsible for supplying the catalytic dinitrogenase with electrons from ferredoxin or other reducing equivalents. One molecule N_2 is reduced to two molecules of ammonia NH_3 within the molybdenum-iron (MoFe) cofactor of the dinitrogenase, which is also called MoFe protein (Fig. 36.6a). This reaction requires eight electrons, 16 molecules ATP, and eight protons and releases one molecule of dihydrogen H_2 as side product. The heterotetrameric dinitrogenase consists of two alpha and two beta subunits and receives electrons from the dinitrogenase reductase. It contains two cofactors, the electron-channeling P-cluster located at the interface of one α- and β-subunit, and the mentioned catalytic MoFe cluster located in the α-subunits. Besides the MoFe cofactor containing nitrogenase (Nif), there is also a vanadium–iron (VFe) cofactor containing isoform (Vnf), as well as an isoform with a Fe-only cofactor (Anf). The expression of Vnf- and Anf-type nitrogenases was shown to be dependent on the availability of the trace element Mo, on temperature and on salinity. Both isoforms are believed to have evolved from the ancestral Nif-type nitrogenase since they only occur in prokaryotes possessing Nif nitrogenase (Mus et al. 2019).

The nitrogenase is irreversibly inactivated by oxygen (O_2). A progenitor of the current Mo-dependent Nif nitrogenase is believed to have emerged around 3.5 billion years ago (Bya) in hydrogenotrophic methanogens, which are archaea. This predates the rise of atmospheric oxygen levels due to the emergence of oxygenic photosynthesis around 2.3 Bya (Allen et al. 2019). Consequently, this explains the susceptibility of nitrogenase to O_2, which did not represent an issue at the time of the evolution of the first nitrogenase. By multiple lateral gene transfer events, Nif nitrogenase was passed on to the domain of bacteria in which the enzyme is broadly distributed. Bacteria with many different lifestyles like phototrophs, chemotrophs, and aerobic as well as obligate and facultative anaerobic bacteria, all possess Nif. Generally, aerobic bacteria have a larger number of *nif* genes encoding the enzyme itself, as well as accessory proteins important for synthesis, regulation, and protection of the nitrogenase. This is likely to be an adaptation to the more efficient metabolism of aerobes. Furthermore, the larger gene number is in accordance with the higher turnover and protection of the nitrogenase from the damage caused by O_2 (Mus et al. 2019).

The sensitivity of the nitrogenase to O_2 poses a major challenge for aerobic N_2-fixing bacteria and even more for photoautotrophs that produce oxygen by oxygenic photosynthesis, like cyanobacteria. Bacteria have evolved different

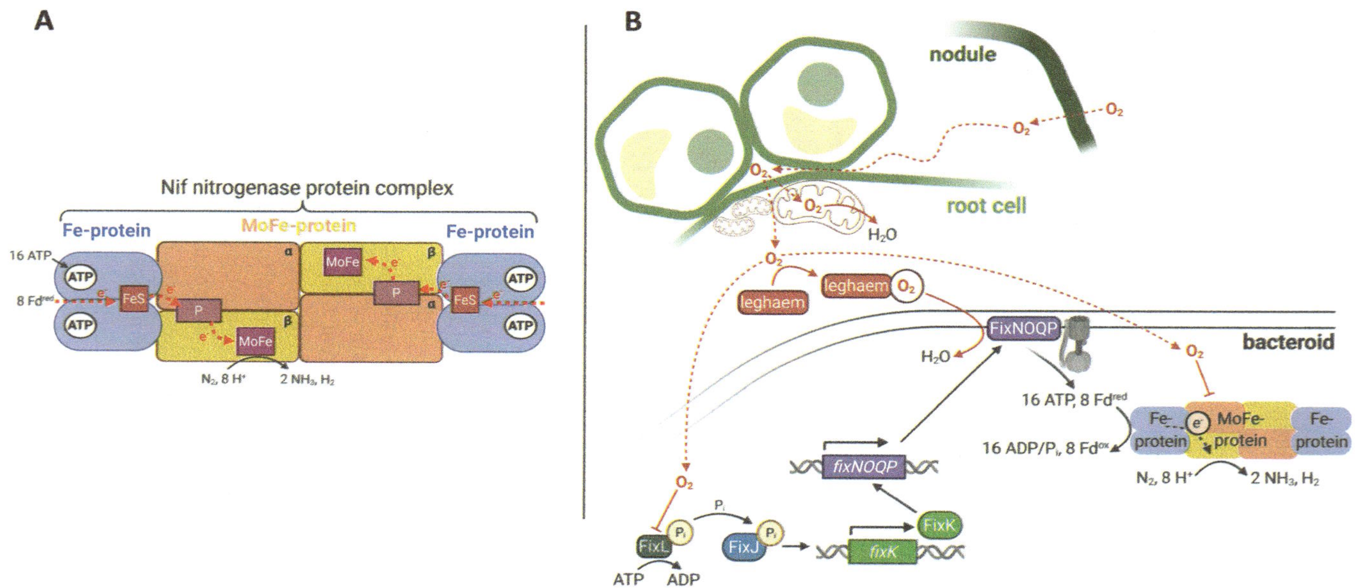

Fig. 36.6 Model for Nif nitrogenase (**a**) and bacteroid differentiation (**b**) to protect the O_2-senstive nitrogenase from O_2 damage

solutions to cope with this challenge. For example, facultative anaerobic bacteria perform N_2-fixation under anaerobic conditions only. This can be achieved by temporally controlling nitrogenase expression and activity to occur only under anaerobic growth conditions. The facultative anaerobe *Azotobacter* was shown to protect its nitrogenase from O_2 by encapsulating its cell with the formation of the polysaccharide alginate under aerobic conditions (Sabra et al. 2000). Furthermore, this bacterium shows an elevated respiratory activity, thereby rapidly consuming intracellular O_2 to form an anaerobic environment around its nitrogenase (Oelze 2000).

Concerning the bacterial genus *Rhizobium*, protection of the nitrogenase from O_2 damage is challenging, because these bacteria are obligate aerobes requiring respiration to cope with the high energy demand of their nitrogenase. *Rhizobium* is proteobacteria that has a symbiotic relationship with legume plants fueling the plant with assimilated nitrogen and receiving carbon assimilates in return. In the process of nodulation, *rhizobia* first migrate into nodules, specialized organs in the plant's root. Subsequently, the bacteria are taken up by plant root cells as endosymbionts into so-called symbiosomes. The bacterial cell within one of the symbiosomes undergoes a differentiation process driven by altered *fix* and *nif* gene expression to become a so-called bacteroid (Fig. 36.6b). This morphologically differs from free-living *Rhizobia* and is specialized in nitrogen fixation within the nodule. The bacteroid is protected from O_2 by three main mechanisms, which together result in nearly anoxic conditions at the nitrogenase expressed within the bacteroid. Firstly, the nodule possesses a cortical O_2 diffusion barrier that allows the plant to regulate O_2 entry into the nodules. Secondly, the mitochondria of the root host cells are relocated to the cell periphery, where they function as an additional barrier by consuming O_2 via respiration. Thirdly, the O_2-binding protein leghemoglobin is expressed in the plant cell cytoplasm. Leghemoglobin is structurally similar to myoglobin; however, it has a much higher affinity and much faster binding kinetics to O_2. Thereby, leghemoglobin concentrates O_2 and guarantees its rapid and even distribution in the nodule. Moreover, leghemoglobin-bound oxygen does not harm the nitrogenase, but it is still available to the bacteroid's respiratory chain. The reason for this is the expression of a high-affinity terminal oxidase, complex IV of the respiratory chain, from the *fixNOQP* operon in bacteroids (Bergersen and Appleby 1981; Rutten and Poole 2019).

Regarding oxygen-producing bacteria, almost all cyanobacteria possess a MoFe cofactor containing nitrogenase. In filamentous diazotrophic cyanobacteria, the nitrogenase is protected from oxygen inside cells specialized in N_2 fixation, so-called heterocyst cells, which possess a thick cell wall and lack O_2-producing photosystem II (PSII). Thereby, nitrogen fixation is spatially separated from O_2-evolving photosynthesis, which occurs in undifferentiated vegetative cells. Hence, the ATP needed for nitrogen fixation is produced by photosystem I (PSI) with cyclic photophosphorylation in vegetative cells and then transferred to the neighboring heterocysts in the form of electron-rich substrates like sucrose. Cell differentiation into heterocyst is tightly regulated and characterized by a series of signaling events via many secondary messenger molecules (Agostoni and

Montgomery 2014). Calcium (Ca^{2+}) is one of the most important second messengers, which is known to operate more broadly in metabolic signaling and/or differentiation processes in cyanobacteria. A significant role for Ca^{2+} has been speculated in the tight regulation of heterocyst differentiation and photosynthesis (Zhao et al. 2005; Shi et al. 2006; Walter et al. 2016, 2019, 2020).

A rare mechanism to protect the nitrogenase from oxygen is the differentiation of vegetative cells to diazocytes observed in the filamentous cyanobacterium *Trichodesmium*. In contrast to heterocysts, diazocytes possess no protective cell envelope and contain both photosystems. However, they have a high respiratory metabolic activity; hence, they consume free oxygen as described above for *Azotobacter* (Sandh et al. 2012). In contrast, some unicellular cyanobacteria temporally separate the two cellular processes of N_2-fixation and oxygen-evolving photosynthesis by only fixating N_2 under anaerobic conditions at night and in microbial mat communities with high respiratory activity (Mus et al. 2019). This protective mechanism is the most common among cyanobacteria, while heterocyst-forming filamentous cyanobacteria are thought to only have evolved about 408 million years ago (Mya) (Allen et al. 2019).

Recently, it was proposed that the late emergence of heterocyst-forming filamentous cyanobacteria in evolution is connected to the rise of the atmospheric oxygen level at the Proterozoic eon, which was about 500 Mya. In the great oxidation event (GOE), which lasted from about 2.4 Bya until 2.3 Bya, the atmospheric O_2 content rose to about 2% due to oxygenic photosynthesis conducted by cyanobacteria. Subsequently, oxygen content remained that low for about 2 billion years until it grew up to 21% by oxygenic photosynthesis performed by terrestrial plants. The low atmospheric oxygen content during the 2 billion years period in the Proterozoic, which is therefore also called "the boring billion," is proposed to be a consequence of the feedback inhibition of nitrogenase by oxygen in cyanobacteria. Only with the increase in atmospheric oxygen content evoked by terrestrial plants did the diverse mechanisms of bacteria to protect nitrogenase from oxygen become necessary (Allen et al. 2019).

The activity of the nitrogenase is tightly regulated. For example, in the diazotrophic bacterium *Azospirillum brasilense*, PII proteins play a key role in the posttranslational regulation of nitrogenase activity. Other main interacting partners of the PII signaling system are the nitrogenase regulatory enzymes DraT (dinitrogenase reductase ADP-ribosyl-transferase) and DraG (di-nitrogenase reductase glycol-hydrolase) (Huergo et al. 2006a,b, 2007, 2009; Rajendran et al. 2011). The mechanism of PII-(GlnB and GlnZ)-dependent activation/inactivation of NifH is summarized in (Fig. 36.7).

36.4.2.2 Adaptation to Nitrogen Starvation in Nondiazotrophic Bacteria

The acclimation response to nitrogen deprivation has been extensively studied in the unicellular cyanobacterial strains *Synechococcus elongatus* PCC 7942 and *Synechocystis* sp. PCC 6803 (Klotz et al. 2016; Selim and Haffner 2020). Unlike

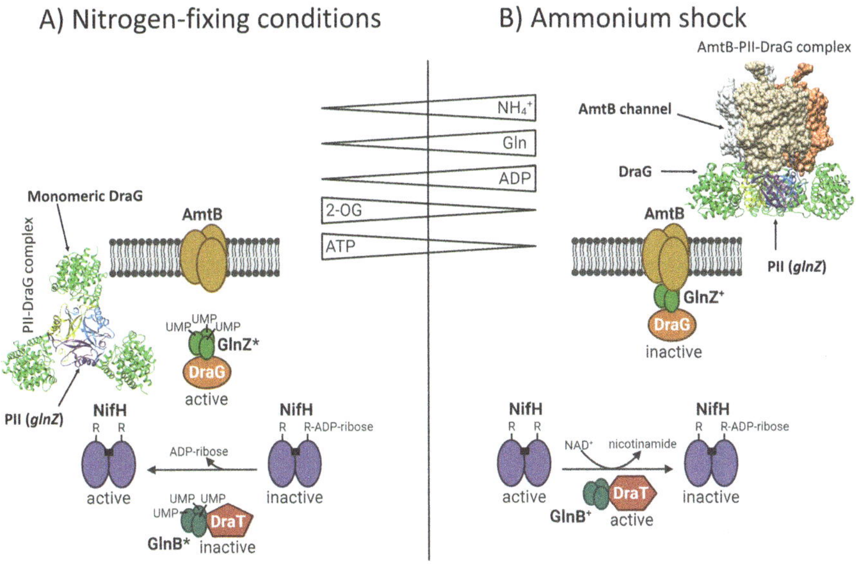

Fig. 36.7 Model for the mechanistic role of PII (GlnB and GlnZ) proteins in the activation/ inactivation of NifH nitrogenase activity, through the interaction with DraT and DraG, modified from Huergo et al. (2009). Under nitrogen-fixing conditions, the cytoplasmic-localized PII (GlnB and GlnZ) proteins are fully uridylylated and DraG is active to remove ADP ribosylation (ADP-R) from NifH, while DraT is inactive, allowing NifH nitrogenase activity. Upon nitrogen excess conditions of NH_4^+ shock, the intracellular increase of Gln due to GS activity leads to deuridylylation of the PII protein. Under this condition, the deuridylylated PII (GlnZ) moves to the membrane to close the NH_4^+ transport channel AmtB, and the DraG is sequestered to the membrane through the formation of the ternary AmtB-PII-DraG complex, while DraT starts again to be active leading to ADP ribosylation of NifH (the nitrogenase inactive form). The membrane localization of DraG within the ternary complex separates the DraG from the cytoplasmic NifH and thereby inhibits the DraG ADP-R removing activity. * indicates fully uridylylated and ATP/2-OG bound PII (GlnB and GlnZ) proteins bound, while + indicates deuridylylated/ADP bound PII (GlnB and GlnZ) proteins. The structures of the PII-DraG complex (PDB: 3O5T) and AmtB-PII-DraG complex, which is modeled based on AmtB-PII structure (PDB: 2NUU), are shown

filamentous cyanobacteria, unicellular strains cannot differentiate specialized cells for nitrogen fixation. Instead, in the absence of a combined nitrogen source, these organisms follow a developmental program that leads to metabolic dormancy and allows them to survive under these starvation conditions for a prolonged period of time (Forchhammer and Schwarz 2019). The most immediate metabolic change caused by nitrogen depletion is a rise of the 2-OG levels, since ammonia assimilation via the GS-GOGAT cycle stops operating. As described in the above Sect. 36.4.1. (*Adaptation to variable nitrogen availability* via *regulation of the glutamine synthetase*), 2-OG is a reporter of the intracellular C/N balance (Fokina et al. 2010; Muro-Pastor et al. 2001), and binds both the signal transduction protein PII and the global nitrogen control factor NtcA, increasing its DNA binding affinity and mediating its interaction with the transcriptional cofactor PipX (Forchhammer 2010). One of the targets of NtcA is *nblA*, a gene encoding for a small protein involved in the degradation of the phycobilisomes (Espinosa et al. 2007; Llácer et al. 2010). When nitrogen assimilation stops, the anabolic pathways involved in amino acid and nucleic acid synthesis are halted, with the consequent intracellular accumulation of ATP and reducing equivalents. Cells respond by adjusting the photosynthetic apparatus to prevent damage due to the extreme reduction of the photosynthetic electron carriers. This adjustment is achieved via degradation of the light-harvesting complexes, the phycobilisomes, which occurs in response to the limitation of various nutrients, but it is particularly rapid under nitrogen deprivation (Forchhammer and Schwarz 2019).

NblA is the main protein involved in phycobilisome degradation. Transcription of the *nblA* gene is induced under nitrogen starvation and is controlled by a very complex regulatory network. As mentioned above, an increase in 2-OG levels and binding of NtcA are required for induction of *nblA* expression under nitrogen-depleted conditions. Moreover, the response regulators NblR, NblC, RpaB, and SrrA are also involved in the regulation of *nblA* expression (Forchhammer and Schwarz 2019). Additionally, *nblA* expression is subjected to redox regulation: Reduction of electron carriers induces *nblA* expression and the initiation of reactions that act as an electron sink represses *nblA* expression (Klotz et al. 2015). This complex system allows a tight regulation of the phycobilisome degradation process, which is essential for survival to environmental changes. In addition to preventing photodamage, phycobilisome disassembly provides amino acids for glycogen synthesis during acclimation to nitrogen starvation. As a result of degrading the light-harvesting complexes, cells experience a color change from blue-green to yellow-orange, gaining a bleached appearance. Therefore, the process of phycobilisome degradation is termed chlorosis (Allen and Smith 1969).

Another immediate metabolic response to nitrogen starvation is the accumulation of glycogen, which has been well investigated in *Synechocystis* (Gründel et al. 2012). When imbalance in the C/N ratio is sensed through elevated levels of 2-OG, the newly photosynthetically fixed carbon is directed toward glycogen synthesis. 3-Phosphoglycerate (3-PGA) is the first stable product of the CO_2 fixation reaction catalyzed by the ribulose-1,5-bisphosphate carboxylase-oxygenase (RuBisCo). 3-PGA can enter the glycolytic route in the catabolic direction, where it is converted to 2-phosphoglycerate (2-PGA) by the phosphoglycerate mutase (Pgam), or in the gluconeogenesis direction, where it is converted to 2,3-bisphosphoglycerate (2,3-PGA) and directed toward glycogen synthesis. The Pgam reaction is a key point in the control of the fate of the photosynthetically fixed carbon. Under nitrogen sufficiency, when 2-OG levels are low, PII binds PirC, a competitive inhibitor of Pgam, and carbon is directed into the catabolic route. When 2-OG levels increase during nitrogen limitation, the PII-PirC complex dissociates and PirC inhibits Pgam, directing carbon into glycogen synthesis (Orthwein et al. 2020). Glycogen accumulation is essential for proper acclimation to nitrogen starvation. Mutants impaired in glycogen synthesis fail to carry out the chlorosis process and do not survive nitrogen depletion (Gründel et al. 2012). Glycogen accumulation starts almost immediately after the onset of nitrogen starvation and reaches a maximum of 60% of the cell's dry weight after 14 hours (Klotz et al. 2016). As glycogen is being synthesized, cells begin the expression of the glycogen catabolic enzymes, which remain inactive until a nitrogen source is again available (Doello et al. 2018). This anticipatory behavior allows cells to rapidly react to the presence of a nitrogen source. Some cyanobacterial strains of the genera *Anabaena, Cyanothece, Microcystis, Nostoc, Oscillatoria, Synechococcus,* and *Synechocystis* accumulate an additional carbon polymer called polyhydroxybutyrate (PHB) (Zilliges 2014), which is produced from glycogen after a few days of nitrogen starvation (Koch et al. 2019). The physiological role of PHB in the survival of periods of nitrogen starvation has, however, not been elucidated (Hauf et al. 2013; Klotz et al. 2016; Koch et al. 2020).

After the first events in the adaptation to nitrogen starvation (i.e., chlorosis and glycogen accumulation) have taken place, cells direct their metabolism into a dormant state that allows prolonged survival under these conditions. The chlorotic state is characterized by growth arrest and reduced metabolic activity. Growth arrest occurs after DNA replication, rendering cells ready for division when they can resume metabolic activity and providing a higher polyploidy to protect them in case of DNA damage. During quiescence, a precise regulation of the residual metabolic processes is essential to ensure cell survival. In *Synechocystis*, upon nitrogen starvation, the intracellular ATP content is reduced to approximately 25% of the levels in vegetative cells and is maintained at this level throughout the entirety of the dormant period (Doello et al. 2018). As cells enter dormancy, they degrade most of their thylakoid membranes. Thus, dormant cells rely on residual photosynthetic and respiratory activity to maintain their ATP content to a minimum level. Chlorotic cells conserve a small proportion of their photosynthetic machinery (Sauer et al. 2001; Spät et al. 2018), but the residual activity they are capable of is required to sustain viability, since treatment with inhibitors of photosynthetic electron transport or prolonged exposure to darkness leads to death (Forchhammer and Schwarz 2019). Due to the vast degradation of the thylakoid membranes and the consequent reduced space for proton storage, the bioenergetics of chlorotic *Synechocystis* cells largely depend on sodium. The plasma membrane is energized by its alternative respiratory chain, which consists of a NAD(P)H dehydrogenase type II (NDHII) and an alternative terminal cytochrome c oxidase (ARTO), creating a sodium motive force that can be employed by the ATP synthases in the plasma membrane to provide dormant cells with ATP. This adaptation strategy seems to extend to other high-salt-adapted cyanobacteria, but not to freshwater species such as *S. elongatus* (Doello et al. 2021).

When dormant nitrogen-starved cells encounter a source of combined nitrogen, they are capable of reverting the metabolic and structural changes described above and restore the vegetative cell cycle. The process of exiting dormancy is termed resuscitation, and it involves a genetically determined program (Klotz et al. 2016; Spät et al. 2018). Immediately after nitrogen availability, the genes encoding for the ATP synthesis, nitrogen assimilation and translation machinery are upregulated (Klotz et al. 2016). The first detected metabolic response of chlorotic cells to the presence of nitrogen is an increase in the ATP levels (Doello et al. 2018), which is necessary to fuel nitrogen assimilating reactions. To synthesize ATP, cells cease the

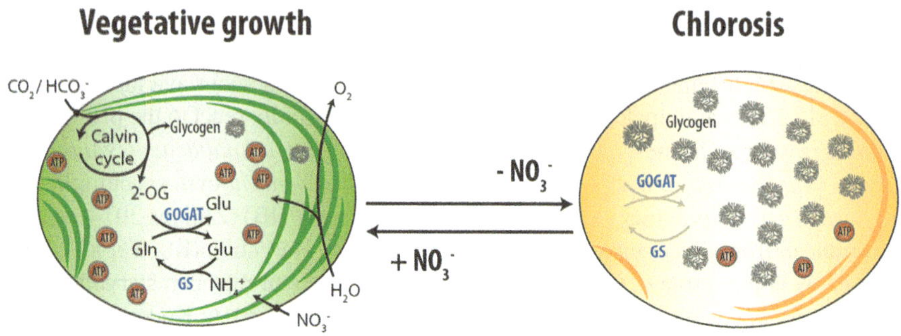

Fig. 36.8 Schematic representation of the adaptation to nitrogen starvation in *Synechocystis* sp. PCC 6803. Under nitrogen deprivation, cells tune down their ATP levels, degrade their photosynthetic machinery and thylakoid membranes, and accumulate glycogen granules, reaching a state of dormancy that allows prolonged survival under these conditions. Upon addition of a combined nitrogen source, cells initiate the resuscitation program: Glycogen degradation is activated, ATP levels rise, and the photosynthetic machinery and thylakoid membranes are rebuilt, restoring vegetative growth within 48 h

residual photosynthetic activity and induce respiration of glycogen, thus switching to a heterotrophic metabolism. Glycogen degradation provides the necessary energy and intermediates to rebuild the photosynthetic machinery, and it is essential for the recovery of dormant cells (Doello et al. 2018). Twenty-four hours after nitrogen availability, cells start to regreen and regain photosynthetic activity, entering a mixotrophic phase. Photoautotrophic growth and cell division resume after 48 h, thereby completing the program (Klotz et al. 2016) (Fig. 36.8).

36.5 Concluding Remarks

Despite all the research efforts, it is necessary to point out that no experimental design truly mimics the complexity of the actual environmental changes experienced by microbes in nature. The seasonal cycles and nutrient scarcity/availability depend on wild or man-made factors, ecosystems, and metaorganisms dynamics. In addition, experiments are time limited, precluding an absolute certainty about the fate of the organisms in question.

Nevertheless, the compilation of information continues to be our strongest tool to prepare for a new norm, as predicted by the intergovernmental panel on climate change, and elaborate accordingly which organisms and locations face the highest degree of risk. This information will serve to inform policy makers on how to control the causes and mitigate the consequences. These can enforce laws and rules for urban sprawl, industrialization, and fuel consumption, but also promote public awareness through campaigns aimed at educating people about waste recycling and the carbon footprint of their own diet.

Acknowledgments The authors gratefully acknowledge Karl Forchhammer for continued support. Furthermore, we would like to acknowledge Libera Lo-Presti for critical scientific and linguistic editing of the manuscript and the infrastructural support by the Cluster of Excellence "Controlling Microbes to Fight Infections" (EXC 2124) of the German research foundation (DFG). We are indebted to the DFG for continuous funding of the work in the authors laboratory. KAS would like to dedicate this chapter to the memory of Dr. Ali Selim, a distinguished father and medical doctor, for his continued support.

Author Contributions KAS conceived and designed this project. All authors analyzed the data and wrote the manuscript, while KSA merged the final version of the manuscript. KAS and EZ prepared the figures (except Fig. 36.8 by SD). All authors but particularly SD commented and edited on the manuscript and approved the final version of the manuscript.

References

Abatenh E, Gizaw B, Tsegaye Z, Tefera G (2018) Microbial function on climate change - a review. Environ Pollut Climate Change 2:1000147

Abduljalil JM (2018) Bacterial riboswitches and RNA thermometers: nature and contributions to pathogenesis. Non-coding RNA Res 3:54–63

Agostoni M, Montgomery BL (2014) Survival strategies in the aquatic and terrestrial world: the impact of second messengers on cyanobacterial processes. Life (Basel, Switzerland) 4 (4):745–769. https://doi.org/10.3390/life4040745

Allen MM, Smith AJ (1969) Nitrogen chlorosis in blue-green algae. Archiv fur Mikrobiologie 69 (2):114–120. https://doi.org/10.1007/bf00409755

Allen JF, Thake B, Martin WF (2019) Nitrogenase inhibition limited oxygenation of Earth's proterozoic atmosphere. Trends Plant Sci 24:1022–1031

Angilletta MJ, Huey RB, Frazier MR (2010) Thermodynamic effects on organismal performance: is hotter better? Physiol Biochem Zool 83:197–206

Bardgett RD, Freeman C, Ostle NJ (2008) Microbial contributions to climate change through carbon cycle feedbacks. ISME J 2:805–814

Beere HM (2004) 'The stress of dying': the role of heat shock proteins in the regulation of apoptosis. J Cell Sci 117:2641–2651

Bergersen FJ, Appleby CA (1981) Leghaemoglobin within bacteroid-enclosing membrane envelopes from soybean root nodules. Planta 152:534–543

Bolay P, Muro-Pastor MI, Florencio FJ, Klahn S (2018) The distinctive regulation of cyanobacterial glutamine synthetase. Life (Basel) 8:52

Bothe H, Schmitz O, Yates MG, Newton WE (2010) Nitrogen fixation and hydrogen metabolism in cyanobacteria. Microbiol Mol Biol Rev 74:529–551

Bradford MA (2013) Thermal adaptation of decomposer communities in warming soils. Front Microbiol 4:333

Brown JR, Masuchi Y, Robb FT, Doolittle WF (1994) Evolutionary relationships of bacterial and archaeal glutamine synthetase genes. J Mol Evol 38:566–576

Calloni G, Chen T, Schermann SM, Chang HC, Genevaux P, Agostini F, Tartaglia GG, Hayer-Hartl M, Hartl FU (2012) DnaK functions as a central hub in the *E. coli* chaperone network. Cell Rep 1:251–264

Cavicchioli R et al (2019) Scientists' warning to humanity: microorganisms and climate change. Nat Rev Microbiol 17:569–586

Chowdhury S, Maris C, Allain FHT, Narberhaus F (2006) Molecular basis for temperature sensing by an RNA thermometer. EMBO J 25:2487–2497

Cimdins A, Klinkert B, Aschke-Sonnenborn U, Kaiser FM, Kortmann J, Narberhaus F (2014) Translational control of small heat shock genes in mesophilic and thermophilic cyanobacteria by RNA thermometers. RNA Biol 11:594–608

Commichau FM, Forchhammer K, Stülke J (2006) Regulatory links between carbon and nitrogen metabolism. Curr Opin Microbiol 9:167–172

Cronan CS (2018) Microbial biogeochemistry. In: Ecosystem biogeochemistry: element cycling in the forest landscape. Springer, Cham, pp 31–40

Dell AI, Pawar S, Savage VM (2011) Systematic variation in the temperature dependence of physiological and ecological traits. Proc Natl Acad Sci U S A 108:10591–10596

Deutsch CA, Tewksbury JJ, Huey RB, Sheldon KS, Ghalambor CK, Haak DC, Martin PR (2008) Impacts of climate warming on terrestrial ectotherms across latitude. Proc Natl Acad Sci U S A 105:6668–6672

Docherty KM, Gutknecht JLM (2012) The role of environmental microorganisms in ecosystem responses to global change: current state of research and future outlooks. Biogeochemistry 109:1–6

Doello S et al (2018) A specific glycogen mobilization strategy enables rapid awakening of dormant cyanobacteria from chlorosis. Plant Physiol 177:594–603. https://doi.org/10.1104/pp.18.00297

Doello S, Burkhardt M, Forchhammer K (2021) The essential role of sodium bioenergetics and ATP homeostasis in the developmental transitions of a cyanobacterium. Curr Biol. https://doi.org/10.1016/j.cub.2021.01.065

Drigo B, Kowalchuk GA, Yergeau E, Bezemer TM, Boschker HTS, Van Veen JA (2007) Impact of elevated carbon dioxide on the rhizosphere communities of Carex arenaria and *Festuca rubra*. Glob Chang Biol 13:2396–2410

Dutta H, Dutta A (2016) The microbial aspect of climate change. In: Energy, ecology and environment, vol 1. Joint Center on Global Change and Earth System Science of the University of Maryland and Beijing Normal University, Beijing, pp 209–232

Espinosa J, Forchhammer K, Contreras A (2007) Role of the Synechococcus PCC 7942 nitrogen regulator protein PipX in NtcA-controlled processes. Microbiology 153:711–718

Fokina O et al (2010) Mechanism of 2-oxoglutarate signaling by the *Synechococcus elongatus* P II signal transduction protein. Proc Natl Acad Sci U S A 107:19760–19765

Forcada-Nadal A, Palomino-Schatzlein M, Neira JL, Pineda-Lucena A, Rubio V (2017) The PipX protein, when not bound to its targets, has its signaling C-terminal Helix in a flexed conformation. Biochemistry 56:3211–3224

Forchhammer K (2010) Network of PII Signalling protein interactions in unicellular cyanobacteria. Recent Adv Phototrophic Prokaryotes 675:71–90

Forchhammer K, Schwarz R (2019) Nitrogen chlorosis in unicellular cyanobacteria – a developmental program for surviving nitrogen deprivation. Environ Microbiol 21:1173–1184

Forchhammer K, Selim KA (2020) Carbon/nitrogen homeostasis control in cyanobacteria. FEMS Microbiol Rev 44:33–53

Franks SJ, Hoffmann AA (2012) Genetics of climate change adaptation. Annu Rev Genet 46:185–208

García-Domínguez M, Reyes JC, Florencio FJ (1999) Glutamine synthetase inactivation by protein-protein interaction. Proc Natl Acad Sci U S A 96:7161–7166

Gionchetta G, Romaní AM, Oliva F, Artigas J (2019) Distinct responses from bacterial, archaeal and fungal streambed communities to severe hydrological disturbances. Sci Rep 9:13506

Grandclement C, Tannieres M, Morera S, Dessaux Y, Faure D (2016) Quorum quenching: role in nature and applied. FEMS Microbiol Rev 40:86–116

Gründel M et al (2012) Impaired glycogen synthesis causes metabolic overflow reactions and affects stress responses in the cyanobacterium Synechocystis sp. PCC 6803. Microbiology (United Kingdom) 158(12):3032–3043. https://doi.org/10.1099/mic.0.062950-0

Hallin S, Bodelier PLE (2020) Grand challenges in terrestrial microbiology: moving on from a decade of progress in microbial biogeochemistry. Front Microbiol 11:981

Hauf W et al (2013) Metabolic changes in Synechocystis PCC6803 upon nitrogen-starvation: excess NADPH sustains polyhydroxybutyrate accumulation. Metabolites 3(1):101–118. https://doi.org/10.3390/metabo3010101

Huergo LF, Dixon R (2015) The emergence of 2-Oxoglutarate as a master regulator metabolite. Microbiol Mol Biol Rev 79:419–435

Huergo LF, Chubatsu LS, Souza EM, Pedrosa FO, Steffens MB, Merrick M (2006a) Interactions between PII proteins and the nitrogenase regulatory enzymes DraT and DraG in Azospirillum brasilense. FEBS Lett 580:5232–5236

Huergo LF, Souza EM, Araujo MS, Pedrosa FO, Chubatsu LS, Steffens MB, Merrick M (2006b) ADP-ribosylation of dinitrogenase reductase in Azospirillum brasilense is regulated by AmtB-dependent membrane sequestration of DraG. Mol Microbiol 59:326–337

Huergo LF, Merrick M, Pedrosa FO, Chubatsu LS, Araujo LM, Souza EM (2007) Ternary complex formation between AmtB, GlnZ and the nitrogenase regulatory enzyme DraG reveals a novel facet of nitrogen regulation in bacteria. Mol Microbiol 66:1523–1535

Huergo LF, Merrick M, Monteiro RA, Chubatsu LS, Steffens MB, Pedrosa FO, Souza EM (2009) In vitro interactions between the PII proteins and the nitrogenase regulatory enzymes dinitrogenase reductase ADP-ribosyltransferase (DraT) and dinitrogenase reductase-activating glycohydrolase (DraG) in Azospirillum brasilense. J Biol Chem 284:6674–6682

Huergo LF, Chandra G, Merrick M (2013) P(II) signal transduction proteins: nitrogen regulation and beyond. FEMS Microbiol Rev 37:251–283

Huey RB, Kearney MR, Krockenberger A, Holtum JAM, Jess M, Williams SE (2012) Predicting organismal vulnerability to climate warming: roles of behaviour, physiology and adaptation. Philos Trans R Soc B Biol Sci 367:1665–1679

Jiang P, Ninfa AJ (2009) α-Ketoglutarate controls the ability of the *Escherichia coli* PII signal transduction protein to regulate the activities of NRII (NtrB) but does not control the binding of PII to NRII. Biochemistry 48:11514–11521

Klahn S, Schaal C, Georg J, Baumgartner D, Knippen G, Hagemann M, Muro-Pastor AM, Hess WR (2015) The sRNA NsiR4 is involved in nitrogen assimilation control in cyanobacteria by targeting glutamine synthetase inactivating factor IF7. Proc Natl Acad Sci U S A 112:E6243–E6252

Klahn S, Bolay P, Wright PR, Atilho RM, Brewer KI, Hagemann M, Breaker RR, Hess WR (2018) A glutamine riboswitch is a key element for the regulation of glutamine synthetase in cyanobacteria. Nucleic Acids Res 46:10082–10094

Klotz A et al (2015) Nitrogen starvation acclimation in *Synechococcus elongatus*: redox-control and the role of nitrate reduction as an electron sink. Life 5(1):888–904. https://doi.org/10.3390/life5010888

Klotz A et al (2016) Awakening of a dormant cyanobacterium from nitrogen chlorosis reveals a genetically determined program. Curr Biol 26(21):2862–2872. https://doi.org/10.1016/j.cub.2016.08.054

Koch M et al (2019) PHB is produced from glycogen turn-over during nitrogen starvation in Synechocystis sp. PCC 6803. Int J Mol Sci 20:1942. https://doi.org/10.3390/ijms20081942

Koch M, Berendzen KW, Forchhammer K (2020) On the role and production of polyhydroxybutyrate (Phb) in the cyanobacterium synechocystis sp. pcc 6803. Life 10. https://doi.org/10.3390/life10040047

Loh E, Righetti F, Eichner H, Twittenhoff C, Narberhaus F (2018) RNA thermometers in bacterial pathogens. Microbiol Spectr 6(2). https://doi.org/10.1128/microbiolspec.RWR-0012-2017

Kortmann J, Narberhaus F (2012) Bacterial RNA thermometers: molecular zippers and switches. Nat Rev Microbiol 10:255–265

Kosaka T, Nakajima Y, Ishii A, Yamashita M, Yoshida S, Murata M, Kato K, Shiromaru Y, Kato S, Kanasaki Y, Yoshikawa H, Matsutani M, Thanonkeo P, Yamada M (2019) Capacity for survival in global warming: adaptation of mesophiles to the temperature upper limit. PLoS One 14:e0218985

Krysenko S, Matthews A, Okoniewski N, Kulik A, Girbas MG, Tsypik O, Meyners CS, Hausch F, Wohlleben W, Bera A (2019) Initial metabolic step of a novel ethanolamine utilization pathway and its regulation in *Streptomyces coelicolor* M145. mBio 10:e00326-19

Lapina T, Selim KA, Forchhammer K, Ermilova E (2018) The PII signaling protein from red algae represents an evolutionary link between cyanobacterial and Chloroplastida PII proteins. Sci Rep 8:790

Llácer JL et al (2010) Structural basis for the regulation of NtcA-dependent transcription by proteins PipX and PII. Proc Natl Acad Sci U S A 107(35):15397–15402. https://doi.org/10.1073/pnas.1007015107

Luque I, Forchhammer K (2008) Nitrogen assimilation and C/N balance sensing, Chapter 13. In: Herrero A, Flores E (eds) The cyanobacteria: molecular biology, genomics and evolution. Caister Academic Press, London

Maleki F, Khosravi A, Nasser A, Taghinejad H, Azizian M (2016) Bacterial heat shock protein activity. J Clin Diagn Res 10:BE01–BE03

Merino N, Aronson HS, Bojanova DP, Feyhl-Buska J, Wong ML, Zhang S, Giovannelli D (2019) Living at the extremes: extremophiles and the limits of life in a planetary context. Front Microbiol 10:780

Miller DJ, Fort PE (2018) Heat shock proteins regulatory role in neurodevelopment. Front Neurosci 12:821

Krysenko S, Okoniewski N, Kulik A, Matthews A, Grimpo J, Wohlleben W, Bera A (2017) Gamma-glutamylpolyamine synthetase GlnA3 is involved in the first step of polyamine degradation pathway in *Streptomyces coelicolor* M145. Front Microbiol 8:726

Muro-Pastor MI, Reyes JC, Florencio FJ (2001) Cyanobacteria perceive nitrogen status by sensing intracellular 2-oxoglutarate levels. J Biol Chem 276:38320–38328

Mus F, Colman DR, Peters JW, Boyd ES (2019) Geobiological feedbacks, oxygen, and the evolution of nitrogenase. Free Radic Biol Med 140:250–259

Narberhaus F, Balsiger S (2003) Structure-function studies of *Escherichia coli* RpoH (σ32) by in vitro linker insertion mutagenesis. J Bacteriol 185:2731–2738

Narberhaus F, Waldminghaus T, Chowdhury S (2006) RNA thermometers. FEMS Microbiol Rev 30:3–16

Neira JL, Ortore MG, Florencio FJ, Muro-Pastor MI, Rizzuti B (2020) Dynamics of the intrinsically disordered inhibitor IF7 of glutamine synthetase in isolation and in complex with its partner. Arch Biochem Biophys 683:108303

Oelze J (2000) Respiratory protection of nitrogenase in azotobacter species: is a widely held hypothesis unequivocally supported by experimental evidence? FEMS Microbiol Rev 24:321–333

Orthwein T et al (2020) The novel PII-interacting regulator PirC (Sll0944) identifies 3-phosphoglycerate mutase (PGAM) as central control point of carbon storage metabolism in cyanobacteria. bioRxiv. https://doi.org/10.1101/2020.09.11.292599

Pantoja-Uceda D, Neira JL, Saelices L, Robles-Rengel R, Florencio FJ, Muro-Pastor MI, Santoro J (2016) Dissecting the binding between glutamine synthetase and its two natively unfolded protein inhibitors. Biochemistry 55:3370–3382

Perez-Mon C, Frey B, Frossard A (2020) Functional and structural responses of arctic and alpine soil prokaryotic and fungal communities under freeze-thaw cycles of different frequencies. Front Microbiol 11:982. https://doi.org/10.3389/fmicb.2020.00982

Rajendran C, Gerhardt EC, Bjelic S, Gasperina A, Scarduelli M, Pedrosa FO, Chubatsu LS, Merrick M, Souza EM, Winkler FK, Huergo LF, Li XD (2011) Crystal structure of the GlnZ-DraG complex reveals a different form of PII-target interaction. Proc Natl Acad Sci U S A 108:18972–18976

Rampelotto PH (2013) Extremophiles and extreme environments. Life 3:482–485

Rexer HU, Schäberle T, Wohlleben W, Engels A (2006) Investigation of the functional properties and regulation of three glutamine synthetase-like genes in Streptomyces coelicolor A3(2). Arch Microbiol 186:447–458

Righetti F, Narberhaus F (2014) How to find RNA thermometers. *Frontiers in Cellular and Infection*. Microbiology 4:132

Rodriguez R, Durán P (2020) Natural Holobiome engineering by using native extreme microbiome to counteract the climate change effects. Front Bioeng Biotechnol 8:568

Ronneau S, Hallez R (2019) Make and break the alarmone: regulation of (p)ppGpp synthetase/hydrolase enzymes in bacteria. FEMS Microbiol Rev 43:389–400

Rousk J, Bengtson P (2014) Microbial regulation of global biogeochemical cycles. Front Microbiol 5:103. https://doi.org/10.3389/fmicb.2014.00103

Rudolph B, Gebendorfer KM, Buchner J, Winter J (2010) Evolution of *Escherichia coli* for growth at high. J Biol Chem 285:19029–19034

Rutten PJ, Poole PS (2019) Oxygen regulatory mechanisms of nitrogen fixation in rhizobia. Adv Microb Physiol 75:325–389

Sabra W, Zeng AP, Lunsdorf H, Deckwer WD (2000) Effect of oxygen on formation and structure of *Azotobacter vinelandii* alginate and its role in protecting nitrogenase. Appl Environ Microbiol 66:4037–4044

Saelices L, Galmozzi CV, Florencio FJ, Muro-Pastor MI, Neira JL (2011) The inactivating factor of glutamine synthetase IF17 is an intrinsically disordered protein, which folds upon binding to its target. Biochemistry 50:9767–9778

Sandberg TE, Pedersen M, Lacroix RA, Ebrahim A, Bonde M, Herrgard MJ, Palsson BO, Sommer M, Feist AM (2014) Evolution of *Escherichia coli* to 42 °C and subsequent genetic engineering reveals adaptive mechanisms and novel mutations. Mol Biol Evol 31:2647–2662

Sandh G, Xu L, Bergman B (2012) Diazocyte development in the marine diazotrophic cyanobacterium Trichodesmium. Microbiol (Reading) 158:345–352

Sauer J et al (2001) Nitrogen starvation-induced chlorosis in Synechococcus PCC 7942. Low-level photosynthesis as a mechanism of long-term survival. Plant Physiol 126:233–243

Selim KA, Haffner M (2020) Heavy metal stress alters the response of the unicellular cyanobacterium *Synechococcus elongatus* PCC 7942 to nitrogen starvation. Life (Basel, Switzerland) 10 (11):275. https://doi.org/10.3390/life10110275

Selim KA, Maldener I (2021) Cellular and molecular strategies in cyanobacterial survival-"In Memory of Prof. Dr. Wolfgang Lockau". Life (Basel, Switzerland) 11(2):132. https://doi.org/10.3390/life11020132

Selim KA, Haase F, Hartmann MD, Hagemann M, Forchhammer K (2018) PII-like signaling protein SbtB links cAMP sensing with cyanobacterial inorganic carbon response. Proc Natl Acad Sci U S A 115(21):E4861–E4869

Selim KA, Haffner M, Watzer B, Forchhammer K (2019) Tuning the in vitro sensing and signaling properties of cyanobacterial PII protein by mutation of key residues. Sci Rep 9(1):18985. https://doi.org/10.1038/s41598-019-55495-y

Selim KA, Ermilova E, Forchhammer K (2020c) From cyanobacteria to Archaeplastida: new evolutionary insights into PII signalling in the plant kingdom. New Phytol 227:722–731

Selim KA, Lapina T, Forchhammer K, Ermilova E (2020b) Interaction of N-acetyl-l-glutamate kinase with the PII signal transducer in the non-photosynthetic alga *Polytomella parva*: co-evolution towards a hetero-oligomeric enzyme. FEBS J 287:465–482

Selim KA, Tremiño L, Alva V, Espinosa J, Contreras A, Marco-Marín C, Hartmann MD, Forchhammer K, Rubio V (2020a) Functional and structural characterization of PII-like protein CutA does not support involvement in heavy metal tolerance in cyanobacteria and hints at a small-molecule carrying/signaling role. FEBS J. https://doi.org/10.1111/febs.15464

Shade A, Peter H, Allison SD, Baho DL, Berga M, Bürgmann H, Huber DH, Langenheder S, Lennon JT, Martiny JBH, Matulich KL, Schmidt TM, Handelsman J (2012) Fundamentals of microbial community resistance and resilience. Front Microbiol 3:417

Shi Y, Zhao W, Zhang W, Ye Z, Zhao J (2006) Regulation of intracellular free calcium concentration during heterocyst differentiation by HetR and NtcA in Anabaena sp. PCC 7120. Proc Natl Acad Sci U S A 103:11334–11339

Singh BK, Bardgett RD, Smith P, Reay DS (2010) Microorganisms and climate change: terrestrial feedbacks and mitigation options. Nat Rev Microbiol 8:779–790

Smith TP, Thomas T, García-Carreras B, Sal S, Yvon-Durocher G, Bell T, Pawar S (2019) Community-level respiration of prokaryotic microbes may rise with global warming. Nat Commun 10(1):5124. https://doi.org/10.1038/s41467-019-13109-1

Spät P et al (2018) Chlorosis as a developmental program in cyanobacteria: the proteomic fundament for survival and awakening. Mol Cell Proteomics 17:1650–1669

Steinchen W, Bange G (2016) The magic dance of the alarmones (p)ppGpp. Mol Microbiol 101:531–544

Tiwari S, Thakur R, Shankar J (2015) Role of heat-shock proteins in cellular function and in the biology of fungi. Biotechnol Res Int 132635

Voolstra CR, Ziegler M (2020) Adapting with microbial help: microbiome flexibility facilitates rapid responses to environmental change. BioEssays 42(7):e2000004

Walter J, Lynch F, Battchikova N, Aro EM, Gollan PJ (2016) Calcium impacts carbon and nitrogen balance in the filamentous cyanobacterium Anabaena sp. PCC 7120. J Exp Bot 67:3997–4008

Walter J, Selim KA, Leganés F, Fernández-Piñas F, Vothknecht UC, Forchhammer K, Aro EM, Gollan PJ (2019) A novel Ca^{2+}-binding protein influences photosynthetic electron transport in *Anabaena* sp. PCC 7120. Biochim Biophys Acta Bioenerg 1860:519–532

Walter J, Leganés F, Aro EM, Gollan PJ (2020) The small Ca^{2+}-binding protein CSE links Ca^{2+} signalling with nitrogen metabolism and filament integrity in Anabaena sp. PCC 7120. BMC Microbiol 20(1):57. https://doi.org/10.1186/s12866-020-01735-5

Wang Y, DeHaseth PL (2003) Sigma 32-dependent promoter activity in vivo: sequence determinants of the groE promoter. J Bacteriol 185:5800–5806

Wooliver R, Tittes SB, Sheth SN (2020) A resurrection study reveals limited evolution of thermal performance in response to recent climate change across the geographic range of the scarlet monkeyflower. Evolution. https://doi.org/10.1111/evo.1404

Xue K, Xie J, Zhou A, Liu F, Li D, Wu L, Deng Y, He Z, Van Nostrand JD, Luo Y, Zhou J (2016) Warming alters expressions of microbial functional genes important to ecosystem functioning. Front Microbiol 7:668. https://doi.org/10.3389/fmicb.2016.00668

Yang Z, Yang S, Van Nostrand JD, Zhou J, Fang W, Qi Q, Liu Y, Wullschleger SD, Liang L, Graham DE, Yang Y, Gu B (2017) Microbial community and functional gene changes in Arctic Tundra soils in a microcosm warming experiment. Front Microbiol 8:1741. https://doi.org/10.3389/fmicb.2017.01741

Zak DR, Blackwood CB, Waldrop MP (2006) A molecular dawn for biogeochemistry. Trends Ecol Evol 21:288–295

Zhao Y, Shi Y, Zhao W, Huang X, Wang D, Brown N, Brand J, Zhao J (2005) CcbP, a calcium-binding protein from Anabaena sp. PCC 7120, provides evidence that calcium ions regulate heterocyst differentiation. Proc Natl Acad Sci U S A 102:5744–5448

Zilliges Y (2014) Glycogen, a dynamic cellular sink and reservoir for carbon. In: Flores E (ed) Cell biology. Caister Academic Press, London

The manufacturer's authorised representative in the EU is Springer
Nature Customer Service Centre GmbH, Europaplatz 3, 69115 Heidelberg,
Germany. If you have any concerns regarding our products, please
contact ProductSafety@springernature.com

Printed and bound by CPI Group (UK) Ltd, Croydon, CR0 4YY
30/09/2026
02225289-0020